ADVANCED TOPICS
IN SCIENCE AND TECHNOLOGY IN CHINA

ADVANCED TOPICS IN SCIENCE AND TECHNOLOGY IN CHINA

Zhejiang University is one of the leading universities in China. In Advanced Topics in Science and Technology in China, Zhejiang University Press and Springer jointly publish monographs by Chinese scholars and professors, as well as invited authors and editors from abroad who are outstanding experts and scholars in their fields. This series will be of interest to researchers, lecturers, and graduate students alike.

Advanced Topics in Science and Technology in China aims to present the latest and most cutting-edge theories, techniques, and methodologies in various research areas in China. It covers all disciplines in the fields of natural science and technology, including but not limited to, computer science, materials science, life sciences, engineering, environmental sciences, mathematics, and physics.

Lanjuan Li *Editor*

Infectious Microecology

Theory and Applications

With 17 figures, 6 of them in color

ZHEJIANG UNIVERSITY PRESS
浙江大学出版社

Springer

Editor
Prof. Lanjuan Li
The State Key Laboratory for Diagnosis and Treatment of Infectious Diseases, the First Affiliated Hospital, School of Medicine, Zhejiang University; Collaborative Innovation Center for Diagnosis and Treatment of Infectious Diseases, Hangzhou, 310003, China.
E-mail: ljli@zju.edu.cn

ISBN 978-3-662-43882-4 ISBN 978-3-662-43883-1 (eBook)
Springer Heidelberg New York Dordrecht London

Library of Congress Control Number: 2014941442

Printed on acid-free paper

Springer is part of Springer Science+Business Media (www.springer.com)

国家科学技术学术著作出版基金资助出版
"十二五"国家重点图书出版规划项目

Lanjuan Li *Editor*

Infectious Microecology

Theory and Applications

With 17 figures, 6 of them in color

ZHEJIANG UNIVERSITY PRESS
浙江大学出版社

Springer

图书在版编目(CIP)数据

感染微生态学：理论与实践 = Infectious microecology: theory and applications：英文 / 李兰娟主编. —杭州：浙江大学出版社，2014.7
ISBN 978-7-308-12435-5

Ⅰ. ①感… Ⅱ. ①李… Ⅲ. ①感染—病原微生物—微生物学—英文 Ⅳ. ①R37

中国版本图书馆 CIP 数据核字(2013)第 255630 号

感染微生态学：理论与实践
李兰娟　主编

责任编辑	张鸽　张凌静
封面设计	俞亚彤
出版发行	浙江大学出版社
	网址：http://www.zjupress.com
	Springer-Verlag GmbH
	网址：http://www.Springer.com
排　　版	杭州理想广告有限公司
印　　刷	浙江印刷集团有限公司
开　　本	710mm×1000mm　1/16
印　　张	42.5
字　　数	1329 千
版 印 次	2014 年 7 月第 1 版　2014 年 7 月第 1 次印刷
书　　号	ISBN 978-7-308-12435-5 (浙江大学出版社)
	ISBN 978-3-662-43882-4 (Springer-Verlag GmbH)
定　　价	268.00 元

浙江大学出版社发行部联系方式 (0571)88925591; http://zjdxcbs.tmall.com

Preface

The first edition of *Infectious Microecology* was published in 2002. After 10 years of basic research and clinical practice, microecology, especially infectious microecology, has made great progress in the world, which confirms the innovative and clinical value of this theory. The development of infectious microecology is based on the progress in molecular biology, metagenomics, metabolomics, proteomics, and it is also a supplement to existing theories and practice in the field of infectious diseases.

With the progress of human civilization and medical technology, the spectrum of diseases has greatly changed. The aging population is increasing in the world. These people have a relatively low immune function and this is often accompanied by one or more underlying diseases, such as hypertension, diabetes, chronic kidney disease, etc. In addition, with the wide use of antibiotics, immunosuppressants, radiochemotherapy, organ transplants and interventional therapy, the life of critically ill patients has been prolonged. Drug-resistant strains, especially multi-drug resistant strains, are prevalent throughout the world. These strains may be the normal flora for healthy people, but they can lead to severe or even fatal infections in the above mentioned populations. The research and development of new antibiotics are far from able to meet the needs of clinical practice, and antibiotics alone cannot solve this problem. Therefore, prevention and treatment of infectious diseases has become a major issue in the new century. It is against this backdrop and in need of new theoretical guidance that infectious microecology has emerged.

Microecology sprouted at the end of the 19th century. Since the 1970s the development of gnotobiology, anaerobic culture techniques, electron microscopy techniques and cellular molecular biology have promoted the development of microecology. In the past 10 years, microecology research in humans has gained extensive attention at home (China) and abroad. Studies have shown that the microecological system is like a human organ with physiological functions, and microecological flora plays an important role in the body's immune system, metabolism and nutrition, especially in the prevention and occurrence of infections. American scholar, Professor Hannah Gordon, said, "To ignore our microbial side would be to ignore an important contributor to our health and our biology." In the book *Infectious Microecology* published in 2002, we proposed the theory of "infectious microecology", which was highly praised by Professor Jeremy K. Nicholson from Imperial College London. Our study "Infectious Microecology: Theory and Application" won second prize nationally for progress in science and technology in 2007, and another study titled "Intestinal Microecology and Infection" was sponsored by the National Basic Research Program of China ("973" Program). The latter has attracted a high degree of international attention, and the magazine *Science* gave a comprehensive introduction to it. The results of this project were published on *Hepatology* and it also attracted Professor Dusko S. Ehrlich to join in, who is in charge of the European human gut metagenomics project. With further research into microecology, people know more about both the useful and harmful effects that microecology brings to the host. What's more, infectious microecology enriches the connotation of the theory of infectious diseases, so that people can look at the incidence, progression and prognosis of infection from the point of view of microecology. It improves the anti-infection strategy, proposing a new idea that the treatment of infection should be to "kill and promote bacteria" rather than only "kill bacteria". In recent years a wider body of evidence has shown that microecology therapy is indeed an effective weapon in the prevention and treatment of bacterial infection.

Utilizing 10 years worth of research and clinical practice, referring to recent literature about the relationship between infection and microecology, and combined with the latest research findings of liver microecology, we updated the theory, knowledge and techniques in the field of infectious microecology. We hope this edition can provide new information for medical students and clinicians.

The book is divided into 23 chapters. Chapters 1 to 5 introduce the origin and development background of the concept "infectious microecology", as well as the composition, physiological and immunological functions of normal microflora. It also details the relationship between normal microbiology host shift, translocation and infection; between normal microflora variation, microecology disturbance and infection, especially nosocomial infection. Chapters 6 to 9 introduce the latest research and technology platforms for infectious microecology. Chapters 10 to 21 systematically dissect the characteristics of infectious microecology, and detail the prevention and treatment of diseases in various systems from the infectious microecology viewpoint. Chapter 22 introduces types and functions of the microecological modulator and its development. Chapter 23 is about the future development of infectious microecology.

Even with the careful contributions of our friends and colleagues, errors of source and misinterpretation may have found their way into the book, so suggestions for improvement will be gratefully received.

Lanjuan Li
Hangzhou, China
May, 2014

Lanjuan Li

Professor Lanjuan Li, a M.D. supervisor, academician of the CAE (Chinese Academy of Engineering) and chief physician of the First Affiliated Hospital of Zhejiang University, is a famous infectious disease specialist in China and she has been engaged in clinical, teaching and scientific research work in the field of infectious diseases for 40 years. She has made great achievements in microecology research by developing new infectious microecology theories that explore the occurrence, development, and outcome of infection from the microecological perspective, and by proposing new strategies for the prevention and control of infections. Prof. Li is also a pioneer in the study of artificial liver in China and established a special and effective artificial liver system (Li's artificial liver system [Li-ALS]), which was a momentous breakthrough in the treatment of severe hepatitis. In addition, she has undertaken more than 10 key research projects such as the national science fund project, National 863 Program and National 973 Program and so on. She has 22 authorized invention patents and has published more than 360 papers, over 150 collected by SCI academic journals including The Lancet and The New England Journal of Medicine. Served as the first completed, she has won several prizes including first prize of National Science and Technology Progress, and second prize of National Science and Technology Progress twice, first prizes of Science and Technology Progress of Zhejiang Province five times, and the second prize of Popularization and Application of Award in Colleges and Universities granted by the Ministry of Education. Presently, she holds the post of director of the Collaborative Innovation Center for Diagnosis and Treatment of Infectious Diseases, director of the State Key Laboratory for the Diagnosis of Infectious Diseases, the Leader of State Key Discipline Department of Internal Medicine (Infectious Diseases), and also director of the Zheijiang Infectious Disease Key Laboratory.

Meanwhile, she is also the chairman of the International Human Microbiome Consortium (IHMC), director of the Department of Bio-Medicine of the Ministry of Education, vice chairman of the Chinese Medical Association (CMA), vice president of the Chinese Health Information Society (CHIA), deputy chairman of the Chinese Society of Biomedical Engineering (CSBE), and director of the National Artificial Liver Training Base, Division Chief of the Microecology Branch of Chinese Preventive Medicine Association (CPMA), division chief of the Infectious Diseases Branch of Chinese Medical Doctor Association (CDMA), a vice-chairman member of the Third Cloud Computing Expert Committee of Chinese Institute of Electronics (CIE), Board trustees of the International Society for Apheresis (ISFA), President of Zhejiang Medical Association, editor-in-chief of the *Chinese Journal of Clinical Infectious*

Diseases, *Chinese Journal of Microecology and Zhejiang Medical Journal*, vice editor-in-chief of the *Chinese Journal of Infectious Diseases* and *International Journal of Epidemiology and Infectious Disease.* She has edited and published 28 monographs including the first edition of "*Artificial Liver*" and "*Infectious Microecology*" of China; and planned textbooks of Epidemiology. Furthermore, she also holds the position of vice chief engineer of the "Twelfth Five-Year Plan" — a major science-technology project titled "Prevention and Treatment of AIDS and Viral Hepatitis and Other Major Infectious Diseases" and is the expert team leader of "Field Study at Comprehensive Prevention and Control Demonstration Area". In 2010, she won the title of "National Excellent Science and Technology Workers" for her great contribution to the diagnosis and treatment of infectious disease.

Contents

Contributors

Hongqi Chen
The Sixth People's Hospital Affiliated to Shanghai Jiao Tong University, 600 Yishan Road, Shanghai, 200233, China

Jianing Chen
The State Key Laboratory for Diagnosis and Treatment of Infectious Diseases, the First Affiliated Hospital, School of Medicine, Zhejiang University; Collaborative Innovation Center for Diagnosis and Treatment of Infectious Diseases, Hangzhou, 310003, China

Nan Chen
The State Key Laboratory for Diagnosis and Treatment of Infectious Diseases, the First Affiliated Hospital, School of Medicine, Zhejiang University; Collaborative Innovation Center for Diagnosis and Treatment of Infectious Diseases, Hangzhou, 310003, China

Yanfei Chen
The State Key Laboratory for Diagnosis and Treatment of Infectious Diseases, the First Affiliated Hospital, School of Medicine, Zhejiang University; Collaborative Innovation Center for Diagnosis and Treatment of Infectious Diseases, Hangzhou, 310003, China

Yu Chen
The State Key Laboratory for Diagnosis and Treatment of Infectious Diseases, the First Affiliated Hospital, School of Medicine, Zhejiang University; Collaborative Innovation Center for Diagnosis and Treatment of Infectious Diseases, Hangzhou, 310003, China

Guangying Cui
The State Key Laboratory for Diagnosis and Treatment of Infectious Diseases, the First Affiliated Hospital, School of Medicine, Zhejiang University; Collaborative Innovation Center for Diagnosis and Treatment of Infectious Diseases, Hangzhou, 310003, China

Hongyan Diao
The State Key Laboratory for Diagnosis and Treatment of Infectious Diseases, the First Affiliated Hospital, School of Medicine, Zhejiang University; Collaborative Innovation Center for Diagnosis and Treatment of Infectious Diseases, Hangzhou, 310003, China

Hong Fang
Department of Dermatology, the First Affiliated Hospital, School of Medicine, Zhejiang University, Hangzhou, 310003, China

Hongchao He
Department of Urology, Ruijin Hospital, School of Medicine, Shanghai Jiao Tong University, Shanghai, 200025, China

Wei He
Department of Urology, Ruijin Hospital, School of Medicine, Shanghai Jiao Tong University, Shanghai, 200025, China

Xinjun Hu
The State Key Laboratory for Diagnosis and Treatment of Infectious Diseases, the First Affiliated Hospital, School of Medicine, Zhejiang University; Collaborative Innovation Center for Diagnosis and Treatment of Infectious Diseases, Hangzhou, 310003, China

Zhou Hua
The First Afflilated Hospital, School of Medicine, Zhejiang University, Hangzhou, 310003, China

Jian Huang
Department of Hematology, the First Affiliated Hospital, School of Medicine, Zhejiang University, Hangzhou, 310003, China

Feng Ji
Department of Gastroenterology, the First Affiliated Hospital, School of Medicine, Zhejiang University, Hangzhou, 310003, China

Jianwen Jiang
Key Lab of Combined Multi-Organ Transplantation, Ministry of Public Health, Department of Hepatobiliary and Pancreatic Surgery, the First Affiliated Hospital, School of Medicine, Zhejiang University, Hangzhou, 310003, China

Jie Jin
Department of Hematology, the First Affiliated Hospital, School of Medicine, Zhejiang University, Hangzhou, 310003, China

Konstantinos Krampis
J. Craig Venter Institute, 9704 Medical Center Drive, Rockville, Maryland, 20850, USA

Lanjuan Li
The State Key Laboratory for Diagnosis and Treatment of Infectious Diseases, the First Affiliated Hospital, School of Medicine, Zhejiang University; Collaborative Innovation Center for Diagnosis and Treatment of Infectious Diseases, Hangzhou, 310003, China

Tao Li
Department of Urology, Ruijin Hospital, School of Medicine, Shanghai Jiao Tong University, Shanghai, 200025, China

Yongtao Li
The State Key Laboratory for Diagnosis and Treatment of Infectious Diseases, the First Affiliated Hospital, School of Medicine, Zhejiang University; Collaborative Innovation Center for Diagnosis and Treatment of Infectious Diseases, Hangzhou, 310003, China

Zongxin Ling
The State Key Laboratory for Diagnosis and Treatment of Infectious Diseases, the First Affiliated Hospital, School of Medicine, Zhejiang University; Collaborative Innovation Center for Diagnosis and Treatment of Infectious Diseases, Hangzhou, 310003, China

Chenyu Mao
Department of Medical Oncology, the First Affiliated Hospital, School of Medicine, Zhejiang University, Hangzhou, 310003, China

Xiaohui Miao
The State Key Laboratory for Diagnosis and Treatment of Infectious Diseases, the First Affiliated Hospital, School of Medicine, Zhejiang University; Collaborative Innovation Center for Diagnosis and Treatment of Infectious Diseases, Hangzhou, 310003, China

Jason Miller
J. Craig Venter Institute, 9704 Medical Center Drive, Rockville, Maryland, 20850, USA

Karen E. Nelson
J. Craig Venter Institute, 9704 Medical Center Drive, Rockville, Maryland, 20850, USA

Huanlong Qin
Shanghai Tenth People's Hospital of Tongji University, 301 Yanchangzhong Road, Shanghai, 200072, China

Zhigang Ren
Key Lab of Combined Multi-Organ Transplantation, Ministry of Public Health, Department of Hepatobiliary and Pancreatic Surgery, the First Affiliated Hospital, School of Medicine, Zhejiang University, Hangzhou, 310003, China

Douglas B. Rusch
J. Craig Venter Institute, 9704 Medical Center Drive, Rockville, Maryland, 20850, USA

Sakaliya
Department of Urology, Ruijin Hospital, School of Medicine, Shanghai Jiao Tong University, Shanghai, 200025, China

Yuan Shao
Department of Urology, Ruijin Hospital, School of Medicine, Shanghai Jiao Tong University, Shanghai, 200025, China

Zhoujun Shen
Department of Urology, Ruijin Hospital, School of Medicine, Shanghai Jiao Tong University, Shanghai, 200025, China

Granger Sutton
J. Craig Venter Institute, 9704 Medical Center Drive, Rockville, Maryland, 20850, USA

Huiru Tang
State Key Laboratory of Magnetic Resonance and Atomic and Molecular Physics, Center for Biospectroscopy and Metabonomics, Wuhan Institute of Physics and Mathematics, Chinese Academy of Sciences, Wuhan, 430071, China

Andrey Tovchigrechko
J. Craig Venter Institute, 9704 Medical Center Drive, Rockville, Maryland, 20850, USA

Baohong Wang
The State Key Laboratory for Diagnosis and Treatment of Infectious Diseases, the First Affiliated Hospital, School of Medicine, Zhejiang University; Collaborative Innovation Center for Diagnosis and Treatment of Infectious Diseases, Hangzhou, 310003, China

Xianjin Wang
Department of Urology, Ruijin Hospital, School of Medicine, Shanghai Jiao Tong University, Shanghai, 200025, China

Yina Wang
Department of Dermatology, the First Affiliated Hospital, School of Medicine, Zhejiang University, Hangzhou, 310003, China

Yulan Wang
State Key Laboratory of Magnetic Resonance and Atomic and Molecular Physics, Center for Biospectroscopy and Metabonomics, Wuhan Institute of Physics and Mathematics, Chinese Academy of Sciences, Wuhan, 430071, China

Yingfeng Wei
The State Key Laboratory for Diagnosis and Treatment of Infectious Diseases, the First Affiliated Hospital, School of Medicine, Zhejiang University; Collaborative Innovation Center for Diagnosis and Treatment of Infectious Diseases, Hangzhou, 310003, China

Jian Wu
Hepatobiliary Pancreatic Surgery, the First Affiliated Hospital, School of Medicine, Zhejiang University, Hangzhou, 310003, China

Nanping Wu
The State Key Laboratory for Diagnosis and Treatment of Infectious Diseases, the First Affiliated Hospital, School of Medicine, Zhejiang University; Collaborative Innovation Center for Diagnosis and Treatment of Infectious Diseases, Hangzhou, 310003, China

Charlie Xiang
The State Key Laboratory for Diagnosis and Treatment of Infectious Diseases, the First Affiliated Hospital, School of Medicine, Zhejiang University; Collaborative Innovation Center for Diagnosis and Treatment of Infectious Diseases, Hangzhou, 310003, China

Xiaorong Xiao
State Key Laboratory of Oral Diseases, West China Hospital of Stomatology, Sichuan University, Chengdu, 610041, China

Yonghong Xiao
The State Key Laboratory for Diagnosis and Treatment of Infectious Diseases, the First Affiliated Hospital, School of Medicine, Zhejiang University; Collaborative Innovation Center for Diagnosis and Treatment of Infectious Diseases, Hangzhou, 310003, China

Ao Xie
Department of Microecology, School of Basic Medical Sciences, Dalian Medical University, Dalian, 116044, China

Liang Xu
Department of Gastroenterology, the First Affiliated Hospital, School of Medicine, Zhejiang University, Hangzhou, 310003, China

Nong Xu
Department of Medical Oncology, the First Affiliated Hospital, School of Medicine, Zhejiang University, Hangzhou, 310003, China

Jing Xue
State Key Laboratory of Oral Diseases, West China Hospital of Stomatology, Sichuan University, Chengdu, 610041, China

Jin Yang
The State Key Laboratory for Diagnosis and Treatment of Infectious Diseases, the First Affiliated Hospital, School of Medicine, Zhejiang University; Collaborative Innovation Center for Diagnosis and Treatment of Infectious Diseases, Hangzhou, 310003, China

Jingyun Yang
Medical College, Jiamusi University, Jiamusi, 154007, China

Xuesong Yang
Medical College, Jinan University, Guangzhou, 510632, China

Shibu Yooseph
J. Craig Venter Institute, 9704 Medical Center Drive, Rockville, Maryland, 20850, USA

Jieli Yuan
Department of Microecology, School of Basic Medical Sciences, Dalian Medical University, Dalian, 116044, China

Chenjing Zhang
Department of Urology, Ruijin Hospital, School of Medicine, Shanghai Jiao Tong University, Shanghai, 200025, China

Qiong Zhang
The State Key Laboratory for Diagnosis and Treatment of Infectious Diseases, the First Affiliated Hospital, School of Medicine, Zhejiang University; Collaborative Innovation Center for Diagnosis and Treatment of Infectious Diseases, Hangzhou, 310003, China

Kekai Zhao
The State Key Laboratory for Diagnosis and Treatment of Infectious Diseases, the First Affiliated Hospital, School of Medicine, Zhejiang University; Collaborative Innovation Center for Diagnosis and Treatment of Infectious Diseases, Hangzhou, 310003, China

Shusen Zheng
Key Lab of Combined Multi-Organ Transplantation, Ministry of Public Health, Department of Hepatobiliary and Pancreatic Surgery, the First Affiliated Hospital, School of Medicine, Zhejiang University; Collaborative Innovation Center for Diagnosis and Treatment of Infectious Diseases, Hangzhou, 310003, China

Shan Zhong
Department of Urology, Ruijin Hospital, School of Medicine, Shanghai Jiao Tong University, Shanghai, 200025, China

Jianying Zhou
The First Afflilated Hospital, School of Medicine, Zhejiang University, Hangzhou, 310003, China

Yu Zhu
Department of Urology, Ruijin Hospital, School of Medicine, Shanghai Jiao Tong University, Shanghai, 200025, China

1

Infectious Microecology

Lanjuan Li
The State Key Laboratory for Diagnosis and Treatment of Infectious Diseases, the First Affiliated Hospital, School of Medicine, Zhejiang University; Collaborative Innovation Center for Diagnosis and Treatment of Infectious Diseases, Hangzhou, 310003, China
E-mail: ljli@zju.edu.cn

Microecology, with information increasing rapidly, has been a new developing subject. In recent years, the discipline of microecology has become a significant subject studied by international scientists. Some scientists proposed the concept of "superorganism", which viewed the human body and the colonized microbiota as a whole and that this whole, composed of the human genome and human microbial genome, be called the "metagenome". Since 2005, many journals, such as *Science* and *Nature*, have reported that gut microbes provide nutrition for humans, regulate the intestinal epithelial development and induce the innate immunity. The functions of these microbes make them as important as an organ in the human body. Thus, the destruction of gut microbes means damage to health.

Infection is the physiological phenomenon which is induced by the interaction that is associated with a microorganism invading its host or macroorganism abnormally. It is not only a basic subject which mainly focuses on the theory of microecology and prevention measures in infection, but also an applied subject revealing the scientific information of infection mechanisms that are associated with the interaction between normal microbiota and the host.

This chapter provides a comprehensive overview of the production, principles, techniques, and prevention of Infectious Microecology.

1.1 Conception of Infectious Microecology

Infectious microecology is an emerging branch of microecology, which is regarded as the inevitable result of the development of microecology. The research into infectious microecology contains the disciplines of infection that are associated with the correlation between normal microflora and their host, which is aimed at investigating the internal rules of the interrelationship between microorganisms and macro organisms to prevent and cure infection and to improve medical fitness diathesis. In this section, we shall take infectious microecology as the focus, to review its conception.

1.1.1 Definition

Infectious microecology is a branch of microecology, which studies the occurrence, development and outcome of infection using the theory and methods of microecology and guides the transformation of infection in the direction of host health. It is conceived as an interaction between microecology, medical microbiology, immunology and infectious diseases. As microecology, it aims to investigate the internal rule of the interrelationship between microorganisms and microorganisms and the manifestation and consequences of the relationship.

1.1.2 Classification of Normal Microbiota a

The human normal microbiota consists of extremely complex microbiota combinations, which are the aggregate of microorganisms that reside on the surface and in deep layers of the skin, in the saliva and the oral mucosa, in the conjunctiva, and in the gastrointestinal tracts. They include bacteria, fungi, and archaea. Some of these organisms perform tasks that are useful for the human host.

1.1.2.1 Normal Microbiota

Microecology is a branch of ecology, which studies in particular the microbial flora's structure, function and the relationship between microorganisms and hosts [1]. It is formed from the long-term study of the symbiotic relationship between microorganisms and macroorganisms. The key focus of microecology is normal microbiota. The normal microbiota is the ecological structure formed by microbes and their hosts in the history of evolution, including bacteria, fungi, viruses and biologically active substances. Macroorganisms can also be divided into different levels, such as human, animal, plant and microorganisms, cell and molecular levels.

The development process of microecology is full of dynamic equilibrium and dynamic disorder. The balance of microecology is the foundation of health. It will cause sickness when the imbalance happens, including imbalance between different microorganisms, microorganisms and hosts, the microorganism-host entirety and the outside environment.

1.1.2.2 Function of Normal Microbiota

To its host, normal microbiota is not harmful but helpful and necessary. There are many kinds of normal flora systems in humans, including intestinal flora, oral nasopharyngeal flora, genitourinary tract flora, skin flora, *etc.* However, intestinal flora is the largest reservoir of human flora and we know more about the function of normal microbiota by studying it. Since 2005, *Science* and *Nature* have reported that bacteria in the gut fulfill a host of useful functions for humans, including providing nutrition, regulating the growth of intestinal epithelial cells and inducing congenital immunity. The metabolic activities performed by these bacteria resemble those of an "organ"; thus, the destruction of gut flora means damage to health.

The human body, consisting of about 10 trillion cells, carries about ten times as many microorganisms in the intestines, which weigh nearly 1.2 kg, as heavy as the human liver [2]. The physiological functions of intestinal flora are listed as follows.

(i) Antagonism: The normal flora in the human body forms a layer of biofilm barrier by adhering to a particular site, planting and breeding, which can prevent species that would harm the host from colonizing the gut through competitive exclusion and adjust the balance between human body and microbiota. Van der Wanij, a microbiologist from Holland, raised the concept of colonization resistance (CR) in the mid 1970s, which is defined as "the resistance to colonization of the alimentary canal by newly ingested microorganisms". The anaerobic bacteria plays a major role in the functioning of the CR. We proposed B/E (bifidobacterium species/enterobacter) in the 1990s and believes that B/E can indicate colonization resistance. In the sterile animal body or after the killing of all intestinal anaerobic bacteria by antibiotics, the colonization resistance decreases, which leads to a weakening of the inhibition ability by the invasion of pathogenic bacteria.

(ii) Immune stimulation: By producing an extensive immunological barrier, gut flora have a continuous and dynamic effect on the host's gut and systemic immune systems, which can be easily proven by aseptic animal experiments [3]. Sterile animals often have poorly developed immune systems, reduced immune cells, a weakened function, *etc.* Research shows that lactobacillus and bifidobacterium could be the effective probiotic for enhancing some aspects of immunity, not only the live microbe but also the bacterial lysate and fermentation solution. The bacteria stimulate the lymphoid tissue associated with the gut mucosa to produce antibodies to pathogens [4]. In recent years, it has been discovered that the segmented filamentous bacteria (SFB), which can promote the differentiation and maturation of intestinal TH17 cells, are closely related to human immunity.

(iii) Metabolic function: In this co-evolutionary procedure, the relationship between gut flora and humans is not merely commensal (a non-harmful coexistence), but rather a mutualistic relationship. The microorganisms perform a host of useful functions for human health. The normal gut microorganisms, such as *Bifidobacterium* and *Lactobacillus*, can synthesize a variety of essential vitamins and nutrients, such as vitamin B, vitamin K, nicotinic acid, pantothenic acidetc, which can also synthesize non-essential amino acid, like aspartic acid, alanine, valine and threonine, using protein residue. Bacteria turn carbohydrates that they ferment into materials that may help the body to absorb essential dietary minerals such as calcium, magnesium and iron. These materials are also in direct relation to diabetes, hypertension, hyperlipaemia, *etc.*

In addition, gut microbes have the effect of detoxification, increasing growth of intestinal epithelial cells, anti-tumor, *etc.*

In summary, the gut microbes provide health protection for the host, preventing invasion of harmful, pathogenic bacteria. To understand the relationship between the normal intestinal flora and the pathogens is also conducive to a better understanding of the normal intestinal flora. Take *Candida albicans* for example. Its infection often occurs in patients with immunodeficiency or long-term antibiotic treatment. Overgrowth of *Candida albicans* in the intestine, which access the bloodstream through the intestinal mucosa, causes systemic *Candida* infection. A large number of animal experiments show that the normal gut flora could prevent the *Candida* overgrowth. In sterile animals and the animal model taking antibiotics, the researchers found that the intestinal colonization force of *Candida albicans* increased, invaded the intestinal epithelium, and prolonged residence time. *Candida albicans* colonization in the intestine in turn also affects the structural changes in the intestinal microbial flora, reducing the diversity of the intestinal microbes. In addition, *Candida albicans* can cause local cytokine changes in the intestine and change the local mucosal immune status.

1.1.2.3 Classification of Normal Microbiota

Normal microflora is a complexity of microbiota. The study of these complexities requires not only the theories and methods of microbiology, but also the special techniques of microecology. According to microecology, normal microbiota can be classified into different categories with different classification methods, such as endogenic and exogenic flora (based on microbial origin); autochthonous and allochthonous flora (based on microbial living environment); resident and transient flora (based on microbial location).

Bacteria are the most various and numerous in the world of macroorganisms, and form the common pathogeny of infection. Therefore, understanding the classification of bacteria has essential theoretical and practical significance for the identification of bacteria, the diagnosis of disease and the prevention and treatment of bacterial infections. Bergey's classification system is commonly used internationally today. The book, *Berger's Manual of Systematic Bacteriology* edited by Boone *et al.* in 2001, classifies bacteria by Phylum, Class, Order, Family and Genus. This book reflects the change in bacterial classification, which transforms from a phenotype system developed by humans to a more natural classification system.

1.2 History of Infectious Microecology

Infectious microecology developed along with microecology and infectious disease when these two disciplines overlapped and merged together. The history of infectious microecology can be divided into four stages.

1.2.1 The Embryonic Stage of Microecology

Before the microscope was invented, humans had already known how to take advantage of the beneficial microorganisms to prevent the growth of harmful microbes although the microorganism could not be studied well in morphology. As early as 2500 B.C., an old covenant recorded the scene of making yogurt on a wall painting. In China in 540 A.D., during the Northern Wei Dynasty, Prof. Sixian Jia wrote a book entitled *Qi Min Yao Shu* in which mould was mentioned repeatedly and divided into beneficial organisms and harmful organisms. Prof. Sixian Jia described the beneficial organisms as colorful as green, red, white, black and yellow, while using the term "meiqmoengx" for obnoxious species. Similar descriptions could be found in the book *Exploitation of the Works of the Nature* written by Yingxing Song in 1637. It was a remarkable leap in the history of microecology or infectious microecology when the first microscope was invented by Antony van Leeuwenhoek in 1676. With the help of this innovation, the Dutch successfully viewed various microorganisms from wells, sewage and samples from humans such as sputum, dental plague, saliva and faeces. At the end of the 19th century, more and more microorganisms had been detected under the microscope such as coccus, bacillus, mycoplasma, spirillum and spirochete, and their morphology and microecology came into sight. The scientists also began to establish the cultural method for microorganisms.

Research into infectious microecology can be traced back to the end of the 19th century. In 1890, research about pathogeny of infant dyspepsia by Dr. Tisser in the Trouseau Children's Hospital of Paris found that intestinal flora of a breast-fed infant was different from that of an infant fed from a baby bottle. The difference in intestinal flora was thought to be one of the causes of the high incidence of diarrhea in the bottle-fed baby. This was of wide concern and made people realize that intestinal dysbiosis was the main cause of diarrhea. It was a big challenge to the academic trend of "one kind of bacterium, one disease" at that time. These discoveries by Tisser were the starting point of microecology.

At the end of the 19th century and the beginning of the 20th century, there were three epoch-making people in bacteriology, Louis Pasteur, Elie Metchnikoff and Robert Koch. These researchers and their theories had a profound impact on the development of microecology. Koch is a very famous etiologist. He promoted the discovery of the pathogen of anthrax, cholera, scarlet fever, diphtheria and pertussis. These findings strengthened the concept of bacteria as pathogen to the public. On the other hand, research into normal flora was neglected. As research continued, people had a greater understanding of the relationship between humans and bacteria. Fermentation theory, the relationship between bacteria and nutrition and the benefit of bacteria theory by Louis Pasteur have been widely accepted by contemporary science and carried forward. The theory of *Lactobacillus* antagonising intestinal spoilage organisms and the theory of health and long life by Metchnikoff are still universally acknowledged by the modern public, and yogurt has already become a worldwide favorite.

1.2.2 The Lag Phase of Microecology

The outbreak of the First World War led to the spread of infectious diseases all over the world during the early part of the 20th century and into the 1940s. Due to the huge disaster those pathogens caused, scientists focused on the research of pathogenetic microorganisms. Thus, microecology lagged behind. Moreover, the limitations inherent in this method of study (*in vitro* culture) became another obstacle in the development of microecology.

1.2.3 The Developmental Stage of Microecology

Penicillin was discovered in 1928 and was put into mass production in 1945, followed by Streptomycin. A large number of antibiotics were introduced to the world, which made many contributions to the struggle between humans and infectious diseases. The wide application of antibiotics also led to interest in the structure and the function of normal flora, which promoted progress in the study of microecology.

In the 1950s, aseptic animal breeding work was a success. The Lobao Laboratory of Notre Dame University, Indiana, USA, and the Sterile Animal Lab of the Swedish Karolinska Institute successfully bred sterile animals. Aseptic animal breeding, as a brand new method, gave strong impetus to research of microecology. Then gnotobiotic biology (Gnotobiology) emerged.

In the 1970s, Dr. Volker Rush from Germany first proposed the word "microecology". He stated that "microecology is ecology on a cellular or molecular level", and established the first microecology institute in the world, in Holborn. This was the first time that microecology appeared as a concept. And then, with the use of anaerobic culture, electron microscopy, cell and molecular biology in the research field of microecology, microecological research maderapid advancement. More and more people come to realize the huge complexity of human body microecology, and put forward the microecological balance theory corresponding to the balance of macroecology. Prof. Kang Bai presented a representative conception of microecological balance based on the summary of previous discourses: "Microecological balance is the dynamics of the physiological combination of the formation of normal microbiota and the host at different developmental stages during long-term historical evolution. This combination is defined in common macro conditions, ecological organizational structure of normal microbial balance and how the corresponding internal and surface ecological spatial structure of its host (human, animal and plant) interact as a physiological unity. The internal structure and state of unity is the microecological balance."

But every coin has two sides. The increasing use of antibiotics, dysbacteriosis, and superinfection lowers host resistance to infection, which causes microbiologists' anxiety in the wide use of antibiotics. Early in the 1950s, microbiologist Prof. Wei Xi once pointed out: "With the use of brilliant antibiotics, we should pay attention to the side effects they bring to us as human beings, the disruption of the normal microbial flora and the cause of dysbacteriosis."

1.2.4 The Establishment and Development of Infectious Microecology

Antibiotics has played an important role in anti-infection studies, but simple application of antibiotics cannot resolve infection problems such as the appearance of drug resistance (especially the multi-drug resistant bacteria), dysbacteriosis, super-infection and so on. In addition, with the negative consequences of the widespread use of immunosuppressive agents, chemoradiation, interventional therapy, *etc.*, we have gradually understood the helpfulness and harmfulness of microecology.

Today, with the wide use of antibiotics, microecological disturbance, and the increasing antibiotic resistance rate, the type of infection has evolved from exogenous and transmissible to endogenous and self-infectious. The protection and cure of infectious diseases are still great issues in the 21st century and we urgently need new theories and methods to instruct and research. With this background, our group first combined microecology with infection, and did in particular a series of research on the relationship between liver diseases and microecology. In 2001, we formally proposed the concept of infectious microecology. In 2002, we released the first publication at home and abroad on infectious microecology, with support from a number of famous experts. Microecology provides new theoretical evidence for the protection and control of infectious diseases, which makes us take a new look at the occurrence, development and transfer process of infection from a microecological viewpoint and updates the tactics of anti-infection. It then provides a new theoretical idea of infectious microecology, from pure sterilization to sterilization along with promoting bacteria. In recent years, the development of genomics transcriptomics, proteomics, metabonomics technology and implementation of the human yuan genome plan provide a variety of optional technologies for human microbes, which meanwhile provide a possibility for deep and comprehensive research on the relationship between the human microbial genome and health and disease. Infectious microecology is expected to make further and systemic development. On October 17, 2008, scientists from around the world in Heidelberg jointly founded the International Human Microbial Groups Union (IHMC), promoting global scientists to cooperate to research the relationship between the microbial genome and human health and disease, of which the human body microbial genome plan (HMP) from NIH and human intestinal macrogenome plan (MetaHIT) from Europe consisted of the core membership.

From 2007 to 2012, as the first scientific department, the national key-laboratory of diagnosis and treatment of infectious diseases in the First Affiliated Hospital of Zhejiang University, undertook a project on microecology and basic research into infection, which belonged to the national basic research program of China (“973” Program).

1.3 The Subject Characteristics of Infectious Microecology

Infectious microecology is an important part of ecology, which is a branch of life science, and its development should be adapted to the development of life science.

1.3.1 Subject Relations

With respect to disease and health the ecological balance and imbalance are controlled by three factors: agent, environment and host. There are several subjects to be studied in relation to these three factors, such as microbiology, immunology, physiology, pathology and clinical medicine. In microbiology the emphasis is on the agent, physiology, pathology, and immunology and in clinical medicine more emphases is put on the host. In microecology the emphasis is on the micro-environment between etiology and host. And micro-environment is the pivot of ecological balance which controls the balance and imbalance. All six subjects have a special focus while each has different degrees of connection to the other two factors.

Therefore, microecology is an independent methodology which has its own theoretical system and uniqueness. As a branch of microecology, infectious microecology makes use of the general rule of microecology to study infection as a new discipline with unique laws.

1.3.2 Subject Core

The core of infectious microecology is the interrelationship between normal microbes and the hosts, whereas pathogeny is focusing on related factor. Different from traditional ideas, microecology not only regards the pathogeny as a specific microrganism, but also considers the quality, quantity and position of microbes and the change in the host, and even regards these as important factors.

The core of infectious microecology is an interdependent and interactive dynamic process of the immunity and nutrition of the normal microbiology group under the control of genetic factors. Microbiology (etiology) and immunology are the auxiliary disciplines of this core research.

1.3.3 Infectious Microecology Is a Basic Subject

Infectious microecology has double disciplinarity. First of all, it is a basic subject. Microecology is the basic discipline of biological research, and an important research fields in biology, physiology, and biochemistry. Biology is about the subject of the structure, function, reproduction, classification and behavior of the individual organism, and physiology is the science of the transformation and manifestation of the physiological and biochemical levels, while microecology is the science of the law between microorganisms and macro biology. Microecology, like other disciplines, constitutes the basic subject of human, animal, plant and microorganism research.

Microecology is a branch of life science that researches the basic rules of organisms taking flora, immunity and nutrition as the center link. The subject occupies an important fundamental position in modern medicine. At present, scientists pay great attention to the research of human body micro-ecology all over the world. We began microecological researches of liver disease in 1994 and has found an intestinal microecological imbalance in patients with liver diseases [5], especially in those with severe hepatitis, and the imbalance degree is correlated with the severity of the liver disease. We consider that the microecological imbalance and the bacterial translocation (including metabolic products, such as endotoxine, bacteria) play important roles in the exacerbation of liver disease.

Through basic and clinical research, it is found that the intervention of probiotics could effectively improve the barrier function of the intestinal wall, reduce the damage to the liver brought about by the unbalanced microecology [6, 7]. A series of research projects has promoted the development of the study of the relationship between liver disease and microecology, and has enriched the study of the pathogenesis of severe hepatitis [8-11]. In 2007, microecology and infection research gained the support of the national key basic research program ("973" program), which was the first large international basic clinical research project on infectious microecology, and a full-scale introduction was made by *Science magazine*. In 2006, Prof. Steven Gill [12] and others at Stanford University explicitly pointed out that disrupting the body's microecology was damaging human health in *Science*. In December 2007, the Human Microbiome Project (HMP) [13], a 5-year project, initiated by the NIH in the United States, aimed to characterize the microbial communities found at several different sites on the human body, including nasal passages, oral cavities, skin, gastrointestinal and urogenital tractsand to analyze the role of these microbes in human health and disease. Now the sequencing of the human body microorganisms of 242 healthy people have been finished, which provides a preliminary outline of the microbial community and its function in healthy humans [14]. MetaHIT (http://www.metahit.eu/) is a project financed by the European Commission under the 7th FP program, which mainly focuses on the research of the composition of gut microbes in the human body, and the relationship between gut microbes and diseases such as inflammatory bowel disease, obesity, and so on.

Infectious microecology, which should be guided by the basic law, is to study and explore the principle and manifestation of infectious microecology. All life phenomena should not be separated from the law of microecology and many subsidiary subjects have gradually developed in this way such as immune microecology, nutritional microecology, gastrointestinal microecology, tumor microecology, and so on.

1.3.4 Infectious Microecology is an Applied Subject

Infectious microecology is also an applied subject. In recent years, infectious microecology has provided new ideas about biological science and laws, which have made significant changes and innovations in respect to medical concepts and medical patterns.

1.3.4.1 Innovation in the cognitive Pattern in Infectious Diseases

The traditional cognitive pattern of infectious diseases is based on an etiological pattern so as to study the reasons, manifestation, development and prognosis of infection. However, the theory that pathogen exposure may cause infection or not has also shown that infection may not lead to disease.

Microecology considers that the human and animal host carry a large number of normal microbial group. Under normal circumstances, the body's biological barrier is formed by normal microbes distributed in specific areas, such as the digestive tract, respiratory tract, genitourinary tract and skin, which resist the foreign invasion by pathogenic microbes.

Animal experiments found that normal intestinal flora showed a certain degree of antagonistic action against foreign agents. Taken into consideration was whether the body was infected, and whether the occurrence and development after infection depend not only on factors such as aggressivity and the toxins of the pathogenic microorganism, but also on the body's normal microbial equilibrium state.

1.3.4.2 Innovation in Biological Aetiology

The traditional biological aetiology says that infection is caused by a pathogenic microorganism. Yet microecology considers that infection is an important content of reciprocal transformation between the ecological balance and imbalance. The microorganism which causes infections is not necessarily pathogenic bacteria or pathogens microorganisms. Normal bacterial translocation may also be the cause of infection. The translocation of the normal bacteria group in the intestinal tract has attracted wide attention.

1.3.4.3 Innovation as a Means to Prevent Infection

The innovation of conception promotes the renovation of anti-infective treatment. Antibiotic treatment for infections has made remarkable progress, but the long-term use of broad-spectrum anti-bacterial drugs causes microecological imbalance and the formation and prevalence of drug-resistant strains. This will lead to uncontrolled infection or even fatal infection. At present, the microecological disorders and drug resistance have already caused global public health problems. Reasonable use of antibiotics is very necessary, while the use of microecological regulating agents in prevention and treatment of infection appears more necessary. Microecological regulating agents include probiotics and synbiotics. The supplementary use of probiotics aims at restoring the intestinal microecological balance, repairing the enterobacteria membrane barrier, improving intestinal planting resistance, inhibiting excessive growth of potential pathogenic bacteria, promoting the secreting mucins of intestinal epithelial cells, secreting sIgA of paneth cells regulating the body's immune function and so on [15,16].

1.4 The Classification of Infection

The classification of infection can be defined in different ways according to the origin of pathogens. Different types of infection have varied clinical manifestations, prognosis, therapeutic effects, so the further understanding of classification of infection helps us provide better prevention and therapeutic measures.

1.4.1 Types of Infection

The infections could be classified into endogenous and exogenous infection according to the origin of pathogens. Endogenous infections are those acquired from normal microbiota (the usually harmless microbes in our bodies) which might become pathogenic when people have compromised immune systems and are more vulnerable to opportunistic infections. Exogenic infection refers to what is caused by organisms not normally present in the body but entering from the environment.

When the moment of infection is taken into account, this can be classified into community acquired infection and nosocomial infection. Community-acquired infection is defined as an infection contracted outside of a health care setting or an infection present on admission. It is often distinguished from nosocomial, or hospital-acquired infection. The WHO defines nosocomial infections (or hospital acquired infections) as infections acquired by patients and their nursing members or hospital staff during patient care, which are not present or incubating at admission, no matter whether the infectious symptoms appear during the duration of hospitalization or not. The average infection rates among the different races and detection technology all over the world were 3.29% – 18.3% (2008). The "National Monitoring Net of Hospital Infection" was founded in order to control the situation of national nosocomial infection by the Ministry of Health of China. In 2003, a cross-sectional nationwide study involving 94,723 hospitalized patients from 159 tertiary hospitals demonstrated that the prevalence rate of nosocomial infections was 4.77%. The top five nosocomial infections included lower respiratory tract infections (33%), upper respiratory tract infections (18%), urinary tract infections (11%), surgical site infections (10%), and gastrointestinal infections (7%) [17-20]. With the widespread use of antibiotics, immunosuppressive agents and invasive technologies, nosocomial infection is becoming more predominant.

The pathogens causing nosocomial infections are as follows:

Bacterial resistance: Bacterial resistance is closely related to nosocomial infection. Since Fleming discovered penicillin in 1928, the application of antibiotic agents and measures has played a vital role in infection control. However, the virulence and drug resistance of bacteria were observed as getting stronger and stronger due to heavy use of antibiotics. These problems have been more significant since the 1980s. The intense use of antibiotic therapy in hospitalized patients promotes the emergence of drug-resistant strains of bacteria. The common multiple drug resistant bacteria and pan-resistant bacteria, such as methicillin-resistant *Staphylococcus aureus* (MRSA), extended-spectrum β-lactamase producing *Enterobacteriaceae*, multiple resistant *Pseudomonas aeruginosa*, pan resistant *Acinetobacter baumannii etc.*, have become the main cause of nosocomial infections. It is necessary to explore a totally new idea for fixing refractory infection due to bacterial resistance. Academician Lanjuan Li put forward a brand new theory about infectious microecology in 2001. She suggested that the infection could be studied using the definition and theory of microecology, so we can understand the infection better from occurrence to progression and outcome.

Mostly normal flora: In the evolution of humans, microorganisms were so intimately related to humans that they compromis a super-organism. In the light of microecology, the host, pathogen and environment are the identity. The definition of normal flora is not absolute. The normal flora can be pathogenic when they leave the regular habitat to another site. If the balance between the host and microorganisms is broken, the ability of the host to restrain the pathogen's colonization will be weakened. As a result, the opportunistic flora that used to be harmless might relocate to another site and cause nosocomial infection. Therefore, the most popular pathogens of nosocomial infection are those that are of normally low virulence, even avirulent. For example, the flora settled in the oral cavity or the intestine can cause hospital acquired pneumonia.

1.4.2 The Evolution of Infection

The traditional understanding of infection focuses on why we get infected, the manifestation, progression and prognosis of the infection. Microecology theory gives us a new insight and shows that infection is the result of the competition between ecological balance and imbalance. How we are infected and how the infection progresses are now thought to be even more dependent on the balance of normal microorganisms than on the invasiveness and virulence of pathogens. Infection is not only the result of invasion by the pathogen, but also the translocation of normal flora.

1.4.3 The Outcome of Infection

Because infection is not an all-or-nothing affair, individual variation in resistance to disease also results in different degrees of reaction to the infectious agents; *i.e.*, the outcome of the interaction between host and parasite is variable in individual cases.

(i) Occult infection: An infection first recognized by secondary manifestations, *e.g.* increased neutrophils in the circulation or fever of unknown origin, often caused by a bacterial infection in an obscure site, *e.g.* a subphrenic or other intra-abdominal region.

(ii) Apparent infection: An infection is readily recognized with the host showing symptoms typical of the disease.

(iii) Latent infection: A lingering infection that may lie dormant in the body for a period of time but may become active under certain conditions.

(iv) Carrier state: The continued presence of an organism (bacteria, virus, or parasite) in the body that does not cause symptoms, but is able to be transmitted and infect other persons.

1.5 Microecological Characteristics of Infection

Studies of human microflora in health and disease and during exposure to professional and ecological factors are traditional problems. The purpose of these studies is to develop methods and means for diagnosing and preventing human microbiocenosis disorders. Fundamental and applied research in cooperation with prophylactic and clinical institutions and departments yielded data contributing to the solution of many pressing problems in the prevention and diagnosis of infectious diseases.

1.5.1 Microecology Helps to Build and Maintain the Host's Immune Systems

The immune system is the most important weapon for animals to detect and fight against infections. It can kill pathogenic microbes and viruses. However, its development and maintenance depend on lots of bacteria as well. Defects in immune system development and function are serious problems for germfree mice to face [21]. These mice have immature lymphoid follicles and an enlarged cecum. Their plasma cells, $CD8^+$ intestinal epithelial cells, and αβ T-cell receptor intestinal epithelial cells are significantly reduced. The production or expression of immunolobulin A (IgA), anti-microbial peptides, major histocompatibility complex II, Toll-like receptor (TLR) 9, and interleukin (IL) 25 is greatly decreased as well. Their Peyer's patches are smaller than conventional animals, and their spleens and MLNs have depletion of lymphocyte zones. Many other defects were also observed, but not listed here.

The germfree animal, which is a type of utopian animal model, plays an important role in the study of human microbiota. By infecting germfree animals with specific microbes, we can investigate the mechanism of how the microbiota works in our body. It is reported that the inherent ingredient of the bacteroides fragilis—polysaccharide (PSA) can correct systemic T-cell deficiencies, direct lymphoid organogenesis, regulate the expression of the Treg cell, and prevent allergic diseases. Beneficial bacteria such as bifidobacteria and lactobacilli can resist pathogenic bacteria and opportunistic infection by reducing the PH value in the intestine and generating colonization resistance. The segmented filamentous bacteria (SFB) were the subject of many studies in recent years. Early research has demonstrated that SFB had an important role in the immunology of their hosts. However, the unculturability of SFB has made many studies of this organism's unique features impossible. Mono-colonized mice with SFB allow the opportunity to further exploit the organism's unique characteristics. Recent reports indicated that SFB could stimulate the development of germinal center reactions in the Peryer's patches, induce the expression of $CD4^+$ T-cell and IgA. The latest study has found that SFB can induce the expression of pro-inflammatory factors by comparing mice from differing sources. The mono-colonized mice with SFB had increased the expression of the Th17 cell and up-regulated the genes associated with inflammation.

1.5.2 Dynamic Balances between Microecology and Host

Prof. Kang Bai pointed out that the balances were formed in a long-term evolutionary process, in which the normal micro-population exists with their hosts in different developmental stages in a dynamic and physiological group. This combination between micro-population and their hosts is a physiogenic entity of normal interaction in the different levels of the ecological organization structure of the normal micro-population and ecosystem spatial structure of hosts (human, animal or plant) in the body or body surface under the effect of the common macro-environment. The internal structure and condition of this entity is the microecological balance.

For individuals, balances between host and microecology have been dynamically established from when the first bacterium enters the host, so that both of them can survive. Microbes and their host share nutrition, tolerate each other, and constitute a "super organism". These balances shift with diets, aging, and diseases *etc.* It is important to disclose which microbes are involved in the host's physiological function, how microbes in microecology respond to each other, and what kind of the host's genes work in their interactions with microbes.

The effect of the microbiota on human health has been best exemplified by studies of many diseases, such as obesity, diabetes, IBD, autoimmune disease and neoplasms. For example, obesity is generally believed to be closely associated with the gut microbiota. Germfree mice can be protected against the development of obesity, even obesity caused by the consumption of a western-style, high-fat, sugar-rich diet. The mechanisms producing this response may be identified with increased fatty-acid metabolism ascribed to the shortage of gut microbiota. In addition, research has also shown that obesity is associated with phylum-level alteration of the gut microbiota. Interestingly, it has been suggested that in humans and mice, higher Firmicutes, lower Bacteroidetes, and decreased bifidobacteria during infancy could very possibly induce obesity.

1.5.3 Many Infections are Correlated with Microecological Imbalance

In the medical field, infection has traditionally been defined as a pathophysiological process of the interaction between the host and invading pathogenic microorganisms. It emphasizes the relationship between pathogen and host. However, much recent research showed that infections were correlated with microecological imbalance. We found that intestinal flora in patients with chronic severe hepatitis was severely disturbed and gut microbiological colonization resistance was impaired [22]. In patients with liver cirrhosis, the proportion of phylum Bacteroidetes was significantly reduced, whereas Proteobacteria and Fusobacteria were highly enriched [23]. The composition of intestinal Bifidobacterium was deeply altered in CHB and HBV cirrhotic patients with a shift from beneficial species to opportunistic pathogens [24]. Iwasaki reported that microbiota could regulate immune defense against respiratory tract influenza A virus infection. Also, Josenhans reviewed the impact of pathogenic and commensal bacteria on IBD-like pathogenesis in mice infected models and summarized important recent developments [25]. David reviewed the relationship between viral infection and gut microbiota [26]. All these studies show that the microecological balance is the foundation of human health. If the microecological balance turns into microecological disturbance, infectious diseases will easily occur. Similarly, when infection happens, the microbiota will be impacted.

From the point of view of microecology, the occurrence, development and ending of any new infection are processes of dynamic imbalance to balance. Great efforts are needed to explain the exact mechanism of interaction between microbiota and the host.

1.6 The Microecological Mechanism of Infection

The occurrence, development and outcome of infection is the process of interaction between pathogen and host, including the pathogen invasion mechanism, the adhesion mechanism of bacteria to the epithelial cell membrane, the antagonism mechanism of the pathogen and other domestic bacteria, as well as the host immune-stimulating mechanism.

1.6.1 Occurrence of Infection

Factors associated with infection occurrence include the following:

1.6.1.1 Quantitative Change

Under normal circumstances, the normal flora which inhabits human skin and the epithelial cells of the cavity communicating with the outside world maintain a relative balance within the limits of fluctuations, and forms a biofilm barrier. The normal flora may be influenced by various factors, such as antibiotics. The reduction in the population of susceptible strains in the original habitat is accompanied by the increase in resistant bacteria, and the opportunistic pathogen can become the cause of the infectious bacteria. In the 1990s, We suggested that B/E value (*i.e.*, ratio of the number of bifidobacteria and Enterobacteriaceae) could be used as the indicators of "colonization resistance" of intestinal microflora of humans.

16.1.2 Qualitative Changes

In susceptible microbiota, the exogenous bacteria, such as passing bacteria and pathogens which cause communicable diseases, can grow and cause infection.

1.6.1.3 The Change of Position

Normal flora has its specific location, but translocation can occur under the influence of antibiotics, trauma and other factors. For example, *Escherichia coli* in the host colonic habitat is normal, but it can cause infection or disease when it translocates into the respiratory tract.

1.6.1.4 The Change in the Host

All kinds of macroscopic organisms have their own particular normal microbial flora. Some microorganisms can cause host disease or infection, if transferred to another macroscopic organism. For example, the normal bacteria of shellfish, *Vibrio*, transferred to the human body, can cause human diarrhea as well as *Campylobacter*, which is normal bacteria of birds, when transferred to humans, can cause gastrointestinal disorders.

These four factors are not isolated and play an important role in etiology, micro-environments and host interactions.

1.6.2 Development of Infection

The development of infections is a phase of infection following the stage of infection occurrence. The main mechanism of infection development is the process of etiology (pathogens) struggling with host immune cells. The etiology might predominate in the early development stage, and the host might do so in the late stage. The immune system is our main defense against infection, including congenital and acquired immunity. In addition to immune factors, the normal flora as a part of host resistance also gradually tends to balance on the new basis. The end stage will come after the peak of the etiological struggle against the host.

1.6.3 Outcome of Infection

The outcome of infection is the result of the etiological struggle against the host. From the population's perspective, the outcome of infection is an infection spectrum of continuous quantitative change combining recovery, illness and death.

Vaccines are also called immunizations because they can take advantage of our natural immune system's ability to prevent infectious illness by artificial infection to establish an artificial immune barrier, which may reduce mortality and morbidity, increase subclinical infection and prevent reinfection. Both the artificial immune barrier and immune barrier have proven the objective fact that infection is physiological, because no matter what kind of immunity, infection is acquired.

1.7 Prevention and Control of Infections Using Infectious Microecology Theories

The use of infectious microecology theories in prevention and treatment of infections has achieved great success in the past decade. This partly is due to a need to find alternatives to traditional therapies, such as antibiotics, as well as the lack of efficient treatment for infections. Although there are increasing reports of the efficacy of probiotics in the treatment of infections, the scientific basis is just beginning to be understood.

1.7.1 The Revolution in Infection Prevention and Control Strategy

The traditional strategy of infection prevention and control is based on killing or inhibiting pathogens. Antibiotics have been frequently and heavily used to combat the majority of bacterial infections. Since antibiotics (especially broad-spectrum ones) can kill beneficial microbes together with pathogens, these anti-infection processes usually lead to microecological imbalance and immune suppression, thus causing other diseases such as diarrhea and multiple organ dysfunction syndromes (MODS). Furthermore, antibiotic resistance often occurs and leads to infections that are difficult to treat and may become more serious diseases [26-29].

Applying theories and methods in microecology, We found that the imbalance in humans may result in severe infections or damage to organs, and then established a new concept of "infectious microecology". In 2001, based on this theory, we proposed preventing and controlling infection from the point of view of keeping or rebuilding the microecological balance. Antibiotics should be used properly and carefully. Before or after killing pathogens with antibiotics, good microbes should be promoted by microecological modulators to reduce the dysbiosis, and to maximize the benefits of the probiotic directly for competitive exclusion and immune stimulation.

1.7.2 Mechanisms for Preventing and Controlling Infections with Microecological Modulators

Microecological modulators are beneficial microbes and/or their growth promoters [30]. They can keep or rebuild the balance of microecology, thus preventing and treating diseases and improving health. Generally, microecology modulators prevent and control infections by the following several mechanisms. Beneficial microbes compete against pathogens by preventing the adhesion and colonization of pathogens through the production of inhibitory substances including acids and hydrogen peroxide and natural antibiotics. They can also enhance the mucous barrier function, simulate host antimicrobial defenses, sequester essential nutrients from invading pathogens, modulate the immune system, and reduce endotoxins and blood cholesterol levels [32].

1.7.3 Frequently Used Microecological Modulators

The microecological modulators can be broadly divided into probiotics, prebiotics and synbiotics. According to the currently adopted definition by FAO/WHO, probiotics are: "Live microorganisms which, when administered in adequate amounts, confer a health benefit on the host". *Bifidobacterium* and *Lactobacillus* are most commonly used as probiotics. Species from other bacterial genera such as *Streptococcus*, *Bacillus*, and *Enterococcus* are also used as probiotics. Also, nonbacterial microorganisms such as yeasts have also been used as probiotics for a long time.

Prebiotics are non-digestible food ingredients that stimulate the growth and (or) activity of probiotics. Most of them are non-digestible oligosaccharides such as fructooligosaccharides, galactooligosaccharides, xylooligosaccharides, and oligosaccharides. Usually, prebiotics increase the number and (or) activity of bifidobacteria and lactic acid bacteria, thus promoting calcium and other mineral absorption, modulating the immune system, reducing colorectal cancer risk, and relieving colon inflammation.

When probiotics and prebiotics are combined, they form a synbiotic. Sometimes, vitamins and trace elements are added. Fermented dairy products, such as yogurt, are considered synbiotic because they contain live bacteria and the fuel they need to thrive.

1.7.4 Prospects of Infection Prevention and Control Using Infectious Microecology Theories

In a sense, mammals have begun to struggle with infection and achieve dynamic equilibrium between host and microbes since they were born. Their embryos are in a sterile environment. After birth, mammals are in contact with the outside world containing a variety of microorganisms. Their immune and metabolism system becomes mature. Microecology and the host mutually adapt to each other. Thus, prevention and treatment of infection from the point of view of microecological balance is not only a deepening of cognition, but also the most fundamental approach and a direction for future development.

Recently, the rapid development of science and technology, especially the study of sterile animals [32] and gene knock mice [33] greatly promoted the research into the relationship between microecology and infectious diseases [34]. Some new microbes with significant biological functions such as segmented filamentous bacteria and Lachnospira have been discovered in the intestine. This supplies potential material for developing novel mocroecological modulators.

Mocroecological modulators such as probiotics have played an important role in infection related diseases including allergies, candidiasis, diarrhea, respiratory infection, inflammatory bowel disease, ulcerative colitis and Crohn's disease [35]. However, there are still many important scientific issues unresolved, such as the role and mechanisms of microecological imbalance in the chronic [36] and severe process of infectious diseases, and also the mechanism of microecological reconstruction [37, 38]. At the same time, the developmental speed of the existing microecological modulators for infection prevention and control is lagging far behind as far as relevant basic research is concerned. Therefore, it is important to pay great attention to the basic research into infectious microecology, as well as to new strategies, new methods, new technologies and new drugs for microecological therapy of infections.

References

[1] Serban D E. The gut microbiota in the metagenomics era: Sometimes a friend, sometimes a foe. Roum Arch Microbiol Immunol, 2011, 70(3):134-140.

[2] Rao Y, Lingamneni B, Reddy D. Probiotics in oral health – a review. JNJ Dent Assoc, 2012, 83(2):28-32.

[3] Hooper L V, Littman DR, Macpherson AJ. Interactions between the microbiota and the immune system. Science, 2012, 8; 336(6086):1268-1273.

[4] Walker J A, McKenzie A. Innate lymphoid cells in the airways. Eur J Immunol, 2012, 42(6):1368-1374.

[5] Wu Z W, Li L J, Ma W H, *et al.* The new means of gut microbial colonization-B/E value. Zhejiang Prev Med, 2000, 12: 4-5.

[6] Mangell P, Nejdfors P, Wang M, *et al.* Lactobacillus plantarum 299v inhibits *Escherichia coli*-induced intestinal permeability. Dig Dis Sci, 2002, 47:511-516.

[7] Chiva M, Soriano G, Rochat I, *et al.* Effect of Lactobacillus johnsonii La1 and antioxidants on intestinal flora and bacterial translocation in rats with experimental cirrhosis. J Hepatol, 2002, 37:456-462.

[8] Chen Y F, Yang F L, Lu H F. Characterization of fecal microbial

Communities in Patients with Liver Cirrhosis. Hepatology, 2011, 54:562-572.
[9] Li Y T, Wang L, Chen Y, *et al*. Effects of gut microflora on hepatic damage after acute liver injury in rats. The Journal of TRAUMA, 2010, 68: 76-83
[10] Wu Z W, Lu H F, Wu J P, *et al*. Assessment of the fecal *Lactobacilli* population in patients with hepatitis B virus-related decompensated cirrhosis and hepatitis B cirrhosis treated with liver transplant. Microb Ecol, 2011, 63:929-937.
[11] Xu M, Wang B H, Fu Y Q, *et al*. Changes of fecal *Bifidobacterium* species in adult patients with hepatitis B virus-induced chronic liver disease. Microb Ecol, 2011, 63:304-313.
[12] Gill S R, Pop M, Deboy R T, *et al*. Metagenomic analysis of the human distal gut microbiome. Science, 2006, 312:1355-1359.
[13] NIH HMP Working Group, Peterson J, Garges S, *et al*. The NIH human microbiome project. Genome Res, 2009, 19:2317-2323.
[14] Qin J, Li R, Raes J, *et al*. A human gut microbial gene catalogue established by metagenomic sequencing. Nature, 2010, 4:59-65.
[15] Prasad S, Dhiman RK, Duseja A, *et al*. Lactulose improves cognitive functions and health-related quality of life in patients with cirrhosis who have minimal hepatic encephalopathy. Hepatology, 2007, 45:549-559.
[16] Kale R A, Gupta R K, Saraswat V A, *et al*. Demonstration of interstitial cerebral edema with diffusion tensor MR imaging in type C hepatic encephalopathy. Hepatology, 2006, 43:698-706.
[17] Wu A H, Ren N, Wen X M, *et al*. One-day prevalence survey of nosocomial infection in 159 hospitals. Chin J Infect Control, 2005, 4:12-21.
[18] Wu A H, Ren N, Wen X M, *et al*. A nosocomial infection point-prevalence survey: results and analysis of 193 hospitals in China in 2001. Chin J Nosocomiol, 2002, 12:561-569.
[19] Lee M K, Chiu C S, Chow V C, *et al*. Prevalence of hospital infection and antibiotic use at a university medical center in Hong Kong. J Hosp Infect, 2007, 65:341-347
[20] CDC. NNIS.National nosocomial infections surveillance (NNIS) system report, data summary from January 1992 to June 2003. Am J Infect Cont, 2003, 31:481-498.
[21] Reading N C, Kasper D L. The starting lineup: Key microbial players in intestinal immunity and homeostasis. Front Microbiol, 2011, 2:148.
[22] Li L, Wu Z, Ma W, Yu Y, Chen Y. Changes in intestinal microflora in patients with chronic severe hepatitis. Chin Med J (Engl), 2001, 114: 869-872.
[23] Chen Y, Yang F, Lu H, *et al*. Characterization of fecal microbial communities in patients with liver cirrhosis. Hepatology, 2011, 54: 562-572.
[24] Xu M, Wang B, Fu Y, *et al*. Changes of fecal Bifidobacterium species in adult patients with hepatitis B virus-induced chronic liver disease. Microb Ecol, 2012, 63:304-313.
[25] Nell S, Suerbaum S, Josenhans C. The impact of the microbiota on the pathogenesis of IBD: lessons from mouse infection models. Nat Rev Microbiol, 2010, 8:564-577.
[26] David R. Viral infection. The gut microbiota: Friend or foe? Nat Rev

Microbiol, 2011, 9:831.
[27] Britton R A, Young V B. Interaction between the intestinal microbiota and host in Clostridium difficile colonization resistance. Trends Microbiol, 2012, 20:313-319.
[28] DuPont AW, DuPont H L. The intestinal microbiota and chronic disorders of the gut. Nat Rev Gastroenterol Hepatol, 2011, 8:523-531.
[29] Ehlers S, Kaufmann S H. Infection, inflammation, and chronic diseases: consequences of a modern lifestyle. Trends Immunol, 2010, 31:184-190.
[30] Isolauri E, Kirjavainen P V, Salminen S. Probiotics: A role in the treatment of intestinal infection and inflammation? Gut, 2002, 50 (Suppl 3):III54-59.
[31] Antoine J M. Probiotics: Beneficial factors of the defence system. Proc Nutr Soc, 2010, 69:429-433.
[32] Reid G, Younes J A, Van der Mei HC, *et al*. Microbiota restoration: natural and supplemented recovery of human microbial communities. Nat Rev Microbiol, 2011, 9:27-38.
[33] Rosenberg E, Koren O, Reshef L, *et al*. The role of microorganisms in coral health, disease and evolution. Nat Rev Microbiol, 2007, 5:355-362.
[34] Stecher B, Hardt W D. The role of microbiota in infectious disease. Trends Microbiol, 2008, 16:107-114.
[35] Nell S, Suerbaum S, Josenhans C. The impact of the microbiota on the pathogenesis of IBD: Lessons from mouse infection models. Nat Rev Microbiol, 2010, 8:564-577.
[36] Chung H, Pamp S J, Hill J A, *et al*. Gut immune maturation depends on colonization with a host-specific microbiota. Cell, 2012,149:1578-1593.
[37] Littman D R, Pamer E G. Role of the commensal microbiota in normal and pathogenic host immune responses. Cell Host Microbe, 2011, 10:311-323.
[38] Lu H, He J, Wu Z, et al. Assessment of microbiome variation during the perioperative period in liver transplant patients: A retrospective analysis. Microb Ecol, 2013, 65: 781-791.

2

Human Microbiota and Its Function

Yongtao Li
The State Key Laboratory for Diagnosis and Treatment of Infectious Diseases, the First Affiliated Hospital, School of Medicine, Zhejiang University, Hangzhou, 310003, China
E-mail: ytli95@163.com

The human body is home to complex communities of microorganisms. Their total number is estimated to be 10^{14}; 10 times the number of human cells per individual [1]. These microbial communities are found on our skin, in the mouth, nose, ears, vagina, and in the intestinal tract. Similar to environmental sources in which microbes are found, such as seawater and soil, the human body could be considered as an ecosystem consisting of different niches, or a meta-community consisting of many local communities. Each anatomical site has its own physiochemical characteristics, and each location is occupied with a specialized set of microbes. The majority of the human-associated microbes and the largest diversity are found in the intestinal tract, where microbial abundance increases from the stomach to the colon, with the highest number of microbes found in stools (10^{11} per mL). This complex ecosystem consists of bacteria, archaea, yeasts and other eukaryotes [1, 2].

Colonisation of the gastrointestinal tract of newborn infants starts immediately after birth and occurs within a few days. Initially, the type of delivery (passage through the birth canal versus caesarean section) and the type of diet (breast versus formula feeding) might affect the colonisation pattern [3]. Other environmental factors also have a major role since differences exist between infants born in developed countries and those born in developing countries, and between infants from different hospital wards [4, 5]. Pioneer bacteria can modulate expression of genes in host epithelial cells, thus creating a favorable habitat for themselves, and can prevent growth of other bacteria introduced later in the ecosystem [2].The initial colonisation is therefore very relevant to the final composition of the permanent flora in adults. Conventional bacteriological analysis of faecal flora requires meticulous techniques for cultivation of bacteria on various growth media and an array of methods for taxonomic identification of the isolates. Results of such studies have shown that anaerobic bacteria outnumber aerobic bacteria by a factor of 100 – 1,000 [6]. The genera bacteroides, bifidobacterium, eubacterium, clostridium, peptococcus, peptostreptococcus, and ruminococcus are predominant in human beings, whereas aerobes (facultative anaerobes) such as escherichia, enterobacter, enterococcus, klebsiella, lactobacillus, proteus, *etc.*, are among the subdominant genera [7]. Every individual has several hundreds of species belonging to these genera, with a particular combination of predominant species that is distinct from that found in other individuals. The species vary greatly between individuals. The composition of the individual's flora can fluctuate under some circumstances, for instance acute diarrhoeal illnesses, antibiotic treatment, or to lesser extent induced by dietary interventions, but individuals' flora composition pattern usually remains constant [8]. Several bacteria that can be seen by direct microscopic examination of diluted faecal specimens cannot be grown in culture media. Unicellular organisms need biodiversity for growth. Thus, 40% – 80% of the total microscopic counts are not recoverable by culture, although estimates vary between individuals and between studies [9]. Molecular biological procedures now can also be used to investigate the microbial ecology in the colon without use of cultures. Results of an analysis of bacterial genes in human faeces showed that many DNA sequences correspond to previously undescribed microorganisms, and some data suggest that every individual has unique strains of bacteria. Quantitative analysis of faecal bacteria shows important differences between individuals and over time within the same individual that are not always detectable by conventional culture techniques. Molecular procedures have shown that aerobes, including *Escherichia coli*, enterococci, and lactobacilli, achieve very high densities and metabolic activity in the human caecum, since 50% of total bacteria ribosomal RNA in caecal contents correspond to these species. By contrast, these species account for only 7% of bacteria ribosomal RNA in faecal samples. Such species could have an important role in caecal fermentations [10–12].

Use of animals bred under germ-free conditions has provided important information about the effect of the microbial community of the gut on host physiology and pathology. Evidence obtained through such studies suggests that microflora have important and specific metabolic, trophic and protective functions (panel) [7].

2.1 Metabolic Functions

A major metabolic function of colonic microflora is the fermentation of non-digestible dietary residue and endogenous mucus produced by the epithelia.

Gene diversity in the microbial community provides various enzymes and biochemical pathways that are distinct from the host's own constitutive resources. Overall outcomes of this complex metabolic activity are recovery of metabolic energy and absorbable substrates for the host, and supply of energy and nutritive products for bacterial growth and proliferation. Fermentation of carbohydrates is a major source of energy in the colon [13]. Non-digestible carbohydrates include large polysaccharides (resistant starches, cellulose, hemicellulose, pectins, and gums), some oligosaccharides that escape digestion, and unabsorbed sugars and alcohols. The metabolic endpoint is generation of short-chain fatty acids. Anaerobic metabolism of peptides and proteins (putrefaction) by the microflora also produces short-chain fatty acids but, at the same time, it generates a series of potentially toxic substances including ammonia, amines, phenols, thiols and indols. Available proteins include elastin and collagen from dietary sources, pancreatic enzymes, sloughed epithelial cells and lysed bacteria [14, 15]. Substrate availability in the human adult colon is about 20 – 60 g of carbohydrates and 5 – 20 g of protein per day. In the caecum and right colon, fermentation is very intense with high production of short-chain fatty acids, an acidic pH (5 – 6), and rapid bacterial growth [16]. By contrast, the substrate in the left or distal colon is less available, the pH is close to neutral, putrefactive processes become quantitatively more important, and bacterial populations are close to static. Colonic microorganisms also play a part in vitamin synthesis and in absorption of calcium, magnesium and iron. Absorption of ions in the caecum is improved by carbohydrate fermentation and production of short-chain fatty acids, especially acetate, propionate, and butyrate. All of these fatty acids have important functions in host physiology. Acetate and propionate are found in portal blood and are eventually metabolised by the liver (propionate) or peripheral tissues, particularly muscle (acetate). Acetate and propionate might also have a role as modulators of glucose metabolism: absorption of these short-chain fatty acids would result in lower glycaemic responses to oral glucose or a standard meal–a response consistent with an ameliorated sensitivity to insulin [17, 18]. In fact, foods with a high proportion of non-digestible carbohydrates all have a low glycaemic index. However, results of one study showed no effect of colonic fermentation of carbohydrates on insulin resistance. Zeneng Wang reported that gut flora metabolism of phosphatidylcholine promotes cardiovascular disease [19]. Their metabolomics studies identify elevated plasma levels of specific analytes that are associated with increased risk of CVD [19].

In order to benefit the host, microflora is bound to participate in the trophic structure: decomposition of nutrients provides energy for the host substrate. Body weight control depends on the long-term mechanism for fine regulation. Over a long period of time (several years), the body's more than the daily consumption of less than 1% of the energy intake can lead to weight gain and metabolic confusion [20]. Therefore, all food affecting the energy mechanism is actually involved in the maintenance of the body's weight balance.

Backhed first discovered intestinal flora to be an environmental factor to regulate the body's energy reserves [21]. When ordinary mice were compared with germ-free mice, the body fat content was higher by 40%, and the gonad fat content was higher by 47%. Interestingly, they also found that ordinary mice fed more than germ-free mice. At the same time, researchers found the weight of sterile mice showed an increase of 60% after colonization of ordinary mouse intestinal flora within two weeks. The authors analyzed three possible mechanisms for gaining weight: firstly, the glucose absorption in the intestine increased; secondly, the energy of non-digestible food ingredients was absorbed (such as the short chain fatty acids by fermentation); thirdly, concomitant occurrence of hyperglycemia and hyperinsulinemia were two key factors to promote the body fat formation. It was interesting that the lipoprotein lipase activity of sterile mice, after normal mice intestinal flora colonization, increased, which prompted the release of fatty acids from triglycerides, lipoprotein complexes, following uptake and utilization by muscle or adipose tissue. The authors pointed out that the increase was the result of suppressing the fasting induced obesity factor (FIAF) in the intestine. Because FIAF can inhibit lipoprotein lipase activity, after colonization in germ-free mice the reduction of FIAF will lead to the accumulation of triglycerides in adipose tissue.

In 2006, Ley *et al.* using 12 obese and non-obese groups, compared the differences between the obese and non-obese distal colonic flora, and found that Bacteroides were significantly reduced, compared with the obese group, in the distal colon, and hard wall strains increased significantly. Moreover, in the obese group showing weight loss after one year, the proportion of hard-walled bacteria became similar to that in the non-obese group [22, 23]. This research from the study of animals and humans concluded that in the obese body there appeared a change in the intestinal flora. However, did the change in the intestinal flora lead to obesity or not? To test this, Ley *et al.* undertook another experiment. The gut floras of obese mice and non-obese mice were planted in two groups of non-obese sterile mice intestines respectively. Two weeks later they found the weight of the sterile mice with the colonization of the gut flora of obese mice had risen, compared to another group of mice [24], which fully explained that the increase in body weight of mice can be attributed to the different intestinal flora. In 2008, Kalliomaki *et al.* reported that in order to clarify whether early intestinal flora related to childhood obesity, they undertook a prospective study of 25 overweight children aged 7, and found the intestinal *Bifidobacterium* reduced and *Enterococcus* increased, compared with normal-weight children of the same age, indicating that the intestinal flora disturbance occurred before the weight gain [25].

However, the view that intestinal flora promotes the changes in body weight was in dispute. Because just as the findings by Ley and other scholars in animal experiments, it was unclear whether the subtle changes in the energy absorbing body could cause significant weight change. In fact, Bajzer pointed out that in Ley's research, when two groups of non-obese sterile mice respectively colonized the intestinal flora of normal obese mice and ordinary non-obese mice, the difference in weight gain was not very obvious, and this difference could also be considered to be the difference in food intake, not necessarily a different energy absorption efficiency [26]. Cani's studies also showed that foods rich in non-digestible but fermentable fiber could reduce body weight and diabetes severity [27]. Fibrous food through full fermentation in the cecum and colon promoted some flora which used fiber as the source of energy, so that colon bacteria increased [28]. This research suggested that some characteristic changes of the intestinal flora had a beneficial effect on the host, even if the characteristics of this flora were not yet fully understood, but do not support the hypothesis that fiber or polysaccharide food digested by intestinal flora increased body weight by increasing energy supply.

2.2 Trophic Functions

Epithelial cell growth and differentiation – Possibly the most important role of short-chain fatty acids on colonic physiology is their trophic effect on the intestinal epithelium. The rate of production of crypt cells is reduced in the colon of rats bred in germ-free environments, and their crypts contain fewer cells than those of rats colonised by conventional flora, which suggests that intraluminal bacteria affect cell proliferation in the colon. Differentiation of epithelial cells is greatly affected by interaction with resident microorganisms. All three major short-chain fatty acids stimulate epithelial cell proliferation and differentiation in the large and small bowel *in vivo*. However, butyrate inhibits cell proliferation and stimulates cell differentiation in epithelial cell lines of neoplastic origin *in vitro*. Moreover, butyrate promotes reversion of cells from neoplastic to non-neoplastic phenotypes. A role for short-chain fatty acids in prevention of some human pathological states such as chronic ulcerative colitis and colonic carcinogenesis has been long suspected although, admittedly, conclusive evidence is still lacking [29, 30].

2.3 Interaction between Gut Bacteria and Host Immunity

The intestinal mucosa is the main interface between the immune system and the external environment. Thus, it is not surprising that gut-associated lymphoid tissues contain the largest pool of immunocompetent cells in the human body. The dialogue between host and bacteria at the mucosal interface seems to play a part in the development of a competent immune system. Animals bred in a germfree environment have low densities of lymphoid cells in the gut mucosa, specialised follicle structures are small, and circulating concentrations of immunoglobulins in the blood are low. Microbial colonisation of the gastrointestinal tract affects the composition of gut associated lymphoid tissue. Immediately after exposure to luminal microbes, the number of intraepithelial lymphocytes expands great, germinal centers with immunoglobulin producing cells arise rapidly in follicles and in the lamina propria, and concentrations of immunoglobulin increase substantially in serum. In mice and rats, a non-pathogenic and non-culturable segmented filamentous bacterium that preferentially attaches to Peyer's patch epithelium stimulates development of mucosal immune architecture and function. Many and diverse interactions between microbes, epithelium and gut-associated lymphoid tissue are involved in modeling the memory mechanisms of systemic immunity. For instance, flora has been implicated in oral tolerance [31]. The systemic response to a specific antigen can be abrogated after ingesting the same antigen. This effect persists for several months in mice with conventional flora, whereas in germfree mice systemic unresponsiveness persists for only a few days. After oral administration of ovoalbumin, germ-free mice maintain a Th2 immune response and produce IgE antibodies against ovoalbumin. Interestingly, the abnormality can be corrected by reconstitution of conventional flora, but this procedure is only effective in neonates and not in older mice. The interaction between gut-associated lymphoid tissue and flora early in life seems to be crucial for appropriate development of complex mucosal and systemic immunoregulatory circuits. In adults, immunity may be constantly reshaped by persistent interactions between the host and its bacteria that take place in the gut. Commensal organisms try to circumvent the immune response. For instance, *Bacteroides fragilis*, a predominant species in the human colon, can change its surface antigenicity by producing distinct capsular polysaccharides. Surface diversity seems to allow the organism to escape immunosurveillance and maintain an ecological niche of predominance in the intestinal tract. However, host defences adapt and keep an active control of bacterial growth. The immune response to microbes relies on innate and adaptive components, such as immunoglobulin secretion. Most bacteria in human faeces are coated with specific IgA units [32, 33]. Innate responses are mediated not only by white blood cells such as neutrophils and macrophages that can phagocytose and kill pathogens, but also by intestinal epithelial cells, which coordinate host responses by synthesising a wide range of inflammatory mediators and transmitting signals to underlying cells in the mucosa. The innate immune system has to discriminate between potential pathogens from commensal bacteria, with use of a restricted number of preformed receptors. Mammalian cells express a series of toll-like receptors, which recognise conserved motifs on bacteria that are not found in higher eukaryotes. The system allows immediate recognition of bacteria to rapidly respond to an eventual challenge. For example, incubation of nonpathogenic bacteria with inflamed human intestinal mucosa elicits different types of immediate cytokine responses, which are transduced to the underlying tissue and promote changes in the phenotype of lamina propria lymphocytes [34, 35].

2.4 Protective Functions: The Barrier Effect

Resident bacteria are a crucial line of resistance to colonisation by exogenous microbes and, therefore, are highly relevant in prevention of invasion of tissues by pathogens. Germ-free animals are very susceptible to infection. Colonisation resistance also applies to opportunistic bacteria that are present in the gut but have restricted growth [36, 37]. The equilibrium between species of resident bacteria provides stability in the microbial population within the same individual under normal conditions. However, use of antibiotics can disrupt the ecological balance and allow overgrowth of species with potential pathogenicity such as toxigenic Clostridium difficile, associated with pseudomembranous colitis. Several mechanisms have been implicated in the barrier effect. *In vitro*, bacteria compete for attachment sites in the brush border of intestinal epithelial cells. Adherent non-pathogenic bacteria can prevent attachment and subsequent entry of pathogen enteroinvasive bacteria into the epithelial cells. Furthermore, bacteria compete for nutrient availability in ecological niches and maintain their collective habitat by administering and consuming all resources – *e.g.*, in the gnotobiotic mouse monocolonised with Bacteroides thetaiotaomicron. The host actively provides a nutrient that the bacterium needs, and the bacterium actively indicates how much it needs to the host. This symbiotic relationship prevents unwanted overproduction of the nutrient, which would favor intrusion of microbial competitors with potential pathogenicity for the host. Finally, bacteria can inhibit the growth of their competitors by producing antimicrobial substances called bacteriocins. The ability to synthesise bacteriocins is widely distributed among microbial collectivities of the gastrointestinal tract. The host can control production of such substances since most of them are protein compounds degradable by digestive proteases [38]. Thus, the role of bacteriocins is mainly restricted to localised niches.

References

[1] Bengmark S. Ecological control of the gastroinstinal tract: the role of probiotic flora. Gut, 1998, 42: 2-7.

[2] Elisabeth M B. Composition and function of the human-associated microbiota Nutr Rev, 2009, 67(suppl 2): 164-171.

[3] Gronlund M M, Lehtonen O P, Eerola E, *et al.* Fecal microflora in healthy infants born by different methods of delivery:permanent changes in intestinal flora after cesarean delivery. J Pediatr Gastroenterol Nutr, 1999, 28:19-25.

[4] Backhed F, Ley R E, Justin L, *et al.* Host-bacterial mutualism in the human intestine. Science, 2005, 307: 1915-1920.

[5] Zoetendal E G Akkermans ADL, Akkermans-van vliet WM, *et al.* The host genotype affects the bacterial community in the human gastro-intestinal tract. Microb. Ecol Health Dis, 2001, 13: 129-134.

[6] Simon G L, Gorbach S L. Intestinal flora in health and disease. Gastroenterology, 1984, 86: 174-193.

[7] Guarner F, Malagelada J R. Gut flora in health and disease, Lancet, 2003, 361: 512-519.

[8] Palmer C, Bik E M, Digiulio D B, *et al.* Development of the human infant intestinal microbiota. PLoS Biol, 2007, 5: e177.

[9] Suau A, Bonnet R, Sutren M, *et al.* Direct rDNA community analysis reveals a myriad of novel bacterial lineages within the human gut. Appl Environ Microbiol, 1999, 65: 4799-4807.

[10] Tannock G W. Molecular assessment of intestinal microflora. Am J Clin Nutr, 2001, 73 (suppl): S410-S414.
[11] Kimura K, McCartney A I, McConnell M A, *et al.* Analysis of fecal populations of bifidobacteria and lactobacilli and investigation of the immunological responses of their human hosts to the predominant strains. Appl Environ Microbiol, 1997, 63: 3394-3398.
[12] Sghir A, Gramet G, Suau A, *et al.* Quantification of bacterial groups within human fecal flora by oligonucleotide probe hybridization. Appl Environ Microbiol, 2000, 66: 2263-2266.
[13] Cummings J H, Beatty E R, Kingman S M, *et al.* Digestion and physiological properties of resistant starch in the human large bowel. Br J Nutr, 1996, 75: 733-747.
[14] Smith E A, Macfarlane G T. Enumeration of human colonic bacteria producing phenolic and indolic compounds: Effects of pH, carbohydrate availability and retention time on dissimilatory aromatic amino acid metabolism. J Appl Bacteriol, 1996, 81: 288-302.
[15] Fallingborg J. Intraluminal pH of the human gastrointestinal tract. Dan Med Bull 1999, 46: 183-196.
[16] Hill M J. Intestinal flora and endogenous vitamin synthesis. Eur J Cancer Prev 1997, 6 (suppl): S43-S45.
[17] Miyazawa E, Iwabuchi A, Yoshida T. Phytate breakdown and apparent absorption of phosphorus, calcium and magnesium in germfree and conventionalized rats. Nutr Res, 1996, 16: 603-613.
[18] Younes H, Coudray C, Bellanger J, *et al.* Effects of two fermentable carbohydrates (inulin and resistant starch) and their combination on calcium and magnesium balance in rats. Br J Nutr, 2001, 86: 479-485.
[19] Wang Z N, Klipfell E, Brian J, *et al.* Gut flora metabolism of phosphatidylcholine promotes cardiovascular disease Nature, 2011, 472 (7341): 57-63.
[20] Hill J. Understanding and addressing the epidemic of obesity: An energy balance perspective. Endocr Rev, 2006, 27: 750-761.
[21] Backhed F, Ding H, Wang T, *et al.* The gut microbiota as an environmental factor that regulates fat storage. Proc Natl Acad Sci USA, 2004, 101: 15718-15723.
[22] Ley R E, Backhed F,Turnbaugh P, *et al.* Obesity alters gut microbial ecology. Proc Natl Acad Sci USA, 2005, 102: 11070-11075.
[23] Ley R E, Turnbaugh P J, Klein S, *et al.* Microbial ecology: Human gut microbes associated with obesity. Nature, 2006, 444:1022-1023.
[24] Turbaugh P J, Ley R E, Mahowald M A, *et al.* An obesity-associated gut microbiome with increased capacity for energy harvest. Nature, 2006, 444: 1027-1031.
[25] Marko K, Collado M C, Salminen S, *et al.* Early differences in fecal microbiota composition in children may predict overweight. Am J Clin Nutr, 2008, 87:534-538.
[26] Bajzer M, Seeley R J. Physiology: Obesity and gut flora. Nature, 2006, 444: 1009-1010.

[27] Cani P D, Neyrinck A M, Maton N, *et al.* Oligofructose promotes satiety in rats fed a high-fat diet: Involvement of glucagon-like peptide-1. Obes Res, 2005, 13: 1000-1007.

[28] Kolida S, Meyer D, Gibson G R. A double-blind placebo-controlled study to establish the bifidogenic dose of inulin in healty humans. Eur J Clin Nutr, 2007, 61: 1189-1195.

[29] Alam M, Midtvedt T, Uribe A. Differential cell kinetics in the ileum and colon of germfree rats. Scand J Gastroenterol, 1994, 29: 445-451.

[30] Gordon J I, Hooper L V, McNevin M S, *et al.* Epithelial cell growth and differentiation. III. Promoting diversity in the intestine: Conversations between the microflora, epithelium, and diffuse GALT. Am J Physiol, 1997, 273: G565-G570.

[31] Krinos C M, Coyne M J, Weinacht K G, *et al.* Extensive surface diversity of a commensal microorganism by multiple DNA inversions. Nature, 2001, 414: 555-558.

[32] Kagnoff M F, Eckmann L. Epithelial cells as sensors for microbial infection. J Clin Invest, 1997, 100: 6-10.

[33] Aderem A, Ulevitch R J. Toll-like receptors in the induction of the innate immune response. Nature, 2000, 406: 782-787.

[34] Borruel N, Carol M, Casellas F, *et al.* Increased mucosal TNF_production in Crohn's disease can be downregulated *ex vivo* by probiotic bacteria. Gut, 2002, 5: 659-664.

[35] Ling Z, Liu X, Luo Y, *et al.* Pyrosequencing analysis of the human microbiota of healthy Chinese undergraduates. BMC Genomics 2013, 14: 390.

[36] Ling Z, Liu X, Wang Y, *et al.* Pyrosequencing analysis of the salivary microbiota of healthy Chinese children and adults. Microb Ecol, 2013, 65: 487-495.

[37] Taguchi H, Takahashi M, Yamaguchi H, *et al.* Experimental infection of germ-free mice with hyper-toxigenic enterohaemorrhagic *Escherichia* coli O157:H7, strain 6. J Med Microbiol, 2002, 51: 336-343.

[38] Lievin V, Peiffer I, Hudault S, *et al.* Bifidobacterium strains from resident infant human gastrointestinal microflora exert antimicrobial activity. Gut, 2000, 47: 646-652.

3

Infectious Microecology and Immunology

Hongyan Diao *, Guangying Cui, Jianing Chen, Yingfeng Wei
The State Key Laboratory for Diagnosis and Treatment of Infectious Diseases, the First Affiliated Hospital, School of Medicine, Zhejiang University, Hangzhou, 310003, China
* E-mail: diao.hy@163.com

Infectious microecology is a branch of microecology that uses microecological theory and methods to investigate the appearance, development and outcome of infection. The main purpose of this chapter is to investigate the pathogenesis of infection and its clinical manifestations, preventions and therapies. Under normal conditions, the gut microflora exists in a state of equilibrium with the host that has been described as a separate "organ" adapted to human physiology. From the moment of birth, the neonatal intestine is confronted with potent immunostimulatory substances while the body surfaces are protected from environmental and microbial exposure during fetal life [1]. Microbial antigenic challenge is required for maturation of several physiological and anatomical functions of the intestinal epithelial barrier (IEB) [2]. Commensal bacteria regulate intestinal innate and adaptive immunity and provide stimuli for ongoing repair and restitution of IEB [3]. Colonization by pathogenic bacteria and (or) dysmature response to microbial antigen may result in flagrant inflammatory response [4].

3.1 Infection and Immunity

When it comes to infectious microecology, it is inevitable to discuss infection. In this chapter, we first discuss infection and immunity.

The development of an infectious disease in an individual involves complex interactions among the host normal flora and pathogen. Microbes can induce disease by killing host cells or liberating toxins, sometimes without extensive colonization of host tissues, and even in some infections the host response is the wrong doer, being the main cause of tissue damage and other disorders [5]. The features of microorganisms determine their virulence and pathogenesis in infectious disease.

3.1.1 Immunity Response to Microbes

Defense against microbes is mediated by the effector mechanisms of innate and adaptive immunity. The early defense of infection mainly depends on the innate immune system, which consists of an anatomical barrier (such as skin and mucous), physiological barrier (such as temperature, pH, dissolved oxygen, normal flora, and a number of dissolution factors) with the characteristics of nonspecific phagocytosis/pinocytosis and nonspecific inflammation. Adaptive immunity mainly consists of antigen-presenting cells, effector lymphoid cell populations, and numerous molecules that mediate cellular interactions. Compared with innate immunity, adaptive immune responses are generally more potent in several responses, such as expansion of the pool of antigen-specific lymphocytes. Its processes include the classical features of antigenic specificity and diversity, immunologic memory and self- /non-self-recognition. Effector cells that eliminate the microbes and memory cells that protect the individual from subsequent infections are the main constituent of adaptive immune responses to microbes.

A microbe differs greatly in patterns of host invasion and colonization, so its elimination requires diverse immune systems. The specialization of adaptive immunity provides the host optimal response to each type of microbe. In many infections, tissue injury and other disorders may be caused by the host immune response to the microbe rather than by the microbe itself [6, 7]. Immunity, like many other defense mechanisms, not only is necessary for host survival, but also induces injury to the host on occasion.

We summarize the main features of immunity to three important pathogenic microorganisms (extracellular bacteria, intracellular bacteria and fungi) and discuss the immune responses to these microorganisms.

3.1.2 Immune Responses to Extracellular Bacteria

Extracellular bacteria can replicate outside host cells. Some different species of extracellular bacteria are pathogenic, but pathogenesis of extracellular bacteria can be classified as two principal mechanisms. First, these bacteria induce inflammation to result in tissue destruction at the site of infection [8]. Second, toxins derived from these bacteria have diverse pathologic effects [9]. Such toxins may be endotoxin or exotoxin. Endotoxin is a component of bacterial cell walls and exotoxin is actively secreted by the bacteria. The mechanisms of innate immunity to extracellular bacteria include: complement activation, phagocytosis, and other inflammatory response. For example, gram-positive bacteria express mannose on their surface which may bind mannose-binding lectin leading to complement activation by the lectin pathway [10]. Further on, complement activation results in opsonization and enhanced phagocytosis of the bacteria. Toll-like receptors of phagocytes induce the activation of the phagocytes as a result of encountering with microbes [11]. These various receptors promote the phagocytosis and stimulate the microbicidal activities of the phagocytes. In addition, activated phagocytes secrete cytokines, which induce inflammation. Cytokines also can induce the systemic manifestations of infection, such as fever and the synthesis of acute-phase proteins. Humoral immunity is the principal protective adaptive immune response against extracellular bacteria, and its functions are blocking infection, eliminating the microbes, and neutralizing their toxins. These inflammatory reactions are usually self-limited and controlled, but there are some injurious effects of immune responses to extracellular bacteria. The principal injurious consequences of host responses to extracellular bacteria including inflammation and septic shock, which is a syndrome characterized by circulatory collapse and disseminated intravascular coagulation. Macrophages that are activated by microbial components produce a "cytokine storm" at the early phase of septic shock [12]. The progression of septic shock is associated with defective immune responses, resulting in unchecked microbial spread. However, certain bacterial toxins called superantigens stimulate all the T cells in a body that express a particular family of V T cell receptor genes [13], with the subsequent production of large amounts of cytokines that can also cause septic shock.

3.1.3 Immune Responses to Intracellular Bacteria

Intracellular bacteria are capable of survivorship and even replication within phagocytes. Because intracellular bacteria can always find a niche where they are inaccessible to circulating antibodies, their elimination requires the mechanisms of cell-mediated immunity. T cell-mediated immunity is the major protective immune response against intracellular bacteria. This form of immunity can be adoptively transferred with lymphoid cells rather than serum from infected or immunized animals.

Cell-mediated immunity consists of two types of reactions: $CD4^+$ T cells that activate macrophages and $CD8^+$ CTLs. $CD4^+$ T cells express CD40 ligand and secrete IFN-γ, and these two stimuli activate macrophage to result in killing of phagocytosed microbes and lysis of infected cells by cytotoxic T lymphocytes (CTLs) [14]. In addition, IFN-γ also opsonizes bacteria for phagocytosis and stimulates the production of antibody isotypes that activate the complement, thus aiding the effector functions for macrophages. When bacteria antigens are transported from phagosomes into the cytoplasm, phagocytosed bacteria could stimulate $CD8^{+}$ T cell response. Thus, the effectors of two types of cell-mediated immunity function cooperatively in defense against intracellular bacteria [15].

Mechanisms of immune evasion by intracellular bacteria include inhibiting phagolysosome fusion or escaping into the cytosol and directly scavenging or inactivating microbicidal substances. Resistance to phagocyte-mediated elimination also means that such bacteria tend to cause chronic infections, often recurring after apparent cure, and difficult to eradicate.

3.1.4 Immune Responses to Fungi

Fungal infections are an important cause of morbidity and mortality in humans. Compromised immunity is an important predisposing factor for clinically significant fungal infections. Neutrophil deficiency as a result of bone marrow suppression is frequently associated with such infections [16].

Different fungi may live in extracellular tissues or within phagocytes. However, less is known about antifungal immunity than anti-bacteria. Neutrophils and macrophages are the principal mediators of innate immunity against fungi. Neutrophils can phagocytose fungi for intracellular killing [17]. Some strains of fungi could inhibit the production of cytokines such as TNF-α and IL-12 by macrophages and stimulate production of IL-10, thus inhibiting macrophage activation [18].

Cell-mediated immunity is an important mechanism of adaptive immunity against fungal infections. In much adaptive immunity, Th1 responses are protective while Th2 responses are detrimental to the host [19]. As is known to all, granulomatous inflammation is an important cause of host tissue injury in some intracellular fungal infections, such as histoplasmosis. Fungi often induce specific antibody responses that have a protective effect. Antibody-dependent cellular cytotoxicity can eliminate some fungi, such as *Cryptococcus neoformans* [20]. The interaction between microbes and the human body will not happen without immunity, especially pathogenic organisms. And also in the human body, this interaction will be thrown into confusion by other microbes, to form a microecology that plays an important role in infection and immunity.

3.2 Infectious Microecology and Immunology

Microbial communities play an important role in the health of human organisms and these highly evolved organisms are bound up with us. It is estimated that

500 – 1,000 species of 10 – 100 trillion organisms settle down in the human intestinal epithelial barrier (IEB) [21] and compose the microflora barrier of IEB [22]. There are ten times more bacterial cells than the total number of cells in the human body [23]. The microbial genomes (microbiomes) outnumber the human genomes 100-fold. Intestinal microecology (IM) mainly consists of bacteria [2]. Although viruses and eukaryotes (*e.g.*, fungi) are also present in IM, they only serve as a minority [21]. Depending on the classification system, the microbiota may contain a larger number of bacteria ranging from 15,000 to 36,000 species. IM performs several important functions thus can be considered to be virtually an essential "organ" [2] which influences absorption and distribution of nourishment [24 - 26], regulates mucosal development of the intestinal epithelial barrier (IEB) [27, 28], and modulates innate and adaptive immunity [29, 30]. The influences of commensal bacteria on the immune and physiological system throughout life are responsible for the proper education of our immune system [31]. The microbiota (collective bacterial population) is responsible for the proper development of immune and inflammatory cells in the healthy gut through the "physiological" or "controlled" inflammation, thus conferring protection against pathogens [32]. However, Mucosal response to abnormal IM in a premature host can result in abnormal inflammatory and immune response that results in disruption of IEB [33]. The task of the IEB is preventing intestinal microbes and their products from translocating into internal milieu. Luminal bacterial presence and their translocations across IEB are proven [2]. On the other hand, it is well established that the colonic microflora plays a crucial role in the development of the intestinal immune system [34]. Animals kept in germ-free conditions lack intestinal immunity [35]. Commensal bacteria exert an anti-inflammatory effect by selectively blocking transcription activation of factor NF-κB in mature adult enterocytes [36]. It has been demonstrated that the commensal bacteria also strengthen tight junction protein assembly increase and alter characteristics of mucus secretions in a manner that inhibits microbial translocation across IEB [2]. In addition, commensal bacteria could produce protective nutrients against inflammation [37]. Commensal bacteria also up-regulate the expression of intermediates that down-regulate the production of inflammatory cytokines and chemokines [38, 39]. Thus, commensal bacteria are considered as the driver for maturation of many innate and adaptive functions and suppress inflammatory signals [2].

3.2.1 *Intestinal Microbes and Intestinal Barrier*

Of many species of bacteria, approximately 300 – 500 different ones have evolved and adapted to living and growing in the human intestine [33]. Interestingly, only a few species of bacteria adhere to the epithelia and some other bacteria transit in the stomach and the small intestine [33]. By contrast, the large intestine with high densities of living bacteria, which achieve concentrations of up to 10^{11} or 10^{12}

organisms/g of luminal contents, contains a complex and dynamic microbial ecosystem [40].

The colonization of the gastrointestinal tract in newborn infants starts immediately after birth and occurs within a few days [41]. Initially, the colonization pattern might be affected by the type of delivery (passage through the birth canal versus caesarean section) and diet (breast versus formula feeding) [42 - 44]. Other environmental factors also have an influence suggested by the differences which exist between infants born in developed countries and those born in developing countries, and among infants from different hospital wards [44]. Pioneer bacteria can create a favorable habitat for themselves, and prevent growth of other bacteria introduced later in the ecosystem, by modulating expression of genes in host epithelial cells [45]. Therefore, a close correlation exists between initial colonization in infants and the final composition of the permanent flora in adults [46]. Individuals' flora composition pattern usually remains constant, but can fluctuate under some circumstances, such as acute diarrhoeal illnesses, antibiotic treatment, to a lesser extent induced by dietary interventions [47].

Bacteria living within the colonic lumen affect host homoeostasis. Some of these bacteria are potential pathogens and can be a source of infection and sepsis when the integrity of the bowel barrier is physically or functionally breached [48]. However, the constant interaction between the host and its microbes can infer important health benefits to the human host, which is drawing particular attention to the functional implications of microflora in host physiology [43, 49].

Normal flora in a body is a crucial line of resistance to colonization by exogenous microbes and, therefore, is highly relevant in preventing pathogens from invasion, for studies have shown that germ-free animals are very susceptible to infection [33, 49]. Colonization resistance also restricts the growth of opportunistic bacteria which are present in the gut [50]. The equilibrium among resident bacteria provides a steady state in the microbial population under normal conditions. However, the ecological balance can be disrupted in some conditions, for instance when we are overusing antibiotics, resulting in overgrowth of species with potential pathogenicity such as toxigenic clostridium associated with pseudomembranous colitis [51]. Several mechanisms have been presumed concerning the barrier effect. *In vitro*, adherent non-pathogenic bacteria can prevent attachment and subsequent entry of pathogen enteroinvasive bacteria from intestinal epithelial cells by competing for the attachment site in the brush border of the epithelial cells [52]. Furthermore, bacteria compete for nutrients and maintain their collective habitat [53]. This symbiotic relationship between host and normal flora prevents unwanted overproduction of the nutrient, which would favor intrusion of potential pathogens for the host [54, 55]. Finally, bacteria can produce antimicrobial substances called bacteriocins to inhibit the growth of their competitors, which can be controlled by the host since most of bacteriocins are protein compounds and degraded by digestive proteases [56].

3.2.2 Intestinal Microecology and Host Immunity

As is well known, the intestinal mucosa is constantly exposed to a vast array of microbes, food antigens, and toxins [57]. The intestinal epithelium must tell pathogenic from nonpathogenic organisms as well as food antigens. It must "tolerate" the commensal as well as sense danger signals of potentially harmful pathogens [58].

Infection of the gastrointestinal tract is a primary health problem for both adults and children worldwide. A healthy gastrointestinal microflora composes a barrier against invading organisms and can heighten the host's defense against pathogens [59]. It can also enhance intestinal immunity by adhering to intestinal mucosa and stimulating local immune responses [60]. Changes in the normal human gut microflora will induce the development of intestinal disorders by altering the intestinal microecology and intestinal colonization resistance [61]. Thus the maintenance of a balanced intestinal microecology is important for preserving intestinal integrity [62].

It is not surprising that gut-associated lymphoid tissues contain the largest pool of immune-competent cells in the human body because the intestinal mucosa is the main interface between the immune system and the external environment [43]. The interaction between host and bacteria at the mucosal interface seems to play a role in the development of a competent immune system [63]. Animals bred in a germfree environment have low densities of lymphoid cells, small specialized follicle structures, and low circulating concentrations of immunoglobulins in the blood [64]. Immediately after exposure to luminal microbes, the number of intraepithelial lymphocytes expands greatly and germinal centers with immunoglobulin producing cells arise rapidly in follicles and in the lamina propria, thus increasing the concentrations of immunoglobulin substantially in serum [65].

Most interactions among microbes, epithelium and gut-associated lymphoid tissue are relevant to the memory mechanisms of systemic immunity [66]. For instance, floras have been implicated in oral tolerance, which persists for several months in mice with conventional flora, whereas only a few days in germfree mice [67]. Interestingly, the abnormality in neonates but older mice can be corrected by reconstitution of conventional flora [68], suggesting that the interactions early in life are crucial for appropriate development of complex mucosal and systemic immunoregulatory circuits.

The immune response to microbes depends on innate and adaptive components [69]. Innate responses are mediated not only by white blood cells but also by intestinal epithelial cells, which adjust host responses by producing a wide range of inflammatory mediators to transmit signals to underlying cells in the mucosa [64, 70]. The innate immune system uses a restricted number of performed receptors to discriminate potential pathogens from commensal bacteria [71]. The system will recognize bacteria and respond to an eventual challenge immediately and rapidly. For example, the coculture of nonpathogenic bacteria and inflamed human intestinal mucosa elicits different types of immediate cytokine responses, which transmit signals to the underlying tissue and promote changes in the phenotype of lamina proprialymphocytes [72].

Some research demonstrated that intestinal inflammation may be due to excessive activity of the innate or adaptive immune system [73]. Inflammatory bowel disease results from pathogenic immune responses to bacterial antigens in genetically susceptible individuals, which may have abnormal mucosal barrier function and dysregulation of immune responses [74].

Mucous, epithelial cells and tight junction compose the initial physical barriers in the gastrointestinal tract. When a pathogenic organism breaks these barriers, the innate immune system provides immediate nonspecific responses. While a pathogen evades the innate system, the adaptive immune system composes an additional layer of protection. Adaptive responses involve antigen-specific humoral and cell-mediated reactions, inducing a memory. The response depends on the cooperation among antigen presenting cells (APC), macrophages and dendritic cells, leading to activation of T cells and participating in the modulation of immune responses.

Bacteria may attach to the luminal side of the mucous layer directly, but to the intestinal epithelium indirectly. The ability to distinguish potentially invasive pathogens from the intestinal microbiota is principally ascribed to pattern recognition receptors (PRRs), which form a vital part of the innate immune system. PRRs, *e.g.*, toll-like receptors (TLRs) and nucleotide-binding oligomerization domains (NODs), recognizing conserved bacterial and viral motifs, allow the interaction between gut epithelium and bacteria [75].

Immune response can not only protect the human body from harm, but it can also induce a lesion, *e.g.*, Crohn's disease. Thus, this tight regulation which is required to maintain homeostasis is achieved through multiple non-immune and immune factors [76].

The apoptosis of active T lymphocytes constitutes a vital control mechanism of inflammatory responses and suppression of inflammation [77]. The mature T lymphocytes are adjusted and controlled tightly by programmed cell death or apoptosis, which can occur throughout the life of a T cell regardless of its resting and activated state, playing a major role after antigen-specific lymphocyte activation and proliferation [78]. After antigen clearance, only a minority of the T cells generated survive and become memory T cells, which prevent a recurrent infection, whereas the vast majority of activated T cells are controlled to undergo apoptosis which is also an important part of the normal immune response [79]. Abnormality of this mechanism can induce the development of autoimmunity (*e.g.*, Crohn's disease) or lymphoma [80, 81]. In Crohn's disease, abnormal activation and apoptosis of intestinal T lymphocytes which induce an exaggerated response to bacterial antigens in the gut lumen play an important role in the development of intestinal mucosal lesions [82].

Regulatory T lymphocytes (Tregs) are another important role in this controlling. They have special ability for immune suppression and are important regulators of the immune response in various settings by suppressing enteroantigen-reactive cells and contributing to the maintenance of intestinal immune homeostasis [83]. There are distinct Treg subsets coexisting in the intestinal mucosa and mesenteric lymph nodes [84]. Disturbances in Treg number and function have a close relationship with immune-mediated disorders [85].

So the gut is the important site for induction of Tregs, which secrete immunoregulatory cytokines such as IL-10 and can regulate Th1 and Th17 responses [86]. Recent findings suggest that some gut commensal bacteria, including *lactobacilli*, *bifidobacteria* and helminths, play an important role in inducing Tregs in gut lymphoid follicles [87]. Such T cell-mediated regulatory pathways are essential controlled mechanisms by which the host can tolerate the massive burden of innocuous antigens in the gut without responding to inflammation [88], and protect the human body from allergy and autoimmune diseases.

In conclusion, we can outline the defense mechanisms of the intestine (Fig. 3.1). Firstly, normal microbiota in the intestine can inhibit the invasion of potential antigens by competing for the attachment site and nutrient and by producing bacteriocins. Secondly, the epithelial cells can distinguish potential antigens from the intestinal microbiota by PRRs, and then induce T cells maturity and activation to eliminate pathogens. Meanwhile, Tregs can be activated and participate in T cells apoptosis after antigens' clearance. Then the memory T cells generated from active T cells are maintained to prevent a recurrent infection. If some of these mechanisms are abnormal, diseases may happen. Set Crohn's disease as an example, abnormal apoptosis of intestinal T lymphocytes will induce an exaggerated response to bacterial antigens in the gut lumen (Fig. 3.1).

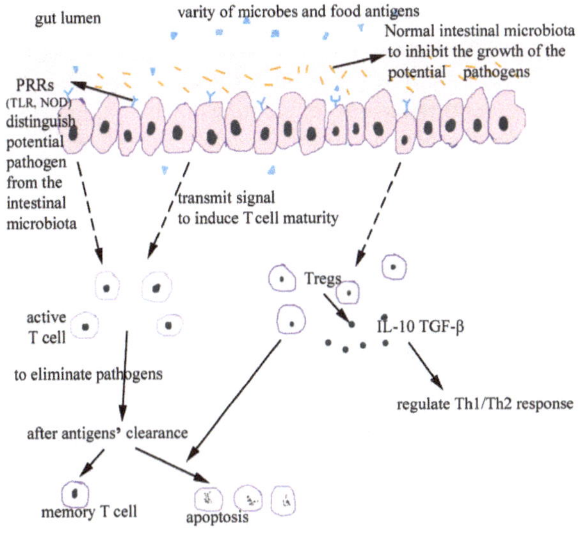

Fig. 3.1. Structure of the mechanism in the intestine to inhabit injury resulting from the pathogens or autoimmunity

3.3 Hepatic Microecology and Immunity

The liver is not only the largest solid organ in the body but also dual inputs for its blood supply. Approximately 80% of its blood supply is provided by the gut through the portal vein, which is abundant in bacterial products, environment toxins, and food antigens. The remaining 20% comes from the hepatic artery. Seventy percent of the cells or 80% of the liver volume consists of hepatocytes, which play an important role in fulfilling the metabolic and detoxifying needs of the body. The remaining cells are composed of nonparenchymal cells, including endothelial cells, stellate cells, Kupffer cells, dendritic cells, and lymphocytes. Much evidence indicates that the liver is a significant part of the body's immune response [89]. Therefore, the liver is considered as an immunologic organ and plays a key role in innate immune defenses against pathogens [89].

The exposure of liver cells to antigens and microbial products from the intestinal bacteria leads to the composition of a distinctive local immune environment [90]. Innate lymphocytes, including NK cells and NKT cells, are plentiful in the liver. Moreover, multiple populations of nonhematopoietic liver cells, including sinusoidal endothelial cells, stellate cells located in the subendothelial space, as well as hepatocytes, serve as non-professional APCs [91]. These cells present antigens in the circumstances of immunosuppressive cytokines and inhibitory cell surface ligands. Thus immune responses to liver antigens often result in tolerance [92]. Some important human pathogens, including hepatitis B virus and the malaria parasite, make use of the liver's environment to induce the abnormality of immunity, and establish persistent infection [90].

The liver has several important functions in innate and adaptive immunity. The major contributions to the innate immune system of the liver include synthesis and secretion of acute phase proteins, nonspecific phagocytosis of particles, nonspecific pinocytosis of molecules, as well as nonspecific cell killing [93]. Hepatic involvement in the adaptive immune system comprises deletion of activated T cells, induction of tolerance to ingested and self-antigens, extrathymic proliferation of T cells, and inactivation or clearance of many of the signaling and effector molecules [94]. Perturbations in the liver's structure or function can induce significant ramifications in both the innate and adaptive immune systems [95].

3.3.1 Liver Involvement in Innate Immunity

Liver innate immunity plays an important role in clearing antigens and protecting the human body by nonspecific phagocytosis and nonspecific cell killing [96].

3.3.1.1 Nonspecific Phagocytosis

The liver has been proved to be an important site for removing antigen and immune complexes. For instance, soluble IgG complexes are eliminated from circulation by the liver, more precisely predominantly by Kupffer cells and, to a lesser degree, endothelial cells [93]. As a general rule, Kupffer cells ingest particulate material through the pathway of phagocytosis and endothelial cells assimilate soluble materials by pinocytosis.

Kupffer cells are plentiful in the liver but the population density, cytologic characteristics and physiologic functions are various in different regions [97]. Kupffer cells are somewhat more abundant in the periportal region [93], and have been shown to be present through zone 1 (periportal), zone 2 (midzonal), and zone 3 (perivenous) of rat liver acinus in a ratio of 4:3:2. Periportal Kupffer cells are the largest and have the highest lysosomal enzyme activities and strongest phagocytic activity of all [98]. These observations suggest the periportal region is the first point of meeting with the incoming blood which might contain potential pathogens [95].

The recognition of the Fc domain of immunoglobulins by Kuffer cells induces nonspecific phagocytosis of immune complexes as well as antibody-coated particles [99]. Thus, Kupffer cells have a significant role in modulation of inflammatory and immunologic processes [100]. Uptake of immunoglobulin complexes is mediated via various subtypes of the Fcγ receptor, primarily the receptors of Fcγ receptor IIB2 (Fcγ RIIB2) and Fcγ receptor III (Fcγ RIII) [101 - 103]. Regulation of Fc receptor expression on Kupffer cells and hepatic sinusoidal endothelial cells appears to play a potential role in the modulation of any process related to the production of circulating immunoglobulins. A series of studies have shown that the level of ligand expression can regulate the level of receptor expression [104].

3.3.1.2 Nonspecific Cell Killing

NK and NKT cells are considered as the primary effector cells involved in nonspecific, intrahepatic cell killing in the liver.

NK cells belong to bone marrow-derived mononuclear cells that have a major role in defense of the liver against invading tumor cells. NK cell killing of target cells is mediated by two major pathways. The Fas/FasL pathway depends on the binding of FasL to Fas and subsequent activation of "death" signaling, resulting in initiating the caspase cascade and apoptosis. The perforin/granzyme pathway makes pores in the cell membrane through perforin and induces granzymes into the cytosol. The perforin/granzyme pathway essentially constitutes a shortcut directly to the downstream caspase cascade [105].

Some signaling molecules might affect the induction of Fas/FasL- or perforin/granzyme-mediated apoptosis. IL-18 is first identified as an IFN-γ-inducing factor that is produced by activated Kupffer cells. IL-18 promotes Fas/FasL-mediated killing by NK cells [106, 107]. Kupffer cells also secrete IL-12 known as NK cell-stimulating factor [106]. These data further prove that Kuffer cells and NK cells have a synergistic cooperation in cell killing [95].

NKT cells are currently considered to be distinguished from NK, which not only possess a series of the phenotypic and physiologic characteristics of NK cells, but also express T cell receptor (TCR). Constitutive expression of CD1 on cell surfaces permits NKT cells to interact with target cells without delay, thus inducing a rapid response. NKT cells are abundant in the liver [108], and can develop extrathymically from liver precursors [109].

T cell-mediated liver diseases including autoimmune hepatitis and viral hepatitis are associated with significant morbidity and mortality worldwide and remain a serious concern in the clinical setting. In autoimmune hepatitis, liver histology shows massive granulocytes accumulation, $CD4^+$ T cells infiltration, influx of a relatively small number of $CD8^+$ T cells, and hepatocytes necrosis/apoptosis. Hepatic NKT cells play a pivotal role in this process (Fig. 3.2). Upon activation, NKT cells secrete various cytokines including IFN-γ that activate resident Kupffer cells and recruit macrophages to produce TNF-α, which subsequently causes liver injury. IL-4 causes NKT cells to express Fas ligand, contributing to Fas/FasL-mediated liver injury. Our recent study indicated a critical link between OPN and NKT cells in the pathogenesis of hepatitis in mice [110]. We found that NKT cells could secrete substantial amounts of OPN upon activation. According to our studies, hepatic NKT cells expressed both OPN and its receptors, $\alpha 9$ and $\alpha 4$ integrins. The strong cellular infiltration occurred at an early stage of Con A-induced hepatitis, preceding the liver tissue damage. In addition, among these infiltrating cells, the major cell populations were neutrophils. Neutrophils also play a pivotal role in liver injury after Con A injection. The distribution of neutrophils as defined by MPO activity correlated well with the area of liver degeneration [110].

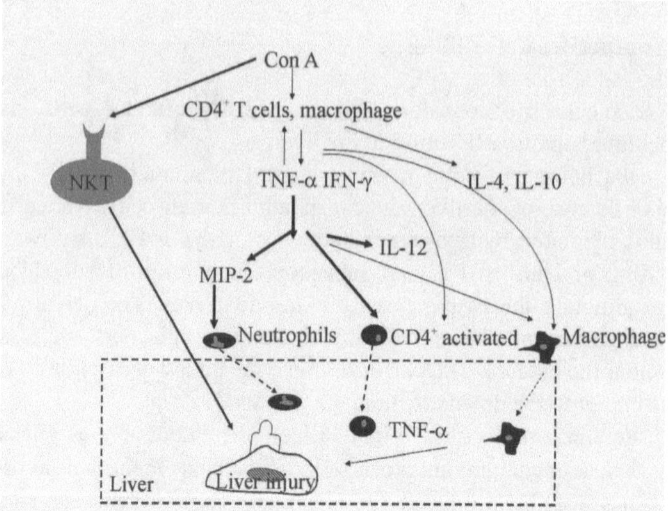

Fig. 3.2. Hepatic NKT cells play a pivotal role in the process of Con A-induced liver injury

Resident intrahepatic iNKT cells clearly expressed OPN. Osteopontin regulates NKT cell development and peripheral NKT cell activation, as evidenced by the inhibition of receptor down-regulation and reduced IL-4 production. The absence of OPN results in impaired NKT cell antigen receptor-mediated signaling pathways, such as activation of NFAT2. IL-4 production of NKT cells due to defective activation of NFAT2 led to reduced FasL expression on NKT cells. Osteopontin also resulted in attenuated cytotoxic activity of iNKT cells against Fas positive hepatocytes [111].

The liver is in a constant state of inflammation suggested by the presence of a series of proinflammatory in the normal liver, on the basis of high hepatic levels of IL-12. Besides the effects on NK cells described above, IL-12 can assist the maturation of $CD8^+$ T cells, double positive $CD4^+$ $CD8^+$ T cells, and NKT cells, all of which have cytotoxic activities [112]. However, the high mortality rate of liver cancer in humans is largely due to liver metastasis in contrast to these effective, redundant systems for nonspecific cell killing, suggesting a failure of defense mechanisms [95].

3.3.2 Liver Involvement in Adaptive Immunity

The normal liver always directly participates in adaptive immunobiology. This involvement essentially consists of (a) removal of activated T cells from any site in the body, (b) induction of tolerance to ingested- and self-antigens, and (c) extrathymic proliferation of T lymphocytes [95].

3.3.2.1 Deletion of Activated T Cells

After an inflammatory reaction subsides, a population of immunologically active cells and molecules exist and require neutralization or disposal [93]. The liver not only has a major role in disposal of circulating macromolecules, as discussed before, but also has a specific role in clearance of activated T cells [95].

Leukocyte emigration from blood vessels is required for activated T cell deletion by the liver. Leukocyte emigration in the liver occurs in sinusoids [113]. Hemodynamic factors, Kupffer cell migration and leukocyte interactions with vessel walls play a synergy to reduce the rate of blood flow through liver sinusoids [114]. These alterations in rate of blood flow are different between regions of the hepatic acinus and species [115]. Thus, leukocytes in hepatic sinusoids are exposed to integrin-type adhesion molecules extensively and slowly. ICAM-1 and VCAM are expressed constitutively and highly in sinusoidal endothelial cells and benefit the integrin-mediated adhesions in the absence of a local inflammatory reaction [113, 116]. The study specificially blocking either ICAM-1 or VCAM-1 can reduce intrahepatic accumulation of murine $CD8^+$ T cells and suggests that these adhesion molecules play a major role in the trapping of activated $CD8^+$ T cells in mouse liver [117].

Further research [118] revealed this trapping selects activated $CD8^+$ T cells, but not resting T cells or T cells that were actively involved in apoptosis. Following trapping there was a temporary 8-fold expansion of the population of activated T cells in the first 48 hours after injection into the portal vein of mice, followed by apoptosis and a reduced population of $CD8^+$ T cells in the next 4 days [119].

The end of expansion of the trapped $CD8^+$ T cell population may be due to the cytokine microenvironment of the liver. Activated T cells must require a co-stimulatory signal to avoid apoptosis [120]. Helper T cell populations are also critical for survival and expansion of activated T cell populations. The liver microenvironment lacks co-stimulatory and helper molecules, resulting in apoptosis of sequestered activated T cells ("death-by neglect") [121].

3.3.2.2 Induction of Tolerance to Ingested and Self-antigens

Professional APC presents antigens to $CD8^+$ T cells in association with MHC molecules and activity of helper T cells, and induces protective immunity against intracellular microorganisms [122]. Liver sinusoidal lining cells can take up antigen, process the antigen and present it to T cells. However, probably owing to the lack of input from helper T cells, the end is tolerance rather than immunity [123 – 125].

3.3.2.3 Extrathymic T Cell Proliferation

It has been proved that extrathymic pathways to T cell differentiation exist in the intestine and the liver [126], where T cell populations may differentiate from their own preexisting precursor cells [127].

Extrathymic T cell differentiation is petty in normal mice, but becomes predominant in some mice with autoimmune diseases, without thymus and aging [128]. These T cells have a number of phenotypic changes, especially including the expression of TCR with an intermediate level named as TCR (int). TCR (int) cells are generated in liver parenchyma and then migrate to the sinusoidal lumen. This is different from the channel of thymus-derived ('normal') T cells which goes from sinusoids into the hepatic parenchyma [129].

3.4 Liver's Immune Privilege

There is a close relationship between the liver and the gastrointestinal tract, not only in anatomy but also in pathology and physiological pathology. The digestive system is the largest microecosystem, as well as the largest reservoir of microbes and endotoxin in the body [130]. Here we discuss that the essential mechanism inducing liver tolerance is likely due to the continuous exposure of various liver cells to endotoxin, generated from the intestinal bacteria. This exposure induces the expression of a series of cytokines, antigen-presenting molecules, and costimulatory signals to impose T-cell inactivation, partly by effects on liver antigen-presenting cells [131].

Liver APCs subsets are important to mediate the liver tolerance [132]. Dendritic cells (DCs) in the liver, because of insufficient signals, stop at a "semimature" stage, so that they are able to present an antigen but fail to deliver a full set of cosignals, inducing T-cell tolerance [133].

In the liver, the sinusoidal endothelial cells, as well as Kupffer cells and hepatocytes, are considered as semiprofessional APCs which can combine the TCRs but express a set of other signals that do not induce full T-cell activation. T cells may be malicious when they run into such APCs.

For instance, with the LSECs expressing not only MHC class Ⅰ but also MHC class Ⅱ along with CD40, CD80, and CD86, but as APCs, the activation of naïve $CD4^+$ T cells induce effector cells to be Th0 like, synthesizing IFN-γ but also IL-4 and IL-10 [134], while the activation of $CD8^+$ T cells causes effector cells to be without cytotoxic function [135]. On the basis of these studies, LSECs have been proposed to be important in the induction of liver tolerance [136 - 138].

During an acute phase or systemic inflammatory response, hepatocytes can be induced to produce high levels of complements and secreted PRRs by a set of proinflammatory cytokines (such as IL-1, IL-6, TNF-α, and IFN-γ). Many of these proteins are secreted into the bloodstream, thereby playing a key role in innate immunity against local and systemic microbial infection. Meanwhile, the liver is also considered as a major source of many other acute phase proteins, which play key roles in innate defenses against infection and in reducing tissue damage through inactivation of proteinases released by pathogens and dead or dying cells [139].

Kupffer cells could recognize the Fc domain of immunoglobulins and induce nonspecific phagocytosis of immune complexes as well as antibody-coated particles.

APCs in the liver also play an important role in liver immunity against antigens from the intestine. Sinusoidal endothelial cells, as well as Kupffer cells and hepatocytes, are considered as semiprofessional APCs. They have the capacity to engage the TCRs but display a set of other signals that do not cause full T-cell activation and play an important role in mediating liver tolerance [96].

NK and NKT cells play an important role in nonspecific, intrahepatic cell killing in the liver by Fas/FasL- or perforin/granzyme-mediated apoptosis with the help of IL-12 and IL-18 secreted mainly by Kupffer cells to eliminate the bacteria from other parts of the human body, especially from the intestine. However, Both NK and NKT cells can make a set of cytokines, including IL-10, which may be important in immune deviation [140]. Therefore, these cells are regulated by the local environment in the liver.

Liver cells are exposed to abundant antigens and LPS from the intestine and plentiful lymphocytes owing to its distinctive vascular architecture. The liver provided a basis for immunological theories of immunity against pathogens and tolerance to some others which are not harmful to the human body. And also they consist of a liver microecosystem.

In this microecosystem, various liver cells perform their functions to eliminate pathogens and ignore some antigens in normal circumstances (Fig. 3.3). Firstly, liver APCs, including LESC, hepatocytes and Kupffer cells, could play a crucial role in eliminating antigens from the intestine by nonspecial phagocytosis. And then they will induce T cells activation but not full activation. For example, $CD4^+$ T cells are prone to Th0 like cells secreting IL-4, IL-10 and IFN-γ, and $CD8^+$ T cells will be activated but without cytotoxic function [141]. Furthermore, NK cells and NKT cells will be induced to carry out the nonspecial killing, by expressing FasL and secreting perforin/granzyme. However, the balance of this system will be destroyed when the intestinal barrier is destroyed, because intestinal microecology and hepatic microecology are an integer and consist of digestive microecology. In this condition, a large number of bacteria and exotoxin will flow into the blood and lead to a set of serious syndromes named systemic inflammatory response syndrome. Or in some conditions, such as HBV infection, liver cells cannot perform a big enough response to some antigens, thus inducing persistent pathogen infection and injury to the body.

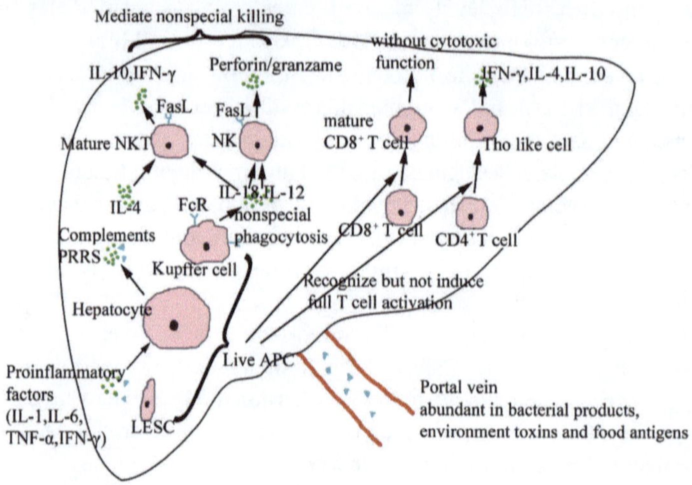

Fig. 3.3. Structure of the mechanism in the liver to eliminate pathogens and keep silent some antigens in normal circumstances

In this chapter, we discuss the relationship between infectious microecology and immunity, especially the immunity in intestinal microecology and hepatic microecology. The gut-liver axis plays an important role in keeping bacteria and endotoxins from the digestive system, and maintaining the immune balance between the human body and the microecosystem.

References

[1] Stockinger S, Hornef M W, Chassin C. Establishment of intestinal homeostasis during the neonatal period. Cell Mol Life Sci, 2011, 68: 3699-3712.

[2] Sharma R, Tepas J J, 3rd. Microecology, intestinal epithelial barrier and necrotizing enterocolitis. Pediatr Surg Int, 2010, 26: 11-21.

[3] Schneeman T A, Bruno M E, Schjerven H, *et al.* Regulation of the polymeric Ig receptor by signaling through TLRs 3 and 4: Linking innate and adaptive immune responses. J Immunol, 2005, 175: 376-384.

[4] Hooper L V, Stappenbeck T S, Hong C V, *et al.* Angiogenins: A new class of microbicidal proteins involved in innate immunity. Nat Immunol, 2003, 4: 269-273.

[5] Dhananjaya B L, D'Souza C J. The pharmacological role of phosphatases (acid and alkaline phosphomonoesterases) in snake venoms related to release of purines – a multitoxin. Basic Clin Pharmacol Toxicol, 2011, 108: 79-83.

[6] Kelly CP, Kyne L. The host immune response to Clostridium difficile. J Med Microbiol, 2011, 60: 1070-1079.

[7] Walsh K B, Teijaro J R, Rosen H, *et al.* Quelling the storm: Utilization of sphingosine-1-phosphate receptor signaling to ameliorate influenza virus-induced cytokine storm. Immunol Res, 2011, 51: 15-25.

[8] Kanno T, Sakaguchi K, Fukuyama M, *et al.* Properties of metabolic substances produced by group A streptococcus from a food-borne epidemic. J Infect Chemother, 2011,17: 462-467.

[9] Shen A. Clostridium difficile toxins: Mediators of inflammation. J Innate Immun, 2012, 4: 149-158.

[10] Takahashi K. Mannose-binding lectin and the balance between immune protection and complication. Expert Rev Anti Infect Ther, 2011, 9: 1179-1190.

[11] Takahashi M, Ishida Y, Iwaki D, *et al.* Essential role of mannose-binding lectin-associated serine protease-1 in activation of the complement factor D. J Exp Med, 2010, 207: 29-37.

[12] Drummond R, Cauvi D M, Hawisher D, *et al.* Deletion of scavenger receptor A gene in mice resulted in protection from septic shock and modulation of TLR4 signaling in isolated peritoneal macrophages. Innate Immun, 2013, 19: 30-41.

[13] Ohashi R, Takaya J, Tsuji S, *et al.* Prognostic usefulness of lymphocyte V beta receptor determination in toxic shock syndrome. Eur J Pediatr, 2005, 164: 703-704.

[14] Schultze J L, Michalak S, Lowne J, *et al.* Human non-germinal center B cell interleukin (IL)-12 production is primarily regulated by T cell signals CD40 ligand, interferon gamma, and IL-10: Role of B cells in the maintenance of T cell responses. J Exp Med, 1999, 189: 1-12.

[15] Hashimoto K, Maeda Y, Kimura H, *et al.* Mycobacterium leprae infection in monocyte-derived dendritic cells and its influence on antigen-presenting function. Infect Immun, 2002, 70: 5167-5176.
[16] Loures F V, Pina A, Felonato M, *et al.* TLR2 is a negative regulator of Th17 cells and tissue pathology in a pulmonary model of fungal infection. J Immunol, 2009, 183: 1279-1290.
[17] Vonk A G, Netea M G, Kullberg B J. Phagocytosis and intracellular killing of Candida albicans by murine polymorphonuclear neutrophils. Methods Mol Biol, 2012, 845: 277-287.
[18] Kaya E G, Ozbilge H, Ustundag M B, *et al.* The effects on immune response of levamisole treatment following infection of U-937 macrophages with Candida albicans. Acta Microbiol Immunol Hung, 2011, 58: 279-288.
[19] Antachopoulos C, Walsh T J. Immunotherapy of Cryptococcus infections. Clin Microbiol Infect, 2012, 18: 126-133.
[20] McClelland E E, Nicola A M, Prados-Rosales R, *et al.* Ab binding alters gene expression in Cryptococcus neoformans and directly modulates fungal metabolism. J Clin Invest, 2010, 120: 1355-1361.
[21] Eckburg P B, Bik E M, Bernstein C N, *et al.* Diversity of the human intestinal microbial flora. Science, 2005, 308: 1635-1638.
[22] Snoek S A, Verstege M I, Boeckxstaens GE, *et al.* The enteric nervous system as a regulator of intestinal epithelial barrier function in health and disease. Expert Rev Gastroenterol Hepatol, 2010, 4: 637-651.
[23] Backhed F, Ley R E, Sonnenburg J L, *et al.* Host-bacterial mutualism in the human intestine. Science, 2005, 307: 1915-1920.
[24] Kadooka Y, Sato M, Imaizumi K, *et al.* Regulation of abdominal adiposity by probiotics (Lactobacillus gasseri SBT2055) in adults with obese tendencies in a randomized controlled trial. Eur J Clin Nutr, 2010, 64: 636-643.
[25] Zhang X, Zhao Y, Zhang M, *et al.* Structural Changes of Gut Microbiota during Berberine-Mediated Prevention of Obesity and Insulin Resistance in High-Fat Diet-Fed Rats. PLoS One, 2012, 7: e42529.
[26] Gauffin Cano P, Santacruz A, Moya A, *et al.* Bacteroides uniformis CECT 7771 Ameliorates Metabolic and Immunological Dysfunction in Mice with High-Fat-Diet Induced Obesity. PLoS One, 2012, 7: e41079.
[27] Shifrin D A, Jr, McConnell R E, Nambiar R, *et al.* Enterocyte microvillus-derived vesicles detoxify bacterial products and regulate epithelial-microbial interactions. Curr Biol, 2012, 22: 627-631.
[28] Vendrig J C, Fink-Gremmels J. Intestinal barrier function in neonatal foals: Options for improvement. Vet J, 2012, 193: 32-37.
[29] Izadpanah A, Dwinell M B, Eckmann L, *et al.* Regulated MIP-3alpha/CCL20 production by human intestinal epithelium: Mechanism for modulating mucosal immunity. Am J Physiol Gastrointest Liver Physiol, 2001, 280: G710-G719.
[30] Adkins B, Contractor N. Immune responses of female BALB/c and C57BL/6

neonatal mice to vaccination or intestinal infection are unaltered by exposure to breast milk lycopene. J Nutr, 2011, 141: 1326-1330.

[31] Lathrop S K, Bloom S M, Rao S M, *et al.* Peripheral education of the immune system by colonic commensal microbiota. Nature, 2011, 478: 250-254.

[32] Prakash S, Rodes L, Coussa-Charley M, *et al.* Gut microbiota: next frontier in understanding human health and development of biotherapeutics. Biologics, 2011, 5: 71-86.

[33] Sekirov I, Russell S L, Antunes L C, *et al.* Gut microbiota in health and disease. Physiol Rev, 2010, 90: 859-904.

[34] Iliev I D, Funari V A, Taylor K D, *et al.* Interactions between commensal fungi and the C-type lectin receptor Dectin-1 influence colitis. Science, 2012, 336: 1314-1317.

[35] Hill D A, Artis D. Intestinal bacteria and the regulation of immune cell homeostasis. Annu Rev Immunol, 2010, 28: 623-667.

[36] Menard S, Candalh C, Bambou J C, *et al.* Lactic acid bacteria secrete metabolites retaining anti-inflammatory properties after intestinal transport. Gut, 2004, 53: 821-828.

[37] Lakhan S E, Kirchgessner A. Gut inflammation in chronic fatigue syndrome. Nutr Metab (Lond), 2010, 7: 79.

[38] Winkler P, Ghadimi D, Schrezenmeir J, *et al.* Molecular and cellular basis of microflora-host interactions. J Nutr, 2007, 137: 756S-772S.

[39] Speca S, Giusti I, Rieder F, *et al.* Cellular and molecular mechanisms of intestinal fibrosis. World J Gastroenterol, 2012, 18: 3635-3661.

[40] Das P, Singh A K, Pal T, *et al.* Colonization of the gut with Gram-negative bacilli, its association with neonatal sepsis and its clinical relevance in a developing country. J Med Microbiol, 2011, 60: 1651-1660.

[41] Ichinohe T, Pang I K, Kumamoto Y, *et al.* Microbiota regulates immune defense against respiratory tract influenza A virus infection. Proc Natl Acad Sci U S A, 2011, 108: 5354-5359.

[42] Hviid A, Svanstrom H, Frisch M. Antibiotic use and inflammatory bowel diseases in childhood. Gut, 2011, 60: 49-54.

[43]Guarner F, Malagelada J R. Gut flora in health and disease. Lancet, 2003, 361: 512-519.

[44] Yatsunenko T, Rey F E, Manary M J, *et al.* Human gut microbiome viewed across age and geography. Nature, 2012, 486: 222-227.

[45] Lutgendorff F, Akkermans L M, Soderholm J D. The role of microbiota and probiotics in stress-induced gastro-intestinal damage. Curr Mol Med, 2008, 8: 282-298.

[46] Rousseau C, Poilane I, De Pontual L, *et al.* Clostridium difficile Carriage in Healthy Infants in the Community: A Potential Pathogenic Strain Reservoir. Clin Infect Dis, 2012.

[47] Moore W E, Moore L H. Intestinal floras of populations that have a high risk of colon cancer. Appl Environ Microbiol, 1995, 61: 3202-3207.

[48] Kobayashi M, Nakamura K, Cornforth M, *et al.* Role of M2b Macrophages in the Acceleration of Bacterial Translocation and Subsequent Sepsis in Mice Exposed to Whole Body [137Cs] Gamma-Irradiation. J Immunol, 2012, 189: 296-303.
[49] Urbaniak C, Burton J P, Reid G. Breast, milk and microbes: A complex relationship that does not end with lactation. Womens Health (Lond Engl), 2012, 8: 385-398.
[50] Volf J, Stepanova H, Matiasovic J, *et al.* Salmonella enterica serovar Typhimurium and Enteritidis infection of pigs and cytokine signalling in palatine tonsils. Vet Microbiol, 2012, 156: 127-135.
[51] Mason K L, Erb Downward J R, Mason K D, *et al.* Candida albicans and bacterial microbiota interactions in the cecum during re-colonization following broad spectrum antibiotic therapy. Infect Immun, 2012.
[52] Barnett A M, Roy N C, McNabb W C, *et al.* The interactions between endogenous bacteria, dietary components and the mucus layer of the large bowel. Food Funct, 2012, 3: 690-699.
[53] Thiennimitr P, Winter S E, Winter M G, *et al.* Intestinal inflammation allows Salmonella to use ethanolamine to compete with the microbiota. Proc Natl Acad Sci U S A, 2011, 108: 17480-17485.
[54] Hooper L V, Xu J, Falk P G, *et al.* A molecular sensor that allows a gut commensal to control its nutrient foundation in a competitive ecosystem. Proc Natl Acad Sci U S A, 1999, 96: 9833-9838.
[55] Acheson D W, Luccioli S. Microbial-gut interactions in health and disease. Mucosal immune responses. Best Pract Res Clin Gastroenterol, 2004, 18: 387-404.
[56] Hassan M, Kjos M, Nes I F, *et al.* Natural antimicrobial peptides from bacteria: Characteristics and potential applications to fight against antibiotic resistance. J Appl Microbiol, 2012.
[57] Neu J, Lorca G, Kingma S D, *et al.* The intestinal microbiome: relationship to type 1 diabetes. Endocrinol Metab Clin North Am, 2010, 39: 563-571.
[58] Vaarala O, Atkinson M A, Neu J. The "perfect storm" for type 1 diabetes: The complex interplay between intestinal microbiota, gut permeability, and mucosal immunity. Diabetes, 2008, 57: 2555-2562.
[59] Wallace T C, Guarner F, Madsen K, *et al.* Human gut microbiota and its relationship to health and disease. Nutr Rev, 2011, 69: 392-403.
[60] Gaboriau-Routhiau V, Lecuyer E, Cerf-Bensussan N. Role of microbiota in postnatal maturation of intestinal T-cell responses. Curr Opin Gastroenterol, 2011, 27: 502-508.
[61] Schwiertz A, Jacobi M, Frick J S, *et al.* Microbiota in pediatric inflammatory bowel disease. J Pediatr, 2010, 157: 240-244, e241.
[62] Salminen S, Isolauri E, Onnela T. Gut flora in normal and disordered states. Chemotherapy, 1995, 41 (Suppl 1): 5-15.
[63] Compare D, Coccoli P, Rocco A, *et al.* Gut-liver axis: The impact of gut microbiota on non alcoholic fatty liver disease. Nutr Metab Cardiovasc Dis,

2012, 22: 471-476.
[64] Pearson C, Uhlig H H, Powrie F. Lymphoid microenvironments and innate lymphoid cells in the gut. Trends Immunol, 2012, 33: 289-296.
[65] Hakansson A, Molin G. Gut microbiota and inflammation. Nutrients, 2011, 3: 637-682.
[66] Brandtzaeg P. Function of mucosa-associated lymphoid tissue in antibody formation. Immunol Invest, 2010, 39: 303-355.
[67] Moreau M C, Gaboriau-Routhiau V. The absence of gut flora, the doses of antigen ingested and aging affect the long-term peripheral tolerance induced by ovalbumin feeding in mice. Res Immunol, 1996, 147: 49-59.
[68] Sudo N, Sawamura S, Tanaka K, *et al.* The requirement of intestinal bacterial flora for the development of an IgE production system fully susceptible to oral tolerance induction. J Immunol, 1997, 159: 1739-1745.
[69] Fagundes C T, Souza D G, Nicoli J R, *et al.* Control of host inflammatory responsiveness by indigenous microbiota reveals an adaptive component of the innate immune system. Microbes Infect, 2011, 13: 1121-1132.
[70] Campeau J L, Salim S Y, Albert E J, *et al.* Intestinal epithelial cells modulate antigen-presenting cell responses to bacterial DNA. Infect Immun, 2012, 80: 2632-2644.
[71] Cieza R J, Cao A T, Cong Y, *et al.* Immunomodulation for gastrointestinal infections. Expert Rev Anti Infect Ther, 2012, 10: 391-400.
[72] Kim J, Hegde M, Jayaraman A. Microfluidic co-culture of epithelial cells and bacteria for investigating soluble signal-mediated interactions. J Vis Exp, 2010.
[73] Matricon J. Immunopathogenesis of inflammatory bowel disease. Med Sci (Paris), 2010, 26: 405-410.
[74] Xavier RJ, Podolsky D K. Unravelling the pathogenesis of inflammatory bowel disease. Nature, 2007, 448: 427-434.
[75] Tennyson C A, Friedman G. Microecology, obesity, and probiotics. Curr Opin Endocrinol Diabetes Obes, 2008, 15: 422-427.
[76] Mizrahi M, Ilan Y. The gut mucosa as a site for induction of regulatory T-cells. Curr Pharm Des, 2009, 15: 1191-1202.
[77] Carol M, Borruel N, Antolin M, *et al.* Modulation of apoptosis in intestinal lymphocytes by a probiotic bacteria in Crohn's disease. J Leukoc Biol, 2006, 79: 917-922.
[78] Sekine Y, Yamamoto C, Kakisaka M, *et al.* Signal-transducing adaptor protein-2 modulates Fas-mediated T cell apoptosis by interacting with caspase-8. J Immunol, 2012, 188: 6194-6204.
[79] Soni C, Karande A A. Glycodelin-A interferes with IL-2/IL-2R signalling to induce cell growth arrest, loss of effector functions and apoptosis in T-lymphocytes. Hum Reprod, 2012, 27: 1005-1015.
[80] Jerez A, Clemente M J, Makishima H, *et al.* STAT3 mutations unify the pathogenesis of chronic lymphoproliferative disorders of NK cells and T cell large granular lymphocyte leukemia. Blood, 2012.

[81] Mitomi H, Ohkura Y, Yokoyama K, *et al.* Contribution of TIA-1+ and granzyme B+ cytotoxic T lymphocytes to cryptal apoptosis and ulceration in active inflammatory bowel disease. Pathol Res Pract, 2007, 203: 717-723.
[82] Santaolalla R, Mane J, Pedrosa E, *et al.* Apoptosis resistance of mucosal lymphocytes and IL-10 deficiency in patients with steroid-refractory Crohn's disease. Inflamm Bowel Dis, 2011, 17: 1490-1500.
[83] de la Fuente H, Cibrian D, Sanchez-Madrid F. Immunoregulatory molecules are master regulators of inflammation during the immune response. FEBS Lett, 2012, 586: 2897-2905.
[84] Veenbergen S, Samsom J N. Maintenance of small intestinal and colonic tolerance by IL-10-producing regulatory T cell subsets. Curr Opin Immunol, 2012, 24: 269-276.
[85] Hormannsperger G, Clavel T, Haller D. Gut matters: Microbe-host interactions in allergic diseases. J Allergy Clin Immunol, 2012, 129: 1452-1459.
[86] Ogino H, Nakamura K, Ihara E, *et al.* $CD4^+$ $CD25^+$ regulatory T cells suppress Th17-responses in an experimental colitis model. Dig Dis Sci, 2011, 56: 376-386.
[87] Ruiter B, Shreffler W G. The role of dendritic cells in food allergy. J Allergy Clin Immunol, 2012, 129: 921-928.
[88] Guarner F. Prebiotics, probiotics and helminths: the "natural" solution? Dig Dis, 2009, 27: 412-417.
[89] Gao B, Jeong W I, Tian Z. Liver: An organ with predominant innate immunity. Hepatology, 2008, 47: 729-736.
[90] Crispe I N. The liver as a lymphoid organ. Annu Rev Immunol, 2009, 27: 147-163.
[91] Lau A H, de Creus A, Lu L, *et al.* Liver tolerance mediated by antigen presenting cells: Fact or fiction? Gut, 2003, 52: 1075-1078.
[92] Mowat A M. Anatomical basis of tolerance and immunity to intestinal antigens. Nat Rev Immunol, 2003, 3: 331-341.
[93] Parker G A, Picut C A. Liver immunobiology. Toxicol Pathol, 2005, 33: 52-62.
[94] Bilate A M, Lafaille J J. Induced $CD4^+Foxp3^+$ regulatory T cells in immune tolerance. Annu Rev Immunol, 2012, 30: 733-758.
[95] Parker G A, Picut C A. Liver immunobiology. Toxicol Pathol, 2005, 33: 52-62.
[96] Crispe I N, Giannandrea M, Klein I, *et al.* Cellular and molecular mechanisms of liver tolerance. Immunol Rev, 2006, 213: 101-118.
[97] LeCluyse E L, Witek R P, Andersen M E, *et al.* Organotypic liver culture models: Meeting current challenges in toxicity testing. Crit Rev Toxicol, 2012, 42: 501-548.
[98] Hallam S, Escorcio-Correia M, Soper R, *et al.* Activated macrophages in the tumour microenvironment-dancing to the tune of TLR and NF-κB. J Pathol, 2009, 219: 143-152.

[99] Parker G A, Picut C A. Immune functioning in non lymphoid organs: The liver. Toxicol Pathol, 2012, 40: 237-247.

[100] Kremer M, Thomas E, Milton R J, *et al.* Kupffer cell and interleukin-12-dependent loss of natural killer T cells in hepatosteatosis. Hepatology, 2010, 51: 130-141.

[101] Smith K G, Clatworthy M R. FcgammaRIIB in autoimmunity and infection: Evolutionary and therapeutic implications. Nat Rev Immunol, 2010, 10: 328-343.

[102]Younes A S, Csire M, Kapusinszky B, *et al.* Heterogeneous pathways of maternal-fetal transmission of human viruses (review). Pathol Oncol Res, 2009, 15: 451-465.

[103] Desjarlais J R, Lazar G A, Zhukovsky E A, *et al.* Optimizing engagement of the immune system by anti-tumor antibodies: An engineer's perspective. Drug Discov Today, 2007, 12: 898-910.

[104]Champsaur M, Lanier L L. Effect of NKG2D ligand expression on host immune responses. Immunol Rev, 2010, 235: 267-285.

[105]Cullen S P, Brunet M, Martin S J. Granzymes in cancer and immunity. Cell Death Differ, 2010, 17: 616-623.

[106]Iannello A, Samarani S, Debbeche O, *et al.* Potential role of interleukin-18 in the immunopathogenesis of AIDS: Involvement in fratricidal killing of NK cells. J Virol, 2009, 83: 5999-6010.

[107]Iannello A, Samarani S, Debbeche O, *et al.* Role of interleukin-18 in the development and pathogenesis of AIDS. AIDS Rev, 2009, 11: 115-125.

[108]Godfrey D I, Uldrich A P, Baxter A G. NKT cells-an early warning system for HBV infection. Nat Med, 2012, 18: 1014-1016.

[109]Stritesky G L, Jameson S C, Hogquist K A. Selection of self-reactive T cells in the thymus. Annu Rev Immunol, 2012, 30: 95-114.

[110]Diao H, Kon S, Iwabuchi K, *et al.* Osteopontin as a mediator of NKT cell function in T cell-mediated liver diseases. Immunity, 2004, 21: 539-550.

[111]Diao H, Iwabuchi K, Li L, *et al.* Osteopontin regulates development and function of invariant natural killer T cells. Proc Natl Acad Sci USA, 2008, 105: 15884-15889.

[112]O'Farrelly C. Immunoregulation in the liver and its extrahepatic relevance. J Pediatr Gastroenterol Nutr, 2004, 39 Suppl 3: S727-S728.

[113]Friedl P, Weigelin B. Interstitial leukocyte migration and immune function. Nat Immunol, 2008, 9: 960-969.

[114]Hickey M J, Kubes P. Intravascular immunity: The host-pathogen encounter in blood vessels. Nat Rev Immunol, 2009, 9: 364-375.

[115]McCuskey R S. The hepatic microvascular system in health and its response to toxicants. Anat Rec (Hoboken), 2008, 291: 661-671.

[116]Deane J A, Hickey M J. Molecular mechanisms of leukocyte trafficking in T-cell-mediated skin inflammation: Insights from intravital imaging. Expert Rev Mol Med, 2009, 11: e25.

[117]John B, Crispe I N. Passive and active mechanisms trap activated $CD8^+$ T

cells in the liver. J Immunol, 2004, 172: 5222-5229.
[118]Cope A P. T cells in rheumatoid arthritis. Arthritis Res Ther, 2008, 10 (Suppl 1): S1.
[119]Kuniyasu Y, Marfani S M, Inayat I B, *et al.* Kupffer cells required for high affinity peptide-induced deletion, not retention, of activated $CD8^+$ T cells by mouse liver. Hepatology, 2004, 39: 1017-1027.
[120]McPherson A J, Snell L M, Mak T W, *et al.* Opposing roles for TRAF1 in the alternative versus classical NF-κB pathway in T cells. J Biol Chem, 2012, 287: 23010-23019.
[121]Stoneman V E, Bennett M R. Role of Fas/Fas-L in vascular cell apoptosis. J Cardiovasc Pharmacol, 2009, 53: 100-108.
[122]Sallusto F, Lanzavecchia A, Araki K, *et al.* From vaccines to memory and back. Immunity, 2010, 33: 451-463.
[123]Castell J V, Castell M. Allergic hepatitis induced by drugs. Curr Opin Allergy Clin Immunol, 2006, 6: 258-265.
[124]Crispe I N. Liver antigen-presenting cells. J Hepatol, 2011, 54: 357-365.
[125]Sallusto F, Impellizzieri D, Basso C, *et al.* T-cell trafficking in the central nervous system. Immunol Rev, 2012, 248: 216-227.
[126]Abo T, Tomiyama C, Watanabe H. Biology of autoreactive extrathymic T cells and B-1 cells of the innate immune system. Immunol Res, 2012, 52: 224-230.
[127]Korn T, Bettelli E, Oukka M, *et al.* IL-17 and Th17 Cells. Annu Rev Immunol, 2009, 27: 485-517.
[128]Roy D, Cai Q, Felty Q, *et al.* Estrogen-induced generation of reactive oxygen and nitrogen species, gene damage, and estrogen-dependent cancers. J Toxicol Environ Health B Crit Rev, 2007, 10: 235-257.
[129]Oo Y H, Adams D H. The role of chemokines in the recruitment of lymphocytes to the liver. J Autoimmun, 2010, 34: 45-54.
[130]Rittler P, Demmelmair H, Koletzko B, *et al.* Effect of elective abdominal surgery on human colon protein synthesis *in situ*. Ann Surg, 2001, 233: 39-44.
[131]Baine I, Abe BT, Macian F. Regulation of T-cell tolerance by calcium/NFAT signaling. Immunol Rev, 2009, 231: 225-240.
[132]Morelli A E, Thomson A W. Tolerogenic dendritic cells and the quest for transplant tolerance. Nat Rev Immunol, 2007, 7: 610-621.
[133]Klein L, Munz C, Lunemann J D. Autophagy-mediated antigen processing in $CD4^+$ T cell tolerance and immunity. FEBS Lett, 2010, 584: 1405-1410.
[134]Milush J M, Long B R, Snyder-Cappione J E, *et al.* Functionally distinct subsets of human NK cells and monocyte/DC-like cells identified by coexpression of CD56, CD7, and CD4. Blood, 2009, 114: 4823-4831.
[135]Limmer A, Ohl J, Kurts C, *et al.* Efficient presentation of exogenous antigen by liver endothelial cells to $CD8^+$ T cells results in antigen-specific T-cell tolerance. Nat Med, 2000, 6: 1348-1354.
[136]Tang L, Yang J, Liu W, *et al.* Liver sinusoidal endothelial cell lectin,

LSECtin, negatively regulates hepatic T-cell immune response. Gastroenterology, 2009, 137: 1498-1508, e1491-1495.

[137]Mehal W Z. The gut-liver axis: a busy two-way street. Hepatology, 2012, 55: 1647-1649.

[138]Holz L E, Warren A, Le Couteur D G, *et al.* $CD8^+$ T cell tolerance following antigen recognition on hepatocytes. J Autoimmun, 2010, 34: 15-22.

[139]Ricklin D, Hajishengallis G, Yang K, *et al.* Complement: A key system for immune surveillance and homeostasis. Nat Immunol, 2010, 11: 785-797.

[140]Sakuishi K, Miyake S, Yamamura T. Role of NK cells and invariant NKT cells in multiple sclerosis. Results Probl Cell Differ, 2010, 51: 127-147.

[141]Levings M K, Sangregorio R, Sartirana C, *et al.* Human $CD25^+CD4^+$ T suppressor cell clones produce transforming growth factor beta, but not interleukin 10, and are distinct from type 1 T regulatory cells. J Exp Med, 2002, 196: 1335-1346.

4

Microecology Disturbance and Infection

Jingyun Yang[1]*, Xuesong Yang[2], Xinjun Hu[3]
[1] Medical College, Jiamusi University, Jiamusi, 154007, China
[2] Medical College, Jinan University, Guangzhou, 510632, China
[3] The State Key Laboratory for Diagnosis and Treatment of Infectious Diseases, the First Affiliated Hospital, School of Medicine, Zhejiang University; Collaborative Innovation Center for Diagnosis and Treatment of Infectious Diseases, Hangzhou, 310003, China
*E-mail: yangjingyun@126.com

To study microecology disturbance, we need to understand microecology balance first. Microecology balance is formed through a long-term evolutionary process, in which the normal micro-population exists with its hosts in different developmental stages in a dynamic and physiological group. This combination between the micro-population and its hosts is a physiogenic entity of normal interaction in the different levels of the ecological organizational structure of the normal micro-population and ecosystem spatial structure of hosts (human, animal or plant) in the body or on the body surface under the effect of the common macro environment. The internal structure and condition of this entity is the ecological balance [1]. And the reverse side of ecological balance is the micro dysbiosis, which includes disturbance among microbes, between microbe and host and between the microbe-host entity and the external environment.

Microecology disturbance is closely related with infection, which is the important content of reciprocal transformation between microecology disturbance and microecology balance. Therefore, microecology disturbance is an important component of infection microecology.

4.1 Microecology Disturbance

Many recent reviews have described the known interactions between the innate and adaptive immune system and the tens of trillions of microbes that live in our body. For example, studies of humans and gnotobiotic mouse models indicate that our mutualistic relations with the gut microbiota influence maturation of the immune system, modulate responses to epithelial cell injury, affect energy balance, and support biotransformations that we are ill-equipped to perform on our own, including processing of xenobiotics [2]. To ignore our microbial side would ignore an important contributor to our health and biology.

4.1.1 Concept of Microecology Disturbance

Early in 1920, Scheunert A, a German microbiologist, pointed out that the chaotic condition of enteroflora was microecological disturbance when he studied enteroflora. This terminology has been used widely and comparably employed in ecological balance since 1960. In 1962, Naebek suggested that it was due to the microecology disturbance that the condition of recomposed micro-colonial was damaged or confused. This concept only emphasized the disturbance of the microbe itself and did not include disturbance between microbe and host, which is deficiency. Obviously the concept mentioned above is only suitable to micro dysbacteria of microecology disturbance [3].

The concept in the book *Microecology*, edited by Kang Bai in 1988, is that "the situation transformed from physiological combination to pathological combination of microecology balance between microbes and between normal microbe and host under the effect of the external environment" is known as microecology disturbance. The concept includes not only the disturbance of the microbe itself but also the disturbance microbe-host and microbe-host to external environment. Therefore, we think that this concept is rather objective compared with previous ones.

4.1.2 Classification of Microecology Disturbance

Microecology disturbance can be classified by ecological classification, clinical classification and synthetic classification.

4.1.2.1 Ecological Classification

Microecology disturbance can be classified into four types as follows by ecological classification.

***Dysbacteria*:** Dysbacteria is an abnormal change in quantity or quality in the micro-colonial or population in the original micro-habitat environment, such as the over-use of antibiotics resulting in suppression of the sensitive bacteria, meanwhile enhancing insensitive bacteria growth, which induces an abnormal change in construction and quantity in certain normal flora. The majority of this change is in quantity. Therefore dysbacteria can also bedisproportionate. Dysbacteria can be divided into three catagories according to the different level of dysbacteria.

(i) Degree 1 dysbacteria (incubation dysbacteria): Degree 1 dysbacteria is light and the change can only be found in bacterium quantity detection with or without few clinical symptoms. The dysbacteria at degree 1 is usually reversible. It could recover automatically after removing the inducing factor, for example degree 1 dysbacteria induced by overusing antibiotics. It will be recovered by itself after stopping antibiotics.

(ii) Degree 2 dysbacteria (localization dybacteria): Degree 2 dysbacteria shows more disproportion in bacterium quantity detection. The change of flora is pathological rather than physiological. Chronic diseases are usually detected, such as chronic enteritis, chronic pyelonephritis, chronic stomatitis and angina *etc.* And this change is nonreversible at this stage.

(iii) Degree 3 dysbacteria (flora chaotic disease, bacterium alternation disease or dual infection): Degree 3 dysbacteria shows that the original flora is suppressed and only few strains grow absolutely dominantly. Acute diseases are usually found and are dangerous. For example, pseudomembranous enterocolitis is mainly induced by clostridium difficile and staphylococci. Clostridium difficile is one of the normal adult enteroflora, above 5%, and not harmful to healthy humans normally. Clostridium difficile infection (CDI) following the administration of antibiotics has been estimated to be the most costly healthcare-associated infection, responsible for an estimated 3 billion dollars in increased healthcare costs annually [4]. If using antibiotics for a long time, some bacteria that usually control clostridium difficile growth will be suppressed. Then clostridium difficile will be activated and produce toxin whose target is the intestinal tract and mainly induces colon diseases and formation of pseudomembrane. Diarrhea, bellyache, fever, hypoproteinemia, toxic megacolon, toxice megacolon and even enterobrosis leading to death are usually found at this time in clinical examination.

Others, such as *Bacillus proteus*, *Bacillus aeruginosus*, *Blastomyces albicans*, pneumobacillus and *Escherichia coli etc.*, can also induce degree 3 dysbacteria, which usually happens after using antibiotics for a long time, immune-depressant cytotoxic drugs, hormones, X-rays and complicated operations or with severe diabetes, cirrhosis, malignant tumors and other diseases.

***Location transfer*:** Location transfer is also called translocation. These are the following reasons for translocation: Firstly, the flora proportion has undergone a grave change and dominant bacteria grow fast and extend everywhere; Secondly, the host immunohypofunction resistance decreases and the original bacteria structure is disturbed because of disease. Immunosuppressant application and radiotherapy can induce location transfer. Location transfer can be divided into transversal transfer and portrait transfer.

(i) Transversal transfer: Transversal transfer refers to normal flora transfer from the original location to the periphery. For example, bacteria of the lower digestive tract transfer to the upper digestive tract, which often takes place in liver disease patients. Coliform bacteria transfer to the small intestine and many live in the small intestine to induce the contaminated small bowel syndrome which is also a good example. These are also usually seen as transversal transfers when bacteria of the upper respiratory tract transfer to the lower respiratory tract, bacteria of the lower urinary tract transfer to the pelvis and those of the vagina transfer to the uterus and oviduct. Schindler D found that depletion of dendritic cells in CD11c-DTR transgenic mice resulted in substantial worsening of infection, as indicated by increased bacterial loads in kidneys and lungs, accelerated mortality, and more severe pathology [5].

(ii) Portrait transfer: Normal flora lives in layers of mucous and skin. For example, oxybiontic bacteria live on the surface layer of oral mucosa, facultative anaerobes in the middle layer and obligate anaerobes in the bottom layer. When a microecology disturbance happens, surface layer bacteria may transfer to the deep layer or even to the under layer. In the circumstances, although disproportion doesn't appear, disease can also be induced and found in many stomatitis. Another example, enteritis shows dysbacteria early, which can induce mucous hyperemia, edema and inflammation. Then bacteria can reach the lymph nodes, liver, spleen through lymph and blood, and result in peritoneum and general infection, which can be clearly seen on electron microscope specimens.

Portrait transfer can be divided into four stages.

i) Body surface stage: Microbes grow abnormally in the microhabitat of the skin, oral cavity, nasopharynx, respiratory tract, small intestine, large intestine and vagina mucous. There are usually no symptoms and signs in the clinic.

ii) Epithelium stage: Microbes grow abnormally in the skin, oral cavity, nose, respiratory tract, digestive tract, vagina, conjunctiva epithelium and dysbacteria occurs obviously. Catarrh, edema and inflammation are shown clinically.

iii) Lymphoid tissue stage: Microbes invade deep lymphoid tissue including the thymus, lymph gland, secondary generative center, bone marrow, liver and spleen. Thymus and lymph gland enlargement, leukocytosis and splenohepatomegalia are usually seen clinically. Sozinov AS found that changes in the intestinal microflora were accompanied by alterations in the morphological structure in the liver [6].

iv) Reticuloendothelial cell stage: Microbes invade joints, pleura, pericardium and blood vessel endothelium and so on. Arthritis, pleurisy and pericarditis are usually seen clinically.

***Hematogenous infection*:** Hematogenous infection takes place before or after location transfer. Location transfer is a pathway for translocation bacteria to spread and is one type of translocation infection too. Hematogenous infection can be divided into bacteremia, septicemia, toxemia and pyemia.

(i) Bacteremia: Bacteremia is the case in which, for any reason, bacteria enter the blood transiently and do not grow in the blood. Bacteremia is very common and happens in nearly 4% to 10% of healthy adults. The inducement factors of bacteremia in the clinic are epithelium and mucosa damage resulting from fire burns, bedsores, ureter detainment and artificial respirators.

It is usual for normal flora to enter the blood, but the infection does not take place in normal healthy conditions, and occurs only in conditions of immunotolerance induced by weakness and infection.

We usually call bacteremia induced by normal flora non-specificity bacteremia, so alternation and cooperative infection of normal flora usually take place. To study the hemoculture of dysbacteria patients, we found bacteria of normal flora or conditional pathogenic bacteria in 43 of 97 patients (44%) and cooperative infection of staphylococci, *Bacillus proteus* and *Blastomyces albicans* in 35 patients (36%). Staphylococci can be separated in nearly half the blood samples of ulcerative colitis, induced by dybacteria patients. Hematogenous infection is important to normal flora translocation and bacteremia exists before the formation of metastatic lesions.

(ii) Septicemias: Septicemias are bacteria that enter the blood and grow dramatically in the blood, which is a severe hematogenous infection. Intestinal epithelial damage alters the capacity to provide nutrient support, thereby contributing to nutritional deficiencies in the setting of sepsis [7]. The bacteria induced septicemias are usually staphylococci, pneumococcus, hemolytic streptococcus, *Escherichia coli* and so on. The clinical symptoms of hyperpyrexia, shiver, rash are usually seen in septicemia. Shock, DIC and multiple organ failure also can take place in severe septicemia.

(iii) Toxemia: Toxemia occurs when toxins secreted by bacteria enter blood circulation from local infection lesions. Toxemia can induce general hyperpyrexia with shock. Anaemia also can be seen, and the toxins in blood can damage blood cells directly.

(iv) Pyemia: Pyemia occurs when pyogenic bacteria transfer to another place through blood circulation to induce severe infection and re-enter the blood to induce more severe infection. A Pyemia patient is too weak to prevent sickness and the prognosis is not good.

***Translocation lesion*:** For some reasons normal microbes transfer to distant organs or tissues to induce infection lesions, such as an abscess on the brain, liver, pancreas and belly cavity. Kazantsev used DNA plasmid to label *Escherichia coli* and found the bacteria inducing an abscess come from enteric *Escherichia coli*.

4.1.2.2 Clinical Classification

Prof. A. Bilibin (1979) divided microecological disturbance into three types.

***Incubation type*:** Incubation type is the term used to describe the microecological disturbance when bacterial flora alternation appears, without clinical observable symptoms. So it is also called a subclinical type.

***Limited type*:** Limited type is also called location type. Normal micropopulation disturbance takes place at the original site and is as similar as the disproportion in microecology classification.

***Suffusion type*:** Suffusion type includes hematogenous infection and translocation lesion.

4.1.2.3 Synthetic Classification

This classification emphasizes that both microecological disturbance and clinical manifestation should be considered. Generally, microecology disturbance can be divided into dominant disturbance and recessive disturbance. There is both clinical manifestation and flora change for dominant disturbance. For recessive disturbance there are two subtypes, subtype A with clinical manifestation and without flora change, and subtype B without clinical manifestation and with flora change.

The information mentioned above suggests that a synthetic classification should be used to separate microecology disturbance.

The manifestation and the flora change in the main habitat of synthetic classification are shown in Table 4.1.

Table 4.1 Clinical symptoms and flora alternation of synthetic classfication

Type	Clinical symptoms	Flora alternation
Incubation type A	+	−
Incubation type B	−	+
Limited type	+	+
Hematogenous infection type	+/−	+/−
Translocation lesion type	+	+

4.1.3 Influencing Factors of Microecology Disturbance

There are many complicated influencing factors in microecology disturbance. Generally speaking, all the factors which can affect ecological balance can result in microecology disturbance. All the factors of physics, chemistry and biotics which can interfere with hosts and normal microbes must induce microecology disturbance, because there are normal microbes on the body surface and in the body, which play an important biological role in the absorption of nourishment and immune antagonism to protect the body. Intestinal macrophages and dendritic cells act in a synergistic fashion with intestinal epithelial cells and microbiota to initiate the triad that governs the intestinal immune responses [8].

Whatever the reasons, diseases can induce microecology disturbance. The relationship between microecology disturbance and disease is the cause and effect as well.

Some therapies of modern medicine, such as long-term antibiotics usage, cytotoxic drugs, hormones, isotopes, immunodepressants, operations, urethral catheterizations, and intubation *etc.*, have harmful effects on normal microbes and hosts so as to destroy the microecological balance and induce microecology disturbance, therefore increase the iatrogenic diseases rate. Otherwise, chronic diseases can create hypoimmunity partly or generally as time passes, so doctors should be aware of preventing microecology disturbance during the therapy.

Below we will only introduce the common factors of microecology disturbance, such as antibiotics, isotopes, operations and chronic diseases.

4.1.3.1 Antibiotics

Alexander Fleming (1888–1955), British, discovered penicillin in 1928. After overcoming several problems, in 1944, antibiotics were thrown into industrial production in the US. It was at that time the antibiotics industry started and now thousands of antibiotics can be produced besides penicillin.

We will never forget the meritorious service of antibiotics for mankind because they have saved the lives of millions of people from various kinds of infection. However, as all the things on the earth have side effects, dysbacteriosis is one of the bad aspects of antibiotics. Early in 1950, Prof. Xi Wei of China and his assistants encountered one case. An 18-year old youth went to the hospital for rat-bite fever. The pathogenic bacteria is rat chain silk bacillus which is sensitive to penicillin and streptomycin. After diagnosis and medicine the patient's condition initially improved but later got worse and the patient finally died. The reason is that after eliminating pathogenic bacteria, double infection was induced and septicemia of pneumobacillus took place, leading to the death of the patient. At that time Prof. Xi Wei showed that "after the brilliant birth of antibiotics we must be aware of the shadow, which is that they interfere with normal microbes to induce dysbacteriosis".

***Inducing dysbacteriosis and destroying microecological balance*:** As we all know, antibiotics can interfere with the ecological balance of normal microbes to induce dysbacteriosis and double infection. However, antibiotics not only induce dysbacteriosis but also decrease field planting resistance. In another word, long-term antibiotics usage can damage the barriers of normal microbes especially to anaerobic bacteria which have an important biological function. The conditioned pathogen or Ectogenesis pathogen, which are inhibited by normal microbes in normal condition, grow fast and induce infection in this case. For example, antibiotics therapy induces alteration of intestinal flora with diarrhea, enteritis, pseudomembranous colitis and so on.

***Bacterial drug resistance induced by abuse of antibiotics*:** Recently, drug resistant bacteria have spread all over the world. The number of deaths from infection by multidrug resistant bacteria has risen yearly. The result of bacterial resistant drug detection in China shows that from 1998 to 1999 hospital using the onset of infection in 13 hospitals in 9 regions, the infection rates induced by MRSA and MRSE were 27.55% and 15.67% and in 2001 the infection rate induced by MRSA was 81.82% and by MRSE was 41.67%.

China is one of the countries for antibacterial abuse. The information given by the Chinese Pharmaceutical Association shows that patients with diarrhea are total more than 0.8 billion yearly. Among them the average usage rate of antibiotics is 84%. In fact 70% of them do not need antibiotics. And, according to the investigation, 30% of hospital infection in patients is induced by drug-resistant bacteria every year. Among all of the bacteria separated in the clinic, ciprofloxacin resistant *Escherichia coli* takes the first place in the world.

The main reason for bacterial drug resistance is antibiotics abuse. Under the effect of antibiotics, mutation and selection of normal flora increase the power of antibiotics. For example, in a study by Knothe, patients take 25 mg tetracycline and an increase in bacterial drug resistance is discovered. In the enterobacteria drug tolerance transmission is very frequent. A stain with R factor can transfer the R factor, which is a determinant group and transmissible drug resistance factor. Under the effect of transposon and other plasmids, the rate of transmission of resistance is above 10-6.

The increase in resistance induced by antibiotics is relative to the R factor strain with F-pili, which can connect donor to receptor. In enterobacteria the transmission function of *E. coli* and *Bacillus aeruginosus* is the strongest. *E. coli* can transfer resistance to *Staphylococcus aureus*, pneumobacillus, *Bacillus influenzae*, Pseudomonas, *Bacillus typhi* and *Bacillus dysenteriae*.

In a word, under the effect of antibiotics, bacteria increase resistance through the mechanism of genetics and biochemistry. Microbial resistance is a natural biological response of microbes to a selective pressure, such as weather conditions, food, oxygen or water availability, or the presence of an antimicrobial drug. When a new class of antibiotics is introduced, it is effective at first, but will eventually only be useful against a small fraction of bacterial populations that have an intrinsic or acquired resistance mechanism [9]. Drug resistance of bacteria is becoming a big problem for scientists all over the world.

Transformation from endogenous to exogenous infection: In hospital, infection the rejection of an antibiotics sensitivity pattern and the selection of a drug resistance pattern (usually R plasmid carrier) are significant. Most of the strains from the intestinal tract of many patients from the countryside or other places with poor medical conditions are sensitive at the beginning of hospitalization. But not long after hospitalization a drug tolerant strain develops. This is because antibiotics kill the sensitive strain so that the drug resistant strain develops.

For many pathogenic bacteria of endogenous infection, features of exogenous infection appear in the hospital onset of infection. Staphylococci, *Bacillus aeruginosus* and others normal microbes are similar to *Bacillus typhi* and *Bacillus dysenteriae* in their epidemic tendency. In epidemiology this is an unwanted case and we should pay much attention to it.

4.1.3.2 Isotopes

An isotope, a radioactive substance, is related to radioactive X-ray application and contributes to human life significantly. It has become a severe social problem that immune function reduction induced by radiation disease results in a threat to human health. The main complication of radiation disease, which was well known after the atomic bomb explosions of Hiroshima and Nagasaki in Japan in the 2nd World War, is ecological disturbance and then a series of infections induced by normal microbes disturbance.

After exposure to an amount of rediogens, the number and function of phagocytes decreased and non-specific sterilizing material in the serum decreased to induce a lower defense ability. However, bacteria have strong resistance to radioactive X-rays. After exposure the drug resistance and toxicity of bacteria increase. So after exposure to radiation the microecological balance of normal microbes and hosts is lost, the sensitivity of the host to microbes increases, unapparent infection becomes apparent infection, the patient's condition is aggravated and the attack rate and death rate increased.

The radiation X-ray effect that induces microecology disturbance can be divided into the host aspect and microbes aspect. The details of both will be introduced below.

***Host aspect*:** After receiving an amount of radiogens and radioactive X-rays, the innate defense mechanism is damaged and the function and the number of phagocytes decreased. The ability of the immunoresponse is obviously destroyed and the biological characteristics of microbes have changed, such as the fermentability of lactose in *E. coli* has disappeared.

(i) The effects on the defense mechanism: The microbes are found in the exposed animal tissues including organs and blood, so we know that one part of the microecological balance of the host has lost the ability to maintain the balance. After being exposed to enteric microbes, *E. coli*, *Bacillus aeruginosus*, *Bacillus proteus*, Monilia, enterococci *etc.* can be found in tissues, but not in the control group. Petrov R V (1957) reported that, after exposure, enteric microbes could reach the lymph nodes after 2 days, alveolus after 3 days and blood after 4 to 5 days respectively, in rats. Likewise, at 48 hours after exposure (11GY/S) *E. coli* was injected subcutaneously into rabbits. After lesion formation the number of bacteria was 2.0×10^9/g in the experimental group and 1.0 – 8.0×10^4/g in the control group. From the above data it can be seen that the exposure affects the host resistance severely.

(ii) The effects on survival time of host: The effects on survival time of the host are relative to carrying or not carrying various kinds of bacteria. Germ-free animals live longer than normal animals. Animals with *Bacillus bifidus* or *Bacterium lacticum* live longer and those with *E. coli* live the shortest (Table 4.2).

Table 4.2 The effects of 20GY/S ray on survival time of SCID mice (1983)

Animal	Quantity	Survival time
GF Mice	10	9.6±1.2
B. breve YIT400g	10	12.2±1.3
Lactobacillus acidophilus XIT0168	10	11.2±1.2
E. Coli 0-26	10	6.9±1.4

Note: The four groups show a significant difference statistically ($P<0.01$)

***Microbes aspect*:** The effect of radiation on microbes is indirect. After the host is exposed to radiation the radiogens secreted in the human body affect microbes. The defensive ability to contain exposure of microbes is obviously bigger than the host's. Several pathological GY that is able to make an effect on humans and animals do not damage the structure of microbes, unless up to several hundred GY.

There are two types of changes after exposure to microbes.

(i) Drug tolerance increases: Before exposure, 92% of the gram positive occurring in dogs are sensitive to penicillin and 30% are sensitive to streptomycin. After exposure the sensitivities are as low as 5% and 19%.

(ii) Toxity reinforcement: In 1941, Bingyang Liu reported that after a mouse was exposed to inoculated Rickettsiae typhi, toxity of Rickettsiae typhi separated from the mouse was reinforced. Furthermore, after exposure the toxity of *Bacillus typhi* is reinforced and staphylococci too. For staphylococci the ability of mannitol resolution, hemolysin and hyaluronidase can become positive.

All the facts told us that isotope exposure has an obvious effect on microecology disturbance and is an inducing factor in microecology disturbance.

4.1.3.3 Operations

All operations can destroy the normal physiological and anatomical structure in varying degrees and are harmful to the ability of normal microbes to induce microecology disturbance. Contaminated small bowel syndrome, colon excision, stomach excision *etc.*, all can damage the intestinal microecosystem to induce steatorrhea, anaemia, water and electrolyte metabolisms and other microecological disturbances. For example, after stomach excision more bacteria can enter the small intestine and induce dysfunction, achlorhydria, abnormal biochemical events and bacteria of the small intestine syndrome. Operations on the low digestive tract induce intestinal flora alternation more obviously, increase the number of aerobes, such as enteric bacilli and enterococci, and decrease the number of anaerobes, such as *Bacillus bifidus*. Then the field planting function is decreased and the rate of postoperative infection is increased.

On the whole, for all the therapies including operation, urethral catheterization and intubation, which can damage the physioanatomic structure of host, we should lighten the damage to the normal microbes as much as possible. After operation, the microeclogical regulator should be used to help recovery of the balance as soon as possible and shorten the convalescent period.

4.1.3.4 Long-Term Chronic Diseases

Diseases such as malignant tumor, leukemia, diabetes, connective tissue disease *etc.*, can all partly lower the general immune function to make the conditioned pathogen grow fast and result in microecology disturbance.

4.2 Infection

Ever since humans appeared on the earth, infectious diseases have existed. Until now, infectious diseases induced by microbes have been one of the main reasons for death and disability. During the long evolutionary process microbes form a symbiotic relationship with humans and are human's lifetime partner, so infection is inescapable and universal. The purpose of infection is to avoid further infection. Infection as immunity is a normal physiological process. Without infection there will be no immunity.

4.2.1 The Concept of Infection

Infection is a very popular topic in social life and literature, so the connotation of infection is well known. With the improvement in science, the development of microecology and especially the study of the relationship between normal microbes and hosts in the medical domain, the concept of infection should be re-evaluated.

The concept of infection in traditional medicine is that "it is a pathophysiological process of the interaction of the host with invading pathogenic microorganisms" or "infection is a process of the host interacting and fighting invading pathogenic microorganisms". In short, all of these emphasize the relationship between pathogen and host and everything about the pathogen is an abstract and vague concept. Thus, to make the concept of infection clear we have to know the entity of the pathogen.

Then what is a pathogen? What is the entity of a pathogen? Why do some pathogens induce diseases and pathogen is the only carrier staying in the host? On the contrary, why not some other pathogens induce various kinds of infection? Using simple etiological ideas we cannot solve these problems clearly. From the view of microecology, microbes have no absolute relation to diseases and it means that there are no absolute pathogenic bacteria. All the so-called pathogenic bacteria now are a microecological manifestation in the host of the transformation and translocation process of normal microbes. For example, if the members of enteric normal microbes translocate to the respiratory tract or blood and induce diseases they become pathogenic microorganisms.

It mismatches the objective fact that calling some kind of microbes is pathogenic bacteria or non-pathogenic bacteria. In the individual microbes because of the difference in structure (capsule *etc.*), toxin production (endotoxin, exotoxin) and bionomics between genera, it is true that the pathogenicities are different. Objectively the differences do exist in that the pathogenicity of some microbes is strong to some hosts and weak to others. This is not up to microbes, but is closely related to the settled host, qualitation, location and quantitation of microbes and the condition and environment of the host. For the same bacterium, it may be normal in plants but pathogenic in animals. Here qualitation can be explained as correctly determining the genus of microbes. This is an important basis to decide the cause of a disease. Here, location refers to determining the microecological space and location where the microbes live in the body. For some bacterium it's normal in the intestinal tract but pathogenic when translocated to the respiratory tract. Quantitation is the process of ascertaining the number of microbes present. For the same bacterium, different number shows different pathogenicity. Thus, when deciding whether one microbe is pathogen of the disease or not, we should consider all of the factors including settled host, qualitation, location and quantitation aspects. Others have noted that avoiding sticking permanent label of pathogenic bacteria to a bacterium, etiologic agent should be used instead of pathogenic bacteria.

In conclusion, from a microecological view, the concept should be "infection is a kind of microecological manifest of the reaction between microbes and hosts induced by microbes abnormally infecting the host and macroorganism. This is the summation of abnormal attack of microbe to host, tissue and blood and defense of the host". The main appearance of infection is that it can induce the host to start the specific and non-specific immunoresponse to original bacteria. For microbes the ends of infection are proliferation, festriction and death and for host the ends are unapparent infections, germ-carrying, health, apparent infection, disease and death.

4.2.2 Types of Infection

Depending on the source of microbes infecting the host, infection can be divided into the two types below.

4.2.2.1 Exogenous Infection

It is the infection induced by microbes from outside of the host, such as typhoid fever, cholera, diphtheria and other infectious diseases.

The exogenous infections are not constant forever. Recently due to the induction of antibiotics some autogenous infection microbes, such as staphylococci, *Bacillus aeruginosus*, *Blastomyces albicans etc.*, have an extreme number of R factors and can be resistant to many antibiotics. For example, recently with the appearance of multidrug resistance bacteria, such as MRSA, penicillin resistant Streptococcus pneumoniae and vancomycin resistant enterococci, they result in the features of onset of infection are similar with those of exogenous infection and spread in population.

4.2.2.2 Endogenous Infection

Endogenous infection is the infection induced by original bacteria of host or other hosts. Recently autoinfection of endogenous infection is very common. With the development of society and improvement in medical conditions, exogenous infection decreased gradually with endogenous infection especially autoinfection mainly now occurring. For example, the original residence bacteria or by-pass bacteria, such as staphylococci, *Klebsiella* and *Blastomyces albicans*, have resistance to antibiotics. When antibiotics kill sensitive bacteria which are barrier and antagonism, the drug-resistance bacteria can grow excessively to induce autoinfection. Additionally when the immune function of the host is weak, autoinfection can be induced by the member of normal flora. Endogenous infection takes place under certain conditions, in clinic endogenous infection it is called opportunistic infection or conditional infection. These bacteria and fungi are the conditioned pathogen. These microbes are usually the cause of hospital infection.

Autoinfection is the infection induced by host original flora disturbance. Some factors make the host's immune function weak to induce infection caused by normal microbe translocation (Table 4.3).

Table 4.3 Microecology dynamic performance of infection

Infection type	Dynamic performance of microecology	
	translocation	change host
Autoinfection	+	–
Endogenous infection	+	+
Exogenous infection	–	+

4.2.3 Etiologic Agent of Infection

From the view of microecology, an etiologic agent is considered as normal flora in which translocation and changing host might induce infection because of certain factors. Frequent etiologic agents in human dysbacteriosis are introduced below.

4.2.3.1 Staphylococci

Staphylococci is a member of normal flora and minority in the intestinal tract. There are 10^4 – 10^6/g stool. The pathogenicity of *Staphylococcus aureus* is the strongest in staphylococci. *Staphylococcus albus* is lower. Both of them can appear in microecology disturbance. Someone observed that color is not a sign of pathogenicity. Generally haemolyticus, plasma-coagulase reaction, mannitol fermentation and activity of Lysozyme are considered as a sign of pathogenicity. Methicillin-resistant *Staphylococcus aureus* (MRSA) emerged in 1960 and was a problem confined largely to the healthcare setting, or hospital-associated MRSA (HA-MRSA). In the 1990s, community-associated MRSA (CA-MRSA) infections appeared [10].

From the view of microecology disturbance, as long as the number of any staphylococci reaching a certain number or planting corresponding site the bacteria make host suffer from diseases. The pathogenicity indices mentioned above are not relative to pathogenicity, sometimes there are many negative cases for coagulate Staphylococcus albus separated from septicopyemia patients. Another feature of staphylococci is its strong drug resistance. The literature shows MRSA is 20% to 66.7% of *Staphylococcus aureus* infection. Ninety seven point six percent of *Staphylococcus aureus* separated from hospital infection is MRSA.

4.2.3.2 *Bacillus proteus*

Bacillus proteus is a type of by-pass bacteria in healthy people. Although *Bacillus proteus* can be detected in 3% to 5% of normal people, the number of bacteria is low. *Bacillus proteus* has certain pathogenicity to rabbit and mice. The positive haemolysis and lecithinase *Bacillus proteus* may indicate pathogenicity.

There is no relationship between species of *Bacillus proteus* and pathogenicity. It can be separated in both severe and mild cases. It can be considered that the alteration of intestinal flora induced by *Bacillus proteus* depends on their number. Additionally if cooperated with the synergistic reaction of other etiologic agent, such as staphylococci and *Blastomyces albicans*, it may be more dangerous.

4.2.3.3 Fungi

Blastomyces albicans and yeast-like fungi are the members of enteric normal bacteria. In normal conditions they do not lead to disease. But under the induction of antibiotics the bacterium number exceeds 10^5/g stool or are dominant so that they can increase the onset of the diseases. Flora alternation disease of *Blastomyces albicans* is widespread. When pathological changes occur in the intestinal tract, infections of the oral cavity and the respiratory tract, such as stomatitis, ulcer of angle of mouth, rose tongue, dermatomycosis *etc.*, are usually seen. The reaction of antibiotics, bacteria and *Blastomyces albicans in vitro* shows that the antibiotics kill or inhibit sensitive bacteria to cause *Blastomyces albicans* to proliferate quickly and finally induce diseases. In light of their characters pathogenic bacteria of deep mycosis can be divided into five groups, the majority of them are *Cryptococcus neoformans* and various kinds of Monilia. And the latter *Blastomyces albicans* is an important infection bacterium in hospitals.

4.2.3.4 *Bacillus aeruginosus*

Bacillus aeruginosus can be detected in healthy people belonging to by-pass bacteria and does not induce disease. In dysbacteriosis, it can induce severe infection and some obstinate disease, especially after burning and scald *Bacillus aeruginosus* has a high infection rate and extensive resistance. So the infections of *Bacillus aeruginosus* are usually seen on the application of antibiotics.

4.2.3.5 *E. coli*

E. coli is a typical member of normal flora, but it can induce dysbacteria in abnormal condition. *E. coli* can be divided into enteropathogenic *E. coli*, enterotoxigenic *E. coli* and normal *E. coli*. The pathogenicity depends on the number, whether it is dominant or not. But we should be pay attention to whether the hemolytic and non-lactose fermenters stains of *E. coli* that separated from patients increased. Some observations prove that this might have some relationship with pathogenicity. Once enteropathogenic *E. coli* and enterotoxigenic *E. coli* grow prevalently, their pathogenicity tend to be very strong.

4.2.3.6 Klebsiella

Klebsiella is bigger Bacterium entericum with thick capsules and without flagella. The representative strain is pneumobacillus, which was first separated in patients with pneumonia by Friedlandel in 1882 and belongs to normal flora. But when dysbacteria takes place for some reasons, it can induce respiratory infection, such as pneumonia, urethra infection, liver and gall disease, septicemia and peritonitis.

4.2.3.7 Clostridia

Anaerobe Clostridia is a frequent etiologic agent of dysbacteria. Clostridia is member of normal flora too, especially *Bacillus aerogenes* capsulatus, although its pathogenicity is true. *Bacillus aerogenes* capsulatus do not induce disease under normal conditions. It is only when the number of clostridia exceeds the normal level that it can induce disease. Recently it is found that Clostridia usually induce severe enteric infection after magnanimous antibiotics and isotope application, especially pseudomembranous enterocolitis induced by Clostridium difficile, the death rate of which is high if no prompt treatment is exerted. Besides many other anaerobes, *Bacteroides*, *Streptococcus anaerobius* and Wei Rong acidilactici *etc.* also can induce dysbacteria. But anaerobic infection is usually diphasic. This means that the secondary anaerobic infection takes place after aerobe infection, which needs oxygen. This just reflects the microecological law of infection.

The frequency of occurrence of etiologic agent in different countries and different areas may vary. The difference mainly lies in antibiotics selection. The extent and type of antibiotics application directly decide which bacterium is predominant and which is in an inferior position. Generally the more extensively and variedly the antibiotics are used, the higher frequency of dysbacteria is, and the bacterium with strong resistance can grow predominately early.

A report from the former Soviet Union Moscow Institute of Medicine shows that aerobe *Bacillus proteus* and staphylococci are predominate in etiologic agents of intestinal dysbacteria while lactose negative *Escherichia* and *Blastomyces albicans* are the next most common. In the dysbacteria induced by these bacteria, synergistic reaction of two or more bacteria is frequent. See the details in Table 4.4.

It is shown in Table 4.4 that *Bacillus proteus* and staphylococci take 11.3% in the total of etiologic agent, lactose negative *Escherichia* is 50%, *Blastomyces albicans* is only 3.5% and synergistic reaction of two or more bacteria with the highest frequency is 68.8%.

Table 4.4 The frequency of occurrence to etiologic agent of intestinal tract dysbacteriosis

Etiologic agent	Research amount	Dysbacteriosis amount	Etiologic agent				
			Proteus	Staphylococcus	lactose (−) *Escherichia*	*Candida albicans*	synergism
Acute dysentery	171	70	10	11	3	2	44
Chronicdysentery	132	71	7	6	7	4	47
Colitis	157	107	10	13	4	3	77
Others	63	34	5	2		1	26
Total	523	282	32	32	14	10	194
(%)			11. 3	11. 3	5. 0	3. 5	68. 8

4.2.4 Epidemic Links of Infection

There are three epidemic links of exogenous infection or communicable diseases, which are infection sources, route of transmission and susceptible population. There are also three links of autoinfection and endogenous infection, reservoir, translocation channel and susceptible habitat. The first three links are the microorganisms spread process among the population; the latter three links are the transfer process of the etiologic agent or pathogenic microorganisms between different habitats in the body.

4.2.4.1 Reservoir

In the macro-epidemiology, "reservoir" is translated as bacteria storage animals or host, while in the micro-epidemiology it can be translated into bacteria storage library or habitats.

In microecology, the reservoir is the storage place or the main source of the etiologic agent. In the human body there are four important reservoirs, oral cavity and upper respiratory tract flora, intestinal flora, urinary tract bacterial flora and vaginal flora. The four reservoirs release etiologic agents to the habitats around. Their relationship was shown in Fig. 4.5.

As can be seen, oral infections, etiologic agents of upper respiratory tract infection, urinary tract infections, pelvic infections and intra-abdominal infection were all from the micro-communities of the nearest microhabitat.

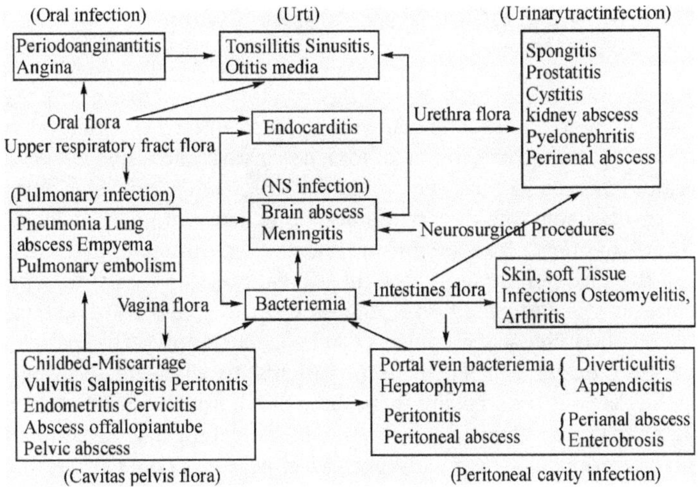

Fig. 4.1. The translocation channel of reservoirs and etiologic agent from autogenous infection

4.2.4.2 Translocation Channel

In addition to the blood and the lymph node, the etiologic agents transfer from the reservoir to the susceptible habitats mainly by trauma, surgery, intubation and other treatment measures.

4.2.4.3 Susceptible Habitats

Susceptible habitats are mostly the injured organs and tissues. The normal organs and tissues in general are not easily infected. Ischemia, trauma or mechanical stimulation can enhance the colonization capacity of etiologic agents. The host's condition such as malnutrition, immune dysfunction, antibiotics, radiation therapy, and aging, also can increase the sensitivity of habitat.

4.3 The Relation between Microecology Disturbance and Infection

According to traditional biological pathogeny theory, infection is caused by pathogenic microorganisms, while in microecology, infection is an important aspect of microecological balance and microdysbiosis transformation. Infections were not necessarily caused by pathogenic bacteria or pathogens, as they may be the results of normal microflora translocation or host.

4.3.1 Traditional Biological Pathogeny Theory

In light of the biological pathogeny theory, infectious disease is caused by infection of pathogenic microorganisms. It starts from the taxonomy of species to study the species and even virulence, toxicity, invasiveness and other biological characteristics between strains. Such research is necessary, but we cannot correctly evaluate the pathogenic role of certain microorganisms without full account of the relationship of the host, microorganisms and the environment, and the microbial biomass. In the past people used "the three conditions of Koch" to set standards for the selection of pathogens: Firstly, the lesions can consistently demonstrate the presence of bacteria; Secondly, bacteria can be isolated only in the patient; Thirdly, transferring the bacteria isolated from patients to the sensitive animals can reproduce the disease. When using these three conditions to measure the members of normal microflora, none can be regarded as pathogens, because all of the pathogenic microorganisms can be separated in healthy people. Therefore we can see that biological pathogeny theory is one-sided and unrealistic.

4.3.2 Ecological Pathogeny Theory

In light of the ecological pathogeny theory, from the view of microecology, pathogenicity of microorganisms depends not only on the characteristics of microbial species, but also on the qualitative and quantitative, location and host settled the host transformation outcome of microecological balance of three aspects, host, environmental and microbiological, which are the performance in the process of microecological imbalance. For example, diseases could result from the significant increase or decrease of the normal flora, the intrusion of non-original bacteria, intestinal flora transferred to the respiratory tract and the normal flora of animals transferred to the human body. It can be seen that ecological pathogeny theory is consistent with the ecological objective laws.

According to traditional biological pathogeny theory the infections are harmful. However, within the ecological pathogeny theory, the infections are common. Incidence is adventives. The infection is the best way to stop infection. Infection as immunity is a physiological function, and only under certain conditions it may shift from physiology to pathology or from microecological balance to microecological imbalance.

4.3.3 The Significance of Infection

Infection is widespread in both the system development and the individual development. From the view of system development, all the organisms and microorganisms in early evolution have experienced the process of contact, interaction, mutual restraint and mutual adaptation. These microbes are an integral part of an organism's existence.

4.3.3.1 The Universality of Infection

From the view of individual development, (*e.g.* from human birth to death) life is endlessly being accompanied by infection. In 1960, Reimann put forward the concept of spectrum infection. He used a large number of evidence to prove that infection is widespread, while the incidence is adventive and death is rarer. Within streptococcal infection, for example, health carriers and latent infection accounted for 86%, the incidence accounted for 12.6%, and deaths, only 1.4%.

Animal experiments show that intestinal epithelial cells and the occurrence of lymph node centers of sterile animals did not develop as well as normal animals did, and compared with the sterile animals, the intestinal epithelial mucosa of normal animals tends to a normal "mild inflammation of signs". This is a sign of infection and also shows that infection is common.

4.3.3.2 Physiology of Infection

From the view of evolution, the universality of infection is a physiological basis of infection. The existence of all living beings is developed in the relationship with the microorganisms and is the necessary outcome of common development of microorganisms with the host. Without infection there would be no immunity, no existence and no continuity of life. Infection is the reason for immunity, while immunity is the result of infection. When an infection has just appeared in the crowd, the spectrum of infection morbidity and mortality accounted for a large proportion, but during a pandemic period, its proportion is shrinking, while the proportion of carriers and the latent infection expand. This is because the infection induces immunity, which can control the infection and thereby protect most of the host from morbidity or death. In addition, compared with normal animals serum the bactericidal effect on the Salmonella of the sterile animal serum is much weaker, which also shows that both the infection and immunity are a result of the physiological phenomenon of living beings.

4.3.3.3 Infections are the Best Measures to Stop Infection

Infection is inevitable. Infection is the best way to prevent and treat infection and is a physiological function to keep the microecological balance of microbes and their hosts. Germ-free animals are exposed to bacteria environment, even if the saprophytic bacterium such as *Bacillus proteus* could also the cause of their death. The reason is that in the past the animals had not received such a bacterial infection. In this sense, we should try to contribute to infection, expand infection, play the physiology of infection, and avoid the pathology of infection. For example, application of artificial infection: the current BCG, polio vaccine, measles vaccine are the typical examples to extend and promote the infection. Use harmless infection to prevent harmful infection and prevent too much control of natural infection. A newborn is sterile at birth. The normal microflora is built up gradually by contacting with the external environment. This is a natural process so that children receive a certain capability to control harmful infection. However, the current health care measures often violate the law of nature. In recent years there is a delivery hyper-health syndrome, under the conditions of a high degree of cleaning and disinfection delivery and nurse newborns, delaying the succession process of the newborns obtaining normal microflora from the natural environment. Once exposed to a normal environment, the newborns are vulnerable to infectious diseases and have indigestion and other syndromes. Based on the above understanding, we should try to contribute to the normal physiological succession.

4.3.4 Microdysbiosis Induces Infection Diseases

In today's society, because of a large number of applications of antibiotics, cytotoxic drugs, immunosuppressive agents, radiotherapy and chemotherapy and surgery and other medical means, incidence rate of infectious disease caused by microecological imbalance is increasing year by year, which attracts widespread concern in the whole society. Below we introduce primitively the diseases of human organs and tissues induced by microecological disturbance.

***Oral cavity*:** Microecological imbalance in the oral cavity expresses itself in the form of periodontitis, oral ulcers, thrush, *etc.* Tongue pain and eating difficulties occur earlier than others. The results of oral secretions, smears and cultures show that the anaerobic bacteria and fungal infections are the most prevalent and the detection rate of *Candida albicans* is the highest.

***Respiratory system*:** Microecological imbalance due to lung infection often appears clinically when the original lung infection of the patient has been controlled gradually. In a large number of antibiotics applications, low heat appears intractably and after the replacement of antibiotics, the disease may turn better for 1 to 2 days and later gradually turn worse after several recurrences of the high fever, which is inconsistent with the signs appearing. In Sputum culture the growth of conditioned pathogen or fungi can be seen. Susceptibility test shows that they are resistant to common antibiotics. The patients died from more severe lung infection due to persistent unhealed, body constitution consumption and organ failure.

***The digestive system*:** The flora in the digestive system is a large and complex ecosystem. The incidence of antibiotic-associated diarrhea (AAD) is common, with a reported incidence of 5% - 30% in critically ill patients [11]. Recently the clinical probability of disease caused by intestinal microecology disturbance is rising year by year, about 2% to 3%. Microecological disturbance in the digestive system can appear as acute and chronic diarrhea, pseudomembranous colitis, small bowel syndrome, and others.

***Urinary system*:** For microecological disturbance, the fungal infections in the urinary system is the most prevalent. An early manifestation is when the urine appears cloudy and with floc. The urinary tract irritation symptoms are often not obvious and the fungi can be seen in urine microscopy.

***The nervous system*:** Dysbacteriosis can induce intracranial infection, in part caused by the conditions of pathogen infection and in the other part by a fungal infection. Fungal infection is characterized by a moderate fever, cerebrospinal fluid slight cloudiness, cell number increasing moderately.

***Multiple organ failure*:** A lot of clinical studies confirm that bacterial translocation in patients with serious disease is an important factor contributing to multiple organ failure. After the intestinal flora disturbance endogenous infection and endotoxemia caused by the bacteria and endotoxin translocation is not only the hotbeds of stimulating the inflammatory response but also the engine of multiple organ failure. Impaired intestinal barrier function is implicated in the pathogenesis of sepsis and multiple organ dysfunction syndrome (MODS) in patients with decreased gut perfusion resulting from surgery, trauma, or shock [12, 13].

4.3.5 Mechanism of Microorganisms and Host

Since the completion of the human genome project, some new subjects, such as genomics, proteomics, infectiomics, have emerged. The so-called "group study" is to study objects generally and comprehensively. Infectriomics is to "group study" infection including two aspects: structure and function, when the body infected by microbes, genotype and phenotype are changed. These changes must be encoded by microorganisms and the host genome. Therefore it includes changes of replication, transcription, translation and the level of post-translational modifications *etc.* in microorganisms and host. Studies of the human microbiome have revealed that even healthy individuals differ remarkably in the microbes that occupy habitats such as the gut, skin and vagina [14]. The diversity of microbes within a given body habitat can be defined as the number and abundance distribution of distinct types of organisms, which has been linked to several human diseases: Low diversity in the gut to obesity and inflammatory bowel disease, for example, and high diversity in the vagina to bacterial vaginosis [15-19].

When microbial infection occurs, different microorganisms within the same host may give rise to inducing different infection groups. Infection groups with different characteristics can be used to study the infection mechanism of microorganisms. Today, DNA microarray and proteome analysis have been used for a comprehensive study of microorganisms' gene expression patterns in a certain environment mode with the same host.

The occurrence and development of infection are determined by the interaction of the host environment – microbe. The balance between these relationships is very important to our health. Once the balance is lost, infectious diseases are likely to occur. However, at present our understanding of the molecular mechanism of microecological balance and disturbance is still very poor.

Therefore, a comprehensive and integrated application of infectiomics combinining the new methods (such as DNA and protein microarray) and traditional methods (such as molecular cloning, PCR, gene knockout, *etc.*) can help us effectively and accurately study the mechanism of interaction between microbes and host.

References

[1] Bai K. Microecological Principle. Dalian: Dalian Publishing House, 1996.
[2] Gill S R, Pop M, Deboy R T, *et al.* Metagenomic analysis of the human distal gut microbiome. Science, 2006, 312: 1355-1359.
[3] Yang J Y. Medical Microecology. Beijing: China Medical Science and Technology Publishing House, 1997.
[4] Proctor L M. The human microbiome project in 2011 and beyond. Cell Host Microbe, 2011, 10: 287-291.
[5] Schindler D, Gutierrez M G, Beineke A, *et al.* Dendritic cells are central coordinators of the host immune response to *Staphylococcus aureus* bloodstream infection. Am J Pathol, 2012, 181: 1327-1337.
[6] Sozinov A S, Abdulkhakov S R, Kiyasov A P, *et al.* Alteration of the liver in rats with experimental dysbiosis. Bull Exp Biol Med, 2003, 1: 19-21.
[7] Ritz M, Fraser R, Tam W, *et al.* Impacts and patterns of disturbed gastrointestinal function in critically ill patients. Am J Gastroenterol, 2000, 95: 3044-3052.
[8] Khan M W, Kale A A, Bere P, *et al.* Microbes, intestinal inflammation and probiotics. Expert Rev Gastroenterol Hepatol, 2012, 6: 81-94.
[9] Soares G M, Figueiredo L C, Faveri M, *et al.* Mechanisms of action of systemic antibiotics used in periodontal treatment and mechanisms of bacterial resistance tothese drugs. J Appl Oral Sci, 2012, 20: 295-309.
[10] El Amin N M, Faidah H S. Methicillin-resistant *Staphylococcus aureus* in the western region of Saudi Arabia: Prevalence and antibiotic susceptibility pattern. Ann Saudi Med, 2012, 32: 513-516.
[11] Szymsnski H, Pejcz J, Jawien M, *et al.* Treatment of acute infectious diarrhea in infants and children with a mixture of three Lactobacillus rhamnosus strains a randomized, double-blind, placebo controlled trial. Aliment Pharmcol Ther, 2006, 23: 247-253.
[12] Derikx J P, Poeze M, van Bijnen A A, *et al.* Evidence for intestinal and liver epithelial cell injury in the early phase of sepsis. Shock, 2007, 28: 544-548.
[13] Holland J, Carey M, Hughes N, *et al.* Intraoperative splanchnic hypoperfusion, increased intestinal permeability, down-regulation of monocyte class II major histocompatibility complex expression, exaggerated acute phase response, and sepsis. Am J Surg, 2005, 190: 393-400.
[14] The Human Microbiome Project Consortium. Structure, function and diversity of the healthy human microbiome. Nature, 2012, 486: 207-214.
[15] Turnbaugh P J, Hamady M, Yatsunenko T, *et al.* A core gut microbiome in obese and lean twins. Nature, 2009, 457: 480-484.

[16] Fredricks D N, Fiedler T L, Marrazzo J M. Molecular identification of bacteria associated with bacterial vaginosis. N Engl J Med, 2005, 353: 1899-1911.
[17] Huang H J, Zhang A Y, Cao H C, et al. Metabolomic analyses of faeces reveals malabsorption in cirrhotic patients. Dig Liver Dis, 2013, 45: 677-682.
[18] Ling Z, Liu X, Chen W, et al. The restoration of the vaginal microbiota after treatment for bacterial vaginosis with metronidazole or probiotics. Microb Ecol, 2013, 65: 773-780.
[19] Ling Z, Liu X, Luo Y, et al. Associations between vaginal pathogenic community and bacterial vaginosis in Chinese reproductive-age women. PLoS One, 2013, 8: e76589.

5

Nosocomial Infections and Bacterial Resistance

Yonghong Xiao
The State Key Laboratory for Diagnosis and Treatment of Infectious Diseases, the First Affiliated Hospital, School of Medicine, Zhejiang University; Collaborative Innovation Center for Diagnosis and Treatment of Infectious Diseases, Hangzhou, 310003, China
E-mail: Xiao-yonghong@163.com

Nosocomial infections and bacterial resistance are two significant healthcare problems, which lead to not only prolonged hospital stays and extra medical costs, but also increased mortality. This chapter covers the epidemiology of nosocomial infections (including incidence, source of infection, route of transmission, risk factors, *etc.*), infection classifications, pathogenic microorganisms, clinical manifestations, diagnosis, treatment, and infection control measures, such as hospital management, infection surveillance, sterilization, isolation, education and preventive measures for common hospital infections. The section about bacterial resistance discusses the global trends of bacterial resistance, especially the "superbugs" (a serious threat to patient safety), including methicillin-resistant *Staphylococcus aureus*, vancomycin-resistant enterococcus, extended-spectrum beta-lactamase producing Enterobacteriaceae, and multiple drug resistant non-fermentative bacteria. As for the mechanisms of bacterial resistance, this chapter mainly describes target-site mutation, antibacterial hydrolyzing or modifying enzyme, and the active drug efflux system in detail. At last, the measures against bacterial resistance are briefly discussed.

Nosocomial infections and bacterial resistance are two significant challenges confronting modern healthcare facilities, Previous research has shown that surgical wound infections could extend patients stay in hospitals 8.2 additional days on average. In the United States, nosocomial infections contribute to about 100,000 deaths a year and $15 billion extra costs. In the UK, nosocomial infections contribute to an average of four additional hospital days and $4.5 billion extra costs yearly [1-3]. Advancing age, higher prevalence of chronic diseases, and increased use of invasive diagnostic and therapeutic procedures in hospitalized patients are risk factors for nosocomial infections and will impose continuing pressure on the control of nosocomial infections in the future.

Bacterial resistance is closely associated with nosocomial infections. Drug-resistant bacteria cause most of the cases of nosocomial infection, which is the most significant characteristic differentiating nosocomial infection from community-acquired infection. The intense antibiotic usage in hospitalized patients, especially those with immune deficiency or receiving invasive procedures, promotes the emergence of drug-resistant strains of bacteria. Antibacterial sensitive microorganisms are inhibited or eliminated, while resistant strains remain and persist, and even become endemic in the hospital. The common multiple drug-resistant bacteria and pan-resistant bacteria, such as methicillin-resistant *Staphylococcus aureus* (MRSA), extended-spectrum beta-lactamase producing Enterobacteriaceae, multiple resistant *Pseudomonas aeruginosa*, pan resistant *Acinetobacter baumannii etc.* have become the main cause of nosocomial infections. Drug-resistant bacterial infections also lead to prolonged hospital length of stay, additional costs and increased mortality. Moreover, bacterial resistance and nosocomial infections interact with each other, making the control of infection trickier [1, 4, 5].

5.1 Nosocomial Infections

With the rapid development of medical technology and the extensive use of medical devices, intervention therapy, tumor chemotherapy and organ transplantation, nosocomial infection has become an increasingly serious health problem, which has a great impact on society, the economy and health care. Strengthening nosocomial infection control, reducing its incidence, relieving the patients' suffering, and cutting down medical costs are challenges confronting the global healthcare system.

5.1.1 Introduction

The WHO defines nosocomial infections (or hospital acquired infections) as infections acquired by the patients and their nursing members or hospital staff during patient care, which are not present or incubating at admission, no matter whether the infectious symptoms appear in the duration of hospitalization or not. In 2002, the Ministry of Health of China adopted the definition and explained nosocomial infections as infections occurring in hospitalized patients, in whom the infection is not present or incubating at the time of admission, including infections which are acquired in the hospital but appear after discharge, and also occupational infections among staff of the facility (revised in 2006). In 2008, CDC/NHSN defines nosocomial infection as a localized or systemic condition that results from adverse reaction to the presence of an infectious agent or its toxin(s), there must be no evidence that the infection was present or incubating at the time of admission to the acute care setting[3, 4, 6].

According to the definitions, the followings are nosocomial infections:

- Infections without specific incubation period occurring at 48 hours or more after hospital admission;
- Infections with specific incubation period occurring after average incubation period counted from the time of admission;
- Infections obviously associated with previous hospital stay;
- New infections (except metastatic focus of sepsis) occurring at a site other than the previous one;
- New infections caused by new pathogens different from the pathogens of previous infection (except contamination and previous co-infection);
- Infections that newborn acquired during or after delivery;
- Infections caused by potential pathogens activated by treatment, such as herpes virus and *Mycobacterium tuberculosis*;
- Infections that the healthcare staff acquired at work.

The followings are not nosocomial infections: Inflammatory symptoms caused by wound or irritation of non-biological factors; infections acquired by newborn via placenta (occurring within 48 hours after birth), such as herpes simplex, toxoplasmosis, chicken pox, hepatitis B; acute attack of previous chronic infections during hospitalization. Patients have topical colonization of microorganisms but without any symptoms of infection.

Nosocomial infections can be classified into two categories depending on the source of pathogens: endogenous infections and exogenous infections.

5.1.2 Epidemiology of Nosocomial Infections

The morbidity and mortality of nosocomial infections are variable in different hospitals, patient population and specific wards. The nosocomial infectious types are also distinct in different areas.

5.1.2.1 Prevalence of Nosocomial Infections

The incidence of nosocomial infections varies greatly in different countries, ranging from 1% to 40% [1, 3]. Early reports showed that the incidence of nosocomial infections was approximately 10% in China. In 2001, a cross-sectional nationwide study involving 107,496 hospitalized patients from 193 tertiary hospitals demonstrated that the prevalence rate of nosocomial infections was 5.52% ranging from 5.73% to 6.39%. The top-five nosocomial infections were respiratory tract infections (48.7%), urinary tract infections (12.8%), surgical site infections (11.9%), intestinal infections (10.9%), and skin infections (6.7%). In 2003, the prevalence rate of nosocomial infections in 159 hospitals nationwide was 4.77%, and the distribution of nosocomial infections had no obvious change. Studies showed larger hospitals (more than 500 beds), public hospitals, and teaching hospitals had higher incidence of nosocomial infections [7, 8]. A cross-sectional study in Hong Kong carried out in 2005 showed that the prevalence rate of nosocomial infections was 4%, in which the rate of pneumonia, blood infections and urinary tract infections were 1.4%, 0.9%, and 0.8%respectively [9].

The WHO reported that the rates of nosocomial infections were 5.1% in Norway, 7% in Spain, and 9.1% in Greece; in the Netherlands, thanks to the effective preventive measures, the rate of nosocomial infections was 1%; in some countries of Asia, South America and Africa, the rate of nosocomial infections was over 40% [1, 3]. In the United States, surveillance data collected by CDC in 2002 indicates that, there are as many as 1.7 million cases of nosocomial infections per year, which contribute to 99,000 deaths and $17–20 billion extra costs. Nosocomial infections are common in intensive care unit and healthcare facilities for patients with serious illness. The most common nosocomial infections are urinary tract infections (36%), surgical wound infections (20%), pneumonia (11%), and bloodstream infections (11%) [10]. In 2006, the rates of nosocomial infections were 8.19% in England, 6.35% in Wales, 5.43% in Northern Ireland and 4.89% in the Republic of Ireland. The most common nosocomial infections in these countries were intestinal infections (20.6%), urinary tract infections (19.9%), surgical wound infections (14.5%), pneumonia (14.1%), skin infections (10.4%) and bloodstream infections (7.0%) [11]. The rate of nosocomial infections in tertiary hospitals in Malaysia was 13.9% [12].

5.1.2.2 Source of Nosocomial Infections

***Exogenous infection*:** Exogenous infection, also called cross infection, is caused by microorganisms acquired from other patients, healthcare staff, visitors as well as hospital surroundings. Pathogenic microorganisms may be present on objects, equipment and in a hospital environment. Gram-negative bacilli survive months or even years in humid conditions, while Gram-positive cocci are present in air and on surfaces of dust and objects. *Klebsiella, Enterobacter, Citrobacter etc.* survive well in glucose infusion. Gram-positive anaerobic bacilli have high resistance and a long lifecycle, and may transmit through air, unsterilized objects, or wound exudates. Pathogenic microorganisms may spread between patient and healthcare staff by hand contact. The outbreak of infections with MRSA, vancomycin-resistant enterococci (VRE) or *Clostridium difficile* are usually exogenous.

***Endogenous infection*:** Endogenous infection, also called autogenic infection, is an infection caused by an infectious agent that is already present on the patient's skin, mouth, throat, stomach, intestines *etc.* When the patient's immunity is impaired and protective barrier is injured, bacterial translocation and flora imbalance may lead to infections. Endogenous bacteria cause most cases of antibiotic-related diarrhea.

5.1.2.3 Transmission Route of Nosocomial Infections

Just like other infectious diseases, nosocomial infections are transmitted through various routes. Most cases are transmitted by direct contact and contaminated blood. Such media as air and devices transmit a few cases [3].

Contact Transmission: Direct contact transmission refers to pathogenic microorganisms from infection sources transmitting to the contacted person, such as neonate infection in the mother's vagina or cervix during delivery. Infections in newborns with group B *Streptococcus hemolyticus*, *Neisseria gonorrhoeae*, *Listeria monocytogenes*, Herpes simplex virus, *Chlamydia trachomatis*, hepatitis B virus *etc.* are all results of direct contact transmission. Indirect contact transmission involves contact of susceptible hosts with a contaminated media, such as contaminated environment, instruments, devices *etc.* The contaminated hands of healthcare staff are the most common indirect contact transmission media [13, 14].

Bloodborne Transmission: Bloodborne transmission refers to infections caused by contaminated blood, blood products or infusion. The common pathogens include hepatitis B virus, hepatitis C virus, HIV, cytomegalovirus, toxoplasma, and plasmodium. If a certain batch of medicinal products is contaminated, an outbreak of infections may occur in a unit or hospital [15].

Airborne Transmission: Pathogens, such as influenza virus, *M. tuberculosis*, herpes virus, and *Aspergillus*, can spread from one patient to other via air, aerosol, *etc.* Air spread staphylococcal and streptococcal infections were reported occasionally, but the cases are much less than those caused by contact transmission. The results showing the air sterilization of the operating room could reduce postoperative wound site infections suggest that some wound infections are caused by airborne microorganisms.

Food Transmission: Contaminated water or food may lead to outbreaks of nosocomial infections, e.g. the outbreak of infectious diarrhea occurs in hospitalized patients after taking *L. monocytogenes* contaminated dairy products or norovirus contaminated foods [16, 17].

Instruments Transmission: Contaminated or inadequately disinfected medical instruments may lead to patient infections, e.g. respiratory tract infections in patients with tracheal intubation or mechanical ventilation caused by *P. aeruginosa*, *Acinetobacter*, and *Klebsiella pneumoniae*. Atypical mycobacterial surgical wound infections caused by contaminated surgical instruments soaked with low concentration disinfectant is another example. The recycle of invasive medical devices (such as tubes, catheters, endoscope, mechanical ventilator, and dialysis device, *etc.*), which is difficult to be thoroughly sterilized after using, is also one route of infection transmission.

5.1.2.4 Risk Factors for Nosocomial Infections

Nosocomial infections mainly occur in patients with impaired immunity and those who received more invasive procedures. The risk factors for nosocomial infections include two categories [1, 3, 13, 14, 18-20]:

(i) Patient factors

i) Patient age: Infants and young children (under 3 years old) have a lower defense to infections due to their underdeveloped immune system. The elderly are susceptible to infections due to their impaired immunity and complication of chronic diseases. The incidence of nosocomial infections in people over 65 years old is twice as high as that in young adults.

ii) Chronic underlying diseases: The incidence of chronic diseases increases with age. Patients with malignant tumors, diabetes, COPD, chronic kidney disease, chronic liver disease, and senile dementia, *etc.* are all exposed to high risk for nosocomial infections.

iii) Immunological status: Patients with congenital or acquired immune deficiency, such as congenital immunity impairment or AIDS patients are vulnerable to nosocomial infections.

iv) Illness condition: Patients suffering from burns or wounds are vulnerable to nosocomial infections, because of the damage to the body's natural protective barrier against pathogens. Unconscious patients are at high risk of aspiration pneumonia.

(ii) Iatrogenic Factors

i) Invasive procedures and indwelling foreign implants. Intravascular catheter, tracheotomy, cardiac catheter, urinary catheter, biliary drainage catheter, mechanical ventilator, peritoneal dialysis, hemodialysis, lumbar puncture, and cerebrospinal fluid shunt *etc.* damage the body's natural protective barrier, introduce pathogens into the human body and increase infection risks. The incidence of hepatitis B infection in hemodialysis patients is up to 13.3%–88.9%. The incidence of urinary tract infections in patients using an indwelling urinary catheter for two weeks is 50%, and the incidence of bacteremia is 5.8 times as high as that of non-catheterizing patients. Implants, such as artificial heart valve, artificial blood vessel, and artificial joint also increase the risk of bacterial infections.

ii) Surgical operations. Prolonged duration of operation, serious damage to tissues, contaminated surgical site, preoperative shaving *etc.* may lead to surgical wound infections.

iii) Use of immunosuppressant. Long-term use of corticosteroids or immunosuppressant may lead to nosocomial infections, especially infections of opportunistic pathogens, such as mycobacterium, fungi, viruses and protozoa.

iv) Chemotherapy and radiotherapy. Patients with malignant tumors are susceptible to nosocomial infections because of their impaired immune systems, non-intact skin and mucosa, and agranulocytosis caused by chemotherapy and radiotherapy. Moreover, the pathogens are usually difficult to identify.

v) Antimicrobial therapy. Antimicrobial agents are often used for treating and preventing bacterial infections, but the long-term and improper use of antimicrobial agents will lead to flora imbalance, super infection and drug-resistant bacterial infections.

5.1.3 Pathogens of Nosocomial Infections

Bacteria, especially Gram-negative bacilli, are the main pathogen of nosocomial infections. In recent years, nosocomial infections with Gram-positive cocci and fungi are increasing. Viruses, mycoplasma, and protozoa are also important pathogens responsible for nosocomial infections. The present incidence of nosocomial infections with fungi is even 2–5 times as high as that in the 1980s. In the United States, nosocomial infection studies in 2006 and 2007 showed that 87% of the nosocomial infections were bacterial infections and 13% were fungal infections. In China, a study in 2003 showed that 75.96% of the nosocomial infections were bacterial infections and 24.04% were fungal infections [1, 7, 21].

5.1.3.1 Bacteria

Over 80% of the nosocomial infections are caused by bacteria, mainly opportunistic pathogens and drug-resistant pathogens, in which Gram-negative bacilli comprise 60%, including Enterobacteriaceae (*E. coli*, *K. pneumoniae*, *Proteus*, *etc.*) , *P. aeruginosa*, *Acinetobacter*, *etc.* In recent years, the incidence of nosocomial infections with *Stenotrophomonas maltophilia* and flavobacterium tend to rise, especially in patients in intensive care units and those with mechanical ventilation. *S. aureus*, coagulase negative staphylococcus (CNS), and enterococcus are common Gram-positive cocci that caused nosocomial infections, especially infections of skin and skin structure infections, surgical wound infections, and primary bloodstream infection. Infections caused by CNS are gaining more attention. *Staphylococcus epidermidis* is the most common pathogen that leads to infections related to an indwelling vascular catheter and ventricle drainage. *S. epidermidis* may also lead to the infections of an artificial joint or artificial heart valve. Enterococci mainly cause urinary infections and surgical wound infections. Group B *S. hemolyticus* is the main pathogen for meningitis and bloodstream infections in newborns. More seriously, over 90% of the isolates of staphylococci are penicillinase positive strain, which makes penicillin inactivation. MRSA is on the rise, accounting for 50%–70% or a higher percentage of isolates of *S. aureus* in some big hospitals. It may also lead to a breakout in some units of the hospital. Vancomycin-resistant enterococci (VRE) are also the pathogen, which comprise 10%–20% in the US but less than 3% in China [22]. *Legionella pneumophila* and other legionella may also cause nosocomial infections. It is reported that *L. pneumophila* pneumonia accounts for 3%–10% of hospital-acquired pneumonia. Fast growing mycobacterium, such as *Mycobacterium avium*, *Mycobacterium chelonae* and *Mycobacterium fortuitum* may lead to surgical wound infections and other infections in patients who have received cardiac surgery or other surgery.

Bacteroides is the most common pathogen of anaerobic bacterial infections. It may cause abdominal infections and pelvic infections following gastrointestinal surgery and gyne-obstetrics surgery. Fusobacterium, peptococcus, and actinomyces may lead to mouth infections and respiratory system infections, such as aspiration pneumonia, necrotizing pneumonia, lung abscess, and pyothorax. There are also cases of bloodstream infections and endocarditis caused by bacteroides and propionibacterium.

Colitis that appears after use of antibiotics caused by *C. difficile*, which is named as antibiotic-associated diarrhea or *C. difficile* associated diarrhea (CDAD), may spread in the hospital. The occurrence of CDAD has tended to rise in recent years. It is estimated that it affects 2 million patients yearly in the US. Besides the use of antimicrobial agents, some cases are also associated with other factors, such as proton pump inhibitor and advanced age. The actual situation is even more serious due to inadequate knowledge of this disease.

In 2006 and 2007, the most common pathogens of nosocomial infections in the United States were CNS (15%), *S. aureus* (15%), enterococcus (12%), *E. coli* (10%), and *P. aeruginosa* (8%). In mainland China, in 2003, the most common pathogens of nosocomial infections were *P. aeruginosa* (12.63%), *S. aureus* (7.47%), *Klebsiella* (6.8%), CNS (6.14%), *Acinetobacter* (5.73%), enterococci (4.5%), *Enterobacter* (4.45%), and *S. maltophilia* (1.89%). In Hong Kong, the most common pathogens of nosocomial infections were *P. aeruginosa* and *S. aureus* [9, 21, 22].

5.1.3.2 Fungi

With the widespread use of broad-spectrum antibiotics, invasive medical instruments, cardiovascular procedures, and organ transplantation, the incidence rate of nosocomial fungal infections is significantly increasing. The rate in the 1990s was 2–5 times as high as that of the 1980s, and outbreaks of fungal infections are more and more common. It is reported that the incidence of fungal infection is 28%–42% in patients that underwent liver transplantation, <5% in kidney transplantation, 10%–35% in heart transplantation, and 10%–35% in bone marrow transplantation. Nosocomial infection surveillance data showed that the most common pathogens of fungal infections are *Candida albicans* (78.3%), *Candida glabrata* (7.3%) and *Aspergillus* (1.3%). Recently, infections with non-albicans *Candida*, such as *Candida tropicalis*, *Candida parapsilosis*, *Candida krusei etc.* tended to increase, and are primarily resistant to fluconazole. *Candida* leads to not only nosocomial lung infections, digestive tract infections, but also bloodstream infections in patients with indwelling intravenous catheters, and even invasive candidiasis in immunocompromised patients. *Aspergillus* is one of the most common infectious pathogens in patients with acute non-lymphocytic leukemia, in which the pulmonary aspergillosis are not rare. Cryptococcal meningitis may also occur in patients with immunodeficiency [23, 24, 25].

5.1.3.3 Viruses

Common nosocomial viral infections include respiratory tract infections caused by respiratory syncytial virus or parainfluenza virus, influenza, rubella and viral hepatitis [15, 17, 19]. Cytomegalovirus infections are common in patients that underwent solid organ transplantation or bone marrow transplantation. Most cases of nosocomial viral hepatitis are hepatitis B or hepatitis C, which are closely associated with blood or blood products transfusion, and hemodialysis. Group B *coxsackievirus* may lead to infections in infants and become epidemic. Diarrhea caused by rotavirus or norovirus is common in infants and the elderly. Herpes simplex virus, cytomegalovirus, and varicella-zoster virus can all lead to endemic infections.

5.1.3.4 Others

C. trachomatis may cause conjunctivitis or pneumonia in newborns. Ureaplasma and Gardnerella vaginalis may parasitize in post kidney transplantation patients, who are also vulnerable to Pneumocystis jiroveci and toxoplasma infections. Blood transfusion may transmit malaria. Infections with amoeba, *Toxocara canis* and *Strongyloides stercoralis* are common in psychotics and children with mental retardation. *S. stercoralis* may also be transmitted through organ transplantation.

5.1.4 Common Nosocomial Infections

The type and incidence of nosocomial infections vary depending on the patients' illness, medical procedures, and use of medical devices in different hospitals and areas. In western countries, it is reported that the most common nosocomial infections are urinary tract infections, lower respiratory tract infections, surgical wound infections, and bloodstream infections. In China, data showed that the most common nosocomial infections are respiratory tract infections (25%–40%), followed by urinary tract infections, wound infections, gastrointestinal infections, and bloodstream infections [2,7,11].

5.1.4.1 Lower Respiratory Tract Infections

Lower respiratory tract infections comprise 15%–20% of nosocomial infections.

The incidence of lower respiratory tract infections in hospitalized patients is 1% – 2%, of which 20% are put on a mechanical ventilator. In China, pulmonary infection is the most common nosocomial infection, accounting for 10% – 30%, the incidence in hospitalized patients is 0.5% – 5% and much higher in patients in intensive care units. According to a CDC report, an average of 6.7 – 24.1 cases of ventilator-associated pneumonia (VAP) occurs per 1000 hospital days. Pulmonary infection is a very common complication in patients in intensive care units and patients with serious underlying diseases, and its mortality is up to 30% – 50%. The mortality of VAP is up to 70% [2, 7, 8, 10, 26].

According to the data of National Nosocomial Infections Surveillance System (NNIS), of the pathogens causing hospital-acquired respiratory infections, 52% are Gram-negative bacilli (mainly *P. aeruginosa*, *Klebsiella*, and *Enterobacter*), 19% are Gram-positive cocci (mainly *S. aureus*). In China, the most common pathogen of hospital-acquired respiratory infections is *P. aeruginosa*, followed by *Klebsiella*, *S. aureus* and *E. coli*.

Pathogens of hospital-acquired pulmonary infections vary with the length of hospital stay. In general, hospital-acquired pneumonia is divided into early onset and late onset pneumonia with a hospitalizing time of 5 days. *S. pneumoniae*, Haemophilus influenzae, staphylococcus, or enteric bacilli often cause early onset pneumonia, while Gram-negative bacilli (> 60%), which includes *P. aeruginosa* (27%), *Acinetobacter* (12%), *Klebsiella* (10%), and *Enterobacter* (7%), *S. aureus* (23%) or fungi (6%) are major pathogens of late onset pneumonia. Multiple drug resistant bacteria, such as MRSA, *P. aeruginosa*, *Acinetobacter*, *S. maltophilia*, and flavobacterium, cause most ventilator-associated pneumonia. Immunocompromised patients are vulnerable to infections with *Aspergillus*, *Candida*, *P. jiroveci*, cytomegalovirus, zoster virus, *C. trachomatis*, and atypical mycobacterium. Patients suffering from coma and shock often have anaerobic bacterial (peptococcus, peptostreptococcus, and fusobacterium) infections or co-infections of anaerobic and aerobic bacteria because of the aspiration of oral secretion. Respiratory syncytial virus is the most ccommon pathogen of lower respiratory infections in infants under two years old, and both its morbidity and mortality rates are high.

Most patients get nosocomial pneumonia by aspiration of oropharynx and gastrointestinal flora, or by cross-infection among patients or by staff's hand transmission. Ventilator, sprayer and humidifier all can be the media for transmission. Medical instruments with a liquid storage device may create appropriate conditions for the growth of Gram-negative bacilli. The aspiration of oropharyngeal and gastrointestinal excretion, substantial proliferation of local colonizing organisms, impaired pharyngeal and pneumobronchial immunodefensiveness and non-sterile or invasive sputum suction in patients undergoing tracheotomy may lead to lung infections. Other risk factors for pulmonary infections include contaminated flushing fluid, antimicrobial therapy, long-term bed rest in the elderly, aspiration of secretion, and contaminated intermittent pressure ventilation, *etc.* In the first 5 days of mechanical ventilation, the incidence of VAP increases by 3% each day, from day 5 to day 10 the incidence of VAP decreases by 2% each day, and after 10 days it decreases by 1% per day. About 20% of patients receiving thoracic surgery or upper abdominal surgery have apparent symptoms of infection; nearly half of the patients indicate radiological evidence of pulmonary infections. Smoking, ongoing lung diseases and more than 2 hours duration of surgery will increase the rate of postoperative infections. Pneumonia usually appears 2 – 4 days after surgery [26, 30].

The pathogenesis of lung infection is associated with colonization of organisms on respiratory mucosa. The fibronectin on the surface of host cells and protease in saliva also plays an important role. The loss of fibronectin on the surface of mucosal cells may facilitate the adhesion, colonization and proliferation of Gram-negative bacilli on the respiratory mucosal cells. Epithelial cell damage caused by some other factors and flora imbalance caused by antimicrobial therapy may promote the colonization of pathogens on respiratory mucosa and thus lead to pulmonary infection.

The clinical manifestations of nosocomial pneumonia lack specificity. The diagnosis mainly depends on the combination of clinical symptoms and chest imageological findings. Patients with fever, leukocytosis, purulent sputum, and newly occurring radiological opacities in pulmonary parenchyma can be diagnosed with pneumonia. To assist the diagnosis and antibacterial therapy, qualified sputum or bronchoalveolar lavage fluid should be collected for microbiological examination.

Nosocomial pneumonia should be timely treated with proper antimicrobial agents; otherwise, the mortality will be high. The choice of antimicrobial agents for empirical therapy should comprehensively take into account the local bacterial resistance and the risk factors for acquiring drug-resistant bacteria, such as a history of antibiotics use within 90 days before admission, hospitalizing duration, local isolation rate of multiple drug resistant (MDR) germs, previous use of intravenous antibiotics within 30 days prior to infection, chemotherapy, wound care, regular hemodialysis, immunity, immunosuppressant application, *etc.* Nosocomial pneumonia without risk factors for MDR bacterial infection can be treated with narrow-spectrum antibiotics, such as third-generation cephalosporins, fluoroquinolones, ertapenem, *etc.*; otherwise, with broad-spectrum antibiotics, such as cephalosporin or carbapenems that act with a potent activity against *P. aeruginosa*, or even a combination of antimicrobial agents [26].

5.1.4.2 Urinary Tract Infections

Urinary tract infection (UTI) is the most common nosocomial infections in western countries. According to an NNIS report, it is the most common nosocomial infection (40%) in the US, and its morbidity ranges from 2% to 5%. However, the incidence of nosocomial UTI has significantly declined recently, accounting for 28% of all nosocomial infections. In patients with hospital-acquired urinary infection, 90% have a history of urinary tract instrument use or urinary catheterization, and 5% – 10% are associated with other urinary procedures such as cystoscopy. Blood spread and unknown original nosocomial UTIs are rare. Being female, aged, with urinary tract obstruction, vesicoureteral reflux, incomplete bladder emptying, and inappropriate use of antimicrobials are all the risk factors of UTI. The most common pathogen of hospital-acquired UTI is *E. coli*, accounting for 30% – 50% of all cases, which is quite lower, nevertheless, than that of community-acquired UTI (80%). Other common pathogens include enterococci (14%), *Klebsiella* (12%), *P. aeruginosa* (6%) and *Proteus* (4%). Ten percent of the hospital-acquired UTI is caused by *Candida*, and its morbidity tends to increase. Besides, a few patients with long-term use of indwelling urinary catheters suffer from co-infection of two or more pathogens. In patients with catheter-associated UTI, 1% – 5% are predisposed to bacteremia [31, 32].

Most nosocomial UTIs are caused by pathogens retrograding transmission through a urinary catheter. Infectious pathogens get to the bladder along with the urinary catheter and urine collection container. Recent study shows that catheter-related UTIs are associated with the bacterial biofilm formation on the contact surface between the urinary catheter and mucous membrane. In female patients, especially those with indwelling urinary catheters, pathogens of stool or bedsore may contaminate and colonize the urethra and lead to ascending infection. Cross infection transmitted by medical staff's hands plays an important role in UTI. In addition, contaminated flushing fluid or inadequately disinfected cystoscopies can transmit pathogens. The morbidity of bacteriuria is 1% – 5% in patients with single catheterization, 90% in patients with an indwelling urinary catheter and without closed drainage. Twenty to twenty-five percent of patients with an indwelling urinary catheter and closed drainage are detected with bacteriuria in 48 hours. The infection rate increases in duration of catheterization; the risk of bacteriuria is 5% – 10% after one day of catheterization, and 50% after two weeks of catheterization [31, 32].

The typical symptoms and signs of hospital-acquired UTI include urinary frequency, urgency, odynuria, lower abdominal pain, fever, *etc.* Catheterization-associated UTI usually has no apparent symptoms, and fever may be the main complaint. UTI is usually diagnosed by urine analysis; the presence of 10 leukocytes per high power field of microscope in centrifuged urine has diagnostic value. Urine culture shall be performed to identify the pathogens and guide the antibacterial therapy. Asymptomatic bacteria may occur in some patients with an indwelling urinary catheter, in which condition urine culture is usually non-obligated. The choice of antimicrobial agents for nosocomial UTI depends on the infection type, pathogen category and bacterial resistance. Asymptomatic bacteriuria usually does not need routine antimicrobial treatment and spontaneous cure after the removal of a urinary catheter is anticipated. However, bacteria in patients with immunodeficiency and patients receiving urinary surgery or artificial implant should be treated with antimicrobials [32].

5.1.4.3 Surgical Wound Infections

Surgical wound infections refer to infections of superficial or deep tissue and cavity in surgical sites. Despite progress achieved in aseptic technique, surgical skill, and infection prevention measures, surgical wound infection is still one of the most common nosocomial infections, which has a proportion ranging from 10% to 19%. The infection morbidities in clean incision, clean-contaminated incision and contaminated incision are 2%, 2% – 10%, and 10% – 20%, respectively. Newborns, the elderly and obesity patients are vulnerable to surgical wound infections. Diabetes, corticosteroids or immunosuppressant application can all lead to such infections. Prolonged preoperative hospital stay, long-term bed rest, hypoproteinemia, increased duration of operation, excessive loss of blood, serious tissue damage and drainage are the risk factors for surgical wound infection. *S. aureus* is the most common pathogen of surgical wound infection, of which the isolating rate ranges from 17% to 20%. In recent years, surgical wound infections caused by CNS and enterococci have tended to increase, accounting for 12.6% and 13.3% respectively. Gram-negative bacilli including *P. aeruginosa* and *E. coli* are responsible for 50% of the surgical wound infections. Anaerobic bacteria, such as bacteroides, are the common pathogen of postoperative gynecological infection [33, 34].

Infective microorganisms may come from the patient's own skin, from other patients or staff members, from contaminated instruments, dressings, *etc.* Most of the surgical wound infections appear 3 – 5 days after surgery. Superficial infections may have such symptoms as redness, swelling, heat, pain and purulent discharge around the wound. Deep infections may have such symptoms as increased wound tenderness, worsened wound pain, heat at site of incision. Persistent infections often lead to delayed healing, wound sinuses formation, and purulent discharge around the wound *etc.*

5.1.4.4 Bloodstream Infection

The morbidities of nosocomial bloodstream infections (BSIs) are 0.3% – 2.8%. There are 30,000 – 50,000 cases of nosocomial BSIs per year in the US, which caused 17,000 deaths. The surveillance and control of pathogens of epidemiologic importance conduced in 49 US hospitals in 1997 – 2002 showed that every 6-hospital patients in 1,000 admissions acquired BSIs, in which 50.5% occurred in intensive care units and 24% were associated with intravenous catheterization, and the mortality was 27%. The Hospitals in Europe Link for Infection Control through Surveillance (HELICS) showed that the incidence of BSIs was 3.3% in patients who stayed in the intensive care units for >2 days, of which 60% was associated with intravenous catheters, and the incidence of catheter associated BSIs was 1.23% – 4.2% every 1,000 catheter-days. In developing countries, the incidence of catheter associated BSIs is as high as 7.8% – 18.5% every 1,000 catheter-days. About 50% of nosocomial BSIs led to 24 additional hospital days on average [10, 35, 36].

BSIs may be either primary or secondary. Primary BSIs with a recognized pathogen cultured from blood and the organism do not relate to an infection at another site. Secondary BSIs result from infection spreading at other sites. Half of primary BSIs do not have infectious origination and another 50% of primary BSIs are associated with intravenous infusion, intravascular operations, or hemodialysis. Secondary BSIs are primarily related to the infections of the urinary tract, surgical wound, the respiratory tract, and skin, *etc.*

Nosocomial BSIs are usually sporadic, and sometimes may cause an outbreak, such as BSIs caused by contaminated intravenous infusion, blood, or blood transfusion apparatus. Sporadic BSIs do not relate to intravascular operations and are often caused by pathogen spread from other sites. Most sporadic BSIs are secondary, and merely one quarter is primary. On the contrary, about 80% epidemic nosocomial BSIs are primary, often outbreak intensively in a certain ward, even in patients with normal immune function.

Newborns, the elderly aged over 60 years old, patients with serious injury, burn, fatal underlying diseases, or agranulocytosis, and patients receiving corticosteroids, immunosuppressant treatment, or anti-tumor chemotherapy are all at high risk of BSIs. Risk factors of BSIs include intravenous infusion, transfusion of blood or blood products, systemic use of antimicrobial agents, intravascular procedures, and hemodialysis. The longer the duration of the catheter (more than 48–72 hours) is, the higher the morbidity of BSIs is. The incidence of BSIs is 0.05% in patients not receiving intravenous infusion, 0.37% in patients with peripheral vascular catheters, and 4.48% in patients with central vascular catheters. According to NNIS surveillance, 3–5 cases of BSIs occur in 1,000 hospitalized patients, and 2–14 cases appear every 1,000 hospital-days on average. The incidence of catheter-related BSIs is also associated with catheter type. The infection rate is low in patients using a peripheral venous catheter or midline catheter, and high in patients using a non-tunneled catheter, central venous catheter and arterial catheter for dialysis [10, 37].

The common pathogens of nosocomial BSIs are Gram-positive cocci (>60%), followed by Gram-negative bacteria (27%) and fungi (8%). In Gram-positive cocci, CNS is the most common (31%), followed by *S. aureus* (15.7%) and *E. faecalis* (11.1%). Catheter-associated BSIs are primarily caused by CNS, and non-catheter-associated BSIs primarily caused by *S. aureus*. Most Gram-negative bacilli BSIs are caused by *E. coli*, *Klebsiella*, *Enterobacter*, and few are caused by *P. aeruginosa* and *Serratia*. *Candida* BSIs comprise 6% – 8% of all BSIs. BSIs caused by *E. coli* occur 12 days after hospitalization on average, *S. aureus* 15 days, and enterococci and *Candida* longer [37, 38].

The diagnosis of BSIs mainly depends on blood culture. CNS detected in a single blood culture shall be evaluated carefully by considering clinical symptoms. CNS in two and more non-consecutive blood cultures from different blood-taking sites usually have diagnostic value. Catheter-related BSIs are usually diagnosed by IDSA criteria: A positive result of semiquantitative or quantitative catheter culture, whereby the same organism is isolated from a venous blood culture taking from another site, and the presence of symptoms of bloodstream infections. For patients without blood culture results, clinical relieving after de-catheterization may indirectly indicate the catheter-related BSI [39].

BSIs need antimicrobial therapy, and catheter-associated BSIs often need de-catheterization or re-catheterization. The mortality of nosocomial BSIs ranges from 20% to 50%. The factors affect the prognosis including advanced age, fatal underlying disease, intra-abdominal infection or lower respiratory tract infection, stay in intensive care unit, metastatic focus of infection, shock, special pathogenic infections such as *P. aeruginosa*, *Serratia*, *Klebsiella*, *Bacteroides*, *Candida*, *etc.*, and improper use of antimicrobial agents for treatment.

5.1.4.5 Gastrointestinal Infections

The most common nosocomial gastrointestinal infections (GIs) are antibiotic associated diarrhea and pseudomembranous enteritis. Other GIs such as viral hepatitis and infectious diarrhea may also occur, and sometimes may even cause an outbreak.

C. difficile is the most important pathogen of antibiotic associated diarrhea. Although *S. aureus* sometimes can be detected in patients' stool, it is just a symbiosis bacterium. As a result, antibiotic associated diarrhea is also named *C. difficile* associated diarrhea (CDAD). CDAD is common in patients who have undergone gastrointestinal surgery, patients with bowel obstruction, uremia, aplastic anemia, or diabetes, and elderly patients using antibiotics. If not treated in time, CDAD may lead to pseudomembranous enteritis, and the mortality is up to 30%. Pathogens can be isolated from medical staff's hands, floor, bathroom, bedding, and furniture. Staff's hands may play an important role in the transmission of *C. difficile* [18].

The morbidity of antibiotic associated diarrhea had been at a low level for a long time, about 20/100,000. However, the incidence of CDAD began to rise since 2000. In Quebec and Canada, its incidence became three times higher and it even became an epidemic in some hospitals. At the same time, the bacterial virulence increased. Apart from toxin A and toxin B, the isolated *C. difficile* also produces binary toxin, and is remarkably resistant to quinolones. With an increase in CDAD cases, not only the elderly but also healthy adults and young children become vulnerable. The disease-related antimicrobial agents are no longer limited to clindamycin and ampicillin. The use of quinolones or proton-pump inhibitors is definitely associated with CDAD [18, 40].

The clinical manifestations of CDAD vary greatly in patients. Mild cases may only have symptoms such as loose stool and increased stool frequency, while serious cases may indicate symptoms such as high fever, severe diarrhea, toxic shock, toxic intestinal tympanites, intestinal perforation *etc.* The diagnosis of CDAD mainly depends on symptoms and a stool bacteriotoxin test. The treatment of CDAD includes suspending previous antimicrobial therapy, wiping out risk factors, and taking metronidazole or vancomycin orally. Metronidazole had been recommended as first choice for CDAD for a long time. However, recent study showed that metronidazole was superior for mild cases but vancomycin for severe ones. Besides, a microecological agent may also be helpful for the treatment of CDAD [40, 41].

Most nosocomial viral hepatitis (hepatitis B and hepatitis C) is caused by blood or blood product transfusion, or hemodialysis. It is reported that the incidence of post-transfusion hepatitis C is 3% – 21% in other countries and 13% – 20% in China. The incidence of post-transfusion hepatitis B is 5% – 10%. Hepatitis C is the most common post-hemodialysis hepatitis in countries other than China, the incidence is 4.6% in France, 5.8% in the US, while in China hepatitis B is the most common. The symptoms, diagnosis and treatment of nosocomial viral hepatitis are the same as routine viral hepatitis [1, 5].

Infectious diarrhea is the most common nosocomial infection, primarily caused by salmonella. Pathogenic *E. coli* or staphylococci may also be the cause. Shigella, campylobacter, yersinia, Entamoeba histolytica, rotavirus and norovirus are all reported to have led to nosocomial infectious diarrhea outbreaks.

5.1.4.6 Other Nosocomial Infections

Hospital-acquired skin and skin structure infections, including pyoderma, scabies, impetigo, bedsore infection and shingles, account for 5% of all nosocomial infections. The most common pathogens of those infections are *S. aureus* (42% – 48%), *P. aeruginosa* (13%), *E. coli* (8%), and enterococci (7%). Skin infections caused by *S. aureus* often can become epidemic. Most of the babies who stayed in the hospital for >4 days will have *S. aureus* colonizing their navels, noses and skin. In baby units of some hospitals, the colonization rate is as high as 25%. Thirty percent of workers in the baby units also have *S. aureus* colonizing in their noses. However, such colonization does not always lead to infections. *S. aureus* can also adhere to bedding, clothes, floor, tables and chairs. It is transmitted by contact in baby units. The staff's hands play an important role in cross infection. Cases transmitted by air are rare.

Nosocomial central nervous system infection is common in patients receiving craniocerebral operation or cerebrospinal fluid shunt. Most of the pathogens are Gram-negative bacteria, accounting for 38%, including *Enterobacter*, *P. aeruginosa*, and *Acinetobacter*. Common pathogens also include *S. aureus*, CNS, and streptococci, accounting for 20%. A few cases are caused by *S. pneumoniae*, *H. influenzae*, enterococci, and *L. monocytogenes* and seldom by *C. albicans*, whose incidence is about 7% and mortality as high as 35% [42, 43].

In recent years, with the increased use of joint diagnosing technology and artificial joint replacement, the incidence of nosocomial joint infection is rising year by year. According to reports, the infection rates of artificial joint replacement are 2.5% – 8.9% in other countries, and 0% – 5% in China. The symptoms of infection include joint redness, joint swelling, joint pain, confined activity, and osteomyelitis in some cases. The most common pathogens are CNS, *P. aeruginosa*, and fungi. Besides antibacterial therapy, removal of artificial devices or implants are often required, so antimicrobial prophylaxis is often recommended for arthroplasty.

Most nosocomial cardiovascular system infections occur in patients receiving angioplasty, heart valve replacement surgery or a heart pacemaker. Reports showed that the incidence of nosocomial cardiovascular system infection is 2.5%. The most common infectious pathogens after post valve replacement are CNS, Gram-negative bacilli and fungi.

Patients undergoing any abdominal surgeries, including surgeries on the liver, gall, pancreas, and spleen for tumor, stone or cyst, are all at risk of postoperative intra-abdominal infection. Patients with necrotizing pancreatitis are even at higher risk of infection, because of the increased use of indwelling drainage tubes. The common pathogens of abdominal infections are Enterobacteriaceae, including *E. coli*, *Citrobacter*, *Klebsiella* and enterococci.

5.1.5 Prevention of Nosocomial Infections

There are many risk factors associated with nosocomial infections. Taking measures to deal with these factors may effectively reduce morbidity of nosocomial infections. Some control measures may be complicated, but most of them are simple, practical and effective. For example, the importance of hand washing has been well demonstrated for over 100 years, but there still is room for further improvement in healthcare staff's hand hygiene. More effort is required in hospital management, organization, regulations, training, supervision, *etc.* so as to control sources of infection, cut down the route of transmission, and reduce nosocomial infections [1, 3, 4, 6].

5.1.5.1 General Measures

Establishing a hospital infection control organization: Hospitals should establish an authorized infection control committee, composed of infection control professionals and administrators, being responsible for nosocomial infection management, training, supervision, monitoring and control.

Formulating rules and regulations for infection control: The Infection Control Committee should formulate written regulations to specify infection control responsibilities and duties of physicians, nurses, pharmacists, microbiologists and administrators.

Enhancing education and training of all staff: All the hospital staff, including medical personnel and all other categories of staff, such as administrators, cooks and cleaners, should be trained and educated for nosocomial infection prevention and control. The compliance of staff with disinfection and isolation practices should be ensured. All staff should follow appropriate practices of hand washing and wash hands before and after contact with patients. Let all staff realize that hand washing is an important way to prevent nosocomial infections [44].

5.1.5.2 Establishing Nosocomial Infection Surveillance System

A nosocomial infection surveillance system aims to monitor the incidence, distribution, risk factors, and pathogens of nosocomial infections, identify deficiencies as early as possible, and work out preventive measures. A computer network should make full use of processing, storing and analyzing surveillance data.

The following formula has been used for years in China to calculate the incidence of nosocomial infections: Incidence (%) = Number of new infections acquired in a period/Number of patients observed in the same period × 100%. The formula widely used internationally is: Incidence/day (‰) = Number of new infections acquired in a period/Total of patient-days for the same period × 1000‰. The latter is more accurate and scientific, but seldom used in China. Besides the incidence of nosocomial infection, risk factors, pathogenic microorganisms and bacterial resistance should also be further investigated. Outbreaks of nosocomial infection should be promptly identified and investigated [3, 45].

New medical staff should undergo comprehensive physical examination. Bacteria culture should be performed regularly for organisms collected from the noses and hands of workers in the wards. Workers who carry *S. aureus* (in nose) should apply topical mupirocin ointment. Persistent carriers of *S. aureus* should stop working in the wards.

Results of nosocomial infection investigation should be reported to relevant departments and clinical offices promptly. Proper preventive intervention should be taken for correcting wrong ideas or actions, and improving staff compliance with infection control measures. Follow-up for significant problems and preventive measures to make further suggestions for infection control are also the compulsory work of nosocomial surveillance.

5.1.5.3 Hospital Construction and Environment

The hospital is a place where patients receive medical treatment and healthcare staff work. Infection control must be considered in hospital design and construction, and all the nosocomial infection associated factors such as hospital layout, functional section, traffic flow, logistics flow, operation room, heating, lighting, air conditioning, disinfection, isolation *etc.* should be considered. During the operation of the hospital, water, food, waste, transportation, and storage of medical supplies, microorganism surveillance, *etc.* should be carefully monitored.

5.1.5.4 Disinfection and Isolation

***Hand hygiene*:** Contaminated healthcare staff's hands are an important means of transmitting pathogens. Hospitals should take measures to improve awareness of medical staff about hand disinfection, monitor and improve their compliance with hand hygiene requirements, and provide necessary disinfectant (such as soap, alcoholic rub) to minimize the contact transmission of hospital acquired infections [3, 44, 45].

***Isolation*:** The isolation of some special infected patients, such as those infected with VRE, is very important for nosocomial infection control. Healthcare staff should take proper protective measures when they are in contact with isolated patients or high-risk patients, such as wearing masks and isolation gowns.

***Disinfection and Sterilization*:** A hospital environment plays a certain role in nosocomial infections. Monitoring wards regularly for bacteria and taking proper measures according to the laboratory bacterial monitoring results are necessary for nosocomial infection prevention and control. The application of disinfectant and modern disinfection technology, the proper disposal of medical waste, and even the optimization of instructions for the cleaning of kitchen and bathroom are all important measures to stop the transmission of nosocomial infections.

5.1.5.5 Proper Use of Antimicrobial Agents

The proper use of antimicrobial agents is a key strategy for bacterial resistance and nosocomial infection control. Nosocomial infection control professionals should cooperate with hospital Drug and Therapeutics Committee to publicize, educate, train and monitor health staff in antimicrobial use to reduce the inappropriate use of antimicrobial drugs.

5.1.5.6 Protection of Patients and Medical Staff

Besides the patient's personal predisposing factors, nosocomial infection is also associated with medical procedures and practices. Healthcare staff should keep patient protection in mind, perform aseptic techniques well, minimize invasive procedures, dispose medical waste in time properly, and nurse tubes and catheters correctly. Patients and medical staff at higher risk should be vaccinated to prevent some nosocomial infections, such as hepatitis B virus.

5.1.5.7 Handle Outbreak of Nosocomial Infection

An outbreak is defined as an unusual or unexpected increase of a known nosocomial infection or the emergence of a new infection. Outbreaks of nosocomial infection usually can be identified by an infection surveillance system.

Once an outbreak occurs, its risk factors, source of infection, route of transmission, *etc.* must be promptly investigated and confirmed, and effective control measures should be taken.

5.1.5.8 Measures for Prevention of Common Nosocomial Infections

***Nosocomial pneumonia and ventilator-associated pneumonia*:** Keeping patients in a semi-recumbent position, avoiding invasive mechanical ventilation if possible, ensuring aseptic intubation and suctioning, and limiting duration of artificial breath support are effective measures for prevention of nosocomial pneumonia. Digestive decontamination and changes of ventilator circuit every 48 or 72 hours are not recommended for all patients. Preventing potentially pathogenic microorganisms from colonizing oropharynx is an important measure. Both hand-washing and taking sterile gloves can prevent exogenous microorganisms from colonizing oropharynx. Patients with gastrorrhagia should avoid taking antacids and H_2 blockers and be treated with sucralfate (it helps prevent gastrorrhagia and does not change intragastric pH) instead. Patients who underwent surgery should get out of bed as early as possible, and take painkillers to relieve wound pain so as not to hinder coughing or deep breathing. Critically ill patients with swallowing difficulty should be fed with a nasogastric tube to reduce the risk of aspiration pneumonia [26, 44].

***Urinary tract infections*:** Avoid urethral catheterization unless there is a compelling indication. To practice aseptic technique at catheter insertion, maintaining closed drainage and limiting duration of catheterization are helpful in preventing UTI. According to some professionals, intermittent catheterization is good in patients with neurogenic bladder, or spinal cord injury for UTI prevention. Prophylaxis use of antimicrobial agents and bladder irrigation of antiseptic solution is proved not effective [32, 44].

***Surgical wound infections*:** Clipping rather than shaving hairs in skin preparation is recommended. Treating nasal cavity *S. aureus* carrier, minimizing the duration of preoperative hospital stay, optimizing surgical techniques, reducing tissue damage, deflating cavity residual, and stopping bleeding instantly will decrease the postoperative infection rate. Ultraviolet radiation in the operating room reduces surgical wound infection rate and a laminar airflow room reduces infections transmitted by air. The appropriate use of preoperative antimicrobial prophylaxis is proven effective in reducing surgical wound infection in some operations [34].

***Bloodstream infections*:** Ensuring hand washing and aseptic techniques, wearing sterile gloves when inserting an indwelling nutrition infusion tube or conducting intravenous fluid infusion for patients who are vulnerable to infections benefit the prevention of bloodstream infections. Using a finer needle for venous puncture, performing peripheral vein instead of femoral venous puncture, limiting an indwelling catheter to as short a duration as possible, and avoiding venesection unless there is a medical indication are valuable measures for BSI prevention. The catheter should be removed immediately, and catheter, pinhead and patient' blood should be collected for bacterial and (or) fungal culture when topical skin infection or phlebitis appears. Topical use of antimicrobial cream has not yet proven effective in preventing such infections [39, 44].

5.2 Bacterial Resistance

Bacterial resistance is becoming a global public crisis. More and more infections are caused by multiple-drug or pan-drug resistant bacteria occurring around the world. WHO appeals to all the member countries to take effective measures against bacterial resistance.

5.2.1 *Prevalence of Bacterial Resistance*

The occurrence and development of bacterial resistance are closely related to the clinical use of antibacterial agents. Back in 1940, when penicillin had not been officially put to clinical application, Abraham and Chain discovered Penicillinase in *E. coli* for the first time in history. In 1956, Newton and Abraham discovered cephalosporinase in *Bacillus cereus*. The problem of bacterial resistance is drawing more and more global attention. Later, β-lactamase stable penicillins (methicillin, oxacillin, *etc.*) were developed and put to clinical application. However, MRSA emerged shortly in 1960. At the turn of the 21st century, vancomycin-resistant *S. aureus* (VRSA) was reported in the USA for the first time. This has also been the case with the drug resistance of Gram-negative bacilli. The β-lactamases generated by the bacilli have developed from a narrow spectrum to a broad spectrum, extended-spectrum β-lactamases (ESBLs), carbapenemase, β-lactamase inhibitor hydrolyzing β-lactamases, *etc.* Bacterial resistance has already become a serious threat to public health world wide.

5.2.1.1 Drug Resistance of Gram-positive Bacteria

***Drug resistance of staphylococci*:** The drug resistant rate of clinically isolated staphylococci against benzylpenicillin has exceeded 90%, mainly due to the generation of penicillinase. Therefore, penicillin is no longer used clinically for treating staphylococcal infection. Moreover, the main clinical problem with the drug resistance of staphylococci is the prevalence of methicillin-resistant strains. Ever since the first MRSA strain was found in the UK in 1960, the drug-resistant isolates have been gradually spreading to the whole world. MRSA isolating rate is now anywhere from 5% to 70% among different countries and areas. Moreover, the rate of methicillin-resistant coagulase-negative staphylococci is even higher. The resistant mechanism of staphylococci against methicillin is mainly due to acquired *mecA* gene that encodes penicillin-binding protein 2a with lower affinity with β-lactams. Such strain often exhibits resistance to multiple drugs, including all β-lactams, aminoglycosides, macrolides, sulfonamides, and quinolones. They are only sensitive to a small number of drugs such as glycopeptides and linezolid. Methicillin-resistant staphylococci are one of the main nosocomial infection pathogens [46, 47].

Due to the disparity in the application of antibacterial agents and measures for infection control, the isolating rate of MRSA differs greatly among different countries or areas. The isolating rate of MRSA in Europe ranges between 2% and 54%, which is relatively low in North Europe, Central Europe, and Scandinavian countries. For example, it is below 5% in the Netherlands, Germany, Belgium, Denmark and Sweden. However, it is 30% – 60% in France, UK, Portugal, Greece, Italy, and Romania. The MRSA isolating rate continuously rose from 22% in 1995 to 57% in 2001, and the present is over 60% in the US hospitals. Even worse, the prevalence rate in ICU rose from 35.9% in 1993 to 64.4% in 2003. Significantly, it is higher in the south of the US than in other regions. By the results of bacterial resistant surveillance in China, the isolating rate of MRSA in hospital-acquired infections was over 60% in 2007. By consecutive resistant surveillance, MRSA has sharply increased by more than twice in the past 10 years. The antibacterial resistant spectrum of MRSA isolated in China is broader than that in other areas and SCC*mec* III is its main drug-resistant genotype. In other regions of Asia, MRSA isolating rates are 67% (Japan), 60% (Taiwan, China), 55% (Hong Kong, China), and 52% (Singapore), respectively, and SCC*mec* II is the main genotype [48-51].

Besides hospital-acquired MRSA, community-acquired MRSA (CA-MRSA) has become an important symbol of the continuous proliferation of drug-resistance in recent years. In 1993, MRSA with a unique genotype was isolated from indigenous people who had never use antibacterial agents in Australia. MRSA infections in a previously healthy population (especially in youngsters) were reported in the US and Europe one after another. These strains belong to type SCC*mec* IV and most of them produce Panton-Valentine Leucocidin (PVL), which causes pyogenic skin infections and necrotizing pneumonia. These strains are resistant to β-lactams but sensitive to sulfanilamide and tetracycline. Due to their unique biological phenotype and origin, they were named CA-MRSA. CA-MRSA outbreaks have been reported all over the world but the detailed prevalence is still unknown in most areas. According to a survey performed by US CDC in three regions (Atlanta, Minnesota, Baltimore) in 2001 and 2002, the isolating rate of CA-MRSA was 8% – 20%. The infection incidence was 25.7/100,000 in Atlanta and 18/100,000 in Baltimore. Recently over 70% of staphylococci from skin infections were CA-MRSA in Atlanta and Houston, US. The molecular epidemiologic assay of CA-MRSA in the US showed USA300 is the dominating type [51-53].

With the prevalence and spread of MRSA, vancomycin is the main agent for the treatment of staphylococcal infections and the resistance to the agent is approaching. In 1997, Hiramatsu *et al.* observed that the sensitivity of clinically isolated MRSA to vancomycin decreased. The minimum inhibition concentration (MIC) of vancomycin against MRSA is 2 – 4 mg/L. The main change in the resistant strain is the thickened cell wall. These bacteria could be further divided into two groups by the subsequent studies; one is vancomycin-intermediate *S. aureus* (VISA) and the other is heterogeneous vancomycin-intermediate *S. aureus* (hVISA). The major difference between the two types is that only a small number of daughter cells in hVISA decrease sensitivity to vancomycin but all the descendants of VISA maintain insensitivity to vancomycin. The mechanism for vancomycin insensitivity is still unclear. Due to higher technical requirements for the detection, the prevalence of VISA and hVISA is difficult to clarify. The ratio of VISA in clinically isolated *S. aureus* in China was 11.1% – 17% according to the report by Sun *et al.*, but it is interesting that this ratio was declining recently [54].

In 2002, vancomycin-resistant *S. aureus* (VRSA) was reported in the US, which was isolated from foot ulcer infection of a diabetic patient in Michigan. This was the first case of its kind in the world. So far, 9 similar bacterial infection cases have been reported. Most of them took place in Michigan, US. The MIC values of vancomycin against those isolates were 16 – 1,024 mg/L. The resistance is mediated by *vanA* cluster, which might originate from vancomycin-resistant enterococci. No similar strain has been isolated in other countries or regions but more attention is still needed [47, 55, 56].

***Drug resistance of streptococci*:** The resistance of *S. pneumoniae* against penicillins and macrolides is outstanding in some countries and regions. For example, the ratio of penicillin-resistant *S. pneumoniae* (PRSP) is over 80% in Korea, about 50% in Hong Kong, China, and 30% in Spain. Surveys in different regions of China have revealed that the ratio of PRSP is about 15%, while the ratio of penicillin-intermediate *S. pneumoniae* (PISP) is about 30%, which is a serious challenge to clinical treatment of *S. pneumoniae* infections. In 2008, the US Clinical and Laboratory Standards Institute (CLSI) amended the sensitivity breakpoint of penicillin against *S. pneumoniae* and raised the drug resistant criteria of non-meningitis *S. pneumoniae*. This may reduce the penicillin resistant rate of *S. pneumoniae* other than those causing central nervous system infections [57, 58].

Resistance of streptococci (including *S. pneumoniae* and β-hemolytic streptococcus) to macrolides is serious around the world but differs greatly with regions. The resistant rates of both *S. pneumoniae* and β-hemolytic *Streptococcus* are between 10% and 70%. Macrolides resistant rate of *S. pneumoniae* is over 40% in European countries like Italy and France, below 10% in Northern European countries like Germany, the Netherlands, Czech, and Poland, about 30% in the US, and over 70% in China. The macrolide-resistant mechanisms of *S. pneumoniae* mainly include target mutation and active efflux. The former leads to resistance to macrolides, clindamycin, and streptogramin (type MLS_B), and the latter only leads to resistance to macrolides (type MefA). The former performs a high level of drug resistance and the latter a low level. MefA drug resistance prevails in North America and MLS_B prevails in China and Europe. Therefore, the clinical value of macrolides differs with regions in the treatment of community-acquired respiratory tract infection. The macrolide resistance of β-hemolytic *Streptococcus* is similar to that of *S. pneumoniae*. The resistance rate may be lower than that of *S. pneumoniae* in some regions such as North America and Europe. According to the surveillance results in China in 2007, erythromycin resistant rate of β-hemolytic *Streptococcus* was 59.2% and the great majority was high-level resistance [57, 59-61].

***Drug resistance of enterococci*:** Enterococci is a normal flora in the microecological environment of the human alimentary tract and one of the common pathogenic bacteria causing nosocomial infections, such as endocarditis, urinary infections, sepsis, and wound infections. According to the investigation by Chinese Nosocomial Infection Surveillance Network, it ranks the fourth among all Gram-positive bacteria. Penicillin and ampicillin used to be the main drugs for treating enterococcal infection, but resistance of enterococci to penicillin was frequently seen in recent years and vancomycin has already become a main antibacterial agent in the treatment of enterococcal infections. The prevalence of vancomycin-resistant enterococci (VRE) varies in different regions of the world. Compared with *E. faecalis*, *E. faecium* is more resistant to antibacterial agents.

Results of national antibacterial surveillance in China (Mohnarin) in 2005 showed that *E. faecalis* isolated from nosocomial infections had a resistant rate of 15% – 23% against penicillin and ampicillin and *E. faecium* was over 80%. The resistant rates of both bacteria against fluoroquinolones were respectively 50% and 90%. However, VRE strains are rare; the isolating rate of VRE was less than 3% in 2007 and most strains were isolated from ICUs. In the US, VRE first appeared in 1989 and expanded quickly thereafter. Data provided by NNIS showed that between 1989 and 1999, the isolating rates of VRE among nosocomial infection patients in ICUs rose from 0.4 % to 25.2%. From 2006 to 2007, 33.3% isolates were VRE. The isolating rate of vancomycin-resistant *E. faecalis* was 6.9% and *E. faecium* 80%. The incidence rate of VRE varies greatly with different regions of the world. TEST surveillance found that the incidence rate of vancomycin-resistant *E. faecium* was below 3% in China, Canada, and Germany, below 40% in the UK, Italy, Swiss, India, and Pakistan, and over 60% in Korea, the US, and Argentina [22, 62-64].

Six glycopeptide resistant phenotypes and genotypes (VanA, B, C, D, E, and G) of VRE have been discovered. VanA and VanB are of high clinical value. The former is resistant to vancomycin, teicoplanin, and the latter only to vancomycin. VanA is the main prevailing type around the world, but infection outbreaks of both types have been reported [63, 64].

Linezolid is a newly launched antibacterial agent against Gram-positive bacteria, which has powerful activity against enterococci including VRE. However, because bacteria can easily get resistance through antibacterial target mutation, and strains being resistant to the agent have been generated in the laboratory and isolated from patients. Moreover, outbreaks of infections by linezolid-resistant enterococci and *S. aureus* in some ICUs have been reported.

5.2.1.2 Drug Resistance in Enterobacteriaceae

***Drug resistance of E. coli and K. pneumoniae*:** To be the main representatives of Enterobacteriaceae, *E. coli* and *K. pneumoniae* are frequent pathogenic isolates in community-acquired and nosocomial infections. They rank as the first and third most common isolates respectively in the national nosocomial infection and resistance surveillance in China and the resistance to β-lactams and quinolones is a major clinical challenge [62].

The main resistant mechanism of *E. coli* and *K. pneumoniae* to β-lactam antibacterial agents lies in the production of extended-spectrum β-lactamases (ESBLs), which leads to the resistance against β-lactams including penicillins, cephalosporins, and aztreonam. Only carbapenems and cephamycins are stable to these enzymes and the germs are sensitive to β-lactams/β-lactamase inhibitor compounds. *E. coli* and *K. pneumoniae* isolated in different regions of the world are greatly diverse in the β-lactamase positive rate. In general, *E. coli* has a lower ESBLs production rate than *K. pneumoniae*. Results of Mohnarin in China show that in 2007, the ESBLs positive rates of both bacteria were 35% and 25%, respectively. Moreover, the rate for bacteria isolated from ICU was higher than 70%. The rate for bacteria isolated from children is higher than that from adults. TEST results in three consecutive years showed that *K. pneumoniae* has an ESBLs positive rate of 44% in South America, 22.4% in the Asian-Pacific Region, 7.5% in North America, and 13.3% in Europe [62, 65-68].

ESBLs from Enterobacteriaceae mainly fall into three types, namely TEM, SHV, and CTX-M. Prevalence of these three types differs with regions. The ceftazidime-hydrolyzing effect of TEM is stronger than that to cefotaxime, while CTX-M mainly hydrolyzes cefotaxime and ceftriaxone. In the early stages of ESBLs prevalence, SHV and TEM mainly prevailed in North America and European countries and CTX-M mainly prevailed in East Asia, India, and South America. In recent years, however, CTX-M seems to be proliferating worldwide. For example, CTX-M-15 and CTX-M-3 have already become the dominating ESBL types in European regions [68, 69].

Besides ESBLs, *E. coli* and *K. pneumoniae* can also produce cephalosporinase represented by AmpC. However, most cephalosporinase-producing bacteria belong to *Enterobacter*. A non-metal carbapenemase firstly found in *K. pneumoniae*, named KPC carbapenemase, is worthy of attention, which can almost hydrolyze all β-lactam antibacterial agents and was first discovered in North Carolina (USA), 2001. Shortly after that, it was reported in many countries around the world such as Israel, China, Brazil, and Greece. Seven types of KPC (KPC-2 to 8) have been discovered up until now. Although the overall positive rate of KPC in *K. pneumoniae* is not very high, it has been detected in other clinical isolates including *E. coli*, *Klebsiella. oxytoca*, *Enterobacter aerogenes*, *Enterobacter cloacae*, *Serratia marcescens*, *Salmonella enteritidis*, and *P. aeruginosa*. This trend is worthy of attention [69, 70].

Resistance of *E. coli* to fluoroquinolones is an outstanding problem in China. Surveillance results show that the resistant rate of *E. coli* to levofloxacin and ciprofloxacin is over 70% and the resistant rate of community-acquired pathogenic bacteria is higher than 50%. The drug resistant rate of *K. pneumoniae* is 35%. In the whole world, Asia has the most severe situation, and Europe and America have relatively lower drug resistant rates. Ciprofloxacin resistant rate of *E. coli* is 35% in Korea, 23% in Taiwan, China, 4% in the UK and 3% in the US. Ciprofloxacin resistant rate of *K. pneumoniae* is 12% in Korea, 11% in Taiwan, China, 11% in UK and 7% in the US [22, 62, 70, 71]. The quinolone resistant mechanism of bacteria mainly results in antibacterial target mutations, especially DNA gyrase subunit A (GyrA) and topoisomerase IV subunit C (ParC) mutations. Recently, plasmid-mediated quinolone resistance has become a focal point of attention around the world. The resistance is mainly caused by the impeded link between quinolones and the target site due to inserting of a protecting protein encoding by plasmid. Bacteria containing this plasmid show low-level resistance to quinolones but are assigned selective advantages for target-mutation drug resistance. The mediating gene in the plasmids is named *qnr*. *qnrA*, *B*, and *S* have been discovered and qnrA is relatively common and has been found in almost all continents where man lives. It is most frequently seen in bacteria of the Enterobacteriaceae, such as *E. coli*, *K. pneumoniae*, *Citrobacter*, and *Providencia stuartii* with a positive rate of 0.4% – 94%. In addition, plasmid-mediated aminoglycoside acetylase mutant *aac (6')-Ib-cr* can hydrolyze the piperazine ring (in ciprofloxacin, norfloxacin) of quinolones and thus deactivate the agents. This plasmid-mediated drug resistance is also widely distributed all over the world. Its positive rate in *E. coli* is 51% in Shanghai, China and 28% in the US [72, 73].

Aminoglycoside resistance to *E. coli* is mainly because of antibacterial agent modifying enzymes. Owing to the different structures of aminoglycoside agents and regional habits in the use of these agents, the bacterial resistance to aminoglycosides may be variable with regions. Mohnarin data of China shows that the resistant rate of *E. coli* is 60% – 80% to gentamycin, tobramycin, and kanamycin but only about 20% to amikacin and isepamicin. Aminoglycoside modifying enzymes identified are mainly *aac(3)-IIb*, which can modify gentamycin and kanamycin. US MYSTIC and SENTRY surveillance results show that *E. coli* from ICU has a < 10% resistant rate to gentamycin and tobramycin and *K. pneumoniae* has a < 15% resistant rate to these agents. In addition, both rates tend to fall year by year. The resistant rate to amikacin in both germs is less than 1% [22, 74, 75].

***Drug Resistance in other Enterobacteriaceae*:** Enterobacteria are common pathogens of nosocomial infections. According to the results of Mohnarin of China, enterobacteria (hereafter, meaning other than *E. coli* and *Klebsiella*) isolated from nosocomial infections have significant resistance to broad-spectrum penicillins, third generation cephalosporin, quinolones, and gentamycin but the resistant rate is lower than that of *E. coli*. Compared with *E. coli*, these bacteria have a broader drug resistant spectrum because of their capability of producing cephalosporinases represented by AmpC. Therefore, carbapenems and fourth generation cephalosporins are main choices for clinical therapy of those bacterial infections. Results of 1998–2003 SENTRY surveillance in North America showed that the resistant rates of enterobacteria to ceftazidime, cefepime, amikacin, imipenem, and ciprofloxacin were 17.5%, 0.4%, 0.2%, 0.1%, and 4.6%, respectively. These rates were far lower than those of China in 2005, which were 26.1%, 0%, 21.7%, 0%, and 26.1%, respectively [22, 74].

Salmonella and *Shigella* are common pathogenic bacteria causing community-acquired intestinal infections. Their antibacterial resistance is weaker than that of other bacteria in the Enterobacteriaceae family. The surveillance results of China in 2005 showed that the drug resistant rates were respectively 5.2% and 6.1% to ceftriaxone, 0% and 15.1% to ciprofloxacin, and 13.3% and 30.3% to gentamycin [22, 62].

5.2.1.3 Drug Resistance of Glucose Non-fermentive Bacteria

***Drug resistance of P. aeruginosa*:** *P. aeruginosa* is a common pathogen in nosocomial infections. Its isolating rate is notably high in ICUs. Results of Mohnarin 2007 showed that *P. aeruginosa* ranked second among all bacteria seen in nosocomial infections. Amikacin, levofloxacin, carbapenem, ceftazidime, and cefepime are usually employed to treat *P. aeruginosa* infections, but its resistance to these drugs and even multi-drug resistant *P. aeruginosa* is already rising year by year. Results of Mohnarin 2007 showed that the resistant rates of *P. aeruginosa* to the above-mentioned antibacterial agents were 21.9%, 31.7%, 33.2%, 29.9%, and 25.3%, respectively. In addition, 2005 US MYSTIC surveillance results were 10.4%, 22.4%, 7.3%, 9.8%, and 4.8%, respectively [62, 74].

The resistance of *P. aeruginosa* to carbapenems is a problem deserving attention, which often displays characteristics of multi-drug resistance and even pan-drug resistance, which means the germ may become resistant to all previously-effective drugs and thus pose a serious challenge to clinical antibacterial therapy. The resistant mechanism of *P. aeruginosa* to carbapenems is mainly attributed to the production of carbapenemases, of which metal carbapenemases play an especially outstanding role in the prevalence of *P. aeruginosa*. The main types of carbapenemases include IMP, VIM, and GIM. According to the 2007 report of European Antibacterial Resistance Surveillance System (EARSS), the resistant rate of *P. aeruginosa* to carbapenems was over 25% in 6 out of 33 participating countries and 51% in Greece. VIM-1 carbapenemase was detected from all these countries. In addition, prevalence of GIM carbapenemase was found in Germany. IMP carbapnemase was found in Mainland China, Taiwan, Japan, Canada, and Australia [69, 75].

The mechanism of the multi-drug resistance of *P. aeruginosa* mainly lies in the activation of bacterial active efflux systems and the loss of outer membrane protein. The major active efflux systems include MexAB-OprM, MexCD-OprJ, and MexEF-OprN, which can be activated by the induction of some antibacterial agents or chemical agents such as meropenem, quinolones, and detergents, *etc.*

***Drug resistance of Acinetobacter*:** In recent years, *Acinetobacter baumannii* has become one of the major pathogenic bacteria responsible for the clinical infections in severely ill patients. Mohnarin results have revealed that *A.*

baumannii is the fourth most common bacteria [62] and the drug resistance of *A. baumannii* is rising rapidly. Outbreaks of infections caused by multi-drug resistant or pan-drug resistant strains have been reported all over the world. Multi-drug resistant *A. baumannii* is called Gram-negative "MRSA" because of its severe resistance and it is hard to treat.

Results of Mohnarin 2007 showed that the resistant rates of *A. baumannii* to amikacin, levofloxacin, cefoperazone/sulbactam, and imipenem were 51.5%, 45.8%, 13.4%, and 23.4%, respectively. The rates of MYSTIC results were tobramycin(10.4%), levofloxacin (22.4%), piperacillin/tazobactam (9%), and imipenem (7.3%) in 2005 in the US, and were imipenem (69.8%), ciprofloxacin (34%), and gentamycin (47.6%) from 2002 to 2004 in Europe, respectively [62, 75, 77].

The resistance of *A. baumannii* to β-lactams is mainly due to that it produces various kinds of β-lactamases (including ESBLs, AmpC, and carbapenemases) and activates the active efflux systems. Carbapenemase is the main reason for the resistance to carbapenems in the germ. Unlike *P. aeruginosa*, *A. baumannii* producing metal β-lactamases (such as IMP, VIM and SIM carbapenemases) or KPC carbapenemase is rare. The main carbapenemases detected in *A. baumannii* is Group D β-lactamase; the representative is OXA β-lactamases with potency to hydrolyze carbapenems. Outbreaks and trans-regional spreading of *A. baumannii* producing OXA-23, OXA-58, and OXA-24 enzymes have been reported in Europe, North America, South America, Asia, and Australia. Lu *et al.* detected that 27 out of 30 clinical isolates of carbapenem-resistant *A. baumannii* strains were OXA-23 positive in China [78, 79].

It should be noted that *A. baumannii* strains being resistant to polymyxin B, an alternative therapeutic agent for multi- or pan-resistant *A. baumannii* infection, have occurred all over the world. Their drug resistant rates were 1.7%, 1.9%, 2.7%, and 1.9% respectively in North America, Asian-Pacific Region, Europe, and South America [68].

5.2.2 Mechanisms of Bacterial Resistance

During their long history of biological evolution, bacteria have developed a complicated drug resistant system with diversiform mechanisms. They may acquire drug resistance through the mutation and regulation of their own chromatin DNA or by obtaining exogenous drug-resistant determinants (*e.g.* drug-resistant plasmids, transposons, integrons). Each bacterium may have several drug resistant mechanisms against different antibacterial agents or several resistant patterns against the same antibacterial agent. For example, staphylococci acquires resistance to β-lactam mainly by target mutation and Gram-negative enterobacteria acquire resistance to β-lactams by β-lactamase production. Most bacteria acquire resistance to quinolones by way of DNA gyrase mutation but active efflux systems play an important role in quinolone resistance of *P. aeruginosa*.

5.2.2.1 Inactivating or Modifying Enzymes against Antibacterial Agents

Inactivating or modifying enzymes encoded by plasmids or chromatin genes against antibacterial agents is one of the major bacterial resistant mechanisms. Those enzymes include β-lactamase, aminoglycoside modifying enzymes (acetyltransferase, phosphotransferase, nucleotidyltransferase), chloramphenicol acetyltransferase, and erythromycin esterase, *etc.*

β-lactamase. β-lactamase is the most common inactivator of resistant bacteria against antimicrobial agents. After stimulating with β-lactam antibacterial agents, most bacteria can produce β-lactamase to hydrolyze and inactivate these agents. β-lactamase acts on the β-lactam rings contained in all β-lactam antibacterial agents, cuts off the peptide bond, opens β-lactam rings, and thus inactivates the agents. Up to now, there are many β-lactamases to be identified and there are several enzyme categorization methods. The universal method is Bush-Jacoby-Medeiros categorization method proposed in 1995 (Table 5.1) [78, 80-82]. By the location of encoding genes, β-lactamases can be classified as plasmid-mediated β-lactamase and chromosome-mediated β-lactamase.

β-lactamase encoding plasmids are transferable between different bacterial species and the plasmids can be descended to the progeny cell. Most of the broad-spectrum β-lactamases and ESBLs are plasmid-mediated.

Broad-spectrum β-lactamases are represented by TEM-1, TEM-2, and SHV-1 enzyme and most of which are produced by Gram-negative enterobacteria and capable of hydrolyzing penicillins and first- and second-generation cephalosporins, but they have no effect on monobactams and third- and fourth-generation cephalosporins. In addition to the agents hydrolyzed by broad-spectrum β-lactamases, plasmid-mediated ESBLs can hydrolyze third- and fourth-generation cephalosporins and monobactams. Therefore, ESBLs producing enterobacteria perform drug resistance to cefotaxime, ceftazidime, and other third-generation cephalosporins and aztreonam at different degrees, but cephamycins or carbapenems are stable to the enzymes. ESBLs are mainly detected in bacteria of Enterobacteriaceae, such as *E. coli* and *K. pneumoniae.* Moreover, their drug-resistant rate is increasing quickly, especially in the isolates from ICUs, where third-generation cephalosporins are extensively used. The sensitivity rate of *E. coli* and *K. pneumoniae* isolated from ICU to β-lactam antibacterial agents other than imipenem is only 70% or even lower in the US and Europe, and the outbreaks of infections caused by ceftazidime-resistant *K. pneumoniae* often occurred [82].

Most of ESBLs discovered in early days were mutants from TEM-1, 2 and SHV-1 β-lactamases including TEM-3 – TEM-26 and SHV-2 – SHV-5. New types of ESBLs keep being identified in recent years, CTX-M β-lactamase is disseminating in various countries around the world and, moreover, it has become the dominant prevailing β-lactamase in some areas such as China, South America, Germany, *etc.* [69, 82].

Table 5.1 Categorization of β-lactamases (modified from Ref. 78, 80–82)

F. type	M. type	Name	Substrates	Source	Inhibitor		Representative enzymes
					CA	EDTA	
1	C	Cephalosporinase	Cephalosporins	C	–	–	ACT-1, CMY-1, FOX-1, AmpC
2a	A	Penicillinase	Penicillins	P	+	–	PC1
2b	A	Broad-spectrum β-lactamase	Penicillins & NS cephalosporins	P	+	–	TEM-1, 2, SHV-1
2be	A	Extended-spectrum β-lactamase	Penicillins, Cephalosporins, monobactams	P	+	–	TEM-3, SHV-2, CTX-M-15, PER -1, VEB-1
2br	A	Enzyme resistant to lactamase inhibitors	Penicillins, NS cephalosporins, BLI	P	–	–	TEM-30, 50, SHV-10
2c	A	Carbenicillinase	Carbenicillin	P	+	–	PSE-1, CARB-3, RTG-4
2d	D	Oxacillinase	Oxacillin	P	+/–	–	OXA-1, 10
2dr	D	Carbapenemase	Carbapenems, oxacillin	P	+/–	–	OXA-23, 48
2e	A	Extended-spectrum Cephalosporinase	Cephalosporins	C	+	–	CepA
2f	A	Non-metal Carbapenemase	Penicillins, Cephalosporins, carbapenems	C	+/–	–	IMI-1, SME-1, KPC-2
3a	B	Metal β-lactamase	Penicillins, cephalosporins, carbapenems	C & P	–	+	CcrA, IMP-1, VIM-1, IND-1, GOB-1, FEZ-1, CAU-1, L1
3b	B	Metal β-lactamase	Carbapenems	C	–	+	CphA, Sfh-1
4	ND	Penicillinase not inhibited by CA	Penicillins	C	–	–	SAR-2

F type: function type; M type: Amble molecular type; CA: clavulanic acid; EDTA: ethylenediamine tetraacetic acid; P: plasmid; C: chromosome; NS: narrow-spectrum; BLI: β-lactamase inhibitor. ND: non-defined.

AmpC is a representative of chromosome-mediated cephalosporinases, which belongs to type Ⅰ β-lactamase in BJM category system and molecule type C in Amble classification, and is often isolated from *E. cloacae*, *E. aerogenes*, *C. freundii*, *S. marcescens* and *P. aeruginosa*, *etc.* The type Ⅰ β-lactamase has very strong inducibility. The above-mentioned bacteria weakly express type Ⅰ β-lactamase in normal physiological status without the presence of antibacterial agents. The β-lactamase output will significantly increase after inducing with β-lactams or regulator de-repression. Therefore, it is also called inducible β-lactamase. In addition, plasmid-mediated cephalosporinase has already been detected recently.

The gene regulation of type Ⅰ β-lactamase has been elucidated. The genomic cluster of AmpC consists of *ampC, ampR, ampD, ampE. ampC* is the structural gene of type Ⅰ β-lactamase which coding apoenzyme AmpC. *ampR, ampD*, and ampE are several regulator genes, which mainly participate in the regulation of

AmpC expression. *ampR* and *ampC* are arrayed side by side within the chromosome and exhibit reversing transcription. *ampR* codes AmpR, a 31 kDa protein. In the absence of β-lactam inducer, AmpR serves as a repressor against ampC transcription, and as an activator to promote the expression of AmpC β-lactamase in the presence of β-lactam antibacterial agents as the inducer. Some hold the view that, without the presence of β-lactam inducer under normal circumstances, AmpR and AmpD exist in the form of a complex and repress the transcription of *ampC. ampD* is the second regulator gene located in a chromosome far away from *ampC. ampD* and *ampE* are arrayed side by side within the chromosome and are parts of the operon. In the presence of an inducer, AmpD, a 21 kDa protein, can interact with inductive β-lactam antibacterial agents to detach AmpR protein from the complex, play the role as an activator, and activate the transcription of *ampC*. In the event of gene mutation, *ampD* will encode defective AmpD protein and lose its original function for forming complex with AmpR. AmpR will provide an activating effect as an activator and cause over-expression of β-lactamase. This process is called depressed β-lactamase expression. If ampR gene has any mutation and causes defective AmpR protein, this defective protein will no longer be able to regulate the transcription of *ampC*. In this case, neither the normal state of AmpD nor the presence of an inducer will have any effect on the defective AmpR protein. *ampE* is also a regulator gene but its function is not as clear as the two regulator genes mentioned above [81, 83, 84].

Carbapenemases is getting to be a focus for the research of antibacterial resistance and a serious challenge for anti-infectious therapy in recent years. These enzymes can hydrolyze almost all the β-lactam antibacterial agents including carbapenems and have several molecule types. Metal β-lactamases (metallo β-lactamases) belong to molecule type B. Their coding genes are located in plasmids or chromosome. The active sites of such enzymes contain bivalent metal ions, with Zn^{2+} as the most common ion. Metallo β-lactamases have the following main features: quickly destroying carbapenems such as imipenem and meropenem, inactivating penicillins, cephalosporins, and β-lactamase inhibitors, and hydrolyzing aztreonam on a limited scale. EDTA, a chelator of bivalent cations, can inhibit the hydrolyzing activity of metallo β-lactamases. *Bacillus cereus* was the first metallo β-lactamase producer detected. After that, metallo β-lactamases were discovered in other bacterial species such as *Bacteroides, Pseudomonas, Xanthomonas, Legionella, Serratia,* and *Acinetobacter*. Up to now, the major metallo β-lactamases identified include IMP, VIM, SPM, GIM, and SIM enzymes. The second type of carbapenemases belongs to the molecule type A. Their coding genes locate in chromosomes or plasmids and the hydrolysis can be inhibited by β-lactamase inhibitors. *P. aeruginosa*, *E. coli*, *K. pneumoniae*, and *Enterobacter* are the producers of molecular type A carbapenemases. KPC enzyme, identified at first in *K. pneumoniae*, is the typical representative and the rest includes SME, NMC, IMI, SFC, and GES enzymes. The third type of carbapenemases includes the molecule type D, also named as oxacillinases (OXA). These enzymes are widely detected in *P. aeruginosa*, *Acinetobacter*, and *K. pneumoniae*. The catalyzing substrates consist of penicillins, first-generation and second-generation cephalosporins, and carbapenems, but third- and fourth-generation cephalosporins and aztreonam are stable to them. β-lactamase inhibitors or sodium chloride can inhibit the activity of OXA enzymes *in vitro*. In addition, Korean investigators found that CMY-10 cephalosporinase could hydrolyze carbapenems and bring out bacterial resistance against carbapenems with synergistic action of bacterial outer membrane permeability decrease [70, 78, 81, 85].

Aminoglycoside modifying enzymes: Aminoglycoside modifying enzymes are usually coded through plasmids or chromosome genes, and the genetic determinants can move between different bacterial species. The modifying targets of these enzymes are amino or hydroxyl moieties of aminoglycosides, the affinity of modified molecules to target sites in bacterial ribosome decreases, and the drug intake promoted by this binding diminishes. Bacterial cells gain resistance against the agents. Aminoglycoside modifying enzymes fall into three types according to their acting mechanisms. N-acetyltransferase (AAC) with acetyl coenzyme A dependent acetylation of amino moieties at aminoglycoside position 1, 3, 6', 2'; O-nucleotidyltransferase (ANT) and O-phosphotransferase (APH), both being ATP-dependent enzymes, ANT leads to nucleotidylation of hydroxyl groups at 2", 4', 3", 6 and APH causes phosphorylation of hydroxyl groups at position 3', 4, 3", 6 of aminoglycosides.

Many types of aminoglycoside modifying enzymes have been identified, and the main prevailing enzymes varied with time and regions. There are four groups of AAC enzyme, namely AAC (2"), AAC (6'), AAC (1), and AAC (3). More than 20 AAC (6') coding genes have been cloned in various bacterial species. Seven groups of APH, namely APH (3'), APH (2"), APH (3"), APH (6), APH (9), APH (4), and APH (7"), are confirmed. All the enzymes share 25% or more homology of amino acids. In addition, APH (3') can modify amikacin and isepamicin. ANT enzymes consist of five groups, *i.e.*, ANT (6), ANT (4'), ANT (3"), ANT (2"), and ANT (9). Besides, there are some bifunction enzymes such as AAC (6')-APH (2") and AAC (6')-ANT (2"). Because of the diversity of aminoglycoside modifying enzymes and versatility of catalyzing sites, the same enzyme could modify different aminoglycoside agents and different enzymes might modify the same agent. Furthermore, some antibacterial agents will not lose antibacterial activity after modification. Therefore, partial cross-resistance among aminoglycosides is their character (Table 5.2) [86-88].

Modifying enzymes of other antibacterial agents: Chloramphenicol acetyltransferase can convert chloramphenicol into metabolite without any antibacterial activity; this enzyme is often detected in staphylococci, group D streptococcus and some Gram-negative bacilli, which is encoded by plasmid or chromosome. Erythromycin esterase is an erythromycin hydrolase but not the main mechanism of bacterial erythromycin resistance, which can break the lactone and deactivate the agent. Aminoglycoside modifying enzyme AAC(6')-Ib-cr variant can hydrolyze the piperazine group of ciprofloxacin and norfloxacin, causing low-level resistance of bacteria to these agents.

Table 5.2 Categorization of aminoglycoside modifying enzymes

Enzymes		Substrate	Enzymes		Substrate
APH	APH(3′)- I	K, N, L, P, R	AAC	AAC(6′)–I	T, A, Ne, D, S, K, I
	APH(3′)- II	K, N, B, P, R		AAC(6′)– II AAC (3)- Ia,-Ib	T, G, Ne, D, S,K, As
	APH(3′)-III	K, N, L, P, R, B, A, I		AAC(3)– II a,- II b,- II c	T, G, Ne,D, S
	APH(3′)-IV	K, N, B, P, R		AAC(3)–IIIa,-IIIb,-IIIc	T, G, D, S, K, N, P, L
	APH(3′)- V	N, P, R		AAC(3)–IV	T, G, Ne, D, S, Ap
	APH(3′)-VI	K, N, P, R, A, I		AAC(3)–VII	G
	APH(3′)-VII	K, N		AAC(1)	P, L, R, Ap
	APH(2″)-Ia	K, G, T, S, D		AAC(2′)–Ia	T, G, Ne, Du, N
	APH(2″)-Ib,-Id	K, G, T, Ne, D	ANT	ANT (2″)-I	T, G, D, S, K
	APH(2″)-Ic	K, G, T		ANT (3″)-I	St, Sp
	APH(3″)-Ia,-Ib	St		ANT (4″)-Ia	T, A, D, K, I
	APH(7″)-Ia	H		ANT (4″)- II a	T, A, K, I
	APH(4)-Ia,-Ib	H		ANT (6)-I	St
	APH(6)-Ia,-Ib,-Ic, -Id	St		ANT (9)-I	Sp
	APH(9)-Ia,-Ib	Sp			

A: amikacin; Ap: apramycin; As: astromicin; B:butirosin; D: dibekacin; Du: duazomycin ; G: gentamycin; H: hygromycin; I: isepamicin; K : kanamycin; L: lividomycin ; N: neomycin; Ne: netilmicin; P: paromomycin; R: ribostamycin; S: sisomicin; Sp: spectinomycin; St: streptomycin; T: tobramycin.

5.2.2.2 Bacterial Outer Membrane Permeability Decrease and Active Efflux

The bacterial outer membrane is a unique cellular organelle with lipid double-layer, which is similar with the cytoplasmic membrane. The outside links to lipopolysaccharides, which can prevent hydrophobic substances from entering the bacterial cell. A great number of proteins embed in the lipid bilayer; some of them are transporting channels for substance (including antibacterial agents) and are named porins. The expression of porins is operated by a bacterial genome and affected by bacterial internal or external environmental alteration. Some porins are special channel for antibacterial agent entrance, the down regulation of porin expression holds back the entrance of antibacterial agents and bacteria get resistance to the agents. For example, porin OprD2 in the outer membrane of *P. aeruginosa* serves as an imipenem-specific channel, the loss or lower expression of OprD2 will create bacterial resistance to imipenem, but remain sensitive to meropenem, which may make use of some other alternative transport channels. This phenomenon is observed in *P. aeruginosa*, *E. coli*, *P. mirabilis*, and *Acinetobacter*, *etc.*, but the resistance is at a low level.

Absence of bacterial porin may lead to drug resistance, but the resistant level is usually low, because the reduced protein expression can only slow down the antibacterial agent influx and prolong the concentration balancing time between the inner and outer bacterial cells. The antibacterial agent can continuously permeate the outer membrane by means of the concentration gradient, finally reach the concentration equilibrium, and provide its due antibacterial effects. Further research has shown that the high-level drug resistance caused by loss of bacterial outer membrane protein often has synergistic actions with the active efflux of antibacterial agents and both of the mechanisms are associated with each other in terms of genetic regulation. An important feature of active drug efflux mechanism is its multi-drug resistance. In other words, the bacteria having active efflux can develop resistance to antibacterial drugs with different chemical structures, or even to antiseptics and detergents.

The widespread existence of active efflux in biological cells is the result of biological evolution which is not only associated with bacterial resistance but also an important physiological functions of cells. According to the composition and action mechanisms, active efflux systems are divided into 5 groups: major facilitation superfamily (MFS) systems using proton motive force as energy, resistance-nodulation division (RND) systems, small multi-drug resistance (Smr) systems, and ATP-binding cassette (ABC) systems, all using ATP hydrolysis to provide energy. Table 5.3 lists the common bacterial active efflux systems [89-91].

MFS systems are often found in Gram-negative and Gram-positive bacteria. They are uni-component efflux pumps located in the bacterial cytoplasmic membrane, but at times they work with membrane fusion proteins and outer membrane proteins to generate active efflux effects. These efflux pumps mainly provide effluent effects on single antibacterial agent. Smr systems are simple in structure, but can be used as the efflux channels for antibacterial agents such as chloramphenicol, tetracycline, and some dyes, cations. Smr and MFS systems lack outer membrane protein constituents. Their pumping proteins can only transfer drugs into periplasmic space, where drugs can quickly permeate through phospholipid bilayers back into the cell. Therefore, they cannot cause drug resistance in Gram-negative bacteria but can cause valuable drug resistance in Gram-positive bacteria. An ABC system related to drug resistance is rare in bacteria.

RND are the main active efflux systems for bacterial resistance. They have already been detected in many bacterial species such as *E. coli*, *salmonella*, *E. aerogenes*, *E. cloacae*, *P. mirabilis*, klebsiella, *N. gonorrhoeae*, *H. influenza*, *P. aeruginosa*, *Acinetobacter*, *Burkholderia*, *S. maltophilia*. Such a system consists of pumping protein, membrane fusion protein (MFP), and outer membrane protein (OMP). They are important physiological structures of bacteria that eliminate metabolic products, harmful substances and toxins, *etc.* In addition, they can evacuate some antibacterial agents matching their transporting sites out of bacteria. The weak substrate specificity of such efflux pumps lays a structural foundation for multi-drug resistanct bacteria. Almost all antibacterial agents can meet with corresponding active efflux systems [89].

Table 5.3 Common active efflux systems of bacteria

Type	Efflux system	Gene	Bacteria	Representative substrate
MFS	EmrB	-	*E. coli*	CCCP, NA, IMP, TR
	QacA	*qac*	*S. aureus*	Monovalent or bivalent organic cations
	Blt	*bltR*	*B. subtilis*	AC, CM, CT, EB, FQ, Rd, tetraphenyl phosphates
	Bmr	*bmrR*	*B. subtilis*	Same as Blt
	NorM	-	*V. parahaemolyticus*	Antibiotics, dyes, lipotropic cations
	NorA	*norA*	*S. aureus*	Same as Blt
	PmrA	*pmrA*	*S. pneumoniae*	FQ
	MefA	*mefA*	*Streptococci*	ML
SMR	Smr	-	*S. aureus*	Monovalent cations such as CV and EB
	QacE	-	*K. pneumoniae*	Same as Smr
	QacEΔ1	-	*G- bacteria*	Same as Smr
	CmlA	-	*P. aeruginosa*	CM
	Tet	-	*G- bacterium*	TC
RND	AcrAB-TolC	*acrR,marA robA,soxS*	*E. coli, E. aerogenes, S. enteritidis, H. influenza*	AC, BL, BS, CM, CV, EB, FA, FQ, ML, NO, OS, RF, SDS, TX
	SdeAB	?	*S. liquefaciens*	EB, CM, FQ, OS
	AcrEF-TolC	*acrS*	*E. coli*	AC, BS, FQ, SDS, TX
	MtrCDE	-	*N. gonorrhoeae*	CV, EB, FA, TX
	CmeABC	?	*C. jejuni*	AP, CM, CT, EB, EM, NA, FQ, TC, SDS
	MexAB-OprM	*mexR*	*P. aeruginosa*	AC, AG, BL, CM, CV, EB, ML, NO, SDS, TC, TM, TR
	AdeABC	*adeT, adeSR*	*Acinetobacter*	AG, CM, EB, FQ, NO, TC, TM
	SmeABC	*smeRS*	*S. maltophilia*	AG, BL, FQ

AC, acriflavine; AG, aminoglycosides; BL, β-lactams; BS, bile salts; CCCP, carbonyl cyanide m-chlorophenylhydrazone; CM, chloramphenicol; CT, cefotaxime; CV, crystal violet; EB, ethidium bromide; EM, erythromycin; FA, fatty acids; FQ, fluoroquinolones; ML, macrolides; NA, nalidixic acid; NO, novobiocin; OS, organic solvents; Rd, rodamine; RF, rifampicin; SDS, sodium dodecyl sulfate; TC, tetracycline; TM, trimethoprim; TR, triclosan; TX, Triton X-100.

At least four RND active efflux systems, namely MexAB-OprM, MexCD-OprJ, MexEF-OprN, and MexXY-OprM were identified in *P. aeruginosa*, which are respectively coded by four chromosome operons. MexB, MexD, MexF, and MexY are inner membrane pumping proteins. MexA, MexC, MexE, and MexX are membrane fusion proteins. OprM, OprJ, and OprN are outer membrane proteins. Pumping proteins on the cytoplasmic membrane take in the substrate from the periplasmic space and the interior of the cell membrane, membrane fusion proteins and outer membrane proteins deliver the substrate out of the bacterial cell. The pre-requirement for molecules being a substrate for the efflux system is hydrophobicity, so that the substrate molecule can insert the lipoid cell membrane. For example, AcrAB efflux system in *S. typhimurium* only expels nafcillin and cloxacillin with a lipotropic side chain but cannot expel cefazolin and cefmetazole without a lipotropic side chain [89, 90].

The protein spatial structure and mode of action of RND efflux systems are still not completely elucidated. Pumping protein MexB is composed of 1,046 amino acids with 12 trans-membrane domains; both amino terminal and carboxyl terminal are anchored on the cytoplasmic side of the cell membrane. 311 and 314 amino acid residues in the first and fourth periplasmic loops, respectively, form a long hydrophilic region that may interconnect with membrane fusion protein MexA and outer membrane protein OprM. Membrane fusion protein connects the inner membrane by the amino terminal, spans the periplasmic space, and reaches the outer membrane. Porin is the channel through which the substrate is discharged from the outer membrane. The hybridizing MexAB-OprJ and MexCD-OprM with changed outer membrane protein and the recombined MexCD-TolC with *E. coli* porin still have active efflux functions. In addition, the types of substrate do not change with the outer membrane protein alternation. However, the hybridizing MexAD-OprM and MexCB-TolC do not hold efflux effect. In other words, in a triad efflux system, the binding between pumping protein and membrane fusion protein has specificity, which determines the substrate types of different systems, and the outer membrane protein can work with multiple inner membrane constituents to create efflux functions.

The genetic regulations of the efflux system are complicated. The local regulating genes *mexR, nfxB*, and *mexT* have been discovered upstream of the mexAB-oprM, mexCD-oprJ, and mexEF-oprN operons of *P. aeruginosa*. *mexR* is highly homogenous with marR gene of *E. coli*, the gene of multiple drug-resistant inhibition protein MarR, a MarA-dependent protein. Gene products of *mexR* can inhibit the expression of mexAB-oprM and provide a negative self-regulation. nalB multi-drug resistant phenomenon is the result of *mexR* mutation, which causes overexpression of mexAB-OprM. Since different mutation sites of *mexR* have variable effects on the expression of mexAB-OprM, *i.e.*, the mutations at the *mexR* 5' end express MexAB-OprM less than mutations at other sites; 5' end of *mexR* may play as a partial promoter. Gene products of *nfxB* can inhibit the transcription of mexCD-oprJ and provide a negative regulation. The mutation of *nfxB* gene will result in overexpression of mexCD-oprJ and thus give rise to *nfxB* multi-drug resistant phenomenon. Unlike *mexR* and *nfxB, mexT* codes LysR transcription activator and up-regulates mexEF-oprN, resulting in overexpression of mexEF-oprN and thus in *nfxC* multi-drug resistant phenomenon. In addition, MexT can reduce OprD expression on the transcription stage, causing bacteria resistance to carbapenems. Besides, the global regulating genes rob, *soxRS*, and *marRAB* also participate in the regulation of active efflux system and associate with the regulation of outer membrane proteins of bacteria. Both of those synergistically cause multi-drug resistance [91-93].

RND active efflux systems of *E. coli* and other Gram-negative bacteria are similar to those of *P. aeruginosa* in composition and are closely associated with the multi-drug resistance of these bacteria.

5.2.2.3 Modifications in the Target Sites of Antibacterial Agents

By way of spontaneous genetic mutation or recombination, bacteria can modify antibacterial targets or protect the target from binding of antibacterial agents and the affinity of antibacterial agents on their target sites decreased; finally, bacteria gain resistance to antibacterial agents. The common and important antibacterial resistant phenomena resulting from target site modifications include mutations of penicillin-binding proteins (PBPs) (*e.g.* MRSA, penicillin-resistant *S. pneumoniae*), DNA gyrase mutations for quinolone-resistant germs, substitution of D-alanine with D-serine in the side chain of cell wall peptidoglycan detected in vancomycin-resistance enterococci, and ribosome mutation for aminoglycoside-resistance.

Drug resistant mechanism of methicillin-resistant staphylococci: Methicillin-resistant staphylococci (MRS) are actually no longer a matter of pure resistance to methicillin. Besides being highly resistant to all kinds of β-lactam antibacterial agents, MRS is also resistant to many structurally irrelevant antibacterial agents such as erythromycin, clindamycin, gentamycin, trimethoprim (TMP), quinolones to different extents. It is only sensitive to a small number of drugs such as vancomycin, teicoplanin, and linezolid [47, 48].

It is confirmed that the resistance of MRS to β-lactams is mainly due to the appearance of a special penicillin-binding protein PBP2a generated in the cell membrane of staphylococci. PBP2a shares the same functions as other PBPs in the synthesis of cell wall peptidoglycan, but its affinity with β-lactams is lower and not inhibited by β-lactams. mecA is the genetic determinant of PBP2a. Apart from *mecA*, there are two *mec* regulator genes in staphylococci, namely *mecI* and *mecRI*. *mecI* encodes MecI protein, which is a repressor of *mecA; mec RI* encodes Mec RI protein, which is a co-inducer of *mecA*. At physiological status, MecI repressor inhibits *mecA* expression and there is no transcription of PBP2a. Once combining with β-lactam antibacterial agents, Mec RI is activated. The activated Mec RI can eliminate the inhibition of Mec I on *mecA*, and PBP2a transcription is initiated and bacteria transform into resistant status. Besides, genes like *fem A* and *femB* also participate in the regulation of the drug resistance of MRS [49, 94, 95].

MecA-mecRI operons are located in a mobile gene segment in staphylococci chromosome, *i.e.*, staphylococcal cassette chromosome *mec* (SCC*mec*). This segment is about 24 – 67 kb in size and inserted into the replication origin of staphylococci chromosome through site attBscc. All SCC*mec* segments contain *mecA* and chromosome intergrase genes (*ccrA/ccrB, ccrC*). Other inserted genes are non-β-lactam antibiotic resistant genes. SCC*mec* segments can be divided into 5 types according to their compositions and sizes. Type Ⅰ, Ⅳ, and Ⅴ SCC*mec* do not contain drug-resistance genes other than mecA. Type Ⅱ and Ⅲ SCC*mec* contains sulfanilamide and tetracycline resistant genes. Epidemiological surveys show that community-acquired MRSA mainly belong to Type Ⅳ and nosocomial MRSA mainly belong to Type Ⅱ, Ⅲ, or Ⅰ. Type Ⅴ is seen in community-acquired MRSA, too. The origin of SCC*mec* is still unknown [47, 94, 96].

There are two phenotypes of MRSA, *i.e.*, homogeneous and heterogeneous MRSA [97]. All individual cells of homogeneous MRSA are highly resistant to antibacterial agents and form a population with the unitary drug resistant level. The cell population of heterogeneous MRSA usually contains two or more sub-populations having different degrees of drug resistance. Only a small number of the cells of these sub-populations are highly resistant to antibacterial agents and most cells have low-level drug resistance. The cultivating temperature (37 – 30 ℃) or medium constitutes for MRSA has an impact on the resistant status.

Resistant mechanism of S. pneumoniae to penicillin: *S. pneumoniae* does not produce β-lactamase and its resistance to β-lactam antibiotics is not plasmid-mediated but mainly caused by the mutations of bacterial PBPs, the action target of penicillins. PBPs are the catalyzing enzymes for the synthesis of the bacterial cell wall at the final-stage; β-lactam antibiotics provide antibacterial effects by covalent-binding through the β-lactam rings with serine activation site of PBPs and thus stop the synthesis of bacterial cell wall peptidoglycan. The penicillin-binding domain of PBPs contains three conserved sequences with serine activation sites: SerXxxXxxLys (SXXK) cassette, SerXxxAsn (SXN) cassette, and LysThr/SerGly (KT/SG) cassette. If these conserved sequences or their adjacent amino acids mutate, β-lactam antibiotics will no longer be able to bind PBPs effectively and will thus result in drug resistance. *S. pneumoniae* has six PBPs whose molecular weights range between 90 – 43 kD. Five of them, namely PBP1a, PBP1b, PBP2a, PBP2b and PBP2x, are high-molecular-weight proteins. PBP3 is a low-molecular-weight protein. PBP2x and PBP2b are the main antibacterial determinants of β-lactam antibiotics. Their reduced affinity will cause low-level drug resistance. PBP2x mutation can cause resistance to cefotaxime, while PBP2b does not react with broad-spectrum cephalosporins like cefotaxime, and therefore has nothing to do with the resistance to such β-lactams. Reduced affinity of PBP1a is frequently detected in highly drug-resistant strains. Low-affinity PBP2a mutant can be detected in clinical and laboratory drug-resistant strains, which is another important drug-resistant determinant. Compared with other PBPs, PBP2a is a low-affinity PBP and is probably a natural drug-resistant form rather than a main target site for β-lactam antibiotics. Effects of PBP1b on drug resistance are still unknown, but its low-affinity mutants have already been detected on drug-resistant mutants. PBP3 mutation has been detected in the laboratory cefotaxime-resistant mutant but its relationship with the drug resistance in clinical strains has not been verified. High-level drug resistance usually results from the co-mutation of several PBPs [98, 99].

Mutative PBP genes contain some highly variable allele from *S. mitis* and *S. viridans*, which are generated by inter-species recombination and thus have mosaic structures occurring in different chromosome segments. The diversity of such exogenous genes and the versatility of recombination sites have decided the complication of mutant PBPs. It was identified that *pbp2x, pbp2b* and *pbp1a* have many allelomorphic gene mutants that differ in mosaic area and sequential relationship. *pbp* genes can move in different drug-resistant bacterial clones and transfer horizontally to sensitive strains through transformation [100] .

There are many point mutations existing in pbp genes among laboratory and clinical drug-resistant bacteria. The single site mutation of *pbp2x* and *pbp2* only leads to low-level drug resistance. *pbp1a* mutation is frequently found in a high-level drug resistant strain and is a part of the high-level drug resistance caused by multi-gene mutations. Due to the mutational site variation, the drug resistance levels and cross-resistance patterns of *S. pneumoniae* are different. For example, Thr550Ala mutation of *pbp2x* will cause bacterial resistance to cefotaxime as well as super-sensitivity to amoxicillin and oxacillin. Meaningful *pbp2x* mutations causing PBP2x amino acid substitute often occur around the three-conserved penicillin-binding domain: Thr550Ala/Gly and Gln552Glu behind K547SG cassette, His394Tyr/Leu near S395SN cassette, and Thr338Ala/Gly/Pro of S337TMK cassette. Pneumococcal resistance to piperacillin involves *pbp2b* mutation, in which the amino acid substitute is at Thr446Ala behind S443SN cassette. In the drug-resistant mutation of pbp1a, amino acid substitute at Thr371Ala/Ser of S370TMK cassette is of the greatest importance, and other substitutes take place at Thr574Asn, Ser575Thr, Gln576Gly and Phe577Thr near K557TG cassette and Ile459Met and Ser462Ala near S428RN cassette [98, 101].

***Resistant mechanism of enterococci to glycopeptides*:** Vancomycin provides its antibacterial effect by interrupting the final synthetic stage of cell wall peptidoglyocan in Gram-positive cocci. Its action target is the D-Ala-D-Ala at the end of the N-acetylmuramylpentapeptid side chain. Once combining with the pentapeptide end, vancomycin will hinder the effects of transpeptidase and carboxypeptidase, block the cross-links of the tetrapeptide or pentapeptide side chain; bacterial cells died from lack of a stout three-dimensional cell wall. In the vancomycin-resistant strain, D-Ala-D-Ala is replaced by D-Ala-D-Lac at the end of the pentapeptide side chain, the affinity between cell walls and glycopeptides reduces, the hydrogen bond between vancomycin and the dipeptide disappears, the cross-links among pentapeptide get on continuously and bacteria change into a drug resistant germ.

Determinants of VRE exist in plasmids or chromosomes. Drug-resistant phenotypes and their genetic operons of VanA, VanB, VanC, VanD, VanE, and VanG have already been identified. VanA and VanB are of clinical value but drug resistance of other types is at a low level, which mostly exists in enterococci other than clinical important pathogenic species. VanA performs resistance to both vancomycin and teicoplanin. However, VanB is resistant only to vancomycin and remains sensitive to teicoplanin [62].

The regulating operon of VanA is contained in transposon Tn1546 and includes a gene cluster of *vanR, vanS*, *vanH*, *vanA*, *vanX*, *vanY*, and *vanZ*. *vanH*, *vanA* and *vanX* are indispensable genes for drug resistance. vanA encodes ligase and catalyzes lactic acid link to the peptide end of the side chain of the cell wall peptidoglycan to form D-Ala-D-Lac. *vanH* locates upstream of *vanA* and encodes α-keto acid reductase which transforms pyruvic acid into lactic acid as the substrate for ligase. vanY encodes DD-carboxypeptidase, hydrolyzes alanine at the pentapeptide end, and breaks down D-Ala-D-Ala. *vanR* and *vanS* are regulatory genes, which sense vancomycin in the environment and activate the resistant operon. *vanY* and *vanZ* encode auxiliary proteins. The comparison of VRE phenotypes and genotypes is listed in Table 5.4 [63, 102, 103].

Table 5.4 A Comparison of VRE phenotypes and genotypes

Characteristics		Phenotypes					
		VanA	VanB	VanC	VanD	VanE	VanG
MIC (mg/L)	Vancomycin	⩾64	⩾4	⩾2	⩾16	16	16
	Teicoplanin	⩾16	0.5–1	0.5–1	⩾2	0.5	0.5
Transferable		+	+	–	–	–	–
Mobile element		Tn1546	Tn1547	Intrinsic	Intrinsic	Acquired	?
Expression		Inducible	Inducible	Inducible/inherent	inherent	Inducible	Inducible
Operon location		Plasmid	Plasmid	Chromosome	Chromosome	Chromosome	Chromosome
Pentapeptide end		D-Ala-D-Ala	D-Ala-D-Ala	D-Ala-D-Ser	D-Ala-D-Ala	D-Ala-D-Ser	D-Ala-D-Ser
Species		*E. faecalis* *E. faecium*	*E. faecalis* *E. faecium*	*E. gallinarum* *E. casseliflavus*	*E. faecium*	*E. faecalis*	*Enterococcus faecalis*

Mechanism of quinolone-resistance: Through combination with bacterial DNA gyrase (composed of GyrA and GyrB subunits) and topoisomerase IV (composed of ParC and ParE subunits), quinolones impede the replication and transcription of bacterial DNA and thus bring out antibacterial effects. Quinolone-resistance of bacteria mainly comes from the subunit gene point mutation of both enzymes. In addition, the mutations have a gradual accumulating process and the accumulated mutations at different points will cause high-level resistance of bacteria to quinolones [104]. The frequent mutation regions located around the target sites are called quinolone resistant determining regions (QRDR).

In addition to changes in action targets, bacteria can also acquire drug resistance through plasmid-mediated quinolone target protection. In 1998, Martinez-Martinez *et al.* reported a plasmid-mediated quinolone gene *qnr* (now *qnrA*), which encodes a protein (QnrA) with topoisomerase-specific binding effect. The binding can bring protection to the target sites of the enzymes and result in bacterial resistance by hampering the formation of the antibacterial triad of topoisomerase, quinolone, and DNA. The binding of QnrA protein to specific sites on topoisomerase IV do not need the presence of DNA, quinolone drugs, or ATP. QnrA contains 218 amino acids and belongs to the pentapeptide-repeat protein family. The Qnr family already has over 90 known members, which have been detected in many bacteria. *qnrS* and *qnrB* are also plasmid-mediated quinolone resistant genes. The plasmid-mediated quinolone resistance performs resistance at low-level but creates a selective advantage for high-level bacterial resistance by target mutations. Bacteria with *qnr* genes are more likely to develop resistance to quinolones. Plasmids bearing *qnr* genes also contain CTX-M gene [71-73].

Other resistant mechanisms of target modifications: Antibacterial resistant mechanisms of macrolide-resistant streptococci include target changes and active efflux (the same as above). Macrolide antibiotics can form a complex with bacterial ribosome 50S subunit, specifically inhibit the synthesis of bacterial proteins, and thus bring about antibacterial effects. Some macrolide resistant streptococci can encode ribosome methylase to catalyze methylation of specific adenine of 23S rRNA in 50S subunit, and thus weaken the binding between macrolide and ribosome. *ermB* (erythromycin resistance methylase), the ribosome methylase gene, is usually located on the transferable transposon Tn154. Its drug-resistance phenotype is MLS and demonstrates cross-resistance among macrolides, lincosamides and streptogramin. According to the resistant performing patterns, MLS drug resistance can be classified as intrinsic resistance (cMLS) and inducible resistance (iMLS). The upstream regulatory region of *ermB* determines the intrinsic or inducible expression of *ermB*. Inducible drug resistance is only low-level resistance to 14-membered and 15-membered macrolides in physiological condition. Once iMLS gene is fully induced, it will exhibit high-level resistance to all macrolides, lincosamides, and streptogramin, which is the same as the intrinsic resistant phenotype. Besides, bacterial ribosome 23S rRNA and/or ribosome protein L4 mutations may also cause drug resistance in *S. pneumoniae*.

Aminoglycoside-producing bacteria usually contain ribosome rRNA methylase, which methylates the specific basic radicals and reduces the affinity with antibacterial agents to create drug resistance. This methylase has also been detected in some bacterial clinical isolates. For example, RmtA of *P. aeruginosa*, RmtB of *S. liquefaciens*, and ArmA of *K. pneumoniae* are methylases causing high-level resistance to all aminoglycoside drugs (MIC>1,024 mg/L). *M. tuberculosis* can also develop streptomycin resistance by modification of ribosome protein.

Linezolid, a novel antibacterial agent of oxazolidone, provides antibacterial action by interfering with the formation of an early-stage 70S ribosome initiation complex in bacteria and blockading bacterial protein synthesis. It has broad antibacterial activity against all of Gram-positive cocci. The main drug-resistant mechanism detected in enterococci and staphylococci is the mutation of G2576T in region V of ribosome 23S rRNA. Bacteria usually have multiple 23S rRNA operons and develop high-level drug resistance through gradual accumulative mutation.

5.2.2.4 Other Drug-Resistant Mechanisms

Bacterial resistance to sulfanilamide is caused by the significantly increased output of aminobenzoic acid in bacteria, which reaches dozens of times as much as that of sensitive bacteria. The increased aminobenzoic acid binds with sulfanilamide antagonistically and bacteria can survive in the presence of the antibacterial agent.

5.2.3 Strategies to Control Bacterial Resistance

It is well known that bacterial resistance causes therapeutic failure of infections, increases patient mortality, and aggravates the socioeconomic burden. To make matters worse, the research and development of new antibacterial agents lag far behind the occurrence of bacterial resistance. In the end, human beings have to enter the so-called "Post-antibiotics Era", in which no more drugs are available and infectious diseases will once again become a public health crisis that threatens the health of humankind. Confronted with the challenge of bacterial resistance, people should rethink the experience in the success and the lessons from the failures in the application of antibacterial drugs, and take measures to contain the further spread of bacterial resistance for the benefit of the whole human race rather than just on the level of the health system. According to recommendations made by the WHO and based on the results of much research, we should take active measures to control bacterial drug resistance in the following aspects [105-107]:

(i) Governments of all countries should lay emphasis on the problem of bacterial resistance and take effective supervising, administrative, and educational measures to promote the reasonable use of antibacterial agents and control the proliferation of drug-resistant bacteria.

(ii) International organizations and governments of all countries should actively collaborate with each other to control the proliferation and spread of drug-resistant bacteria in different regions, countries, and even continents.

(iii) Educate the public and professionals to help them learn about the threat of bacterial resistance and reduce the improper use of antibacterial agents.

(iv) We should be encouraged to establish a bacterial resistance surveillance system, carry out fundamental and practical researches in the field of bacterial resistance, and explore effective strategies to control drug-resistant bacteria.

(v) Encourage the research and development of new antibacterial agents and actively seek alternative drugs and approaches in replacement of antibacterial agents for the treatment of infections.

(vi) Develop and research vaccines for the prevention of infectious diseases to reduce the incidence of all kinds of infectious diseases.

(vii) Regulate industrial behavior in the sale of antibacterial agents, enforcing knowledge and patent protection, striking out at fake and poor-quality drugs, and strictly carrying out the prescription drug policies regarding the sales of antibacterial agents.

(viii) Forbid the feedstuff addition of antibacterial agents for animal growth and reserve the crucial antibacterial agents to human beings.

References

[1] McFee R B. Nosocomial or hospital-acquired infections: An overview. Dis

Mon, 2009, 55: 422-438.

[2] CDC, NNIS. National nosocomial infections surveillance (NNIS) system report, data summary from January 1992 through June 2004. Am J Infect Cont, 2004, 32: 470-485.

[3] WHO. Prevention of hospital-acquired infections, A practical guide. 2nd ed. Geneva, 2002.

[4] Horan T C, Andrus M, Dudeck M A. CDC/NHSN surveillance definition of health care-associated infection and criteria for specific types of infections in the acute care setting. Am J Infect Control, 2008, 36:309-332.

[5] Burke JP. Infection control–a problem for patient safety. N Engl J Med, 2004, 348: 651-656.

[6] MOH, People's Republic of China. Nosocomial infection administrative regulations. Beijing, China, 2006.

[7] Wu A H, Ren N, Wen X M *et al.* One-day prevalence survey of nosocomial infection in 159 hospitals. Chin J Infect Control, 2005, 4(1): 12-21.

[8] Wu A H, Ren N, Wen X M *et al.* A nosocomial infection point-prevalence survey: results and analysis of 193 hospitals in China in 2001. Chin J Nosocomiol, 2002, 12(8): 561-569.

[9] Lee M K, Chiu C S, Chow V C *et al.* Prevalence of hospital infection and antibiotic use at a university medical center in Hong Kong. J Hosp Infect, 2007, 65: 341-347.

[10] CDC, NNIS. National nosocomial infections surveillance (NNIS) system report, data summary from January 1992 to June 2003. Am J Infect Cont, 2003, 31: 481-498.

[11] Smyth ETM, McIlvenny G, Enstone J E *et al.* Four country healthcare associated infection prevalence survey 2006: Overview of the results. J Hosp Infect, 2008, 69: 230-248.

[12] Hughes A J, Ariffin N, Huat T L *et al.* Prevalence of nosocomial infection and antibiotic use at a university medical center in Malaysia. Infect Cont & Hosp Epidemiol, 2005, 26: 100-104.

[13] Tacconelli E, Cataldo M A. Vancomycin-resistant enterococci (VRE): Transmission and control. Int J Antimicrob Agent, 2008, 31:99-106.

[14] Harbarth S. Nosocomial transmission of antibiotic-resistant microorganisms. Curr Opin Infect Dis, 2001, 14:437-442.

[15] Wang N S, Sheng X H, Zhang X G *et al.* Hepatitis C virus infection in uremic patients on maintenance hemodialysis: A follow-up study for 126 months. Chin J Blood Puri, 2009, 8(11): 593-597.

[16] Gikas A, Kritsotakis E I, Maraki S. A nosocomial, foodborne outbreak of Salmonella enterica serovar enteritidis in a university hospital in Greece: The importance of establishing HACCP systems in hospital catering. J Hosp Infect, 2007, 66(2): 194-196.

[17] Simon A, Schildgen O, Maria Eis-Hübinger. A Norovirus outbreak in a pediatric oncology unit. Scand J Gastroenterol, 2006, 41(6): 693-699.

[18] Ashu N J, Tompkins D, Wilcox M H. Comparative analysis of prevalence,

risk factors and molecular epidemiology of antibiotic associated diarrhea due to Clostridium difficile, Clostridium perfringens, and *Staphylococcus aureus*. J Clin Microbiol, 2006, 44: 2785-2791.
[19] Goldmann D A. Epidemiology and prevention of pediatric viral respiratory infections in health care institutions. Emerg Infect Dis, 2001, 7: 249-253.
[20] Jarvis W R. Infection control and changing health care delivery systems. Emerg Infect Dis, 2001, 7(2): 170-3.
[21] Edwards J R, Peterson K D, Andrus M L *et al.* National Healthcare Safety Network (NHSN) Report, Data summary for 2006. Am J Infect Control, 2007, 35: 290-301.
[22] Xiao Y H, Wang J, Li Y. Bacterial resistance surveillance in China: A report from Mohnarin 2004–2005. Eur J Clin Microbiol Infect Dis, 2008, 27: 697-709.
[23] Clark T A, Hajjeh R A. Recent trends in the epidemiology of invasive mycoses. Curr Opin Infect Dis, 2002, 15: 569-574.
[24] Pfaller M A, Diekema D J. Epidemiology of invasive candidiasis: A persistent public health problem. Clin Microbiol Rev, 2007, 20(1): 133-163.
[25] Menzin J, Meyers J L, Friedman M *et al.* Mortality, length of hospitalization, and costs associated with invasive fungal infections in high-risk patients. Am J Health Syst Pharm, 2009, 66(19): 1711-1717.
[26] ATS/IDSA. Guidelines for the management of adults with hospital-acquired, ventilator-associated, and healthcare-associated pneumonia. Am J Respir Crit Care Med, 2005, 171: 388-416.
[27] Athanassa Z, Siempos II, Falagas ME. Impact of methicillin resistance on mortality in Staphylococcus aureus VAP: A systematic review. Eur Respir J, 2008, 31: 625-632.
[28] Fiel S. Guidelines and critical pathways for severe hospital-acquired pneumonia. Chest, 2001, 119: 412S-418S.
[29] Hu B J, Wei L, Zhang X Z *et al.* A retrospective cohort study of the influence of time of hospital-acquired pneumonia onset on pathogen constitution. Chin J Tuberc Respir Dis, 2005, 28(2): 112-116.
[30] Valenciaa M, Torres A. Ventilator-associated pneumonia. Current Opinion in Critical Care, 2009, 15: 30-35.
[31] Tenke P, Kovacs B, Johansen TEB *et al.* European and Asian guidelines on management and prevention of catheter-associated urinary tract infections. Int J Antimicrob Agent, 2008, 31S: S68-S78.
[32] Lo E, Nicolle L, Classen D *et al.* Strategies to prevent catheter-associated urinary tract infections in acute care hospitals. Infect Control Hosp Epidemiol, 2008, 29(Suppl 1): S41-S50.
[33] Olsen M A, Lefta M, Dietz J R *et al.* Risk factors for surgical site infection after major breast operation. J Am Coll Surg, 2008, 207(3): 326-335.
[34] Haridas M, Malangoni M A. Predictive factors for surgical site infection in general surgery. Surgery, 2008, 144(4): 496-501.
[35] Wisplinghoff H, Bischoff T, Tallent S M *et al.* Nosocomial bloodstream infections in U S hospitals: Analysis of 24, 179 cases from a prospective

nationwide surveillance study. Clin Infect Dis, 2004, 39(3): 309-317.
[36] Hansen S, Schwab F, Behnke M *et al.* National influences on catheter-associated bloodstream infection rates: practices among national surveillance networks participating in the European HELICS project. J Hosp Infect, 2009, 71(1): 66-73.
[37] Maki DG, Kluger D M, Crnich C J. The risk of bloodstream infection in adults with different intravascular devices: A systematic review of 200 published prospective Studies. Mayo Clin Proc, 2006, 81(9): 1159-1171.
[38] Wang J, Xiao Y H. Mohnarin report 2006–2007: Bacterial distribution and resistance in bloodstream infections. Chin J Nosocomiol, 2008, 18(9): 1238-1342.
[39] Mermel L A, Allon M, Bouza E *et al.* Clinical practice guidelines for the diagnosis and management of intravascular catheter-related infection: 2009 Update by the Infectious Diseases Society of America. Clin Infect Dis, 2009, 49(1): 1-45.
[40] McFarland L V, Beneda H W, Clarridge J E *et al.* Implications of the changing face of Clostridium difficile disease for health care practitioners. Am J Infect Control, 2006, 35: 237-253.
[41] Kelly C M, LaMont J T. Clostridium difficile – more difficult than ever. N Eng J Med, 2008, 359: 1932-1940.
[42] Hao F L, Wang J, Xiao Y H. Mohnarin report 2006-2007: Bacterial distribution and resistance in central nervous system infections. Chin J Nosocomiol, 2008, 18(9): 1243-1247.
[43] McClelland S, Hall W A. Postoperative central nervous system infection: Incidence and associated factors in 2,111 neurosurgical procedures. Clin Infect Dis, 2007, 45(1): 55-59.
[44] Yokoe D S, Mermel L A, Anderson D L *et al.* A compendium of strategies to prevent healthcare-associated infections in acute care hospitals. Infect Control Hosp Epidemiol, 2008, 29: S12-S21.
[45] Tablan O C, Anderson L J, Besser R *et al.* Guidelines for preventing health-care–associated pneumonia, 2003: Recommendations of CDC and the Healthcare Infection Control Practices Advisory Committee. MMWR Recomm Rep, 2004, 53(RR-3): 1-36.
[46] Chambers H F. The changing epidemiology of *Staphylococcus aureus*? Emerg Infect Dis, 2001, 7(2): 178-182.
[47] Karchmer A W, Bayer A S. Methicillin-resistant *Staphylococcus aureus*: an evolving clinical challenge. Clin Infect Dis, 2008, 46: S342–S343.
[48] Boucher H W, Corey G R. Epidemiology of methicillin-resistant *Staphylococcus aureus*. Clin Infect Dis, 2008, 46: S344-S349.
[49] Klevens R M, Morrison M A, Nadle J *et al.* Invasive methicillin-resistant *Staphylococcus aureus* infections in the United States. JAMA, 2007, 298(15): 1763-1771.
[50] Zhao C Y, Xiao Y H, Wang S *et al.* An investigation of *staphylococcal chromosomal* cassette mec typing of methicillin-resistant *Staphylococcus aureus* clinical isolates. Chin J Infect Dis, 2007, 25(10): 611-615.

[51] Aiello A E, Lowy F D, Wright L N *et al.* Meticillin-resistant *Staphylococcus aureus* among US prisoners and military personnel: Review and recommendations for future studies. Lancet Infect Dis, 2006, 6: 335-341.
[52] Chambers H F. Community-associated MRSA: resistance and virulence converge. N Engl J Med, 2005, 352: 1485-1487.
[53] Fridkin S K, Hageman J C, Morrison M *et al.* Methicillin-resistant *Staphylococcus aureus* disease in three communities. N Engl J Med, 2005, 352: 1436-1444.
[54] Sun W J, Chen H B, Liu Y D *et al.* Prevalence and characterization of heterogeneous vancomycin-intermediate *Staphylococcus aureus* (hVISA) from 14 cities in China. Antimicrob Agent Chemother, 2009, doi:10.1128/AAC.00206-09.
[55] Finks J, Wells E, Dyke T L *et al.* Vancomycin-Resistant *Staphylococcus aureus*, Michigan, USA, 2007. Emerg Infect Dis, 2009, 15(6): 943-945.
[56] Hiramatsu K, Aritaka N, Hanaki H *et al.* Dissemination in Japanese hospitals of strains of *Staphylococcus aureus* heterogeneously resistant to vancomycin. Lancet, 1997, 350: 1670-1673.
[57] Pallares R, Fenoll A, Linares J *et al.* The epidemiology of antibiotic resistance in *Streptococcus pneumoniae* and the clinical relevance of resistance to cephalosporins, macrolides and quinolones. Int J Antimicrob Agent, 2003, 22: S15-S24.
[58] CLSI. Performance Standards for Antimicrobial Susceptibility Testing; Seventeenth Informational Supplement. Pennsylvania, USA, 2008, M100-S17, 27(1).
[59] Yao C, Yu Z, Xiao Y H. Study on drug resistance and molecular epidemiology of Streptococcus pneumoniae isolated in Chongqing. Chin J Epidemiol, 2005, 26(6): 431-434.
[60] Xiao Y H, Liu J, Wang Z *et al.* Macrolides resistant phenotype and genotype of Streptococcus pyogenes. Chin J Antibiot, 2007, 32(4): 323-326.
[61] Leclercq R. Mechanisms of resistance to macrolides and lincosamides: nature of the resistance elements and their clinical implications. Clin Infect Dis, 2002, 34: 482-492.
[62] Xiao Y H, Wang J, Zhao C Y *et al.* Mohnarin bacterial resistance surveillance 2006–2007. Chin J Nosocomiol, 2008, 18(8): 1051-1056.
[63] Pootoolal J, Neu J,, Wright GD. Glycopeptide antibiotic resistance. Ann. Rev Pharmacol Toxicol, 2002, 42: 381-408.
[64] Werner G, Coque T M, Hammerum A M *et al.* Emergence and spread of vancomycin resistance among enterococcus in Europe. Eurosurveillance, 2008, 13(47): 1-11.
[65] Xiao Y H, Wang J. Mohnarin Report 2006–2007: Bacterial Distribution and Resistance in Intensive Care Units. Chin J Nosocomiol, 2008, 18(9): 1223-1227.
[66] Li X Y, Xiao Y H. 2006–2007 Mohnarin report: Bacterial resistance in patients under 14 years old. Chin J Antibiot, 2008, 33(10): 579-589.

[67] Reinert R R, Low D E, Rossi F *et al.* Antimicrobial susceptibility among organisms from the Asia/Pacific Rim, Europe and Latin and North America collected as part of TEST and the *in vitro* activity of tigecycline. J Antimicrob Chemother, 2007, 60: 1018-1029.

[68] Isturiz R. Global resistance trends and the potential impact on empirical therapy. Int J Antimicrob Agent, 2008, 32(S4): S201-S206.

[69] Hawkey P M, Jone A M. The changing epidemiology of resistance. J Antimicrob Chemother, 2009, 64(Suppl 1): i3-i10.

[70] Patel J B, Rosheed J K, Kitchel B. Carbapenemases in Enterobacteriaceae: Activity, Epidemiology, and Laboratory Detection. Clin Microbiol Newsl, 2009, 31(8): 55-63.

[71] Livermore D M, James D, Reacher M *et al.* Trends in fluoroquinolone (ciprofloxacin) resistance in Enterobacteriaceae from bacteremias, England and Wales, 1990–1999. Emerg Infect Dis, 2002, 8: 473-478.

[72] Martinez-Martinez L, Cano L E, Rodriguez-Martinez J *et al.* Plasmid-mediated quinolone resistance. Expert Rev Anti Infect Ther, 2008, 6(5): 685-711.

[73] Robicsek A, Jacoby G A, Hooper D C. The worldwide emergence of plasmid-mediated quinolone resistance. Lancet Infect Dis, 2006, 6: 629-640.

[74] Sadera H S, Fritschea T R, Jones R N. Potency and spectrum trends for cefepime tested against 65746 clinical bacterial isolates collected in North American medical centers: Results from the SENTRY Antimicrobial Surveillance Program (1998–2003). Diag Microbiol Infect Dis, 2005, 52: 265-273.

[75] Rhomberga P R, Jones R N. Contemporary activity of meropenem and comparator broad-spectrum agents: MYSTIC program report from the United States component. Diag Microbiol Infect Dis, 2005, 57: 207-215.

[76] Souli M, Galani I, Giamarellou H. Emergence of extensively drug-resistant and pandrug-resistant Gram-negative bacilli in Europe. Eurosurveillance, 2008, 13: pii: 19045.

[77] Unal S, Garcia-Rodriguez J A. Activity of meropenem and comparators against Pseudomonas aeruginosa and Acinetobacter spp. isolated in the MYSTIC Program, 2002–2004. Diagn Microbiol Infect Dis, 2005, 53: 265-271.

[78] Poirel L, Pitout J D, Nordmann P. Carbapenamases: Molecular diversity and clinical consequences. Future Microbiol, 2007, 2(5): 501-512.

[79] Lu C Y, Zhang Z. A study on mechanism for resistance of carbapenems in Acinetobacter baumannii. Chin J Clin Lab Sci, 2006, 24(4): 295-298.

[80] Bush K, Jacoby G A, Medeiros A A. A functional classification scheme for b-lactamases and its correlation with molecular structure. Antimicrob Agent Chemother, 1995, 39(6): 1211-1233.

[81] Bush K, Jacoby G A. An updated functional classification of beta-Lactamases. Antimicrob Agent Chemother, 2009, doi:10.1128/AAC. 01009-09.

[82] Paterson D L, Bonomo R. Extended-spectrum β-lactamases: A clinical

update. Clin Microbiol Rev, 2005, 18(4): 657-686.
[83] Hanson N D, Sanders C C. Regulation of inducible AmpC beta-lactamase expression among Enterobacteriaceae. Curr Pharm Des, 1999, 5(11): 881-894.
[84] Jacoby G A. AmpC beta-Lactamases. Clin Microbiol Rev, 2009, 22(1): 161-182.
[85] Lee, S H, Jeong S H, Park Y M. Characterization of blaCMY-10 a novel, plasmid-encoded AmpC-type β-lactamase gene in a clinical isolate of Enterobacter aerogenes. J Appl Microbiol, 2003, 95: 744-752.
[86] Wright G D. Aminoglycoside-modifying enzymes. Curr Opin Microbiol, 1999, 2: 499-503.
[87] Kotra L P, Haddad J, Mobashery S. Aminoglycosides: Perspectives on mechanisms of action and resistance and strategies to counter resistance. Antimicrob Agent Chemother, 2000, 44(12): 3249-3256.
[88] Jana S., Deb J K. Molecular understanding of aminoglycoside action and resistance. Appl Microbiol Biotechnol, 2006, 70: 140-150.
[89] Kumar A, Schweizer HP. Bacterial resistance to antibiotics: Active efflux and reduced uptake. Adv Drug Del Rev, 2005, 57: 1486–1513.
[90] Pagès J M, Amaral L. Mechanisms of drug efflux and strategies to combat them: challenging the efflux pump of Gram-negative bacteria. Biochem Biophys Acta, 2009, 1794(5): 826-833.
[91] Li X Z, Nikaido H. Efflux-mediated drug resistance in bacteria. Drugs, 2004, 64(2): 159-204.
[92] Aeschlimann J R. The role of multidrug efflux pumps in the antibiotic resistance of Pseudomonas aeruginosa and other gram-negative bacteria. Insights from the Society of Infectious Diseases Pharmacists. Pharmacotherapy, 2003, 23(7): 916-924.
[93] Chen H, Hu J, Chen P R. The Pseudomonas aeruginosa multidrug efflux regulator MexR uses an oxidation-sensing mechanism. Proc Natl Acad Sci USA, 2008, 105(36): 13586-13591.
[94] Domínguez M A, Liñares J, Martín R. Molecular mechanisms of methicillin resistance in *Staphylococcus aureus*. Microbiologia, 1997, 13(3): 301-308.
[95] Berger-Bächi B, Rohrer S. Factors influencing methicillin resistance in staphylococci. Arch Microbiol, 2002, 178(3): 165-171.
[96] Deurenberg R H, Stobberingh EE. The molecular evolution of hospital-and community-associated methicillin-resistant *Staphylococcus aureus*. Curr Mol Med, 2009, 9(2): 100-115.
[97] Ryffel C, Strässle A, Kayser F H *et al.* Mechanisms of heteroresistance in methicillin-resistant *Staphylococcus aureus*. Antimicrob Agents Chemother, 1994, 38(4): 724-728.
[98] Hakenbeck R. β-Lactam-resistant Streptococcus pneumoniae: Epidemiology and evolutionary mechanism. Chemotherapy, 1999, 45: 83-94.
[99] Grebe T, Hakenbeck R. Penicillin-binding proteins 2b and 2x of Streptococcus pneumoniae are primary resistance determinants for different class of β-lactam antibiotics. Antimicrob Agents Chemother, 1996, 40(4):

829-834.
[100] Hakenbeck R, KÖnig A, Kern I *et al.* Acquisition of five high-Mr Penicillin-binding protein variants during transfer of high-lever β-lactam resistance from Streptococcus mitis to Streptococcus pneumoniae. J. Bacteriol, 1998, 180(70): 1831-1840.
[101] Rogers P D, Liu T T, Barker K S *et al.* Gene expression profiling of the response of Streptococcus pneumoniae to penicillin. J Antimicrob Chemother, 2007, 59(4): 616-626.
[102] Woodford N, Johnson A P, Morrison D *et al.* Current perspectives on glycopeptide resistance. Clin Microbiol Rev, 1995, 8(4): 585-615.
[103] Courvalin P. Vancomycin resistance in Gram-positive cocci. Clin Infect Dis, 2006, 42: S25-S34.
[104] Hawkey P M. Mechanisms of quinolone action and microbial response. J Antimicrob Chemother, 2003, 51 Suppl 1: 29-35.
[105] Cosgrove S E, Carmeli Y. The impact of antimicrobial resistance on health and economic Outcomes. Clin Infect Dis, 2003, 36: 1433-1437.
[106] Cosgrove S E. The relationship between antimicrobial resistance and patient outcomes: mortality, length of hospital stay, and health care costs. Clin Infect Dis, 2006, 42: S82-S89.
[107] WHO. WHO global strategy for containment of antimicrobial resistance, executive summary. Geneva, 2001.

6

Microbial Culture and Its Clinical Application

Xiaohui Miao, Kekai Zhao, Nan Chen, Yu Chen *
The State Key Laboratory for Diagnosis and Treatment of Infectious Diseases, the First Affiliated Hospital, School of Medicine, Zhejiang University; Collaborative Innovation Center for Diagnosis and Treatment of Infectious Diseases, Hangzhou, 310003, China
* E-mail: chenyu6812@sina.com

Infectious disease is a common disease which affects almost everyone at least once or more during their lifetime. Infections may be caused by endogenous or exogenous microorganisms including bacteria, chlamydiae, mycoplasmas, rickettsiae, viruses, fungi, or parasites. Diagnosis of infection depends on clinical assessment of symptoms and signs, but more often it is necessary to identify a specific etiologic agent by utilizing microbiologic laboratory methods, among which microbial culture is generally accepted as the gold standard.

6.1 Conventional Microbial Culture and Clinical Application

Microbial culture is a way of growing and enriching microbes in a laboratory setting in order to find a certain etiologic microorganism or harvest more specific microbes. Some kinds of bacteria are prone to causing infection in special sites. It is necessary to provide proper cultural conditions for bacteria recovery from different infected sites.

6.1.1 Clinical Blood Culture

Bacteremia may be continuous, intermittent, or transient. If infection occurs in the blood vessel (for example, endocarditis), organisms are continuously in the blood. And in some infections bacteremia is intermittent. For example, bacteria from extra vascular infected sites can translocate into the blood via the lymphatic system. These bacteria may be from gastrointestinal tract infections, urinary tract infections, respiratory infections, meningitis, abscesses, and so on. On the other hand, transient bacteremia could be caused by normal bacterial flora after various clinical manipulations, such as tooth extraction, barium enema, cystoscopy, and so on. Mostly bacteria can be cleared from blood in minutes to hours. Sometimes bacteria that cannot be cleared causes overwhelming or intravascular infection.

After bacteria enter the bloodstream, the patient's body temperature begins to rise, possibly accompanied by chills. Unfortunately, by the time the patient has an elevated temperature; organisms may already be cleared from the blood. Optimally, the blood specimen should be collected just before a fever spike. Actually most bacteremia is intermittent, so it is necessary that the blood culture be performed several times. For adult patients, two sets of cultures should be collected per febrile episode to help distinguish probable pathogens from possible contaminants. In general, contaminants are more likely to be recovered from a single set of cultures whereas pathogens typically are recovered from more than one set. The majority of bacteremia should be detected with three sets of blood cultures. Generally, blood specimens should be collected before any antimicrobial agents are given, although it is difficult in clinical therapy, particularly for patients with persistent symptoms of sepsis.

During blood culture collection, the skin should be decontaminated first, and blood obtained by venipuncture. The blood specimen's volume is critical because the number of bacteria in blood is extremely small (1 – 10 CFU/mL). For adults, 20 – 30 mL of blood is usually collected in each venipuncture [1]. Increasing the volume of blood specimens helps with bacteria recovery [2]. Less blood is required from infants (1 – 5 mL), because the number of bacteria in blood is more than that in adults' blood [3]. Blood-to-medium ratios from 1:5 to 1:10 can be acceptable [4]. Lower dilutions may result in bacterial inhibition by serum factors, and higher dilutions prolong detection time.

Conventional blood culture systems that use nutritionally enriched liquid media can recover most bacteria, including anaerobes. Bacteria growth is indicated by turbidity, hemolysis, gas production, or the presence of colonies. The subculture of the contents from aerobic bottles after 6 – 18 h of incubation is required. Routine subculture of the content from anaerobic bottles is not necessary.

The isolator system is a lysis-centrifugation system that is used to concentrate bacteria or fungi present in blood. This system contains several agents, including saponin, which lyses the red blood cells. Organisms are pelleted during centrifugation, and the sediment is plated to solid media. This system provides improved recovery of certain organisms, including fungi. The main disadvantages of this system are that it is labor-intensive and, because of the increased manipulation during processing, is easily contaminated.

The fully automated continuously monitored blood culture systems are the newest type of systems developed for the detection of bacteria and fungi in blood. The systems also have culture bottles incubated in an instrument, where they are continuously monitored for the production and(or) consumption of gas. The data collected are transmitted to a computer and analyzed to allow rapid detection of microbial growth. Each system utilizes a noninvasive method (*e.g.*, colorimetric, fluorescent, or manometric methods to detect CO_2 or other gases) to monitor growth. The system offers a variety of media, which could be used for the recovery of aerobes or anaerobes. It could also be used for patients taking antibiotics. These systems and available media have been extensively evaluated, and the data from these studies indicate that these systems are a reliable and rapid alternative to conventional systems for the detection of bacteria and yeast in blood.

All blood cultures used for bacteria detection should be incubated at 35℃. Manual blood culture systems should be incubated for 7 days. Manual broth-only systems should include a terminal subculture of the aerobic bottle on the final day. For automated systems, data from several studies have shown that incubation for 5 days is sufficient to recover most bacteria. Terminal subculture of automated blood culture bottles is unnecessary [5]. When subculturing positive blood culture vials, the aerobic media should be selected based on the Gram stain results. Anaerobic media and culture conditions should be used if the Gram stain result is suggestive of an anerobic organism or if the organism is recovered from an anaerobic blood bottle only.

Most microorganisms can cause septicemia, therefore positive isolate should be reported. The most common and significant isolates are *Escherichia coli*, *Staphylococcus aureus*, *Streptococcus pneumoniae*, and anaerobic bacteria (usually 5% – 15% of all isolates). And 6% – 15% of positive blood cultures indicate polymicrobial infection.

Blood cultures may be contaminated with bacteria from skin flora or environmental organisms. Most common bacteria include *Staphylococcus epidermidis*, *Propionibacteria*, *Corynebacteria*, *Bacillus* spp. and so on. Isolation of these organisms from a single bottle is not important for clinical diagnosis. Also *Viridans streptococci* and *Enterococci* found in one bottle are usually not significant and are probably the result of transient bacteremias. Occasionally all of these bacteria can cause significant infection; therefore the diagnosis of infection should be based on both the clinical laboratory examination and the clinical assessment of symptoms.

6.1.2 Clinical Body Fluids Culture

Body fluid specimens are all collected by fine needle aspiration after skin decontamination. The specimen is delivered in the syringe or in a sterile screw-capped tube. It is better to collect as much fluid as possible because the number of organisms in it may be small. Bacteria are then concentrated by centrifugation or filtration. For cerebrospinal fluid (CSF), centrifugation at 1500×g for 15 minutes is sufficient [6]. Filtration cannot offer any advantage and may inhibit bacteria if antimicrobial agents are in the field.

Bacterial meningitis is life threatening. And CSF should be processed immediately. Gram stain smear results are helpful for media selection. Otherwise blood agar, chocolate agar, and enriched thioglycolate broth are selected at the same time. Plates are incubated in 5% – 10% CO_2 for 2 – 3 days.

The most common pathogens of bacterial meningitis are *Haemophilus influenza*, *Neisseria meningitides*, and *Streptococcus pneumoniae*. *Streptococcus agalactiae* and *E. coli* are the major causes of meningitis in the newborn. Other pathgens of bacterial meningitis include *Mycobacterium tuberculosis*, *Staphylococci*, *Listeria monocytgenes*, *Leptospira interrogans*, *Citrobacter spp.*, and other gram-negative rods. But anaerobic bacteria are rarely isolated from CSF. *Staphylococcus* is often isolated from shunt infections. These strains are frequently multiple resistant and difficult to eradicate.

Nonbacterial meningitis may be caused by viruses, fungi, or free-living amoebae. The latter two can be detected by direct CSF examination.

Several rapid techniques have been developed for examination of common bacterial meningitis pathogens. Co-agglutination, latex agglutination, or counter immunoelectrophoresis can be performed on direct CSF examination. If organisms resembling *H. influenzae* or *S. pneumoniae* are seen on a smear, a quellung test with specific antiserum provides an immediate identification. Latex agglutination can detect cryptococcal antigen in CSF or serum and is more sensitive than direct detection by India ink.

Meningitis can be caused after other infections, so it is necessary to collect other kinds of specimens. Blood culture is often performed with CSF examination. Besides blood, specimens may be taken from the ears, sinuses and so on. Examination of pathogens in pleural, pericardial, peritoneal, or synovial fluid is performed in the same way as CSF examination. Pleural and peritoneal fluid and synovial fluid culture should include anaerobic agar media.

6.1.3 Clinical Urine Culture

In patients with symptoms of urinary tract infection (painful urination, urgent micturition,frequent micturition), urine is collected for culture at the onset of symptoms and may be repeated 48 – 72 hours after institution of therapy. If bacteriuria is asymptomatic, two or three specimen's cultures are repeated to confirm an initial infection. In suspected renal tuberculosis three consecutive first-morning urine specimens should be submitted. Pooled 24-hour collection of urine is unacceptable.

Urine is collected in a sterile screw-cap container. Most bacteria grow well in urine at room temperature; therefore specimens must be processed within 2 hours or refrigerated to prevent overgrowth of contaminating bacteria. Alternatively, the urine may be aspirated into commercially available devices that contain a preservative fluid for at least 24 hours at room temperature [7].

Urine tract infection includes the lower tract (cystitis, acute urethral syndrome) and upper tract (pyelonephritis) infection. Microscopic examination of unspun urine can provide helpful information. In women with asymptomatic bacteriuria or pyelonephritis, colony counts equal to or greater than 10^5 CFU/mL are considered significant bacteriuria, and contaminating bacteria often has colony counts of less than 10^3 CFU/mL. Women with symptoms of urinary tract infection and pyuria but colony counts of less than 10^5 CFU/mL are designated as having acute urethral syndrome.

The most common bacteria that cause urinary tract infection are *E. coli* and other Gram-negative rods, *Staphylococcus saprophyticus*, *S. aureus*, yeasts, enterococci, and β-hemolytic streptococci. No evidence supports the pathogenicity of lactobacilli or diphtheroids in urinary tract infection. Even though colony counts are greater than 10^5 CFU/mL, their isolation is more likely an indication of improper collection or transport.

Patients with long-term catheterization may be at risk of polymicrobial infection when two or more organisms' colony counts are greater than 10^5 CFU/mL. Repeat urine culture is recommended whenever two or more organisms are isolated.

6.1.4 Culture of Specimens from Gastrointestinal Tract

Specimens from feces, vomitus, and duodenal contents are collected mainly for isolation of pathogen causing diarrhea. Feces are passed directly into a clean, leakproof container or may be collected from a bedpan.

Temperature or pH change can greatly affect bacterial recovery such as *Shigella* spp., so ideally specimens should be plated immediately. If specimens cannot be plated within two hours after collection, they should be mixed with a transport medium and refrigerated. Buffered glycerol saline is more appropriate for *Shigella* than Cary-Blair medium [8]. But glycerol saline is not suitable for *Vibrio* spp. or *Campylobacter jejuni*.

For conventional stool culture, the media inoculated should allow the detection of *Salmonella*, *Shigella*, and *Campylobacter* spp. Media for the recovery of *Salmonella* and *Shigella* spp. are incubated in air at 35℃ for 2 days. Media for the recovery of *Campylobacter* spp. are incubated in a micro-aerophilic environment at 42℃ for up to 3 days.

6.2 Requirements for Collection of Cultural or Non-Cultural Specimens

The specimens to be collected for culture depend on signs and symptoms of the disease. In other words, different types of specimens should be taken at the appropriate time from the locus of the host where the pathogen is most probably present according to the clinical manifestation of the patient. For example, when septicemia is suspected, blood for bacterial culture should be extracted when the patient is shivering just before emerging from fever or is suffering a sustained high fever. For microbiological diagnosis of typhoid fever, culture of blood or bone marrow would be the first choice during the early stage of disease while a stool would be the optimal sample during the late stage.

6.2.1 Conventional Principles for Collection and Transportation of Specimens Used for Culture

The types of specimen that may be used for culture can be:

- Body fluid: Blood, urine, cerebrospinal fluid, pleural effusion, ascites, *etc.*
- Secretions or discharges: Sputum, stool, urethral discharge, vaginal or cervical discharge, respiratory secretions, vomitus, drainage after operation, *etc.*
- Swabs: Throat swabs, urine tract swabs, rectal swabs, decubitus ulcer swabs, *etc.*
- Bone marrow aspirates.
- Exudates or pus from wound, ears, eyes, abscess puncture, *etc.*
- Endoscopic washings.
- Biopsy or tissue.
- Skin scrapings.

There are some points needed to be followed for collection and transportation of specimens for culture:

- Enough amount of the specimen should be collected before starting treatment with the antimicrobial agent.
- Deliver specimen to the laboratory as soon as possible after collection. If transport is delayed, the following specimens should be refrigerated: Urines (within 30 minutes), stool (within 1 hour), and respiratory specimens. Specimens for viral culture must be transported to the laboratory immediately in ice.
- Special precautions should be taken to avoid contamination of the specimen by organisms in the surrounding skin, mucus membrane and air during collection.
- Specimens must be appropriately labeled or enclosed with enough information of the patient, such as the patient name, ID number, date and time of collection, specimen type and tests requested, etc.
- Specimens should be stored and transported in tightly sealed, leak proof containers.

Specific guidelines for collecting specimens to be used for bacterial culture:

- Specimens can be transported at room temperature within two hours unless otherwise specified.
- Blood: Enough volume of blood (0.5 – 2 mL for infants, 2 – 5 mL for children, and at least 10 mL for adults) should be collected. It is recommended to collect two initial sets of blood cultures sequentially followed by a third or a fourth set at 4 – 6 hours intervals. The higher the volume of blood cultures, the higher the yield of detection of bloodstream infections [9].
- Sputum: Sputum should be collected in a sterile container, preferably early in the morning before eating or drinking anything. The mouth should be rinsed with water to rinse out bacteria from the mouth and dilute the saliva which may contaminate the specimen. With a forceful cough, the sputum should be spat out into the sterile container immediately, avoiding prolonged collection in the mouth cavity. Three consecutive samples may have to be collected if testing for *M. tuberculosis*.

- Stool: Submit 10 – 20 g in sterile container to the lab within 1 hour. Refrigerate it if transport is delayed.
- Urine: Collect 1 – 10 mL of fresh urine in a sterile specimen container. Random fresh midstream urine is the preferred type of specimen for culture and 24-hour urine is unacceptable. Patients are required to first cleanse the urethral area, and void the first portion of the urine stream into the toilet, and then the midstream urine is then collected into a clean container. Transport urine specimen to the microbiology laboratory immediately or refrigerate it within 30 minutes.
- Cerebrospinal Fluid (CSF): Aseptically collect CSF from a lumbar puncture into sterile tubes and send the second tube (>3 mL) in sterile, leak-proof container to the laboratory within 15 minutes. Cerebrospinal fluid for bacterial culture should never be refrigerated.
- Other body fluids: Submit as much fluid as possible to the laboratory in sterile, leak-proof container or anaerobic transport vial within 15 minutes. NEVER submit a swab dipped in fluid. One aerobic blood culture bottle inoculated at bedside (up to 10 mL) is highly recommended if an adequate sample is available. If the blood culture bottle is inoculated, submit separate aliquot in anaerobic vial or sterile container for preparation of cytocentrifuged Gram stain and inoculation of solid media.
- Swabs/Exudate: Swabs or exudate can be collected from ear, eye, nose, throat, urethra, vagina, cervix, catheter tip, wound, *etc.* Moisten swab with sterile normal saline, then swab the surface of the diseased region (throat, eye, *etc.*), or insert the swab 2 – 4 cm into the canal or cavity (urethra, vagina, cervix, etc.) and gently rotate swab for 10 – 30 seconds. Place specimen collection swab into transport tube and carefully break swab shaft at scoreline. Send specimen immediately after collection. Dry swabs are not suitable for culture.
- Tissue: Submit in anaerobic collection jar or sterile screw-cap container within 15 minutes. Add drops of sterile saline to keep small pieces of tissue moist.
- Abscess: Tissue or aspirates are always superior to swab specimens in the abscess culture. Remove surface exudate by wiping with sterile saline or 70% alcohol. Aspirate with needle and syringe, and inject all abscess material on top of agar in the tube. If a swab must be used, pass the swab deep into the base of the lesion to firmly sample the fresh border.

6.2.2 Special Conditions for Collecting Specimens Used for 'Non-Routine' Culture

Specimens used for special microbes' detection are usually collected under special conditions.

6.2.2.1 Mycobacterial Culture (AFB Culture)

Following the procedure of collection and transport for bacterial culture, swabs are suboptimal for recovery of mycobacteria due to limited material and the hydrophobicity of the mycobacterial cell envelope.

6.2.2.2 Anaerobic Cultures

Tissues or aspirates are preferred rather than swabs. Fluid collections should be aspirated through disinfected tissue or skin. For superficial ulcers, collect material from below the surface (after surface debridement or use a needle and syringe). Submit specimens using anaerobic transport media. Deliver specimens to the laboratory immediately after collection. Anaerobic flora is prevalent on mucosal surfaces of the oral cavity, upper respiratory, gastrointestinal and genital tracts. Therefore, specimens collected from these sites should not ordinarily be cultured for anaerobic bacteria.

6.2.2.3 Fungal Culture

For diagnosis of fungi infection in the skin, scrape the periphery of the lesion border using a scalpel blade, and transport the specimen in a sterile container as soon as possible after collection. Collection and transport of all other specimen types may follow the same procedure as that of bacterial culture. However, swabs are unacceptable for fungal studies. If fungi infection in addition to bacteria is suspected, double sets of specimen should be collected both for bacterial culture and fungal culture.

6.2.2.4 Viral Culture

Collect specimens for culture early in illness when viral shedding is maximal. Swabs and other types of specimen are unacceptable. Specimen should generally be placed in a sterile, leak-proof container with viral transport medium (VTM). Transport the specimen to the laboratory immediately in ice or refrigerate.

6.2.3 Inoculating Samples in an Optimal Media: Selection of the Culture Media

Culture media are nutrient mixtures made specifically for the growth, storage, or transport of microorganisms or other types of cells under laboratory conditions. Microorganisms need nutrients, a source of energy and certain environmental conditions in order to grow and reproduce. In the environment, microbes have adapted to the habitats most suitable for their needs. In the laboratory, however, these requirements must be met by a culture medium.

There are three main physical forms of culture media: Liquid/broth media, semisolid media and solid media. The solid or semisolid media are distinctive from liquid media in the additional solidifying agent, which is usually agar. Liquid media, such as nutrient broth, brain heart infusion broth, or tryptic soy broth, are often used as enrichment media to encourage the growth of microorganisms from small inoculants. Microbial growth is observed to see how quickly the broth becomes cloudy. A cloudier broth typically means a greater number of microbes. Liquid cultures can often contain multiple microbial species, so they tend to be less useful than solid cultures for diagnosis of bacteria and fungi infection. Semisolid media are often used in determining bacterial motility, and in promoting anaerobic growth. Solid media, such as nutrient agar or blood agar, are the most frequently used culture media for: (i) Pure culture isolations; (ii) Observing the appearance of specific microorganism colonies; and (iii) Observing specific biochemical reactions or color. Colonies are made up of clones in the solid culture media, in which all cells are identical to each other. This feature is what makes solid culture so useful for microbial identification. Different kinds of colonies from various species will have distinct traits and characteristics (*e.g.*, color, size, shape and growth rate of the colony), which help microbiologists identify the microbe.

Based on the type and combination of nutrients, different categories of media can be made, such as natural or complex media, synthetic or defined media, differential or selective media, enrichment media, reducing media and living media. Selection of the optimal culture media should be taken into account before inoculating a specimen depending on the main purpose of the microbial culture. Therefore, it is critical to establish effective cooperation and communication between clinicians and microbiologists.

***Natural or complex media*:** Natural or complex media are rich in nutrients, they contain water soluble extracts of plant or animal tissue (*e.g.*, enzymatically digested animal proteins such as peptone and tryptone). Usually a sugar, often glucose, is added to serve as the main carbon and energy source. The combination of extracts and sugar creates a medium which is rich in minerals and organic nutrients, but since the exact composition is unknown, the medium is called complex.

***Defined or synthetic media*:** Defined or synthetic media are media composed of pure ingredients in carefully measured concentrations dissolved in double distilled water, *i.e.*, the exact chemical composition of the medium is known. Typically, they contain a simple sugar as the carbon and energy source, an inorganic nitrogen source, various mineral salts and if necessary growth factors (purified amino acids, vitamins, purines and pyrimidines).

***Selective or differential media*:** Selective or differential media are media based on either of the two categories above supplemented with growth-promoting or growth-inhibiting additives. The additives may be species- or organism-selective (*e.g.*, a specific substrate, or an inhibitor such as cyclohexamide which inhibits all eucaryotic growth and is typically used to prevent fungal growth in mixed cultures).

Enrichment media: Enrichment media are media which promote the growth of a particular organism by providing it with the essential nutrients and rarely contain certain inhibitory substance to prevent the growth of normal competitors. For example, Selenite-F- broth media favor the growth of *Salmonella* and also prevent the growth of normal competitors like *E. coli*. The latter do not perish in the medium but they will not flourish like *Salmonella* [10, 11].

Reducing media: Reducing media are employed for growing obligate anaerobes, which particularly contains chemicals (reducing agents) that deplete molecular oxygen.

Living media: Living media are groups of living cells such as tissues, callus or an organ used for growing viruses, rickettsias *etc.* Chick embryos are commonly used for cultivation of viruses. Yeasts, molds and bacteria, if they enter as contaminants, are able to multiply in the fluids of the chick embryos as in a culture tube.

6.3 Clinical Choice of Microbial Culture or Non-Culture

Traditional types of laboratory microbial tests include microscopy, culture and immunological tests (agglutination tests, precipitation tests, complement fixation tests, enzyme immunoassays, immunofluorescence, Western blot, and immuno-histological techniques). In past decades, many molecular methods based on detection of microbial nucleic acids have been developed and applied in the clinical diagnosis of infection, including polymerase chain reaction (PCR) and reverse-transcriptase PCR (RT-PCR), branched DNA (bDNA), nucleic acid sequence-based amplification (NASBA) and gene microarrays. In recent years, many attempts have been made to develop new rapid, sensitive and reliable methods to promote a microbe positive rate, such as flow cytometry or microbial biosensors.

6.3.1 Overview of Microbial Culture and Non-Culture Based Methods

Microbial culture is normally the gold standard for identification of the pathogenic microorganisms for an infection. According to the Koch's postulates [12], to establish the link between a specific pathogen and infection, the pathogen can be isolated from the diseased host and grown in pure culture, and the specific disease must be reproduced when a pure culture of the pathogen is inoculated into a healthy host. Furthermore, when a pathogen is culturable and identified, the laboratory can also evaluate its susceptibility to antimicrobial drugs, which may facilitate the selection of sensitive anti-microorganism agents and control of infection. However, microbial culture is time-consuming and results of the culture may not be available until several days (most bacteria) or even weeks (such as *Mycobacteria tuberculosis*) later. Therefore, patient management still relies primarily on clinical diagnosis and empiric estimates of the most likely organisms causing infections. Other limitations of conventional culture techniques include low sensitivity. Moreover, not all pathogens can be cultured, making alternative methods other than microbial culture necessary.

Microscopy for microbes can be done quickly, and the causative agent of infection could be clarified immediately if microorganisms with typical morphous or staining is observed, such as eggs or larva of some parasites, acid-fast bacillus, Gram-negative diplococcus in the cerebrospinal fluid, certain fungi and so on. But accuracy depends on the experience of the microscopist, equipment as well as quality control. Enough microbes are required for microscopy so that they are less likely to be neglected, which make it necessary for concentration of fluid samples or even multiplying pathogens by microbial culture so that microbes reach a sufficient population density allowing visual identification. Yet not all pathogens are visible by routine microscopy. For example, most viruses can only be observed by electronic microscopy.

Immunologic methods for diagnosis of infectious diseases are increasing rapidly. There are two basic types of immunodiagnostic tests for infectious disease, *i.e.*, test for antigen produced by the infectious agent, and test for antibody response to such antigen. These tests are based on a simple concept that the antigen-antibody reaction is specific. The antigen used for immunoassay can be either components of the microbes or the microbial metabolites such as envelope protein or bacterial toxins. Specific antigen detected from serum can be interpreted by definite existence of corresponding pathogen in the host, while a detectable specific antibody may indicate that the immune system of the host has responded to certain microbial agent infection, but could not clarify whether the pathogen is present in the host at the particular point of time. However, dynamically ascending titers of serological IgM antibody may contribute to establishing the diagnosis of current or latest infection. Moreover, the specific antibody will not be detectable until the pathogen has invaded the host for a certain period, namely "window phase", particularly in chronic viral infections. The immune state of the host may also interfere with the detection of the antibody, especially in those patients with autoimmune or immunodifficient disease, leading to false-positive or false-negative results. Immunological tests are usually more rapid than direct microbial culture in the detection of pathogens, and are more sensitive, but are less specific than a culture because some microbes may share similar or the same antigen determinants. Therefore, immunological tests are relatively less used in the detection of most bacteria which can be culturable, and are more frequently applied in the diagnosis of microorganisms difficult to be isolated or cultured such as rickettsia, chlamydia, virus, etc.

The nucleic acid-based molecular method is playing an increasingly important role in the rapid detection and identification of pathogenic organisms in clinical samples. Organism-specific DNA or RNA sequences are extracted from the samples, and are detected based on sequence amplification (such as PCR or NASBA) or signal magnification (such as bDNA). They are generally highly specific and highly sensitive and can be used for all categories of microbes. Compared to the microbial culture, the nucleic acid-based method needs less volume of samples and concludes far more rapidly [13], and results are rarely affected by the antibiotics in samples or the existing state of pathogens in the host (dead or alive). Moreover, with the development of technology, the quantity of a certain pathogen in the host may be accurately assessed by the molecular method, which is hard to be realized by a microbial culture while it is very important for evaluating the effect or prognosis of some kinds of infection such as chronic hepatitis B. Similar to conventional tests, the first-generation nucleic acid-based assays determined only a single organism. Recent improvements in technologies have paved the way for high-throughput detection of more microorganisms in a single test based on multiplex PCR, microarrays or macroarrays [14], which is particularly useful for diseases of unknown origin. Use of new molecular technologies is not restricted to detection and identification of microbial pathogens but also can be used for genotyping, allowing one to determine antimicrobial resistance or to perform microbial fingerprinting. In recent years, molecular diagnostics have gradually replaced or complemented culture-based, biochemical and immunological assays in routine microbiology laboratories.

Biosensor technology is the fastest growing technology for pathogen detection when compared to the cultural, immunological or molecular method. A biosensor is a compact analytical device incorporating a biological or biologically derived sensing element (such as an enzyme, antibody, microbe or DNA) either integrated within or intimately associated with a physicochemical transducer. The transducer can be electrochemical, optical, piezoelectric or thermal devices. Advantages of microbial biosensors include the ability to provide continuous data with respect to a specific organism, targeted specificity, fast response time, the capability for mass production, the elimination or simplification of sample preparation steps, and the ability to obtain measurements with minimal perturbation of the sample. Although much progress has been made in this area, it is noted that very few of these devices have achieved commercial success until now [15].

Flow cytometry (FCM) was discovered in the late 1960s, and has been extensively applied to mammalian cells and chromosomes, such as cell cycle analysis and medical diagnostic studies. However, its application in the field of microbiology has not yet been fully developed. One of the main obstacles is the size of microbes. For example, the diameter of a bacterial cell is around one-tenth of that of a mammalian blood cell, resulting in a smaller surface area for staining, and the DNA content of a bacterial cell is about 10^{-3} times that of a diploid human cell. Therefore, the bacterial cells require more sensitive instruments and more bright fluorescent dyes. FCM has been utilized in serological discrimination of bacteria, fungi, viruses and parasites.

6.3.2 Clinical Indication of Microbial Culture and Non-Culture

Since microorganisms can live or exist in proper environments *in vitro* or *in vivo*, it can be expected theoretically that a microbial culture may be suitable for diagnosis of infections caused by all pathogenic microbes. However, this has never been realized in a clinical microbiology laboratory probably for two reasons: First, it is hard to find out all details of the living environment of every microorganism and accurately simulate it under an experimental condition. Second, convenience and cost-effect should be taken into account when selecting a method for the laboratory diagnosis of infection, so a microbial culture will inevitably be inferior to, or be replaced by, an alternative method if it is time-consuming, complicated or expensive.

Bacterial culture is the most often utilized and essential method in the diagnosis of bacterial infection, and is often incorporated with other methods for further identification of pathogens. Of note is that a viable but non-culturable (VBNC) state of bacteria has been reported [16], which means that bacteria fail to grow on the routine bacteriological media on which they would normally grow and develop into colonies, but are alive and capable of renewed metabolic activity so that they will be culturable again after resuscitation. Bacteria may enter the VBNC state as a response to some form of environmental stress, such as starvation, incubation outside the temperature range of growth, elevated osmotic concentrations, oxygen concentration, or exposure to white light [16].

Mycobacteria are difficult to culture. Specimens containing normal flora (*e.g.*, sputum) must first be decontaminated and concentrated. Mycobacterium tuberculosis and some other mycobacteria grow slowly. Growth of *M. tuberculosis* is typically faster in liquid than in solid media. Routine use of automated systems with liquid media can result in growth within 2 weeks *vs.* $\geqslant$ 4 weeks on solid media such as Lowenstein-Jensen agar [17]. Specimens should be allowed to grow for 8 weeks before being discarded.

Fungal culture is also a routine laboratory method, which attempts to grow and identify any fungus originating from a patient's specimen when fungal infection is suspected. Furthermore, the goal is to determine whether the isolated fungus is clinically significant, which is the causative agent of the patient's disease. It should be noted that 90% of fungal infections are due to dermatophytes (fungi that infect skin, hair and nails), which do not normally need a fungal culture for diagnosis. Another contraindication for fungal cultures is that the patient has already been treated with antifungal medication. Moreover, fungi are common microorganisms in the environment, skins or body cavity open to the outside, and so whether the positive fungal culture is the normal flora, opportunistic microbe or clinically significant fungi, should be evaluated before coming to a conclusion.

Viral culture is more often used in basic research rather than in clinical diagnosis, but it may be particularly useful in clinical cases with a broad viral differential, as this technique provides a relatively unbiased approach to the identification of viral pathogens. Viral culture utilizes primary cell lines or continuous cell lines to support the replication of viruses. Specimens are inoculated onto cell culture and monitored by light microscopy for cytopathic effect (CPE), the visible cellular changes that occur in response to viral infection. Based on the specimen source, the time to CPE, the quality of the CPE, and the cell line(s) showing CPE, a preliminary identification can be made [18]. The presence of a specific virus can be further confirmed by immunofluorescent staining using virus-specific, fluorescently-labeled antibodies, or by molecular detection. However, not all viruses can grow on cell culture, or even be visualized by electronic microscope.

Detection of mycoplasma by culture is the reference method of detection and has a theoretical level of detection of one colony-forming unit (CFU). The culture method uses a combination of selective growth media and incubation conditions to positively enrich any mycoplasma present in a sample. Most mycoplasma contaminants can be detected by growth on standardized agar, with the exception of certain strains of *M. hyorhinis*. Colonies of mycoplasma exhibit distinctive "fried egg" morphology when viewed under a plate microscope. However, there are some strains of mycoplasma that are non-cultivable. Moreover, pathogenic mycoplasmas are slow growing, fastidious organisms so that the assay requires a 28-day test interval before a definitive result can be obtained. Isolation might be impaired by overgrowth of saprophytic, non-pathogenic mycoplasmas and contaminant bacteria and fungi. Serologic tests have long been the basis for mycoplasma surveillance, but are of less value for the detection of early infections. Real-time PCR has gained a promising role in the diagnosis of mycoplasma infection in recent years for its higher specificity and superior sensitivity to that of culture or single point serology [19, 20].

Chlamydial infection is the most commonly reported bacterial sexually transmitted infection [21]. Up until the early 1980s, the optimum laboratory test to confirm the presence of chlamydial infection was inoculation of clinical material in cell culture and demonstration of characteristic chlamydial inclusions. However, this method necessitated good transport and cold-storage facilities in order to maintain the viability of the organism prior to inoculation. Moreover, chlamydial cell culture facilities are available only in very few clinical laboratories and this technique was, at best, only 60% – 80% sensitive. It seems to have been almost replaced by other methods, such as chlamydial antigen detection tests, or detection of chlamydial nucleic acid. Nucleic acid-based methods generally offer superior sensitivity and specificity to a culture [22].

Culture of rickettsia is difficult, laborious and dangerous (tissue culture or isolation on embryonated eggs), for there is a significant risk of laboratory infection. Serological tests continue to be the main method in the laboratory diagnosis of rickettsia infection. Recently, PCR technology became very important in identifying rickettsial species and strains.

In the diagnosis of parasites infection, microbial culture is an option inferior to microscopy, immunological tests or molecular biological assay. *In vitro* or *ex vivo* culture is sometimes used in protozoans such as malaria parasites. There are a number of issues involved in the culture of parasites that make it highly complex and subject to many variables, such as the complex life cycles. *In vitro* culture of parasites at any one stage within the life cycle involves a tremendous number of variables, including parasite stage, host site, host temperature, host immune responses, parasite species and(or) strain, and parasite-protective mechanisms. However, *in vitro* cultivation of parasites that cause human disease is invaluable, as it provides not only information on the development of the parasite but also ways for new approaches to the containment and (or) eradication of the parasite [23].

6.4 Interpretation of the Microbial Culture Results

The result of microbial culture is very useful for the clinicians when managing infectious disease from many aspects: Identifying the pathogen of the infectious disease, optimizing the antimicrobial regimen, evaluating the effect and prognosis, deciding the endpoint of therapy, or even deciding when the patient can be dismissed from the isolation state in case of the pathogen spreading.

It is obvious that two kinds of results may be expected for microbial culture: A positive result, which means that a certain microorganism had been observed or obtained by culture; or a negative result, which means that no microbe is detected by the microbiological amplification. What can be obtained from two kinds of results is very important for clinical diagnosis.

6.4.1 Interpreting the Positive Results of the Microbial Culture

A positive microbial culture is the ideal result expected both by the clinician and the microbiologist, which will favor or support the clinical decision. However, the following questions should be taken into equal consideration when evaluating the culture results and making a clinical decision.

6.4.1.1 Has the Specimen Been Contaminated during Collection and Transportation?

Contamination of the specimen is the most common factor which may result in false-positive microbial culture. To sterilize the site for obtaining a specimen and putting the specimen into sterile containers are often emphasized and rarely neglected, but other details, for example exposure of the specimen in the air for a long time, delayed transport without refrigerated management (especially for urines and stool), or collecting the stool directly from the chamber pot or toilet, are often ignored and may result in false-positive microbial culture. Therefore, the reliability of a positive result should be prudently evaluated before a clinical decision is made depending on it.

6.4.1.2 Is it the Microbe from a Normal or Resident Flora?

Some microorganisms, such as *Shigella dysenteriae, Mycobacterium tuberculosis, Coccidioides immitis*, and influenza virus, are always considered clinically significant. Others that ordinarily are harmless components of the indigenous flora of the skin and mucous membranes, or that are common in the environment, may or may not be clinically significant, depending on the specimen source from which they are isolated. For example, coagulase-negative staphylococci are normal inhabitants of the skin, the gastrointestinal tract, vagina, urethra, and the upper respiratory tract. Therefore, their isolation from superficial ulcers, wounds and sputum cannot usually be interpreted as clinically significant. However, they do commonly induce infections associated with intravascular devices and implanted prosthetic materials, and can be interpreted as clinically significant when the microorganism is isolated in large numbers from the surface of an intravascular device, from each of several sites surrounding an implanted prosthetic device or, in the case of prosthetic valve endocarditis, from several separately collected blood samples. Physicians must also consider that the composition of microbial species on the skin and mucous membranes may be altered by disease, administration of antibiotics, endotracheal or gastric intubation, and the hospital environment. For example, potentially pathogenic bacteria can often be cultured from the pharynx of seriously ill, debilitated patients in the intensive care unit, but may not cause infection.

6.4.1.3 What is the Cause of Conflicting Results of Serial Specimen at Intervals?

Serial microbial culture of the same kind of specimen at intervals is common in the clinic to increase the chance of a positive rate and evaluate the effect of therapy. Unfortunately, conflicting results often occur in serial microbial culture, which is that different microbes may be obtained at intervals and may be susceptible to various or contrastive antimicrobial agents, which often puzzle the doctor in the selection of antibiotics. Two possible factors may account for such conflicting results: Shift of the dominating flora in the host arising from the use of antimicrobial agents, and contamination of the specimen at a certain point in time.

6.4.1.4 Can a Susceptibility Test Predict the Outcome of Treatment?

Susceptibility tests determine a microbe's vulnerability to antimicrobial drugs by exposing a standardized concentration of an organism to specific concentrations of antimicrobial drugs. Results can be qualitatively reported as susceptible (S), intermediate (I), or resistant (R), or can be semi-quantitatively indicated by the minimum inhibitory concentration (MIC). Minimal killing bactericidal concentration (MBC) can also be determined but is technically difficult, and standards for interpretation have not been agreed on. For some organisms, results obtained with one drug predict results with similar drugs. Thus, not all potentially useful drugs need to be tested simultaneously.

However, a susceptibility test occurs *in vitro* and only represents the effect of the free drug on the isolated microbes. Many *in vivo* factors, such as pharmacodynamics and pharmacokinetics, site-specific drug concentrations, host immune status site-specific host defenses may influence the effect of antimicrobial agents and the treatment success. Thus, susceptibility test results do not always predict treatment outcome, and should not be the only evidence for selection of agents.

6.4.2 Interpreting the Negative Results of the Microbial Culture

The negative result of microbial culture can be interpreted as: No pathogens exist in the patient (true negative); or pathogens exist in the patient, but have not been successfully isolated or cultured for various reasons (false negative).

6.4.2.1 What Do the Negative Results Mean to a Physician?

The negative result of microbial culture is a disappointed and frustrated response to the physicians' effort to identify the pathogen of infectious disease, which may delay the diagnosis and appropriate decision of treatment. In contrast, such a result may be also encouraging and expected by the physicians after antimicrobial therapy, which may justify the selection of the antimicrobial agents and mean forthcoming recovery of patients from infection. However, microbial cultures are fraught with false-negative test results so that the medical staff must interpret culture results in light of the patient's status and response to the therapy.

6.4.2.2 What is the Cause of the False-Negative Results?

The false-negative result of microbial culture can be confirmed by repeated culture or other methods such as immunological test or nucleic acid detection. A typical procedure for a clinical microbial culture includes collecting samples from patients with suspected infection, transport of the specimen to the microbiological laboratory, inoculating the specimen into the culture medium and maintaining the culture under predefined conditions. So it can be reasonably inferred that any inhibiting factor or improper management during this procedure may induce a false-negative result, which includes:

***Status of the patient*:** The most common reason for the false-negative result is that the patient has been pretreated with an antimicrobial agent before collection of the specimen, so that growth of the microbe may be inhibited both *in vivo* and *in vitro*. In addition, if the patient is in the prodromal or recovery stage of an infection, pathogens will be less multiplied and released so that not enough pathogens can be obtained for culture.

***Improper collection and transport of specimen*:** Not enough volume of a specimen, collecting samples from the wrong site of the body, delayed transport, no VTM for viral specimen and so on will inevitably result in false-negative culture.

***Inoculating the specimen into the non-optimal medium*:** For example, *Mycobacterium tuberculosis* grows very slowly in solid media and may appear as a negative result within a period as long as 8 weeks. Use of some automated systems with liquid media such as TB BACTAC can yield results of primary cultures within the shortest periods (3 – 21 days) [24].

***Improper condition for maintaining culture*:** Improper moisture and temperature in the incubating chamber may significantly inhibit the growth of most microbes. Anaerobic bacteria will not grow if oxygen has not been deprived from the incubating atmosphere. The concentration of carbon dioxide is also very important for maintaining a viral culture. Contamination of mycoplasma is common in the cell culture and may make it not suitable for supporting the multiplication of a virus [25, 26].

In conclusion, microbial culture plays a very important role in clinical infection diagnosis and therapy. At the same time, consideration of many kinds of possibilities will help to correct interpretation of microbial culture results and give a correct direction for clinical diagnosis and therapy.

References

[1] Wilson M L, Weinstein M P. General principles in the laboratory detection of bacteremia and fungemia. Clin Lab Med, 1994, 14: 69-82.
[2] Li J, Plorde J J, Carlson L G. Effects of volume and periodicity on blood cultures. J Clin Microbiol, 1994, 32: 2829-2831.
[3] Paisley J W, Lauer B A. Pediatric blood cultures. Clin Lab Med, 1994, 14: 17-30.
[4] Auckenthaler R, Ilstrup D M, Washington J A. Comparison of recovery of organisms from blood cultures diluted 10% (volume/volume) and 20% (volume/volume). J Clin Microbiol, 1982, 15: 860-864.
[5] Reimer L G, Wilson M L, Weinstein M P. Update on detection of bacteremia and fungemia. Clin Microbiol Rev, 1997, 10: 444-465.
[6] Murray P R, Hampton C M. Recovery of pathogenic bacteria from cerebrospinal fluid. J Clin Microbiol, 1980, 12: 554-557.
[7] Lauer B A, Reller L B, Mirrett S. Evaluation of preservative fluid for urine collected for culture. J Clin Microbiol, 1979, 10: 42-45.
[8] Wells J G, Morris G K. Evaluation of transport methods for isolating *Shigella* spp. J Clin Microbiol, 1981, 13: 789-790.
[9] Bouza E, Sousa D, Rodríguez-Créixems M, *et al.* Is the volume of blood cultured still a significant factor in the diagnosis of bloodstream infections? J Clin Microbiol, 2007, 45: 2765-2769.
[10] Parry C M, Wijedoru L, Arjyal A, *et al.* The utility of diagnostic tests for enteric fever in endemic locations. Expert Rev Anti Infect Ther, 2011, 9: 711-725.
[11] Wain J, Hosoglu S. The laboratory diagnosis of enteric fever. J Infection in Developing Countries, 2008, 2: 421-425.
[12] Ani A E. Advances in the laboratory diagnosis of mycobacterium tuberculosis. Ann Afric Med, 2008, 7: 57-61.
[13] Polisena J, Chen S, Cimon K, *et al.* Clinical effectiveness of rapid tests for methicillin resistant *Staphylococcus aureus* (MRSA) in hospitalized patients: A systematic review. BMC Infect Dis, 2011, 11: 336-348.
[14] Millar B C, Xu J, Moor J E. Molecular diagnostics of medically important bacterial infections. Curr Issues Mol Biol, 2006, 9: 21-40.
[15] Deisingh A K, Thompson M. Biosensors for the detection of bacteria. Can J Microbiol, 2004, 50: 69-77.

[16] Oliver J D. The viable but nonculturable state in bacteria. J Microbiol, 2005, 43: 93-100.
[17] Cruciani M, Scarparo C, Malena M, *et al.* Meta-analysis of BACTEC MGIT 960 and BACTEC 460 TB, with or without solid media, for detection of mycobacteria. J Clin Microbiol, 2004, 42: 2321-2325.
[18] Leland D S, Ginocchio C C. Role of cell culture for virus detection in the age of technology. Clin Microbiol Rev, 2007, 20: 49-78.
[19] She R C, Thurber A, Hymas W C, *et al.* Limited utility of culture for *Mycoplasma pneumoniae* and *Chlamydophila pneumoniae* for diagnosis of respiratory tract infections. J Clin Microbiol, 2010, 48: 3380-3382.
[20] Kashyap B, Kumar S, Sethi G R, *et al.* Comparison of PCR, culture & serological tests for the diagnosis of *Mycoplasma pneumoniae* in community-acquired lower respiratory tract infections in children. Indian J Med Res, 2008, 128: 134-139.
[21] Centers for Disease Control and Prevention, Workowski KA, Berman SM. Sexually transmitted diseases treatment guidelines, 2006. MMWR Recomm Rep, 2006, 55: 1-94.
[22] Johnson R E, Newhall W J, Papp J R, *et al.* Screening tests to detect Chlamydia trachomatis and Neisseria gonorrhoeae infections. MMWR Recomm Rep, 2002, 51: 1-38.
[23] Visvesvara G S, Garcia L S. Culture of protozoan parasites. Clin Microbiol Rev, 2002, 15: 327-328.
[24] Dinnes J, Deeks J, Kunst H, *et al.* A systematic review of rapid diagnostic tests for the detection of tuberculosis infection. Health Technol Assess, 2007, 11: 1-196.
[25] Jun-Jie Li, Zi-Ke Sheng, Mei Deng, *et al.* Epidemic of klebsiella pneumoniae ST11 clone coproducing KPC-2 and 16S rRNA methylase RmtB in a Chinese university hospital. BMC Infectious Diseases, 2012, 12: 373.
[26] Min Xu, Baohong Wang, Yiqi Fu, *et al.* Changes of fecal bifidobacterium species in adult patients with hepatitis B virus-Induced chronic liver disease. Microb Ecol, 2012, 63: 304-313.

7

Molecular Microecological Techniques

Zongxin Ling, Charlie Xiang *
The State Key Laboratory for Diagnosis and Treatment of Infectious Diseases, the First Affiliated Hospital, School of Medicine, Zhejiang University; Collaborative Innovation Center for Diagnosis and Treatment of Infectious Diseases, Hangzhou, 310003, China
* E-mail: cxiang@zju.edu.cn

Recent researches have shown that microbes associated with the human body are approximately 10 times more numerous than our own cells and contain, in aggregate, about 100 times more genes. This has led to the viewpoint that humans and our microbial symbionts should be considered as "superorganisms". Some microbes cause disease, but the overwhelming majority are either innocuous or play a vital role in human development, physiology, immunity, and nutrition. However, the overall extent of bacterial diversity in different microhabitats, such as the oral cavity, urogenital tract, skin, gastrointestinal tract, nasal passages and so on, has not been studied extensively yet. The traditional microbiological techniques, such as bacterial cultivation in combination with accurate molecular identification, could not keep pace with our demands for understanding the real world of the field. Despite the rapid development in microbiology, the basic requirements for microecological studies have tremendous limitations. However, these limitations can be overcome with the advent of molecular ecology techniques based on sequence comparisons of nucleic acids (DNA and RNA) and can be used to provide molecular characteristics, while also providing a classification scheme that predicts phylogenetic relationships at the same time. Either direct or in combination with microbial cultivation, the benefits of molecular methods for microecological studies are apparent, which can acquire new and more detailed information about the microbial communities in different microhabitats. The molecular techniques used widely today provide powerful approaches for comprehensive analyses of the microbial communities. This also can be achieved by optimally combined use of different molecular methods that complement each other. For example, fluorescent *in situ* hybridization (FISH) that cells labeled by group-specific probes can be enumerated using flow cytometry and be simultaneously sorted and collected. Subsequent diversity screenings are possibly used by other methods such as molecular fingerprinting methods or 16S rDNA sequence analysis. Among molecular fingerprinting methods, PCR-denaturing gradient gel electrophoresis (PCR-DGGE) / PCR-temperature gradient gel electrophoresis (PCR-TGGE), terminal restriction fragment length polymorphism (T-RFLP) represent rapid and reliable techniques that have been used successfully to evaluate the microbial compositions of different microhabitats. Sequencing of 16S rDNA by constructing clone libraries has revolutionized our understanding of bacterial diversity. Further improvements in molecular monitoring of microbial communities can be expected when new high-throughput techniques such as DNA-microarray technology and next-generation sequencing techniques will become applicable for molecular ecology studies. All these will give us a deeper understanding of the microbial ecology in different microhabitats and, consequently, about the possibilities for modulating the microbiota in specific microhabitat to the benefit of host health.

7.1 Introduction

The human body is home to many indigenous microorganisms, with distinct communities at different anatomical sites. An enormous number of microorganisms, the vast majority of which are bacterial species, are known to colonize and form complex communities, or microbiota, at various sites within the human body [1-3]. Microbial cells that thrive on and within the human body are approximately 10 times more numerous than our own cells and contain, in aggregate, about 100 times more genes, leading to the suggestion that humans and our microbial symbionts be considered "superorganisms" [4]. A growing body of evidence suggests that the composition and function of microbiota in different human body habitats play a vital role in human development, physiology, immunity, and nutrition [1, 5-9]. For example, in the gastrointestinal tract, the taxonomic composition of the microbial community may affect the propensity to develop obesity [7, 10], inflammatory bowel disease [11], type 1 diabetes [12], cardiovascular diseases [13], even chronic diseases [14] and so on. The expansion of these studies paralleled the development of various high throughput-omics technologies [15, 16].

Methodological developments have played crucial roles in our conceptual understanding of the microecological related diseases of the specific microhabitat of the human body mentioned above. Traditional microbiology has focused on the study of individual species as isolated units. In fact, uncultured microorganisms comprise the majority of the planet's biological diversity [17]. Many microbes have never been successfully isolated as viable specimens for analysis, presumably because their growth condition is dependent upon a specific microenvironment that has not been, or cannot be, reproduced experimentally. Among those isolated species, analyses of genetic structure, gene expression modes, and metabolic characteristics have rarely extended to inter-species interactions or microbe-host interactions. A reliance on culture-based enumeration methods apparently underestimates of true microbial diversity. There is a growing opinion that the term "unculturable" is inappropriate, while we have not found the correct cultural conditions. With our inability to cultivate most of the microbial species colonizing in the specific microenvironment, however, the taxonomic composition, its community structure, and ultimately its function, have not been fully explored. With the advent of molecular techniques, the dynamics of bacterial diversity in different microhabitats have been investigated using molecular fingerprinting methods and sequence analysis of microbial small subunit ribosomal RNA genes (16S rRNA) and other universal targets (such as *cpn60*) [18, 19]. Among these molecular fingerprinting methods, T-RFLP and PCR-DGGE represent rapid and reliable techniques that have been used successfully to identify the bacterial compositions of different microhabitats. The sequencing of 16S rRNA genes from different samples by constructing clone libraries (typically at most a few thousand clones from a low number of individuals) has revolutionized our understanding of microbial systematics and diversity [20]. However, this cloning and sequencing method only identifies the abundant members of the communities, which can still not reach the real word of the microbial communities in the specific microhabitat. Recently, next-generation high-throughput sequencing techniques, especially ultra-high-throughput pyrosequencing, have been developed for the monitoring of microbial communities in different microhabitats. The improvements in pyrosequencing (*e.g.*, barcoded pyrosequencing on the Genome Sequencer FLX/454 Life Sciences platform) enable a great increase in throughput via parallel in-depth analysis of large scale samples with limited sample processing and lower costs [21, 22]. This technique has been successfully used in various ecosystems including deep mines [23], soil [24], fermented seafood [25], skin [26], chronic wounds [27], and oral microbiota [28, 29]. Of course, other molecular approaches such as amplified rDNA restriction analysis (ARDRA), FISH, internal transcribed spacer (ITS) and single-stranded conformation polymorphism (SSCP), are all widely used for analyzing the microbial diversity in the specific microhabitat of the human body. Recently, a custom-designed, gut microbiome specific phylogenetic microarray (CombiMatrix, CustomArray 4X2K Microarray) termed the Aus-HIT (human intestinal tract) chip was used for identifying the gastrointestinal microbiota. Fig. 7.1 shows the common approaches used for microbial diversity analysis in microecology. All of these approaches produce specific patterns or profiles of nucleic acids amplified from samples and that patterns reflect the structure of microbial community.

Recent development of metagenomics has shown that uncultured microorganisms represent the vast majority of organisms in most microhabitats. This evidence is derived from analyses of 16S rRNA gene sequences amplified directly from the samples, which can avoid the bias imposed by culturing and lead to the discovery of vast new phylotypes of microbial life. In fact, the 16S rRNA genes analysis has revolutionized microbial world, however, such studies only obtain a phylogenetic description of community membership, while providing little insight into the genetics, physiology, and biochemistry of the members. Metagenomics provides a new technical innovation that facilitates study of a more direct and unbiased access to uncultured organisms. The application of metagenomic sequence information will be helpful to understand the physiology and ecology of environmental microorganisms and facilitates the design of better culturing strategies to link genomic analysis with pure culture studies [17].

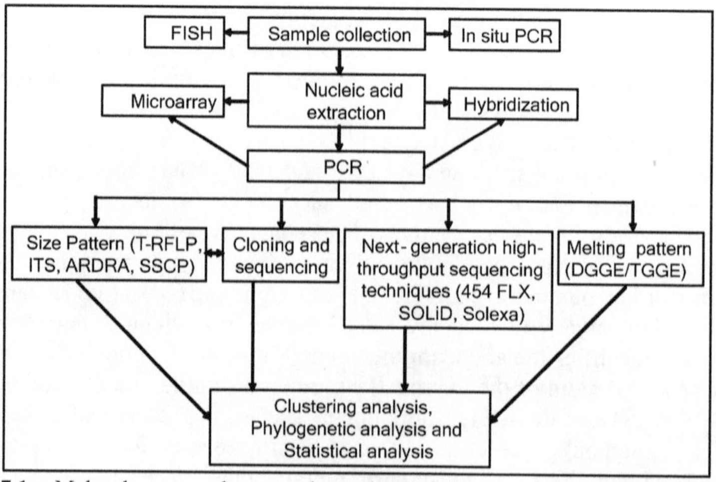

Fig. 7.1. Molecular approaches to analyze microbial diversity in the specific microhabitat

7.2 Size Pattern Analysis — T-RFLP Polymorphism Analysis of 16S rRNA Genes

Terminal restriction fragment length polymorphism (T-RFLP) analysis is one of the popular fingerprinting techniques used to monitor changes in the structure and composition of microbial communities [30]. T-RFLP has been successfully used in numerous studies to explore microbial diversity of the predominant populations in various habitats and offers the advantage that it is more amenable to high throughput and more comprehensive than cultivation-dependent methods [31-36]. With a compromise between the information gained and labor intensity, T-RFLP has been widely used for microecological analysis. In fact, the size polymorphism of terminal restriction fragments from a PCR amplified marker is measured by T-RFLP, which is implied by its name. It is the combination of at least three technologies including comparative genomics/RFLP, PCR, and nucleic acid electrophoresis (Fig. 7.2). According to the results of comparative genomics, a specific DNA sequences including phylogenetic marker is chose as the target sequence. The universal primer is designed and PCR amplifies the signal from a high background of unrelated markers. Subsequent digestion with suitable restriction endonucleases produces terminal fragments appropriate for sizing on high resolution sequencing gels. The latter step is conveniently performed on automated systems such as the ABI gel or capillary electrophoresis systems that can provide digital output. A fluorescently tagged primer use limits the analysis to only the terminal fragments of the digestion. Because size markers bearing a different fluorophore from the samples can be included in every lane, the sizing is extremely accurate [37].

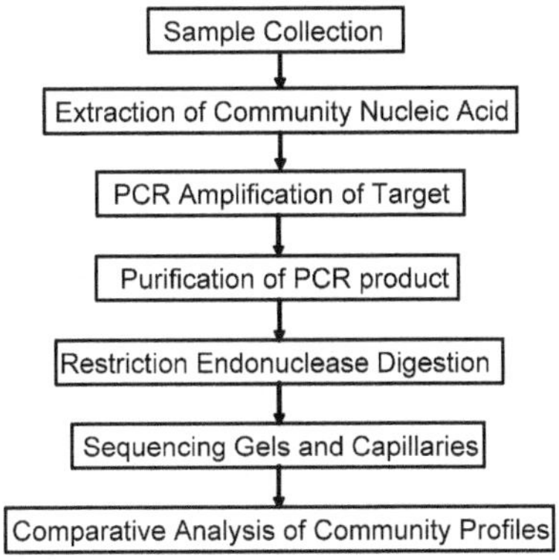

Fig. 7.2. The workflow of the T-RFLP

After community DNA is extracted from the sample, bacterial 16S rRNA gene is specifically amplified by PCR using primer pairs designed from its conserved regions. The bacterial 16S rRNA gene can be used to investigate the all predominant bacteria in the microbial communities. Then, the PCR product is digested by a restriction enzyme, and the length profile of terminal restriction fragment (TRF) labeled by the fluorescent dye is detected, which can be used for the identification of species composition of the bacterial communities. The ratios of each PCR amplicon that estimate by measuring fluorescence emission intensity, can indicate the relative abundance of bacterial species. This method can be used to estimate the relative abundance of predominant bacterial species (relatively quantitative) and to study the pattern shift (in terms of predominant bacterial groups and their relative abundance) of bacterial community structures.

For T-RFLPs analysis, one primer is labeled at the 5' terminus with a fluorescent dye (such as NED, VIC, 6-FAM and so on). Generally, one species will contribute one terminal fragment of a given size, although several species may have terminal fragments of identical size. T-RFLP is a relative high-throughput, reproducible method that can be used to carry out both qualitative and quantitative analyses of a particular gene in a community. The 16S rRNA gene fragments are usually targeted, which fragments can be separated by gel electrophoresis in non-denaturing polyacrylamide gels or by capillary electrophoresis, and discriminated by laser-induced fluorescent detection. These fluorescence data are converted into electrophoregrams, in which the areas under the peaks indicate the relative proportions, and the peaks represent fragments differing in size of the fragments. The advantage of T-RFLPs is its capability to detect even rarer members of a microbial community, but it is greatly affected by the amount of the total bacterial genomic DNA. In addition, phylogenetic assignments can be deduced from the sizes of the terminal restriction fragments (T-RFs) using web-based resources that predict T-RF sizes for known bacteria, such as such as the Ribosomal Database Project (http://rdp.cme.msu.edu) or the Microbial Community Analysis (MiCA) website at the University of Idaho (http://hermes.campus.uidaho. edu) [38, 39].

The clustering analysis algorithms for T-RFLP data described by Abdo *et al.* [40] are used for identifying the threshold for defining peaks and for the cluster analysis of T-RFLP data. First, true peaks are determined with a defined threshold. Second, hierarchical clustering is carried out to identify those fragments with lengths close enough to justly group them in the same length category. Third, the Euclidean distances between T-RFLP profiles are calculated and these are hierarchically clustered based on average linkage and a dendrogram is constructed. Finally, three cluster criteria were employed to identify a statistically meaningful number of groups in the data [40]. Recently, a free, online software — T-REX — was developed for the processing and analysis of T-RFLP data. With this software it was able to: (i) Label raw data with attributes related to the experimental design of the samples; (ii) Distinguish true peaks over noise with a baseline threshold; (iii) Align T-RFs in all samples; (iv) Construct a two-way data matrix from labeled data and process the matrix in a variety of ways; (v) Produce several measures of data matrix complexity, including the distribution of variance between main and interaction effects and sample heterogeneity; and (vi) Analyze a data matrix with the additive main effects and multiplicative interaction (AMMI) model. In fact, T-REX can provide a free, platform-independent tool to the research community that allows for an integrated, rapid, and more robust analysis of T-RFLP data [41].

T-RFLP is rapid and sensitive molecular technique, which is a highly reproducible and robust technique that yields high-quality fingerprints consisting of fragments of precise sizes. In principle, with an appropriate database constructed, these T-RFLP data could be phylogenetically assigned. As a powerful molecular technique, it has been used for strain identification, comparative community analysis and to derive estimates of the diversity of a phylogenetic group within a community. Because of its high-throughput capacity and the supporting sequence databases, T-RFLP will be proved most valuable in comparative community analysis. With systematic phylogenetic specific primers used in T-RFLP, increasing levels of community dissection can be attained. When T-RFLP data combines with complementary data on the diversity and distribution of fundamental physiological markers as well as physicochemical data describing the particular microhabitat, they will provide new insight into the structure and function of microbial communities. However, T-RFLP data of the bacterial community profile can be considered as "semi-quantitative" according to the number of peaks in each sample (*i.e.* number of distinguishable bacterial types) as well as qualitative according to the position of peaks (*i.e.* occurrence of unique bacterial types) [42]. T-RFLP data can only be regarded as "semi-quantitative" since one peak may represent many species of bacteria that share the same cutting sites for the restriction enzyme of experimenters' choice [43]. The same as for all other molecular tools, minority bacterial populations may not be detected by

T-RFLP analysis since template DNA from these populations represent a small fraction of the total extracted DNA and may not be amplified by PCR due to kinetic bias [30]. Engebretson and Moyer [44] have shown that T-RFLP seems to be very useful for estimating diversity in communities characterized by low-to-intermediate species richness, but is not suitable for complex microbial populations. In addition, it cannot differentiate closely related DNA sequences, which are likely to have the same terminal restriction site, and thus may reduce the number of detectable operational taxonomic units (OTUs). However, the identified limitations and pitfalls for this technique include the formation of pseudo-terminal restriction fragments, which can result in the overestimation of microbial diversity, which was reported by Egert and Friedrich [45]. Besides, the choice of the primers and restriction endonucleases also seems to be very important for obtaining an accurate evaluation of the microbial diversity [44, 46].

7.3 Melting Pattern Analysis — PCR-DGGE Analysis of 16S rRNA Genes

Denaturing gradient gel electrophoresis (DGGE), an electrophoretic method to identify single base changes in a segment of DNA, was originally developed in the 1980s to detect single mutations in genomic DNA [47-49]. Separation techniques on which DGGE is based are first described by Fischer and Lerman [50]. Double-stranded DNA (dsDNA) is subjected to an increasing denaturant environment and will melt in discrete segments called “melting domains” in a denaturing gradient acrylamide gel. It is sequence-specific for the melting temperature (T_m) of these domains. When it reaches the T_m of the lowest melting domain, the dsDNA will become partially melted and then create branched molecules. Partial melting of the dsDNA also reduces its mobility in the gel. Since the T_m of a particular melting domain is sequence-specific, the presence of a mutation will alter the melting profile of that DNA when compared to the wild-type. DNA containing mutations will encounter mobility shifts at different positions in the gel to the wild-type. If the fragment completely denatures, then migration again becomes a function of size [51].

Recently, microbial ecologists take advantage of this technique for analyzing whole bacterial communities and for studying the microbial diversity in various bacterial populations. As a molecular fingerprinting method, PCR-DGGE can separate PCR-generated DNA products. The environmental DNA PCR can generate templates of different DNA sequences that represent many of the predominant microorganisms. However, since PCR products from a given reaction are of similar size (bp), conventional separation results only in a single DNA band by agarose gel electrophoresis that is largely non-descriptive. DGGE can overcome this limitation by separating PCR products based on sequence differences that results in differential denaturing characteristics of the DNA. During DGGE, an approximately 40 bp GC-rich sequence (GC clamp) is added into the 5’-end of the forward primer, which facilitates migration of PCR products of similar size but varying sequence, to different locations within a polyacrylamide gel containing a denaturing gradient of urea and formamide. Upon reaching a threshold denaturant concentration, the weaker melting domains of the double-stranded PCR product will begin to denature at which time migration slows dramatically. Different sequences of PCR products will denature at different denaturant concentrations resulting in different patterns of bands. Each band theoretically represents a different bacterial phylotypes present in the community. Once generated, fingerprints can be uploaded into databases in which fingerprint similarity can be assessed to determine microbial structural differences between environments or among treatments. Furthermore, with the breadth of PCR primers available, DGGE can also be used to investigate broad phylogenies or specific target organisms such as pathogens.

The small subunit (SSU) or 16S rRNA gene and its rRNA counterpart are the most widely used target molecules in the field of bacterial identification, phylogeny, and detection. Compared to the most common housekeeping genetic marker, there are a number of reasons to use the 16S rRNA gene sequences to study bacterial phylogeny and taxonomy, which include (i) Its presence in almost all bacteria; (ii) Unchanged function of the 16S rRNA gene over time, suggesting that random sequence changes are a more accurate measure of time; and (iii) The sequences (1,500 bp) is large enough for informatics purposes [52]. Because the 16S rRNA gene is ubiquitous and composed of alternating conserved (C) and hypervariable (V) regions, it represents the prototype genetic marker in molecular microbial ecology (Fig. 7.3).

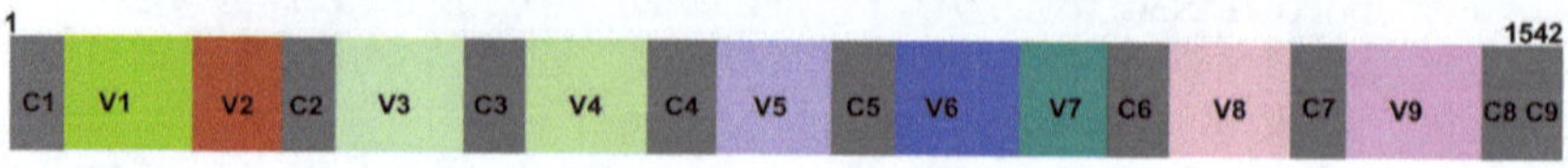

Fig. 7.3. The structure of SSU or 16S rRNA gene
V1: 61–106 bp; V2: 121–240 bp; V3: 436–500 bp; V4: 588–671 bp; V5: 734–754 bp; V6: 829–857 bp; V7: 990–1,045 bp; V8: 1,118–1,160 bp; V9: 1240–1,298 bp

As a result, 16S rRNA gene sequences have provided the basis for the design of oligonucleotide probes and primers for analysis of complex bacterial ecosystems by means of clone libraries, and population fingerprinting (*e.g.*, using DGGE or TGGE, FISH, and real-time PCR). Huys et al. has summarized the universal and group-specific PCR primers used in sequence dependent electrophoresis-based profiling of human microbial communities [53]. However, the hypervariable region(s) chosen for amplification can affect the DGGE profiles and diversity indices produced from community DNA samples and even subtle differences in primer sequences can result in substantially different profiles and the downstream assessment of microbial diversity. By comparing different hypervariable regions of 16S rRNA genes for PCR-DGGE, Yu and Morrison have shown that the DGGE profiles of the V3 region were the most reliable [54]. Likewise, Amp and co-workers reported the co-detection of chloroplast and mitochondrial DNA of plants (Zea mays) by DGGE fingerprinting in the analysis of maize fermentations using a frequently applied prokaryotic V3 – 16S rRNA gene primer set [55].

Generally, different PCR strategies are used for the following DGGE analysis. For total bacteria detection, the "touchdown" PCR in combination with reconditioning PCR is performed, which offers a simple and rapid means to optimize PCRs, increasing specificity, sensitivity and yield, without the need for lengthy optimizations and/or the redesigning of primers. The most common used primers are 341F (5'-ACTCCTACGGGAGGCAGCAGT-3') and 518R (5'-GTATTACCGCGGCTGCTGGCAC-3') that targeting on the 16S rRNA hypervariable V3 region. With the GC-clamp and primers, the PCR products are approximately 250 bp and are suitable for DGGE analysis. However, for group-specific DGGE analysis, such as the group of *Lactobacillus* and *Bifidobacterium*, the common PCR strategy is nested PCR. In the first PCR round, nearly full length 16S rRNA gene sequences are amplified using primer sets such as 8F (5'-AGAGTTTGATCCTGGCTCAG-3') and 1492R (5'-GGTTACCTTGTT ACGACTT-3'), and in the second PCR round, group specific primer sets were used. One advantage of using nested PCR with specific primers is the increased sensitivity, which makes it possible to visualize also those species that are present in lower numbers [56]. However, the use of nested PCR should be evaluated with caution. A possible disadvantage of applying two successive PCR reactions is the introduction of an even greater bias due to preferential amplification [57]. In practice, DGGE-based community profiling comprises four steps: (i) Total bacterial genomic DNA extraction from the samples; (ii) PCR-controlled amplification using specific oligonucleotide primers; (iii) DGGE analysis; and (iv) Fingerprint processing and analysis.

PCR with GC-clamp attached primers. Generally, the commonly used DGGE primers were the universal bacterial primers 341F and 534R targeting on the V3 regions of the 16S rRNA genes [29, 58, 59]. Enough bacterial genomic DNA is extracted from environmental samples and used as a template for PCR amplification of 16S rRNA genes originating from resident prokaryotic populations. In one of our studies of vaginal microbiota analysis, a pair of universal bacterial primers (341F 5'- GTATTACCGCGGCTGCTGG-3', 534R 5'-CGCCCGCCGCGCGCGGCGGGCGGGGCGGGGGCACGGGGGGACTCCT ACGGGAGGCAGCAG-3') is used to amplify PCR products for DGGE analysis. In fact, the 16S rRNA-V3 PCR amplification uses the hot-start touchdown protocol described by Muyzer *et al.* [60], and the reaction mixture contained 1.25 unit of Hot Start *Taq* polymerase (Takara, Dalian, China), 1× PCR buffer (2.5 mmol/L $MgCl_2$ included), 3 pmol of each primer, 200 mmol/L each deoxynucleoside triphosphate (dNTP) and 20 ng of extracted bacterial DNA in a total volume of 50 μL. The thermal cycling program consists of the following time and temperature profile: 95 ℃ for 5 min; 20 cycles of touchdown PCR: denaturizing at 95 ℃ for 30 s, annealing at 65 ℃ for 30 s which was decreased by 0.5 ℃ every second cycle, and extending at 72 ℃ for 30 s; then additional 5 cycles with annealing temperature at 55 ℃ for 30 s were performed; followed by final extension at 72 ℃ for 7 min. In order to minimize heteroduplex formation and single-stranded DNA (ssDNA) contamination during PCR amplification that might cause sequence heterogeneity in a single DGGE band, an additional 5 cycles of reconditioning PCR and PCR products are purified by electrophoresis on a 1%

agarose gel and eluted with QIAquick Gel extraction kit (QIAGEN) before DGGE analysis. All PCRs are performed in a thermocycler PCR system (DNA Engine Tetrad 2, Bio-Rad, Hemel Hempstead, Herts, UK). Samples (5 μL) of the amplified products (approximately 200 bp) are checked by electrophoresis on 1% (*wt/vol*) agarose gel and visualized by ethidium bromide staining, and the concentrations are measured by using a NanoDrop ND-1000 spectrophotometer (Thermo Electron Corporation). All amplified products are stored at −20 ℃ before DGGE analysis.

Optimal denaturant gradient determination. In DGGE, the denaturing environment is created by a combination of uniform temperature, typically between 50 and 65 ℃ and a linear denaturant gradient formed with urea and formamide. A solution of 100% chemical denaturant consists of 7M urea and 40% formamide. The denaturing gradient may be formed perpendicular or parallel to the direction of electrophoresis. A perpendicular gradient gel, in which the gradient is perpendicular to the electric field, typically uses a broad denaturing gradient range, such as 0 – 100%. In parallel DGGE, the denaturing gradient is parallel to the electric field, and the range of denaturant is narrowed to allow better separation of fragments. Generally, different samples from different microhabitats should be used different optimal denaturant gradient. With the uniform temperature maintaining in 60 ℃, the optimal denaturant gradient for samples from the vagina is 30% – 55%, for samples from oral cavity is 28% – 58%, and for samples from skin, gastrointestine and nasal passages is 30% – 60%. For other environmental samples, the optimal denaturant gradient should be determined before formal experiments. The optimal denaturant gradients will differentiate the predominant bacteria in the specific microhabitat.

DGGE analysis. Our parallel DGGE analysis is performed using the D-Code universal mutation detection system apparatus (Bio-Rad, Hercules, CA) with 16-cm by 16-cm (or 20-cm by 20-cm) by 1-mm gels according to the manufacturer's protocol. The sequence-specific separation of the PCR fragments is obtained in 8% (*wt/vol*) polyacrylamide [acrylamide-N, *N′* bisacrylamide; 37.5:1 (*wt/vol*)] gels in 1× TAE buffer (40 mmol/L Tris, 20 mmol/L glacial acetic acid, 1 mmol/L EDTA, pH 8.0). The denaturing gels with optimal denaturant gradients are increasing in the direction of electrophoresis. A 100% denaturing solution contained 40% (*vol/vol*) formamide and 7 mol/L urea. A stacking gel containing 8% (*wt/vol*) polyacrylamide was applied onto the denaturing gel. A volume of 13 – 16 μL of PCR samples is loaded onto the stacking gel. Electrophoresis is conducted at a constant voltage of 200 V and a temperature of 60 ℃ for approximately 4 h. The addition of a 40 bp GC clamp to one of the PCR primers insures that the region screened is in the lower melting domain and that the DNA will remain partially double-stranded. In other words, the 40 bp GC clamp can improve the resolution of DGGE. Following electrophoresis, the gel is stained by SYBR green I (Amresco, Ohio, USA) and photographed with UVI gel documentation system (UVItec). Each set of samples was independently PCR amplified and analyzed by DGGE twice to confirm the reproducibility of banding profiles.

DGGE bands identification. In order to identify phylogenetic affiliation, fragments of interest were excised from denaturing gradient gels with a sterile scalpel and placed into a single Eppendorf tube. Gel pieces were washed once in 1× PCR buffer and incubated in 20 μL of the same buffer overnight at 4 ℃. Five microliters of the buffer solution were used as a template for PCR re-amplification with universal bacterial primers, as described above for DGGE, but without GC clamps. PCR products were excised from 1.0% agarose gel and purified with QIAGEN, then ligated with pGEM-T Easy Vector (Promega), transformed into competent *Escherichia coli* DH5α cells. The positive clones were verified and sequenced. The sequences of excised DGGE bands were submitted to RDP II release 10 database to determine their closest isolate relatives with length <1,200 bp. Sequences similarity searches were used to assign each clone to major bacterial phylotypes.

DGGE image analysis. The digitized gel images can be analyzed using Quantity One® 1-D analysis software or Bionumerics software program. These software are used to detect bands by normalizing against the total intensity data for each lane. Bands with a minimum density of 5% were detected in each lane and bands were matched using a match tolerance of 2%. Bands occupying the same position in the different lanes of the gels were identified. A similarity matrix was constructed using Dice's similarity coefficient. This is defined as $[2j/(a+b)] \times 100$, where j is the number of bands in common between two lanes, and $(a+b)$ is the total band number of both lanes. Dendrograms were constructed by the unweighted pair group method, using arithmetic averages (UPGMA).

In our studies, we found that PCR-DGGE was a useful tool for detecting changes in predominant microbiota in specific microhabitats, and this has been widely applied for comparative analysis of parallel samples. In one studies, our PCR-DGGE profiles were obtained by amplifying bacterial DNA from vaginal swabs with and without bacterial vaginosis (BV). Each lane represented one subject which was selected from its group at random. As shown in Fig. 7.4(a), the PCR-DGGE profiles of BV and CN reveal significant differences in the overall structure and composition of the vaginal community by targeting the V3 region of 16S rDNA. Bacterial diversity is higher in the BV group than that in the healthy control (CN) group. DGGE profiles are significantly different from one another and varied with the participants. Fig. 7.4(b) depicts the results of Ward's analysis in which the Dice coefficient for measuring similarity in banding patterns was applied. The BV and CN groups display a statistically significant clustering of profiles, cluster I (BV group) and cluster II (CN group). 11 dominant fragments that can represent the pattern of the DGGE profiles are excised from the DGGE lanes, reamplified, sequenced and identified by BLAST with the 16S rRNA V3 region sequences. *Lactobacillus* was the predominant genus in the CN group and bacterial diversity of the BV group was far more complex and was dominated by *A. vaginae*, uncultured *Sneathia sp.*, *Fusobacterium nucleatum subsp.*, uncultured *Eggerthella sp.*, uncultured *Megasphaera sp.*, *Clostridium acetobutylicum* and *Clostridium thermocellum*. From these results, we propose that PCR-DGGE analysis can be used to monitor the dramatic shift of bacterial transition and routinely defined BV in laboratory. PCR-DGGE can be one useful tool for BV diagnosis with microecological viewpoints.

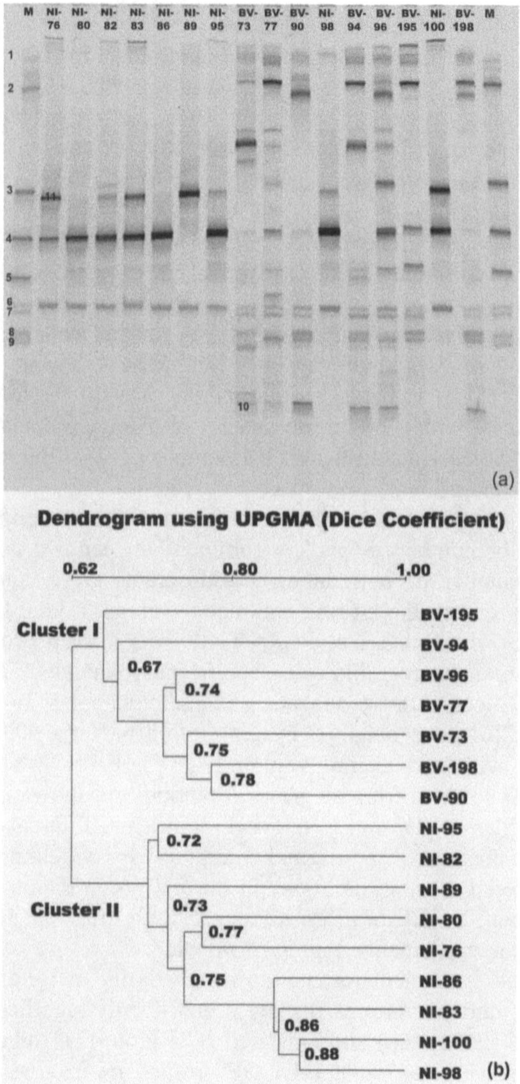

Fig. 7.4. (a) PCR-Denaturing Gradient Gel Electrophoresis (DGGE) analysis of the predominant bacterial communities in vaginal swabs from bacterial vaginosis (BV group) and healthy women (CN group). Each lane represented one subject which was selected in its group at random. M represents a marker constructed in this study with the identified bands to facilitate the interpretation of the figure. Bands: 1: Uncultured Sneathia sp.; 2: Fusobacterium nucleatum subsp. nucleatum ATCC 25586; 3: Clostridium thermocellum ATCC 27405; 4: Lactobacillus iners; 5: Clostridium acetobutylicum; 6: Lactobacillus iners; 7: Clostridium thermocellum ATCC 27405; 8: Atopobium vaginae; 9: uncultured Eggerthella sp.; 10: uncultured Megasphaera sp.; 11: Lactobacillus crispatus. (b) Dendrogram of the DGGE profiles shown in panel (a)

As one of the most promising molecular techniques in microbial ecology, PCR-DGGE can be used to evaluate the complex microbial communities and the dynamics of community structure in the specific microhabitat. In a word, high resolution, relative long fragment length detection, non-isotopic, easy manipulation and fast are the major advantage of DGGE. That's why DGGE analysis has been used widely in microbial ecology. Of course, although PCR-DGGE is a useful tool to detect the predominant microbiota of the specific microhabitat, Li *et al.* also points several drawbacks existing in the technique such as: (i) Unrelated bacterial species may have similar or identical migration distances [61]; (ii) Two or more phylogenetically related bacterial species may display close or overlapping band positions on the DGGE gel [62]; (iii) Levels of certain microbes in the sample may have been at or below detectable levels for PCR; (iv) Although many more bacterial species have been identified by PCR-DGGE [63, 64], not all the species have been detected in a single individual's specific microhabitat; (v) PCR-DGGE can only detect the structure of the predominant bacteria in the specific microhabitat, but not quantify these bacteria. In combination with real-time quantitative PCR (qPCR), the structure and abundance of the predominant bacteria will be clearly clarified [58, 59]. It might be the limitations of the molecular approach for the study of microbial ecology [65].

7.4 FISH

In microbial ecology, FISH is a culture-independent approach for the *in situ* analysis of the composition of microbial communities and their dynamics. Since its introduction in the late 1980s, FISH has become a widely used approach for the bacterial identification, quantification and — in combination with other techniques — characterization of phylogenetically defined microbial populations in complex environments [66, 67]. In fact, the new microscopic method for simultaneously determining *in situ* the identities, activities, and specific substrate uptake profiles of individual bacterial cells within complex microbial communities was developed by combining FISH performed with rRNA-targeted oligonucleotide probes and microautoradiography. In 1989, fluorescent labeled oligonucleotides are first used for the detection of single microbial cells. When compared to the radioactive probes, fluorescent probes are much safer; they offer better resolution and do not need additional detection steps. Moreover, fluorescent probes can be labeled with dyes of different emission wavelength thus enabling detection of several target sequences within a single hybridization step [68]. Its sensitivity and speed have made FISH a powerful tool for evaluating the phylogenetic identity, morphology, number, and spatial arrangements of microorganisms in environmental settings [67].

Generally, rRNAs are the main target molecules for FISH. There are several reasons: (i) rRNAs can be found in all living organisms; (ii) They are relatively stable and occur in high copy numbers (usually several thousand per cell); and (iii) They include both variable and highly conserved sequence domains [67, 69]. Signature sequences unique to a chosen group of microorganisms, ranging from whole phyla to individual species, can therefore be identified by comparative sequence analysis. Bacteria and archaea contain 5S, 16S, and 23S rRNAs with lengths of approximately 120, 1500 and 3000 nucleotides, respectively. In the vast majority of applications, the probes used in FISH are often targeting on 16S rRNA. In FISH probe design, the major steps are identifying short regions (usually 15 – 30 nucleotides in length) in a sequence alignment unique to the target group of interest, centralizing mismatches to nontarget organisms (where possible), and modifying the sequence to meet probe design criteria such as a minimum melting temperature [70]. The relative short probes have an easier access to their target in the whole cells. The public databases now include 16S rRNA sequences for most cultured microbial species, as well as numerous sequences directly retrieved from the environment. Probes are designed using sequence information from these databases and program packages such as ARB (www.arb-home.de) [71, 72]. PROBER (http://prober.cshl.edu/) is an oligonucleotide primer design software application that designs multiple primer pairs for generating PCR probes useful for FISH [73].

FISH detects nucleic acid sequences by a fluorescently labeled probe that hybridizes specifically to its complementary target sequence within the intact cell. After specific probes designed, a typical FISH protocol includes the following steps: (i) Fixation of the specimen; (ii) Preparation of the sample, possibly including specific pretreatment steps; (iii) Hybridization with the respective probes for detecting the respective target sequences; (iv) Washing steps to remove unbound probes; (v) Mounting, visualization and documentation of results. Detailed descriptions of this procedure, which can be completed within a few hours, are shown in Fig. 7.5. In fact, FISH is fully compatible with direct count methods [74]. The oligonucleotide probes used in FISH are covalently linked at the 5'-end to a single fluorescent dye molecule. Fluorochromes with different excitation and emission maxima allow simultaneous detection of two or more microorganisms. Dyes commonly used for FISH in microbiology are fluorescein-derivates (Fluorescein-Isothiocyanate (FITC), 5-(-6-) carboxyfluorescein-N-hydroxysuccimide-ester (FluoX)) and rhodamine-derivates (Tetramethyl-Rhodamine-Isothiocyanate (TRITC), Texas Red), and more recently cyanine dyes like Cy3 and Cy5 [68, 75]. The carbocyanine dyes have greatly increased the sensitivity of FISH, but further improvements are still needed. The microorganisms living in oligotrophic environments, such as the open ocean, are typically small with low cellular rRNA content. The profound influence of cellular growth rate and nutritional status on cell detection by FISH has been described [69]. Of course, the optimal FISH conditions still influence the results of the detection. Recently, Yilmaz *et al.* modified the standard rRNA-targeted FISH protocol by removing the fixation steps to allow recovery of unmodified nucleic acids. With this method, hybridized cells could be visualized in two different samples by epifluorescence microscopy. And then they target one bacterial and one archaeal population with group-specific oligonucleotide probes using in-solution fixation-free FISH and sort hybridized populations using fluorescence-activated cell sorting (FACS). They find that sorted populations are highly enriched for the target organisms based on 16S rRNA gene sequencing, thus confirming probe specificity using the modified FISH protocol. This approach should facilitate subsequent genomic sequencing and analysis of targeted populations as DNA is not compromised by crosslinking during fixation [76].

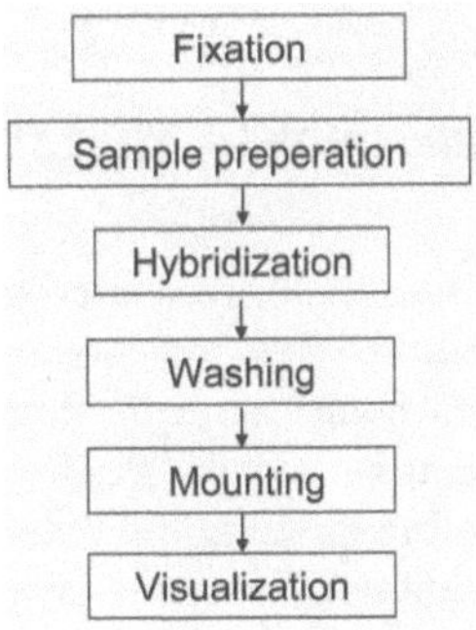

Fig.7.5. Flow chart of a typical FISH protocol

With its advantage of rapid and direct single-cell identification of microbes by detecting signature regions in their rRNA molecules, FISH has been widely used for exploring the microbial diversity in environment, and the complex microbial communities from different microhabitats of human body such as oral cavity, gastrointestine, vagina [77], and so on. So far, FISH has become an indispensable tool for the study of microbial ecology. When compared to microbiological culture or in vitro amplification of nucleic acids, FISH is fast, cheap and easy to carry out. It should be especially valuable for the detection of slow-growing, fastidious or unculturable bacteria. However, there are also several problems and pitfalls we should mentioned. (i) False positive results that are caused by autofluorescence from microorganisms themselves and low specificity of the probes; (ii) False negative results that might be associated with insufficient probe penetration, higher order structure of target or probe, low rRNA content of the whole cells and photobleaching over time [68]. The techniques development will help to solve these problems mentioned above. Recent advances include new web-based tools for assisting probe design and optimization of experimental conditions, easy-to-implement signal amplification strategies, innovative multiplexing approaches, and the combination of FISH with transmission electron microscopy (TEM) or extracellular staining techniques. Further emerging developments focus on sorting FISH-identified cells for subsequent single-cell genomics and on the direct detection of specific genes within single microbial cells by advanced FISH techniques employing various strategies for massive signal amplification [78].

In summary, FISH using rRNA-targeted probes is the method of choice for all studies in which exact cell numbers and cellular locations need to be determined. The methodology is being continuously improved. The developments of FISH technology used in microbial ecology will include more advanced and sophisticated ways of performing multicolor FISH analyses in a high-throughput format, to provide a deeper level of detail concerning the structure and composition of the specific microhabitat. However, accurate quantification still remains a challenging task and each new study needs careful controls. Further method development is therefore needed with respect to FISH sensitivity and automation [79].

7.5 Microarray Applications in Microbial Ecology Research

Researches have begun to investigate the incredible diversity of microorganisms in the human body and natural environments over the last two decades. The development of technologies such as PCR fingerprinting techniques (DGGE, T-RFLP and so on), reverse transcriptase PCR, real-time qPCR, reporter genes and FISH, have made it possible to study the dynamics of simple communities or small groups of predominant microorganisms in the specific microhabitat. However, many microorganisms in the specific microhabitat present in low numbers, which can not be understand with the above techniques. In order to fully understand the ecology of a complex microhabitat such as the gastrointestinal tract, it is necessary to analyze the dynamics and/or activity of hundreds to thousands of different microbial populations simultaneously. As specific, sensitive, quantitative and high-throughput tools for microbial detection, identification and characterization in the specific microenvironment of the human body, microarrays have the unprecedented potential to achieve this objective [80]. Microarrays can now be produced that contain thousands to hundreds of thousands of probes with the rapid advances in printing technology. Since 1995, DNA microarrays comprising hundreds or thousands of DNA fragments arrayed on small glass slides are originally developed for gene expression profiling [81]. These are subsequently applied to the study of different aspects of microbial ecology, including methane cycling, total microbial diversity and a range of biogeochemical functions [82].

DNA microarrays for use in microbial ecology have been developed utilizing different types of probes: Oligonucleotides, cDNA and microbial genomes. Oligonucleotide microarrays offer a fast, high-throughput alternative for the parallel detection of microbes from virtually any sample. So far, high-density 16S rRNA gene-targeting microarray that comprised nearly 30,000 oligonucleotide probes for the investigation of bacterial and archaeal diversity [83, 84]. In fact, phylogenetic oligonucleotide array (POA) is designed based on a conserved marker such as the 16S rRNA gene, which is used to compare the relatedness of communities in different environments. Phylogenetic microarrays are among the leading comprehensive molecular techniques that enable high-throughput analysis of microbial ecology in the specific microhabitat and can be used for strain typing, the determination of diversity and the analysis of the functionality of microbial ecosystems [80, 85, 86]. These phylogenetic microarrays are in most cases based on the small subunit ribosomal RNA (SSU rRNA) gene [87]. Such microarrays can be useful for the characterization of human intestinal microbiota composition and dynamics as more than 60% of the currently known diversity of gut microbiota phylotypes has only been detected by 16S rRNA gene sequences, whereas less than 40% was found by cultivation [88]. This situation resembles the situation reported for many other complex microbial ecosystems [89]. A recent application of a comprehensive SSU rRNA gene-based phylogenetic microarray showed that this technology provides superior diagnostic power for the analysis of the microbial community structure when compared with the clone library approach [80, 85].

cDNA microarrays, initially fabricated using randomly generated 1kb gene fragment probes to reveal taxonomic relationships among Pseudomonas species with high sensitivity, allow higher sensitivity and better resolution than oligonucleotide microarrays [84]. In fact, cDNA microarrays is generally applicable to nonmodel organisms at the current time, as it requires only that a large library of cDNAs be available as a source of clones to be arrayed . Community genome array (CGA) or genome probing microarrays (GPM) have been successfully applied to microbial ecology research. Community genome array (CGA) contains the whole genomic DNA of cultured organisms and can describe a community based on its relationship to these cultivated organisms. CGA that employ microbial genome probes can circumvent PCR artifacts and bias and also enhance their specificity and sensitivity [82]. CGA have been used to characterize the complex microbial composition of soil, river and marine sediments [77], as well as fermented vegetable food monitored over the course of the fermentation process [76].

Functional gene array (FGA) are designed for key functional genes that code for proteins catalyzing various biogeochemical processes, such as the carbon, nitrogen and sulfur cycles and may also provide information on the microbial populations and communities controlling these processes in natural environments [80, 90, 91]. Both oligonucleotides and cDNA derived from functional genes can be used for fabricating FGA.

As far as the microbial genomes arrays are concerned, metagenomic array (MGA) is a potentially powerful technique because, unlike the other arrays, they contain probes produced directly from environmental DNA itself and can be applied with no prior sequence knowledge of the community. MGA was fabricated for rapid characterization of metagenomic libraries with whole microbial and community genomes. This MGA is different from other microarrays in terms of the concept of probe and target. Microarray probes are generally spotted onto a glass slide, whereas the MGA format contains microarray targets arrayed on a glass slide and uses a labeled, specific gene as a probe [82]. The reverse approach used with MGA has made it possible to rapidly screen metagenomic libraries that comprise fosmid clone libraries from marine sediment and cosmid clone libraries derived from a groundwater microcosm [82, 92, 93]. Whole-genome open reading frame (ORF) array (WGA) contains probes for all of the ORFs in one or multiple genomes. These arrays have traditionally been used for functional genomic analyses of individual organisms, but they can also be used for comparative genomic analyses or to investigate the interactions of multiple organisms at the transcriptional level [80].

One study of human gastrointestinal microbiota focused on MGA was conducted with a phylogenetic microarray — referred to as the human intestinal tract chip (HITChip) by Rajilic-Stojanovic *et al.* recently [94]. In the HITChip, over 4,800 oligonucleotide probes are designed based on V1 and V6 hypervariable regions of the 16S rRNA gene targeting 1,140 unique microbial phylotypes (<

98% identity) following analysis of over 16,000 human intestinal 16S rRNA sequences. The number of intestinal phylotypes is steadily increasing and the currently known number of uncultured phylotypes has been estimated to amount to around 1,800 [95]. These unique microbial phylotypes included different taxonomic level from orders, genera to species. The rational designed HITChip can be used to analyze the predominant bacteria in human gastrointestinal tract. With this microarray, they have revealed temporal dynamics of fecal microbiota between children and elderly adults and also confirmed the adult intestinal microbiota is an individual-specific and relatively stable ecosystem. Despite these quantitative differences with respect to the shared microbiota, however, the overall phylogenetic composition of the microbiota core appeared to be similar in the young microbiota and elderly adults. This core consists of members of *Actinobacteria*, *Bacteroidetes* and *Firmicutes*. Although the fecal samples of young and elderly adults contain different marked microbiota, the different groups of bacteria have a highly similar phylogenetic position and may perform similar functions. In this regard, the hypothesis that a functional core microbiota existed in the specific microbiota is raised and is also confirmed by metagenome sequence analysis of obese and lean twins [10]. Used as a phylogenetic fingerprinting tool with the possibility for relative quantification, the HITChip has the potential to bridge the gaps in our knowledge in the quantitative and qualitative description of the human gastrointestinal microbiota composition [94].

So far, it is still not enough to only understand the structure and composition of the microbiota in the specific microhabitat. Understanding the functions of microbiota is critical for basic science discovery, biotechnology, agriculture, energy, environment and human health. However, the majority of microorganisms in the specific microhabitat are not cultured yet [69]. In order to overcome such obstacles for studying microbial communities in natural settings, a novel comprehensive microarray, termed GeoChip functional gene arrays, has been developed, used and updated by Zhou *et al.* recently [96, 97]. Zhou's lab focuses on the development of new microarrays using in the filed of microbial ecology. GeoChip contains 24,243 oligonucleotide (50 mer) probes and coveres 410,000 genes in 4150 functional groups which involved in nitrogen, carbon, sulfur and phosphorus cycling, metal reduction and resistance, and organic contaminant degradation. With so many functional genes covered, GeoChip is the first comprehensive microarray currently available for studying biogeochemical processes and functional activities of microbial communities important to human health, agriculture, energy, global climate change, ecosystem management, and environmental cleanup and restoration. It is particularly useful for providing direct linkages of microbial genes/populations to ecosystem processes and functions [96]. So far, the GeoChip has been updated to version 3.0, which contains more probes and covers more functional genes [97]. At the same times, Zhou's lab also develops another new functional gene array (the PathoChip), which is helpful to understand the population dynamics of pathogens and their pathogenic potential in the environment. PathoChip constructed with key virulence genes related to major virulence factors such as adherence, colonization, motility, invasion, toxin, immune evasion, and iron uptake. A total of 3,715 best probes were designed from virulence factors, covering 7,417 coding sequences from 1,397 microbial species (2,336 strains) [98]. This array has been applied to community samples from soil, oil plume, and saliva to assess the occurrence of virulence genes in natural environments. It has been confirmed that the PathoChip provides a useful tool to identify virulence genes in microbial populations, examine the dynamics of virulence gene in response to environmental perturbations, and determine the pathogenic potential of microbial communities [98].

With the development of microarrays used in microbial ecology, development of standardized methods for data analysis and interpretation are one of the greatest needs for microarray analysis of microbial communities. More and more probes targeting on huge genes are developed in new microarrays. It is necessary to develop new statistical methods for understanding the complex communities in those comprehensive microarrays. Those statistical methods developed for functional genomics may not be appropriate for analyzing the complex data sets often produced from microarray analysis of environmental samples. New statistical methods need to be devised and/or existing methods adapted to meet the specific challenges posed by these types of microarrays. The development of improved universal standards would also enhance data analysis and enable comparison of array data between experiments and laboratories [99].

However, like most other techniques, microarrays currently still detect only the dominant populations in many environmental samples. In order for microarray technology to reach its full high-throughput potential and provide real-time information of microbial populations in environmental samples, it will be necessary for the technology to eventually be automated and field deployable [100]. With advances in microfabrication and microfluidic technologies, it is now becoming possible to assemble all of the chambers, pumps, valves, mixers, heaters, and detectors that are required for microarray analysis on a single chip [101, 102]. In addition, PCR-induced artifacts and biases such as chimeras, mutations and heteroduplex molecules and skewed template-to-product ratios have been well documented, which can lead to over- or underestimation of microbial community diversity [82]. Furthermore, probe specificity is strongly affected by probe length in microarrays: as probe length increases, specificity decreases [103]. When considering the choice of the probe length in terms of both specificity and sensitivity, a suitable compromise is to utilize probes with well-established, non-equilibrium, thermal dissociation for real-time hybridization analysis, thereby allowing the discrimination of perfect-match and mismatch duplexes in microarray platforms [82, 104, 105]. These "lab-on-a-chip" still face the same analytical challenges as encountered with manual microarrays and are just in early stages of development, but they have the potential to revolutionize microarray analysis of environmental microbial populations. Ultimately, for whichever array format is used, more comprehensive, broad-scale applications are necessary to further validate and demonstrate the analytical power of microarrays for investigating various biological questions.

7.6 Cloning Library Construction and Sequencing

As one of the most important molecular techniques, cloning library construction and sequencing have been used widely for analyzing the overall structure and composition of microbiota of the specific microhabitat and combined with other molecular techniques such as PCR-DGGE and T-RFLP [20]. Detailed information of the microbial community composition in natural systems can be gained from the phylogenetic analysis of 16S rRNA gene sequences obtained directly from samples by PCR amplification, cloning, and sequencing, although this procedure may be biased as well [57, 106, 107]. 16S rRNA gene cloning and sequencing has been applied to analyze the bacterial community in humans [20] and in a pig [108] and compared to culture-based methods. The results showed that the microbial community is much more complex and that the bacterial diversity cannot be comprehended by culturing. In fact, 16S rRNA gene cloning library and sequencing has become a powerful tool for the analysis of the specific microhabitat. 16S rRNA gene can be amplified from both cultured and uncultured microorganisms. By constructing a cloning library, cultured and uncultured bacteria from the samples can be analyzed with the same detection efficiency. After sequencing, the number of clones will represent the relative abundance of the specific phylotypes in the specific microhabitat. There are incomparable advantages to understand the structure of the microbial communities, even the next generation high throughput sequencing techniques introduced.

Currently, 16S rRNA gene cloning and sequencing has become a standardized procedure for bacterial identification. So far, more than 5 million 16S rRNA gene sequences can be accessed in GenBank, while the number is still increasing. With the large 16S rRNA gene library, the cloned sequences can be submitted to the nucleotide sequence databases to perform comparative analysis and construct phylogenetic tree by Clustal W and TreeView. The common used nucleotide sequence databases are as follows: the GenBank from U.S. National Institutes of Health (NIH, http://www.ncbi.nlm.nih.gov/BLAST) and the Ribosomal Database Project (RDP, http://rdp.cme.msu.edu/) from Michigan State University [109]. In addition, ARB from Technical University of Munich (Germany) can provide automatic sequence alignment of comparison, secondary structure testing and phylogenetic tree building (http://www.arb-home.de/) [72]. By comparing with nucleotide sequence databases, it is easy to construct phylogenetic tree by the calculated from the sequence differences. With this method, 16S rDNA cloning library is established from the human feces, ileum and colon samples, which find a large numbers of bacteria that have been ignored in the previous studies. Another study by constructing large scale 16S rRNA gene cloning library from feces of healthy Chinese, 7255 16S rRNA gene sequences represent 476 operational taxonomic units (OTUs), which demonstrated the overall structure of gastrointestinal microbiota. It is the first study to understand the structure and composition of gastrointestinal microbiota from Chinese in a deeper level [20].

The 16S rRNA gene cloning library construction and sequencing is performed as follows (Fig. 7.6). It is generally including the several steps: (i) Sample collection and total bacterial genomic DNA extraction; (ii) 16S rRNA gene amplification and purification; (iii) 16S rRNA gene fragments cloning and transforming; (iv) Sequencing; (v) Data analysis and phylogenetic tree construction.

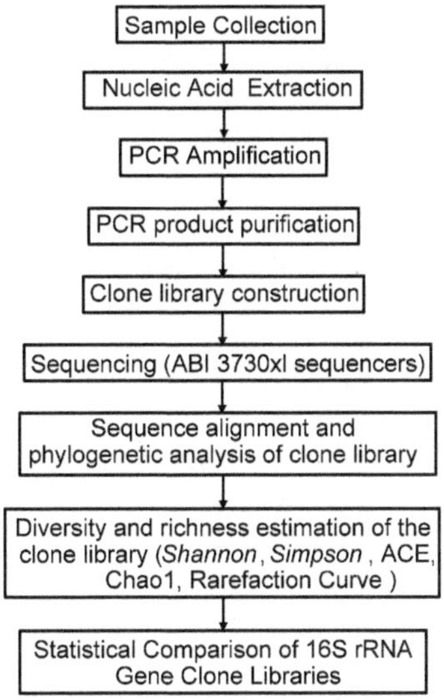

Fig. 7.6. Flow scheme showing cloning library construction and sequencing

There are unique advantages for bacterial 16S rRNA gene clone library analysis, which can reveal the presence of bacteria in the sample more accurately, especially for those uncultured bacteria. Therefore, cloning library construction and sequencing demonstrated the most extensive bacterial diversity of the specific microhabitat rather than culture dependent techniques before the advent of next-generation high throughput sequencing approaches. However, this technique could not also overcome the drawbacks of the other techniques that are mentioned above and detected only the predominant microbiota of the specific microhabitat of the human body. It still could not give the full extent of bacterial diversity of the specific microhabitat. And it was also significantly influenced by the choice of primer pairs [110]. There was clearly a need to complement the earlier culture-based data and the data generated with techniques based on PCR amplification with data that are independent of amplification bias, such as *in situ* hybridization. In addition, cloning library construction and sequencing is difficult to use in clinical practice as it is time- and cost-consuming, difficult to standardize operating conditions and require great amount of bacterial genomic DNA. Therefore, 16S rRNA gene-based molecular biology techniques can still not substitute other techniques, such as bacteria culture and isolation, optical and electron microscopy observation, biochemical identification, and so on. As technology continues to improve, 16S rRNA gene sequence analysis will gradually become a powerful tool for studies of the structure and composition of the microbial communities, and are likely to be widely used in clinical diagnosis.

7.7 Next-Generation Sequencing Techniques for Microbial Ecology Research

Advances in DNA sequencing technologies have created a new field of research, called metagenomics, allowing comprehensive examination of microbial communities, even those comprised of uncultivable organisms since 2005. Metagenomics (also referred to as community genomics, environmental genomics, and population genomics) is the genomic analysis of microorganisms by direct extraction and cloning of DNA from an assemblage of microorganisms [111]. The metagenomic approach allows analyzing the genetic material derived from complete microbial communities harvested from natural environments, instead of examining the genome of an individual bacterial strain that has been cultured. A large-scale sequencing project that is initiated by Craig Venter for the metagenome of the Sargasso Sea resulted in the identification of numerous novel genes and is a very famous example of sequence-based metagenome analyses [112]. So far, the human microbiome has been considered as the second genome of the human body, which is a source of genetic diversity, a modifier of disease, an essential component of immunity, and a functional entity that influences metabolism and modulates drug interactions [113]. Metagenomics has certainly become the research hotspots in the field of microbial ecology. Two large-scale projects such as the NIH Human Microbiome Project and the European MetaHIT consortium have promoted the progress of metagenomics greatly [114, 115]. However, characterization and analysis of the human microbiome have been greatly depended on the advances in genomic technologies, especially the high throughput sequencing tecniques.

With the advent of next-generation sequencing (NGS) platforms, it have made possible to recover DNA sequence data directly from environmental samples [116]. The traditional Sanger DNA-sequencing method can only sequence specimens individually and, therefore, is inadequate for processing complex environmental samples, especially for large-scale studies [117]. Conventional DNA-sequencing technique has provided the most efficient method for the development of large DNA barcode reference libraries, the number of individuals in an environmental sample is beyond the scope of its ability [118]. NGS can make up its shortcomings and can read DNA sequences data from multiple templates in parallel from the thousands of specimens present in an environmental bulk sample, which it works effectively, and with costs ever-lowly [117]. Using of specific gene markers such as species-specific DNA barcodes, NGS technologies have been applied to analyze the environmental DNA in ecological and environmental research. Different from conventional Sanger DNA-sequencing technology, this massively parallel throughput sequencing techniques may require only one or two instrument runs to complete an experiment., A series of high-throughput sequencing devices have been commercially introduced based on different chemistries and detection techniques in the last few years. In all cases, NGS reads are produced from fragment "libraries" without the need of a conventional vector-based cloning and *E. coli.*-based amplification stages used in capillary sequencing. These NGS technologies can potentially generate several hundred thousand to tens of millions of sequencing reads in parallel. As such, some of the cloning bias issues that impact genome representation in sequencing projects may be avoided, although each sequencing platform may have its own associated biases. This massively parallel throughput sequencing capacity can generate sequence reads from fragmented libraries of a specific genome (*i.e.* genome sequencing); from a pool of cDNA library fragments generated through reverse transcription of RNA molecules (*i.e.* RNAseq or transcriptome sequencing); or from a pool of PCR amplified molecules (*i.e.* amplicon sequencing) [117]. And these methods can largely be grouped into three main types: sequencing by synthesis, sequencing by ligation, and single-molecule sequencing [119]. Although the available NGS devices use quite diverse chemistry and base incorporation/detection tools, they share two main steps: Library fragmentation/amplicon library preparation and detection of the incorporated nucleotides [117, 120, 121].

Sequencing by synthesis includes three NGS platforms such as Roche 454 pyrosequencing, Illumina sequencer and Ion Torrent system, which differ by read length and how templates are amplified and immobilized. Consistent with Sanger DNA-sequencing, NGS techniques largely determine base composition through the detection of chemiluminescence created by nucleotide incorporation during synthesis of the complementary DNA strand by DNA polymerase. DNA is fragmented to the appropriate size, ligated to adaptor sequences, and then clonally amplified to enhance the fluorescent or chemical signal in sequencing by synthesis. Templates are then separated and immobilized in preparation for flow-cell cycles [117].

Roche 454 pyrosequencing is widely used for metagenomic analysis since 2005. So far, 3,082 papers were published with the 454 pyrosequencing and 1,260 papers among them were focused on metagenomics and microbial diversity (http://www.454.com/). And we think that more and more papers that use this technique in microbial ecology will be published in the future. In 454 pyrosequencing, primed DNA template is adhered to a microbead and amplified using emulsion PCR. Each nucleotide incorporated by DNA polymerase results in the release of a pyrophosphate molecule, which initiates a series of downstream reactions to produce light by the action of the enzyme luciferase. The amount of generated light is directly proportional to the number of nucleotides incorporated [122]. The workflow of the 454 pyrosequencing is shown in Fig. 7.7. After genomic DNA extraction, it is fragmented, ligated to adapters and separated into single strands. The library fragments are immobilized on either sepharose or styrofoam beads whose surfaces carry oligonucleotides complementary to the 454-specific adapter sequences ligated or PCR-generated onto both ends of the fragmented library. Under specific conditions that favor one fragment per bead, fragments are amplified in a PCR-reaction-mixture-in-oil emulsion and PCR amplification occurs within each droplet, resulting in beads each carrying ten million copies of a unique DNA template. The amplified beads are enriched and prepared as singlestranded and annealed to a specific sequencing primer. These beads are then arrayed into a picotiter plate (PTP) that is designed to have more than one million wells per plate. Each of the wells can hold only one amplified DNA bead. Four layers of engineered beads are deposited into the PTP. From bottom to top, diluted pyrosequencing enzyme beads, DNA amplified beads, pyrosequencing enzyme beads and, finally, PPiase beads. All bead layers are deposited by centrifugation [117, 123]. The PTP is sequenced in the 454 pyrosequencing devices. The steps for 454 pyrosequencing include the flow of repetitive cycles of nucleotide solutions (T, C, A and G). The 454 sequencing instrument consists of the following major subsystems: a fluidic assembly, a flow cell that includes the well-containing fiber-optic slide, a CCD camera-based imaging assembly with its own fiber-optic bundle used to image the fiberoptic slide, and a computer that provides the necessary user interface and instrument control [123]. With the advance of the 454 pyrosequencing, its read length and output in a run have been improved steadily, which can generate up to 800 bp sequencing reads and nearly 1 million reads in a single run. Due to its longer reads, 454 pyrosequencing is also used in genomic or transcriptomic sequencing when de novo assembly is involved [119].

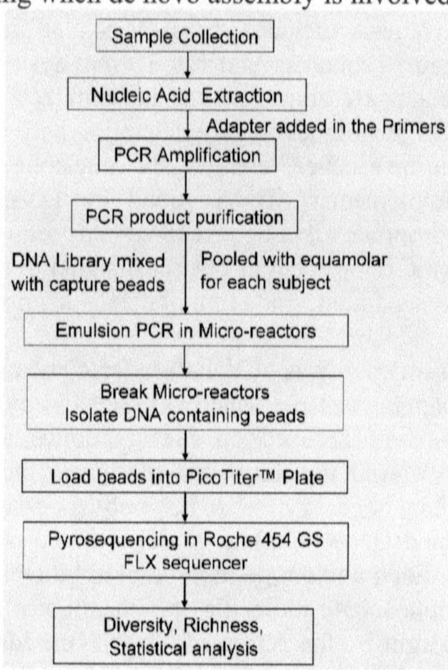

Fig. 7.7. Overview of the 454 sequencing technology

Illumina sequencer, formerly known as Solexa, utilizes a sequencing-by-synthesis coupled with solid-phase bridge amplification in which 5' and 3' adapters are ligated to each end of a DNA template on the surfaces of a flow cell (http://www.illumina.com). One end of the fragment is attached to the substrate. The adapters hybridize to immobilized forward or reverse primers by active heating and cooling steps, creating a bridge that facilitates amplification, generating amplicons that remain attached to the substrate, thus forming clusters of identical templates, which enhances chemiluminescent detection. Millions of such clusters of the library fragments are generated within each lane of the flow cell. Following amplification, the denatured and primed DNA amplicons are supplied with polymerase and a mixture of four nucleotides, which labeled with a different fluorophore and inactivated their 3'-OH chemically. In per flow cycle, only a single base will be incorporated, which is ensured by blocking modification. After each nucleotide is incorporated, an excitation followed by an imaging step takes place to identify the incorporated nucleotide in each cluster. A chemical deblocking treatment removes the fluorescent group and allows the incorporation of the following nucleotide during the next flow cycle. This cycle is repeated until the DNA fragment has been synthesized to its target length [117, 119]. So far, HiSeq 2000, HiSeq 1000 and Genome Analyzer IIx have sequencing outputs of up to 600, 300 and 95 Gb, respectively. In 2012, HiSeq 2500 is the upgrade of HiSeq 2000, which can generate up to 120 Gb of data in 27 h. Recently, MiSeq system is introduced by Illumina and can generate up to 150 bp sequencing reads with a total throughput of 1.5 – 2 Gb per run. Because of its high capacity, it has been widely used for whole-genome resequencing, including human and model organisms projects [117].

Ion Torrent system is a unique NGS technology, which is not based on fluorescent dyes but relied on the real-time detection of hydrogen ion concentration, released as a by-product when a nucleotide is incorporated into a strand of DNA by the polymerase action [124]. By sequentially adding nucleotides, Ion Torrent system is able to detect which nucleotide has been incorporated into the growing strand. Today, two versions of Ion Torrent system are commercially available. The Ion PGM can utilize different ion chips: 314, 316 or 318, which can generate up to 10 Mb, 100Mb or 1Gb, respectively. Ion Proton will provide higher throughput, which will deliver the human genome or human exome in just a few hours [117, 119]. Currently, the Ion Torrent system is mainly focused on the shotgun sequencing of microbial genomes [124, 125].

Sequencing by ligation is quite different from sequencing by synthesis, which uses the mismatch sensitivity of DNA ligase to determine the sequence of nucleotides in a given DNA strand. Different lengths of oligonucleotide probes are labeled with fluorescent tags, depending on the nucleotide(s) to be determined. DNA templates are fragmented and primed with a short, known anchor sequence, which allows the probes to hybridize. DNA ligase is added to the flow cell and joins the fluorescently labeled probe to the primer and template. Fluorescence imaging is performed to determine which probe was incorporated. This process is repeated using different sets of probes to query the DNA template and assess the sequence of nucleotides [119]. Two platforms, such as Applied Biosystems SOLiD sequencer and Polonator G.007 system, are worked based on sequencing by ligation.

Applied Biosystems (Life Technologies) introduced SOLiD platform that utilizes sequencing by ligation to determine DNA sequence composition in 2007. This specific process couples oligo adaptor-linked DNA fragments with 1-μm magnetic beads that are decorated with complementary oligos and amplifies each bead-DNA complex by emPCR. After amplification, the beads are covalently attached to the surface of a specially treated glass slide, which is placed into a fluidics cassette within sequencer. The ligation-based sequencing process starts with the annealing of a universal sequencing primer that is complementary to the SOLiD-specific adapters ligated to the library fragments. The addition of a limited set of semi-degenerate 8mer oligonucleotides and DNA ligase is automated by the instrument. When a matching 8mer hybridizes to the DNA fragment sequence adjacent to the universal primer 3' end, DNA ligase seals the phosphate backbone. After the ligation step, a fluorescent readout identifies the fixed base of the 8mer, which corresponds to either the fifth position or the second position, depending on the cycle number. A subsequent chemical cleavage step removes the sixth through eighth base of the ligated 8mer by attacking the linkage between fifth and sixth bases, thereby removing the fluorescent group and enabling a subsequent ligation round. A second round of sequencing initiates with the hybridization of an n-1 positioned universal primer, and subsequent rounds of ligation-mediated sequencing, and so on. The same process is repeated with n-2, n-3 and n-4 positioned universal primers. The generated fluorescence from the five universal primers is decoded with a two-base calling processing software [117, 126]. Each SOLiD run requires approximately 5 days and produces 3 – 4 Gb of sequence data with an average read length of 25 – 35 bp. SOLiD platform has been successfully used in resequencing studies, transcriptomics, or in genomic sequencing in combination with other technologies [127, 128].

Introduced by the same team as SOLiD system, the Polonator G.007 system is available through Azco Biotech. On this platform, library preparation for sequencing is accomplished using emPCR for amplification of template DNA, loading of the beads onto the flow cells and fully automated polymerase colony sequencing by ligation [129]. This platform can combine a high-performance instrument at very low price and the freely downloadable, open-source software and protocols in this sequencing system. Today, the output of a full eight-flow-cell Polonator-sequencing run is up to 240 million mappable reads of sequence with a read length of 40 bases accomplished in nearly 80 h [119]. So far, the Polonator G.007 system is mainly used in the Personal Genome Project.

With the increasing usage and new modification in NGS, the third generation sequencing (mainly single-molecule sequencing, SMS) is coming out with new insight in the sequencing. When compared with NGS, third-generation sequencing has two main characteristics: (i) PCR is not needed before sequencing, which shortens DNA preparation time for sequencing; (ii) The signal is captured in real time, which means that the signal, no matter whether it is fluorescent (Pacbio) or electric current (Nanopore), is monitored during the enzymatic reaction of adding nucleotide in the complementary strand [130]. Currently, three technology platforms comprise this third generation. Two achieve SMS by incorporating and detecting fluorescently labeled nucleotides: Helicos Genetic Analysis System and Pacific Biosciences Single Molecule Real Time (SMRT) System are already commercially available. The third, Oxford Nanopore's nanopore sequencing, is not ready for commercial prime time, but is likely to be the cheapest of the three (Oxford Nanopore is also considered as the fourth generation sequencing technology by Ku *et al.* [131]). So far, SMS technologies are relatively new to the market. As these methods become more readily available and further developed, its applications in microbial ecology are underway with publications soon to follow.

The Helicos Genetic Analysis System is the first commercially available SMS on the market in 2008, which utilizes sequencing-by-synthesis on a single DNA molecule [132]. The preparation of Single-stranded DNA fragments is the key step for library construction. No amplification step is required. During the sequencing cycles, repetitive cycles of the DNA polymerase and one of the four fluorescently labeled nucleotides are flowed in, resulting in template-dependent extension of DNA strands according to the flowed nucleotide. The fluorescent nucleotides are modified to stop the polymerase extension until the incorporated nucleotide's fluorescence is captured and the images are recorded with a highly sensitive CCD camera connected to a fluorescent microscope. A washing step then takes place to wash off the unincorporated nucleotides as well as the by-products of the previous cycle. After washing, fluorescent labels on the extended strands are chemically cleaved and removed. Another cycle of single-base extension, label-cleaving and imaging follows [117]. Helicos produces average read lengths of 35 bp across 600 million to 1 billion reads, totaling 21 – 35 Gb per run at a rate of > 1 Gb/h [119]. Using Helicos's proprietary true single-molecule sequencing (tSMS) technology, Helicos BioSciences has demonstrated the first single-molecule sequencing of the M13 viral genome and confirmed its perfect accuracy [132].

Pacific Biosciences SMRT System is commercially available in 2010. SMRT is based on the natural DNA synthesis by a DNA polymerase with phospholink nucleotides as it occurs in a continuous, processive manner. This approach is enabled by two key innovations: nanophotonic visualization chambers (SMRT chips) and phospholinked nucleotides. Each SMRT chip, containing thousands of zero-mode waveguides (ZMWs) [133], consists of tens of thousands of subwavelength, ten nanometre diameter holes, which are fabricated by perforating a thin metal film supported by a transparent substrate. During the sequencing workflow, the complimentary DNA strand is synthesized from the single-stranded template by the action of DNA polymerase planted at the bottom of each waveguide with four different-coloured phosphor-linked nucleotides. Different from other technologies, the fluorescence label is attached on the terminal phosphate group rather than the nucleotide base, leading to the release of the fluorescence moiety with the nucleotide incorporation [117, 134]. Although SMRT sequencing has several advantages over NGS and Helicos tSMS technology, the error rate of the raw read is high (> 5%), and its throughput is lower than that of NGS and tSMS. With their new SMRTbell protocol in which templates are circularized and sequenced repeatedly, SMRT has increased its accuracy to 99.999% for 30 × coverage [119].

With the introduction of and advancement in high throughput sequencing techniques mentioned above, more and more novel information has been unearthed in the field of microbial ecology, especially the overall structure of the specific microbiota and the functions of those uncultured microorganisms. New developed methods will provide huge sequencing reads that contain information approaches the real world of the specific microhabitat. Modern molecular techniques are facilitating improved understanding of host-microbe dialogue in human health and in disease processes. Rapid progress will provide optimism for a bright future for the field of next-generation environmental DNA analysis.

7.8 Conclusion

The molecular techniques available today provide a powerful toolkit for a comprehensive analysis of the microbiota. This is optimally achieved by combined use of different methods that complement each other. For example, cells labeled by group-specific probes in fluorescent *in situ* hybridization (FISH) can be enumerated using flow cytometry and simultaneously sorted and collected. Subsequent diversity screening is then possible by other methods such as molecular fingerprinting methods or 16S rDNA sequence analysis. Among these molecular fingerprinting methods, PCR-DGGE/PCR-TGGE, ARDRA and T-RFLP represent rapid and reliable techniques that have been used successfully to identify the bacterial compositions of different ecological niches. Sequencing of 16S rRNA genes from different samples by constructing clone libraries has revolutionized our understanding of microbial systematics and diversity (metagenomics). Great improvements in molecular monitoring of microbiota can be expected when new high-throughput techniques such as DNA-microarray technology, NGS techniques and SMS techniques become applicable for molecular ecology studies. This will give us a deeper understanding of the microbial ecology of the specific microhabitat and, consequently, about the possibilities to modulate the microbiota in specific niches to the benefit of host health.

References

[1] Dethlefsen L, McFall-Ngai M, Relman D A. An ecological and evolutionary perspective on human-microbe mutualism and disease. Nature, 2007, 449: 811-818.
[2] Turnbaugh P J, Ley R E, Hamady M, *et al.* The human microbiome project. Nature, 2007, 449: 804-810.
[3] Ling Z, Liu X, Luo Y, *et al.* Pyrosequencing analysis of the human microbiota of healthy Chinese undergraduates. BMC Genomics, 2013, 14: 390.
[4] Gill S R, Pop M, Deboy R T, *et al.* Metagenomic analysis of the human distal gut microbiome. Science, 2006, 312: 1355-1359.
[5] Cash H L, Whitham C V, Behrendt C L, *et al.* Symbiotic bacteria direct expression of an intestinal bactericidal lectin. Science, 2006, 313: 1126-1130.
[6] Ley R E, Peterson D A, Gordon J I. Ecological and evolutionary forces shaping microbial diversity in the human intestine. Cell, 2006, 124: 837-848.
[7] Ley R E, Turnbaugh P J, Klein S, *et al.* Microbial ecology: Human gut microbes associated with obesity. Nature, 2006, 444: 1022-1023.
[8] Mazmanian S K, Liu C H, Tzianabos A O, *et al.* An immunomodulatory molecule of symbiotic bacteria directs maturation of the host immune system. Cell, 2005, 122:107-118.
[9] Turnbaugh P J, Ley R E, Mahowald M A, *et al.* An obesity-associated gut microbiome with increased capacity for energy harvest. Nature, 2006, 444:

1027-1031.
[10] Turnbaugh P J, Hamady M, Yatsunenko T, *et al.* A core gut microbiome in obese and lean twins. Nature, 2009, 457: 480-484.
[11] Frank D N, St Amand A L, Feldman R A, *et al.* Molecular-phylogenetic characterization of microbial community imbalances in human inflammatory bowel diseases. Proc Natl Acad Sci USA, 2007, 104: 13780-13785.
[12] Wen L, Ley R E, Volchkov P Y, *et al.* Innate immunity and intestinal microbiota in the development of Type 1 diabetes. Nature, 2008, 455: 1109-1113.
[13] Ordovas J M, Mooser V. Metagenomics: The role of the microbiome in cardiovascular diseases. Curr Opin Lipidol, 2006, 17: 157-161.
[14] Zhao L, Shen J. Whole-body systems approaches for gut microbiota-targeted, preventive healthcare. J Biotechnol, 2010, 149: 183-190.
[15] Kurokawa K, Itoh T, Kuwahara T, *et al.* Comparative metagenomics revealed commonly enriched gene sets in human gut microbiomes. DNA Res, 2007, 14: 169-181.
[16] Nicholson J K, Holmes E, Wilson I D. Gut microorganisms, mammalian metabolism and personalized health care. Nat Rev Microbiol, 2005, 3: 431-438.
[17] Riesenfeld C S, Schloss P D, Handelsman J. Metagenomics: Genomic analysis of microbial communities. Annu Rev Genet, 2004, 38: 525-552.
[18] Hill J E, Goh S H, Money D M, *et al.* Characterization of vaginal microflora of healthy, nonpregnant women by chaperonin-60 sequence-based methods. Am J Obstet Gynecol, 2005, 193(3 Pt 1): 682-692.
[19] Schellenberg J, Links M G, Hill J E, *et al.* Pyrosequencing of the chaperonin-60 universal target as a tool for determining microbial community composition. Appl Environ Microbiol, 2009, 75: 2889-2898.
[20] Li M, Wang B, Zhang M, *et al.* Symbiotic gut microbes modulate human metabolic phenotypes. Proc Natl Acad Sci USA, 2008, 105: 2117-2122.
[21] Meyer M, Stenzel U, Hofreiter M. Parallel tagged sequencing on the 454 platform. Nat Protoc, 2008, 3: 267-278.
[22] von Bubnoff A. Next-generation sequencing: The race is on. Cell, 2008, 132: 721-723.
[23] Edwards R A, Rodriguez-Brito B, Wegley L, *et al.* Using pyrosequencing to shed light on deep mine microbial ecology. BMC Genomics, 2006, 7: 57.
[24] Roesch L F, Fulthorpe R R, Riva A, *et al.* Pyrosequencing enumerates and contrasts soil microbial diversity. ISME J, 2007, 1: 283-290.
[25] Roh S W, Kim K H, Nam Y D, *et al.* Investigation of archaeal and bacterial diversity in fermented seafood using barcoded pyrosequencing. ISME J, 2010, 4: 1-16.
[26] Fierer N, Hamady M, Lauber C L, *et al.* The influence of sex, handedness, and washing on the diversity of hand surface bacteria. Proc Natl Acad Sci USA, 2008, 105: 17994-17999.
[27] Dowd S E, Sun Y, Secor P R, *et al.* Survey of bacterial diversity in chronic wounds using pyrosequencing, DGGE, and full ribosome shotgun sequencing. BMC Microbiol, 2008, 8: 43.

[28] Keijser B J, Zaura E, Huse S M, *et al.* Pyrosequencing analysis of the oral microflora of healthy adults. J Dent Res, 2008, 87: 1016-1020.

[29] Ling Z, Kong J, Jia P, *et al.* Analysis of oral microbiota in children with dental caries by PCR-DGGE and barcoded pyrosequencing. Microb Ecol, 2010, 60: 677-690.

[30] Liu W T, Marsh T L, Cheng H, *et al.* Characterization of microbial diversity by determining terminal restriction fragment length polymorphisms of genes encoding 16S rRNA. Appl Environ Microbiol, 1997, 63: 4516-4522.

[31] Field K G, Bernhard A E, Brodeur T J. Molecular approaches to microbiological monitoring: Fecal source detection. Environ Monit Assess, 2003, 81: 313-326.

[32] Kirk J L, Beaudette L A, Hart M, *et al.* Methods of studying soil microbial diversity. J Microbiol Methods, 2004, 58: 169-188.

[33] Lukow T, Dunfield P F, Liesack W. Use of the T-RFLP technique to assess spatial and temporal changes in the bacterial community structure within an agricultural soil planted with transgenic and non-transgenic potato plants. FEMS Microbiol Ecol, 2000, 32: 241-247.

[34] McCartney A L. Application of molecular biological methods for studying probiotics and the gut flora. Br J Nutr, 2002, 88 Suppl 1: S29-S37.

[35] Sakamoto M, Hayashi H, Benno Y. Terminal restriction fragment length polymorphism analysis for human fecal microbiota and its application for analysis of complex bifidobacterial communities. Microbiol Immunol, 2003, 47: 133-142.

[36] Sakamoto M, Rocas I N, Siqueira J F, Jr., *et al.* Molecular analysis of bacteria in asymptomatic and symptomatic endodontic infections. Oral Microbiol Immunol, 2006, 21: 112-122.

[37] Marsh T L. Terminal restriction fragment length polymorphism (T-RFLP): An emerging method for characterizing diversity among homologous populations of amplification products. Curr Opin Microbiol, 1999, 2: 323-327.

[38] Dorigo U, Volatier L, Humbert J F. Molecular approaches to the assessment of biodiversity in aquatic microbial communities. Water Res, 2005, 39: 2207-2218.

[39] Kent A D, Smith D J, Benson B J, *et al.* Web-based phylogenetic assignment tool for analysis of terminal restriction fragment length polymorphism profiles of microbial communities. Appl Environ Microbiol, 2003, 69: 6768-6776.

[40] Zhou X, Brown C J, Abdo Z, *et al.* Differences in the composition of vaginal microbial communities found in healthy Caucasian and black women. ISME J, 2007, 1: 121-133.

[41] Culman S W, Bukowski R, Gauch H G, *et al.* T-REX: Software for the processing and analysis of T-RFLP data. BMC Bioinformatics, 2009, 10: 171.

[42] Qian P Y, Thiyagarajan V, Lau S C K, *et al.* Relationship between bacterial community profile in biofilm and attachment of the acorn barnacle Balanus amphitrite. Aquat Microb Ecol, 2003, 33: 225-237.

[43] Zhang R, Thiyagarajan V, Qian P Y. Evaluation of terminal-restriction fragment length polymorphism analysis in contrasting marine environments. FEMS Microbiol Ecol, 2008, 65: 169-178.
[44] Engebretson J J, Moyer C L. Fidelity of select restriction endonucleases in determining microbial diversity by terminal-restriction fragment length polymorphism. Appl Environ Microbiol, 2003, 69: 4823-4829.
[45] Egert M, Friedrich M W. Formation of pseudo-terminal restriction fragments, a PCR-related bias affecting terminal restriction fragment length polymorphism analysis of microbial community structure. Appl Environ Microbiol, 2003, 69: 2555-2562.
[46] Lueders T, Friedrich M W. Evaluation of PCR amplification bias by terminal restriction fragment length polymorphism analysis of small-subunit rRNA and mcrA genes by using defined template mixtures of methanogenic pure cultures and soil DNA extracts. Appl Environ Microbiol, 2003, 69: S320-S326.
[47] Borresen A L, Hovig E, Brogger A. Detection of base mutations in genomic DNA using denaturing gradient gel electrophoresis (DGGE) followed by transfer and hybridization with gene-specific probes. Mutat Res, 1988, 202: 77-83.
[48] Cariello N F, Scott J K, Kat A G, *et al.* Resolution of a missense mutant in human genomic DNA by denaturing gradient gel electrophoresis and direct sequencing using in vitro DNA amplification: HPRT Munich. Am J Hum Genet, 1988, 42: 726-734.
[49] Sheffield V C, Cox D R, Lerman L S, *et al.* Attachment of a 40-base-pair G + C-rich sequence (GC-clamp) to genomic DNA fragments by the polymerase chain reaction results in improved detection of single-base changes. Proc Natl Acad Sci USA, 1989, 86: 232-236.
[50] Fischer S G, Lerman L S. DNA fragments differing by single base-pair substitutions are separated in denaturing gradient gels: correspondence with melting theory. Proc Natl Acad Sci USA, 1983, 80: 1579-1583.
[51] Yamamoto M, Kameda A, Matsuura N, *et al.* A separation method for DNA computing based on concentration control. New Generat Comput, 2002, 20: 251-261.
[52] Janda J M, Abbott S L. 16S rRNA gene sequencing for bacterial identification in the diagnostic laboratory: Pluses, perils, and pitfalls. J Clin Microbiol, 2007, 45: 2761-2764.
[53] Huys G, Vanhoutte T, Vandamme P. Application of sequence-dependent electrophoresis fingerprinting in exploring biodiversity and population dynamics of human intestinal microbiota: What can be revealed? Interdiscip Perspect Infect Dis, 2008, 2008: 597-603.
[54] Yu Z, Morrison M. Comparisons of different hypervariable regions of rrs genes for use in fingerprinting of microbial communities by PCR-denaturing gradient gel electrophoresis. Appl Environ Microbiol, 2004, 70: 4800-4806.
[55] Amp F, Miambi E. Cluster analysis, richness and biodiversity indexes derived from denaturing gradient gel electrophoresis fingerprints of bacterial communities demonstrate that traditional maize fermentations are driven by

the transformation process. Int J Food Microbiol, 2000, 60: 91-97.
[56] Boon N, Windt W, Verstraete W, *et al.* Evaluation of nested PCR-DGGE (denaturing gradient gel electrophoresis) with group-specific 16S rRNA primers for the analysis of bacterial communities from different wastewater treatment plants. FEMS Microbiol Ecol, 2002, 39: 101-112.
[57] Suzuki M T, Giovannoni S J. Bias caused by template annealing in the amplification of mixtures of 16S rRNA genes by PCR. Appl Environ Microbiol, 1996, 62: 625-630.
[58] Ling Z, Kong J, Liu F, *et al.* Molecular analysis of the diversity of vaginal microbiota associated with bacterial vaginosis. BMC Genomics, 2010, 11: 488.
[59] Ling Z, Liu X, Chen X, *et al.* Diversity of cervicovaginal microbiota associated with female lower genital tract infections. Microb Ecol, 2011, 61: 704-714.
[60] Muyzer G, de Waal E C, Uitterlinden A G. Profiling of complex microbial populations by denaturing gradient gel electrophoresis analysis of polymerase chain reaction-amplified genes coding for 16S rRNA. Appl Environ Microbiol, 1993, 59: 695-700.
[61] Muyzer G. DGGE/TGGE a method for identifying genes from natural ecosystems. Curr Opin Microbiol, 1999, 2: 317-322.
[62] Hayes V M, Wu Y, Osinga J, *et al.* Improvements in gel composition and electrophoretic conditions for broad-range mutation analysis by denaturing gradient gel electrophoresis. Nucleic Acids Res, 1999, 27: e29.
[63] Moore W E, Holdeman L V, Cato E P, *et al.* Comparative bacteriology of juvenile periodontitis. Infect Immun, 1985, 48: 507-519.
[64] Paster B J, Boches S K, Galvin J L, *et al.* Bacterial diversity in human subgingival plaque. J Bacteriol, 2001, 183: 3770-3783.
[65] Li Y, Ku C Y, Xu J, *et al.* Survey of oral microbial diversity using PCR-based denaturing gradient gel electrophoresis. J Dent Res, 2005, 84: 559-564.
[66] DeLong E F, Wickham G S, Pace N R. Phylogenetic stains: ribosomal RNA-based probes for the identification of single cells. Science, 1989, 243: 1360-1363.
[67] Amann R I, Krumholz L, Stahl D A. Fluorescent-oligonucleotide probing of whole cells for determinative, phylogenetic, and environmental studies in microbiology. J Bacteriol, 1990, 172: 762-770.
[68] Moter A, Gobel U B. Fluorescence in situ hybridization (FISH) for direct visualization of microorganisms. J Microbiol Methods, 2000, 41: 85-112.
[69] Amann R I, Ludwig W, Schleifer K H. Phylogenetic identification and in situ detection of individual microbial cells without cultivation. Microbiol Rev, 1995, 59: 143-169.
[70] Hugenholtz P, Tyson G W, Blackall L L. Design and evaluation of 16S rRNA-targeted oligonucleotide probes for fluorescence in situ hybridization. Methods Mol Biol, 2002, 179: 29-42.
[71] Amann R, Ludwig W. Ribosomal RNA-targeted nucleic acid probes for studies in microbial ecology. FEMS Microbiol Rev, 2000, 24: 555-565.
[72] Ludwig W, Strunk O, Westram R, *et al.* ARB: a software environment for

sequence data. Nucleic Acids Res, 2004, 32(4): 1363-1371.
[73] Navin N, Grubor V, Hicks J, *et al.* PROBER: Oligonucleotide FISH probe design software. Bioinformatics, 2006, 22: 2437-2438.
[74] Maruyama A, Sunamura M. Simultaneous direct counting of total and specific microbial cells in seawater, using a deep-sea microbe as target. Appl Environ Microbiol, 2000, 66: 2211-2215.
[75] Southwick P L, Ernst L A, Tauriello E W, *et al.* Cyanine dye labeling reagents--carboxymethylindocyanine succinimidyl esters. Cytometry, 1990, 11: 418-430.
[76] Yilmaz S, Haroon M F, Rabkin B A, *et al.* Fixation-free fluorescence in situ hybridization for targeted enrichment of microbial populations. ISME J, 2010, 4: 1352-1356.
[77] Fredricks D N, Fiedler T L, Marrazzo J M. Molecular identification of bacteria associated with bacterial vaginosis. N Engl J Med, 2005, 353: 1899-1911.
[78] Wagner M, Haider S. New trends in fluorescence in situ hybridization for identification and functional analyses of microbes. Curr Opin Biotechnol, 2012, 23: 96-102.
[79] Amann R, Fuchs B M, Behrens S. The identification of microorganisms by fluorescence in situ hybridisation. Curr Opin Biotechnol, 2001, 12: 231-236.
[80] Gentry T J, Wickham G S, Schadt C W, *et al.* Microarray applications in microbial ecology research. Microb Ecol, 2006, 52: 159-175.
[81] Schena M, Shalon D, Davis R W, *et al.* Quantitative monitoring of gene expression patterns with a complementary DNA microarray. Science, 1995, 270: 467-470.
[82] Roh S W, Abell G C, Kim K H, *et al.* Comparing microarrays and next-generation sequencing technologies for microbial ecology research. Trends Biotechnol, 2010, 28: 291-299.
[83] Wilson K H, Wilson W J, Radosevich J L, *et al.* High-density microarray of small-subunit ribosomal DNA probes. Appl Environ Microbiol, 2002, 68: 2535-2541.
[84] Brodie E L, Desantis T Z, Joyner D C, *et al.* Application of a high-density oligonucleotide microarray approach to study bacterial population dynamics during uranium reduction and reoxidation. Appl Environ Microbiol, 2006, 72: 6288-6298.
[85] DeSantis T Z, Brodie E L, Moberg J P, *et al.* High-density universal 16S rRNA microarray analysis reveals broader diversity than typical clone library when sampling the environment. Microb Ecol, 2007, 53: 371-383.
[86] Wagner M, Smidt H, Loy A, *et al.* Unravelling microbial communities with DNA-microarrays: challenges and future directions. Microb Ecol, 2007, 53: 498-506.
[87] Bodrossy L, Sessitsch A. Oligonucleotide microarrays in microbial diagnostics. Curr Opin Microbiol, 2004, 7: 245-254.
[88] Rajilic-Stojanovic M, Smidt H, de Vos W M. Diversity of the human gastrointestinal tract microbiota revisited. Environ Microbiol, 2007, 9: 2125-2136.

[89] Schloss P D, Handelsman J. Status of the microbial census. Microbiol Mol Biol Rev, 2004, 68: 686-691.

[90] Frias-Lopez J, Shi Y, Tyson G W, *et al.* Microbial community gene expression in ocean surface waters. Proc Natl Acad Sci USA, 2008, 105: 3805-3810.

[91] Wu L, Thompson D K, Li G, *et al.* Development and evaluation of functional gene arrays for detection of selected genes in the environment. Appl Environ Microbiol, 2001, 67: 5780-5790.

[92] Sebat J L, Colwell F S, Crawford R L. Metagenomic profiling: microarray analysis of an environmental genomic library. Appl Environ Microbiol, 2003, 69: 4927-4934.

[93] Park S J, Kang C H, Chae J C, *et al.* Metagenome microarray for screening of fosmid clones containing specific genes. FEMS Microbiol Lett, 2008, 284: 28-34.

[94] Rajilic-Stojanovic M, Heilig H G, Molenaar D, *et al.* Development and application of the human intestinal tract chip, a phylogenetic microarray: analysis of universally conserved phylotypes in the abundant microbiota of young and elderly adults. Environ Microbiol, 2009, 11: 1736-1751.

[95] Zoetendal E G, Rajilic-Stojanovic M, de Vos W M. High-throughput diversity and functionality analysis of the gastrointestinal tract microbiota. Gut, 2008, 57: 1605-1615.

[96] He Z, Gentry T J, Schadt C W, *et al.* GeoChip: A comprehensive microarray for investigating biogeochemical, ecological and environmental processes. ISME J, 2007, 1: 67-77.

[97] He Z, Deng Y, Van Nostrand J D, *et al.* GeoChip 3.0 as a high-throughput tool for analyzing microbial community composition, structure and functional activity. ISME J, 2010, 4: 1167-1179.

[98] Lee Y J, van Nostrand J D, Tu Q, *et al.* The PathoChip, a functional gene array for assessing pathogenic properties of diverse microbial communities. ISME J, 2013, 7: 1974-1984.

[99] He Z, Van Nostrand J D, Zhou J. Applications of functional gene microarrays for profiling microbial communities. Curr Opin Biotechnol, 2012, 23: 460-466.

[100]Chandler D P, Jarrell A E. Automated purification and suspension array detection of 16S rRNA from soil and sediment extracts by using tunable surface microparticles. Appl Environ Microbiol, 2004, 70: 2621-2631.

[101]Liu R H, Yang J, Lenigk R, *et al.* Self-contained, fully integrated biochip for sample preparation, polymerase chain reaction amplification, and DNA microarray detection. Anal Chem, 2004, 76: 1824-1831.

[102]Liu W T, Zhu L. Environmental microbiology-on-a-chip and its future impacts. Trends Biotechnol, 2005, 23: 174-179.

[103]Jayaraman A, Hall C K, Genzer J. Computer simulation study of molecular recognition in model DNA microarrays. Biophys J, 2006, 91: 2227-2236.

[104]Liu W T, Mirzabekov A D, Stahl D A. Optimization of an oligonucleotide microchip for microbial identification studies: A non-equilibrium dissociation approach. Environ Microbiol, 2001, 3: 619-629.

[105]Pozhitkov A E, Stedtfeld R D, Hashsham S A, *et al.* Revision of the nonequilibrium thermal dissociation and stringent washing approaches for identification of mixed nucleic acid targets by microarrays. Nucleic Acids Res, 2007, 35: e70.

[106]von Wintzingerode F, Gobel U B, Stackebrandt E. Determination of microbial diversity in environmental samples: pitfalls of PCR-based rRNA analysis. FEMS Microbiol Rev, 1997, 21: 213-229.

[107]Wang G C, Wang Y. The frequency of chimeric molecules as a consequence of PCR co-amplification of 16S rRNA genes from different bacterial species. Microbiology, 1996, 142 (Pt 5): 1107-1114.

[108]Pryde S E, Richardson A J, Stewart C S, *et al.* Molecular analysis of the microbial diversity present in the colonic wall, colonic lumen, and cecal lumen of a pig. Appl Environ Microbiol, 1999, 65: 5372-5377.

[109]Wang Q, Garrity G M, Tiedje J M, *et al.* Naive Bayesian classifier for rapid assignment of rRNA sequences into the new bacterial taxonomy. Appl Environ Microbiol, 2007, 73: 5261-5267.

[110]Zhu X Y, Zhong T, Pandya Y, *et al.* 16S rRNA-based analysis of microbiota from the cecum of broiler chickens. Appl Environ Microbiol, 2002, 68: 124-137.

[111]Handelsman J. Metagenomics: Application of genomics to uncultured microorganisms. Microbiol Mol Biol Rev, 2004, 68: 669-685.

[112]Venter J C, Remington K, Heidelberg J F, *et al.* Environmental genome shotgun sequencing of the Sargasso Sea. Science, 2004, 304: 66-74.

[113]Grice E A, Segre J A. The human microbiome: Our second genome. Annual review of genomics and human genetics, 2012, 13: 151-170.

[114]Nelson K E, Weinstock G M, Highlander S K, *et al.* A catalog of reference genomes from the human microbiome. Science, 2010, 328: 994-999.

[115]Qin J, Li R, Raes J, *et al.* A human gut microbial gene catalogue established by metagenomic sequencing. Nature, 2010, 464: 59-65.

[116]Sogin M L, Morrison H G, Huber J A, *et al.* Microbial diversity in the deep sea and the underexplored "rare biosphere . Proc Natl Acad Sci USA, 2006, 103: 12115-12120.

[117]Shokralla S, Spall J L, Gibson J F, *et al.* Next-generation sequencing technologies for environmental DNA research. Molecular ecology, 2012, 21: 1794-1805.

[118]Hajibabaei M, Shokralla S, Zhou X, *et al.* Environmental barcoding: A next-generation sequencing approach for biomonitoring applications using river benthos. PloS one 2011, 6(4): e17497.

[119]Egan A N, Schlueter J, Spooner D M. Applications of next-generation sequencing in plant biology. American Journal of Botany, 2012, 99: 175-185.

[120]Zhang J, Chiodini R, Badr A, *et al.* The impact of next-generation sequencing on genomics. Journal of genetics and genomics = Yi chuan xue bao, 2011, 38: 95-109.

[121]Glenn T C. Field guide to next-generation DNA sequencers. Molecular ecology resources, 2011, 11: 759-769.

[122]Margulies M, Egholm M, Altman W E, *et al.* Genome sequencing in

microfabricated high-density picolitre reactors. Nature, 2005, 437: 376-380.
[123]Rothberg J M, Leamon J H. The development and impact of 454 sequencing. Nat Biotechnol, 2008, 26: 1117-1124.
[124]Rothberg J M, Hinz W, Rearick T M, *et al.* An integrated semiconductor device enabling non-optical genome sequencing. Nature, 2011, 475: 348-352.
[125]Howden B P, McEvoy C R, Allen D L, *et al.* Evolution of multidrug resistance during Staphylococcus aureus infection involves mutation of the essential two component regulator WalKR. PLoS Pathog, 2011, 7: e1002359.
[126]Mardis E R. The impact of next-generation sequencing technology on genetics. Trends Genet, 2008, 24: 133-141.
[127]Ashelford K, Eriksson M E, Allen C M, *et al.* Full genome re-sequencing reveals a novel circadian clock mutation in Arabidopsis. Genome Biol, 2011, 12: R28.
[128]Shulaev V, Sargent D J, Crowhurst R N, *et al.* The genome of woodland strawberry (Fragaria vesca). Nat Genet, 2011, 43: 109-116.
[129]Shendure J, Porreca G J, Reppas N B, *et al.* Accurate multiplex polony sequencing of an evolved bacterial genome. Science, 2005, 309: 1728-1732.
[130]Liu L, Li Y, Li S, *et al.* Comparison of next-generation sequencing systems. J Biomed Biotechnol, 2012, 2012: 251364.
[131]Ku C S, Roukos D H. From next-generation sequencing to nanopore sequencing technology: paving the way to personalized genomic medicine. Expert Rev Med Devices, 2013, 10: 1-6.
[132]Harris T D, Buzby P R, Babcock H, *et al.* Single-molecule DNA sequencing of a viral genome. Science, 2008, 320: 106-109.
[133]Eid J, Fehr A, Gray J, *et al.* Real-time DNA sequencing from single polymerase molecules. Science, 2009, 323: 133-138.
[134]Flusberg B A, Webster D R, Lee J H, *et al.* Direct detection of DNA methylation during single-molecule, real-time sequencing. Nat Methods, 2010, 7: 461-465.

8

Metabonomic Phenotyping for the Gut Microbiota and Mammal Interactions

Huiru Tang *, Yulan Wang
State Key Laboratory of Magnetic Resonance and Atomic and Molecular Physics, Center for Biospectroscopy and Metabonomics, Wuhan Institute of Physics and Mathematics, Chinese Academy of Sciences, Wuhan, 430071, China
* E-mail: Huiru.tang@wipm.ac.cn

8.1 Mammals are 'Superorganisms'

All mammals consist of two distinct but integrated parts including hosts themselves and some symbiotic microorganisms [1-3]. Their symbiosis is established interactively through co-evolution and mutual selections [3-5]. Therefore, mammals are regarded as 'superorganisms' and their physiology and health in entirety have to be understood by taking into consideration hosts, symbiotic microbes and their interactions [1-4]. The symbiotic microorganisms are living mostly in the mammals' gut and also known in different contexts as the gut microbiota, microparasites and microbiomes. It is now known that mammals harbor trillions of symbiotic microbes mainly in their gastrointestinal tract (GIT) with many different microbial species [2-7]. In normal adult human GIT, for instance, there is more than one kilogram of microbes with over ten times more cells than hosts and several thousands of species [2-7]. These symbiotic gut microbiota are co-developed with their hosts' growth playing essential roles in many aspects of mammalian physiology and thus have profound effects on the hosts' health [3-7]. For this reason, microbiomes are now considered collectively as an 'essential organ' or extended genomes, transcriptomes, proteomes and metabonomes [4, 7, 8] for their mammalian hosts. However, it is nontrivial at the moment to completely define the genomes of these microbiomes as has been done for human and rodent hosts. Neither can their composition, transcriptomes and proteomes be defined in detail, since many species cannot be cultured *ex vivo*.

The compositional structure of gut microbiota is often dynamic, varying with both the host's endogenous factors such as genotypes, age and immunological capabilities, and environmental factors such as diets and living conditions [2-10]. Functions of gut microbiota include assistance to the development of hosts' immune system [11-15], food digestion and nutrient absorption [16-24], and participation in the host metabolic regulations such as drug and bile acid metabolisms [25-29]. More and more evidence suggests that gut microbiota are associated with the pathogenesis of many non-infectious diseases such as colorectal cancer [30, 31], inflammatory bowel disease [12], metabolic disorders [16-23], atherosclerosis [32] and even neurological disorders [33, 34]. As a result, the composition and functions of mammalian microbiomes have aroused outstanding interest across many different fields including microbial ecology [2-10], pathophysiology [16-23, 30-36], drug metabolism [37-39], and mammal-microbiome interactions [3-10, 24, 40, 41].

8.2 Co-Metabolisms and the Mammal-Microbiome Interactions

Co-metabolisms are important aspects of interactions between the mammal hosts and symbiotic gut microbiota. Such interactions are reflected by biotransformation of bile acids in enterohepatic circulation [25-28] together with the presence of many aromatic and choline co-metabolites in mammalian urine [35-37, 42-47]. For example, urinary phenylacetylglycine and hippurate are formed in the hosts' liver through glycine conjugation of phenylacetic acid and benzoic acid, respectively, which are intestinal bacterial metabolites of aromatic compounds [35, 36, 42-46]. In fact, other urinary aromatic metabolites have also been reported with similar significance including indoxyl and *p*-cresyl metabolites [42-46]. Furthermore, organic amines such as dimethylamine (DMA) and trimethylamine (TMA) in urine are associated with the gut microbiota metabolisms of choline [35, 36, 44-46].

More extensive host-microbiome co-metabolisms are highlighted by bile acid metabolisms since about 95% bile acids are re-cycled through co-metabolisms by the hosts' liver and gut microbiota in intestines [24-28]. Mammals convert cholesterols from *de novo* biosynthesis and dietary sources into chenodeoxycholic acid and cholic acid in their liver which are further converted into dozens of bile acids *via* hydroxylation, epimerization and conjugations with glycine and taurine [25-28]. These bile acids are then excreted into small intestines to fulfill their normal functions of neutralization and fatty acid absorptions through emulsifications. On the other hand, microbes in intestines conduct a series of reverse reactions of deconjugations and dehydroxylations together with some further reactions to produce so-called secondary bile acids such as deoxycholic acid and lithocholic acid [25]. These bile acids are then re-absorbed from the gastrointestinal tract, returned to the liver *via* the hepatic port vein and re-secreted through the well-known enterohepatic circulation process. Since the co-metabolites of bile acids have critical roles to play in absorption of lipophillic vitamins, mammalian endocrine functions and preventing intestinal microbial translocations [25, 26], gut microbes may be implicated in transgenomic regulations of the mammalian physiology [25, 26].

Under normal physiological conditions, 85% of bile acids are recovered in the small intestine whilst about 10% of such acids are further absorbed in the large intestine with about 5% of bile acids excreted in feces [24]. It is now known that about 80% of bile acids are re-absorbed in jejunum and the proximal part of ileum with about 5% of them recovered from the distal ileum [24]. Although only about 10% of bile acids are re-absorbed from colon to rectum, such a small amount of bile acids in large intestines have important implications in many pathological processes such as colorectal cancers and inflammatory bowl diseases. However, the details of such implications remain to be fully understood.

In fact, enterohepatic circulation is crucially important for the co-metabolic processes of many other compounds including both endogenous metabolites and xenobiotics such as drugs. For example, absorption, (co)metabolism and excretion of many drugs are associated with the essential contributions of gut microbiota to drug activations or detoxications [25, 26]. The effects of dietary cholines on mammals [48] are likely related to the gut microbial degradation of cholines into DMA and TMA. Enterohepatic circulation may also be involved with mammal-microbiome co-metabolisms of aromatic compounds which are important for the fates of aromatic amino acids, bioactive polyphenols from diets or functional foods and degradation of unwanted xenobiotics. Such circulation appears to provide a unique and safe route for the hosts' liver and gut microbiota in GIT to exchange materials and convert metabolites collaboratively. It is worth-noting that epithelium cells in the colon also interact with bacteria living there through the utilization of short-chain fatty acids (SCFAs) produced by the microbial fermentation of resistant polysaccharides [49-51]. Therefore, it becomes obviously important to analyze such co-metabolism and its contributions to the hosts' metabonomic phenotypes.

8.3 Metabonomic Phenotyping for Mammals

Metabolic phenotyping is an essential part of molecular phenotyping processes offering vital information for understanding the health and diseases of mammals [52-57] especially when genomes, transcriptomes and proteomes of gut microbiota cannot be easily defined. This is particularly true since co-metabolisms are important aspects of the mammal-microbiome interactions underpinning the gut microbiota functions [8, 25, 58-60]. However, the metabolite compositions (metabonomes) of all biological samples are complex for mammals [43, 61]. Such complexity is reflected in the sheer number, dynamic concentration ranges, diversity of physicochemical properties of encountered metabolites, and their background biological matrices. Therefore, complete quantifications of mammalian metabonomes of biofluids, cellular and tissue samples remain a formidable task and a huge analytical challenge.

Currently, two mainstream analytical technologies are widely employed, namely nuclear magnetic resonance (NMR) spectroscopy and mass spectrometry (MS) offering complementary information [56, 62-65]. With the former, a variety of techniques has been developed for both liquid samples (biofluids and tissue extracts) and semi-solid samples (cells and tissues). It is worth-mentioning that high-resolution magic-angle spinning NMR (HRMAS NMR) methods are useful for non-destructive metabonomic analysis of cells [66] and tissues [67-70]. NMR techniques are also available for metabolite analysis *in vivo*, which are particularly useful for clinical settings although special facilities are required.

In the case of the latter, methods based on both liquid chromatography-mass spectrometry (LC-MS) and gas chromatography-mass spectrometry (GC-MS) have been developed for biofluids and cell extracts [56, 63-65]. Direct-injection MS analyses are also possible [71] although care has to be taken if salts are present in samples. More recently, MS analysis and imaging of cross-sectioned tissue samples have also been reported [72-75] and such methods are expected to have particular importance in molecular pathological investigations. However, one has to be clearly aware of the advantages and disadvantages of both technologies before considering employment of them. The choice of methods obviously is critically dependent on the scientific questions one aims to answer.

Nevertheless, it is obvious that the combination of both NMR- and MS-based technologies will offer some combined advantages [76, 77]. The simplest such combination is the "logical combination" of the separately acquired NMR and MS data, which can further be combined mathematically such as in the case of STOCSY [78-84]. In fact, NMR and LC-MS technologies can also be combined by hyphenation. In this case, samples can be separated by chromatography followed with simultaneous analyses with ultraviolet-visible spectroscopy (with a DAD), NMR and MS [85-91]. Low abundant metabolites with small chromatographic peaks can be enriched with multiple trappings on micro-scale solid-phase columns prior to NMR and MS analyses. Such hyphenated LC-DAD-SPE-NMR/MS methods are flexible and versatile offering rich information and thus powerful for identification and structural determinations of novel metabolites including both endogenous and xenobiotic metabolites.

Multivariate data analysis of NMR data started decades ago [92-94] and has now been used routinely to recover metabolite information. Such information is the basis for constructing metabolic networks of the metabonomic phenotypic variations following the changes in both endogenous factors (*e.g.*, development and gene alterations) and environmental factors (*e.g.*, stressors and interventions by xenobiotics). In such analysis, signal normalization and scaling are normally employed prior to multivariate statistics. However, extra care has to be taken there since data from different techniques have completely different properties and implications. In general, whilst different metabolites have a similar NMR response-coefficient, similar metabolites may have drastically different MS response-coefficients. Therefore, special attention is needed for quantitative or semi-quantitative analysis of the MS data. Furthermore, for supervised analysis such as PLS-DA and OPLS-DA, modelling have to be critically validated and rigorously assessed [95-97] as widely documented and discussed in the literature.

For metabonomic phenotyping of mammals, two inseparable aspects have to be taken into consideration, namely, mammalian metabolic features and mammal-microbiome co-metabolic features. With most microbes living in the gut, attention has naturally been focused on the topographical metabolic features of the gastrointestinal tract of mammals and their variations with growth development. As a technique with no invasiveness and destruction of tissues, high-resolution magic-angle-spinning NMR spectroscopy can be used to define the metabonomic features *in situ* and has been employed for metabonomic phenotyping of rat [98], mice [60] and human intestines [81, 99]. Results have revealed that tissues from different regions of GIT have their own unique metabolic fingerprints for both animal models [60, 98] and human beings [81, 99] with the greatest differences observed between the small and large intestines. Such intestinal metabolic fingerprints are dependent on the animal developmental process as well [98]. It is likely that the topographic functions for GIT are also related to the communal compositions of gut microbiota varying with growth development and topographic regions of GIT [60, 100].

Information for the compositional structure of gut microbiota, their metabolic features and variations in different intestinal regions is essential for understanding the topographical or compartmentational aspects of the gut microbiota functions. Such information is also vital for understanding the mechanistic aspects of microbiome contributions to the hosts' physiology and for developing new therapeutic strategies for maintaining health and disease preventions. However, both the microbial composition and the gut microbiota metabolisms in different intestinal regions cannot be fully understood without the possibility of isolating and culturing all gut microbes externally.

Nevertheless, a potentially effective solution to this problem is to comprehensively analyze the metabolic profiles of intestinal contents "as a whole" since such profiles carry metabolic information for hosts, gut microbiota and their co-metabolisms. In fact, some metabonomic analyses of mammalian feces have already been documented in a number of published reports [21, 24, 30, 101-106]. Optimized analytical methods have been reported for fecal metabolite composition analysis [24, 101]. Some significant interspecies differences have been defined for humans, mice and rats in their fecal metabolite compositions [102]. NMR analysis has shown that the metabolic phenotypes of the rat GIT contents are age and topographically dependent being related to the topographic functions of the rat GIT [24]. Applications of fecal metabonomic analysis have further been explored for the diagnosis of diseases, such as inflammatory bowel disease [21, 103], liver cirrhosis and hepatocellular carcinoma [104], chronic pancreatitis [105] and colorectal cancer [30, 106]. For example, metabonomic analysis of the patient fecal metabolite composition has shown as a rapid and non-invasive diagnosis of gut microbiota dysfunctions in inflammatory bowel diseases [21, 103]. This is because significant changes in bacterial metabolites, such as MA and TMA, are expected when dysfunctions of gut microbiota occur. SCFAs in fecal samples are produced by the gut microbiota *via* fermentation of complex carbohydrates [24] and their significant changes may also provide prominent indications for problems related to gut microbiota and nutritional status [59, 107].

Urinary metabonomic analysis is yet another effective way to understand the symbiotic co-metabolisms since there is a close correlation between the symbiotic gut microbiota and urinary metabonomes [40]. Mammalian urine carries extremely rich metabolic information of the hosts and symbiotic micriobials as well as the host-microbiome co-metabolisms for both endogenous metabolites and xenobiotics [40]. Urinary metabonomic analyses have already suggested that the symbiotic gut microbiota interact with mammalian metabolisms to make contributions not only to the hosts' biology but also to mammalian pathology and efficacy of xenobiotic (*e.g.* drug) interventions [8, 42, 108-110]. For example, these co-metabolic interactions are implicated in the effects of nutrients and phytomedicines [111] and parasitic infections [44, 45, 112, 113] as well as in the development of glucose intolerance and insulin resistance [35].

The combination of urinary metabotyping and microecological analysis is an effective way of investigating the microbiome and mammalian metabolism interactions [40]. Such combination has already revealed correlations between the host metabolic features and the microbial species [40]. However, the direct contributions of gut microbiota to urinary metabolic details remain to be fully elucidated. The effectiveness and usefulness remain to be exploited for the potential of gut microbiota modifications in disease treatments and health management.

8.4 Future Perspectives

Although there are some methods for metabonomic analysis with both main stream technologies [114-117], methodology developments remain as the major tasks in the foreseeable future. More complete coverage of metabonomes is further required in detailed metabolite quantifications especially for the low-abundance metabolites. The implications of metabolite changes in a more systematic and mechanistic way remain to be fully understood, especially in terms of the metabolite bioactivities related to their modulations to gene regulations, transcriptions and protein expressions for both hosts and gut microbiota. With such information available, the roles of metabonomic phenotyping of mammals as super-organisms may become more effectively explored in health management and many clinical applications such as disease diagnosis, prognosis and preventions.

References

[1] Lederberg J. Infectious history. Science, 2000, 288: 287-293.

[2] Qin J J, Li R Q, Raes J, *et al.* A human gut microbial gene catalogue established by metagenomic sequencing. Nature, 2010, 464: 59-65.

[3] Xu J, Bjursell M K, Himrod J, *et al.* A genomic view of the human-bacteroides thetaiotaomicron symbiosis. Science, 2003, 299: 2074-2076.

[4] Nicholson J K, Holmes E, Wilson I D. Gut microorganisms, mammalian metabolism and personalized health care. Nat Rev Microbiol, 2005, 3: 431-438.

[5] Ley R E, Hamady M, Lozupone C, *et al.* Evolution of mammals and their gut microbes. Science, 2008, 320: 1647-1651.

[6] Backhed F, Ley R E, Sonnenburg J L, *et al.* Host-bacterial mutualism in the human intestine. Science, 2005, 307: 1915-1920.

[7] Yatsunenko T, Rey F E, Manary M J, *et al.* Human gut microbiome viewed across age and geography. Nature, 2012, 486: 222-227.

[8] Nicholson J K, Holmes E, Kinross J, *et al.* Host-gut microbiota metabolic interactions. Science, 2012, 336: 1262-1267.

[9] Xu J, Chiang H C, Bjursell M K, *et al.* Message from a human gut symbiont: sensitivity is a prerequisite for sharing. Trends Microbiol, 2004, 12: 21-28.

[10] Hooper L V, Gordon J I. Commensal host-bacterial relationships in the gut. Science, 2001, 292: 1115-1118.

[11] Hooper L V, Littman D R, Macpherson A J. Interactions between the microbiota and the immune system. Science, 2012, 336: 1268-1273.

[12] Macpherson A, Khoo U Y, Forgacs I, *et al.* Mucosal antibodies in inflammatory bowel disease are directed against intestinal bacteria. Gut, 1996, 38: 365-375.

[13] Peterson D A, McNulty N P, Guruge J L, *et al.* Iga response to symbiotic bacteria as a mediator of gut homeostasis. Cell Host & Microbe, 2007, 2: 328-339.

[14] Kau A L, Ahern P P, Griffin N W, *et al.* Human nutrition, the gut microbiome and the immune system. Nature, 2011, 474: 327-336.

[15] Garrett W S, Gordon J I, Glimcher L H. Homeostasis and inflammation in the intestine. Cell, 2010, 140: 859-870.

[16] Sanz Y, Santacruz A, Gauffin P. Gut microbiota in obesity and metabolic disorders. Proc Nutri Soc, 2010, 69: 434-441.

[17] Turnbaugh P J, Hamady M, Yatsunenko T, *et al.* A core gut microbiome in obese and lean twins. Nature, 2009, 457: 480-487.

[18] Ley R E, Backhed F, Turnbaugh P, *et al.* Obesity alters gut microbial ecology. Proc Natl Acad Sci USA, 2005, 102: 11070-11075.

[19] Ley R E, Turnbaugh P J, Klein S, *et al.* Microbial ecology-human gut microbes associated with obesity. Nature, 2006, 444: 1022-1023.

[20] Turnbaugh P J, Ley R E, Mahowald M A, *et al.* An obesity-associated gut microbiome with increased capacity for energy harvest. Nature, 2006, 444: 1027-1031.

[21] Marchesi J R, Holmes E, Khan F, *et al.* Rapid and non-invasive metabonomic characterisation of inflammatory bowel disease. J Proteome Res, 2007, 6: 546-552.

[22] Zhang X Y, Wang Y L, Hao F H, *et al.* Human serum metabonomic analysis reveals progression axes for glucose intolerance and insulin resistance statuses. J Proteome Res, 2009, 8: 5188-5195.

[23] Backhed F, Ding H, Wang T, *et al.* The gut microbiota as an environmental factor that regulates fat storage. Proc Natl Acad Sci USA, 2004, 101: 15718-15723.

[24] Tian Y, Zhang L M, Wang Y L, *et al.* Age-related topographical metabolic signatures for the rat gastrointestinal contents. J Proteome Res, 2012, 11: 1397-1411.

[25] Martin F P J, Dumas M E, Wang Y L, *et al.* A top-down systems biology

view of microbiome- mammalian metabolic interactions in a mouse model. Mol Systems Biol, 2007, 3: article 112,
[26] Swann J R, Want E J, Geier F M, *et al.* Systemic gut microbial modulation of bile acid metabolism in host tissue compartments. Proc Natl Acad Sci USA, 2011, 108: 4523-4530.
[27] Li J V, Ashrafian H, Bueter M, *et al.* Metabolic surgery profoundly influences gut microbial-host metabolic cross-talk. Gut, 2011, 60: 1214-1223.
[28] Ridlon J M, Kang D J, Hylemon P B. Bile salt biotransformations by human intestinal bacteria. J Lipid Res, 2006, 47: 241-259.
[29] Jones B V, Begley M, Hill C, *et al.* Functional and comparative metagenomic analysis of bile salt hydrolase activity in the human gut microbiome. Proc Natl Acad Sci USA, 2008, 105: 13580-13585.
[30] O'Keefe S J D, Ou J H, Aufreiter S, *et al.* Products of the colonic microbiota mediate the effects of diet on colon cancer risk. J Nutr, 2009, 139: 2044-2048.
[31] Hope M E, Hold G L, Kain R, *et al.* Sporadic colorectal cancer – Role of the commensal microbiota. FEMS Microbiol Lett, 2005, 244: 1-7.
[32] Stepankova R, Tonar Z, Bartova J, *et al.* Absence of microbiota (germ-free conditions) accelerates the atherosclerosis in apoe-deficient mice fed standard low cholesterol diet. J Atheroscl Thromb, 2010, 17: 796-804.
[33] Rhee S H, Pothoulakis C, Mayer E A. Principles and clinical implications of the brain-gut-enteric microbiota axis. Nat Rev Gastroenterol Hepatol, 2009, 6: 306-314.
[34] Parracho HMRT, Bingham M O, Gibson G R, *et al.* Differences between the gut microflora of children with autistic spectrum disorders and that of healthy children. J Med Microbiol, 2005, 54: 987-991.
[35] Xu W X, Wu J F, An Y P, *et al.* Streptozotocin-induced dynamic metabonomic changes in rat biofluids. J Proteome Res, 2012, 11: 3423-3435.
[36] Dumas M E, Barton R H, Toye A, *et al.* Metabolic profiling reveals a contribution of gut microbiota to fatty liver phenotype in insulin-resistant mice. Proc Natl Acad Sci USA, 2006, 103: 12511-12516.
[37] Clayton T A, Lindon J C, Cloarec O, *et al.* Pharmaco-metabonomic phenotyping and personalized drug treatment. Nature, 2006, 440: 1073-1077.
[38] Clayton T A, Baker D, Lindon J C, *et al.* Pharmacometabonomic identification of a significant host-microbiome metabolic interaction affecting human drug metabolism. Proc Natl Acad Sci USA, 2009, 106: 14728-14733.
[39] Wilson I D, Nicholson J K. The role of gut microbiota in drug response. Curr Pharmaceut Design, 2009, 15: 1519-1523.
[40] Li M, Wang B H, Zhang M H, *et al.* Symbiotic gut microbes modulate human metabolic phenotypes. Proc Natl Acad Sci USA, 2008, 105: 2117-2122.
[41] Hooper L V, Wong M H, Thelin A, *et al.* Molecular analysis of commensal

host-microbial relations hips in the intestine. Science, 2001, 291: 881-884.

[42] Zheng X J, Xie G X, Zhao A H, *et al.* The footprints of gut microbial-mammalian co-metabolism. J Proteome Res, 2011, 10: 5512-5522.

[43] Nicholson J K, Wilson I D. Understanding 'global' systems biology: Metabonomics and the continuum of metabolism. Nat Rev Drug Discov, 2003, 2: 668-676.

[44] Wang Y L, Holmes E, Nicholson J K, *et al.* Metabonomic investigations in mice infected with *schistosoma mansoni*: An approach for biomarker identification. Proc Natl Acad Sci USA, 2004, 101: 12676-12681.

[45] Wang Y L, Utzinger J, Saric J, *et al.* Global metabolic responses of mice to *trypanosoma brucei brucei* infection. Proc Natl Acad Sci USA, 2008, 105: 6127-6132.

[46] Zhang L M, Ye Y F, An Y P, *et al.* Systems responses of rats to aflatoxin b1 exposure revealed with metabonomic changes in multiple biological matrices. J Proteome Res, 2011, 10: 614-623.

[47] Martin F P J, Wang Y L, Sprenger N, *et al.* Top-down systems biology integration of conditional prebiotic modulated transgenomic interactions in a humanized microbiome mouse model. Mol Systems Biol, 2008, 4:article 205.

[48] Wang Z N, Klipfell E, Bennett B J, *et al.* Gut flora metabolism of phosphatidylcholine promotes cardiovascular disease. Nature, 2011, 472: 57-82.

[49] Fukuda S, Toh H, Hase K, *et al.* Bifidobacteria can protect from enteropathogenic infection through production of acetate. Nature, 2011, 469: 543-791.

[50] Scheppach W. Effects of short-chain fatty-acids on gut morphology and function. Gut, 1994, 35: S35-S38.

[51] Wong J M W, de Souza R, Kendall C W C, *et al.* Colonic health: Fermentation and short chain fatty acids. J Clin Gastroenterol, 2006, 40: 235-243.

[52] Nicholson J K, Connelly J, Lindon J C, *et al.* Metabonomics: A platform for studying drug toxicity and gene function. Nat Rev Drug Discov, 2002, 1: 153-161.

[53] Wijeyesekera A, Selman C, Barton R H, *et al.* Metabotyping of long-lived mice using h-1 nmr spectroscopy. J Proteome Res, 2012, 11: 2224-2235.

[54] Kinross J M, Holmes E, Darzi A W, *et al.* Metabolic phenotyping for monitoring surgical patients. Lancet, 2011, 377: 1817-1819.

[55] Holmes E, Loo R L, Stamler J, *et al.* Human metabolic phenotype diversity and its association with diet and blood pressure. Nature, 2008, 453: 396-400.

[56] Nicholson J K, Lindon J C. Systems biology-metabonomics. Nature, 2008, 455: 1054-1056.

[57] Holmes E, Wilson I D, Nicholson J K. Metabolic phenotyping in health and disease. Cell, 2008, 134: 714-717.

[58] Martin FPJ, Collino S, Rezzi S. ^{1}H NMR-based metabonomic applications to decipher gut microbial metabolic influence on mammalian health. Magn

Reson Chem, 2011, 49: S47-S54.
[59] Martin F P J, Sprenger N, Montoliu I, *et al.* Dietary modulation of gut functional ecology studied by fecal metabonomics. J Proteome Res, 2010, 9: 5284-5295.
[60] Martin F P J, Wang Y, Yap I K S, *et al.* Topographical variation in murine intestinal metabolic profiles in relation to microbiome speciation and functional ecological activity. J Proteome Res, 2009, 8: 3464-3474.
[61] Nicholson J K, Holmes E, Lindon J C, *et al.* The challenges of modeling mammalian biocomplexity. Nat Biotechnol, 2004, 22: 1268-1274.
[62] Tang H R,Wang Y L. Metabonomics: A revolution in progress. Prog Biochem Biophys, 2006, 33: 401-417.
[63] Tian J, Shi C Y, Gao P, *et al.* Phenotype differentiation of three e-coli strains by gc-fid and gc-ms based metabolomics. J Chromat Anal Technol Biomed Life Sci, 2008, 871: 220-226.
[64] Wilson I D, Plumb R, Granger J, *et al.* HPLC-MS-based methods for the study of metabonomics. J Chromat Anal Technol Biomed Life Sci, 2005, 817: 67-76.
[65] Lenz E M, Wilson I D. Analytical strategies in metabonomics. J Proteome Res, 2007, 6: 443-458.
[66] Humpfer E, Spraul M, Nicholls A W, *et al.* Direct observation of resolved intracellular and extracellular water signals in intact human red blood cells using ^{1}H MAS NMR spectroscopy. Magn Reson Med, 1997, 38: 334-336.
[67] Cheng L L, Ma M J, Becerra L, *et al.* Quantitative neuropathology by high resolution magic angle spinning proton magnetic resonance spectroscopy. Proc Natl Acad Sci USA, 1997, 94: 6408-6413.
[68] Cheng L L, Chang I W, Louis D N, *et al.* Correlation of high-resolution magic angle spinning proton magnetic resonance spectroscopy with histopathology of intact human brain tumor specimens. Cancer Res, 1998, 58: 1825-1832.
[69] Ding L N, Hao F H, Shi Z M, *et al.* Systems biological responses to chronic perfluorododecanoic acid exposure by integrated metabonomic and transcriptomic studies. J Proteome Res, 2009, 8: 2882-2891.
[70] Yang Y X, Li C L, Nie X, *et al.* Metabonomic studies of human hepatocellular carcinoma using high-resolution magic-angle spinning ^{1}H NMR spectroscopy in conjunction with multivariate data analysis. J Proteome Res, 2007, 6: 2605-2614.
[71] Beckmann M, Parker D, Enot D P, *et al.* High-throughput, nontargeted metabolite fingerprinting using nominal mass flow injection electrospray mass spectrometry. Nature Protocols, 2008, 3: 486-504.
[72] Fonville J M, Carter C, Cloarec O, *et al.* Robust data processing and normalization strategy for maldi mass spectrometric imaging. Anal Chem, 2012, 84: 1310-1319.
[73] Nemes P, Woods A S, Vertes A. Simultaneous imaging of small metabolites and lipids in rat brain tissues at atmospheric pressure by laser ablation electrospray ionization mass spectrometry. Anal Chem, 2010, 82: 982-988.
[74] Koizumi S, Yamamoto S, Hayasaka T, *et al.* Imaging mass spectrometry

revealed the production of lyso-phosphatidylcholine in the injured ischemic rat brain. Neuroscience, 2010, 168: 219-225.

[75] Cooks R G, Ouyang Z, Takats Z, *et al.* Ambient mass spectrometry. Science, 2006, 311: 1566-1570.

[76] Dai H, Xiao C N, Liu H B, *et al.* Combined NMR and LC-MS analysis reveals the metabonomic changes in *salvia miltiorrhiza* bunge induced by water depletion. J Proteome Res, 2010, 9: 1460-1475.

[77] Dai H, Xiao C N, Liu H B, *et al.* Combined NMR and LC-DAD-MS analysis reveals comprehensive metabonomic variations for three phenotypic cultivars of *salvia miltiorrhiza* bunge. J Proteome Res, 2010, 9: 1565-1578.

[78] Holmes E, Loo R L, Cloarec O, *et al.* Detection of urinary drug metabolite (xenometabolome) signatures in molecular epidemiology studies via statistical total correlation (NMR) spectroscopy. Anal Chem, 2007, 79: 2629-2640.

[79] Cloarec O, Campbell A, Tseng L H, *et al.* Virtual chromatographic resolution enhancement in cryoflow LC-NMR experiments via statistical total correlation spectroscopy. Anal Chem, 2007, 79: 3304-3311.

[80] Smith L M, Maher A D, Cloarec O, *et al.* statistical correlation and projection methods for improved information recovery from diffusion-edited NMR spectra of biological samples. Anal Chem, 2007, 79: 5682-5689.

[81] Wang Y L, Cloarec O, Tang H R, *et al.* Magic angle spinning NMR and ^{1}H-^{31}P heteronuclear statistical total correlation spectroscopy of intact human gut biopsies. Anal Chem, 2008, 80: 1058-1066.

[82] Maher A D, Fonville J M, Coen M, *et al.* Statistical total correlation spectroscopy scaling for enhancement of metabolic information recovery in biological NMR spectra. Anal Chem, 2012, 84: 1083-1091.

[83] Cloarec O, Dumas M E, Craig A, *et al.* Statistical total correlation spectroscopy: An exploratory approach for latent biomarker identification from metabolic ^{1}H NMR data sets. Anal Chem, 2005, 77: 1282-1289.

[84] Crockford D J, Lindon J C, Cloarec O, *et al.* Statistical search space reduction and two-dimensional data display approaches for UPLC-MS in biomarker discovery and pathway analysis. Anal Chem, 2006, 78: 4398-4408.

[85] Lommen A, Godejohann M, Venema D P, *et al.* Application of directly coupled HPLC-NMR-MS to the identification and confirmation of quercetin glycosides and phloretin glycosides in apple peel. Anal Chem, 2000, 72: 1793-1797.

[86] Duarte I F, Godejohann M, Braumann U, *et al.* Application of NMR spectroscopy and LC-NMR/MS to the identification of carbohydrates in beer. J Agri Food Chem, 2003, 51: 4847-4852.

[87] Corcoran O, Spraul M. LC-NMR-MS in drug discovery. DDT, 2003, 8: 624-631.

[88] Spraul M, Freund A S, Nast R E, *et al.* Advancing NMR sensitivity for LC-NMR-MS using a cryoflow probe: Application to the analysis of acetaminophen metabolites in urine. Anal Chem, 2003, 75: 1536-1541.

[89] Holmes E, Tang H R, Wang Y L, *et al.* The assessment of plant metabolite profiles by NMR-based methodologies. Plant Med, 2006, 72: 771-785.
[90] Tang H R, Xiao C N, Wang Y L. Important roles of the hyphenated HPLC-DAD-SPE-MS/NMR technique in metabonomics. Magn Reson Chem, 2009, 47: S157-S162.
[91] Duarte I F, Legido-Quigley C, Parker D A, *et al.* Identification of metabolites in human hepatic bile using 800 MHz ^{1}H NMR spectroscopy, HPLC-NMR/MS and UPLC-MS. Mol Biosys, 2009, 5: 180-190.
[92] Holmes E, Bonner F W, Sweatman B C, *et al.* Nuclear-magnetic-resonance spectroscopy and pattern-recognition analysis of the biochemical processes associated with the progression of and recovery from nephrotoxic lesions in the rat induced by mercury(II) chloride and 2-bromoethanamine. Mol Pharmacol, 1992, 42: 922-930.
[93] Ghauri F Y K, Nicholson J K, Sweatman B C, *et al.* NMR spectroscopy of human postmortem cerebrospinal-fluid-distinction of alzheimers-disease from control using pattern- recognition and statistics. NMR Biomed, 1993, 6: 163-167.
[94] Gartland K P R, Beddell C R, Lindon J C, *et al.* A pattern-recognition approach to the comparison of PMR and clinical chemical-data for classification of nephrotoxicity. J Pharm Biomed Anal, 1990, 8: 963-968.
[95] Lindon J C, Nicholson J K, Holmes E, *et al.* Summary recommendations for standardization and reporting of metabolic analyses. Nat Biotechnol, 2005, 23: 833-838.
[96] Eriksson L, Trygg J, Wold S. CV-ANOVA for significance testing of PLS and OPLS (r) models. J Chemometr, 2008, 22: 594-600.
[97] Trygg J, Wold S. Orthogonal projections to latent structures (O-PLS). J Chemometr, 2002, 16: 119-128.
[98] Wang Y L, Tang H R, Holmes E, *et al.* Biochemical characterization of rat intestine development using high-resolution magic-angle-spinning ^{1}H NMR spectroscopy and multivariate data analysis. J Proteome Res, 2005, 4: 1324-1329.
[99] Wang Y L, Holmes E, Comelli E M, *et al.* Topographical variation in metabolic signatures of human gastrointestinal biopsies revealed by high-resolution magic-angle spinning ^{1}H NMR spectroscopy. J Proteome Res, 2007, 6: 3944-3951.
[100] Hopkins M J, Sharp R, Macfarlane G T. Age and disease related changes in intestinal bacterial popular-ions assessed by cell culture, 16S rRNA abundance, and community cellular fatty acid profiles. Gut, 2001, 48: 198-205.
[101] Wu J F, An Y P, Yao J W, *et al.* An optimised sample preparation method for NMR-based faecal metabonomic analysis. Analyst, 2010, 135: 1023-1030.
[102] Saric J, Wang Y, Li J, *et al.* Species variation in the fecal metabolome gives insight into differential gastrointestinal function. J Proteome Res, 2008, 7: 352-360.
[103] Le Gall G, Noor S O, Ridgway K, *et al.* Metabolomics of fecal extracts detects altered metabolic activity of gut microbiota in ulcerative colitis and

irritable bowel syndrome. J Proteome Res, 2011, 10: 4208-4218.

[104] Cao H C, Huang H J, Xu W, *et al.* Fecal metabolome profiling of liver cirrhosis and hepatocellular carcinoma patients by ultra performance liquid chromatography-mass spectrometry. Anal Chim Acta, 2011, 691: 68-75.

[105] Naruse S, Ishiguro H, Ko S B H, *et al.* Fecal pancreatic elastase: A reproducible marker for severe exocrine pancreatic insufficiency. J Gastroenterol, 2006, 41: 901-908.

[106] Hu S, Dong T S, Dalal S R, *et al.* The microbe-derived short chain fatty acid butyrate targets miRNA-dependent p21 gene expression in human colon cancer. Plos One, 2011, 6: e16221.

[107] Jacobs D M, Deltimple N, van Velzen E, *et al.* ^{1}H NMR metabolite profiling of feces as a tool to assess the impact of nutrition on the human microbiome. NMR Biomed, 2008, 21: 615-626.

[108] Holmes E, Kinross J, Gibson G R, *et al.* Therapeutic modulation of microbiota-host metabolic interactions. Sci Transl Med, 2012, 4: 137-142.

[109] Wang X N, Wang X Y, Xie G X, *et al.* Urinary metabolite variation is associated with pathological progression of the post-hepatitis B cirrhosis patients. J Proteome Res, 2012, 11: 3838-3847.

[110] Cheng Y, Xie G X, Chen T L, *et al.* Distinct urinary metabolic profile of human colorectal cancer. J Proteome Res, 2012, 11: 1354-1363.

[111] Wang Y L, Tang H R, Nicholson J K, *et al.* A metabonomic strategy for the detection of the metabolic effects of chamomile (*matricaria recutita* L.) ingestion. J Agri Food Chem, 2005, 53: 191-196.

[112] Wu J F, Holmes E, Xue J, *et al.* Metabolic alterations in the hamster co-infected with *schistosoma japonicum* and *necator americanus*. Int J Parasitol, 2010, 40: 695-703.

[113] Wu J F, Xu W X, Ming Z P, *et al.* Metabolic changes reveal the development of schistosomiasis in mice. PLOS Negl Trop Dis, 2010, 4: e807.

[114] Dunn WB, Broadhurst D, Begley P, *et al.* Procedures for large-scale metabolic profiling of serum and plasma using gas chromatography and liquid chromatography coupled to mass spectrometry. Nature Protocols, 2011, 6: 1060-1083.

[115] Chan E C Y, Pasikanti K K, Nicholson J K. Global urinary metabolic profiling procedures using gas chromatography-mass spectrometry. Nature Protocols, 2011, 6: 1483-1499.

[116] Beckonert O, Coen M, Keun H C, *et al.* High-resolution magic-angle-spinning NMR spectroscopy for metabolic profiling of intact tissues. Nature Protocols, 2010, 5: 1019-1032.

[117] Beckonert O, Keun H C, Ebbels T M D, *et al.* Metabolic profiling, metabolomic and metabonomic procedures for nmr spectroscopy of urine, plasma, serum and tissue extracts. Nature Protocols, 2007, 2: 2692-2703.

9

Bioinformatics for Genomes and Metagenomes in Ecology Studies

Douglas B. Rusch, Jason Miller, Konstantinos Krampis, Andrey Tovchigrechko, Granger Sutton, Shibu Yooseph, Karen E. Nelson *
J. Craig Venter Institute, 9704 Medical Center Drive, Rockville, Maryland, 20850, USA
* E-mail: KNelson@jcvi.org

Major technological developments in the field of microbial ecology are redefining the science, moving the focus of research away from studies of individual isolates and species that are studied under carefully controlled conditions in the laboratory, towards the study of entire communities of organisms in their natural environments. Ever more efficient sequencing technologies mean that we can generate huge volumes of sequence data — shifting the cost burden from sequence generation to sequence analysis. The bioinformatic techniques for managing and analyzing both the new types of data and the vastly increased volumes of data are transforming our understanding of life and its interdependencies. These data sets, in conjunction with bioinformatics are enhancing our understanding of microbial diversity and microbial ecology in many different environments. In this chapter, we provide an overview of some of the genomic, metagenomic and informatics approaches currently being used and or being developed for the study of microbial diversity and ecology.

9.1 Introduction to Advances in Microbial Ecology

Only a few years ago, we, as a scientific community did not have a true appreciation for the extent of microbial diversity in nature. Things have changed significantly, and we can expect that we will continue to make new discoveries about the microorganisms in our world. While it has long been known that microbes are ubiquitous and play a crucial role in energy, nutrient cycling, agriculture, disease and health, our understanding of these organisms has been grounded in, and limited to traditional approaches of observation (phenotype) and cultivation dating back to the times of Leeuwenhoek and Pasteur. With the development of molecular biology and DNA sequencing technologies, the identification of conserved phylogenetic markers, the launch of the genomics era, as well as tremendous advances in the science of metagenomics, we have made the use of sequencing approaches to evaluate diversity in environmental settings widely accepted. Advanced sequencing technologies have allowed for increased base-pair (bp) throughput per run while reducing costs significantly [1]. The newer sequencing technologies such as 454-pyrosequencing and Illumina allow for massive volumes of sequence data; for example, the Illumina GAIIx system produces more than 50 Giga-base pair (Gbp) per run, and the updated HiSeq 2000 technology from Illumina has a 4-fold higher volume. The reduction in cost of the technologies has democratized sequencing such that it is widely accessible to most laboratories – both big and small. Smaller and less expensive sequencers have also recently become available (for example GS Junior-454 Sequencing / http://www.gsjunior.com/research-applications.php). The result has been that sequencing of 16S rDNA genes, complete bacterial, fungal and viral genomes as well as environmental metagenomes are now commonplace, and large-scale sequencing and data analysis have become far more routine in microbial ecology studies. In parallel, the informatics challenges have increased, and new tools have had to be developed to assemble and analyze the datasets generated by these ecology studies.

In this chapter, we give an overview of the science along with developments for informatics as it relates directly to microbial ecology studies. This will include an overview of sequencing, assembly, tools for data analysis inclusive of measurements of abundance and diversity, as well as the analysis of gene and pathway function. Finally we will present our thoughts on where the field might be headed and some of the challenges that remain.

9.2 16S rDNA in Ecology Studies

16S rDNA gene analysis has been used widely to identify microbial species and interrogate the diversity of microbes in a range of environments. The value in using the 16S rDNA gene as a phylogenetic marker derives from its being a necessary component of the ribosomal apparatus of microbial cells, and hence being present in all prokaryotic genomes. It also has a relatively slow rate of evolution coupled with highly conserved motifs, which are ideal for primer design. Several faster evolving (variable) segments within the 16S rDNA gene sequence provide useful phylogenetic information and are compatible with some of the next generation sequencing technologies. Lastly, the 16S rDNA gene rarely moves by lateral gene transfer and thus tracks closely with the protein coding genes that define a particular microbial lineage. However, The 16S rDNA gene can be less than perfect as it is often found in multiple copies in many microbial species, with copy numbers ranging anywhere from one to a dozen copies per cell. As such, 16S rDNA gene sequencing is an unreliable indicator of the actual abundance of the species being investigated. Intraspecies differences in ribosomal operons have also been described [2-5] again leading to possible misinterpretation of the results from surveys. Furthermore, while 16S rDNA has largely been successful at classifying organisms down to the species level it evolves too slowly to capture the diversity within species. Finally, the primer design and amplification processes that are used to obtain the 16S rDNA gene sequences are also known to introduce biases and error in the process including the formation of chimeric products, biased rates of amplification, and failure to amplify some targets due to primer mismatch.

Regardless of these limitations, 16S rDNA sequencing continues to produce a wealth of survey data. The update from the Ribosomal Database Project (http://rdp.cme.msu.edu/; RDP Release 10, Aug 30, 2010) includes a total of 1,418,497 16S rRNAs from species that inhabit many different environments. Combined with the absence of better resources for detailed comparative purposes, 16S rDNA gene-based approaches to characterizing microbial diversity have become powerful tools for microbial ecology. Moving past a heavy reliance on Sanger based full length 16S rDNA gene sequencing where the ~1,500 bp sequence was the goal, advances in sequencing technology have made it possible to generate hundreds of thousands of 16S genes from a single environment allowing the community to be studied in unprecedented detail. The newer generations of sequencing technologies generate shorter read-lengths and as such have been used primarily in large surveys and for inference on microbial diversity in ecological samples [6]. Most of the studies that employ the use of next-generation sequencing technologies tend to focus on one or more variable regions for inference of the population diversity [7]. These new technologies offer the ability to sample genomes of organisms that occur at low abundances in communities at much lower cost. This has resulted in a high discovery rate of previously uncultured and novel organisms.

9.3 16S rDNA Gene Analysis

In addition to taxonomic composition, 16S sequences data can also be used to study community structure and biodiversity at different levels (α, β, and γ) [8]. The identification of organisms from 16S rDNA gene samples is facilitated by reference databases (for example [9-11]) containing 16S sequences derived from both cultured and previously unseen uncultured microbes. Both phylogenetic and non-phylogenetic methods have been developed for assigning taxonomies to sequences in a given 16S sample. In a phylogenetic method, an evolutionary tree containing sample and reference sequences is constructed, and the placement of a sample sequence in the tree relative to the reference sequences is used to infer taxonomic origin. STAP [12] is an example of a phylogenetic method. Non-phylogenetic methods, on the other hand, do not construct evolutionary trees, but rather use features computed from sequences data to associate the sample sequences with reference sequences, and subsequently infer their taxonomy; machine learning approaches (both unsupervised and supervised) for taxonomic inference fall into this category, with RDP's Naïve Bayes classifier [11] an example of a widely used program.

A complementary analysis approach involves the identification of operational taxonomic units (OTUs) from 16S sequences data. OTUs are clusters of sequences where the sequences within a cluster meet a preset distance threshold. The choice of distance threshold affects the taxonomic granularity of the OTUs; for instance, 3% sequence distance (or equivalently, 97% sequence identity) is used as a threshold to identify bacterial species [13, 14]. Agglomerative clustering approaches, including single-, complete- and average-linkage, are used typically for the identification of OTUs [15]. Given that the available 16S reference databases almost never capture the full range of microbial taxonomic diversity in an environment, it can often happen that a non-trivial fraction of 16S sequences from a given sample remain unclassified. An OTU-based analysis can be used in these situations to quantify the novelty and diversity of microbes in the community. In addition, rarefaction curves based on OTUs can be used to assess the extent of coverage of the community by the given 16S dataset, and to decide whether additional sequencing is required.

Prior to any taxonomic or OTU-based analysis of 16S samples, an important step is the detection and removal of chimeric 16S sequences. As mentioned above, a chimeric 16S sequence is a PCR amplification artifact and is derived from more than one DNA template [16-18]. Failure to remove chimeric sequences can result in incorrect estimation of taxonomic content and an artificial inflation of community diversity estimates. Several methods have been proposed in the literature to detect chimeras (for example [18-21]) and these are routinely applied to clean 16S datasets. As mentioned above, when using traditional Sanger sequencing technology, it is possible to obtain full-length 16S sequences averaging 1,500 bp. However, cost considerations along with greater sequencing depth have resulted in 454-pyrosequencing becoming a technology of choice for 16S sequencing (for example [22-24]), even though this technology produces shorter reads. Recently, 16S sequencing using Illumina technology has also been reported [25]. With these newer technologies, selected variable regions on the 16S gene are targeted for sequencing [11, 26]. However, the use of short reads results in lower accuracies in taxonomic classification [11]. In addition, it has also become clear that sequencing error rates associated with 454-pyrosequencing can result in an over-estimation of community diversity [27, 28]. Various informatic approaches have been proposed to alleviate this problem, including methods that work directly on 454 flowgrams [29] and those that work on sequence reads [30].

The identification of taxonomic groups and OTUs sets the basis to study ecological diversity within and across microbial communities. Ecological diversity measures that can be computed from these data include species richness indices, evenness and dominance indices, and species abundance models [31]. Many of these indices, including several non-parametric estimators for species richness, are available in 16S analysis programs such as Mothur [21] and QIIME [32]. These programs also include methods for 16S sample comparison and ordination.

9.4 Metagenomics

The new generation of sequencing technologies is also being used in the area of metagenomics — a paradigm shift for traditional microbiologists that enables cultivation independent approaches to the study of microbial communities. Strategically metagenomics involves the consideration of three categories of data: (i) environmental (metadata); (ii) the community of organisms it supports; and (iii) the genes they contain. Each of these categories encompasses a complex set of variables. In a macroscopic environment, the community is typically a census of all the distinct species present. In a microbial community, the very concept of a species is poorly defined and often debated, because it does not clearly reflect the overall genomic or functional equivalence of two organisms. Despite this, closely related organisms do tend to share a substantial amount of genomic sequence along with a great number of physiological and biochemical capabilities lending a community census based approach of great value while also providing a well-established nomenclature to work with.

Most microbial communities are dynamic such that the current population is altering the environment making it possible for new organisms to move in or become abundant in turn altering the functional repertoire of genes present. Organisms often contain a larger repertoire of genes than would be necessary for their survival in any specific environment. The flexibility and adaptability provided by these genes presumably out weighs the cost of maintaining these genes in the genome allowing them to persist perhaps indefinitely according to the lifestyle of the microbe in question. This bundling argues against assuming that a metagenomic sample can be treated like a bag of genes effectively ignoring the fact that genes are functionally organized into operons, genomes, and cells in ways the can profoundly alter their activity. Assuming these categories are dependent on one another implies that measuring environment, community, or the genes should provide considerable information about the other categories. Determining the rules that relate the environment, community, and genes however is far from trivial and will dictate the techniques and approaches that will be used in future metagenomic studies.

Given a sufficient quantity of genomic sequencing reads, it is possible to directly compare communities without relying on marker genes such as 16S. Similarity is measured using the number of overlapping sequencing reads at some percent identity cutoff. By normalizing the intra and inter-overlap frequencies, the similarity between two samples can be calculated. Given the correct cutoff, this genomic approach should be more sensitive than a marker based approach. Because the comparisons are carried out in nucleotide space only, the closely related populations will contribute to the similarity score.

Though it would be expected that the most abundant microbial populations would dictate community similarity, it can be shown that by excluding the sequencing reads from these organisms the pattern of similarity persists even among the less abundant microbial populations. It remains to be seen whether genomic approaches to comparing community will provide sufficient information to be cost effective compared to marker based approaches.

9.5 Recent Applications of Environmental Metagenomic Sequencing

The Sorcerer II Global Ocean Sampling (GOS) Expedition to study marine microbial diversity is the largest dataset to date from a complex environment and provides the best opportunity yet for understanding the relationship between the environment, community, and genes it contains [33, 34]. When the GOS data from the first phase of the expedition were published [35], that study more than doubled the number of proteins in the public databases. The GOS studies have also resulted in a number of spin-off studies that have been focused on the datasets that were initially generated [36-38].

Other large-scale metagenomic projects include the National Institutes of Health (NIH) funded Human Microbiome Project (HMP) to study human associated microbial communities. Through the HMP and other similar projects around the globe [39], accessing the genomes of all the species (bacteria, viruses, phages and eukaryotes) associated with the human body has become a priority. The HMP has by now generated metagenomic sequence data from 15 – 18 body sites of 300 "normal" individuals, many of whom have become repeat sample donors. In parallel to the metagenomic work, culture-based, non-culture based and single cell approaches are allowing us to access the genomes of reference species that will provide insight into the genetic diversity of these microorganisms as well as acting as scaffolding for the metagenomic datasets [40]. The HMP aims to generate more than 3,000 bacterial reference genomes as well as several other phage, viral and eukaryotic genomes—a tiny portion of the total number of microbes that inhabit the human body (http://www.hmpdacc.org/reference_genomes.php).

The first large-scale paper from the HMP consortium describes 178 reference genomes that were generated as a result of our sequencing and annotation efforts [41]. From these genomes, we identified approximately 547,968 polypeptides that were greater than 100 amino acids in length, of which 29,987 were unique; *i.e.* they had not been seen before when we compared our datasets against all publicly available data. Although at first sight this might appear to be a large dataset, we know that the genomes included in this initial study — although the largest conglomeration of sequenced genomes in a single publication — represent but a subset of the species associated with the human body based on interrogation of available 16S surveys of the human body. Continued data-mining of these and other datasets that are being generated by the HMP and international consortia [42] as well as other smaller groups is anticipated to reveal additional significant gene clusters, antibiotic markers, plasmids, phagess *etc.*

Many other large-scale metagenomic studies are underway world-wide, and include but are not limited to studies on ruminants [43], canines [44], poultry [45] and non-human primates [46], as well as various insect species [47, 48].

9.6 Analysis of Viral Communities

The study of viral populations in ecological studies has been challenging at best when compared to bacterial, archaeal and eukaryotic communities. There are no reliable phylogenetic markers for viruses, and they are more numerous and more diverse than the proks and the euks. In addition, the viruses come in many different forms (DNA, RNA, single stranded *vs.* double stranded). Regardless, viruses are key to most microbial communities, driving the evolution and diversity of various ecosystems. High throughput has presented a major advantage for the analysis of viral populations in environmental samples, with 454-pyrosequencing for example proving to be an efficient method for detecting novel viruses among mixtures of highly diverse sequences [49]. Metagenomics approaches are perhaps among the best methods for broad viral surveys and have been used successfully in a number of environments including the oceans, various insect vectors, animal species and human fecal material with a goal to generate a broad characterization of the diverse populations that might be present [49-52]. Here, invariably, the sequences that are generated are characterized based on sequence homology to available sequences in public databases [50]. Viral sequences also assemble and can be described without enriching solely for the viral fractions as a result of general metagenomic surveys [43, 44]. Challenges with using metagenomics to the studying of viral populations remain, and are usually associated with linking specific viral genomes to hosts, the inability to completely assemble viral genomes from mixed populations, and the frequent presence on unknown gene sequences.

Finally, high throughput pipelines are available for sequencing many viruses as in the event of a pandemic. Informatic tools allow for comparisons with close relatives such that differences between these isolates can be readily identified.

9.7 Assembly of Sequence Data

All sequencing technologies generate reads that are much shorter than genomes. Some sequencing technologies generate reads that are shorter than most genes and nearly all sequencers generate reads shorter than operons. Assembly is an informatics work-around for this fundamental limitation of sequencing technology and involves the deduction of long sequences given the short DNA sequences provided by sequencing machines. Compared to bare reads, assembled sequence provides a better substrate for downstream annotation and analysis. Assemblies are more likely than reads to contain complete gene structures that can be recognized by gene-finding software. Assemblies that span gene-pairs provide evidence of their co-existence within cells, and these findings can support inference of linked function, expression, and regulation. Assemblies that capture operons provide additional evidence of co-regulation and co-expression of genes. Assembly complements the shotgun sequencing method whereby DNA is fragmented and sequencing fragments randomly [53]. Shotgun sequencing by nature sacrifices the positional or contextual information that could be provided by directed sequencing. In return, shotgun delivers vast quantities of sequence in high-throughput mode. If the sequence data over-samples the DNA, then assembly software should find shotgun sequences that match each other over long stretches. Long pair-wise sequence alignments imply the sequences are redundant reads of the same DNA locus. Assembly software combines such sequences and generates a representative consensus sequence. The consensus sequence is usually longer and more accurate than any of the component sequences. Thus, assembly collapses redundant sequence, reduces the data quantity, overcomes some sequencing error, and generally restores context to individual shotgun reads. The accuracy of shotgun followed by assembly has been demonstrated on small individual genomes [54, 55], large genomes such as fruit fly [56, 57] and human [58], and complex environmental communities such as ocean water [33].

Assembly usually requires sophisticated algorithms and software [59, 60]. The computational process begins with 'reads' from a sequencing machine. Each read is typically a record in a computer file that lists a few hundred bases of sequence in order. The sequence is derived from one strand of one part of one fragment of one DNA molecule, but none of that information is captured. 'Fragment sequencing' delivers at most one read per fragment. 'Mate pair sequencing' and 'paired end sequencing' deliver two reads per fragment. Most sequencing data includes numerical expectations of the distance between each pair, as well as their relative strands, and assembly software can exploit this extra information. Assembly software combines reads that share sequence. Assemblers generate 'contigs', which are multiple sequence alignments (MSA) of individual reads, and provide a 'consensus sequence' that represents the MSA. Assemblers also generate 'scaffolds' or 'supercontigs' that put contigs in groups, where the relationship between every pair of contigs is derived from mate pairs that have one read in each contig. Unlike contigs, scaffolds can include long gaps that have a predicted length but no sequence. Gaps are typically represented in computer files by a specific number of repetitions of the non-DNA character, 'N'.

The ideal metagenomic assembly would reconstruct one scaffold, or even one contig, that captures all of one genome, for every genome in the sample under study. In practice, metagenomic assembly always falls short due to confounding factors. The most challenging issue is the tremendous diversity found even within microbial populations that are closely placed and closely related. Different levels of divergence between related organisms, at the sequence and structural levels, make it difficult to determine proper thresholds for assembly. Repeat elements, such as ribosomal genes, induce false-positive alignments during assembly. Differing abundance levels per genome complicate the utility of read coverage statistics as indicators of repetitive sequence. It will be some time before closed genomes can be routinely assembled from even moderately complex environments.

Within the metagenomics community, there is some resistance to assembly based on its propensity to generate sequences absent from the sample. Chimeras, *i.e.* joins of unrelated sequences, can be induced by elements such as ribosomal RNA genes that are repeated and conserved across genomes. Whereas an assembly's consensus sequence should "average out" the random error within individual reads, it can 'average in' incorrect base calls when the reads derive from slightly divergent genomes. An inaccurate consensus may be better than no information at all about sequences that are divergent and represented by few reads. However, metagenomics assemblies can contain frame shifts that disrupt true open reading frames. Assembly as an amplifier for metagenomics needs higher fidelity to become widely adopted.

9.8 Assembly: Strategies

Here we describe a model community for which shotgun sequencing and assembly would recover most of the genomic sequence; this model community would consist of genomes that are sufficiently diverged that they do not share any subsequences. It would contain many cells per genome such that each cell would have a nearly identical copy of that genome. If such a community were sequenced to high coverage, assembly would be straight-forward, complicated only by the repeats within individual genomes. Real communities are never so simple. Variable abundance per genome, and variable divergence per genome are components of community diversity that complicate metagenomic assembly. Because diversity differs from community to community, there is not one assembly strategy that works for all.

The microflora of an individual human stool sample is likely to contain large and nearly clonal populations of a few dissimilar bacterial species. Even this environment is far from the model community described above. The MetaHIT consortium studied microbial sequence in 124 human stool samples [42]. Sequencing yielded 8 billion Illumina reads, mostly 75 bp each, with mate pairs separated by 135 bp, 200 bp, or 400 bp. The SOAP *de novo* assembly software, developed for single genomes, was applied with almost no special parameterization. However, the reads were assembled in bins representing individual samples. The assemblies incorporated less than half the reads and the contigs were quite small. The investigators found that most of the unassembled reads could be aligned to the contigs, indicating that differences between similar genomes limited assembly. An assembly of the previously unassembled reads generated 4% more contig sequence, indicating that some genomic sequence was present at low abundance in any one sample but conserved across samples.

The GOS study [33, 34] involved a different type of community and a different assembly strategy. DNA had been taken from ocean surface waters around the globe and the bacterial fraction extracted by filtration. Sequencing by Sanger chemistry on ABI 3730 machines produced 8 million reads, approximately 800 bp each. This data set presented high diversity at all levels of bacterial taxonomy and very low coverage of individual bacterial genomes. Some genomes seemed to be represented by just a few reads while others had been sequenced to 100-fold redundancy or more. The most abundant genome was represented by 10% of all the reads. Assembly was applied to the combined data from all sites without binning. The assembly used low-stringency alignment (minimum 86% nucleotide identity) to encourage collapse of closely related genomes. The single-genome Celera Assembler software [57] was modified to operate with very aggressive alignment parameters while ignoring conflicting data. The assembly strategy was risky but it did generate much larger sequences than would have otherwise been possible. Subsequent analysis indicated that the aggressive assemblies were representative of specific sub-populations in the metagenome and that chimeric sequences could be easily identified. This assembly strategy sought contigs that characterized whole clades where individual genome reconstruction was precluded. The multiple sequence alignments within such contigs can reveal information about abundance and diversity in the community.

Technical replicates in reads can confound estimation of abundance and can mislead assembly. Some assembly software can identify and remove technical replicates. Recent versions of the Celera Assembler software [61] do this. Other software can do this without assembly [62, 63].

9.9 Assembly: Future Directions

Perhaps because assembly software development is a protracted process, there is almost no assembly software specifically for the young field of metagenomics. Scientists have so far made due with single-genome assembly software, possibly altered with parameter adjustments. As metagenomics grows in importance, this void should be filled with multiple offerings. Existing assembly software is clearly confounded by metagenomics data that includes closely related genomes. With manual exploration, we have identified putative complete genomes that were present at high coverage in the GOS data and had sequence similarity to other genomes in the GOS data and had not assembled automatically. The assembly software probably terminated contigs at loci where the similar genomes transitioned between above and below our thresholds for alignment. The lack of contigs larger than 500 kbp indicates a barrier to assembly, but the existence of contigs longer than 100 kbp indicates that coverage was not it.

New data types might help assemblers meet the metagenomics challenge. One potential new data type involves groupings other than mate pairs. The Pacific Biosciences machine will offer a "strobe sequencing" option that generates an arbitrary number of reads, at arbitrary distance from each other, from the same DNA fragment (There are limits on the base call accuracy, the number of bases per fragment, and the precision of the distance estimates). This technique provides a generalization of the mate pair data type. If strobe sequence data were assembled in combination with long and accurate reads, the combination might help assemblers reconstruct longer genomic sequences from metagenomics reads.

One promising data type is DNA sequence derived from the amplified genome of a single cell. While an environmental sample is being subjected to metagenomic sequencing, selected cells are isolated, amplified, and sequenced individually. Cell sorting can assist in the selection of a range of cell types and therefore organisms. The amplification generates micrograms of DNA ready for sequencing from the picograms of DNA in the cell. The amplification is typically uneven (amplifying portions of the genome more than others) and inaccurate (amplifying "chimera" that concatenate different sequences). Some cells resist lysis and amplification. Nevertheless, the resulting sequence data provides conclusive evidence of the co-occurrence of certain sequences within one cell. Each set of single-cell data can be assembled separately. Each assembly can then be used as a substrate for mapping the metagenomics reads. Reads that map at high stringency, and their mates, can be assumed to derive from a closely related genome. This sort of read annotation could assist automatic assembly of the full metagenomics shotgun data set.

Another promising data type is read alignment to reference genomes. In a study of *Prochlorococcus* using the GOS data [64], a reference genome helped identify the related reads which were used to estimate diversity and to refine the assembly. Of course, using references to assemble metagenomics data risks circular reasoning. Careful application of these techniques and others should lead to improved metagenomic assemblies in the future.

One of the ideal outcomes of a metagenomic study would be to generate complete genomes for all the organisms in the community. Genes would thus have biological context and be organized in organismal units (cells). In turn, this would facilitate the annotation of genes and pathways, the determination of the role of the organism, and ultimately lead to an understanding of how the community behaves. As we have discussed above, this is unlikely to occur in the near future given the technical issues associated with metagenomic assembly and the challenges associated with single cell genomics. Interestingly, individual genomes possess intrinsic signals that help us bin unlinked scaffolds or contigs into genomic units. One of these signals is the frequency with which specific oligonucleotides (length 3 – 6 bp) are found within any given sequence. Principal component analysis of the oligonucleotide frequencies can then reveal clusters of sequences that correspond roughly to individual organisms (Fig. 9.1). This approach was firstly used in a metagenomic context to study the microbial community associated with the gutless worm *Olavius algarvensis* [65]. Binning by oligonucleotide frequency requires fairly long sequences (3 kb or more) to provide sufficient signal but given a reasonable assembly of a low to moderately complex community it can be very successful (Fig. 9.1). This functionality allowing the direct analysis of assembled metagenomes has been combined into an online tool known as the Multi-Dimensional Scatter Plot Viewer which provides capabilities for analyzing genomic sequences and annotating the resulting plots (http://gos.jcvi.org/openAccess/scatterPlotViewer2.html). The input is a metagenomic dataset (in fasta format) and the output is a web page showing the input sequences plotted as points in 3 dimensions. The approach can be improved by sub-sampling larger sequences using shorter overlapping subsequences. In these plots, individual sub-sequences are organized into clouds whose spread shows the variability in oligonucleotide frequencies across the larger sequence. Variation within a genome highlights unusual genomic segments such as those recently acquired by horizontal gene transfer. This approach is valuable for studying simpler metagenomes and in principle could be used during the assembly process to produce longer and more accurate initial assemblies.

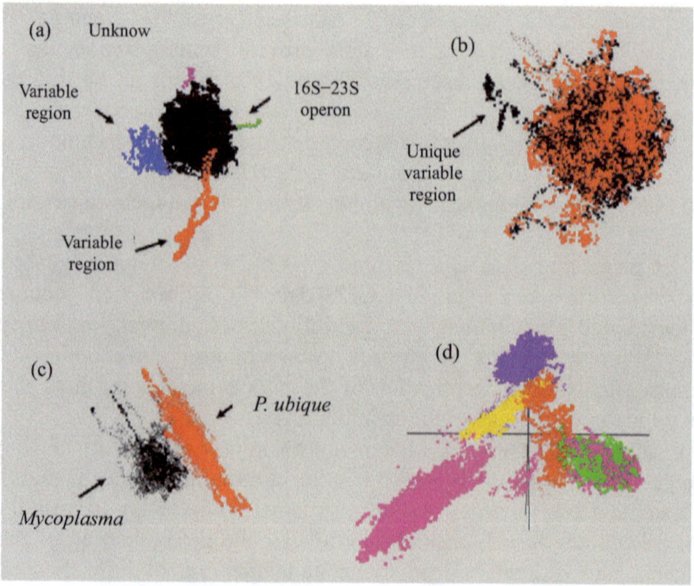

Fig. 9.1. Examples from the Multi-Dimensional Scatter Plot Viewer of three dimensional oligonucleotide frequency plots used to analyze individual genomes and compare genomes and metagenomes. In each instance, the input genomic sequence has been chopped into 5 kb overlapping subsequences (overlap of 4.9 kb) prior to analysis. (a) Analysis of *Prochlorococcus marinus* AS9601 genome. Tendrils representing portions of the genome with unusual oligonucleotide frequencies extend from the central core of the genome. These tendrils correspond to the 16S – 23S operon, hyper-variable segments associated with viral mediated horizontal gene transfer and regions that are to date unexplained. (b) Comparative analysis of two closely related *Prochlorococcus marinus* genomes (one colored red, one colored black). Note that in most cases, red and black points are closely paired except for a single tendril (as indicated) that is exclusively black. (c) Comparison of two distantly related genomes (*Peligabacter ubique* in red and *Mycoplasma genitalium* in black) indicates how well oligonucleotide frequencies can distinguish two different organisms. (d) Five metagenomic assemblies each from a different hot spring in Yellowstone National Park. Each point is colored to represent the sample from which it is derived.

9.10 Fragment Recruitment

Genomic and metagenomic studies have exposed the amazing diversity within microbial species. Complete reference genomes are powerful tools for exploring the metabolic and functional capabilities of an organism. As multiple strains of bacteria from the same species have been sequenced it has become clear how different even very closely related organisms can be. Because metagenomic studies do not require cultivation they can reveal the true extent of the diversity in a community. The technique of fragment recruitment is a powerful approach for both measuring the abundance and the diversity of organisms in a community relative to known reference sequences [34].

The fragment recruitment approach is technically simple, sequencing reads are aligned at low stringency to complete and draft genomes using your favorite sequence alignment tool (typically BLASTN). A simple filter removes poorly aligned reads and the remaining alignments are plotted showing the position (on the genome) and the identity of the alignment. These simple percent identity plots can then be annotated with additional information further enhancing the utility of the plots. This additional information is typically environmental metadata showing the sample and/or environment characteristics associated with a specific read. Alternatively, if mate-paired reads were recruited, the additional information can be in the form of metadata describing the relative placement of the paired reads highlighting structural differences between the genomes in the environment and the reference genome. For example, structural information can indicate whether the reads are closer or further away than expected indicating a possible deletion or insertion relative to the reference respectively. Though the fragment recruitment concept is straight-forward, these plots contain a depth of information that belies their simple origins.

Recruitment plots can be both an attractive and elegant means of organizing metagenomic information. Generally, recruitment provides useful information only when the metagenome contains reads from the same species or genus as one or more of the references genomes. Given a suitable reference, recruited reads will be more or less evenly distributed along the length of the reference and the density (number of reads per kbp) will be indicative of the abundance of that genus or species in the environment. In general, identity indicates how closely related and therefore how representative a genome is of the environmental organisms. Typically, there are gaps in the recruitment where tens or hundreds of kilobases of the reference will have few if any recruited reads. These correspond to hyper-variable segments that appear to be strain specific and contain largely uncharacterized proteins [35]. In some cases, there will be segments of the genome that recruit reads from only a subset of samples highlighting potential geographically or environmentally adaptive genes. Plots can be very complicated as the abundance of related organisms goes up and as the number of related strains in the community increases. There are artifacts due to the recruitment process. Virtually, every microbial genome will have some recruitment, not because they are present in a particular metagenomic sample, but because conserved genes and motifs will align to even very distantly related genes from other organisms.

The Advanced Recruitment Viewer available through the JCVI website provides a mechanism to view pre-computed recruitment plots for GOS, HMP, and other metagenomic studies. It provides a number of basic capabilities for assessing abundance, examining recruitment in the context of existing annotation, viewing structural metadata, and exporting recruited reads of interest.

9.11 Taxonomic Classification

Two main categories of methods are used to assign taxonomy to metagenomic reads or scaffolds that do not contain marker genes such as the 16S rDNA gene. Methods in the first category search the databases of taxonomically characterized proteins such as NCBI NR for the homologs of proteins predicted on the metagenomic sequences. Typically, this is done with BLAST. These methods then try to taxonomically label a given metagenomic protein in a way that fits best the distribution of its homologous genes between taxa. In some methods, this is done by choosing as label the lowest common ancestor taxonomic node for all homologs, possibly taking into an account additional BLAST match characteristics. In other methods, phylogenetic inference methods are used to either construct a tree from those homologs together with the target protein, or to place the metagenomic protein onto an existing tree.

The second category of methods explores a signal that is encoded in the poly-nucleotide (*k-mer*) composition of the sequences and is independent of homology. Either *k-mer* frequencies or more complicated derivatives of those are extracted from both metagehomic and database sequences. Various supervised or semi-supervised machine learning methods are applied in vector space to classify metagenomic data points based on taxonomically labeled database data points. Alternatively, probabilistic models for the sequence composition are built based on the database sequences and applied to classify the metagenomic samples.

McHardy *et al.* [66] demonstrated phylogenetic classification of 1 – 50 kbp genome fragments based on the Support Vector Machines trained on the *k-mer* frequency vectors. At the genus-level, specificity was consistently above 80% and sensitivity was so high, or better on sequences$\geqslant$10 kbp. Note, these types of predictors will gain accuracy with time as reference databases grow.

9.12 MGTAXA

In the JCVI open-source MGTAXA project (http://andreyto.github.com/mgtaxa/), we extended the approach used by PhyloPythia algorithm to use training sequences automatically collected from NCBI WGS datasets and introduced a tunable classification threshold into one-against-rest Support Vector Machine (SVM) classification scheme. That allowed an order of magnitude increase in the taxonomic coverage without loss of accuracy as compared to PhyloPythia, which restricted training to the completely sequenced genomes from NCBI RefSeq.

The major barrier for composition approaches is the length of a query sequence. Recently, project used Interpolated Context Models (ICMs) for the purposes of taxonomic binning in order to handle query sequences as short as 100 bp [67]. We integrated the ICM based methodology into our parallel high-throughput implementation, introducing the assignment directly to higher-order taxa by building the additional models from a mixture of underlying sequences.

One of the primary questions during the analysis of a viral metagenomic fraction is establishing which host(s) a particular virus infects. Due to a high diversity of viruses, it is often very difficult to make such assignment based on the sequence homology to the viral or bacterial databases. As a component of MGTAXA project, we have created the first method that is able to predict the putative hosts for bacteriophages in a metagenomic sample based on the sequence composition. The input is a set of scaffolds at least 5 kbp long assembled from a separately sequenced viral fraction. The hosts are picked from reference database sequences such as NCBI RefSeq optionally expanded with 100 kbp bacterial scaffolds from the metagenomic sample. The method explores a long-standing observation that bacteriophages tend to adopt the compositional signatures of their hosts [68]. Thus, the viral sequences are scored against ICM models built from the bacterial sequences, and the resulting scores are used to assign the probable hosts.

Within the MGTAXA project, we have also created a pipeline for the analysis of Clustered Regularly Interspaced Short Palindromic Repeats (CRISPR) subsystem [69, 70] in the genomes and metagenomes. CRISPR is primarily an immune mechanism of bacteria against viruses. Because the CRISPR "spacer" sequences are acquired from the past invading viruses, finding matches to spacers within the viral metagenomic fraction represents an alternative way to establish host range for viruses in the metagenome. The pipeline has been applied to the GOS datasets, as well as to the collection of marine genomes sequenced by the Moore Foundation project [71]. The presence of CRISPR system demonstrated a clear correlation with the known lifestyle of the sequenced microbes. The genomes where a well-defined CRISPR system was detected were typically associated with particles, surfaces or host organisms, or formed dense colonies in mats or blooms. On the contrary, most of the genomes lacking CRISPR/CAS were free-floating surface picoplankton. Densely populated monocultures can easily be destroyed by a single type of virus, so investing in a defense system makes sense. By contrast, planktonic communities are very diverse and are preyed upon by a diverse set of viruses making a CRISPR style-defense cost-ineffective.

9.13 High Performance Computing

In the scope of MGTAXA project, we have created two open-source applications designed to run on High Performance Computing (HPC) architecture. The first one addresses a problem of adapting typical bioinformatics algorithms to HPC environments dominating large federated resources such as NSF TeraGrid. We created a location-aware scheduler for the MapReduce-MPI library (http://www.sandia.gov/ ~sjplimp/mapreduce.html) and used this library to build a parallel BLAST implementation that calls the methods of unmodified NCBI C++ Toolkit [72]. We demonstrated scaling for up to 2,000 cores on TACC Ranger cluster. The second application is a Self-Organizing Map machine-learning algorithm, popular in the metagenomic binning domain. Our implementation of the "batch SOM" uses a mix of MapReduce-MPI and direct MPI calls and scales to 2,000 cores as well. This parallel implementation allows building of large ("emergent") SOMs from dataset which were out of reach for sequential implementations.

9.14 Functional Annotation

The process of assigning function to a predicted protein from a metagenome involves comparing the sequence against reference databases. In addition, similarity comparison of 16S rDNA from metagenomic samples to sequences of annotated microorganisms, is used to identify the species in the sample [73]. At JCVI, we have developed pipelines for annotating large metagenomic datasets generated using Sanger and 454 technologies, and we continue to make our pipelines more efficient using algorithmic techniques such as clustering [35]. Sequence similarity searches, forming the bulk of the computes, will continue to create a bottleneck as data volumes continue to increase at a rapid pace. For example, analysis of the GOS sequence data consumed one million CPU hours in two large compute clusters [35]. Therefore, the sequence similarity search component of the pipelines will require fundamentally different implementations so as to scale with the data volume.

9.15 Analysis of Eukaryotes in Ecology Studies

Microeukaryotes in microbial communities have been studied but to a lesser extent than the bacterial, archaeal and viral communities. This is probably because of difficulty in cultivation of many of these species, larger genome sizes and lack of data on conserved phylogenetic markers that could be used for comparative purposes. As with the analysis of viral populations described below, information on eukaryotic communities can often be recovered *via* large-scale metagenomic analysis of different environments [43]. As a follow up to the Sargasso Sea study [33], Piganeau and Moreau [74] used sequences from the genome of *Ostreococcus tauri* to identify two new strains of *Ostreococcus*. They were also able to show that the *Ostreococcus* nuclear sequence data derived from the Sargasso metagenome was divided onto 731 scaffolds, which equated to approximately 23% of the complete nuclear genome. In a follow up study [75], they surveyed the picoeukaryotic diversity of the Sargasso dataset by searching for homologs of eight nuclear anchor genes that are conserved throughout the eukaryotic lineage, as well as one chloroplastic and one mitochondrial gene [75]. Sequence similarity as inferred from BLAST twice was the method of choice in this analysis. The gene "anchors", included 18S rRNA, 28S rRNA, and the genes encoding elongation factor 1a (EF1a), elongation factor 2 (EF2), the large subunit of RNA polymerase II (RPB1), actin, α-tubulin and β-tubulin. They described 41 broadly spread distinct eukaryotic scaffolds from this follow up analysis of the Sargasso Sea dataset.

9.16 Challenges Presented by Data Volume (Computational and Storage Requirements, Cloud Computing Solutions)

Most sequencing technologies available on the market today generate massive volumes of data, with large part of the output composed by "raw" images recording the light signals emitted by the chemical reactions during sequencing. The Illumina GAIIx system generates approximately 100,000 image files per run with total size 2.8 TeraByte [1 TB=1,000 GB, (GigaByte)]. ABI's SOLiD has a similar data yield, while other instruments can range from 0.5 to 600 GB depending also whether paired-end fragments are sequenced (for a detailed list see http://www.politigenomics.com/next-generation-sequencing-informatics). After base calling the raw images are reduced to files containing only the sequence read, but these still can range from 300 GB up to 1.2 TB in size.

Even if the image data are disregarded, 10 – 20 sequencing runs can still add up to tens of TB in size and overwhelm available disk systems in laboratories or institutes that lack significant informatics infrastructure.

The Human Microbiome Project (HMP) consisting of four sequencing centers is an example of microbial ecology project generating large scale metagenomic datasets: the Baylor College of Medicine Human Genome Sequencing Center, the Broad Institute, the JCVI, and the Genome Center at Washington University. Under this project, 20,864 million Illumina sequence reads were generated, while comparison of the reads to protein databases alone is estimated to exceed 12 TB in data size. For projects of such scale, data management plans are required which take into account the type of bioinformatic analysis that will be performed at each phase of the experiment, define whether and for how long images, quantified intensities, base pair calls, and sequence assemblies need to be kept. Significant planning is also required for appropriate cost allocation for storage, taking in to account issues such as the proportion of expenditure for high-cost disk systems for running the computational analysis, versus commodity, low-cost data archiving storage.

Investment to sequence technology must be also accompanied by an almost equal or greater investment in high-end computing servers, since running downstream bioinformatic analysis with large-scale sequence datasets requires significant computational capacity. Sequencing instruments are typically bundled with only a minimal compute resource for supporting the data capture during runs, such as the IPAR (Integrated Primary Analysis and Reporting) system by Illumina Solexa which consists of a preconfigured 4-core server with 3 TB of usable storage.

For studying microbial populations through metagenomic sequencing, a common type of analysis performed is alarge-scale homology comparison of the sequence reads from against annotated genomes, in order to identify the species in the sample.

Publicly available online software tools such as those from NCBI [76] are not option this type of analysis, since they offer limited computational capacity (For example, NCBI-BLAST cannot accept input data files of 0.5 GB size for sequence homology search). During the data analysis phase of the *Sorcerer II* Global Ocean Sampling Expedition [35], 588,298 CPU hours were used on a computing cluster with 128 nodes for performing all-against-all BLAST comparison of the sequence reads. Beside high-end compute servers, an informatics infrastructure for working with sequencing data also requires hiring trained bioinformaticians competent to install, configure and uses specific software to analyze the data.

An alternative option to investing in informatics infrastructure, is obtaining computational capacity from a cloud service provider [77]. Cloud services provide researchers with the ability to perform data analysis on a practically unlimited pool of Virtual Machines (VMs), without owning or maintaining any computer hardware. The charge model used by cloud service providers is similar to utilities such as electricity, and customers are billed based on amounts of compute resources consumed. This can work better for smaller research laboratories instead of investing to computer hardware and data center infrastructure, for which the cost cannot be justified for only a handful of experiments. Using as an example the Amazon Elastic Compute Cloud service (EC2, http://aws.amazon.com/ec2), prices range from 0.085$ US per hour per small server VMs (1.7 GB memory, 1 CPU core, 160 GB storage), up to 2$ US per hour for the large server VMs (68.4 GB memory, 26 CPU cores, 1,690 GB storage). In our experiments with small microbial genome assemblies on the cloud, we used intermediate types of VMs on Amazon EC2 (15 GB memory, 4 CPU cores), and were able to successfully complete the assemblies in 3 h at a cost of $0.68/h.

Both public and commercial offerings of sequence analysis suites are currently available on the Amazon EC2 cloud. On the commercial side, DNAnexus (http://www.dnanexus.com) provides support for different sequencing platforms including Illumina and SOLiD, and available tools include sequence statistics and quality metrics, sequence homology comparison and mapping assembly, in addition to ChiPseq, RNAseq and 3'-end sequencing for expression quantification (3SEQ). A recent commercial offering that is targeted specifically to large-scale sequence homology comparison, was made available by SeqCentral (https://www.seqcentral.com/). Despite being a new offering, SeqCentral is a promising option since it provides large-scale computational capacity for comparison of sequences using BLAST against public databases of annotated genomes, which is one of the most common types of analysis performed with metagenomic data from microbial populations.

The Galaxy bioinformatics workbench (http://main.g2.bx.psu.edu/) is a public offering that also runs on the Amazon EC2 cloud, and includes a similar range of tools with DNAnexus, in addition to a set of phylogenetic tools for metagenomic analysis. Galaxy is a self-contained platform which can run on a researcher's own servers, and expanded with additional software packages through simple configuration scripts. Another public offering for next-generation sequencing computing on the cloud is available through our own work on JCVI Cloud Biolinux (http://www.jcvi.org/cms/research/projects/jcvi-cloud-biolinux/overview/), which is a virtual high performance server publicly available through Amazon EC2. Our solution bundles a large set of sequence analysis tools including BLAST, Glimmer, HMMER, Phylip, RasMol, Genespring, ClustalW, and the EMBOSS analysis suite. Users can start the virtual servers with a few clicks on the EC2 console (http://aws.amazon.com/console), and access the interface through a remote connection from their desktop computer. Furthermore, the JCVI Cloud Biolinux virtual servers are open-source, can be downloaded and modified, while advanced users have the option to run the tools on a private installation of the open-source Eucalyptus cloud platform (http://www.eucalyptus.com/).

An important concern for exchanging metagenomic data in microbial ecology projects among many participating centers, submitting data to public databases, or when using cloud-based bioinformatic tools, is the data transfer bottleneck from the local to remote storage. According to results published by the Amazon cloud service (http://aws.amazon.com/importexport), 600GB of data would require approximately one week to be transferred over the network when using an average broadband connection (10Mbp), while a faster connection (T3, 40Mbp) would require approximately 2 d. A commercial software solution that increases data transfer over the network using intelligent routing compared to traditional File Transfer Protocol (FTP) has been developed by Aspera (http://www.asperasoft.com). This technology has been integrated with the NCBI servers, and researchers can download a free software client which allows for increased upload and download speeds from/to the Short Read Archive. Another option offered by the Amazon cloud service provider is to physically ship disk drives to the company's offices, and have the data copied directly to the cloud servers. The cost for disk drives up to 4TB of data is $80 US, which makes it the most cost-efficient method when taking into account the expense required for obtaining a high-bandwidth Internet connection.

9.17 Future Directions

Microbial Ecology will continue to grow as a science and will continue to reward us with a diversity of microbial species. Genomics is one tool, but culture based, FISH and other types of ecology studies continue to give us insight on the diversity of these species. Metabolomic and proteomic studies allow us to see which gene predictions are real under different environmental conditions.

We can also expect that new tools will be developed for ecological studies. These includes sequencing and informatic tools. New genomes, metagenomes and uncultured species will be sequenced. We will continuously need to update the information that we are getting from studies, and large databases should be created to collate information from different labs such that the community as a whole will benefit. These databases should allow us to access sequencing, informatics, microarray, proteomics transcriptomic, and metabolomic data on single species, similar environments or on a single environment. They should also allow for an integration of analytic tools being used to understand the ecology of a system.

The importance of microbial diversity, the challenges in understanding this incredibly complex topic, and the rapidity of technological advance are pushing researchers to develop new and ever more creative approaches to analyzing microbial communities. The high throughput random shotgun sequencing of a microbial community provides a new and powerful perspective on the underlying diversity. It does not suffer from the limitations of indirect approaches that measure the environment or the communities using marker genes and from which the biological capabilities can only be inferred. As new informatic and molecular techniques are developed, *e.g.* single cell sequencing [65, 78], as sequencing technologies continue to improve and as the number of environments explored with metagenomic techniques continues to increase, the potential to develop a theory based understanding of diversity and ecology seems close at hand.

References

[1] Metzker M L. Sequencing technologies - the next generation. Nature Reviews Genetics, 2009, 11: 31-46.

[2] Mylvaganam S, Dennis P P. Sequence heterogeneity between the two genes encoding 16S rRNA from the halophilic archaebacterium Haloarcula marismortui. Genetics, 1992, 130: 399-410.

[3] López-López A, Benlloch S, Bonfá M, *et al.* Intragenomic 16S rDNA divergence in Haloarcula marismortui is an adaptation to different temperatures. Journal of molecular evolution, 2007, 65: 687-696.

[4] Pei A Y, Oberdorf W E, Nossa C W, *et al.* Diversity of 16S rRNA genes within individual prokaryotic genomes. Applied and environmental microbiology, 2010,76: 3886-3897.

[5] Ray A E, Connon S A, Sheridan P P, *et al.* Intragenomic heterogeneity of the 16S rRNA gene in strain UFO1 caused by a 100 - bp insertion in helix 6. FEMS microbiology ecology, 2010, 72: 343-353.

[6] Unno T, Jang J, Han D, *et al.* Use of barcoded pyrosequencing and shared OTUs to determine sources of fecal bacteria in watersheds. Environmental science & technology, 2010, 44: 7777-7782.

[7] Thompson F L, Bruce T, Gonzalez A, *et al.* Coastal bacterioplankton community diversity along a latitudinal gradient in Latin America by means of V6 tag pyrosequencing. Arch Microbiol, 2011, 193: 105-114.

[8] Whittaker R H. Evolution and measurement of species diversity. Taxon, 1972, 21: 213-251.

[9] DeSantis T Z, Hugenholtz P, Larsen N, *et al.* Greengenes, a chimera-checked 16S rRNA gene database and workbench compatible with ARB. Applied and

environmental microbiology, 2006, 72: 5069-5072.

[10] Pruesse E, Quast C, Knittel K, *et al.* SILVA: A comprehensive online resource for quality checked and aligned ribosomal RNA sequence data compatible with ARB. Nucleic acids research, 2007, 35: 7188-7196.

[11] Cole J, Wang Q, Cardenas E, *et al.* The ribosomal database project: Improved alignments and new tools for rRNA analysis. Nucleic acids research, 2009, 37: D141-D145.

[12] Wu D, Hartman A, Ward N, *et al.* An automated phylogenetic tree-based small subunit rRNA taxonomy and alignment pipeline (STAP). PloS one, 2008, 3: e2566.

[13] Bond P L, Hugenholtz P, Keller J, *et al.* Bacterial community structures of phosphate-removing and non-phosphate-removing activated sludges from sequencing batch reactors. Applied and Environmental Microbiology, 1995, 61: 1910-1916.

[14] McCaig A E, Glover L A, Prosser J I. Molecular analysis of bacterial community structure and diversity in unimproved and improved upland grass pastures. Applied and Environmental Microbiology, 1999, 65: 1721-1730.

[15] Schloss P D, Handelsman J. Introducing DOTUR, a computer program for defining operational taxonomic units and estimating species richness. Applied and environmental microbiology, 2005, 71: 1501-1506.

[16] Shuldiner A R, Nirula A, Roth J. Hybrid DNA artifact from PCR of closely related target sequences. Nucleic acids research, 1989, 17: 4409.

[17] Hugenholtz P, Huber T. Chimeric 16S rDNA sequences of diverse origin are accumulating in the public databases. International Journal of Systematic and Evolutionary Microbiology, 2003, 53: 289-293.

[18] Ashelford K E, Chuzhanova N A, Fry J C, *et al.* At least 1 in 20 16S rRNA sequence records currently held in public repositories is estimated to contain substantial anomalies. Applied and Environmental Microbiology, 2005, 71: 7724-7736.

[19] Komatsoulis G A, Waterman M S. A new computational method for detection of chimeric 16S rRNA artifacts generated by PCR amplification from mixed bacterial populations. Applied and Environmental Microbiology, 1997, 63: 2338-2346.

[20] Huber T, Faulkner G, Hugenholtz P. Bellerophon: A program to detect chimeric sequences in multiple sequence alignments. Bioinformatics, 2004, 20: 2317-2319.

[21] Schloss P D, Westcott S L, Ryabin T, *et al.* Introducing mothur: Open-source, platform-independent, community-supported software for describing and comparing microbial communities. Applied and environmental microbiology, 2009, 75: 7537-7541.

[22] Sogin M L, Morrison H G, Huber J A, *et al.* Microbial diversity in the deep sea and the underexplored “rare biosphere”. Proceedings of the National Academy of Sciences, 2006, 103: 12115-12120.

[23] Hamady M, Knight R. Microbial community profiling for human microbiome projects: Tools, techniques, and challenges. Genome research, 2009, 19: 1141-1152.

[24] Turnbaugh P J, Hamady M, Yatsunenko T, *et al.* A core gut microbiome in obese and lean twins. Nature, 2008, 457: 480-484.

[25] Caporaso J G, Lauber C L, Walters W A, *et al.* Global patterns of 16S rRNA diversity at a depth of millions of sequences per sample. Proceedings of the National Academy of Sciences, 2011, 108: 4516-4522.

[26] Liu Z, Lozupone C, Hamady M, *et al.* Short pyrosequencing reads suffice for accurate microbial community analysis. Nucleic acids research, 2007, 35: e120.

[27] Reeder J, Knight R. The "rare biosphere": A reality check. Nature Methods, 2009, 6: 636-637.

[28] Kunin V, Engelbrektson A, Ochman H, *et al.* Wrinkles in the rare biosphere: Pyrosequencing errors can lead to artificial inflation of diversity estimates. Environmental microbiology, 2010, 12: 118-123.

[29] Quince C, Lanzén A, Curtis T P, *et al.* Accurate determination of microbial diversity from 454 pyrosequencing data. Nature methods, 2009, 6: 639-641.

[30] Huse S M, Welch D M, Morrison H G, *et al.* Ironing out the wrinkles in the rare biosphere through improved OTU clustering. Environmental Microbiology, 2010, 12: 1889-1898.

[31] Magurran A E. Ecological diversity and its measurement. Princeton: Princeton university press, 1988.

[32] Caporaso J G, Kuczynski J, Stombaugh J, *et al.* QIIME allows analysis of high-throughput community sequencing data. Nature methods, 2010, 7: 335-336.

[33] Venter J C, Remington K, Heidelberg J F, *et al.* Environmental genome shotgun sequencing of the Sargasso Sea. Science, 2004, 304: 66-74.

[34] Rusch D B, Halpern A L, Sutton G, *et al.* The Sorcerer II global ocean sampling expedition: Northwest Atlantic through eastern tropical Pacific. PLoS biology, 2007, 5: e77.

[35] Yooseph S, Sutton G, Rusch D B, *et al.* The Sorcerer II Global Ocean Sampling expedition: Expanding the universe of protein families. PLoS biology, 2007, 5: e16.

[36] Sharon I, Alperovitch A, Rohwer F, *et al.* Photosystem I gene cassettes are present in marine virus genomes. Nature, 2009, 461: 258-262.

[37] Comeau A M, Arbiol C, Krisch H. Gene network visualization and quantitative synteny analysis of more than 300 marine T_4-like phage scaffolds from the GOS metagenome. Molecular biology and evolution, 2010, 27: 1935-1944.

[38] Sorokin V A, Gelfand M S, Artamonova II. Evolutionary dynamics of clustered irregularly interspaced short palindromic repeat systems in the ocean metagenome. Applied and environmental microbiology, 2010, 76: 2136-2144.

[39] Peterson J, Garges S, Giovanni M, *et al.* The NIH human microbiome project. Genome research, 2009, 19: 2317-2323.

[40] Yeoman C J, Yildirim S, Thomas S M, *et al.* Comparative genomics of Gardnerella vaginalis strains reveals substantial differences in metabolic and virulence potential. PLoS One, 2010, 5: e12411.

[41] Nelson K E, Weinstock G M, Highlander S K, *et al.* A catalog of reference

genomes from the human microbiome. Science (New York, NY), 2010, 328: 994.

[42] Qin J, Li R, Raes J, *et al.* A human gut microbial gene catalogue established by metagenomic sequencing. Nature, 2010, 464: 59-65.

[43] Brulc J M, Antonopoulos D A, Miller MEB, *et al.* Gene-centric metagenomics of the fiber-adherent bovine rumen microbiome reveals forage specific glycoside hydrolases. Proceedings of the National Academy of Sciences, 2009, 106: 1948-1953.

[44] Swanson K S, Dowd S E, Suchodolski J S, *et al.* Phylogenetic and gene-centric metagenomics of the canine intestinal microbiome reveals similarities with humans and mice. The ISME Journal, 2010, 5: 639-649.

[45] Qu A, Brulc J M, Wilson M K, *et al.* Comparative metagenomics reveals host specific metavirulomes and horizontal gene transfer elements in the chicken cecum microbiome. PLoS One, 2008, 3: e2945.

[46] Yildirim S, Yeoman C J, Sipos M, *et al.* Characterization of the fecal microbiome from non-human wild primates reveals species specific microbial communities. PLoS One, 2010, 5: e13963.

[47] Allen H K, Cloud-Hansen K A, Wolinski J M, *et al.* Resident microbiota of the gypsy moth midgut harbors antibiotic resistance determinants. DNA and cell biology, 2009, 28: 109-117.

[48] Suen G, Scott J J, Aylward F O, *et al.* An insect herbivore microbiome with high plant biomass-degrading capacity. PLoS genetics, 2010, 6: e1001129.

[49] Bishop-Lilly K A, Turell M J, Willner K M, *et al.* Arbovirus detection in insect vectors by rapid, high-throughput pyrosequencing. PLoS neglected tropical diseases, 2010, 4: e878.

[50] Bench S R, Hanson T E, Williamson K E, *et al.* Metagenomic characterization of chesapeake bay virioplankton. Applied and Environ-mental Microbiology, 2007, 73: 7629-7641.

[51] Day J M, Ballard L L, Duke M V, *et al.* Metagenomic analysis of the turkey gut RNA virus community. Virol J, 2010, 7: 313.

[52] Reyes A, Haynes M, Hanson N, *et al.* Viruses in the faecal microbiota of monozygotic twins and their mothers. Nature, 2010, 466: 334-338.

[53] Sanger F, Coulson A R, Barrell B G, *et al.* Cloning in single-stranded bacteriophage as an aid to rapid DNA sequencing. J Mol Biol, 1980, 143: 161-178.

[54] Fleischmann R D, Adams M D, White O, *et al.* Whole-genome random sequencing and assembly of Haemophilus influenzae Rd. Science, 1995, 269: 496-512.

[55] Sutton G G, White O, Adams M D, *et al.* TIGR Assembler: A new tool for assembling large shotgun sequencing projects. Genome Science and Technology, 1995, 1: 9-19.

[56] Adams M D, Celniker S E, Holt R A, *et al.* The genome sequence of Drosophila melanogaster. Science, 2000, 287: 2185-2195.

[57] Myers E W, Sutton G G, Delcher A L, *et al.* A whole-genome assembly of Drosophila. Science, 2000, 287: 2196-2204.

[58] Istrail S, Sutton G G, Florea L, *et al.* Whole-genome shotgun assembly and

comparison of human genome assemblies. Proceedings of the National Academy of Sciences of the United States of America, 2004, 101: 1916-1921.
[59] Pop M. Genome assembly reborn: Recent computational challenges. Briefings in bioinformatics, 2009, 10: 354-366.
[60] Miller J R, Koren S, Sutton G. Assembly algorithms for next-generation sequencing data. Genomics, 2010, 95: 315.
[61] Miller J R, Delcher A L, Koren S, *et al.* Aggressive assembly of pyrosequencing reads with mates. Bioinformatics, 2008, 24: 2818-2824.
[62] Niu B, Fu L, Sun S, *et al.* Artificial and natural duplicates in pyrosequencing reads of metagenomic data. BMC bioinformatics, 2010, 11: 187.
[63] Teal T K, Schmidt T M. Identifying and removing artificial replicates from 454 pyrosequencing data. Cold Spring Harbor Protocols, 2010, 2010: prot5409.
[64] Rusch D B, Martiny A C, Dupont C L, *et al.* Characterization of Prochlorococcus clades from iron-depleted oceanic regions. Proc Natl Acad Sci USA, 2010, 107: 16184-16189.
[65] Woyke T, Tighe D, Mavromatis K, *et al.* One bacterial cell, one complete genome. PLoS One, 2010, 5: e10314.
[66] McHardy A C, Martin H G, Tsirigos A, *et al.* Accurate phylogenetic classification of variable-length DNA fragments. Nat Methods, 2007, 4: 63-72.
[67] Brady A, Salzberg S L. Phymm and PhymmBL: Metagenomic phylogenetic classification with interpolated Markov models. Nat Methods, 2009, 6: 673-676.
[68] Lucks J B, Nelson D R, Kudla G R, *et al.* Genome landscapes and bacteriophage codon usage. PLoS Comput Biol, 2008, 4: e1000001.
[69] Haft D H, Selengut J, Mongodin E F, *et al.* A guild of 45 CRISPR-associated (Cas) protein families and multiple CRISPR/Cas subtypes exist in prokaryotic genomes. PLoS computational biology, 2005, 1: e60.
[70] Barrangou R, Fremaux C, Deveau H, *et al.* CRISPR provides acquired resistance against viruses in prokaryotes. Science, 2007, 315: 1709-1712.
[71] Yooseph S, Nealson K H, Rusch D B, *et al.* Genomic and functional adaptation in surface ocean planktonic prokaryotes. Nature, 2010, 468: 60-66.
[72] Camacho C, Coulouris G, Avagyan V, *et al.* BLAST+: Architecture and applications. BMC Bioinformatics, 2009, 10: 421.
[73] Wooley J C, Godzik A, Friedberg I. A primer on metagenomics. PLoS computational biology, 2010, 6: e1000667.
[74] Piganeau G, Moreau H. Screening the Sargasso Sea metagenome for data to investigate genome evolution in Ostreococcus (Prasinophyceae, Chlorophyta). Gene, 2007, 406: 184-190.
[75] Piganeau G, Desdevises Y, Derelle E, *et al.* Picoeukaryotic sequences in the Sargasso sea metagenome. Genome Biol, 2008, 9: R5.
[76] Johnson M, Zaretskaya I, Raytselis Y, *et al.* NCBI BLAST: A better web interface. Nucleic Acids Res, 2008, 36: W5-W9.
[77] Sansom C. Up in a cloud? Nat Biotechnol, 2010, 28: 13-15.
[78] Lasken R. Genomic DNA amplification by the multiple displacement amplification (MDA) method. Biochemical Society Transactions, 2009, 37: 450.

10

Ecology of Oral Infectious Diseases

Jing Xue, Xiaorong Xiao *
State Key Laboratory of Oral Diseases, West China Hospital of Stomatology, Sichuan University, Chengdu, 610041, China
* E-mail: 1135364841@qq.com

Oral infectious diseases are general terms for infectious diseases which occur in the oral cavity and maxillofacial region, which mainly include dental caries, dental pulp and periapical diseases, periodontal diseases, maxillofacial odontogenic and non-odontogenic infections, oral mucosal infections, and peri-implant inflammation.

For a long time, the understanding of human oral infectious diseases was mostly restricted to maxillofacial infections, such as lip carbuncle, cellulitis, traumatic infection, *etc.*, while the infectious diseases in the oral cavity were not fully understood or given sufficient attention, and even often neglected. One important reason is that oral infectious diseases are mostly caused by the resident microflora, or normal microflora of the oral cavity. Due to the slow progression of the disease, early clinical symptoms are not easily detectable, clinical symptoms are less evident than other clinical infectious diseases even in the acute phase. Along with social progress and development, particularly in developed countries, a high priority has been placed on oral health.

The World Health Organization (WHO) lists dental caries, periodontal disease and cardiovascular disease, along with cancer, as diseases which threaten human health, and "teeth with no holes, no pain and no bleeding gums", is one of the 10 criteria for healthy people.

As the old saying goes, "disease comes from the mouth". The mouth is not only a gateway to the outside of the body; it is also one of the body's major organs. The initial processes of human digestion are: food intake, chewing, and swallowing which all begin in the mouth; while external harmful substances and pathogenic microorganisms can also enter the body through the mouth, these foreign invaders or substances may be partially or fully degraded, eradicated, or inhaled into the body.

The mouth is not only an important entry to the digestive system of the body, but also provides an important safeguard to the health and defense of the body. At the same time, a number of local organs infectious diseases or systemic diseases may also correlate with oral flora or oral infectious diseases, or manifest some clinical oral symptoms, *e.g. Viridans streptococci* from bacterial endocarditis and leukemia patients with bacteremia, *Helicobacter pylori* from chronic gastritis, those pathogenic bacteria were found in the mouth. The author had isolated the same serotype C-type *Streptococcus mutans* of dental caries from blood culture samples of a leukemia patient. HIV infection of certain oral infectious diseases such as *Candida* infections, necrotizing ulcerative gingivitis and angular cheilitis are considered as important early manifestations of early diagnosis of AIDS. In traditional Chinese medicine, clinical observation of the tongue is used as a routine examination method which helps to diagnose other organ diseases. Thus, oral health is closely related to the human body health, paying more attention to oral health and understanding the etiology and ecological control methods of oral infectious diseases are very important in maintaining body health [1].

10.1 Ecological Basis

Ecology is the study of the interactions between organisms and their environment. Microecology is the study of the interactions between the normal microflora and their host. Human body ecology is the study of the interactions between organisms and the human body. Oral microecology is the study of the interactions between oral microorganisms and their biotic environment and is an important ingredient in human ecology [2].

There are a variety of favorable conditions for oral microorganisms to adapt to adhesion, growth and reproduction, so great quantities and types of complex microorganisms inhabit all parts of the mouth: teeth, gingival sulcus, mucosa, saliva, dental prosthesis and orthodontic appliances. Previous studies found that ecological relationships between oral microorganisms and between oral microorganisms and humans are exemplified in the oral cavity. With the development and progress of anaerobic bacteriology, molecular biology, gnotobiology and microecology in the 1980s, as an important branch of medicine, the establishment and development of oral microbiology and oral microecology have attracted wide attention in the relevant medical profession and medical microbiology field. Researchers not only discovered and named a large number of new species of oral microflora, but also had a more in-depth knowledge and understanding of the cause of oral infectious diseases.

Oral resident microorganisms were found to be the pathogens of the majority of oral infectious diseases, and the characteristics of mixed bacterial infection of the mouth indicated that the majority of oral infectious diseases came from the destruction of the ecological balance between each of the oral microorganisms and between oral microorganisms and the host in the oral cavity, namely the oral ecological imbalance. This is a brand new etiology theory of oral infectious diseases, and it is a new field based on ecology theory to study the prevention and control of oral infectious diseases. Normal microflora of the oral cavity, oral ecology, oral ecosystem, dental plaque biofilm, ecological plaque, oral beneficial bacteria, balance and imbalance of the oral ecosystem, and oral infectious diseases ecology, and other new terms and new concepts are being understood and identified by more and more people.

The study of the relationships of oral microorganisms with one another and with their oral environment is called "oral ecology". Oral ecology is an important component of human ecology. It is also a dental basic science closely related to oral anatomy, oral microbiology, oral biochemistry, oral immunology, oral molecular biology and oral clinical diseases. Oral ecology is based on ecological theory, discussing the relationship between balance and imbalance of the oral ecosystem and oral health and disease, in order to establish an effective way to maintain oral health and to prevent and control oral diseases.

The ecosystem is a complete system including the bio-complex with all the environmental factors. The oral ecosystem is formed in the long-term evolution process between oral resident microflora and oral ecological areas of the host (including tissue cells and various ecological factors). It is a unified biological system which can exchange material, energy and information independently, including three major components: oral biotic areas, oral microorganisms and oral ecological factors.

10.1.1 Oral Biotic Area

The biotic area is the survival environmental area of organisms; it is the space level and basic component unit of the ecosystem. In ecological research, in order to better understand the relationship between organisms and their living environment, environmental areas are artificially divided into different biotic areas, habitat, biotopes and niches [3].

Composition and characteristics. The oral biotic area includes a number of biotic areas, as well as many habitats, biotopes and niches. The biotic area, habitat, biotope and niche have their own similar or dissimilar characteristics, and thus constitute the complex mouth biotic area. Lip, tongue, cheek, palate, teeth, gingival, and alveolar bones are considered small ecological zones of the oral biotic area. In fact, these small ecological areas include different habitats and ecological sites. Lips can be divided into vermilion and outer lip; the tongue can be divided into the dorsal of the tongue and ventral surface of the tongue, apex of the tongue, and root of the tongue. Teeth are hard tissue and they can be divided into anterior teeth and posterior teeth by location. They can be divided into incisors, canines, molars, upper teeth and molars, the mandibular incisors and molars by anatomical classification; for the same tooth, it also includes the crown, root, pulp, enamel, dentin, cementum. These ecological zones or habitats, niches, sites are important components of the oral ecosystem, and each of them has its own characteristics and provides a variety of colonization conditions for microbial colonization.

Teeth biotic areas are important components of the mouth. There are two sets of teeth during a person's life, a set of deciduous tooth and a set of permanent teeth. Deciduous teeth start to develop from the first 2 months of the embryos and erupt at about 6 mon after birth. Eruption will finish at 2.5 years old with 20 total deciduous teeth. Permanent teeth start to develop from the first 5 months of the embryos and erupt at 6 years of age. Eruption will finish at 20 years of age, with the total permanent teeth numbering 28 – 32.

Each tooth is formed by the crown and root. The crown is exposed to the mouth; the root is embedded in the alveolar bone. Several different surfaces of the crown surface form the ecological communities with different environments. Crown smooth surfaces of the tooth include the labial surface, buccal surface, lingual surface and palatal surface, *etc.* The labial or buccal surface is close to the lip and buccal mucosa, lingual surface and palatal surface are close to the tongue and upper palate. The smooth surface of the crown is a self-cleaning area; it is vulnerable to the cleaning effect from saliva, food friction and oral hygiene measures (mouthwash, tooth brushing, *etc.*). The microbes that can colonize the smooth surface must possess a special anti-out force. There are various shapes of pits, fissures, grooves and cracks on the occlusal surface of the tooth, making the tooth surface present an uneven like shape. There are a lot of prominence and depression areas. Prominence areas include, for example, ridge, cingulum; depression areas include, for example, pit, fissure, groove and crack. These areas of the crown are oral non-self-cleaning areas and self-cleaning by a variety of physical or mechanical actions are not easy to achieve. The special ecological conditions create a good environment for bacterial colonization, making it an aggregation site of oral bacteria.

Both of the gingival and alveolar bone belong to supporting tissue of the tooth – periodontal tissue, whose characteristics will be described in detail in section III of this chapter concerning the related contents of periodontal disease.

Due to the special anatomical structure and position of the oral biotic area, except for the general specificity of ecological zones, such as relatively stable temperature (36.5 – 37.0 ℃), humidity, pH (5.6 – 7.0) and rich source of nutrients, they also have their own features. These features are closely related to types and quantity of oral microbial colonization and the balance and the disorders of the oral ecosystem.

Except for metabolic death, detachment and regeneration of normal tissue cells in the oral biotic area, food chewing, swallowing, the scouring effect of saliva and gingival crevicular fluid flow, and conventional oral hygiene measures, such as tooth brushing, mouthwash and other characteristics, it is obvious that they are of great impact on oral microbial colonization, and play an important role in maintaining the balance of the oral ecosystem. The eruption of teeth begins approximately 6 months after a person's birth; stagnation areas and the types and quantities of microbial colonization are not too many in a toothless mouth. Usually all deciduous teeth will erupt at 2 – 3 years old and at about 7 year-old deciduous teeth begin to fall out. Then permanent teeth begin to erupt, commonly known as "teeth change" and permanent teeth almost complete eruption at about 12 year-old. Such normal physiological changes, namely, the eruption and falling out of deciduous teeth, the eruption of permanent teeth, and the changes in human dentition from no teeth to a mouth full of teeth, are important characteristics of the oral biotic area. The dentition changes not only help chewing and digestion of food, more importantly they provide more habitat and ecological sites for microbial colonization, while the formation of several stagnation areas and the existence of the varied oxidation-reduction potential (Eh) are important reasons for the complex of microbial colonization. Tooth eruption not only provides a smooth surface for the tooth colonization habitat, but also provides proximal tooth surface, occlusal fissure, and sulcus *etc.* stagnation areas. Stagnation areas are areas where it is easy to store food and bacteria. In the stagnation areas, not only the flushing effect of saliva is weak, but also the conventional oral hygiene measures are difficult for providing an effective cleaning. As a result, stagnation areas provide useful protection areas for microorganisms, especially anaerobic microorganisms, and they also become the hidden dangers of oral infectious diseases, for example, dental caries, endodontic diseases, periodontal diseases and pericoronitis, while the microorganisms colonized in these stagnation areas are likely to be the initial factor in oral infectious diseases.

People will wear the appliance due to abnormality of dentition or occlusion, or need to use the prosthesis or implant for various reasons of missing teeth. The wearing of prosthesis, implants, and appliances are important characteristics of changes in the oral ecological areas, which not only provide a new oral microbial colonization habitat, but the resulting stagnation area also promotes colonization and proliferation of *Candida albicans*, *S. mutans* and a variety of anaerobic bacteria, and thus this increases the occurrence and development of oral infectious diseases of gingivitis, periodontitis, caries and peri-implant inflammation and denture stomatitis *etc.* [4].

10.1.2 Normal Oral Microflora

Oral microorganisms were among the first to be observed by humans. In 1683, Anton van Leeuwenhoek observed a variety of spherical, rod and spiral-shaped "small animals" in saliva and in material about the teeth with a simplified microscope, and these were what we now recognize as the cocci, bacilli and spirochetes.

The oral ecological areas might be considered an ideal microbial residence, which favor the established location and growth of a great variety of microorganisms. The oral cavity provides a suitable temperature of approximately 35 – 36℃, an abundance of moisture, differences in Eh, pH and a rich source of nutrition including the gingival crevice fluid, materials around the teeth, the epithelial cells undergoing degradation, and the salivary components (whole saliva has been found to contain 18 free amino acids) [4].

10.1.2.1 Definition

The oral microflora, like other microflora of the body, may be divided into two groups: (i) The resident microflora also referred to as normal or indigenous microflora; and (ii) the transient microflora.

Normal microflora of the oral cavity is an important ingredient representing normal microflora of the human body. The relation of the oral normal microorganism to human oral and systemic health and disease is a scientific study. The normal oral microflora is an important component of the oral ecosystem, and it is also an important ecological factor which affects the eubiosis and dysbiosis of the oral ecosystem.

Although in 1885 Pasteur pointed out that life cannot exist in germ free conditions, there is a mutualistic relationship with bacteria in 90% of the cells in the human body. The human survives as a symbiont interdependent with microbial colonization *in vivo*, but the understanding of human body normal microflora is still controversial. The focus of the problem is that these microorganisms can cause host endogenous infection in certain conditions. Since the 1970s, the significance of research of normal microflora has received more attention and interest. Chinese microecologist professor Kang Bai indicated that normal microflora is a necessary ecological structure in the host body, which formed with their host in the same evolution process and is harmless to the host, being beneficial, useful and necessary. In 2002, the US professor Savage clearly pointed out that humans or higher animals are a complex consisting of the body's tissues, eukaryotic cells and bacterial prokaryotic cells; normal microflora is an essential organ for the survival of humans and higher animals in nature.

A definition of normal oral microflora is: (i) They are formed in the same evolution process of mankind and through natural selection; (ii) They have a close symbiotic relationship with the human oral cavity; (iii) They are not pathogenic but beneficial under normal physiological conditions.

Normal oral microflora is also called resident oral microflora, or indigenous oral microflora, or autochthonous oral microflora, or natural oral microflora.

10.1.2.2 Population

The microflora of the oral cavity consists of bacteria, fungi, mycoplasms, protozoa and viruses. Research of oral normal microflora suggest that bacteria is the major member of oral normal microflora, and also of the oral microorganisms that people know the most about at present.

Bacteria. Bacteria are considered as one class of the most common and important microorganisms in infectious diseases of the human oral cavity receiving extensive attention. Aas *et al.* [5] reported that more than 700 bacterial species or phylotypes had been detected in the oral cavity, of which over 50% had not been cultivated, and 114 predominant species were detected. Commonly detected species in all sites belonged to the genera *Gemella*, *Granulicatella*, *Streptococcus* and *Veillonella*. Other species are *Neisseria*, *Rothia*, *Actinomyces*, *Eubacterium*, *Campylobacter*, *Fusobacterium*, *Peptostreptococcus*, *Haemophilus*, *Abiotrophia defectiva*, *Prevotella*, *porphyromonas*, *Capnocytophaga*, *etc.*

Streptococcus constitute the most numerous and complex group of bacteria and can be isolated from all areas of the oral cavity, 48% in plaque, 45% in the tongue, 29% in the gingival sulcus [4, 6].

Fungus. Candida is the most common fungus in the human oral cavity. Lav and Russell reported that the prevalence of Candida species was 82% in the mouths of 140 infants at the age of 1 mon. More than 30 species of the genus *Candida* have been described and the frequent species among the *Candida* species in the mouth are *Candida albicans*, *Candida tropicalis*, *Candida krusei*, *Candida parapsilosis* and *Candida glabrata*.

Mycoplasms. The oral cavity and the oropharynx are the ecological niches for a number of *mycoplasms* species. These *mycoplasms* have been isolated from dental plaque, dental calculus and the caries of teeth. The common *mycoplasms* species are *mycoplasms salivarium* and *mycoplasms orale*. They may be considered as members of normal oral microorganisms that were commonly isolated from the gingival sulcus, dental calculus in the human oral cavity and oropharynx, especially among healthy young adults. *M. orale* is frequently found in the oral cavity and throat.

Protozoa. Protozoa are unicellular microscopic animal parasites on the human body. *Trichomonas tenax* and *Entamoeba gingivalis* are common protozoa species in the human oral cavity. *T. tenax* species are typically pear shaped, with anterior flagella, an undulating membrane, and a protruding axostyle.

T. tenax is the only parasitic flagellate inhabitant found in the human oral cavity. The organism can be isolated from the saliva in a normal mouth, but it is most common in the diseased gingival sulcus, being found in acute gingivitis.

E. gingivalis is an inhabitant of the human oral cavity that occurs only as a trophozoite which is 8 – 11 μm in diameter. *E. gingivalis* occurs in small numbers in a healthy mouth, but it was thought to play an important role in periodontitis. It is found most commonly in acute necrotizing ulcerative gingivitis, chronic suppurative periodontal lesions, gingival pockets, ulcers of the buccal mucosa and tongue. *E. gingivalis* and *T. tenax* occurred together in acute necrotizing ulcerative gingivitis.

Viruses. Viruses are the smallest known infection agents; hepatitis virus has a diameter of about 25 nm. Herpes simplex virus (HSV) can be found in a healthy mouth, but it occurs the most frequently in herpes simplex type I infection.

10.1.2.3 Sources and Succession

The establishment and succession of the resident microorganisms in an ecosystem are important forms of ecosystem dynamics expression.

Sources. Researchers did not detect bacteria in the mouth of caesarean newborns, thus it is believed that the mouth of the newborn may be sterile. The mouth of a neonatus with normal delivery may be contaminated with a few microorganisms of the mother's vagina during childbirth, including *Enterococcus*, *Lactobacillus*, and *Staphylococcus epidermidis*. These vaginal bacterial floras may be considered as a transient oral flora.

The source of the oral microflora of the newborn is introduced by contact with the outside world after birth, and a great number and wide variety of microorganisms in the environment may be introduced into the newborn's mouth. Selective colonization theory suggests that not all introduced microorganisms can colonize in the oral cavity and form the oral microbial colonization populations, and only those ecological microbial populations associated with the host can adhere to and colonize the oral cavity, for short periods of time from hours to days. The microflora is well established in the oral cavity of the newborn and becomes the oral resident microflora.

Experts believe the oral resident microorganisms are mainly coming from their parents or relatives, but for newborns who did not contact their parents and relatives, the introduction of oral microbial colonization comes from the mouth of their close contact person. In summary, the composition of human oral microorganisms may be influenced by the introducers. A study of oral microflora at 1 – 2 d neonatus shows the homologous serotype of *Streptococcus salivarius* comes from the oral cavity of the neonatus and their mother. In addition, the homologous serotype *Porphyromonas gingivalis* and *Prevotella intermedius* were also found in the oral cavity of the father and son (or daughter) or mother and son (or daughter). These findings indicated that the bacteria came from the newborn's mother or a close contact person and as well as the possibility of oral bacteria spreading among their family members and relatives.

Succession. The quantitative change in phenotype and microbial communities in the ecological environment of the host in different developmental stages and periods are called "succession". Under normal physiological conditions, such as saliva secretion, chewing food and regular oral hygiene measures, as well as the host at different developmental stages, changes of dentition and hormone secretion, the changes in the number and the type of the normal microorganism community in the host, is a characteristic of physiological dynamic change, known as the physilogical succession. On the other hand, due to non-physiological factors such as external environmental factors or pathological factors, changes were induced in

the number and type of the normal microorganism community and manifested in excessive growth of certain species and transformation of dominant species, known as the pathological succession. Pathological succession is the important cause and manifestation which leads to oral infectious diseases.
Birth. During birth the mouth of the newborn may be sterile and the mouth of the neonatus with normal delivery may be contaminated with a few microorganisms from the mother's vagina.
Neonatus. The early oral microflora after birth is mainly and facultatively anaerobic, such as *Streptococcus*, *Staphylococcus*, *Neisseria*, *Bacillus* and *Lactobacillus*. *S. salivarius* establishes itself early in the oral cavity of the baby within a few days after birth.
Infancy. The common species are *Streptococcus mitis*, *Neisseria*, *Veillonella*, *S. epidermidis* and *Lactobacillus*. Fungi, such as *C. albicans*, *C. krusei* and *C. tropicalis* also can be detected in the oral cavity of infants. With the eruption of deciduous teeth, some hidden areas began to emerge in mouth, such as the gingival sulcus, pit, fissure and grooves of tooth occlusal surface, proximal contact spot of a tooth, adjacent place of malposition tooth. These are places where dental hygiene measures are not easily achieved, the normal flow of saliva is blocked, and food debris is easily squeezed in. At the same time, these places are relatively anoxic, thus the succession of communities from aerobic bacteria of the neonatal period to facultative anaerobic bacteria and anaerobic bacteria which are the main dominant bacteria.

S. sanguis and *S. mutans* were found in the mouth with erupted teeth, but were not found in a mouth without erupted teeth. Irregular gram-positive bacteria, such as *Actinomyces* and *Rothia* began to increase. Due to dentition and occlusion differences between individuals, or influence of eating habits and other factors, the type and quantity of microbial colonization show the differences between individuals. Researchers believe that the incidence of early childhood caries relates to genetic and dietary habits and other factors. Great amounts of *S. mutans* can be isolated in the mouths of children who are caries-sensitive and love to have sucrose.

Since the eruption of permanent teeth, tooth structure and dentition has been relatively stable, the succession of adolescent oral microbial colonization is in a relatively stable succession climax, which is called "climax community". Due to the formation of more stagnation areas, including gingival sulcus that is difficult to clean, fissures and proximal surfaces, as well as the rise of sex hormone levels, increase of gingivitis and other factors, make the types and quantities of oral microbial colonization increase significantly, especially an increase of anaerobic bacteria, which is the main characteristic of adolescent oral flora. These anaerobic bacteria, including: *Prevotella oralis*, *Leptothrix buccalis*, *Capnocytophaga*, *Fusobacterium nucleatum*, melanin producing gram-negative anaerobic bacilli and *Treponema, etc.* An increase in hormone secretion and wearing of appliances are also the characteristics of the adolescent mouth, thus affecting the composition and quantity of local oral flora.
Adulthood. The main features of the composition and succession of adult oral flora

are the complexity of the flora and increasing differences between each individual. In adults, as the incidence of chronic periodontal disease increases significantly, so the number and type of oral microbial colonization increase combined with the caries, bad prosthesis or implants and many other factors, also increase microbial colonization differences among individuals. Most studies indicated that there are changes in the composition of oral flora and differences among individuals in the adult period. A variety of different facultative anaerobic and obligate anaerobic bacteria can be detected in the adult mouth. The most common ones are *S. sanguis*, *S. oralis*, *Actinomyces*, *Capnocytophaga*, *P. intermedius* and *P. gingivalis*, *etc.* *Treponema* is the most common spirochete specie in plaque of gingival sulcus and gingival margin. In adult patients with periodontitis, an increased number of *P. gingivalis* and *Treponema* will be found in gingival sulcus and gingival margin.

Distribution. In the oral cavity the microflora differ basically in the oral anatomy. The characteristics and differences between ecological areas in the oral cavity, such as Eh, saliva washed force, stagnation area or non-stagnation area, lead to differences in the distribution of the normal oral microflora in different habitats and ecological sites. The microflora population living on the surface of teeth differs from that found on the gingival crevice, tongue and mucous membrane of the cheek. It is very important to understand the distribution of the normal microflora and influencing factors for the maintenance of oral health, prevention and treatment of oral infectious diseases.

Lip. Due to location of the lip which is most vulnerable to the influence of the outside world, skin and saliva, so the main colonized flora are Oral *Streptococci* and *S. epidermidis*, In addition, *Propionibacterium*, *Neisseria* may also be detected. *Candida albicans* may be detected in Angular cheilitis and cheilitis keratosis, while the pathogens of lip carbuncle are mostly *Staphylococcus aureus* and pyogenic hemolytic *Streptococcus*.

Buccal. The dominant colonization bacteria of buccal mucosa are Viridans *Streptococcus* group, including *S. mitis*, *S. oralis*, *S. sanguis*, and *S. salivarius*. *Candida* may also be detected in the buccal mucosa. Due to saliva cover and washing, remaining saliva flora is often detected in buccal mucosa, such as *Neisseria* and *Haemophilus*, *etc.*

Palate. Streptococcus is the dominant flora of the hard palate. As there is physical contact between saliva, tongue tip and palate, so the flora from the saliva and tongue tip can be detected in the hard palate, such as *Neisseria*, *Haemophilus*, *Lactobacillus* and *Veillonella parvula*. It is more difficult for anaerobic bacteria to proliferate greatly on the palatal mucosal surfaces. For the mouth with removable dentures, due to stagnation areas formed by contact between denture base and palate, *C. albicans* and *Lactobacillus* will show an increase in detection. Due to the close relationship between the soft palate and the respiratory tract, apart from colonization of saliva bacteria, there is often colonization of respiratory tract bacteria, such as *Neisseria*, *Haemophilus*, *Rothia* and Oral *Streptococci*, *etc.* β-Hemolytic *Streptococcus* is often detected in uvula, tongue palate folds and palate and pharyngeal folds. The main anaerobic bacteria that may be detected on the soft palate are *V. parvula* and *Actinomyces*.

Tongue. The dominant floras of the tongue are *S. salivarius*, *S. mitis*, *S.*

epidermidis and *Neisseria*. Other common species are *Haemophilus*, *Rothia mucilaginous* and *Candida albicans*. The cornification back side of tongue is an ideal anaerobe microbial inhabit, *V. parvula*, *Lactobacillus* and *Actinomyces* are frequently isolated from this inhabit.

Gingival crevice. It is the most abundant stagnation area and it is an ideal anaerobe habitation in the oral cavity after eruption of the teeth; the Eh of the gingival crevice is about –180 mv. There are 1,010 – 1,011 CFU (colony form unit) of organisms per gram in specimens of the gingival crevice, including facultative anaerobic and aerobic gram-positive or gram-negative coccus, bacilli, and spirochaetes, and the anaerobe is about 70%, such as *Prevotella oris*, *Fusobacterium* and melanin producing anaerobic rods. Microaerophilic species such as *Capnocytophaga* are also frequently isolated in the gingival crevice.

Tooth. Bacteria were found on all surfaces of teeth and their colony forms an association with plaque. The inhabitants of the teeth include the pit and fissure of the occlusal surface, smooth surface, and interproximal space of the tooth. *Streptococcus* is a common bacterial species on teeth surface. *S. sanguis* and *S. mutans* could establish itself in the mouth of infants only after the eruption of teeth. The high isolation rate of gram-positive bacilli and gram-negative anaerobic bacilli is found in the interproximal space of teeth.

10.1.2.4 Microflora Relationship

For a long time, due to our poor knowledge of the relationship between oral microflora and the dynamics of the oral ecosystem, a confusing issue facing the doctor at the dental clinic is that caries-active patients rarely suffer from periodontitis. In contrast, periodontitis patients seldom suffer from dental caries. With in-depth research on the relationship between oral microorganisms continuing, this issue already has a preliminary answer. Oral ecology scholars have pointed out that the reasons for this situation are closely related with the bacteria interaction of dental plaque in the oral ecosystems, in the dental plaque biofilm. The microorganisms are both symbiotically interdependent and provide antagonistic competition. The interplay between microorganisms affects the eubiosis of the mouth, thereby affecting oral health and disease.

Our knowledge of oral ecosystem dynamics is poor. The interactions between microorganisms are an important content of oral microbial ecology. Population, guilds and communities are basic concepts to discuss relationship between microorganisms. Population is the formation of individual cell growth and proliferation, guilds are the populations relevant to the metabolism, and communities are microbial populations having complementary physiological effect. Interaction of various guilds in the microbial communities is the closest factor affecting system ecology.

Symbiosis is different individual organisms that live together in any form, including mutualism, symbiotic commensalism and parasitism. Symbiosis exists widely in dental plaque, such as the consumption of oxygen by facultative anaerobes and aerobes in the plaque promotes the growth of obligate anaerobes; Veillonella does not use carbohydrates as a carbon source, but uses organic acids from other bacteria in plaque, such as *S. mutans*, to alleviate the acidic environment within the plaque and in favor of the growth of non-acid-resistant bacteria. Plaque ecology is the focus of the study; researchers are constantly exploring the interaction between oral bacteria in order to clarify the role of the relationship between oral bacteria and the development of oral infectious diseases, looking for strong evidence for ecological prevention and treatment of diseases.

Of course, as bacterial types are too complex and with too many influencing factors in the plaque ecosystem, so there are many technical difficulties to fully understand the interaction between bacteria. Present studies on the relationship between oral microorganisms include mutual gathering, mutual nutrition, mutual communication and mutual competition and antagonism.

Aggregation relationship. In studies on the dental plaque ecosystem, the impact of the bacteria aggregation and adhesion to the ecological system cannot be ignored. The aggregation of bacteria is a bacterium attaching to another bacterium's surface; aggregation may exist between the same kinds of bacteria and may also exist between the different kinds of bacteria. Some researchers call the adsorption of the same types of bacteria cells as the aggregation, and the adsorption of two different types of bacteria cells as the co-aggregation. The researchers confirmed that *Prevotella loescheii* VPI 12530 could serve as a bridge-mediated aggregation for *A. israelii* ATCC100485 and *S. sanguis* 34. In addition, *Capnocytophaga ochracea* serve as the bridge-mediated aggregation for *Actinomyces israelii* and *Rothia dentocariosa* has also been confirmed.

Aggregation of bacteria may be directly mediated through the adhesion of bacterial surface structures, such as pili, surface proteins, or their synthesis of extracellular polymers such as dextran, *etc.* For example, the adhesion between *S. mutans* relates to extracellular glucan synthesized by *S. mutans*; those bacteria that cannot aggregate directly will be mediated through the so called coaggregation bridge of other bacteria to complete aggregation. Aggregation or co-aggregation of bacteria is influenced by the nutritional substances that can be used in environments, such as bacterial extracellular polysaccharides, salivary glycoproteins and cell-cell direct binding, for example adhesion and colonization of *S. mutans* on the dental surface need to rely on dextran and para-amino benzoic acid (PABA) provided by early colonization of *S. sanguis*, and *S. mitis*, interact with glycoprotein, mucin, albumin and other polymer in saliva, promoting the aggregation of bacteria.

Adhesin-receptor theory put forward the specific adhesion mechanism of oral bacteria, namely it is the result of specific binding between adhesin on the oral bacteria surface and tooth surface receptor molecules in the form of stereochemical manner. Oral bacterial adhesin and its receptor are an important manifestation between microorganisms interaction. Co-aggregation between dental plaque bacteria is beneficial to the adhesion of bacteria in the dental plaque ecosystem to oral tissues' surface. In particular, it is very important for colonization of a number of weak adhesion bacteria such as Veillonella in the dental plaque ecosystem. Researchers have observed from *in vitro* experiments that there exists co-aggregation between *S. mutans* and *Neisseria*, and among *S. sanguis*, *A. viscosus*, *P. buccalis*, *P. denticola*, *P. loescheii*, *P. intermedius*, *P. oris*, *P. veroralis* and *A. viscosus*, *A. naeslundii*, *A. odontolyticus*, *R. dentocariosa* and *S. sanguis*.

Nutrition relationship. Oral bacteria not only have their own metabolic activities, but also have complex metabolic inter-relationships between microorganisms. The relationship between nutrition and growth of microorganisms is an important form of microbial interaction, and it is an important influencing factor on bacteria growth and proliferation. The trophic relationship between oral microorganisms can explain various microorganisms detected in the same ecological niche. *S. sanguis* can provide the required PABA for *S. mutans* growth, PABA is the important growth factor to promote the growth of *S. mutans*, which can explain the reason why *S. mutans* must proliferate on tooth surfaces with *S. sanguis* colonization. Some bacteria such as *Actinomyces* can provide growth required naphthoquinone (such as vitamin K substances) for melanin producing gram-negative anaerobic bacteria; *Treponema denticola* provide the growth necessary acetic acid for *P. gingivalis*, and *P. gingivalis* provide the growth necessary butyric acid for *T. denticola*; *Prevotella meninogenica* and *Actinomyces* can provide proliferation required formic acid for *Campylobacter sputigena*; the production of vitamin K from *Corynebacterium* can promote the growth of *P. meninogenica*.

When *S. mutans* and *Candida albicans* were cultivated together, the increase in the total amount of plaque indicated that there exists a nutritional relationship between two kinds of microorganisms. Table 10.1 lists the nutritional relationships of current parts of oral bacteria.

Table 10.1 the nutritional relationships between oral bacteria

Providing bacteria	Accepting bacteria	Nutrients
Neisseria	*S. oralis* *Lactobacillus*	Pullulan
Actinomyces *T. denticola*	*P. gingivalis*	Succinic acid
P. gingivalis	*T. denticola*	Iso-butyric
Prevotella *Fusobacterium*	*C. sputigena* *C. rectus*	H_2 and formic acid
F. nucleatum	*P. gingivalis* *T. forsythensis*	N-acetylmuramic acid
C. rectus	*P. gingivalis*	Heme substances
S. sanguis	*S. mutans*	Para-amino-benzoic acid
S. mutans	*V. parvula*	Lactic acid

Communications relationship. Communication and information interchange between microorganisms in a microbial ecosystem is known as the information flow. The exchange of information is closely related to growth, proliferation, genetic diversity and mutual relationships (symbiosis, promotion, restriction and exclusion) of the microflora of the ecosystem, and it is the initiator and the influence of the mechanism of microbial interactions, manifested as genetic diversity. The exchange of information includes nutritional information, chemical information and genetic information. The food chain is considered as the most persuasive model of nutrition information delivery systems in the ecosystem; chemical information is the initiator, and the influences of the energy flow, various enzymes, vitamins, metabolic acid, auxin *etc.* of the ecological system are chemical information transfer substances; transmission of genetic information is known as gene flow, and gene flow is the most popular field of present study. It reveals the interrelationship between the various populations, behaviors, pathogenic mechanism of the ecosystem and the nature of the existence of life. Genetic information includes quorum sensing systems (QS), resistance plasmid transfer, toxic and non-toxic clone expression, regulation and gene mutation *etc.*

The current study is focusing more on QS systems mediated by N-acyl homoserine lactone (AHL), which is present in the signal transduction system of gram-negative bacteria, mainly composed by the AHL signal molecule and its receptor protein LuxS. AHL belongs to autoinducer-1 (AI-1). It can freely access the cell membrane of bacteria in order to complete signal transduction. luxS gene can regulate the adhesion of bacteria by influencing the synthesis of AI-2, block or promote the signal transduction between bacteria. At present, luxS gene is detected in cells of *P. gingivalis*, *F. nucleatum* and other periodontal pathogens [7].

Plasmid is the genetic material outside the chromosome of a bacterial cell. It possesses the role and function of replication in the cell, maintenance and transfer between the cells, as well as acquiring genetic information from other cells. Transmission of drug resistant plasmids among a variety of oral bacteria has been confirmed. Studies on expression and transmission of virulence factors of periodontal pathogens toxic clone found that ions (Fe^{2+}) concentration of gingival sulcus and periodontal pockets may be an important initiating factor of the *P. gingivalis* toxic clone.

Competition and antagonistic relationship. Competition or antagonism between microorganisms is an important factor affecting the ecological eubiosis and dysbiosis of the ecosystem. The competition and antagonism between bacteria in dental plaque is an important factor affecting the ecological balance of the plaque ecosystem and also one of the important factors affecting pathogenicity of the dental plaque. Taking advantage of this principle to support the competitiveness of oral beneficial bacteria may be of great significance in the maintenance of the eubiosis of the oral ecosystem, prevention of both occurrence and development of oral infectious diseases.

The competition between microorganisms mainly refers to competition for colonization space and nutrient. It is the instinct of microbial growth requirements for microorganisms to compete for colonization sites and nutrients in the same habitat. Competitive microorganisms usually grow fast, and it may be the dominant microflora of habitats, except for the impact of microbial populations on competitiveness. It is also influenced by environmental factors. For example, the proliferation ability of oral competitive group of *Streptococci* will be reduced in deep plaque or gingival sulcus, which is affected by environmental Eh effects.

The antagonistic relationship between oral microorganisms is a more complex mutual relationship than their competition. It means that some of the substances produced by microorganisms are harmful to other microorganisms; this kind of antagonism can prevent the excess proliferation of these microorganisms. One kind of microorganism in plaque may produce a metabolic substance that is not beneficial to the survival of another kind of microorganism, so that the growth environment of microorganisms changes. For example, one kind of microorganism interferes with metabolism activity and inhibits the growth and reproduction of another kind of microorganism by changing the osmotic pressure, oxygen and carbon dioxide tension, or by producing bacteriocins and other products within its microecological environment.

Lactobacillus is a powerful acid producing bacteria; it can produce large amounts of lactic acid, making the pH of plaque drop as low as 4.5, inhibiting the growth of acid-sensitive bacteria.

The toxic product of bacteria (antibacterial substance) is the basis of this antagonistic relationship. Antibacterial substances of microorganisms include bacteriocin and other antimicrobial substances. The Sanguicin and H_2O_2 produced by *S. sanguis* can inhibit most periodontal pathogens such as *P. gingivalis*, *F. nucleatum*, *Capnocytophaga*, *P. intermedius*, *S. mutans* and other oral *Streptococci*, *etc.* H_2O_2 is a non-specific antibacterial substance. At low concentration, it may be inhibitory; at high concentrations it can be bactericidal. H_2O_2 in plaque is a metabolic byproduct of *Streptococcus* and *Lactobacillus* under aerobic conditions. H_2O_2 produced by *S. sanguis* can inhibit the growth of *Lactobacillus* and *Actinomyces*. *S. salivarius* and *S. mitis* in plaque can also produce H_2O_2, and can competitively inhibit the growth of *S. mutans*.

Bacteriocin is a bactericidal protein synthesized by the bacteria with the gene outside chromosomes. It could kill a bacteria cell by inhibiting protein synthesis, affecting DNA metabolism and energy metabolism, *etc.* There are both bacteriocin producing bacteria and bacteriocin-susceptible bacteria in dental plaque. *In vitro* experiments found that various serotypes strains of *S. mutans* were able to synthesize a bacteriocin called "mutacin", and serum c-type *S. mutans* produce most of bacteriocin. Mutacin can antagonize *S. sanguis* and *A. viscosus*. With the increase of *S. mutans*, it is often accompanied by a decrease of *S. sanguis* in the plaque. Sanguicin produced by *S. sanguis* can inhibit other *Streptococcus* species and gram-negative anaerobic bacteria.

Melaninocin, which is a bacteriocin-like material produced by *Porphyromonas* and melanin producing *Prevotella*, can also inhibit gram-negative non-spore and not produce melanin anaerobic bacilli, such as *P. oralis* and *Fusobacterium* in the plaque.

10.1.2.5 Affecting Factors

The adhesion, growth, reproduction and colonization process of oral microorganisms in different oral habitats will be affected by a variety of factors including host, microorganisms, and the external environment.

Adherence and retention. Adherence and retention are the primary conditions for microbial colonization. The retention and adherence of oral microorganisms have been given widespread attention all the time, but there are still many questions that are not entirely clear. Experts have found that some bacteria can adhere to oral soft tissue. For example, *S. salivarius* can adhere to dorsum mucosa and buccal mucosa of other soft tissues, while *S. sanguis* and *S. mutans* adhere to enamel, and neither of them could be found in a toothless mouth. There are many mechanisms of oral adherence, such as the calcium bridges, hydrogen bonds, hydrophobic interactions, and the adhesin-receptor theory and so on. Factors affecting adhesion are difficult to fully clarify, but the following aspects can be affirmed and recognized: (i) A certain bacterial surface structure and its product is the specific adhesion or media of bacteria adhesion in the oral cavity. (ii) Host salivary proteins contribute to bacterial adhesion. *In vitro* experiments showed that the early basis of formation of dental plaque — the formation of acquired pellicle was the result of salivary glycoprotein selectively adhering to hard tissues (*e.g.* teeth or restorations) and formed a structureless and acellular pellicle. Bacteria adhered or agglomerated on the pellicle and formed dental plaque. Salivary proline-rich proteins and Statherin are supposed to have bacterial specific adhesion receptor components (adhesin — receptor theory). (iii) The role bacteria interactions play in bacterial adhesion is also affirmative. The difference in the selective adsorption has been confirmed due to the differences in different bacterial adhesion or other support structures and products. The difference in bacteria distribution in the oral cavity also proved this point, for example *S. mutans* and *S. sanguis* can only be detected in the oral cavity after tooth eruption. *Porphyromonas gingivalis* and other obligate anaerobes were first found in gingival sulcus of erupted teeth. Researchers also found that *Actinomyces* adhere to tooth surfaces or gingival sulcus through type I pili (bacteria-specific adhesin), and its type II pili (agglutination bridge matrix) could help bacteria agglomeration. The adhesion of *Actinomyces* in the oral cavity not only relied on their own cell structure pili, but also it relied on hyluronic acid-mediated adhesin to gingival sulcus or the tooth surface. One of the important reasons for *S. sanguis* and *S. mutans* tooth surface adhesion is that they rely on their production of extracellular polysaccharide. Some bacteria in the oral cavity often adhere and colonize through agglomerating with other bacteria, for example adhering to other bacteria through the cell structure, or adhering through the extracellular products produced by other bacteria. The bacteria that can mediate other bacteria's adhesion are known as a coaggregation bridge. Some oral gram-positive and gram-negative bacilli such as *Actinomyces*, and *Capnocytophaga* were often used as the coaggregation bridge involved in the adhesion and colonization of oral bacteria [8]. The wheatear-like structure and bottle brush-like structure of dental plaque clearly showed the role of the accumulation of bacteria in the adhesion and colonization. (iv) Existence of retention areas or protected areas. The researchers observed that some weak adhesive bacteria like *Veillonella* could settle on depressed areas of enamel, such as pit and fissure of teeth. One reason is that these places are a stagnant area. A stagnant area is a colonization habitat that oral microorganisms prefer; it is also known as a protected area. Microorganisms in the stagnant area can be protected from the scavenging effect of colonization due to the scouring force of saliva or other oral friction forces. The gingival sulcus is the biggest stagnant area in the oral cavity and it is also the colonization habitat with the most number and types of bacteria, and therefore it is also known as the largest protected area in the oral cavity; (v) Elimination force and defense force are important factors of the host which affect microbial adhesion and retention. Mechanical eliminating forces from the host include saliva flow, gingival crevicular fluid flow, swallowing of tongue and soft tissue, friction of food chewing and the metabolism of the oral tissue cells (*e.g.*, epithelial cell shedding), *etc.*; Other defense forces of the host include saliva lysozyme, secretary immunoglobulin A (SIgA), lactoferrin, peroxidase, gingival crevicular fluid and phagocytosis. Elimination and defense forces of the host not only prevent exogenous bacterial colonization and adhesion, but also regulate the amount of oral microbial colonization, and they are normal physiological adjustment factors.

Multiplication. A microorganism must multiplicate after adhesion to finally colonize. Factors affecting multiplication include: the use of matrix, pH and oxidation-reduction potential (Eh) of habitats and the interaction between microorganisms.

Nutrients and growth factors. Microbial multiplication must metabolise the available matrix. As the sources of personal nutrients and energy, these matrices may be derived from food intake of host and microbial products from the same colonization area or neighborhood areas. The biggest possible role of the increase in food carbohydrates is considered to be promoting an increase in oral bacteria, especially the number of *Streptococci*; for example, a high intake of sucrose significantly increased the number of *S. mutans* in the oral cavity, which may lead to dental caries.

pH and Eh. In addition to nutrients required for basic conditions of microbial growth, it also requires a suitable environment for growth and multiplication, including pH and Eh. pH and Eh are important factors affecting microbial growth and multiplication.

Most oral microorganisms require environmental pH to be around 7.0, but some bacteria are found to grow and multiplicate under a low pH environment, *e.g.* The tolerable pH of *Lactobacillus* and *Candida albicans* is around 5.2. However, some other bacteria such as *Prevotella melaninogenica*, *Porphyromonas gingivalis* and *Veillonella* cannot tolerate the environment whose pH is under 5.5. Their multiplication was inhibited, *Veillonella* prefer to grow in an alkaline environment such as with a pH around 8.5. The pH of healthy saliva is about 7.0. It is a suitable pH environment for the majority of microbial multiplication. Saliva pH buffer systems can maintain relatively stable pH, and bicarbonate is the most important buffer substance in the saliva buffer system.

$$pH=PK_a+\lg L[(HCO_3)|[H_2CO_3]$$

PK_a is the negative logarithm of weak acid ionization constants in this buffer system.

The existence of different Eh in the oral cavity is one of the important reasons for complex colonization of bacterial populations; it is also one of the important affecting factors of flora distribution differences. Facultative anaerobic bacteria usually multiplicate in an Eh-high environment; However, *Actinomyces*, *Capnocytophaga* and *Campylobacter* require a micro-aerobic environment; Obligate anaerobic bacteria such as *Treponema*, *Fusobacterium*, *Porphyromonas gingivalis*, *Peptostreptococcus* and *Eubacterium* require a redox low Eh environment to multiplicate.

Low Eh environments in the oral cavity are the gingival sulcus whose Eh is about +100 mV, the periodontal pocket whose Eh is as low as about –300 mV, and deep plaque whose Eh is as low as –140 mV.

10.1.3 Saliva and Dental Plaque Biofilm

Saliva and dental plaque are not only the most important ecological factors in the oral ecosystem, but also are important influencing factors for normal oral microorganisms. Saliva and dental plaque are closely related to oral microorganisms, oral health and diseases, the relationship between saliva and dental plaque, and the occurrence and development of oral infectious diseases have always been important research subjects in oral medicine.

10.1.3.1 Saliva

Saliva is secreted and formed by the submandibular gland, sublingual gland, parotid gland, as well as numerous small salivary glands; it is the most important body fluid in the oral cavity, with infiltration, buffering, cleansing, digestion, sterilization and other physiological functions maintaining human health. Saliva function disorders will lead to oral and body diseases.

Saliva microorganisms. Saliva is an important component of the oral ecosystem; it is one of the sources of nutrition for oral microorganisms. The mineral content, organic content, ion concentration, fluoride content, buffering capacity, Eh, vitamins changes of saliva influence the number and types of oral bacteria, thus affecting the oral eubiosis.

Saliva is closely related to oral microorganisms. It not only provides the necessary conditions for adhesion, proliferation and colonization of oral microorganisms, but also it is the carrier and reservoir of oral microorganisms which affect the types and quantities of microbial colonization. It is closely related to the occurrence and development of oral infectious diseases. Relations between saliva and oral microorganisms include saliva and oral microbial adhesion, growth and proliferation. Because of suitable temperature, humidity, pH and the rich source of nutrients such as protein, amino acids, vitamins, and inorganic salts *etc.* of saliva, it provides a basic and necessary condition for the survival of oral microorganisms. Microorganisms can be detected in saliva in all parts of colonization in the mouth. This is related to the covering of the entire mouth and constant movement of saliva. Therefore saliva is also known as the carrier and reservoir of oral microorganisms.

The relationship between saliva and microorganisms also include the regulation and maintenance of types and quantities of normal oral microbial colonization and prevention of the exogenous microorganisms' invasion by saliva ion concentration, buffering capacity, scouring action and antimicrobial proteins, and enzymes, *etc.*

Saliva microorganisms can reflect the status of oral infections, thus it is used as the measurement indicator of oral infectious diseases. Patients susceptible to dental caries or with high incidence of caries can be found to have an increased level of *S. mutans* which is close to dental caries, while patients with gingivitis and periodontitis are often detected with a marked increase in the number of gram-negative anaerobic bacilli which is a periodontal pathogen, such as *P. gingivalis*, *P. intermedius*, *F. nucleatum* and so on.

The dominant microorganisms of normal oral saliva include *S. salivarius*, *S. mitis*, *Neisseria*, *Hemophilus* and *V. parvula*, as well as β-hemolytic *Streptococcus*, *Lactobacillus*, *Actinomyces*, *Capnocytophaga* and so on. *Candida albicans* is the most common fungus in saliva. In studies on the development of dental antibacterial drugs and preparations, the effects of the agents or drugs on oral normal flora in saliva can be tested by detecting the dynamic changes of flora.

Oral *Streptococci* are the dominant flora in saliva; *S. salivarius* and *S. mitis* are the most common ones. In addition, saliva also contains certain amount of *Corynebacterium*, *Actinomyces*, *Veillonella*, *Neisseria*, *Lactobacillus*, *Fusobacterium*, *Prevotella* and other bacteria, as well as fungi, yeast, *Mycoplasma*, virus, Protoza and other microorganisms. The proportions of various bacteria in saliva are listed in Table 10.2.

Table 10.2 Proportion of bacteria in saliva

Bacteria name	Proportion
Streptococcus	41%
Gram-positive bacilli	16.6%
Gram-negative cocci	15.9%
Facultative anaerobic gram-negative bacilli	12%
Anaerobic gram-negative bacilli	3.7%

Saliva protein. There are about 40 types of salivary proteins; some proteins are beneficial to oral microbial adhesion, agglutination and proliferation. The thin film formed on a tooth surface or dental prosthesis surface by saliva glycoprotein is called acquired pellicle, which is the foundation of bacterial adhesion and plaque formation. *In vitro* experiments demonstrate that oral bacteria can quickly adhere to saliva-coated hydroxyapatite (HA), while saliva mucins, proline-rich proteins (PRPs), histatins (histidin-rich proteins, HRPs), and statherin were considered to be more closely related to bacterial adhesion.

The selective affinity of high molecular weight glycoproteins (MG1) to HA may be related to a large number of hydrophobic structures, while the low molecular weight glycoprotein (MG2) can promote bacteria copolymerization and may be related to a large number of neuraminic acids.

Salivary proteins with antibacterial activity include proline-rich proteins, histatins, statherin, lysozyme, peroxidase system, and lactoferrin and so on.

PRPs. The PRPs in saliva can be divided into APRPs, BPRPs and GPRPs, and their contents were 30%, 23% and 17% of total salivary protein respectively. APRPs were secreted mainly by parotid and submandibular glands; their contents are stable in the saliva and are more close to bacteria adhesion. The functions of APRPs on bacterial adhesion include: (i) Adsorption effect to Ca^{2+} and HA; (ii) Promotion effect on bacterial adhesion; (iii) With similar primary structure of the collagen; (iv) Hidden receptors. Researchers found that salivary protein and APRPs adsorbed on the HA accounted for 42%. The binding of HA and APRPs may exist in the N terminal of the 3 – 25 amino acid residues, while within the binding of Ca^{2+} and APRPs there exists in the N terminal 30 amino acid residues. The binding dissociation constant of Ca^{2+} and APRPs is 2×10^{-4} mol/L, the binding force is unaffected by the impact of trypsin and collagenase, but can be affected by pH, ionic strength and Ca^{2+} concentration. The total amount of Ca^{2+} adsorbed in APRPs will normally decrease as the ionic strength increases and pH decreases. A substantial increase in serine phosphorylation can promote adsorption of APRPs to Ca^{2+}.

Gibbons *et al.* [9], found that both APRPs and statherin can promote the adhesion of bacteria to HA. These bacteria include *A. viscosus*, *S. mutans*, *A. israllii*, *A. odontolyticus*, *P. gingivalis*, *P. loescheii*, *P. meninogenica*, and *S. gordonii*. *A. viscous* with type I pili can adhere to HA with adsorbed APRPs and statherin, but *A. viscous* with type II pili cannot adsorb to HA. Such type II pili are considered to play an important role in bacteria agglomeration. Researchers thought that the mutual adhesion of non-carbohydrate receptor type I pili as an adhesin and non-glucosyl APRPs and statherin as receptor are models of adhesin-receptor theory.

The characteristics of similarity between the primary structure of APRPs and collagen structure are thought to be helpful for bacterial adhesion with a special affinity to collagen matrix in the mouth, such as *P. gingivalis*, *A. viscous* and *S. mutans*. They all have a high affinity to collagen matrix.

In 1989, Gibbons found that *A. viscous* cannot adsorb to APRPs with concentration as high as 1,000 μg/mL, but it can absorb to HA coated with APRPs. The results suggested that there exist eptitopes on the APRPs molecule. The folding polypeptide chain of the hidden molecular fragments can be opened and exposed on the surface of HA. The recognition ability of bacteria to a hidden receptor of APRPs molecule is the mechanism of specific adhesion.

(Histatins, HRPs). Histatins are small molecules of polypeptide which are rich in histidine. 12 kinds of HPRs have been isolated from the saliva, and they are respectively named as HRPs 1 – 12. It was found that they may also be involved in the formation of acquired films, and HRPs 1 has been proven to selectively adsorb on the HA and enamel powder.

Statherin. Statherin is a salivary protein which is rich in tyrosine. It not only can absorb HA the same as APRPs and HPRs, but also can promote bacterial adhesion to HA, the same as APRPs.

The antibacterial role of saliva is an important function to maintain saliva. Early *in vitro* studies pointed out that saliva of the submandibular gland, sublingual gland and parotid gland can selectively inhibit the transmission of the type I herpes simplex virus (HSV-1) and further evidence indicated that PRPs have the ability to inhibit viral replication, in which the effect of the BPRPs is more obvious. BPRPs interfere with the penetration and cell processes of virus target cell to achieve the anti-viral effect. In 1998, Heineman and Grecnberge proved that the whole saliva can prevent infection of HSV-1 in epithelial cells.

Lysozyme. Lysozyme is a low molecular weight basic protein derived from big or small salivary glands and sublingual glands. It can hydrolyze the main structure — peptidoglycan of a gram-positive bacteria cell wall, it also has a bacteriolytic effect on gram-negative bacteria as *Klebsiella* and *Escherichia coli*. The bactericidal effect of lysozyme on *S. mutans* has also been confirmed, but experts believe that the antibacterial effect of lysozyme on exogenous bacteria will be stronger, and the effect on the normal oral microflora of oral ecosystem belongs to ecological opsonization. Researchers found that saliva lysozyme of those who suffer from dental caries are lower than in those without caries; this proved the antibacterial effect of lysozyme on cariogenic bacteria.

Peroxidase system. Peroxidase system is an important antimicrobial system in saliva; it consists of peroxidase, H_2O_2 and SCN^-. Peroxidase can catalyze H_2O_2 and SCN^- into $OCSN^-$.

$$H_2O_2 + SCN^- \xrightarrow{} \text{peroxidase} \quad OCSN^- + H_2O$$

$OCSN^-$ is the active ingredient of salivary peroxidase system, anti-microbial factors, through its oxidation of sulfhydryl and inactivation of bacteria sugar metabolic enzymes such as hexokinase, aldolase, phosphofructokinase and so on, and thus interferes with the metabolism of bacteria and achieves an antibacterial effect. H_2O_2 production of *S. sanguis* and *S. oralis* is one of the reasons why they are periodontal beneficial bacteria. The antibacterial mechanism of *S. sanguis* and *S. oralis*、to most gram-negative anaerobic bacteria such as *P. gingivalis*, *F. nucleatum*, and *S. mutans* relates to their H_2O_2 production.

Lactoferrin. Lactoferrin is a kind of glycoprotein bonded with iron and synthesized by neutrophil and glandular epithelial cells in saliva. The protein can chelate the iron atom, thus affecting the growth and metabolism of bacteria. *In vitro* experiments have proved that lactoferrin can inhibit the growth of *S. mutans*.

SIgA. SIgA is the main immunoglobulin of saliva and the concentration of SIgA changes with saliva flow rate. The local role of SIgA in the mouth includes:

(i) Interfering with adhesion and proliferation of microorganisms on the mouth surface and inhibiting the formation of dental plaque; (ii) Inhibiting bacterial enzymes, such as inhibition of glucosyltransferase of *S. mutans*, thereby interfering with synthesis ability of glucan of *S. mutans* in dental plaque formation; (iii) Activating the opsonization of complement C3 bypass; effectively agglutinating bacteria and neutralizing bacterial toxins and viruses.

10.1.3.2 Dental Plaque Biofilm

In 1989, Williams proposed the term dental plaque, but a deeper understanding of plaque began after the 1970s. A lot of studies discussed the formation, composition, metabolism of plaque, and its relationship to caries and periodontal diseases, as well as a variety of factors affecting the plaque. Plaque is a sticky bacterial plaque on tooth surfaces and local produced acid can cause cavities; plaque is a micro ecological system, and bacteria as the main body live on teeth and oral hard tissue structures or surface of prosthesis. Plaque is the ecological environment for the bacteria's survival, metabolism and pathogenesis.

Based on ecological theory, researchers in the 21st century explore the relationship between dental plaque and oral infectious diseases, emphasizing the concept of plaque biofilm; plaque is the bacterial biofilm adhering to teeth or intraoral restoration and appliance surfaces. Adhesion and colonization of various oral microorganisms on the teeth and different parts of tissues around teeth exist in the form of plaque biofilm. The commensalism, competition and antagonism of the microorganisms in the plaque biofilm work in the form of population, community and guilds, establishing a dynamic ecological relationship between each other and the teeth.

Plaque ecology is the study of the correlation between tooth body — oral organs and oral microorganisms, as well as the correlation between microorganisms. It elucidates the mechanism of physiological balance and pathological disorders of the interaction between tooth body and microorganisms, studies the incidence mechanism of caries — one of the three diseases recognized by WHO in a new perspective, lays the foundation of ecological prevention and control of caries.

In the discussion of the etiology of the most common bacterial oral infectious diseases — endodontic diseases and periodontal diseases, researchers clearly pointed out that plaque is the initiating factor for the occurrence and development of caries and periodontal diseases. With oral microbiology, biochemistry and molecular biology development, more in-depth understanding of plaque also helps us to understand the "plaque disease" in oral infection, the cause and mechanism of endodontic disease and periodontal disease, thus establishing effective control methods.

Classification, Formation. According to location, plaque is divided into supragingival plaque and subgingival plaque. Supragingival plaque refers to the plaque above the gingival margin of the cervical tooth surface: smooth surface plaque locates on the smooth tooth surface; grooves and cracks plaque locate in the pits, fissures, grooves and cracks tooth surface points of the ditch fissure. Subgingival plaque is the plaque below the gingival margin of the cervical teeth surface and is covered by gingival. It is divided into attached subgingival plaque and non-attached subgingival plaque.

Acquired pellicle is the basis of formation of a plaque micro ecological system. It is a thin layer of salivary glycoprotein selectively adhering to the surface of the tooth, and it is cell-free, non-structural organic film with a thickness of 1 – 10 μm. Once the acquired pellicle forms, it has changed the ecological environment of the tooth surface, and performs its biological functions: (i) The formation of the acquired pellicle change the properties of the original tooth surface, affecting the colonization and adhesion of bacteria on the tooth surface. Some constituents of the acquired pellicle are specific receptors of some bacteria; they are beneficial to the adhesion of some bacteria, but also can shelter a number of bacterial receptors, reducing the adhesion of these bacteria. For example, the acquired pellicle is beneficial to the adhesion of *S. sanguis*, but it is not beneficial to the adhesion of *S. mutans*. (ii) The acquired pellicle is rich in protein and carbohydrates and it can be used as nutrients for adhesion of bacteria on its surface. Deep bacteria of plaque can also make use of the acquired pellicle to maintain normal metabolism; (iii) The protein and amino acids contained in the acquired pellicle can combine calcium, phosphorus, fluoride and other minerals, which are beneficial to tooth mineralization and remineralization. (iv) The acquired pellicle can reduce the sensitivity of enamel to acid and protect the teeth's enamel surface. Some people think that the early emergence of subsurface enamel damage and the special pathological changes of a relatively complete surface may be related to the protective effects of the acquired pellicle.

After the acquired pellicle forms, the bacteria quickly adhere in the form of populations, communities and guilds, gathering, growing and reproducing, constituting the necessary conditions of formation and maturity of plaque.

The adhesion of bacteria to the tooth surface through the acquired pellicle is a complex process, and the adhesion mechanism may be the following: (i) The role of calcium bridges: the bacteria adhere to the surface of the acquired pellicle through Ca^{2+} electrostatic effect. The $-SO_3$ and $-COOH$ radical of glycoprotein contained in the acquired pellicle are negatively charged, and the terminal $-PO_3$ radical of teichoic acid of the bacterial cell wall is negatively charged, with the help of Ca^{2+} bridge link, bacteria adhere to the tooth surface. (ii) Hydrogen bonding interaction: the bacteria produced extracellular polysaccharides connect with the hydroxyl of glycoproteins in the acquired pellicle by hydrogen bonds. (iii) Hydrophobic interaction: Bacterial surface hydrophobic protein bond to the hydrophobic radical of the acquired pellicle, the stronger the surface hydrophobicity, the stronger the bacterial adhesion. Hydrophobicity of the acquired pellicle relates to lipid content, the more the lipid content, the stronger the surface hydrophobicity. (iv) Adhesin-receptor effect: Adhesin refers to the bacterial surface components related to the adhesion, while the receptor is the component of the host surface reacting with adhesin. Many bacteria surfaces have adhesions, and they specifically and stereochemically bond to complement molecules or receptors of the tissue surface. This is a specific binding and bacteria firmly adhere to the tooth surface. Most adhesin are plant lectin and they mainly exist in the appendix of the bacterial surface, such as pili. They also exist in the binding proteins and a variety of surface enzymes such as GTF. Different bacteria have different adhesin, and different adhesin regulate with bacterial adhesion to different parts of tooth surfaces [10].

Mature plaque. Plaque maturing is a dynamic process. Plaque structure, composition and proportion of bacteria are constantly changing to adapt to environmental changes. Generally it will take 7 d for dental plaque to mature and develop into a mature ecosystem. An important feature of mature plaque is the increase of anaerobes and the decrease of aerobes with increasing age of plaque; early plaque has a larger number of aerobes.

The basic structure of mature plaque is divided into three parts, including the basal layer, the middle layer and the surface layer. The basal layer is the acquired pellicle; it is a layer of uniform acellular structure close to the tooth surface. Its thickness ranges from 0.1 to 1.0 μm. The basal layer staining is relatively constant, hematoxylin-eosin (HE) stained red. The middle layer is the main part of the plaque, also known as body of plaque. It accounts for the largest portion of plaque and it is composed of filamentous bacteria adhered to the acquired pellicle as the main part, the filamentous bacteria parallel to each other, and vertical to the tooth surface. There are a large number of cocci and bacillus distributed or agglomerating between filamentous bacteria, forming a typical palisade structure. A palisade structure is the basic structure of dental plaque, and is also the channel of deep plaque bacteria access to nutrition. The bacteria of the middle layer mainly consist of anaerobes and aerobes.

The surface layer is the layer away from the tooth surface, also known as the outer layer; its structure is relatively loose, with many food residues and shedding of epithelial cells. The composition of surface bacteria changes a lot; their colonized bacteria are mainly facultative anaerobic cocci and rods. The "corncob" and "bristlebrush" structure can be observed in different parts of dental plaque under the scanning electron microscope. The surface layer of groove and crack plaque is thin and filamentous fungi centered, "corncob" structures of bacteria adhesion around its surface are more common, and the "bristlebrush" structure surrounded by adhesion of long rods is more common in proximal dental plaque and the surface layer of subgingival plaque.

Microorganisms of dental plaque. The microbial composition characteristics of plaque are: (i) With the largest number of bacteria and the most complex types; (ii) Different parts are different, with the dynamic variation in its composition; (iii) There is similarity of bacteria species between pathogenic plaque and non-pathogenic plaque, but its composition may vary greatly. Researchers found differences in composition between caries active plaque and caries inactive plaque. The detection rate and amount of *S. mutans* and *A. viscous* of the former were significantly higher. We also can see the difference in subgingival plaque between healthy humans and patients with periodontal diseases. The former was dominated by Oral *Streptococci* and other facultative anaerobic bacteria, and the latter was mainly dominated by gram-negative anaerobic bacteria [9].

Ecological plaque hypothesis. Although plaque is regarded as the initiating factor of caries and periodontal disease, dentists and researchers observed that dental plaque present in each individual would not always cause dental caries and periodontal disease. The pathogenicity of dental plaque has differences in different individuals and different parts of the same individual. Although, according to its pathogenicity, we simply divide the dental plaque into cariogenic plaque, non-cariogenic plaque, and periodontal disease causing plaque, *etc.*, in-depth study found that the pathogenicity of dental plaque relates to a variety of factors, and is influenced by many factors, so it is not easy to define it as pathogenic or non-pathogenic plaque. Based on ecological theory, with the use of modern molecular biology techniques, research into the relationship between dental plaque and oral infectious diseases has become one of the studies of researchers [11].

Basic concept. In 1991, Marsh *et al.* [11] proposed ecological plaque hypothesis: The onset of caries and periodontal disease attributed to environmental factors changes, leading to disorder of the eubiosis of the plaque ecosystem. The significance of ecological plaque hypothesis lies in defining caries and periodontal disease as the dysbiosis diseases and the basis of dysbiosis are environmental factors, including the changes in host, bacteria and external factors. The concept is very similar to the definition of infection by Professor Kang Bai, a microecologist in China. He said: Infection is an ecological phenomenon of interaction between microorganisms and host caused by abnormal invasion of the host. For infectious diseases, especially the study of the etiology of infections caused by the resident microbial flora, simply discussing one or a few bacterial virulence factors and invasion ability makes it difficult to evaluate its pathogenic effect, so as to reveal the nature of infection and find effective ways for prevention and treatment.

The author thinks that the basis of ecological plaque hypothesis is the consideration of the host, microorganisms, environment and other factors, and the eubiosis and dysbiosis are the core.

Dental plaque ecosystem. Actually, dental plaque is the ecological system attaching to the surface of oral tissues and then the "plaque ecosystem" concept proposed by latecomers.

The main body of the dental plaque ecosystem consists of microorganisms and ecological factors. With metabolism, energy and information exchange, it contains the dynamic balance between microorganisms of dental plaque and the host, and between microbes, and the impact on the performance of the external environment.

Discussion of the plaque ecosystem includes plaque formation, maturation and variation, and the relationship between dental plaque microflora and various factors influencing the dental plaque ecosystem. The purpose of it is to allow more scholars to understand the basis of ecological theory and to explore the relationship between dental plaque and oral health and disease, and ecological prevention and treatment of oral infectious diseases.

In dental plaque, the interrelationship among microorganisms affects the ecological balance of dental plaque. The oxygen consumption of facultative anaerobes and aerobic bacteria of plaque promotes the growth of anaerobic bacteria; *Veillonella* does not use carbohydrates as a carbon source, but uses organic acids produced by other plaque bacteria, which will help relieve the acidic environment within the plaque and be beneficial to the growth of non-acidic bacteria. When *S. mutans* and *Candida albicans* are a mixed culture, the total amount of plaque will increase; however, when *Lactobacillus*, *S. mitis*, *Neisseria* and *S. mutans* are a mixed culture, the total amount of plaque will decrease.

H_2O_2 is a non-specific and anti-bacterial substance. It is bacteriostat at low concentration and bactericide at high concentration. H_2O_2 produced by *S. sanguis* can inhibit the growth of *Lactobacillus* and *Actinomyces*, *S. salivarius* and *S. mitis* of plaque can produce H_2O_2 too, and can competitively inhibit the growth of *S. mutans*. Bacteriocin is a bactericidal protein synthesized by the bacteria with the gene outside chromosomes. It could kill the bacteria cells by inhibiting protein synthesis, affecting DNA metabolism and energy metabolism, *etc.* There are both bacteriocin producing bacteria and bacteriocin-susceptible bacteria in dental plaque. Various serotypes strains of *S. mutans* were able to synthesize a bacteriocin called "mutacin". They can antagonize *S. sanguis*, *A. viscosus*, and serum c-type *S. mutans* produce most of the bacteriocin.

Plaque pH is an important parameter to measure the eubiosis of dental plaque, and pH can affect the bacterial enzyme activity and cell division, thus affecting the growth and reproduction of bacteria. At the same time, the changes in pH also affect the solubility of dental hard tissue and transfer of calcium, phosphorus and fluoride, *etc.* In the dental plaque microecosystem, pH has significant changes; acid-producing bacteria produce organic acids and reduce plaque pH.

The pH which can cause dissolution of hard dental tissue is called critical pH 5.2 – 5.5. It is the important indicator to decide demineralization, remineralization and mineral transfer of hard tooth tissue at the plaque—enamel interface. Enamel will dissolve when it is below critical pH, and calcium and phosphorus will release into plaque. Acid tolerance of different bacteria is listed in Table 10.3.

Table 10.3 Acid tolerance of different species

bacteria	pH
Lactobacillus *Yeast*	< 4.5
Streptococcus	5.0
Veillonella	9.0 – 9.5
Other plaque bacteria	6.0 – 8.0

Bacteria can use carbohydrates to produce acid, and the place where acid accumulates is the ecologically dominant area of aciduric bacteria in dental plaque. *S. mutans* use carbohydrates to produce acid, sugar is the major metabolic substrate of *S. mutans*, sucrose makes the number of *S. mutans*, *Lactobacillus*, *Actinomyces* of plaque increase significantly, but high concentration of sucrose on *S. mutans* will have "sugar kill" effect. Zhou *et al.*[4] indicated that the growth of *S.*

mutans was inhibited significantly at a sucrose concentration of 10% – 20%; xylitol, maltose can reduce the number of *S. mutans* in plaque; when glucose is used as a carbon source, xylitol and sorbitol can inhibit the growth of *S. mutans*. If fructose is used as carbon source, there will not be such inhibition.

Fluoride is an effective anti-caries element, it can increase the strength of enamel caries resistance, inhibit bacterial action, and thereby regulates the ecological balance of dental plaque. Fluoride can protect the enamel by influencing the nature and structure of enamel, including the interruption of the Calcium Bridge, removing adsorption of bacteria and protein on the enamel surface.

10.2 Oral Infectious Diseases

Oral infectious diseases include dental caries, dental pulp and periapical disease, periodontal disease, maxillofacial infections, mucosal disease, and secondary infections of teeth restoration.

10.2.1 Dental Caries

Dental caries is one of the common frequently occurring human diseases and it is regarded as one of the three highest incidence diseases by WHO, but due to its slow course, it does not threaten patients' lives under normal circumstances, so it does not gain much attention.

Dental caries is a chronic progressive destruction disease; it is influenced by many factors, and bacteria are the main factor, with the demineralization of inorganic minerals and decomposition of organic matter in dental hard tissue. Although dental caries is a bacterium infectious disease of dental hard tissue, the harm of dental caries is not confined to dental hard tissue. It can develop into deep parts of the teeth, causing dental pulp disease, periapical disease, jaw inflammation and a series of complications, even having serious impact on general health. With the continuous damage to tooth hard tissue, it can cause crown defects and gradually become a residual root problem, eventually leading to tooth loss, damaging the integrity of the masticatory organ. This not only affects the digestive function, but also affects growth and development of the dentognathic system in childhood, so that the quality of health declines. In addition, lesions, dental caries and its secondary diseases can cause bacterial endocarditis and other infectious diseases [12].

The clinical features of dental caries are changes in color, shape, quality and all other aspects of dental hard tissues. Demineralization of hard tissue in carious teeth occurs at the initial stage, the microcrystalline structure changes, teeth transparency declines, causing enamel to become chalk in color. Then there is pigmentation in the lesion, a local part becomes yellowish-brown or dark brown in color. With the continuous demineralization of inorganic ingredients and decomposition of organic ingredients, enamel and dentin will loosen and soften, and ultimately disintegrate into a tooth defect, forming a carious cavity (Fig. 10.1). Once a carious cavity forms, it will not have a self repair capacity [13].

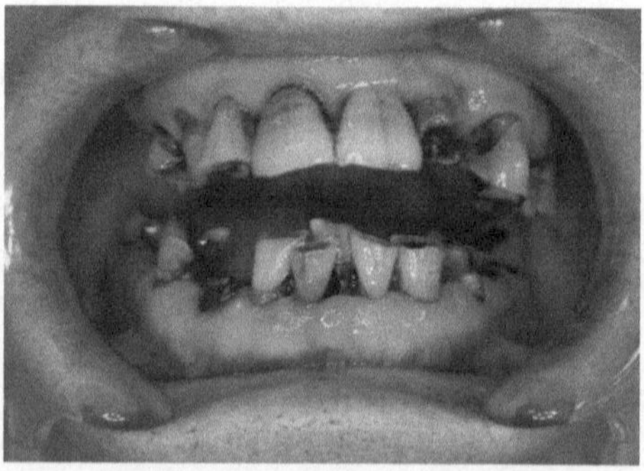

Fig. 10.1. Clinical signs of caries

10.2.1.1 Cariogenic Evidence of Microorganisms

The mouth is a complex environment and the incidence and development of dental caries closely relate to the microorganisms of the oral ecosystem. And like most oral infectious diseases, dental caries is not caused by a single bacterial infection, a variety of bacteria on the tooth surface relate to the occurrence and development of caries.

In 1881, Underwood and Miller pointed out that caries was caused by bacteria and the acid produced by bacteria. Miller isolated more than 30 kinds of bacteria from saliva and caries lesions. He pointed out that bacteria producing acids and decompose proteins could cause dental caries. Through a large number of studies on human and animals, scholars confirmed the role of bacteria in the development of dental caries. Before eruption, a tooth does not have contact with the oral environment, so caries will not occur.

Sterile animal caries test. In 1954, Orland *et al.* observed the relationship between bacteria and dental caries by inoculation of oral bacteria in sterile animals. The results indicated that without inoculation of bacteria, even if they were fed with high-sucrose cariogenic food, sterile mice did not develop caries, while sterile mice inoculated with gram-positive bacillus and *enterococci*, or only *enterococci* would develop caries.

Bacteria can cause caries-like lesions. In the early part of the 19th century, Miller *et al.* confirmed that bacteria can cause caries-like lesions by *in vitro* experiment, incubating teeth, saliva and bread together. Tooth demineralization will occur, because of acid produced by bacterial fermentation of carbohydrates in saliva. A large number of bacteria were observed in carious dentin and enamel under light microscope and electron microscope. When incubating teeth, bread and heated saliva (bacteria were killed) together, because there is no bacterial involvement, even if there are carbohydrates and no tooth demineralization will occur.

Antibiotics can reduce the incidence and severity of caries. In 1946, Meclure and Hewitt through rat caries experiments, after adding penicillin into beverages and water, showed that antibiotics will significantly reduce the incidence and severity of caries in rats. Epidemiological surveys of human dental caries found that the incidence of caries in patients with long-term use of antibiotics was significantly lower than those who did not receive antibiotics.

10.2.1.2 The Etiologic Theory of Caries

The etiology of dental caries has been the discussion focus of experts, including the theories of specific bacterial infection and non-specific bacterial infection, single factor and multiple factor pathogenesis theory of dental caries, and the ecological pathogenesis theory.

Hypothesis of specific and non-specific bacterial infection. The mouth is a complex ecological environment. In this environment, a great variety of bacteria exist in the complex symbiosis relationship and antagonistic relationship with each other. Bacteria have their own metabolic activities, and generate complex metabolites. Although there is no doubt of the role of bacteria in the occurrence of dental caries, however, only 1/10 of oral bacteria can be cultivated *in vitro*. In a complex system of oral bacteria, to determine the relationship between specific bacteria and dental caries is still the direction in which researchers are working.

The theory of specific bacteria. Through a series of experiments on human and animals, some researchers sought to confirm caries was caused by specific bacterial infection. In 1946, Meclure and Hewiit carried out animal caries experiments. Adding penicillin could inhibit the incidence of caries in rats. As a result, it was thought that the incidence of dental caries was related to antibiotics sensitive bacteria, the cariogenicity of plaque related to the composition of bacteria. Two commonly used animal models of experimental caries are: one with its own normal flora, but specific pathogen free animal, the other is the gnotobiotic animal. The researchers believe that bacteria induced caries experiments in animal models show that the occurrence of dental caries relates to specific pathogens, and has the following characteristics: (i) Different bacteria have different cariogenicity; (ii) Different types of bacteria can cause different parts of caries (smooth surface caries, pit and fissure caries and root caries) (Table 10.4); (iii) All cariogenic bacteria can degrade sugar and produce acid. *S. mutans* is a major pathogen of dental caries, followed by *S. sobrinus*, *Actinomyces* and *Lactobacillus*.

Table 10.4 The relationship between oral bacteria and dental caries

Bacteria	Smooth surface caries	Pit and fissure caries	Root caries
S. mutans	+	+	+
S. sobrinus	+	+	+
S. salivarius	+	–	–
S. sanguis	–	+	–
S. mitis	–	+	–
Peptostreptococcus	–	+	–
A. viscosus	–	+	+
A. naeslundii	–	+	+
A. israelii	–	+	+
L. casei	–	+	–
L. acidophilus	–	+	–

Non-specific bacterial hypothesis. The theory of a chemical parasitic proposed by Miller is representative of non-specific bacterial infection hypothesis. At first, Miller tried to find one or several bacteria in samples that were closely related to caries, but he isolated more than 30 kinds of bacteria from saliva and caries-like lesions, one of the many bacteria that can ferment carbohydrates. Some can decompose protein, but he still did not identify a specific bacterial cause of dental caries. Accordingly, he considered that the bacteria causing dental caries were non-specific. Any bacteria that could produce acid and decompose protein can lead to dental caries.

Triple and quadruple etiologic theory of caries. The triple etiologic theory of caries regards host, microorganism and diet as the main factors causing dental caries; three kinds of factors interact and play an important role in the occurrence of the disease. The quadruple etiologic theory of caries thought that the time factor must be added, that the occurrence of caries involves the caries-sensitive host, the role of oral cariogenic bacteria and the appropriate substrates (including food carbohydrates), and these substrates must remain in the oral cavity for enough time (Fig. 10.2).

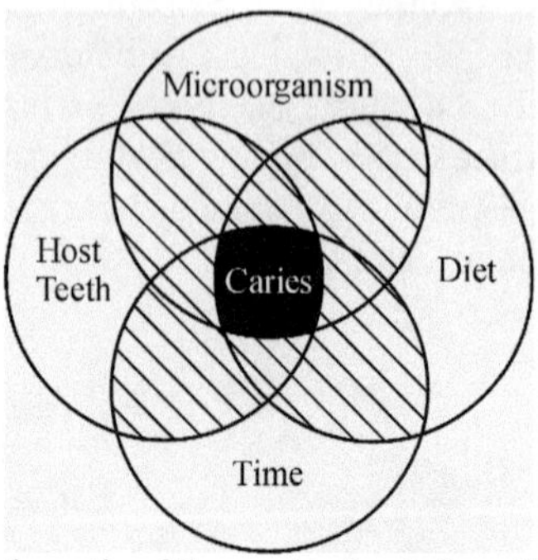

Fig. 10.2. Schematic of the quadruple etiology theory of caries

Ecological etiology theory of caries. The mouth is a complex ecological environment, and in this environment, a great variety of bacteria forms a complex symbiosis relationship and antagonistic relationship with each other. Although there is no doubt about the role of bacteria in dental caries, however, only 1/10 of bacteria in the oral cavity can be cultivated *in vitro*. In a complex oral bacteria system, the specific pathogen of dental caries has not yet been definitively identified. With the ever deepening development of molecular biology and oral ecology, ecological etiology theory of oral infectious diseases including dental caries, dental pulp and periapical infections, periodontal disease and odontogenic maxillofacial infections are being given more attention and recognition by scholars. In 1986, Xiaorong Xiao indicated that lots of normal oral flora existing in the mouth were pathogens of oral caries, dental pulp and periapical infections, periodontal disease and odontogenic maxillofacial infections. The triple etiology theory of caries which has long been recognized as the cause of caries and the current proposed quadruple etiology theory of caries is in essence caries etiology theory based on ecological theory. Dental caries is a result of the interaction of the microbial, the host's and their related factors (such as diet). It is a manifestation of oral ecological imbalance. In 1991, Marsh *et al.* [11] proposed ecological plaque hypothesis. It understands dental plaque and interaction, balance and imbalance of various ecological factors in dental plaque from an ecological perspective, and reaches a dynamic understanding of the pathogenicity of dental plaque. A dental plaque microecological system is an ecological system with certain functions consisting of microbiota and their environment. The teeth acquired pellicle, plaque bacteria and the matrix, saliva factors and exogenous food are the main ecological factors of the system. The occurrence of dental caries is attributed to changes in environmental factors and disorder of plaque ecological balance.

10.2.1.3 The Pathogenic Microorganisms of Dental Caries

Through the sampling inspection of bacteria quantity and type in the mouth of patients with caries or in caries lesions, to understand the relationship between the existence of certain bacteria and the dental caries, we find that *S. mutans* in caries lesions are more than those on a normal healthy tooth surface. The numbers of them are increasing with the occurrence of dental caries. However, *S. sanguis* and *Veillonella* are decreasing. *S. mutans* are positively correlated with dental caries. In addition, we also isolated many *Lactobacillus* and *Actinomyces* from caries lesions [14].

Further *in vivo* experimental study of cariogenic bacteria showed that: (i) The incidence of caries is closely associated with *S. mutans*, *S. mutans* is the main cariogenic bacterium, and other cariogenic bacteria include *A. viscosus*, *A. naeslundii* and *Lactobacillus* strains; (ii) Cariogenic bacteria are not a single bacterium, but are several or many acidogenic and aciduric bacteria. With the incidence of dental caries, the bacteria ratio in dental plaque can change constantly. As the number of certain species increases, the number of other bacteria may decrease (Table 10.5).

Table 10.5 Plaque bacteria changes of caries

↑	↓
S. mutans	*S. sanguis*
Actinomyces	*Veillonella*
Lactobacillus	

***Streptococcus mutans*.** This is a Gram-positive, facultative anaerobic coccus, arranged in chains. *S. mutans* and *S. sobrinus*, *S. cricetus*, *S. rattus*, *S. ferus*, and *S. macacae* make up the *mutans Streptococcus* group, and according to differences in antigen composition in the bacterial wall, they are divided into 8 serotypes of variants: a, b, c, d, e, f, g, h, and the serotype c of *S. mutans* and *S. sobrinus* closely relate to human dental caries [15].

S. mutans are usually found in plaques from individuals with high caries activity, and the cariogenicity is mainly attributed to its biological characteristics such as strong adhesion, strong acidogenic and aciduric ability and production of water-insoluble polysaccharides *etc.* [16].

Adhesion ability. The adhesion of *S. mutans* to the tooth surface occurs with high selectivity and affinity. The lipoteichoic acid on the cell surface of *S. mutans* could attach to the tooth surface through the calcium bridge connection with acidic saliva glycoprotein in the acquired pellicle. Bacterial cell constituents can interact with serum reactive salivary mucin and produce adhesion. *S. mutans* have strong cell surface hydrophobicity (CSH). This may be relevant to the specific adhesion of bacteria. Adhesin-receptor theory explains the oral bacterial adhesion mechanism at the molecular level. Adhesin is a protein-like component on the bacterial surface. It could specifically bond to complementary molecules in a stereo chemical way, and these complementary molecules located on the surface of tissues are called receptors. Adhesins are also known as attachments, they are the fimbriae or pili of the fiber-like protrusions on the bacterial surface, and there are some non-protruding adhesin. Most of adhesins are plant lectin-like substances, they are mainly bonded to glycoprotein and glycolipid receptors, and some adhesin, can bond to non-glycosylated glycoproteins.

The cell surface components that relate to adhesion of *S. mutans* include surface protein PS1 with molecular weight 185 kD, glycoprotein with molecular weight of 115 kD, 120 kD and 135 kD and lipids teichoic acid (LTA). Surface protein P1 and P1-like protein are adhesin on the cell surface of *S. mutans* which are being recognized and studied more. They involve non-sucrose-dependent adhesion between *S. mutans* and teeth surfaces, and they have strong affinity to APRPs.

Acidogenic and aciduric ability. The strong acidogenic and aciduric abilities are one of the important mechanisms of cariogenicity of *S. mutans*. *S. mutans* can produce a large amount of lactic acid by fermentation of carbohydrates, resulting in enamel demineralization. Lactate-dehydrogenase (LDH) is a key enzyme for lactic acid synthesis by *S. mutans*. It is the end enzyme of the Embden-Meyerhor way, and it can catalyze the conversion of pyruvate to lactate, known as the leader of the "lactic acid gate".

LDH activity of *S. mutans* depends on the presence of 1,6-diphosphate (D-fructose-1, 6-diphosphate, FDP) High concentrations of FDP could enhance its activity and it is called the FDP-dependent lactate dehydrogenase (FDP-dependent-LDH). When sugar is adequate in the environment, the bacterial cells can produce high levels of FDP and can activate LDH activity, making the "lactic acid gate" open, metabolizing pyruvate and generating large amounts of lactic acid, reducing the environmental pH value. When the pH is lower than the critical pH of teeth demineralization, enamel demineralization and caries occur.

The strong aciduric ability of *S. mutans* is that it can grow and metabolize in the acidic environment with a pH value of around 5. The growth and glycolysis aciduric ability of *S. mutans* mainly depend on the regulation of proton-translocating membrane adenosine triphosphatase (H^+-ATPase), its acid tolerance is mainly achieved by improving the H^+-ATPase activity and maintaining the cell transmembrane pH.

Li *et al.*[17] found that comC, comD or comE are aciduric-related genes of *S. mutans*. The aciduric ability of the comCDE gene deletion strains decrease. The RopA gene may be one of the major regulatory genes of aciduric ability and biofilm formation of *S. mutans*. In a low pH environment, the tolerance of RopA gene deletion *S. mutans* strains to acid killing and oxidative stress will reduce.

Glucosyl transferase (GTFs). It is the important virulence factor of *S. mutans*; it can mediate bacterial adhesion and agglomeration. In 1986, Loesche first reported *S. mutans* could produce extracellular GTFs gene. *S. mutans* could produce three GTFs: non-water soluble GTF (GTF-1), non-dependent water-soluble GTF enzyme (GTF-SI) and the dependent water-soluble GTF (GTF-S). Their coding genes were *gtfB*, *gtfC* and *gtfD*. *GtfB* encoded GTF-1 can synthesise IG; *gtfC* encoded GTF-SI can synthesise a mixture of IG and SG; *gtfD* encoded GTF-S can synthesise SG. *gtfC* gene is close to downstream of the *gtfB* gene, and their genes are highly homologous. *GtfB* and *gtfC* genes closely relate to the sucrose-dependent adhesion of *S. mutans*, the inactivation of any one of these two genes would cause the *in vitro* sucrose dependent adhesion ability of *S. mutans* to reduce significantly. Munro *et al.* built the *gtfB* and *gtfC* missing *S. mutans* deficient strain, and the *in vitro* sucrose dependent adhesion ability of them were lost. In the gnotobiotic rat model, the ability to cause buccal, proximal and fissure caries was significantly lower than in the wild strain.

Lactobacillus. Gram-positive, nonacid-fast, nonmotile, facultative or obligate anaerobic rods are divided into two sets: (i) Obligate homofermentative species produce mainly lactic acid by fermentation of glucose and are closely related to dental caries and the represented strain are *L. casei* and *L. acidophilus*; (ii) Facultative heterofermentative species produce lactic acid and relatively large amounts of acetic acid, ethanol and carbon dioxide and the represented strain is *L. fermentum*.

Lactobacillus is known as a major pathogen of caries and relates more to dentine caries. *L. acidophilus* is frequently isolated in saliva, but *L. fermentum* and *L. casei* are frequently isolated in dental plaque.

Actinomyces. Gram-positive, non-sporation, nonmotile, irregular rods have branch or slender filaments, 0.2 – 1.0 in diameter and the length changes significantly. *Actinomyces* can produce major lactic acid and a small quantity of acetic acid, succinic acid and a trace amount of formic acid by fermentation of glucose. The subculture of *A. naeslundii* and *A. viscosus* can grow in a facultative condition, but *A. israelii*, *A. meyeri* and *A. odontolyticus* only grow in an anaerobic condition.

Actinomyces is usually found in plaque of root caries and subgingival plaque and the most frequently isolated gram-positive bacillus in subgingival flora and human root surface caries plaque is *Actinomyces*. In all supragingival plaque, you can find *Actinomyces*, the amount of it accounts for about 50% of total bacteria. *A. naeslundii* is often isolated from dorsum lingual, saliva and dental plaque in children, and *A. viscosus* is one of the nonage establishment floras in supragingival plaque.

The inoculation of *A. viscosus* and *A. naeslundii* can cause root caries, pit and fissure caries and periodontal tissue destruction in experimental animals. *A. viscosus*, *A. naeslundii* and *A. odontolyticus* have a strong affinity to the main component of the dentin and cementum matrix-collagen on the tooth surface. It is clearly beneficial to the adhesion of the root surface. In dental plaque biofilm, *A. viscosus* and *A. odontolyticus* gather as bridges, working as adhesion scaffold for other bacteria, forming corn-like or bottle brush-like structure, accelerating plaque formation and maturation.

Adhesins of *A. viscosus* mainly exist in the fimbriae on the cell surface. They are fimbriae I and II. Fimbriae II mainly mediate agglomeration and epithelial cell adhesion of *S. mutans*, *S. sanguis*, *Veillonella*, *P. gingivalis* and *F. nucleatum*. *A. viscosus* forms extracellular levans and heteropolysaccharides, the main components of which are hexosamine and hexose. These polysaccharides only have low-cariogenicity.

A. viscosus, *A. naeslundii* and *A. odontolyticus* have better affinity to facing the collagen component of dentin and cementum and in favor of adhering on the root face. *A. viscosus*i and *A. odontolyticus* as a coaggregation bridge in plaque biofilm can become adhere support and form an ear or brush structure and advance the plaque form and mature. Pilus and pilus II are adhesin of *A. viscosus* and pilus II mediate in coaggregation with other bacteria, such as *S. mutans*, *S. sanguis*, *Veillonella*, *P. gingivalis* and *F. nucleatum*, and adhere to epithelium. *A. viscosus* can form ecto-levans and heteropolysaccharides and the main component is hexosamine and hexose which has lower cariogenicity.

Ecological relationship between *S. mutans* and *S. sanguis*. *S. sanguis* is one of the resident floras inhabiting early on the tooth surface. It is also the frequently isolated *Streptococcus* species in the mouth. It is mainly distributed on the tooth surface and accounts for half of the *Streptococcus* species in plaque. Like *S. mutans*, *S. sanguis* can make use of sucrose and insoluble syntheses and soluble extracellular polysaccharides. These polysaccharides play an important role in plaque formation and aggregation of bacteria on hard tissue. Although some scholars believed that *S. sanguis* might relate to fissure caries, the detection rate of *S. sanguis* in patients with caries was not high. Some scholars pointed out that the number of *S. sanguis* was negatively correlated to the appearance of caries lesions. The ecological relationship between *S. sanguis* and the main cariogenic bacteria *S. mutans* is the content that is of interest to researchers.

Adhesion and agglomeration. *S. sanguis* is one of the bacteria that first colonize in the mouth. The villus structure of *S. sanguis* is similar to cilia of *Streptococcus*. It contains M protein and is resistant to phagocytic cells. It can attach to epithelial cells; the main adhesin of *S. sanguis* are surface lipoteichoic acid (LTA), P1 and P1-like protein of SR and SpaA, they can bond to salivary protein and S-IgA *etc.*, they have a special affinity to the acquired pellicle of the tooth surface with the protein binding to the acquired pellicle, and they can first adhere to the precursor of dental plaque-saliva acquired pellicle; H_2O_2 produced by *S. sanguis* can inhibit the growth of a variety of oral microorganisms, which suggests that in the early stages of plaque formation, *S. sanguis* lives in an environment with high oxygen pressure. It can first adhere to tooth surfaces by competitive inhibition of other bacteria. Carlsson proposed that teeth were colonization places of *S. sanguis* and *S. mutans*, but they could not be detected in baby's mouth before teeth eruption, and large numbers of these two bacteria would appear after teeth eruption, but they would disappear when all teeth were lost. *S. sanguis* was detected earlier than *S. mutans* in the oral cavity. It suggests that the early adhesion and colonization on the tooth surface of *S. sanguis* provide the necessary conditions for adhesion of *S. mutans* to the tooth surface.

Symbiosis. The mutual relationship of nutrition can be the metabolites of some bacteria which become the necessary nutrients or growth factors of other bacteria. It is also called the symbiotic ecological factor. The para-aminobenzoic acid, (PABA) produced by *S. sanguis*, was a specific growth factor for *S. mutans*. It might affect the growth, colonization and agglomeration of *S. mutans* on the tooth surface. In 1999, Guo and Zhou *et al.*[4] used the adhesion method with the 3H-thymidine label. They found that a certain concentration of PABA might interact with the composition of the bacterial cell surface adhesin or change its configuration, and thus interfere with the interaction between bacteria and receptors in the salivary acquired pellicle.

Antagonism. According to characteristics of *S. sanguis* colonized in the oral cavity over a certain period of time (about 8 – 12 mon after birth, average 9 mon), research found that after *S. sanguis* colonized in a baby's mouth, the number kept rising, but when *S. mutans* colonized in the mouth, the number of *S. sanguis* started to decline. It also found that if there was a high proportion of *S. sanguis* in the baby's mouth, the colonization of *S. mutans* would be delayed at least 6 mon.

H_2O_2 produced by *S. sanguis* can inhibit the growth of a variety of oral microorganisms including *S. mutans*. The bacteriocin and bacteriocin-like substance is a hot spot of current research, bacteriocin and bacteriocin-like substances are a general designation of proteins or complexes of proteins and fat or sugar with strong bactericidal activity and produced by bacteria. As the oral dominant bacteria and the resident flora, both *S. sanguis* and *S. mutans* can produce corresponding bacteriocin, and have significant antibacterial activity on the corresponding bacteria. The sanguicin produced by *S. sanguis* has antagonistic effects on *S. mutans*, and the mutacin produced by serotype c-type and e-type *S.*

mutans can inhibit *S. sanguis*, *A. viscous* and other oral bacteria. It has been shown that with the increase in *S. mutans* in dental plaque, there will be a reduction of *S. sanguis*. For the bacteriocins produced by *S. sanguis* and *S. mutans* and their antagonistic bacteria see Table 10.6 [18].

Table 10.6 Bacteriocin of *S. sanguis* and *S. mutans*

Bacteria	Bacteriocin	Antistatic bacterium
S. sanguis	sanguicin	*P. gingivalis* *Aggregatibacter actinomycetemcomitans* *F. nucleatum* *Capnocytophaga* *S. mutans*
S. mutans	mutacin	*S. sanguis* *A. viscosus*

The relationship with the formation and ecological balance of plaque. Adhesin on the cell surface of *Streptococcus* can specifically bond to the composition of a salivary acquired pellicle; it makes a good affinity with the bacteria of the acquired pellicle. H_2O_2 produced by *S. sanguis* can inhibit the growth of a variety of oral microorganisms, which suggests that in the early stages of plaque formation, *S. sanguis* lives in the environment with high oxygen pressure; it can firstly adhere to tooth surfaces by competitive inhibition of other bacteria. In addition, *S. sanguis* also can promote the formation of the acquired pellicle, the mechanism of which is neuraminidase produced by *S. sanguis* decomposed neuraminic acid, and makes the sialic acid molecule of mucopolysaccharide dissociate and promotes adhesive glycoprotein precipitation and adsorption to the tooth surface, forming the acquired pellicle. Thus it can be seen that *S. sanguis* has a good foundation for the first adhesion and colonization on the teeth surface. It is known as the pioneer bacteria. For the ecological relationships of *S. sanguis* and *S. mutans* detailed in Section I of this chapter, the early colonization of *S. sanguis* provides the necessary conditions for the adhesion and colonization of *S. mutans* on tooth surfaces. A variety of cell surface proteins of *S. mutans* such as GTF, fructosyltransferase (FIF), glucan-binding protein (GBP), and cell surface antigen P1, PAC antigen can also directly mediate the adhesion of bacteria and salivary glycoprotein of the acquired pellicle on the tooth surface.

The relationship between *S. sanguis* and the development and maturity of the dental plaque in a number of studies show that, *S. sanguis* is the earliest colonized bacterium on the tooth surface. *S. sanguis* begins a fast-growing trend after the initial colonization, the number of *S. sanguis* increases significantly in 2 – 6 h of plaque formation and it decreases after 6 – 24 h. The ecological relationship between *S. sanguis* and *S. mutans* not only provides the necessary conditions for the adhesion and colonization of *S. mutans* on the tooth surface, but also promotes the growth of *S. mutans* in dental plaque and provides the necessary growth factor – PABA for its growth.

The maturity of dental plaque is a dynamic process. In the process, the composition and proportion of bacteria and the structure of plaque are constantly changing; the reduction of oxygen content of plaque and the decrease in redox potential are beneficial to the growth of anaerobic bacteria. *S. sanguis* and *S. mutans* are both facultative anaerobes, they have ecological relationships not only mutually symbiotic but also mutually antagonistic in the plaque, and they are carrying out a complex metabolism in the plaque and produce water-soluble glucan, extracellular water-insoluble glucan, sanguicin and mutacin *etc.*, which accelerate the development and growth speed of dental plaque. In dental plaque, *S. sanguis* and *S. mutans* are residential and dominant bacteria, the ecological relationship of mutual symbiosis and antagonism of the two bacteria are closely related to the ecological balance of dental plaque. The symbiosis of bacteria is the basis of the microecological balance of plaque, while the antagonism of bacteria lies in regulating the microecological balance, and neither one of them is dispensable.

During the past 30 years, many scholars carried out broad and in-depth studies on the relationship between *S. sanguis*, *S. mutans* and caries. Animal experiments and epidemiological investigation studies have both found that the number of *S. mutans* is positively correlated with dental caries damage, and the number of *S. sanguis* is negatively correlated with the appearance of caries damage. There is no significant difference between cariogenic plaque and non-cariogenic plaque on the type of the bacteria composition, but the proportion of different bacteria varies. *S. mutans* cariogenic plaque, however, *S. sanguis* takes up a high proportion of non-cariogenic plaque (Table 10.7). The bacterial composition of early stages dental plaque also has significant differences between active caries patients and inactive caries patients. A high percentage of *S. sanguis* colonizes in plaque of inactive caries patients. On the contrary, less *S. sanguis* colonizes in active caries patients

Table 10.7 Bacteria species of non-cariogenic and cariogenic plaque

Sort of bacteria composition	Non-cariogenic plaque	Cariogenic plaque
high	*Actinomyces*	*S. mutans*
|	*S. sanguis*	*Actinomyces*
|	*S. mutans*	*S. sanguis*
low	*Lactobacillus*	*Lactobacillus*

Ecological plaque hypothesis theory indicated that: dental caries is the bacterial infectious disease occurring on dental hard tissues. It is caused by disorder of plaque ecology and the direct reason for the demineralization of dental hard tissue and caries lesion formation are the effects of cariogenic bacteria. However, many factors can affect the ecological balance, including bacteria composition of plaque, Eh, pH, caries susceptibility of host, type and concentration of intake sugar and heredity *etc.*

Scholars have carried out a great deal of research on the role of *S. mutans* with strong cariogenicity and *S. sanguis* and the close ecological relationship *S. mutans* play in the ecological imbalance of dental plaque and development of caries. The detection rate of *S. mutans* in plaque is positively related to dental caries in the mouth; the proportion of *S. mutans* in caries lesion is higher than that on a caries-free tooth surface; the distribution of *S. mutans* in a caries lesion is more in the caries lesion and less in the tooth surface far away from caries; using the vaccine made from *S. mutans* can produce an anti-caries effect. Despite the fact that there is some cariogenicity of *S. sanguis* in the animal model, the detection rate of *S. sanguis* in patients with caries is higher than in those without caries. There is no evidence to prove that *S. sanguis* is the main cariogenic bacterium of human dental caries.

In the studies on the relationship between colonization of *S. sanguis* and *S. mutans* on a tooth surface and caries prevalence, they found that despite the early colonization of *S. sanguis* on the tooth surface, it provides the necessary conditions for colonization of *S. mutans* on the tooth surface, but the early colonization of *S. sanguis* in a baby's mouth will lead to a time delay in colonization of *S. mutans*, and the time delay in colonization of *Streptococcus mutans* will in turn lead to caries reduction. As to the close relationship between delayed colonization of *S. mutans* and the reduction of the caries prevalence rate, it can be assumed that the caries risk is based on the level of *S. sanguis* or initial infection time of *S. sanguis*. This view has been approved by more and more scholars. Because the cariogenicity of *S. sanguis* is lower than that of *S. mutans*, many scholars have suggested that the ratio of *S. mutans*/*S. sanguis* is used as a sign of caries risk. The smaller the ratio, the lower the risk of caries. More in-depth studies on the relationship between the ecological relationship of *S. sanguis*, *S. mutans* and caries caused by ecological imbalance of dental plaque will promote studies on ecological control of dental plaque.

10.2.1.4 Ecological Control of Dental Caries

Ecological control of dental caries is one of the important areas of applied research in caries microecology. It refers to a variety of medical interventions that can promote the ecological system, which consists of microorganisms, teeth and their environment, to change from a state of pathological imbalance to a state of physiological balance, and thus meet ecological principles. Dental caries is a disease of symbiotic disorder; it is possible to do comprehensive prevention by way of ecological control.

Basic principles. Basic principles of ecological control are to remove the pathological state that causes the ecological imbalance. The key point of ecological control of dental caries is to maintain or restore the ecological balance of dental plaque, which must remove the internal and external environmental factors causing ecological imbalance of dental plaque. The basic principles are as follows.

Ecological imbalance can cause pathological conditions or diseases, and pathological conditions or diseases can also cause ecological imbalance. When conducting ecological control of dental caries, the role of these diseases should be excluded.

The abnormalities of anatomical tooth structures such as irregular dentition and poor tooth development *etc.*, is either physiological or pathological and can cause ecological imbalances and thus effective orthodontic treatment is necessary to improve abnormal dentition.

To improve the adaptability of the teeth is to improve the tooth resistance to adverse microbial flora effects, Including immunization, nutrition and all health interventions.

Important approaches. Microbial studies of dental caries consider that the prevention of dental caries can not only be achieved by controlling the pathogen but also by preventing multiple factors of plaque microorganisms to take effect. The main approaches of the ecological control of caries are as follows.

Fluoride. The primary role of fluoride is to enhance enamel resistance to demineralization, promote remineralization and inhibit bacteria growth. The anti-microbial effect of fluoride is very clear in dental caries, and is also affected by environmental pH. Under low pH conditions, the antibacterial and anti-metabolism effect of sub-inhibitory concentration of fluoride increase significantly, slowing down the change in pH can reduce the plaque bacteria turning into a disease state, which means that low concentrations of fluoride can maintain the acid sensitive bacteria at a high level by slowing the speed of acid production, while there will be no such phenomenon when there is no fluoride. Therefore, the fluoride can stabilize the composition of bacteria. Applications of fluoride against dental caries include systemic and topical application. Systemic fluoride applications can enhance the caries resistance of teeth by changing the morphology of teeth and creating a state of fluoride hydroxyapatite, reducing the solubility, and increasing the caries lesion remineralization. They also include water fluoridation, food fluoridation, milk fluoridation, fluoride tablets, fluoride drops and other systemic methods; topical applications include fluoride toothpaste, fluoride mouth rinse, local fluoride and fluoride gel *etc.* Topical fluoride can inhibit the growth of *S. mutans*. The sub-inhibitory concentration of fluoride could reduce the glycolysis of *S. mutans* at low pH conditions and reduce the acid production rate, preventing excessive growth of it at low pH. Long-term use of fluoride toothpicks will inhibit the growth of *S. mutans*.

Food regulation. Low pH caused by repeated intake of fermentable carbohydrates in the oral cavity will lead to microecological imbalance. Ecological control is to reduce the frequency and amount of acid production, using the following measures: (i) Inhibiting acid production, such as the use of sub-inhibitory concentration of chlorhexidine *etc.*; (ii) Simply avoiding snacks between meals; (iii) Using sugar substitutes such as xylitol which are only utilized by a small number of bacteria and do not produce acid; (iv) Stimulating saliva secretion after a meal; such as chewing gum after meals *etc.*; (v) Eating more vegetables, fruits and tubers; (vi) Eating beans or their products.

Stimulating saliva secretion. The role of mechanical saliva washing can eliminate residual sugar in the mouth. Buffer systems, urea, ammonia, peptides and other alkali contained in saliva can increase the local pH, and abundant mineral in saliva can promote remineralization. Methods for stimulating saliva secretion are: (i) By using sugar substitutes, stimulating saliva secretion by sweets and promoting remineralization. (ii) Chewing gum can stimulate saliva. Regular use of sugar substitutes-sorbitol gum is not only non-cariogenic; it also has some therapeutic effects.

Improving the adaptability of the host. Improve the adaptability of the host to enhance the resistance to adverse conditions. For example, carry out extensive and in-depth oral health education, and supply timely and appropriate amounts of calcium for infants to promote normal development of teeth.

Antibacterial agents. Antibiotics, chlorhexidine, natural medicine and some trace elements can inhibit the growth and metabolism of *S. mutans*. The use of antibiotics can inhibit the growth of *S. mutans*, but no report has been made on the specific role of antibiotics on *S. mutans*. In addition to fluoride, certain trace elements also can inhibit the growth and acid production of *S. mutans*; it has been proved that zinc can inhibit the growth of *S. mutans* significantly in the artificial mouth. Gallagher and Cutres have proved that: Fluoride, manganese, copper, zinc, selenium, silver, cadmium, antimony, barium etc can strongly inhibit growth and acid production of *S. mutans* group. Titanium, nickel and other elements in prostheses are found to inhibit the growth and acid production of *S. mutans*. There has been much literature recently on studies about natural medicines such as tea polyphenol, scutellaria, Galla Chinensis, Nidus Vespae *etc*. related to growth of *S. mutans*, acid production and GTF influence. Galla Chinensis, Nidus Vespae and polyphenol extracted from green tea (a type of polyphenol) were found to significantly inhibit the growth of *S. mutans* (MIC approximately 4 mg/mL).

Chlorhexidine can control plaque, interfere with bacteria metabolism and inhibit acid production. Studies found that sub-inhibitory concentration of chlorhexidine had this effect too. Using sub-inhibitory concentration of chlorhexidine can reduce the influence of rapid changes of pH on dental plaque microorganisms and demineralization. It prevents the disease without destroying the existing oral ecological environment.

Anti-caries vaccine. Active immunization against dental caries can be accomplished by developing a specific vaccine against caries. The body will receive specific anti-caries antibodies by appropriate vaccination. Surface antigens of *S. mutans*, such as AgI/II (a protein of relative molecular weight 1.9×10^5), AgIII (a protein of relative molecular weight 0.39×10^5), GTFs, (glucan binding proteins, GBPs), *etc*., can be used to construct a caries vaccine through genetic alterations. TianJia Liu's research group had already inserted an entire gene of PPC41 Pac into the eukaryotic PCDNA3 expression vector, and carried out detection of the expression in mammalian cells.

Genetic engineering recombinant antigen vaccine inserts gene encoding fragments of designated immune antigen into genes of bacteria, yeast or mammalian cells with continuous conversion. Preparing purified vaccine containing only antigen immunity is complex. Köhler and Milstein created a new technology: Passive immunization of monoclonal antibody (McAb) showed significance in the prevention of dental caries. McAb may interfere with some important metabolism aspects of the growth of *S. mutans*, resulting in inhibited growth of *S. mutans*. In addition, it is reported that after the use of GTFase strong expression strains of *S. mutans* provided immunity in cows and hens, and specific antibodies against *S. mutans* of IgG and IgA were found in milk and egg yolk, and continued for several months but, if boiled, both antibodies lost immunological activity.

Biological replacement therapy. In studies of ecology prevention and treatment, the ecologist is mostly expected to use biological replacement therapy, known as a new anti-microbial strategy, namely using antagonistic bacteria (effect bacteria) to replace pathogens, also known as replacement therapy. How to select effect bacteria and improve the colonization stability of effect bacteria in the mouth is critical. Tanzer thought that the effect of bacteria in replacement therapy must meet the following requirements: (i) Have the ability to compete with oral inherent bacteria; (ii) Have the bonding force on the tooth surface, bonding to teeth surfaces in the presence of sugar; (iii) Rapid growth in the mouth; (iv) Adapt to the mouth environment in which it locates, especially to adapt to rapid changes in pH; (v) Adapt to the microecological environment in which in which it locates to resist attack from the plaque matrix.

Currently there are two forms of replacement therapy: (i) Pro-emptive colonization, which is using the effective bacteria to occupy ecological sites in dental plaque before the pathogen colonization, establishing the microecological system, thus eliminating pathogens; (ii) Competitive displacement, using the more competitive bacteria to exclude pathogens from the plaque. *S. sanguis*, *S. salivarius* and GTFs gene knockout mutant strain of *S. mutans* are currently being studied more [19].

10.2.2 Pulpal and Periapical Diseases

Pulpal and periapical diseases and dental caries are the most common diseases in a dental clinic. Bacteria are the main pathogenic factor leading to these diseases.

10.2.2.1 The Characteristics of Pulp and Periapical Tissues

Dental pulp tissue is the only soft tissue in the tooth tissues. It locates in the pulp cavity surrounded by tooth tissue, connects to apical and periapical tissues through a narrow apical foramen. Dental pulp is a loose connective tissue; it is composed of cells, intercellular matrix and intercellular fluid. Under the microscope, pulp is artificially divided into four layers: Odontoblast cell layer, cell-free layer, multi-cell layer and the central area. Pulp has 4 basic functions: (i) The formation function; pulp odontoblasts continue to form dentin throughout the life of the teeth. (ii) Nutritional function; pulp provides oxygen, nutrients and dentin fluid for the odontoblast and odontoblast process to maintain the vitality of dentin. (iii) Sensory function; nerve-rich distribution of pulp is the basis for the exercise of function of pain. (iv) Defense function; pulp odontoblasts and connective tissues components can respond to external stimulation or injury and produce a protective defense function, including pain, reparative dentin formation and inflammation.

Periapical tissue is the apical periodontal tissue, including cementum, periodontium and the alveolar bone. 2/3 of the Cementum at the root coronal is a thin lamellar structure, and at the apical 1/3 of the root is a thick and irregular lamellar structure. The basic function of cementum is to attach the principal fiber to the root surface, and it can also repair pathological root absorption caused by inflammation and physiological root absorption caused by translocation of teeth. Apical periodontium locates in the gap between cementum and the alveolar bone. It connects the pulp through the apical foramen. Periodontium is composed of bundles of collagen fibers and the intervening loose connective tissues, connective tissues containing nerves, blood vessels and various cell components (such as fibroblasts, histocyte and undifferentiated mesenchymal cells). Periodontium has the function of feeling and pain conduction, as well as nutrition of cementum and differentiated cells. Undifferentiated mesenchymal cells in periodontium can differentiate into cementoblasts, osteoblasts or osteoclasts, *etc.* in the inflammatory process.

The alveolar bone is composed of the inherent alveolar bone and supporting bone; the inherent alveolar bone is a thin and dense bone forming the inner wall of the socket. There are many holes in an inherent alveolar bone, they are the channels of blood vessels and nerves and these holes give the inherent alveolar bone a cribriform appearance, which is called the laminae cribriformis.

10.2.2.2 Microbial Infection Approaches

Pathogenic bacteria and their products cause pulp and periapical tissue infections through tooth caries, periodontal pockets, or *via* blood circulation.

Via tooth infection. The most common approach of pulp and periapical infections is via tooth infection. The predisposition of the incidence of infection comes from caries, tooth wear, tooth fracture, wedge-shaped defects, cracked teeth and dental malformations and other tooth diseases existing in the oral cavity; exposure of dentinal tubules or direct exposure of pulp are the main reasons for pulp infection. Bacteria and their toxic products invade exposed dentinal tubules, causing demineralization of the dentin wall, expansion of the tubular cavity and finally injury, inflammation and necrosis of pulp; bacteria and metabolic products in the infected root canal can spread to periapical tissue via apical foramen or lateral root canals, then causing periapical abscess and other diseases.

Via periodontal infection. The bacteria in the deep periodontal pocket can go into the pulp through the apical foramen or apical lateral root canals and cause pulp infection; the pulp infection caused by this is called retrograde infection, and the pulpitis caused by periodontal infection is called retrograde pulpitis. The approach of this kind of infection is related to patients with periodontal diseases, in particular the elderly patients. The prevalence and severity of periodontal disease in elderly patients increase with age; when there is periodontal disease, the bacteria in deep periodontal pockets go into the pulp through the apical foramen or apical lateral root canals, or directly infect the periapical tissues, resulting in the infection of dental pulp and periapical tissues. Bacteriological studies found that bacteria causing pulp and periapical infections are similar to the bacterial composition of deep periodontal pockets, which suggests the correlation and possible cross-infection of pathogenic microorganisms between pulp and periapical infection and periodontal disease.

Via blood infection. The blood routes of infection are extremely rare in clinical practice; it is known as anachoresis. The predisposition can be derived from transient bacteremia caused by extractions, scaling and root canal treatment *etc.*, Bacteria can go into the traumatic pulp or periapical tissues through the bloodstream, colonizing and proliferating in the metabolic disordered or traumatic pulp and then causing pulp infection.

10.2.2.3 Type and Ecological Characteristics of Pathogens

In 1890, Milier *et al.* found the bacteria in human necrotic pulp tissue for the first time. In 1965, the bacteria were the initiator of pulp and periapical tissues infection in classic sterile mice experiments under sterile conditions (Table 10.8).

Table 10.8 Sterile mice experiments of pulp and periapical infection

Drilling to open pulp cavity of ordinary mouse and free mouse , and let it be exposed in the oral cavity in the asepsis condition	
↓Ordinary mouse	**↓Sterile mouse**
The experimental tooth develops with serious pulpitis, pulp necrosis and periapical periodontitis in 8 d	The pulp of the experimental tooth does not have progressive inflammation and the surgical wound heals quickly in 42 d

Type of pathogenic microorganisms. In the infected dental pulp and periapical tissues, the type of isolated bacteria ranges from several to dozens, due to infection approaches and clinical symptoms in different ways. As well as individual differences, the pathogenic microorganisms of pulp and periapical infection are different. Common pathogenic microorganisms are as follows:

Black-pigmented anaerobic buds. They are a large group of anaerobic bacteria that can produce black or brown pigment on the blood agar plate. They are the general terms of gram-negative anaerobic bacilli whose growth requires hemin and vitamin K, and they include *Porphyromonas* and black-pigmented *Prevotella*. Black-pigmented anaerobic buds belong to the *Bacteroides* group. As the variation of the classification in the Bergey's taxonomy of bacteria, the middle-fermented sugar *Bacteroides* group becomes a new genus, named *Prevotella*, including *P. intermedius etc.*; non-fermented sugar and melanin-producing *Bacteroides* group is renamed *Porphyromonas*, including *P. gingivalis etc.*.

Melanin-producing group of gram-negative anaerobic bacteria are the dominant bacteria in infected root canal. The bacteria closely related to pulp and periapical tissue infection are mainly *P. meningogenica*, *P. intermedius*, *P. asaccharolytica*, *P. endodontalis* and *P. gingivalis*, *etc*.

Melanin-producing gram-negative anaerobic bacillus can be cultured from the root canal of teeth with periapical periodontitis. The most frequently detected bacteria in clinical samples are *P. intermedius*, *P. endodontalis* and *P. gingivalis*. *P. endodontalis* can strongly induce purulent infection and it is considered as the special pathogen of pulp infection. *P. endodontalis* is isolated more from the root canal and apical abscess; it is not easily to detect in the general area of the mouth and the reason may be that it lacks pilus and is difficult to attach to oral soft tissues. *P. gingivalis* with pilus can firmly attach to host tissue cells and Gram-positive bacteria in dental plaque. It is the dominant bacteria in the infected root canal.

Non-melanin-producing Prevotella spp. is gram-negative anaerobic bacillus. The detection rates of *P. buccae*, *P. oris* and *P. oralis etc.* in the root canal are as high as that of melanin-producing anaerobic bacillus. A variety of non-melanin-producing anaerobic bacteria were isolated from root canals with necrotic pulp, among which *P. oris* and *P. buccae* were dominant bacteria.

Tannerella forsythensis. It is a gram-negative anaerobic bacillus; early studies found that it is closely related to rapidly progressive periodontitis and refractory periodontitis, and recent studies found that the detection rate *T. forsythensis* is higher in pulpitis and periapical periodontitis; it is related to pain, percussion pain, swelling and other clinical symptoms of acute exacerbation of chronic periapical periodontitis. Huang *et al.* used a Polymerase chain reaction technology in 38 cases of chronic periapical periodontitis, and found the detection rate of *T. forsythensis* in infected root canals to be 26.3%.

Campylobacter rectus. Former name is *Wolinella rectus*; it is a lively motility gram-negative obligate anaerobic bacteria. It is often isolated from infected root canal and chronic periapical granuloma, and with the infected root canal is often accompanied by periapical bone destruction.

Peptostreptococcus. It is gram-positive obligate anaerobic coccus; it is often isolated from infected pulp and periapical tissues. *P. micros* is the most common bacteria, the hyaluronidase produced by it plays an important role in damage of periapical tissues. At the same time, it generates a lot of hydrogen sulfide, and is closely related to stink of the root canal. In 1993, Siqueira *et al.* studied the bacteria composition of 30 cases of chronic periapical periodontitis using the 16S rDNA-PCR method; they found that *P. micros* was the major pathogen of the infected root canal [20].

Actinomyces. They are closely related to long-unhealed fistula and periapical bone destruction. *A. israelii*, *A. naeslundii*, *A. odontolyticus*, *A. meyeri* and *A. viscosus* were often detected in pulp and periapical infection.

Enterococcus faecalis. It is a gram-positive facultative anaerobic coccus; it is one of the research hot spots of pathogenic microorganisms of pulp and periapical infection. It is the important pathogen of root canal persistent infection and reinfection, the detection rate of it is higher in the infected root canal after treatment failure. *E. faecalis* cells can secrete cytolysin, gelatinase and other toxic substances to damage host cells, tolerate host non-specific immune response, and spread pathogenicity by plasmid conjugation among *Enterococcus* species.

Ecological characteristics of pathogens. Pulp and periapical diseases are secondary infections mainly caused by sick teeth or periodontal tissue inflammation; the pathogens detected from the lesions of pulp and periapical diseases are not a single bacterium but are mixed bacteria and they include caries pathogens and periodontal pathogens. Normal oral flora which causes endogenous mixed infection and anaerobic bacteria is the main pathogen, which is one of the ecological characteristics of pathogenic microorganisms, while the number of pathogenic microorganisms, types, and infection approaches which are closely related to clinical symptoms are another ecological characteristic of pathogenic microorganisms.

Normal oral microorganisms are the main pathogens. From infection approaches and pathogens types, it can be seen that pulp and periapical infection is the endogenous infection caused by opportunistic infection of normal oral microorganisms, and thus the dominant pathogens detected in lesions are normal oral microorganisms.

Anaerobic bacteria are the main pathogens. Bacteria are important pathogens of pulp and periapical diseases, including a variety of facultative anaerobes and obligate anaerobes, gram-positive or gram-negative cocci, bacilli, *campylobacteria etc.* Among many bacteria, the number and types of anaerobic bacteria are the most common, including melanin-producing gram-negative anaerobic bacteria such as *P. gingivalis*, *T. forsythensis*, *F. nucleatum*, *Peptostreptococcus etc.*

Mixed bacteria and multiple species infection. There is no report about single pathogen detection in lesions of pulp and periapical diseases, but are all mixed bacteria and multiple strains. Sousa *et al.* analyzed the bacteria composition of 30 cases of acute periapical periodontitis accompanied by pain and swelling using a culture method; a total of 117 bacteria were isolated, 75 of them were obligate anaerobes, 3.9 species were detected in each root canal on average [21]. Other bacteriological studies on pulp and periapical infection also pointed out that a variety of bacteria were isolated, which indicates that the pulp and periapical infections are mixed bacteria and multiple strain infections.

Pathogens are related to the infection approaches. As different infection approaches in the oral cavity, the number and types of pathogens is different and the main pathogens of pulp and periapical diseases are closely related to pathogenic organisms of dental caries and periodontitis. Bacteria species of pulpitis resulting from caries are similar to bacteria species of deep dentin caries; bacteria species of retrograde pulpitis resulting from deep periodontal pockets of periodontitis are similar to those of periodontal pockets. The types of inflammatory bacteria species in pulp are related to pulp infection approaches and whether the pulp cavity is exposed or not, pulpitis is mostly the result of caries; the types of bacteria are similar to those of deep dentin caries. Pathogens of exposed pulp cavity pulpitis are mostly dominated by gram-positive facultative anaerobes, such as *Streptococcus*, *Actinomyces* and *Lactobacillus* species *etc.*, *Candida* may also be detected; pathogens of closed pulp cavity pulpitis are less and mostly dominated by gram-negative anaerobic bacteria, including *P. gingivalis*, *T. forsythensis*, *P. micros*, *P. melaninogenica* and *F. nucleatum etc.*

Pathogens are associated with the clinical symptoms. The studies found that the number and types of pathogens of pulp and periapical diseases have a close relationship with clinical symptoms. (i) The co-existence and high detection rate of *P. gingivalis* and *P. micros* are closely related to inflammatory necrosis of pulp and irreversible pulpitis. (ii) Infected root canal is the root canal with necrotic pulp; *Prevotella* species and *Peptostreptococcus* species are often found in the root canal with exudation in patients with acute symptoms, but it is difficult to find these bacteria in dry root canals; anaerobic bacteria are the dominant pathogens of the infected root canal and periapical infection; they account for 2/3 or more of the total number of bacteria, and gram-negative anaerobic bacteria are more common especially. Results of bacteriological analysis of the infected root canal by a lot of researchers suggest that obligate anaerobic bacteria are the main pathogens in the infected root canal, and the most common ones are *P. micros*, *P. gingivalis*, *P. intermedius* and *F. nucleatum*. *Prevotella* species and *P. micros* are common in acute infection of root canals. Bacteria isolated from infected root canals are mixed bacterial infections. The bacteria of root canal mixed infection also often detect *P. melaninogenica*, *P. oralis*, *T. forsythensis*, *P. micros*, *F. nucleatum* and *E. faecalis etc.*; Sundqvist *et al.* found that *P. micros*, *F. nucleatum*, *P. intermedius* and *Peptostreptcoccus anaerobius* were usually detected in the same root canal [22]. (iii) Extensive periapical lesions are closely related with anaerobic bacteria. Periapical infections mostly result from endodontic infections, including periapical periodontitis and periapical abscess; the main pathogens are melanin-producing by anaerobic bacillus such as *P. gingivalis*, *P. intermedius*, *T. forsythensis*, *Fusobacterium*, *P. micros*, *Eubacterium* species and *Actinomycetes* strains. The pathogens related to acute periapical periodontitis, apical black secretion of closed pulp, pain, swelling, pus discharge, odor and fistula formation include *P. intermedius*, *P. gingivalis*, *T. forsythensis*, *P. melaninogenica*, *F. nucleatum*, *P. micros*, *Eubacterium* species and *Actinomycetes* strains. *Prevotella* bacteria and *Peptostreptococcus* bacteria are highly correlated with pain and swelling of patients, while the percussion pain of teeth are closely related to *Porphyromonas*, *Fusobacterium* and *P. micros*. Branner found that the detection rate of obligate anaerobic bacteria in acute periapical infection was 82.3%. The strains of the detection rate above 10% are: *P. intermedius*, *P. micros*, *A. odontolyticus*, *Capnocytophaga*, *Arachnia propionica*, *F. nucleatum*, *A. meyeri*, and *E. lentum*. The strains of the detection rate below 10% are: *P. oralis*, *P. buccae*, *Veillonella*, *Bacteroide* species, *Bifidobacterium* species, *Gemella morbillorum*, *Actinomyces*, *E. limosum*, *Staphylococcus*, *F. varium*, *F. necrophorum*, *P. meninogenica*, *Propionibacterium propionicum*, *Peptostreptococcus anaerobius*. de Souse *et al.* isolated 117 species of bacteria from 30 cases of periapical abscess. The most common bacteria were: *Prevotella* species, *P. micros* and *F. necrophorum*. Jung *et al.* reported that the common pathogens of 79 cases of periapical infection were *P. gingivalis*, *T. forsythensis*, *Treponema maltophilum* and *Treponema socranskii*. (iv) The related pathogens of refractory periapical lesions and persistent fistula are *E. faecalis*, *Actinomyces* and *P. micros*. (v) The main pathogens of periapical bone destruction are *P. intermedius* and *P. micros*, the characteristic of which is that two types of bacteria often exist in combined form. In addition, the bacteria often co-existing with these two bacteria are *Streptococcus* and *Actinomyces* species. (vii) There are differences in the type of pathogen between individuals.

10.2.3 Periodontal Disease

Periodontal disease is a devastating disease of periodontal supporting tissues; the incidence of periodontal disease in our country is as high as 70%. Periodontal disease is one of the oldest and the most common human oral diseases; analysis of primitive skull data indicates the absorption of alveolar bone and the resulting loss of gingival attachment. Periodontal disease is not only one of the reasons leading to tooth loss, but also it is a risk factor in cardiovascular disease, cerebrovascular accidents, pulmonary infections, gastrointestinal infections and preterm delivery.

Periodontal disease, dental caries and pulp and periapical infection are the most common human chronic oral infectious diseases; they not only harm oral health, but also work as a source of clinical infection causing systemic diseases, especially more common in the elderly.

According to the new classification of periodontal disease, periodontal diseases can be divided into two major categories, gingival disease and periodontitis. Periodontal disease has a higher prevalence around the world, and the prevalence of periodontal disease in adolescents and adults is higher than that of dental caries; gingivitis prevalence in our country is as high as 70% – 90%, the earliest of gingivitis can be found in 3 – 5 years old children, and age increases the prevalence and severity of gingivitis which will gradually increase to reach peak at adolescence. Periodontitis is mainly the periodontal disease in adults; most are mild or moderate periodontitis patients, only a small number of people develop severe periodontitis (prevalence rate ranges from 5% to 20%).

The main symptoms of periodontal disease are bleeding gums and inflammation, or the formation of a periodontal pocket. Bleeding gums are the chief complaint symptoms of most patients with periodontal disease. It happens when patients brush their teeth or bite hard food, and occasionally spontaneous bleeding can also be seen. Loose teeth or odontoptosis is the manifestation of patients with severe periodontitis. Changes of gingival in color, texture, shape, and increased volume of gingival crevicular fluid (GCF) and sulcus bleeding on probing can be seen in periodontal clinical examination. When the depth of the sulcus is more than 3 mm, it can form gum bags (false periodontal pocket) or a periodontal pocket, epithelial attachment loss and alveolar bone loss and other typical symptoms.

The bleeding gums and halitosis of patients with periodontitis not only bring pain to patients butperiodontitis is also one of the important causes of tooth loss. The relationship between periodontitis and systemic health has been paid extensive attention; recent studies found that the translocation of bacteria and spread of toxins in periodontitis lesions are related to certain systemic diseases.

These related diseases include diabetes, acute or sub acute infective endocarditis, coronary heart disease, digestive, respiratory, arthritis and kidney disease, *etc.* Researchers detected periodontal pathogens in the atherosclerotic plaques: *P. gingivalis* and *T. denticola.* Plaque of periodontitis patients is suspected to be the "reservoir pool" of *Helicobacter pylori*, the pathogen which causes chronic gastritis, gastric ulcer, even gastric cancer. In addition, it is noteworthy that the clinical symptoms of periodontitis may show signs of systemic disease; the most typical one in early symptoms of AIDS patients can be manifested as ulcerative periodontitis or periodontal candidiasis. Emphasis on periodontal health and prevention of periodontal disease are not only beneficial to oral health, but also conducive to general health [23].

10.2.3.1 Periodontal ecosystem

Periodontium is an important part of the oral cavity. Periodontal tissue is composed of gingival, pericementum, cementum and alveolar bone. Periodontium firmly attach the teeth to alveolar bone, connecting oral mucosal and dental hard tissue and form a well closed state. It not only prevents the impact of external factors, but also bears the bite force. Thus the periodontal tissue is also known as periodontal support tissue. From an ecological point of view, periodontal tissue should be one of the mouth's ecological zones. Due to the characteristics of anatomical structure and physical and chemical properties *etc.*, it provides a good condition for adhesion and proliferation of oral microorganisms. A periodontal ecosystem is composed of periodontal microorganism, periodontal tissue and a variety of ecological factors such as GCF, gingival crevicular plaque and subgingival plaque. The dynamic change in the periodontal ecosystem, ecological balance and ecological imbalance are closely related to occurrence of periodontal health and disease.

Gingival crevice and periodontal pocket. The gingival crevice is the most important ecological zone of the periodontal ecosystem; it is also the largest stagnation zone and protection zone of oral microorganisms. Gingival is the oral epithelium and underlying connective tissue that covers the alveolar surface and the vicinity of the tooth cervical; it is composed of free gingival tissue, attached gingival and gingival papilla. The gap between free gingival and the tooth surface is called the gingival crevice. Crevicular flora has the most abundant types of oral bacteria in a healthy mouth; among these facultative anaerobic bacteria and obligate anaerobic bacteria, melanin-producing anaerobic bacillus, Spirochete, *F. nucleatum*, *Haemophilus*, *Aggregatibacter actinomycetemcomitans etc.* can colonize in healthy sulcus and are considered to have a close relationship to the occurrence and development of periodontal disease, but the number of them is fewer. In inflammation, the depth of the gingival sulcus and the number of suspected periodontal pathogens in crevicular microflora will increase, and toxic clones will be expressed.

Gingival crevicular fluid, infiltrated from gingival connective tissue into the gingival sulcus through the junctional epithelium of the sulcular epithelium, is the important ecological factor of the periodontal ecosystem; the main component of it comes from serum, and contains epithelial cells shed from the host and products of the microorganism in gingival crevicular *etc.* The amount of gingival crevicular fluid and its content relate to gingival inflammation; when inflammation occurs, the amount of gingival crevicular fluid increases, the level of some cytokines such as interleukin, IL, and TNF increase.

A periodontal pocket is the pathological deepening of gingival sulcus under the ecological imbalance condition of a periodontal ecosystem; it is also one of the most important pathological changes. When suffering from periodontitis, the gingival junctional epithelium begin to migrate to cementum, the coronal part of it separates from the tooth surface and forms the periodontal pocket; periodontal pockets are the pathological stagnation zone and protection zone bigger than the normal gingival sulcus. With the depth in the pocket increasing, Eh will decrease to as low as –300 mV.

A periodontal pocket is the new bacterial colonization zone formed by the ecological imbalance of the periodontal ecosystem; the formation of it relates to gingival connective tissue inflammation, collagen destruction caused by inflammation and migration of junctional epithelial to cementum. The fundamental reason is the result of bacterial action and host inflammatory response. Collagenase and hyaluronidase produced by bacteria, increase of inflammation and exudation produced by host reaction, and lysosomal enzymes released from neutrophils and macrophages *etc.*, may dissolve or destroy collagen of connective tissue in the gingival sulcus or the vicinity of the periodontal pocket.

The increase in periodontal pocket depth and aggravation of gingivitis swelling are more beneficial to the proliferation of suspected periodontal pathogens and the accumulation and retention of dental plaque, and thus worsen inflammation and deepen periodontal pockets. The ecological characteristics and low Eh of a periodontal pocket are beneficial to colonization of obligate anaerobic bacteria and the expression of toxic clones.

Supragingival plaque and subgingival plaque. Dental plaque is the initiating factor of the occurrence of gingival disease and periodontal disease, and the plaques which have a close relationship to gingival disease and periodontitis are supragingival plaque and subgingival plaque located above or below the gingival margin. As the plaque accumulation is affected by many factors, so the microbial composition of plaque is different between individuals, or between the different teeth in the same individual. The occurrence of periodontal disease is the result of the ecological imbalance caused by an abnormal increase in the number and types of pathogens in the gingival margin plaque and subgingival plaque; their influencing factors include the interaction between bacteria, decrease of host defense force, expression of periodontal pathogen toxic clones.

Supragingival plaque is the plaque above the marginal gingival. What damages gingival tissue is mainly the marginal gingival plaque located in the vicinity of the marginal gingival, and relates to the plaque-related gingivitis. Marginal gingival plaque is mainly composed of facultative anaerobic gram-positive cocci and bacilli, such as Oral *streptococci* and *Rothia etc.* Gram-negative and gram-positive anaerobic bacilli, cocci and spirochetes are often detected in marginal gingival plaque from gingivitis patients.

Subgingival plaque is the plaque located below the gingival margin; it is distributed in the normal gingival sulcus, pathological gingival pocket and periodontal pocket. The amount of plaque in healthy periodontal tissue is little because of the shallow gingival sulcus; the depth probe of gingival sulcus beyond the normal depth is about 3 mm, but there is no attachment loss. However, the periodontal pocket not only has the depth probe > 3 mm, but also has attachment loss. When the depth of the periodontal pocket is ⩾ 6 mm, due to a significant decrease in Eh (as low as − 300 mV), the amount of plaque will increase with multiplication of a large number of gram-negative anaerobic bacilli [24].

Subgingival plaques are divided into attached subgingival plaque and unattached subgingival plaque. Attached subgingival plaque extends from the supragingival plaque to periodontal pockets and attaches to the root surface, while the unattached subgingival plaque locates on the surface of attached subgingival plaque, directly in contact with gingival sulcus epithelial or pocket epithelium.

Due to ecological differences of attached subgingival plaque and unattached subgingival plaque, there are differences including location, composition, number and types of microbes. Attached plaque is mainly composed of gram-positive cocci, bacilli, and a few gram-negative bacilli, spirochetes. They relate to subgingival calculus, root caries and periodontitis; the types of bacteria in unattached subgingival plaque are less than those in attached subgingival plaque; unattached subgingival plaque is mainly composed of gram-negative obligate anaerobic bacilli, mobile bacteria and spirochetes. They relate to progressive periodontitis and rapid damage of the alveolar bone.

Subgingival plaque hides in the gingival sulcus and periodontal pockets, which provides a favorable habitat for periodontal bacteria, in particular the obligate anaerobic bacteria, as well as bacteria not easy to adhere to the tooth surface or with weaker adhesion ability. Protection of gingival sulcus and periodontal pocket makes subgingival plaque free from the clearance effect of saliva flushing and defensive function, combining with the low Eh to provide a favorable growth environment for anaerobic bacteria.

The matrix between plaques may reduce or prevent the host's role, such as the effect of leukocyte, antibodies, and drugs. Therefore, removal of subgingival plaque or calculus of partial subgingival scaling in prevention of periodontitis are effective measures for the maintenance of periodontal ecological balance, correcting ecological imbalances and prevention of periodontal disease. Ecological characteristics of marginal gingival plaque and subgingival plaque are listed in Table 10.9.

Table 10.9 Characteristics of marginal gingival and subgingival plaque

	Anatomy characteristics	Microorganism	Disease
Marginal gingival plaque	Locate on gingival margin, more influenced by saliva, other friction and drug. Eh is higher.	Gram-positive facultative anaerobes are major bacteria, some gram-negative anaerobic bacteria and spirochetes.	gingivitis
Subgingival plaque	Locate on sulcus or periodontal pocket stagnation area, less influenced by saliva, other friction and drug. Eh is lower.	Gram-negative obligate anaerobic bacteria and anaerobic gram-positive cocci and spirochetes	Periodontitis

Influencing factors of periodontal ecosystem. The basic condition of occurrence of periodontal disease is the accumulation of subgingival plaque in the gingival sulcular pocket, while bacteria in dental plaque and its products are the initiating factor of periodontal tissue inflammation and damage. The main influencing factors of the periodontal ecosystem are the host and subgingival plaque.

Host factors. The host factors influencing periodontal ecosystem include anatomic factors, such as abnormal tooth position and malocclusion, bad habits such as smoking, bruxism, unilateral chewing, mouth breathing and poor brushing *etc.*; iatrogenic factors such as wear of prosthesis and orthodontic appliances; genetic factors such as hyperkeratosis of palms and soles-premature periodontal destruction of teeth syndrome. Some systemic diseases such as diabetes, AIDS can also manifest symptoms of periodontal disease. In addition, increase in sex hormones and the increase in stress are also possible host factors.

The gingival epithelial cell not only has a barrier function, but also has a defensive function through cell surface receptor-mediated secretion of specific antimicrobial peptide, protein and cytokines *etc.* Toll-like receptor (TLR) is one of the family members of the pathogen pattern recognition receptor; it is an important component of the reaction of gingival epithelial cells to the outside world and its signal transduction. TLR can widely recognize endotoxin lipopolysaccharide, peptidoglycan (PGN), lipoteichoic acid, bacterial lipoprotein and other bacterial ligands, and mediates production of anti-microbial factors and chemokines of gingival epithelial cells through different signaling pathways. In addition to directly killing microorganisms by secreted antimicrobial peptide of gingival epithelial cells, they can also activate a-defensins, secrete IL-8, intercellular adhesion molecular (ICAM-1) and other cytokines of chemotactic leukocyte and neutrophile granulocyte. In addition, calprotectin and matrix metalloprotease (MMP) *etc.* are also defensive factors for protection against invasion risk factors and maintenance of epithelial tissue balance of gingival epithelial cells.

Periodontal clinical and experiment studies found that certain genetic factors or genetic diseases may increase host susceptibility to periodontal disease. These genetic factors include abnormal tooth position and malocclusion, familial transmission or susceptibility of periodontal pathogens. Certain genetic diseases, such as hyperkeratosis of palms and soles-premature periodontal destruction of teeth syndrome, could affect the periodontal ecosystem and manifest severe damage in different tissues.

Individual tooth dislocation, excessive eruption or insufficient eruption induced by genetic factors are beneficial to plaque accumulation, or causing occlusal trauma, food impaction *etc.*; long-term occlusal trauma associated with partial stimulate such as periodontitis may exacerbate periodontal pocket and alveolar bone loss. Food impaction is the most common reason for local periodontal tissue inflammation and destruction. Food impaction not only produces mechanical stimulation and oppression on the local gingival, but also increases bacterial colonization, which could result in gingival tissue inflammation and bleeding or lead to gingival recession, periodontal acute pericementitis, bad breath and so on. Obesity may be the second largest risk factor only after smoking in patients suffering from periodontitis, and its biological mechanisms still need to be made clear. Levels of IL-6, TNF-α, C-reactive protein and other inflammatory molecules increased in the fat of obese people, thus it is thought that inflammation may be associated with obesity. The mechanism may be the increase of serum IL-6 and TNF-α exacerbating inflammation.

Familial aggregation performance of early-onset periodontitis (EOP) occurring in adolescence suggests it relates to genetic factors. Transmission of aggregatibacter actinomycetemcomitans among family members was diagnosed through analysis of its serotypes, biotypes and restriction fragment length polymorphism. Genetic analysis of 631 patients in 149 families diagnosed EOP by Maraziat *et al.* indicated that EOP patients may have an autosomal main locus point, whose autosomal dominant is more likely; the explicit rate was 70%.

Systemic diseases such as diabetes are considered as a risk factor for periodontal disease. Diabetic patients are susceptible to periodontitis; increased level of inflammatory cytokines IL-1β and TNF-α could be detected in crevicular fluid of diabetic patients. The mechanism of periodontal disease may relate to defects of leukocyte chemotaxis and phagocytosis, decrease in immune ability and decrease in infection resistance of diabetes patients. Leukemia patients with decreased immune function and AIDS patients with damaged immune function tend to suffer from periodontal diseases.

The relationship between hormone secretion and periodontal diseases is expressed as pregnancy gingivitis in pregnant women and adolescent gingivitis. The influence of some bad habits on the periodontal ecosystem is evident. Long-term bruxism can produce abnormal occlusal force, thereby increasing the affordability of the periodontal tissue and lead to damage. Air flow of mouth breathing is a local stimulating factor to gingival; it may cause gingival swelling or chronic inflammation. Bad oral hygiene habits or methods, including non-standard toothbrush and toothpaste, and incorrect brushing methods, such as applying too much pressure causing tooth abrasion or gingival damage, or a brushing angle not correct to clean teeth may lead to increased plaque accumulation and calculus formation to stimulate gingival, and then cause gingival swelling and bleeding. Smoking is considered a high risk factor for periodontitis; smokers have more dental plaque accumulation and calculus, with recession of lingual gingival finally causing tooth attachment loss and bone resorption. It is reported that smoking may promote proliferation of suspected periodontal pathogen *T. forsythus* and *P. gingivalis*, but the mechanism is not clear.

The influences of prosthesis wear or treatment and after treatment of abnormal dentition orthodontics on the oral ecosystem are evident. Prosthesis and orthodontic appliances become new ecological zones, and thus form a number of stagnation zones, for example the edge of the base and gingival contact position of orthodontic appliances can oppress the gingival margin of abutment teeth. Bad prosthesis with problems of denture design and manufacture not only increase the burden and loosen abutment teeth, but also increase plaque accumulation causing inflammation of gingival and periodontal tissue, and aggravating illness. Orthodontic treatment requires wearing the appliance for a long time. It will also increase the stagnation of plaque accumulation, and could easily lead to food impaction, promoting the proliferation of gingival tissue or inflammation, aggravating the original gingival or periodontal inflammation.

Microbial factors. There are mass and complex microfloras in the health periodontal tissue and they are resident flora of the oral cavity. The microbial influencing factors include the invasive power and the interrelation of microorganisms. The invasive power of microorganisms is connected with the surface structure, as pilus, outer membrane protein (OMP), capsule, endotoxin, and virulent products such as protease, histaminase, hyaluronidase, collagenase, *etc.* The interrelations of periodontal microorganisms are important influencing factors including the co-gathering, co-nutrition, mutual communication and mutual competition and antagonism.

10.2.3.2 Pathogenic Microorganisms

The pathogen species exist in complexes in subgingival plaque of periodontal disease and as many as 700 – 1,000 bacteria species are isolated from periodontal lesions. It has been recognized for the main periodontal pathogens species currently including *P. gingivalis*, *A. actinomycetemcomitans*, *P. intermedius*, *T. forsythus*, *T. denticola*, *P. micros*, and other common bacteria in the periodontal lesion place that they should also be pathogens in the development and progression of periodontal disease; at least they exist as a collaborative bacteria in the disease process, but their role needs to be explored, such as *F. nucleatum*, *Campylobacter rectus*, *Eikenella corrodens*, *Selonomonas* species and *Eubacterium* species *etc.*

P. gingivalis and *A. actinomycetemcomitans* are the bacteria studied most, and they are considered as the most closely suspected periodontal pathogens with destructive periodontal disease.

Porphyromonas gingivalis. It is gram-negative, anaerobic, non-spore, nonmotile and asaccharolytica bacilli. Cells of cocci and bacillus form a black stench colony on the surface of blood agar; they do not ferment carbohydrates and their main phenotype identification features are indole producing. *P. gingivalis* is the major pathogen in chronic periodontitis, and it is often detected from lesions of pulp and

periapical infection, pericoronitis and infection after tooth extraction.
Porphyromonas gingivalis has a variety of virulence factors.
Fimbrillin. Fimbriae are composed by a single subunit. Fimbrillin is a subunit of fimbriae, and it is the component with polypeptide but without polysaccharide. *Porphyromonas gingivalis* strain 381 is composed of at least 1,000 fimbrillin, whose N-terminal is involved in the process of fimbrillin forming fimbriae. fimA is the encoding gene of fimbrillin, existing as a single copy in the chromosome. Fimbriae genotype relates to the periodontal state. Fujiwara *et al.* divided fimA genes into genotype I – V according to nucleotide sequence differences of fimA genes open reading frame. Most patients with periodontitis are genotype II, followed by genotype IV, while periodontal health people are genotype I [25-28].
Gingipains. These are Cysteine proteases, also known as TLP. Gingipains with a close relationship to periodontal disease are mainly Gingipains R (arginine-specific cysteine proteinase, Rgp) and Gingipains K (lysine-specific cysteine proteinase, Kgp). Rgp exist in the form of RgpA and RgpB. The main toxicity of Gingipains are: (i) Affecting the biosynthesis of fimbriae; (ii) As the adhesion factor, directly involving in or regulating bacterial adhesion; (iii) Strong proteolytic effect and degradation of type I type II collagen; (iv) Reducing neutrophil activity; (v) Degradation of IgG, IgA and C3.
Aggregatibacter actinomycetemcomitans. It is gram-negative, capnophilic, glycolytic, nonmotile round end small bacillus and called *Actinobacillus actinomycetemcomitans* or *Haemophilus actinomycetemcomitans*. There is a high detection rate in localized juvenile periodontitis lesions. *A. actinomycetemcomitans* can invade gingival epithelial cells and induce experimental animals to suffer disease. *A. actinomycetemcomitans* can be detected from adult destructive periodontitis lesions, and the detection rate and amount of it are lower than that in local juvenile periodontitis. Serum-type analysis indicates that type A is more common; however, type B is more common in local juvenile periodontitis. It can produce a large number of potentially destructive toxic factors, including leukotoxin (LTX) and cytolethal distending toxin (CDT) and so on [23].
Leukotoxin. LTX is sensitive to heat and protease; molecular weight is 115 kD, encoded by a multi-gene operon, and there are four known genes: *ltxA*, *ltxB*, *ltxC* and *ltxD*. *ltxA* relates to toxin structure encoding and LTX activation. 530 bp deletion of promoter sequence relates to high level production of LTX. LTX can destroy neutrophils and monocytes.
Cytolethal distending toxin. CDT is a thermally unstable protein, encoded by three adjacent genes: *cdtA*, *cdtB* and *cdtC*. CDT can induce cell expansion, arresting the cell cycle not to enter mitosis, affecting the immunosuppressive factor of T cell activity, inducing apoptosis of lymphocytes, inducing secretion of cytokines (IL-1β, IL-6, IL-8).

Aetiology of *P. gingivalis* and *A. actinomycetemcomitans* are listed in Table 10.10.

Table 10.10 Aetiology of *P. gingivalis* and *A. actinomycetemcomitans*

	P. gingivalis	*A. actinomycetemcomitans*
disease	Chronic periodontitis lesions periodontitis active lesions.	Local juvenile periodontitis juvenile periodontal disease
virulence factor	Gingipains,collagenase,endotoxin, plasminogen, fimbrillin, phospholipase A, H_2S, NH_3, fatty acids, degradated immunoglobulins, fibroblast inhibitory factor, inducing bone resorption factor, inducing host cells to produce cytokine, enhanced chemotactic activity. Invading gingival epithelial cells and periodontal membrane cells	Leukotoxin,collagenase, endotoxin, epithelial cells toxins, fibroblast inhibitory factor, bone resorption inducing factor, inducing macrophages to produce cytokines, modifying neutrophil function, degradation of immunoglobulins, invading epithelial cells
animal	Laboratory single or mixed infection plays an important role in induction of disease in gnotobiotic mice	Forming abscess under the skin of gnotobiotic mice induced disease

Socransky and Haffajee (1991)

***Tannerella forsythus* (*Bacteroides forsythus*).** It is gram-negative, has no spores, nonmotile, asaccharolytica, and anaerobic bacilli. *T. forsythus* grows slowly, it requires 7 – 14 d culture to form small colonies; the growth of it often requires N-acetylmuramic acid. The studies found that *T. forsythus* is often isolated from adult periodontitis, refractory periodontitis or active period of periodontal disease, it is one of the most important periodontal pathogens, it relates to periodontal attachment loss. *T. forsythus* strains are mainly isolated from subgingival plaque of periodontitis patients; with periodontal pocket depth increase, the number of them increases significantly, the detection rate in the active portion of periodontal disease is 16%, but the detection rate in the supragingival plaque, healthy sites, gingivitis patients and non-active periodontitis lesions is low.

Virulence factors of *T. forsythus*: (i) Surface protein S-Layer and BspA. A layer of the unique surface structure of T. forsythus is S-Layer glycoprotein; it is composed of jagged glycoprotein subunit, including two proteins encoded by *tfsA* and *tfsB* genes with relative molecular weight of 2.00×10^5 and 2.10×10^5. S-Layer is not only a protective layer of bacterial cells, but also has lectin activity and can mediate adhesion and invasion of *T. forsythus* to histocyte. S-Layer has immune activity and can induce humoral immunity of rapidly progressive periodontitis patients, promoting the formation of an abscess. BspA protein is encoded by the *bspA* gene and the relative molecular weight is 9.8×10^4. BspA protein can mediate adhesion and biofilm formation of Tannerella forsythus to extracellular matrix of histocyte, inducing immune response of the host to produce TNF-α and IL-1β. (ii) Lipopolysaccharide is another virulence factor of *T. forsythus*; it may stimulate the host cells to produce inflammatory mediators and produce antagonism on other bacteria. (iii) *T. forsythus* can produce a variety of enzymes, including tryptase, sialidase, saliva glucosidase, glucosidase *etc.* Tryptase has hemolytic activity and can increase hemoglobin in a habitat, promoting the growth of *P. gingivalis*, *P. intermedius* and other periodontal pathogen.

Treponema denticola. Spirochetes are firstly found to proliferate in biopsy tissue of lesions in patients with acute necrotizing ulcerative gingivitis. It has been confirmed that spirochetes is the pathogen of destructive periodontal disease, its role in other types of periodontal disease is not clear. *T. denticola* is the most discussed spirochete related to destructive periodontal disease. *T. denticola* is more common in lesions of severe periodontitis compared with healthy sites and gingivitis sites; it is more common in subgingival plaque compared to supragingival plaque; in-depth studies prove that these active *T. denticola* can adhere to epithelial fibroblast of human gingival, its virulence factors include outer membrane proteins and a variety of enzymes produced by it such as plasmin, trypsin-like protease, hyaluronidase, acid phosphatase, chondroitin sulfate and peptidase.

***Prevotella intermedius* and *Prevotella nigrescens*.** They are gram-negative, non-spore, nonmotile, melanin producing, glycolytic anaerobic bacilli. A study of periodontal pathogens found that there was a high detection rate of *P. intermedius* in pregnancy gingivitis, which is considered the main pathogen of pregnancy gingivitis. The reasons may be related to the use of estradiol and progesterone by the bacteria. In addition, it is found that the detection rate and amount of *P. intermedius* increase in acute necrotizing ulcerative gingivitis can be isolated in other types of periodontitis, gingivitis and a variety of oral mixed infections, such as pericoronitis, pulp and periapical infection, dry socket after tooth extraction *etc.*

Pathogenicities of *P. intermedius* have certain similarities to *P. gingivalis*, including the induction of mixed infection in experimental animals; elevated serum antibody levels in patients; producing fatty acids, indole, hydrogen sulfide, producing a variety of proteases such as gelatinase, phospholipase (*P. nigrescens* does not produce), inhibiting neutrophil chemotaxis and macrophage phagocytosis ability.

***Peptostreptococcus micros*.** It is gram-positive anaerobic asaccharolytica cocci. It relates to anaerobic mixed infection in the oral cavity and other parts of the bodysuch as infection after tooth extraction, pericoronitis, pulp and periapical infection, vaginitis, abdominal infections *etc.* The detection rate of *P. micros* in periodontal disease sites is higher than that in healthy sites and gingivitis sites; it can be detected in high levels of bacteria in active periodontal lesions, and successful treatment can lead to a reduction of such bacteria. Results of systemic antibody levels test of suspected periodontal pathogen show that there are more elevated levels of specific antibodies in patients with severe periodontitis than in healthy or local juvenile periodontitis.

***Fusobacterium nucleatum*.** It is a gram-negative, non-spore, asaccharolytica or weak fermentation of carbohydrates spindle anaerobic bacilli. *F. nucleatum* is one of the common oral bacteria and mainly colonizes in the gingival margin and subgingival plaque. The detection rate and amount of it is higher in periodontal lesions than in healthy periodontal tissue. It can be detected in adult periodontitis, juvenile periodontitis, gingivitis and pulp and periapical infections, maxillofacial odontogenic infections.

Virulence factors of *F. nucleatum* include endotoxin, fatty acids, NH_3, indole and H_2S *etc.*; it can inhibit fibroblast proliferation, induce macrophages to produce

IL-1.

The role *F. nucleatum* plays in the occurrence and development of periodontal disease is of special concern recently. The ability of the extensive copolymerization force and production of N-acetylmuramic acid of *F. nucleatum* play an important role in the colonization and growth of *P. gingivalis*, *T. forsythus* and *T. denticola*. In an environment with a high concentration of oxygen, *F. nucleatum* cells become elongated, forming an interwoven network structure, parceling *P. gingivalis* in it, which will be beneficial to its adhesion, colonization and growth on oral tissues. In addition, there is the extensive copolymerization force of *F. nucleatum* predominant in the process of the maturation of subgingival plaque and dental plaque, as the copolymer bridge to make *P. gingivalis*, *T. forsythus*, *T. denticola* and other bacteria colonized on the root surface, becoming a part of subgingival biofilm. *F. nucleatum* can also provide the required growth factor of N-acetylmuramic acid for *P. gingivalis* and *T. forsythus*. *T. forsythus* are often isolated with *F. nucleatum* in deep periodontal pockets; the co-culture of it with *F. nucleatum* can enhance the strength of its growth, so researchers think that there may exist a symbiotic relationship between the two bacteria, because they provide the growth required N-acetylmuramic acid for *F. nucleatum*.

Eikenella corrodens. It is a gram-negative, capnophilic, asaccharolytica small regular bacillus; the growth of it requires 5% – 10% CO_2 supply in an anoxic environment. *E. corrodens* is the pathogen of periodontal disease, root canal infection, osteomyelitis, central nervous system infection; the detection rate in periodontal destruction sites is higher than that in healthy parts and it is often found and detected at a high level in active lesions and lesions of refractory periodontal disease. *E. corrodens* and *A. actinomycetemcomitans* are detected together in lesions of local juvenile periodontitis.

The pathogenic mechanism of *E. corrodens* is still not clear; it may be synergistic bacteria of periodontal mixed infection.

10.2.3.3 Ecology of Periodontal Disease Microorganism

The dispute about specific and non-specific periodontal pathogens is a problem that has long troubled clinical dental doctors and researchers. Although supporters of the theory of specific pathogens proposed that the specific type of periodontal disease caused by specific pathogens, for example the pathogen of adult periodontitis is *P. gingivalis*, the pathogen of local juvenile periodontitis is *A. actinomycetemcomitans*, and the pathogen of pregnancy gingivitis is *P. intermedius*, but clinical bacteriology studies have shown that periodontal disease is definitely not a single bacterium infection; to identify specific pathogen in dozens of isolated bacteria in periodontal lesions is considerably difficult. In 1992, the well known US experts of periodontology and periodontal microbiology Socransky and Haffajee, proposed a new concept of suspected pathogens and beneficial bacteria in a paper "modern concept of bacterial etiology of destructive periodontal disease". It is very enlightening and helpful for us to know and understand ecology and the mechanism of periodontal disease with ecological theory. The study of red-orange flora of periodontal pathogens completed in 1998 by Socransky *et al.*[24] is the best annotation of microbial ecology of periodontal disease.

Periodontal suspected pathogens and beneficial bacteria. Socransky *et al.* indicated that although the search for specific pathogens of destructive periodontal disease is more than 100 years old, due to the complexity of microorganisms and technical difficulties, as well as individual differences of host response and other factors, this increases difficulties in relevant research; thus the pathogen and mechanism of periodontal disease has not yet been fully understood. Based on long-term study on periodontal pathogens, Socransky *et al.* proposed the concept of "periodontal suspected pathogens" and "periodontal beneficial bacteria" to represent the bacterial species under different periodontal conditions.

The bacteria which has a high detection rate in periodontal lesion, and is more closely related to the occurrence of periodontal disease is known as periodontal suspected pathogens, such as *P. gingivalis* and *A. actinomycetemcomitans*.

Socransky *et al.* thought that the periodontal suspected pathogens must have the following characteristics: (i) The detection rate and amount in a lesion is higher than in healthy sites; (ii) Eliminating or reducing their number can lead to successful treatment; (iii) Stimulating the host's specific immune response, such as elevated serum antibody levels; (iv) Causing periodontal tissue destruction in experimental animals; (v) Must be a toxic clone and possessing a chromosome or outer chromosome gene factor to initiate disease; (vi) Colonization in lesion, the host must be sensitive to such pathogens, the quantity exceeding the host threshold (refers to the number of normal colonization); (vii) During the proliferation process, other bacteria strains must be the helping bacteria, the local environment being conducive to the expression of the toxic clone.

The detection rate in healthy periodontal tissues in periodontal lesions detected in low places, and periodontal suspected pathogens with antagonistic bacteria is known as periodontal beneficial bacteria, such as *Streptococcus sanguis* and *Veillonella*.

Red flora and orange flora. In 1998, Socransky carried out a bacteriology study and measured a clinical index of 160 cases of periodontitis patients with periodontal attachment loss and 25 healthy people; PD and CAL measured 6 sites per tooth per person, a total of 13,261 copies of subgingival bacteria plaque samples, using DNA-DNA hybridization. 40 species of bacteria were detected and cluster analysis was carried out; based on a similarity greater than 60%, they were divided into 5 categories: red flora, orange flora, yellow flora, green flora, purple flora and blue flora. The red flora and orange flora are considered to be more closely related to periodontal disease, while the relationship between other color flora and periodontal disease is to be discussed.

Red flora bacteria. They are considered the most closely related bacteria to periodontitis; they are the main periodontal pathogens, including *P. gingivalis*, *T. forsythus* and *T. denticola*. The relationship between red bacteria and the periodontal pocket depth (PD) is strong, the detection rate and number of *P.*

gingivalis, *T. forsythus* and *T. denticola* increases with the increase in PD. The periodontal pocket with three kinds of red bacteria is the deepest one, and PD without red bacteria is the shallowest one. The PD in sites with separately detected *P. gingivalis* and detected *P. gingivalis* combined with the other two kinds of red bacteria is the maximum. Red bacteria are also closely related to BOP.

Orange flora bacteria. The relationship to periodontal disease is considered secondary to the red flora bacteria, but they have a close relationship to red flora bacteria; two floras exist in complex form. Orange flora bacteria include *F. nucleatum*, *P. intermedius*, *P. nigrescens*, *P. micro*, *Campylobacter rectus*, *Capnocytophaga sputigena*, *Capnocytophaga gingivalis*, *Eubacterium nodatum etc.*; although the relationship between orange flora bacteria and periodontitis is not as close as red flora bacteria, they are essential in the occurrence and development of periodontal disease. Red bacteria are rare in the absence of orange bacteria, the more the orange bacteria, the more the red bacteria. The number of orange bacteria, *F. nucleatum*, *P. intermedius*, *A. viscosus* increases with an increase in PD. The sites without red bacteria mostly have colonization of orange bacteria. In conclusion, orange bacteria may colonize before red bacteria. *F. nucleatum* plays an important role in growth and colonization of *P. gingivalis*.

Other color flora. Yellow flora (*streptococci* belong to this column); Green flora (including three kinds of capnophilic bacteria, namely *Campylobacter concisus*, *E. corrodens* and serotype a of *A. actinomycetemcomitans*; Purple flora (including *A. odontolyticus* and *Veillonella parvula*); Blue flora (including *A. viscosus*, *Selenomonas noxia* and serotype b of *A. actinomycetemcomitans*.

10.2.4 Maxillofacial Infectious Diseases

Oral and maxillofacial infections are common clinical oral infectious diseases, including pericoronitis, orofacial space infections (such as the infraorbital space infection, buccal space infection, temporal space infection, submandibular space infection, multi space infection of mouth floor *etc.*), mandibular osteomyelitis, face and neck lymphnoditis, swelling pain, facial furuncle and carbuncle, and the specific pathogen maxillofacial infections such as tetanus, maxillofacial bone tuberculosis, maxillofacial actinomycosis, maxillofacial syphilis, HIV infection (AIDS) secondary to maxillofacial infections.

In the modern concept of maxillofacial infections classification, maxillofacial infections are divided into two categories by infection routes; they are odontogenic maxillofacial infections and non-odontogenic maxillofacial infections. This classification is extremely beneficial for the understanding of infection routes, pathogens, and selecting effective treatment. Maxillofacial infection micro-organisms are characteristic due to their location and the type of infection. The etiology of odontogenic infections relates to the translocation and abnormal proliferation of oral normal flora; very few infections are single infection; however, non-odontogenic infections are mostly caused by infection of exogenous bacteria invasion and a single infection is more common, including the specific pathogen infection.

10.2.4.1 Orofacial Odontogenic Infection

Infections caused by pathogens entering into the maxillofacial regions through the diseased teeth or diseased periodontal diseases are collectively called odontogenic maxillofacial infections or maxillofacial odontogenic infections. Maxillofacial odontogenic infections are the most common type requiring maxillofacial clinical surgery; it is also one of the most common diseases of maxillofacial clinical surgery.

The most common odontogenic maxillofacial infection of maxillofacial clinical surgery is pericoronitis of the wisdom tooth; other common odontogenic maxillofacial infections include the pterygomandibular space infection and submandibular space infection caused by mandibular third molar pericoronitis and mandibular molar periapical inflammation. In addition, post extraction infection has also been included in the odontogenic infection, since the hole formed after the teeth extraction is subject to bacteria retention and contributes to growth and reproduction of bacteria, leading to a dry socket after tooth extraction, and may cause other facial secondary infections.

Odontogenic infection is the maxillofacial infection currently studied the most; most maxillofacial infection are odontogenic, the etiology of which is generally considered as causing the oral bacteria. It is typical of many bacteria.

The characteristics of pathogenic microorganisms of odontogenic infection are: (i) The oral normal flora and bacteria are the most common pathogenic microorganisms. (ii) There are mixed species infections. (iii) Anaerobic bacteria especially non-sporing anaerobic bacteria are the dominant pathogens.

Tianzhong Run isolated 184 anaerobic bacteria from the jaw periostitis, cellulitis, and other maxillofacial pyogenic infections, of which 178 were the non-sporing anaerobes. Xiao *et al.*[6] reported a 95.85% anaerobes detection rate from 217 cases of odontogenic maxillofacial infections, of which the anaerobic bacteria separation rate from 150 cases of acute pericoronitis clinical samples was 100%, and the dominant anaerobes were *Peptostreptococcus*, melanin-producing anaerobic bacilli, *Prevotella*, *Capnocytophaga*, and *Fusobacterium*.

10.2.4.2 Non-Odontogenic Facial Infection

The maxillofacial infections caused by non-teeth or non-periodontal tissue are collectively called non-odontogenic maxillofacial infections, including all maxillofacial infections except odontogenic infections. Non-odontogenic maxillofacial infections mainly include the following five categories:

Gland-derived maxillofacial infection are caused by the spread of infection of the maxillofacial lymph nodes, parotid, submandibular duct, such as maxillofacial lymphadenitis, sialadenitis, parotitis *etc.*

Injury-derived maxillofacial infections Result from maxillofacial infections causing by maxillofacial injury such as knife injury, firearms injury, bite injury, jaw open fracture or foreign matter embedded deeply into maxillofacial regions.

Blood-borne maxillofacial infections are maxillofacial soft tissue or jaw infections caused by blood circulation of purulent lesions in other parts of the body or bacteremia.

Iatrogenic maxillofacial infections are secondary infections caused in the course of medical practices (such as surgery, puncture).

Maxillofacial specific pathogens infections. The maxillofacial infections are caused by specific pathogens and the major pathogenic microorganisms are single species, such as maxillofacial actinomycosis caused by *Actinomyces* spp., maxillofacial tuberculosis caused by mycobacterium tuberculosis, tetanus caused by *Clostridium tetani* and AIDS caused by human immunodeficiency virus (HIV).

10.2.5 Oral Mucosal Infections

Oral mucosal infection is also common in oral infections; the infections understood most at present are mainly mucous infections caused by bacteria, virus and fungus; most mucosal infections are thought to be caused by exogenous pathogens except for a few infections caused by normal oral bacteria.

Bacteria and oral mucosa infections

Oral mucosa is an important colonization habitat for oral normal microorganisms; it is influenced more by saliva microbiota, so damaged oral mucosa will be induced diseases due to translocated microbial colonization or dysplasia, and the occurance of diseases will also affect the indigenous microbial colonization, causing flora disequilibrium, or interact as both cause and effect. In a study of 70 patients with chronic oral mucositis, it was found that 65.7% of patients had flora disequilibrium, shown in Table 10.11.

S. aureus and *Streptococcus* can cause coccigenic stomatitis, mucosal congestion, edema, or formations of pseudomembrane (so called pseudo. membranous stomatitis), visible in clinical examination. Gonococcal stomatitis and pharyngitis caused by *Neisseria gonorrhoeae*, full mouth mucosa congestion, superficial ulceration, or the formation of pseudomembrane can be seen. Oral mucosa infections caused by *Treponema pallidum* manifest as mucous patch or chancre. The resident oral bacteria of *Borrelia vincentii* and *Fusobacterium* can collaboratively cause Vincent's angina, acute necrotizing ulcerative gingivitis, stomatitis, Cancrum oris and other endogenous infection. *Mycobacterium tuberculosis* can cause an oral mucosa tuberculosis ulcer. *Helicobacter pylori* relates to recurrent aphthous ulcers and oral lichen planus infection [29, 30].

Table 10.11 Oral flora dysbiosis of patients with chronic oral mucosa infection*

Cases (number)	Normal	flora disequilibrium			
		Compensation	Ⅰ – Ⅱ degree	Ⅲ degree	Ⅳ degree
Recurrent aphthous stomatitis (18)	2	8	4	4	0
Recurrent herpetic stomatitis (10)	4	0	0	4	2
Necrotizing ulcerative gingivitis (8)	2	2	0	4	0
Red lichen planus (10)	4	2	2	2	0
Oral mucosa dysfunction (10)	4	2	2	2	0
Tongue or mouth pain, burning sensation, taste disorders etc					

* Compensation: it is the slight change in the composition of oral flora, including the increase of one kind of opportunistic pathogen; Ⅰ – Ⅱ degree is the decrease in *Lactobacillus*; there are 2 – 3 species of opportunistic pathogens; Ⅲ degree is a significant decrease or disappeance of *Lactobacillus* and non-hemolytic *Streptococcus*; Ⅳ degree is similar to Ⅲ degree, but it has *Candida* detection.

10.2.5.1 Virus and Oral Mucosa Infection

The pathogens discussed most in oral mucosal disease are viruses, such as Herpes simplex virus (HSV), Epstein-Barr virus (EBV), Human Immunodeficiency Viruses (HIV), Human papilloma virus, (HPV) and so on. HSV may cause herpes stomatitis; EBV may cause oral mucosa hairy leukoplakia. Most HIV infections have a manifestation of oral symptoms and appear in the early stages of the disease and receive treatment in dental clinics. In 1992, the oral manifestations of HIV infection of the WHO collaborating centre developed the standards of classification and diagnosis of oral manifestations in HIV infection; it showed the value of oral manifestations.

The aim of it is also based on early diagnosis and treatment of HIV infected patients, strengthening self-protection awareness of dentists and avoiding cross infection caused by medical practice. There is a high prevalence (11% – 96%) of oral candidiasis in HIV-infected people, and it is often the first symptom of HIV infection. The main clinical symptoms are: (i) Erythematosus type: it is common on tongue and palate; (ii) Pseudomembranous type: it can occur in various parts of oral mucosa, but the most common parts are in the tongue, hard palate and buccal mucosa; (iii) Hyperplastic type: it is common in buccal mucosa, and nodular hyperplasia are usually observed on upper palate and tongue; (iv) Angular stomatitis type: *S. aureus* infections are often accompanied.

Hairy leukoplakia is considered as one of the major oral lesions of HIV infection, and it is important to indicate the occurrence of AIDS. Patients with hairy leukoplakia often appear with white stripe damage at the edge of the tongue, and occasionally on cheek, lips, mouth floor, soft palate or pharynx. Of course, it is worth saying that oral candidiasis or hairy leukoplakia may occur in the mouth of non-HIV infected patients, for example hairy leukoplakia is caused by opportunistic infections of EB virus, and it is an oral mucosa expression of EB virus infection. Oral ulcers, herpes stomatitis or cheilitis, necrotizing ulcerative gingivitis are also common in the mouth of HIV-infected persons.

10.2.5.2 Fungus and oral mucosa infection

The fungus relating to oral mucosa infection first comes as *Candida* and the strains detected in the mouth are *Candida albicans*, *Candida tropicalis*, *Candida krusei*, *Candida parapsilosis* and *Candida glabrata.* The above strains can be isolated in a healthy mouth, especially in a neonatal mouth; only the detection amount and rate is less than that of bacteria, and there are differences between individuals. *C. albicans* is the most common pathogenic fungus; it can directly cause acute and chronic oral mucosal lesions, causing *Candida* infection in a mouth with dentures. Long-time use of broad-spectrum antibiotics, steroid hormones, or radiotherapy in oral cancer patients *etc.* will cause oral candidiasis. Pathogenic mechanism of *C. albicans* include: (i) Adhesion to the host cell through a variety of adhesion mechanisms, including the protein-protein interaction and the role of Phytohaemagglutinin (PHA). Agglutinin-like sequence protein (AlSP) and the hyphal wall protein-1 (hwp-1) are both adhesion-related factors of *Candida* adhesion; they can identify the extracellular matrix adhesion protein of the adhesion object, mediating the adhesion on epithelial cells, endothelial cells or microbial cells. (ii) *C. albicans* is a dimorphic fungus containing pseudohypha and blastospore. The growth of hypha and the germ tube of cylindrical protrusions have a resistance effect on phagocytic cells, and they are beneficial to the adhesion of *C. albicans* on epithelial cells. A germ tube can promote *Candida* cell aggregation, working as the connection between adjacent hypha. Pseudohypha not only has strong adhesion on buccal epithelial cells, but also can still change in the cells after phagocytosis; the hypha formed by it can penetrate macrophages and lead to death. (iii) Pathogenicity can be strengthened due to transformation of colonial morphology, smooth colony — sprouting spores — mycelial phase, and the transformation of *C. albicans* from the spore to hypha is the important inducing factor of its pathogenicity. (iv) *C. albicans* can produce a variety of enzymes, including secreted aspartic proteinase, phospholipase, hexosaminase *etc.* and toxins, acid metabolites and other virulence factors.

The main identification features of *C. albicans* are the detection of the germ tube and chlamydospore.

Germ tube formation experiment is achieved by inoculating the isolated bacteria in 0.5 – 1 mL of normal human or sheep serum, 2 – 4 h 37 ℃ incubation, direct microscopic examination or microscopic examination after staining by crystal violet staining or lactic acid phenol Medan staining. *Candida albicans* blastospore and germ tube formation can be observed under an oil immersion lens.

Chlamydospore formation test. Inoculating the isolated in 1% Tween-corn meal agar, after 25 ℃ 1 – 2 d culture will form translucent, sparkling and crystal-clear colonies. After taking a few colonies to smear with crystal violet staining or lactic acid phenol Medan staining and observing under an oil immersion lens, clear chlamydospore can be seen on the top and side edge of *C. albicans* hypha.

10.2.6 Secondary Infection from the Wearing of Dentures

Missing teeth or dentition deformities and other reasonsrequires patients to wear dentures or dental implants, or braces, thus causing changes in oral ecotope: including the formation of a new colonization habitat and stagnation zone. The oral flora disequilibrium will cause dental caries and periodontal secondary infection.

10.2.6.1 Secondary Infection of the Denture wears

Possible secondary infection after the intervention of oral prosthesis has always been the concern of prosthetic clinical practice; secondary caries and periodontal disease are important issues puzzling prosthetic clinicians and patients after the use of denture material, teeth preparation and wearing of dentures. Removable partial dentures (RPDs) are the most common way for dentition restoration; studies of the potential relationship between RPDs and dental caries found that bacterial composition varied a lot after 7 d, 14 d, 21 d and 3 mon wear, *S. mutans* and other cariogenic bacteria were detected, *S. mutans* was the dominant bacteria and rose gradually. Studies found that after 3 months of wearing all-ceramic fixed bridges restoration, the number of bacteria in the butment gingival sulcus and gingival surface of the bridge increased significantly; after 14d of RPDs wear, dental plaque of the butment increased, the plaque index (PI), gingivitis index (CI) and gingival bleeding index (GBI) increased, thus increasing the incidence of periodontal disease.

10.2.6.2 Peri-Implant Inflammation

Dental implants are denture prosthetics used as implant abutment. Peri-implant inflammation is peri-implant tissue inflammation caused by oral microorganisms; it is also an important reason for implant failure. Peri-implant inflammation is a disease of typical oral ecological imbalance, the pathogenic microorganisms of which are normal oral flora. The microorganisms in the vicinity of a successful implant are dominated by gram-positive facultative anaerobes (*Streptococcus* accounted for more than 80%). Periodontal pathogens are a major cause, leading to peri-implant inflammation; the vicinity of the failure implant is dominated by gram-negative anaerobic bacteria such as *F. nucleatum*, *P. gingivalis*, *P.*

intermedius, and Spirochete. Other researchers reported that *Peptostreptococcus*, *Campylobacteria rectus*, *A. actinomycetemcomitans*, *Capnocytophaga*, and *C. albicans* may also be detected in lesions from the patients with peri-implant inflammation, and more research suggests that gram-negative anaerobic bacteria are important pathogens of occurrence and development of peri-implant inflammation.

References

[1] Nolte W A. Oral microbiology: With basic microbiology and immunology: CV Mosby, 1982.
[2] Mackenzie A, Ball A S, Virdee S R. Instant notes in ecology. Recherche, 1998, 67: 2.
[3] Madigan M T, Martinko J M, Yang W B. Microbial biology. 8 ed. Beijing: Science Press, 2001.
[4] Zhou X D, Xiao X R. Oral Microbiology. Chengdu: Sichuan University Press, 2002.
[5] Aas J A, Paster B J, Stokes L N, *et al.* Defining the normal bacterial flora of the oral cavity. *Journal of Clinical Microbiology*, 2005, 43: 5721-5732.
[6] Xiao X R. Oral microbiology and practical techniques. Chengdu: United Press of Peking Union Medical College, Beijing Medical University, 1993 .
[7] Xavier K B, Bassler B L. LuxS quorum sensing: More than just a numbers game. Current opinion in microbiology, 2003, 6: 191-197.
[8] Palmer Jr R J, Gordon S M, Cisar J O, *et al.* Coaggregation-mediated interactions of *streptococci* and *actinomyces* detected in initial human dental plaque. Journal of bacteriology, 2003, 185: 3400-3409.
[9] Gibbons R J, Van Houte J. Selective bacterial adherence to oral epithelial surfaces and its role as an ecological determinant. Infection and immunity, 1971, 3: 567-573.
[10] Rosan B, Lamont R J. Dental plaque formation. Microbes and infection, 2000, 2: 1599-1607.
[11] Marsh P D. Are dental diseases examples of ecological catastrophes? Microbiology, 2003, 149: 279-294.
[12] ZhangW Z, Hao Y Q. Relation between the surface protein of *Streptococcus mutans* and infective endocarditis. International Journal of Stomatology, 2010, 37: 287-290.
[13] Fan M W. Operative dentistry and endodontics. Beijing: People's Medical Publishing House, 2008.
[14] Peterson S N, Snesrud E, Schork N J, *et al.* Dental caries pathogenicity: A genomic and metagenomic perspective. International Dental Journal, 2011, 61: 11-22.
[15] Fitzgerald R J, Adams B O, Sandham H J, *et al.* Cariogenicity of a lactate dehydrogenase-deficient mutant of *Streptococcus mutans* serotype c in

gnotobiotic rats. Infection and immunity, 1989, 57: 823-826.

[16] Rolerson E, Swick A, Newlon L, *et al.* The SloR/Dlg metalloregulator modulates *Streptococcus mutans* virulence gene expression. Journal of bacteriology, 2006, 188: 5033-5044.

[17] Li Y H, Lau P C Y, Tang N, *et al.* Novel two-component regulatory system involved in biofilm formation and acid resistance in *Streptococcus mutans*. Journal of Bacteriology, 2002, 184: 6333-6342.

[18] Li L, Zhou X D. Bacteriocins produced by caries-related bacteria. International Journal of Stomatology, 2009, 36: 32-37.

[19] Hillman J D, Brooks T A, Michalek S M, *et al.* Construction and characterization of an effector strain of *Streptococcus mutans* for replacement therapy of dental caries. Infection and Immunity, 2000, 68: 543-549.

[20] Siqueira Jr J, Rôças I, Paiva S, *et al.* Cultivable bacteria in infected root canals as identified by 16S rRNA gene sequencing. Oral microbiology and immunology, 2007, 22: 266-271.

[21] de Sousa E L R, Ferraz C C R, Gomes B P F A, *et al.* Bacteriological study of root canals associated with periapical abscesses. Oral Surgery, Oral Medicine, Oral Pathology, Oral Radiology, and Endodontology, 2003, 96: 332-339.

[22] Sundqvist G. Associations between microbial species in dental root canal infections. Oral Microbiology and Immunology, 1992, 7: 257-262.

[23] Armitage G C.Development of a classification system for periodontal diseases and conditions. Annals of Periodontology, 1999, 4: 1-6.

[24] Socransky S S, Haffajee A D, Cugini M A, *et al.* Microbial complexes in subgingival plaque. Journal of clinical periodontology,1998, 25: 134-144.

[25] Ploeg J R, Giertsen E, Lüdin B, *et al.* Quantitative detection of *Porphyromonas gingivalis* fimA genotypes in dental plaque. FEMS microbiology letters, 2004, 232: 31-37.

[26] Nakagawa I, Amano A, Kimura R K, *et al.* Distribution and Molecular Characterization of *Porphyromonas gingivalis* Carrying a New Type of fimA Gene. Journal of clinical microbiology, 2000, 38: 1909-1914.

[27] Amano A, Kuboniwa A M, Nakagawa I, *et al.* Prevalence of specific genotypes of *Porphyromonas gingivalis* fimA and periodontal health status. Journal of dental research, 2000, 79: 1664-1668.

[28] Wu Y F, Guo Y H, Liu T J, *et al.* Distribution of f imA genotype of *Porphyromonas gingivalis* in Chinese periodontitis patients and its relationship with chronic periodontitis. Journal of sichuan university medical science edition, 2006, 37: 101-104.

[29] Kleinegger C L, Lockhart S R, Vargas K, *et al.* Frequency, intensity, species, and strains of oral Candida vary as a function of host age. Journal of clinical microbiology, 1996, 34: 2246-2254.

[30] Sullivan D J, Westerneng T J, Haynes K A, *et al.* Candida dubliniensis sp. nov.: Phenotypic and molecular characterization of a novel species associated with oral candidosis in HIV-infected individuals. Microbiology, 1995, 141: 1507-1521.

11

Gastrointestinal Infectious Microecology

Liang Xu, Feng Ji *
Department of Gastroenterology, the First Affiliated Hospital, School of Medicine, Zhejiang University, Hangzhou, 310003, China
* E-mail: jifeng1126@sina.com

Many species of microbes live and grow in the human gastrointestinal tract. Under normal circumstances, they adapt to their environment and contribute to an ecological balance in the host, which is important for the gastrointestinal function. Imbalance of the gastrointestinal mucosa homeostasis leads to a multitude of diseases.

11.1 Microbiota in Health

Colonization of the human gut with microbes begins immediately at birth. After 1 year, the intestinal microbiota starts to resemble that of a young adult, and stabilizes [1]. There exists a possible co-evolution of the host and its indigenous microbiota. The healthy host has intricate mechanisms that allow local control of the resident microbiota without the induction of concurrent damaging systemic immune responses [2].

11.1.1 Normal Microbiota in the Stomach

Anatomically, the human stomach can be divided into cardia, fundus, gastric body, antrum and pylorus. Different parts have different ecological environments [3].

According to acid secretion, the stomach can be divided into a secreting and non-secreting area. The former mainly refers to the gastric body and fundus, and the latter mainly refers to the antrum. The secreting area is the habitat for a large amount of yeast, and lactobacillus lives in the non-secreting area. Since gastric acid has a bactericidal effect, most of the foreign bacteria have been killed. Other than yeast and *lactobacillus*, there are some other bacteria that can be isolated, such as *Streptococcus*, *Bifidobacteria*, *E. coli*, *Peptostreptococcus*, *Bacteroides*, *Staphylococcus*, *Actinomyces*, and *Candida*, *etc.*, but the concentrations are often very low, usually less than 10^3/mL. The normal flora may change with a change in the human physiological state, resulting in species transformation, variation or disappearance. Recently, it has been strongly argued that the overuse of antibiotics causes flora imbalance.

It is worth mentioning that the most studied foreign bacteria in the stomach is *Helicobacter pylori* (*H. pylori*) [4]. Since Warren and Marshall successfully isolated *H. pylori* in 1983, it has caused widespread interest in the medical profession, and a lot of intensive studies have been made. It has been accepted that *H. pylori* infection is an important pathogenic factor in gastritis, peptic ulcers, gastric cancer, MALT lymphoma and other diseases.

11.1.2 Normal Microbiota in the Intestine

The number of bacterial cells present in the mammalian gut shows a continuum that goes $10 - 10^3$ /g of contents in the stomach and duodenum, progressing to $10^4 - 10^7$ /g in the jejunum and ileum and culminating in $10^{11} - 10^{12}$ /g in the colon. Most of the bacteria in the small intestine are gram-negative bacteria. The lack of bacteria in the upper tract seems to be due to the composition of the luminal medium (acid, bile, pancreatic secretion), which kills most ingested microorganisms, and due to the phasic propulsive motor activity towards the ileal end, which impedes stable colonisation of bacteria in the lumen. By contrast, the large intestine contains a complex and dynamic microbial ecosystem with high densities of living bacteria. A large proportion of the faecal mass consists of bacteria (around 60% of faecal solids). Studies have shown that anaerobic bacteria outnumber aerobic bacteria by a factor of 100 – 1,000. The genera bacteroides, bifidobacterium, eubacterium, clostridium, enterobacteriaceae, peptococcus, peptostreptococcus, and ruminococcus are predominant in human beings, whereas aerobes (facultative anaerobes) such as escherichia, enterobacter, enterococcus, klebsiella, lactobacillus, and proteus, *etc.*, are among the subdominant genera. The microbiota present in the intestinal lumen differs significantly from the microbiota attached and embedded in the mucus layer as well as the microbiota present in the epithelium. So some microbial types are only found in faeces but cannot be isolated from the large intestine. The constant interaction between the host and its microbial guests can bring important health benefits to the human host. However, some of those bacteria are potential pathogens and can be a source of infection in some circumstances – for instance when the integrity of the bowel barrier is physically or functionally breached [3].

11.1.3 Physiological Functions of Gastrointestinal Microbiota

Animal studies have provided important information about the effect of the microbial community of the gut on host physiology and pathology. Evidence obtained through such studies suggests that microflora have important and specific metabolic, trophic, immune modulatory and protective functions [2, 5].

11.1.3.1 Nutrition and Metabolism

A major metabolic function of colonic microflora is the fermentation of non-digestible dietary residue and endogenous mucus produced by the epithelia. Fermentation of carbohydrates is a major source of energy in the colon. Non-digestible carbohydrates include large polysaccharides, some oligosac charides that escape digestion, and unabsorbed sugars and alcohols. Those metabolic processess are beneficial to the host in either nutrient acquisition or xenobiotic processing [6]. Microorganisms also anaerobically metabolize peptides and proteins including elastin and collagen from dietary sources, pancreatic enzymes, sloughed epithelial cells and lysed bacteria. The metabolic endpoints of carbohydrates and protein are both associated with the generation of short-chain fatty acids. These fatty acids have important functions in host physiology, such as being a source of energy for colonocytes, facilitating the absorption of salt and water by the colon, stimulating mucosal growth in the colon, playing a role as modulators of glucose metabolism and so on. Colonic microoganisms play a part in vitamin synthesis (vitamin K, vitamin B_{12}, biotin, folic acid and thiamine B complex vitamins) and in absorption of calcium, magnesium and iron. Intestinal bacteria are also involved in the metabolism of host endogenously synthesized compounds, for instance cholesterol, bile acids, bile pigments, androgens, estrogens and so on.

Some natural and synthetic compounds that are used as drugs have been shown to be metabolized by the intestinal microflora [3]. DOPA is used for the treatment of Parkinson's disease. The bacterial modification involves a dehydroxylation, which generates an inactive product in treating this disease. Azulfidine is beneficial for the treatment of ulcerative colitis. The drug structurally has sulfapyridine and aminosalicylate moieties attached *via* an azo bond which is reductively cleaved by bacteria in the colon to release the aminosalicylate, the active component for treating ulcerative colitis. The role of the intestinal bacterial metabolism is important in the action of the cardiac glycoside drug digoxin. In order to form a pharmacologically active drug, the bacterial flora has to remove a trisacchride from the parent compound, releasing digoxigenin. The bacterial intestinal flora can further reduce the double bond in the lactone ring to form dihydrodigoxigenin, which is pharmacologically inactive. Consequently, a portion of the population receiving digoxin will not achieve predicted serum levels resulting from the action of the intestinal microflora.

11.1.3.2 Immune Modulation by the Intestinal Microbiota

The central role of gut microbiota in the development of mucosal immunity is not surprising, considering that the intestinal mucosa is the main interface in contact with the antigens of the external environment. Despite the constant presence of a great amount of antigens from the food and microorganisms, the commensal microbiota stimulates and then coordinates the gastrointestinal associated immune systems to achieve a disease-free state. The commensal microbiota is crucial for stimulating the Peyer's patches, immunoglobulin or IgA producing cells, the dendritic cells. It coordinates pro- and anti-inflammatory signals to regulate the number of microbes and promote the efficacy of the immunological barrier provided by the intestinal mucosa [7].

11.1.3.3 Protective Functions

Gut microbiota provides its host with a physical barrier to incoming pathogens by several mechanisms involving competitive exclusion [2, 3]. Firstly, non-pathogenic bacteria adhere to the epithelial cells of intestinal mucosa, which prevents a pathogen attaching to and entering the epithelial cells. Secondly, bacteria compete for available nutrients in ecological niches. The bacteria can actively indicate how much nutrients they need in the host which supplies the product, so this prevents overproduction of the nutrient, which will favour the intrusive microbial competitors with potential pathogenicity. Finally, bacteria can produce bacteriocins, acids and hydrogen peroxide to inhibit the growth of their competitors. They also stimulate the host to produce various antimicrobial compounds.

11.1.4 Factors Affecting Gastrointestinal Microecological Balance of the Host

The host posseses strong regulatory mechanisms to maintain microecological balance, such as the effect of gastric acid and bile, peristalsis, *etc.* In addition to those, there are other factors affecting the microbiota homeostasis. Mucus is the environment which most gut flora live in, so it regulates the constitution of flora. The bacteria separated from the intestine have the enzyme hydrolyzing mucus which is the carbon and energy source of these strains. If mucus is destroyed, the balance of normal intestinal flora will be breached. The immune system also can impact the gut flora balance. The mucosal immune system needs to fulfill two functions. It needs to be tolerant of the overlying microbiota to prevent the induction of an excessive and detrimental systemic immune response, yet it needs to be able to control the gut microbiota to prevent its overgrowth and translocation to systemic sites. Both innate immunity and adaptive immunity perform to protect indigenous microbes and restrain non-indigenous microbes.

11.1.4.1 Environment

Some obvious environmental changes will lead to the change in the body bacterium group. For instance, it was found that there was a decrease in *Pepto strep to coccus* and an increase in *B. thetaiotaomicron* in aerospace personnel. The young and those eating more fiber and vegetables have more bifidobacterium than the old and those eating more meat and fat. Hunger may change the colonization state of indigenous intestinal flora in the epithelial cells.

11.1.4.2 Microbes

In the normal gastrointestinal tract, indigenous microbes exclude intrusive bacteria, and maintain stable bacterial populations. In this process, the factors at work include bacteriocin, antibiotics, nutrition competition, toxic metabolic end products, low redox potential and so on. Bacteriocin can rule out the intraspecific bacteria, while antibiotics can exclude the interspecific ones; volatile fatty acids inhibit the growth of foreign bacteria and facultative anaerobic bacteria; H_2S produced by the anaerobic bacteria controls growth of *E. coli*.

11.1.4.3 Antibiotics

Administration of spectrum antibiotics can destroy the ecological balance and lead to overgrowth of some species with potential pathogenicity, such as Clostridium difficile (associated with pseudomembranous colitis) and mould. Repeated exposure of the microbial inhabitants of the GIT to the same antibiotics is likely to result in the development of various resistance patterns. Furthermore, it can promote the spread of antibiotic resistance among pathogenic bacteria that will be in contact with the antibiotic-resistant microbiota.

11.1.4.4 Translocation of Bacteria

The passage of living bacteria from the gastrointestinal tract through the epithelial mucosa is called bacterial translocation [8]. Dysfunction of the gut mucosal barrier can cause translocation of many living microorganisms, usually gram-negative aerobic genera (escherichia, proteus, klebsiella). After crossing the epithelial barrier, bacteria can travel *via* the lymph to extra-intestinal sites, such as the mesenteric lymph nodes, liver and spleen. Subsequently, enteric bacteria can disseminate throughout the body producing sepsis, shock, multisystem organ failure, and even death of the host. The systemic presence of the intestinal microbiota and resulting associated complications can occur by two mechanisms. The first one relies on an internal cause such as impaired microvilli function, which leads to bacterial overgrowth and disruption of gut homeostasis resulting in the induction of initial systemic complications and onset of disease. The second one relies on an external injury or inflammatory reaction to cause stress in the body, leading to changes in intestinal permeability promoting bacterial translocation that results in further systemic complications and more serious disease.

11.2 Helicobacter Pylori and Gastroduodenal Disease

Helicobacter pylori (*H. pylori*) is a motile, spiral shaped, gram-negative bacterium that resides in the gastric mucus layer overlying the epithelium. Evidence is that *H. pylori* may induce some gastroduodenal diseases, including chronic gastritis, peptic ulcer and gastric cancer. Host genetics, host immune response, and bacterial virulence appear to play important roles.

11.2.1 Helicobacter Pylori and Chronic Gastritis

H. pylori possesses numerous unipolar flagella that enables it to move and makes it penetrate the mucin layer. The bacterium contains some adhesins which allow it to adhere to the epithelial cells. The adherence increases gastric secretion of IL-8 which, together with bacterial antigens, attracts monocytes and polymorphs. Antigen-presenting cells activate lymphocytes and other mononuclear cells that are attracted to the inflamed mucosa, initiating a cytotoxic or an antigen-producing Th response and causing chronic gastritis.

H. pylori can elaborate a large amount of urease, which neutralizes acid, enables *H. pylori* to infect and colonize the gastric mucosa. The bacterium also produces phospholipases A, and C, catalases and proteases which impair integrity of the gastric epithelial.

11.2.2 Helicobacter Pylori and Peptic Ulcer

Eradicating *H. pylori* may promote ulcer healing and reduce the recurrence. Duodenal ulcers are related to the hyper-secretion of acid and hyper-gastrinaemia. Recent studies have reported increased serum gastrin concentrations in persons infected with *H. pylori* while reductions in serum gastrin concentrations after *H. pylori* eradication. Production of alkaline ammonia by bacteria prevents D cells in the glands from sensing the true level of acidity, leading to inappropriate release of somatostatin and an increase in gastrin, and consequently excess acid secretion.

H. pylori isolated from patients with ulcers carries a high virulence. It usually includes a strong adhesive property and increased production of enzymes with toxic potential. Urease catalysis producing high concentrations of ammonia is followed by formation of toxic complexes such as NH_4Cl. Bacterial phospholipases A and C damage cell membranes and degrade gastric mucus. Most prevalent *H. pylori* genotypes in patients with peptic ulcerations are vacuolating cytotoxin positive (vacA^+) and cytotoxin-associated gene A positive (cagA^+). VacA causes vacuolar degeneration in cultured gastric-cell preparations and gastric ulceration in laboratory animals. The cagA is a marker of increased virulence and enhances the local inflammatory response [9].

11.2.3 Helicobacter Pylori and Gastric Cancer

It is generally accepted that *H. pylori* infection is the most recognized etiological risk factor for gastric cancer (GC). CagA^+ *H. pylori* is associated with an increased risk of GC. CagA^+ strains inject the cagA protein into host cells where it undergoes tyrosine phosphorylation. Translocated cagA forms a physical complex with the SRC homology domain (SH2)-containing tyrosine phosphatase (SHP-2) and stimulates phosphatase activity. Deregulation of SHP-2 by cagA induces cytoskeletal rearrangements, proliferation and increased motility of gastric epithelial cells. Moreover, the cagA^+ strains infection up-regulates the expression of COX-2 in gastric mucosa and GC. COX-2 is usually involved in inflammatory responses and is over-expressed in GC [10].

Recently, evidence has confirmed the importance of genetic polymorphism in the pathogenesis of GC. Functional polymorphisms of toll-like receptor 4, a lipopolysaccharide receptor involved in *H. pylori* recognition and host response, have been associated with increased risk of noncardia GC [11].

However, *H. pylori* infection is not considered a sufficient cause for cancer development. Host and environmental factors act synergistically in this multi-factorial disease.

11.2.4 Diagnosis for H. Pylori Infection

Diagnostic methods for *H. pylori* infection are categorized into two groups: invasive and non-invasive tests. The former includes histology, rapid urease test, culture and PCR. urea breath test and antibody test are widely used as the non-invasive tests.

11.2.4.1 Histology

H. pylori infection is easily recognized on histological examination. Giemsa and hematoxylin-eosin are the most widely used strains. The sensitivity and specificity of the strains are limited by several factors, such as the number and quality of biopsies, and the expertise of the pathologists.

11.2.4.2 Rapid Urease Test

The rapid urease test is a convenient method to detect *H. pylori* directly in gastric biopsies based on the enzymatic activity of urease. The urease activity elevates the pH of the solution when urea is converted to ammonia and it rapidly detects in a color change of the pH indicator.

11.2.4.3 Culture

A culture is probably the best method to identify an infectious agent and to detect the infecting strains' sensitivity to antibiotics. However, the peculiar growth characteristics of *H. pylori* make it difficult to use in a clinical setting.

11.2.4.4 PCR

The *H. pylori* DNA amplification is becoming one of the most practical tests to identify *H. pylori* in gastric mucosa. Real-time PCR-based kits available have great sensitivity, and are easy to use.

11.2.4.5 Urea Breath Test

The UBT identifies *H. pylori* indirectly in exhaled breath. It is simple, rapid and inexpensive. UBT is based on the enzymatic activity of bacterial urease that splits CO_2 labeled ^{13}C or ^{14}C isotopes from urea contained in a capsule ingested by the patients. The ^{12}C : ^{13}C ratio can be measured by a mass spectrometer or by a scintillation counter if ^{14}C is used. The UBT is currently the diagnostic method universally accepted to confirm the eradication of *H. pylori* after treatment ends.

11.2.4.6 Antibody Test

Detection of *H. pylori* antibodies has been widely used in epidemiological studies. The test detects IgG quantitatively by ELISA. The specificity is low because of the slow decline of the antibody titer after *H. pylori* eradication.

11.2.5 Treatment

The eradication of *H. pylori* infection is considered mandatory in peptic ulcer and gastric malignancies, such as gastric adenocarcinoma and mucosa-associated lymphoid tissue lymphoma. Furthermore, eradication is recommended in patients with nonulcer dyspepsia, especially in those with evidence of macroscopic or microscopic mucosal abnormalities (erosions, intestinal metaplasia and atrophy), in chronic NSAID users, in naive NSAID users, in first-degree relatives of gastric cancer patients, as well as in unexplained iron deficiency anemia and in idiopathic thrombocytopenic purpura.

The first-line option for treating *H. pylori* infection includes a proton pump inhibitor (PPI) or ranitidine bismuth, with any two antibiotics among clarithromycin, amoxicillinand metronidazole, given for 7 – 14 d. However, the failure rate for first-line therapy is high in actual clinical practice because of the indiscriminate use of antibiotics. Bismuth-based quadruple therapy and levofloxacin-based regimes have been shown to be effective as by second-line treatment [12, 13].

11.3 Inflammatory Bowel Disease

Inflammatory bowel disease comprises two major diseases, ulcerative colitis (UC) and Crohn's disease (CD). These two diseases share similar features of gut mucosal inflammation. It has been long supposed that IBD has a genetic basis and involves a response of the immune system to some environmental agents. The current opinion is that the immune system and its relationship to the intestinal microbiota are considered to play a central role in the initiation and persistance of IBD.

11.3.1 Role of Microbiota

The intestinal microbiota profoundly affects host immune composition in physiological conditions and is likely to be the most important environmental factor in IBD as the target of the inflammatory response [14]. IBD is mainly localized in the intestines in which most of the bacteria are congregated. This is consistent with the bacterial hypothesis that the commensal microbiota is the major environmental driver of IBD. Specifically, the bacterial hypothesis states that IBD represents the inappropriate response of the mucosal-associated immune system to the commensal microbiota in a genetically susceptible host, such that removal of the commensal microbiota will prevent the development of IBD as supported by model systems for mice.

Although bacteria are important in the pathogenesis of IBD, no specific microorganism has been involved in causing the intestinal inflammation. Some strains of bacteria have been isolated or cultured from the intestine of patients with CD, such as adherent-invasive *E. coli* and *M. paratuberculosis*. However, a direct cause and effect relationship has not been established for any of these organisms. In fact, conditions favoring transmission of infection (low socio-economic status, overcrowding, poor sanitation) appear to protect against IBD, rebutting the infectious etiology. While a single microorganism responsible for IBD has not been identified, numerous studies have shown an alteration to the composition of the gut microbial community in IBD. The majority of these studies have reported a relative abundance of *Enterobacteriaceae* in IBD patients and, recently, low counts of *Faecalibacterium prausnitzii* were shown to be associated with, and highly indicative of, CD localized to the ileum. However, it is still unclear whether the IBD-associated alterations in the gut microbiota are the cause or the consequence of the disease. Some studies of mice suggested that abnormal microbiota was able to initiate an inflammatory response in individuals not genetically predisposed and highlight the fact that colitis can be communicable.

11.3.2 Immune Response

It has been well documented that IBD is associated with the dysregulated mucosal immune response to the commensal microbial antigens in a genetically susceptible host [15, 16]. The immune response can be divided into two basic components: Innate and adaptive, respectively. In IBD, innate immune cells (such as macrophages and dendritic cells) within the lamina propria are inappropriately responding to bacterial components in individuals who have polymorphisms such as the NOD2 gene. This abnormal innate immune signaling from dendritic cells would be predicted to result in the secretion of inflammatory cytokines (such as TNF and IL-6) as well as other important cytokines that modify the adaptive immune response leading the genetically susceptible host to an exaggerated degree of adaptive immunity. Antibodies (IgG especially) production in active IBD are increased, both in the circulation and at the mucosal levels, which is likely to be inflammatory. T cells secret highly pro-inflammatory cytokines as well as an inappropriately low number of T-regulatory cells or the effector T cells resist the regulatory effects of the T-regulatory cells.

11.3.3 Clinical Findings and Diagnosis

The most common symptoms seen in both ulcerative colitis and Crohn's disease are diarrhea, rectal bleeding, abdominal cramps and pain, fever and weight loss. In Crohn's disease, symptoms can result from complications of the disease. A stricture can lead to intestinal obstruction with symptoms of filling up quickly after meals, nausea and vomiting. Fistulas can lead to openings in the skin and around the anal region that drain stool and infected material. An abscess can lead to symptoms of severe pain and fever.

In addition, organs other than the intestinal tract can be involved in the underlying inflammation of IBD. These organs include the eyes (symptoms of red eye or blurred vision), the mouth (symptoms of sores in the mouth), joints (symptoms of joint pain with or without joint swelling and redness), and skin (symptoms of rashes or skin ulcers most commonly involving the lower legs).

Young and middle-aged patients should be examined with the diagnosis of Crohn's disease with manifestations as follows: Intermittent bouts of low fever, diarrhea, right lower quadrant pain, mass and tenderness, perianal disease (abcess, fistulas, *etc.*), and ulceration, structuring or fistuals of the small intestine or colon revealed by radiography or endoscopy.

As for the diagnosis of ulcerative colitis, after ruling out infectious colitis and the other non-infectious colitis, the diagnosis can be made when the following evidence is found: intermittent bouts of bloody diarrhea, lower abdominal cramps, fecal urgency and characteristic endoscopic performance or histological appearance.

11.3.4 Treatment

Optimal treatment of inflammatory bowel disease depends on the disease patterns and level of severity. For example, mesalazine is more useful in ulcerative colitis than in Crohn's disease. For severe IBD' it may require immunosuppression to control the symptom, such as prednisone, TNF inhibition, azathioprine (Imuran), methotrexate, or 6-mercaptopurine. Mesalazine is required more commonly in treatment of IBD. Steroids are used to control disease flares and were once acceptable as a maintenance drug. Biologicals such as TNF inhibitors have been used in Crohn's disease patients for several years. Severe cases may require surgery, such as bowel resection, strictureplasty or a temporary or permanent colostomy or ileostomy [15, 16].

Drugs mentioned above primarily act to suppress the enhanced immune response in IBD and do not affect the microbiota. Antibiotics have some efficacy, especially in active CD and pouchitis, but cannot be used to maintain remission because of lack of long-term efficacy and side-effects. A therapy of probiotics and prebiotics that selectively manipulate the GI microbiota presents another treatment option with a low side-effect burden [17].

Probiotics are living microbes that are beneficial to the host in the gut. They include bacteria such as *Lactobacilli*, *Bifidobacteria*, gram-positive Cocci, *Enterococci* and yeast species such as *Saccharomyces boulardii*. Prebiotics are selectively fermented short-chain carbohydrates, including fructo-oligosaccharides and galacto-oligosaccharides, which allow specific changes both in the composition and/or activity of the GI microbiota that confer benefits upon host health. Combinations of probiotics and prebiotics are termed synbiotics. Animal studies of IBD indicate that probiotics can alter the intestinal microbiota and improve disease. Prebiotics also have shown the efficacy of reducing the activity of pro-inflammatory transcription factors and attenuate inflammation by enhancing gut immunoregulatory bacteria. In addition, prebiotics produce SCFA, such as acetate and butyrate, which inhibit mucosal inflammation, impacting both epithelial and dendritic cell functions. At present, there is some evidence to support the use of probiotics and prebiotics in IBD, but larger clinical trials are required using standard methodology to confirm this evidence.

11.4 Infectious Diarrhea

Acute diarrheal illness, caused by bacterial, viral, protozoal, and parasite pathogens, varies from mild bowel dysfunction to fulminant, life-threatening diseases. In general, pathogens or microbial toxins that cause acute diarrhea are ingested. Pathogenic microorganisms can pass through the hostile environment of the stomach if they are acid resistant or ingested with food. People with decreased gastric acidity are at increased risk of acute diarrheal disease. In the small bowel, the organisms either colonize or invade the mucosa. Organisms that do not have special colonization properties pass into the terminal ileum and colon, where they may compete with the naturally residing microorganisms. The ability of the colonic enteropathogens to invade intestinal mucosa allows these organisms to multiply preferentially.

Microbes can induce diarrhea directly by invasion of the gut mucosa as well as indirectly by producing one of three classes of microbial toxins: secretory enterotoxins, cytotoxins, or neurotoxins. Toxins may be elaborated after microbial replication in the gut or ingested directly [18].

11.4.1 Diarrhea Caused by Toxins

Patients infected with secretory toxin-producing pathogens seldom have fever or other major systemic symptoms, and little inflammatory response occurs. The large numbers of bacteria ingested with contaminated food or water colonize the small bowel. After multiplying to larger numbers, the bacteria produce enterotoxins that bind to mucosal cells, causing hypersecretion of isotonic fluid that far exceeds the reabsoptive capacity of the colon. The diarrhea caused by *V. cholerae* or *E. coli* enterotoxins can result in massive intestinal fluid loss, sometimes more than 1 L/h in adults.

Cytotoxins directly destroy mucosal epithelial cells. Shigella dysenteriae produces a toxin (Shiga toxin) that causes destructive colitis. A similar cytotoxin is produced by enterohemorrhagic *E. coli* strains that are associated with hemorrhagic colitis and hemolytic uremic syndrome. Other bacteria which can elaborate cytotoxins include Clostridium perfringens and Vibrio parahaemolyticus.

11.4.2 Diarrhea Caused by Invasive Pathogens

Diarrheas caused by invasive pathogens are usually accompanied by fever and other systemic symptoms, such as headache and myalgia. Cramping abdominal pain may be prominent, and small amounts of stool which contains pus cells, large amounts of protein and often gross blood are passed at frequent intervals. Epidemiologic characteristics are more helpful than signs or symptoms in determining the etiologic agent in invasive diarrheal illnesses. Shigella dysenteriae and salmonella are common invasive enteropathogens.

Acute shigellosis occurs when susceptible individuals ingest fecal-contaminated water or food. The bacteria are relatively resistant to gastric acid, and the diarrhea occurs after ingestion of only 10 – 100 microorganisms. Largely for this reason, direct person-to-person transmission is more common with shigellosis than it is with other bacterial enteric infections. The organism initially multiplies in the small intestine, producing watery, noninflammatory diarrhea. Later the organisms invade the colonic epithelium, causing the characteristic bloody stool.

Acute salmonellosis results from ingestion of contaminated meat, dairy, or poultry products. Unlike shigella, salmonella is remarkably resistant to desiccation. The nontyphoidal salmonellae invade primarily the distal ileum. The organisms typically cause a short-lived (2 – 3 d) illness characterized by fever, nausea, vomiting and diarrhea.

Other invasive enteropathogens include *Campylobacter Jejuni*, *Yersinia enterocolitica*, enteroinvasive *E. Coli* and *V. parahaemolyticus*. Although most diarrhea-causing pathogens produce either invasive or enterotoxic diarrhea, both processes contribute to the illness in some situations.

11.4.3 Diarrhea Caused by Viruses

Both Rotavirus and Norovirus invade and damage villous epithelial cells. The degree of injury ranges from modest distortion of epithelial cells to sloughing of villi. They may cause diarrhea by interfering with the absorption of normal intestinal secretions, possibly through selective destruction of absorptive villous tip cells with the sparing of secretory crypt cells. The stool is usually watery with few inflammatory cells, probably because of a lack of damage to the colon.

11.4.4 Diagnosis

Discerning the epidemiologic features of the illness is often more helpful than laboratory techniques in identifying patients. The examination of a methylene blue-stained stool preparation for erythrocytes and pus cells may help distinguish acute diarrheal illnesses caused by invasive pathogens from those caused by noninvasive pathogens. The precise diagnosis of any diarrheal illness lasting longer than 4 – 5 d is important because these illnesses may be responsive to specific antimicrobial therapy. Furthermore, among patients with negative stool examinations and cultures, endoscopy may yield a diagnosis of a noninfectious disease.

11.4.5 Treatment

All acute diarrheal diseases respond to a similar fluid repletion regimen because voluminous infectious diarrhea in adults consistently produces the same pattern of electrolyte loss. The fluid losses of massive diarrhea can rapidly be corrected by infusing fluids intravenously. Lactated Ringer's solution is readily available and provides uniformly good results. Almost every patient with diarrhea caused by toxigenic bacteria will recover after administration of adequate fluids intravenously throughout the illness. Fluid repletion can also be achieved through the oral route, using isotonic glucose-containing electrolyte solutions. Patients with fluid deletion caused by invasive microbial agents also respond well to oral glucose-electrolyte therapy.

Most acute infectious diarrheas do not require antibiotic therapy. Of the noninvasive bacterial diarrhea, antibiotics (such as Doxycycline) dramatically decrease the volume of diarrhea only in cholera. Of the invasive bacterial diarrheas, short-term antimicrobial treatment decreases the duration and severity of shigellosis, enteritis caused by Yersinia and Campylobacter. A quinolone such as ciprofloxacin is effective.

Prophylactic antimicrobial agents (such as doxycycline, trimethoprim-sulfamethoxazole and ciprofloxacin) are effective in preventing traveler's diarrhea, which is caused most often by ETEC.

11.5 Irritable Bowel Syndrome

Irritable bowel syndrome (IBS) is a functional gastrointestinal disorder characterized by altered motility, visceral hypersensitivity, abnormal brain-gut interaction, autonomic dysfunction, and immune activation. There are four subtypes of IBS: constipation-predominant IBS, diarrhea-predominant IBS, mixed IBS and unclassified IBS, which may change from one to another in many patients. Epidemiologic studies show prevalence of IBS to be 10% – 20% and an incidence of 1% – 2% per year.

11.5.1 Etiology

IBS is a multifactorial disorder, and the precise etiology is still unknown. There is evidence regarding the contribution of intestinal motility alteration, visceral hypersensitivity, disturbed intestinal reflexes, psychological disorders, food intolerance, GI infection and imbalance of gut flora and genetic factors. Over recent decades, growing evidence suggests that gut flora play an important role in the development of IBS.

11.5.1.1 Gut Flora

Evidence of gut microbes playing a role in the pathogenesis of IBS is supported by three main lines of reasoning. (i) A cause and effect relationship has been documented between the GI microbiota and a specific form of IBS, post-infectious IBS (PI-IBS). Some prospective studies showed IBS symptoms developed in 7% – 32% of patients after they recovered from bacterial gastroenteritis. The potential mechanisms include: difficulty in down-regulating intestinal inflammation in the colonic mucosa, increased gut permeability, increased mucosal enterochromaffin cell production, and increased concentration of mast cells and T lymphocytes in the gut mucosa. (ii) The GI microbiota is altered in IBS patients. The differences in the intestinal microbiota between IBS patients and healthy controls have mostly been studied using faecal material. Both an increase and a decrease in variation have been presented to characterize the GI microbiota in IBS. Ecologically, the abnormal variation most likely reflects a loss of homeostasis, in which the community is unable to maintain its normal structure. (iii) IBS symptoms can be improved by antibiotics, probiotics and prebiotics which target the microbiota [19, 20].

Small intestinal bacterial overgrowth (SIBO) has been proposed as a potential etiologic factor in IBS [21, 22]. It leads to fermentation and production of excess gas, which may induce some of the symptoms of IBS, including discomfort and bloating. Migrating motor complex was found to be an important mechanism controlling bacterial growth in the upper small bowel. Its disruption could promote duodenal bacterial overgrowth and bacterial translocation. However, whether or not the presence of SIBO is related to IBS is still controversial.

11.5.1.2 Dietary Factor

Food intolerance was recently considered as a pathogenesis factor in IBS. Certain foods, such as coffee, alcohol, spices, some raw fruits, vegetables, and milk, may cause the gut to malfunction. It was reported in a study that the IBS symptoms were relieved 8 weeks after eliminating the allergic food, which strongly suggested that food intolerance played a role in IBS.

11.5.1.3 Psychological Factor

The psychological factor plays an important role in visceral hypersensitivity in IBS. Most IBS patients have psychological disorders. It can alter the rhythm of motility and migrating motor complex (MMC). Some reports also showed that it may be a psychological factor which increased the perception of stimuli and induced more GI symptoms.

11.5.1.4 Brain-Gut Axis

Current evidence suggests that an altered brain-gut axis is the key mechanism associated with disordered motility, visceral hypersensitivity and autonomic dysfunction. Regulation of these connections occurs *via* numerous neurotransmitters such as cholecystokinin(CCK), VIP, 5-HT. Recent studies have also shown the involvement of the corticotropin-releasing hormone (CRH) in stress-related pathophysiology of IBS.

11.5.1.5 Genetic Factors

Gene polymorphism might be another factor inducing IBS. The genes included IL-10, serotonin transporter (5-SERT) gene, TRPV1. Individuals with a family history of the disorder are more susceptible to depression when they are exposed to more stress. Genetic factors may directly be linked to GI sensory and motor functions or initiate the modifications underlying the symptoms in the presence of exogenous factors.

11.5.2 Clinical Findings and Diagnosis

The primary symptoms of IBS are abdominal pain or discomfort in association with frequent diarrhea or constipation, and a change in bowel habits. It may also be urgent for bowel movements, tenesmus, bloating or abdominal distention. Some non-GI symptoms could also be present in IBS patients, such as headache, fatigue, backache, depression, anxiety and so on.

The diagnosis of IBS often can be suspected just by a review of the patient's medical history, and the need to rule out other diseases of the bowel. Today the diagnosis of IBS can be made *via* Roma III criteria. A patient should have abdominal discomfort or discomfort (not described as pain) at least once weekly for at least 2 months. The discomfort should be associated with two of the following three features:

(i) Can obtain relief with a bowel movement;

(ii) Onset is associated with a change in the frequency of the bowel movement;

(iii) Onset is associated with a change in the form (appearance) of stool;

There should be no evidence of an inflammatory, anatomic (obstructive), metabolic, or neoplastic (tumorous) cause of the symptoms.

11.5.3 Treatment

IBS is a long-term but manageable condition, and medication affects people differently. It depends on the symptoms and severity.

As mentioned above, the dietary factor and psychological factor can lead to IBS, so changing diet or lifestyle and managing stress are useful to relieve symptoms sometimes.

As for medication, fiber supplements, antispasmodic agents, laxatives and antidiarrheal agents are often used clinically to relieve symptoms. Probiotics are also very common in medical therapy. Probiotics are live bacteria that, when ingested, result in a beneficial response in the individual. The most common probiotic bacteria are bifidobacteria, which are found in the intestine of normal individuals. The explanation as to how probiotics benefit individuals is still unclear. There are two possible mechanisms implicated. The one is that the probiotic bacteria suppress the other bacteria in the intestine which cause symptoms, and the other is that the probiotic bacteria act on the host's intestinal immune system to suppress inflammation. Furthermore, therapy by antibiotics is reported in some research to play a role in treating irritable bowel syndrome by preventing the overgrowth of intestinal bacteria and is believed to improve the symptoms of bloating, diarrhea, abdominal pain, and constipation [23].

11.6 Antibiotic-Associated Diarrhea

Antibiotic-associated diarrhea (AAD) refers to otherwise unexplained diarrhea following the use of antibiotics. Usually, the diarrhea is caused by altered composition and function of the intestinal flora, and the specific pathogens cannot often be identified. Most patients respond to discontinuation of antibiotics and supportive treatment. Risk factors mainly include advanced age, types and prolonged use of antibiotics, the length of hospitalization, and compromised immune status. All groups of antibiotics may cause AAD, but those with broad-spectrum coverage, in particular cephalosporins, extended-coverage penicillins, and clindamycinare are the most common culprits [24, 25].

11.6.1 Pathogenesis

The prolonged use of multiple antibiotics, especially broad-spectrum agents with poor intestinal absorption or high biliary excretion, induces a change in the composition and function of the intestinal flora and therefore results in a higher incidence of AAD.

A decrease in the colonic anaerobic flora interferes with carbohydrate and bile acid metabolism. Osmotic or secretory diarrhea may occur. Overgrowth of opportunistic pathogens takes place as a result of microbiologic and metabolic alterations.

C. difficile, an anaerobic gram-positive rod, accounts for 15% – 20% of all AAD cases. It has been clear that *C. difficile* is the cause of diarrhoea in the setting of antibiotic-associated pseudomembranous colitis, but its role is more questionable in benign self-limited cases of AAD in the community [26]. *C. difficile* is widely present in the environment, transmitted by the fecal-oral route to susceptible individuals. It is considered as part of the normal flora of infants and can be isolated in about 5% of healthy adults and in up to 1/3 of asymptomatic or colonized hospitalized patients.

Toxins A (enterotoxin) and B (cytotoxin) are the best known virulence factors of *C. difficile*. The mechanism of action is by toxin binding on intestinal receptors, leading to disruption of the cytoskeleton and intracellular junctions. Protein synthesis and cell division are inhibited. Important inflammatory mediators will attract neutrophils and monocytes, increasing capillary permeability, tissue necrosis, hemorrhage and edema.

Serum antibodies to *C. difficile* infection are detected in many infected patients. The host's immune response has a role in the clinical outcome. The colonized patients who become asymptomatic carriers have significantly more serum IgG antitoxin A than those with *C. difficile* diarrhea. An elevated IgG amount is also measured in patients who have no relapse of *C. difficile* diarrhea. It suggests that antibodies provide a protective function.

Furthermore, ADD is also supposed to be associated with the direct effects of antibiotics. Some antibiotics have been found to cause small bowel enteropathy with malabsorption and epithelial transport dysfunction.

11.6.2 Clinical Findings and Diagnosis

The clinical manifestations of AAD may vary from mild diarrhea to fatal colitis. The cardinal symptom of the disease is diarrhea, which commonly develops during treatment but may appear as late as 8 weeks after discontinuation of antibiotics. Typical cases of *C. difficile* infection manifest themselves by a profuse, mucous, foul-smelling diarrhea accompanied by cramps and tenesmus. Other clinical manifestations of antibiotic-associated colitis include fever, leukocytosis, fecal leukocytes, hypoalbuminemia, colonic thickening on computed tomography, and characteristic changes by endoscopic inspection or biopsy. In severe cases, toxic megacolon may occur along with the deceptive findings of "improved diarrhea". Then, obvious colonic distention, peritoneal irritation, fever and elevated white blood count are commonly present.

The diagnosis of AAD should be considered in any patient presenting new-onset diarrhea after being treated with antibiotics recently. Exposure can be up to 8 weeks before any antibiotics, including antifungal agents. Clinical presentation, laboratory data, imaging studies, and endoscopic examinations are all useful.

The cytotoxin assay using tissue culture for detecting toxin B has been the gold standard for diagnosis. It is the most sensitive test. However, most laboratories do not offer tissue-culture assays. Alternatives include enzyme immunoassays and stool culture.

11.6.3 Treatment

On one hand, in the majority of patients, AAD is a mild and self-limited illness that responds to the discontinuation of antibiotics, supportive care, and fluid and electrolyte replacement. On the other hand, in cases in which signs and symptoms of colitis develop, the use of effective oral antimicrobial agents against *C. difficile* is often necessary [24].

Oral metronidazole (500 mg three times daily or 250 mg four times daily) and oral vancomycin (125 mg four times daily) have similar rates of efficacy, with response rates of 90% - 97%. Metronidazole is preferred because it is less expensive than vancomycin, and avoids the potential risk of promoting vancomycin-resistant enterococci in nosocomial cases. Indications for oral vancomycin, as opposed to metronidazole, are pregnancy, lactation, intolerance of metronidazole, or failure to respond to metronidazole after 3 - 5 d of treatment. Most relapses respond to another course of antibiotics in standard doses for 10 d. For the repeated replases, the administration of the use of pulsed doses of vancomycin, anion-exchange resins to absorb *C. difficile* toxin, the use of agents to antagonize *C. difficile*, or intravenous immune globulin may be useful.

Probiotics, live microorganisms which confer a health benefit on the host, have been tried in AAD. Several meta-analyses indicated that probiotics may protect against AAD. A prospective, double-blind, controlled study in 1994 showed *Saccharomyces boulardii*, a live nonpathogenic yeast, had some benefits in the treatment of AAD by binding to the glycoprotein receptor site for toxin A at the intestinal brush border. But there is insufficient evidence to recommend probiotic therapy as an adjunct to antibiotic therapy for *C. difficile* colitis.

11.7 Colorectal Cancer

Colorectal cancer (CRC) is a cancer characterized by neoplasia in the colon, rectum, or vermiform appendix. Globally, cancer of the colon and rectum is the third leading cause of cancer in males and the fourth leading cause of cancer in females. The frequency of colorectal cancer varies around the world. It is common in the western world and is rare in Asia and Africa. In countries where people have adopted western diets, the incidence of colorectal cancer is increasing.

11.7.1 Pathogenesis

The pathogenesis of CRC is still unclear. However, many studies have shown that some factors do increase a person's risk of CRC, including high fat intake, a family history of CRC, the presence of polyps in the large intestine, and chronic ulcerative colitis.

Nowadays a lot of evidence from a wide range of sources supports the view that the colonic microbiota is involved in the etiology of cancer. In animals, the presence of the intestinal microbiota has a major impact on colonic tumor formation. One study showed that the rate of tumor formation was much more rapid in conventional than in germ-free rats treated with the tumor initiator 1,2-dimethylhydrazine (DMH) [27]. Bacteria of the bacteroides and clostridium genera increase the incidence and growth rate of colonic tumors induced in animals, whereas other genera such as *lactobacillus* and *bifidobacteria* prevent tumorigenesis. A descriptive human study compared the composition of the faecal flora of people with different risks of CRC. A high risk of CRC was associated with the presence of *Bacteroides vulgatus* and *Bacteroides stercoris*. A low risk was associated with presence of *Lactobacillus acidiphilus*, *Lactobacillus* S06 and *Eubacterium aerofaciens*.

Intestinal bacteria may play a part in initiation of CRC through production of carcinogens, cocarcinogens, or procarcinogens. Epidemiological evidence suggests that diet is a major influence in the development of colorectal malignancies and, consequently, it was proposed that microbial metabolic by-products of dietary compounds could produce either a cytoprotective or a cytotoxic effect in the intestinal mucosa [28]. One of the mainstream hypotheses proposed it that it is associated with the concentrations and composition of microbiota-produced SCFAs, which play a part in maintaining integrity of the epithelial layer and providing energy to epithelial cells. A study showed SCFAs induced different levels of apoptosis, cell cycle arrest and differentiation in animal models of colon cancer, as well as in CRC cell lines compared to normal cells. Another theory postulates that increased production of toxic compounds, such as hydrogen sulfide by the microbiota could produce a cytotoxic and consequently carcinogenic effect. Hydrogen sulfide can be produced through metabolism of amino acids, which are the products of digestion of dietary proteins, and a high-protein diet has been linked to an increased incidence of colon cancer. Populations at low and high risk for CRC have been shown to harbor methanogenic and nonmethanogenic microbiota, respectively. Methanogenic microbiota preferentially produces harmless methane as an end-product of amino acid metabolism, while non-methanogenic microbiota that is enriched with sulfate-reducing bacteria will result in excessive elaboration of highly toxic hydrogen sulfide. Diets rich in fat and meat but poor in vegetables also increase the faecal excretion of N-nitroso compounds, a group of genotoxic substances that are known initiators and promoters of colon cancer [29]. Another group of carcinogens of dietary origin are the heterocyclic aromatic amines that are formed in meat when cooked. Some intestinal microorganisms strongly increase damage to DNA in colon cells induced by heterocyclic amines, whereas other intestinal bacteria can uptake and detoxify such compounds.

There is also increasing evidence supporting the role of inflammation in the pathogenesis of CRC, with evidence indicating that commensal colonic bacteria are important in influencing this process [30]. The microbiota provides a major stimulus in the activation and development of the normal intestinal immune system. Persistent inflammation can result in continued tissue damage and even tumorigenesis. One piece of evidence is that individuals with inflammatory bowel disease are at an increased risk of developing IBD-associated cancer.

Some other factors may also be implicated in CRC development, such as fecal water activities, bacterial enzymes and so on.

11.7.2 Symptoms

The symptoms of colorectal cancer depend on the location of the tumor in the bowel, and whether it has spread elsewhere in the body.

Most patients may have a change in bowel habits and a feeling of incomplete defecation. The right colon is spacious, and the mass of the right colon can grow to a large size before causing any abdominal symptoms. Typically, right-sided tumors cause iron deficiency anemia due to the slow loss of blood over a long period of time. The left colon is narrower than the right colon. Therefore, tumors of the left colon are more likely to cause partial or complete bowel obstruction, resulting in symptoms of constipation, narrowed stool, diarrhea, abdominal pains, cramps, and bloating. Bright red blood in the stool may also indicate a growth near the end of the left colon or rectum. CRC may also lead to weight loss, generally due to a decreased appetite. More unusual constitutional symptoms are an unexplained fever and several paraneoplastic syndromes. The most common paraneoplastic syndrome is thrombosis, usually deep vein thrombosis.

11.7.3 Diagnosis

Screening for the disease is recommended in individuals who are at an increased risk. There are several different tests available for diagnosing cancer and finding out if it has metastasized.

Digital rectal exam (DRE): It only detects tumors large enough to be felt in the distal part of the rectum, but is useful as an initial screening test.

Colonoscopy: A lighted probe called a colonoscope is inserted into the rectum and the entire colon to look for polyps and other abnormalities that may be caused by cancer. This is the only screening test that allows the removal of polyps, which can also prevent CRC.

Double contrast barium enema (DCBE): It is a method for patients who cannot have a colonoscopy. But the barium enema has a lower likelihood of detecting precancerous polyps than a colonoscopy, or CT colonography.

Carcinoembryonic antigen (CEA): High levels of CEA may indicate that a cancer has spread to other parts of the body. But CEA is frequently false positive or false negative. Therefore, CEA is more useful to assess disease recurrence rather than screening it.

CT scan: A CT scan creates a three-dimensional picture of the inside of the body with an *X*-ray machine. In a patient with CRC, a CT scan can check for the spread of cancer in the lungs, liver, lymph node, and other organs. It is often done before surgery.

11.7.4 Treatment

The treatment of CRC depends on the size and location of the tumor, whether the cancer has spread, and the person's overall health. The three primary treatment options are: surgery, chemotherapy and radiation.

Surgery can be categorised into curative, palliative, bypass, fecal diversion, or open-and-close. The most common treatment for CRC is surgery to remove the tumor, called surgical resection. Some patients may need surgery on the liver or lungs to remove tumors that have spread to those organs.

Systemic chemotherapy is delivered through the bloodstream, targeting cancer cells throughout the body. Chemotherapy may be given after surgery to eliminate any remaining cancer cells. In some situations, the chemotherapy and radiation therapy are given before surgery to reduce the size of a rectal tumor and lower the chance of the cancer returning.

Radiation therapy is commonly used for treating rectal cancer because this tumor tends to recur locally. Radiation may be used in conjunction with surgery as definitive therapy, or may be used to reduce, or palliate the symptoms of colorectal cancer such as pain, bleeding, or blockage.

11.8 Gastrointestinal Tuberculosis

Tuberculosis (TB) in the GI tract is common in developed countries. Its incidence in developing countries has been rising in immigrant groups and patients with AIDS. It is caused by both mycobacterium tuberculosis and M bovis. The most frequent site of involvement is the ileocecal region. However, any region of the GI tract may be involved. Modes of GI infection include the following: (i) Spread by means of the ingestion of infected sputum, in patients with active pulmonary TB and especially in patients with pulmonary cavitation and positive sputum smears;

(ii) Hematogenous spread from a primary lung focus that reactivates later or miliary tuberculosis, spread *via* lymphatics from infected nodes; (iii) Local spread from adjacent organs involved in primary tuberculous infection (*e.g.*, renal TB causing fistulas in the duodenum or mediastinal TB lymphadenopathy involving the esophagus). Intestinal tuberculosis may cause granuloma formation, caseation, mucosal ulceration, or scarring and fibrosis with narrowing of the lumen.

Patients often complain of chronic abdominal pain, obstructive symptoms, weight loss, diarrhea, and fever. Dysphagia is seen in esophageal TB. Gastric TB may present similarities to peptic ulcers or gastric carcinoma. Complications include intestinal obstruction, hemorrhage, fistula formation, and bacterial overgrowth with malabsorption. The PPD skin test may be negative, especially in patients with weight loss or AIDS. The differential diagnosis includes Crohn's disease, carcinoma, and intestinal amebiasis. The diagnosis is established by either an endoscopic or surgical biopsy revealing acid-fast bacilli within involved tissue. Computed tomography and/or ultrasonography of the abdomen are helpful for diagnosis. Conventional antituberculous therapy for a minimum of 6 months is effective [31, 32].

References

[1] Mackie R I, Sghir A, Gaskins H R. Developmental microbial ecology of the neonatal gastrointestinal tract. Am J Clin Nutr, 1999, 69: 1035S-1045S.

[2] Guarner F, Malagelada J R. Gut flora in health and disease. Lancet, 2003, 361: 512-519.

[3] Ouwehand A C, Vaughan E E. Gastrointestinal Microbiology. New York: Taylor & Francis Group, 2006: 29-45, 147-150.

[4] Andersen L P. Colonization and infection by Helicobacter pylori in humans. Helicobacter, 2007, 12: S12-S15.

[5] Sekirov I, Russell S L, Antunes L C, *et al.* Gut microbiota in health and disease. physiol Rev, 2010, 90: 859-904.

[6] Tappenden K A, Deutsch A S. The physiological relevance of the intestinal microbiota—contributions to human health. J Am Coll Nutr, 2007, 26: 679S-683S.

[7] Mason KL, Huffnagle GB, Noverr MC, *et al.* Overview of gut immunology. Adv Exp Med Biol, 2008, 635: 1-14.

[8] Berg RD. Bacterial translocation from the gastrointestinal tract. Adv Exp Med Biol, 1999, 473: 11-30.

[9] Malfertheiner P, Chan F K, McColl K E. Peptic ulcer disease. Lancet, 2009, 374: 1449-1461.

[10] Konturek P C, Konturek S J, Brzozowski T. Helicobacter pylori infection in gastric cancerogenesis. J Physiol Pharmacol, 2009, 60: 3-21.

[11] Fuccio L, Eusebi L H, Bazzoli F. Gastric cancer, Helicobacter pylori infection and other risk factors. World J Gastrointest Oncol, 2010, 2: 342-347.

[12] Suzuki H, Nishizawa T, Hibi T. Helicobacter pylori eradication therapy.

Future Microbiol, 2010, 5: 639-948.
[13] Costa F, D'Elios M M. Management of Helicobacter pylori infection. Expert Rev Anti Infect Ther, 2010, 8: 887-892.
[14] Sartor R B. Microbial influences in inflammatory bowel diseases. Gastroenterology, 2008, 134: 577-594.
[15] Blumberg R S. Inflammation in the intestinal tract: Pathogenesis and treatment. Dig Dis, 2009, 27: 455-464.
[16] Kaser A, Zeissig S, Blumberg R S. Inflammatory bowel disease. Annu Rev Immunol, 2010, 28: 573-621.
[17] Campierei M, Gionchetti P. Probiotics in inflammatory bowel disease: New insight to pathogenesis or possible therapeutic alternative. Gastroenterology, 1998, 116: 1246-1260.
[18] Carpenter C C J. Cecil Essential of Medicine. Beijing: Peking university press, 2008.
[19] Salonen A, de Vos W M, Palva A. Gastrointestinal microbiota in irritable bowel. Microbiology, 2010, 156: 3250-3215.
[20] Si J M, Yu Y C, Fan Y J, *et al.* Intestinal microecology and quality of life in irritable bowel syndrome patients. World J Gastroenterol, 2004, 10: 1802-1805.
[21] Lin HC. Small intestinal bacterial overgrowth: A framework for understanding irritable bowel syndrome. JAMA, 2004, 292: 852-858.
[22] Peralta S, Cottone C, Doveri T, *et al.* Small intestine bacterial overgrowth and irritable bowel syndrome-related symptoms: Experience with Rifaximin. World J Gastroenterol, 2009, 15: 2628-2631.
[23] Christopher W Hammerle, Surawicz C M. Updates on treatment of irritable bowel syndrome,World. Gastroenterol, 2008, 14: 2639-2649.
[24] Bartlett J G. Antibiotic-associated diarrhea. N Engl J Med, 2002, 346: 334-339.
[25] Beaugerie L, Petit J C. Antibiotic-associated diarrhoea. Best Practice & Research Clinical Gastroenterology, 2004, 18: 337-352.
[26] American College of Gastroenterology, Practice Parameters Committee. Guidelines for the diagnosis and management of Clostridium difficile – associated diarrhea and colitis. Am J Gastroenterol, 1997, 92: 739-750.
[27] Reddy B S, Weisburger J H, Narisawa T, *et al.* Colon carcinogenesis in germ-free rats with dimethylhydrazine, and N-nitrosamines in health and gastroduodenal disease. Lancet, 1974, ii: 550-552.
[28] O'Keefe S J. Nutrition and colonic health: The critical role of the microbiota. Curr Opin Gastroenterol, 2008, 24: 51-58.
[29] Hughes R, Cross A J, Pollock J R, *et al.* Dose-dependent effect of dietary meat on endogenous colonic N-nitrosation. Carcinogenesis, 2001, 22: 199-202.
[30] Scanlan P D, Shanahan F, Clune Y, *et al.* Culture-independent analysis of the gut microbiota in colorectal cancer and polyposis. Environ Microbiol, 2008, 10: 789-798.
[31] Rashed K A, Jader S, Karla P, *et al.* Intestinal tuberculosis. Int J Tuberc Lung Dis, 1998, 2: 70.
[32] Lazarus AA, Thilagar B. Abdominal tuberculosis. Dis Mon, 2007, 53: 32-38.

12

Infectious Microecology in Liver Disease

Lanjuan Li
The State Key Laboratory for Diagnosis and Treatment of Infectious Diseases, the First Affiliated Hospital, School of Medicine, Zhejiang University; Collaborative Innovation Center for Diagnosis and Treatment of Infectious Diseases, Hangzhou, 310003, China
E-mail: ljli@zju.edu.cn

More than 100 years ago, Pavlov discovered that the liver can remove enteric toxins. Now we think that, under normal circumstances, the liver can also remove bacteria, fungi, and their metabolites such as toxins, endotoxin, ammonia, indoles, phenols, short-chain fatty acids (C4-C6), pseudo-nerve delivery mass precursors. Gnotobiote research reveals that intestinal flora and endotoxin in the liver increase the number of Kupffer cells and play an important role in the improvement. When the functioning of the liver suffers damage (including chronic hepatitis, alcoholic and non-alcoholic fatty liver disease, cirrhosis, severe hepatitis, liver transplantation, *etc.*), the system will incur a gut microflora imbalance in varying degrees, damage to the intestinal barrier and allowing intestinal bacteria and its metabolites to enter extra-intestinal organs (including blood) , resulting in abdominal infection, endotoxemia and so on. Meanwhile, the inflammatory stimuli activate the immune system excessively, which can cause an abnormal immune response, leading to systemic inflammatory response syndrome (SIRS) or multiple organ failure, such as gastrointestinal dysfunction or failure. Endotoxemia, and gastrointestinal dysfunction can also increase the imbalance in gut microflora, and further aggravate liver injury, a vicious cycle. Through a series of basic and clinical microbial research of liver diseases, we found that probiotics can adjust the gut microflora, repairing and improving the intestinal barrier function and reducing intestinal bacterial translocation, decreasing endotoxemia, reducing infection and improving the functioning of the liver. Therefore, the intestinal microflora has a close association with the liver in the anatomical location, and other aspects of physiology and pathophysiology. The balance of the microecology in humans plays an important role in the protection of vital organ functions.

12.1 An Overview of Infectious Microecology in Liver Disease

There is a strong relationship between the liver and the gut through the portal system and enterohepatic recycling. The liver receives blood rich in nutrients and certain toxic contents from the gut and affects intestinal functions through bile secretions into the intestinal lumen.

Alterations of intestinal microbiota seem to play an important role in induction and promotion of liver damage progression, in addition to direct injury resulting from different causal agents (*e.g.*, viral, toxic, metabolic) [1], through mechanisms such as activation of Kupffer cells by bacterial endotoxins. Gut microbiota participate in the pathogenesis of complications of liver cirrhosis, such as infections, spontaneous bacterial peritonitis, hepatic encephalopathy, and renal failure.

12.1.1 Liver Anatomy and Enterohepatic Recycling

The liver is the largest visceral organ in the body, receiving approximately a quarter of the cardiac output. The normal liver has a dual blood supply, deriving 70% to 80% of its blood, rich in nutrients and certain toxic contents absorbed in the gut, from the portal vein, and the other 20% to 30%, rich in oxygen, from the hepatic artery [2].

There is a strong relationship between the liver and the gut: The portal system receives blood from the gut, and intestinal blood content activates liver functions. The liver, in turn, affects intestinal functions through bile secretion into the intestinal lumen [1]. In enterohepatic recycling, foreign chemicals entering the alimentary tract are absorbed into portal venous blood by enterocytes, removed from blood by uptake into hepatocytes, secreted into the bile, and then deposited back into the intestinal lumen where they may be reabsorbed by intestinal cells and available for recycling [3].

12.1.2 Gut-Liver Axis

The gut is a habitat for billions of microorganisms. The intestinal microbiota form is a complex ecological system that participates, under physiological conditions, in the production of vitamins, degradation of bile acids, digestion of nutrients, and local and general immunity [4]. Finally, together with the intestinal mucosa, the endogenous gut flora forms an important barrier against pathogens.

12.1.3 The Role of Kupffer Cells in Liver Disease

The liver consists of the hepatic parenchyma and a large proportion of nonparenchymal cells including sinusoidal endothelial cells, ito cells, and the dedicated hepatic macrophage known as the Kupffer cells. The majority of the metabolism functions associated with the liver's detoxification function takes place within the hepatocytes, whereas the nonparenchymal cell population provides physical and biochemical structure to the liver. The sinusiodal endothelia line the sinusiodal space and form a barrier, which serves to divide the liver into functional compartments. However, Kupffer cells are able to traverse this barrier and are able to pass in and out of the hepatic space, facilitating their signaling function. Central to this signaling role is the ability of Kupffer cells to respond to local changes by the release of cytokines and other signaling molecules such as reactive oxygen species that induce hepatocyte cell proliferation and may enhance clonal expansion of preneoplastic cells leading to neoplasia [5].

Kupffer cells are macrophages that reside in the hepatic sinusoids and constitute the largest number of resident macrophages of the reticuloendothelial system in the body. In normal liver, Kupffer cells comprise 30% of the total hepatic nonparenchymal cell population and 3% of the weight of the liver. Kupffer cells are located in the lumens of the sinusoids, with long processes anchored to the fenestrae on the endothelium, which facilitates the Kupffer cells' filtration and removal of harmful substances from microcirculation. As an important component in the host's defense system, Kupffer cells remove infective and foreign substances from circulation by endocytosis. These cells also release toxic mediators involved in liver injury and cirrhosis[2].

Kupffer cells play an important role in the liver's normal physiology and homeostasis as well as participating in the acute and chronic responses of the liver to toxic compounds. Activation of Kupffer cells directly or indirectly by toxic agents results in the release of an array of inflammatory mediators, growth factors, and reactive oxygen species. This activation appears to modulate acute hepatocyte injury as well as chronic liver responses including hepatic cancer. Understanding the role Kupffer cells play in these diverse responses is the key to understanding mechanisms of liver injury. As with the other roles of Kupffer cells in chronic injury and carcinogenesis, the mechanisms by which Kupffer cells contribute to acute liver injury are varied. In general, the mechanisms involve release by Kupffer cells of mediators including cytokines such as tumor necrosis factor alpha (TNF-α) and interleukins. Reactive oxygen, nitrogen species, proteases, and lipid metabolites such as prostaglandins and thromboxane are also released [5].

12.1.4 Endotoxin Activates Kupffer Cells in Liver Disease

One substance that can effectively activate Kupffer cells is bacterial endotoxins (*e.g.*, bacterial lipopolysaccharide (LPS), peptidoglycan, lipoproteins, and various lipopeptides). The LPS, which is thought to be an important endotoxin, is a component of the cell walls of some of the bacteria (especially Gram-negative bacteria) that normally inhabit the intestine. When the bacteria die, the LPS is released into the intestine, from which some of it can cross the intestinal wall and enter the bloodstream. Higher than normal amounts of LPS entering the bloodstream or tissues can cause fever, chills, shock, and various other symptoms and can often lead to more severe conditions, such as endotoxemia or adult respiratory distress syndrome (ARDS) [5].

Gut flora alterations consist of bacteria overgrowth and release in the circulation of bacterial endotoxins. LPS normally penetrate the mucosa only in trace amounts, enter the portal circulation, and become cleared in the liver to maintain the control of immune homeostasis. Resident macrophages (Kupffer cells) and hepatocytes both contribute to this process through different LPS recognition systems. There is a positive correlation between liver dysfunction and the occurrence of bacterial translocation, and the clearance of LPS from circulation is decreased in states of hepatic dysfunction, such as cirrhosis. It has been proposed that LPS initially is taken up by Kupffer cells and then by hepatocytes. LPS is removed via several mechanisms, including molecules that bind LPS and prevent it from activating TLR4, enzymes that degrade the lipid, a moiety to decrease its activity, inactivation of LPS following uptake into the liver and the spleen, and cellular adaptations that modify target cell responses. Another mechanism for LPS neutralization involves serum lipoproteins, HDL, LDL, VLDL, and chylomicrons, apolipoproteins apoE and apoA-I LPS. All of these mechanisms can chaperone endotoxin to hepatocytes, Kupffer cells, or sinusoidal endothelial cells, resulting in clearance of LPS without significant inflammatory cell activation [6].

LPS activates Kupffer cells via two pathways: One is membrane attachment from mCD14 which is the classical CD14 dependent pathway, which requires the LPS binding protein (LBP) as a co-factor carrying LPS to the membrane of Kupffer cells bound to the receptor CD14; the soluble compound formed by combining LBP with LPS increases affinity of LPS with CD14. Whereas the other pathway-sCD14 may not require LBP, combined with corresponding receptor on Kupffer cells by aid of other proteins (such as HDL, LDL *etc.*), these two pathways finally activate the signal transduction system and trigger the synthesis and release of cytokines and inflammatory mediators [7].

Kupffer cells are macrophages that reside in the liver. Their main role – removing bacteria and foreign proteins from the blood – is essential to the liver's primary function, which is cleansing the blood of foreign materials and toxic substances. When no foreign materials are present, Kupffer cells are in a resting state. They can be activated by numerous molecules, including bacterial endotoxins. Various causal agents can lead to increased endotoxin levels in the blood and the liver. When activated, Kupffer cells secrete a variety of cytokines, including a molecule called tumor necrosis factor alpha (TNF-α), reactive oxygen species (ROS), and several types of interleukins. All of these molecules can act as inflammatory cytokines – that is, they induce an inflammatory response necessary to remove the offending toxic or foreign molecules, damage liver cells and initiate the healing process [5].

12.1.5 Pathogenesis of Bacterial Translocation in Liver Disease

Bacterial translocation is defined as the migration of viable bacteria or bacteria products from the intestinal lumen to mesenteric lymph nodes or other extra-intestinal sites, such as liver, spleen, kidney, and blood. Gram-negative members of the Enterobactereaceae family (such as Escherichia coli and Klebsiella spp), enterococci and other streptococci species are frequently seen in bacterial translocation[8]. Factors that promote bacterial translocation include overgrowth with Gram-negative bacteria, impaired host immune defenses, and injury to the intestinal mucosal barrier resulting in increased intestinal permeability. These mechanisms can act in concert to promote synergistically the systemic spread of indigenous translocating bacteria to cause lethal complications including sepsis.

12.1.5.1 Intestinal Bacterial Overgrowth

Intestinal bacterial overgrowth (both Gram-negative and Gram-positive bacteria) is a major factor promoting bacterial translocation. Impaired intestinal motility, as well as other factors, such as suppressed secretion of gastric acid and intestinal enzymes and reduced bile flow, is important in facilitating bacterial translocation in liver disease. Increasing evidence suggests that small intestinal bacterial overgrowth (SIBO) occurs in approximately 60% of the patients with liver cirrhosis [10]. and its prevalence correlates directly with the severity of liver disease [11].

SIBO is also believed to be a predictor of spontaneous bacterial peritonitis [10, 12].

12.1.5.2 Dysfunction of Intestinal Mucosal Barrier

The host defence against invasion of microbes from the gut consists of numerous local factors, such as mucus, gastric acid, pancreatic enzymes, bile, the epithelial cell barrier with its intracellular junctions, and bowel motility. The most critical barrier against uptake of intra luminal microbes/microbial products, however, is the epithelium per se. Most clinical studies performed to investigate structural changes of intestinal mucosa in patients with cirrhosis have focused on the small intestine. Shorter and thicker microvilli have been described. Thick-walled, dilated capillaries along with oedema of the lamina propria, fibromuscular proliferation, a reduced villus/crypt ratio and thickened muscularis mucosa in the small bowel have been found in cirrhotic patients with portal hypertension and it has been proposed that an increased potential for bacterial translocation may exist in this setting [9, 13].

12.1.5.3 Luminal Factors Contributing to Intestinal Barrier Function

Other factors apart from the epithelium per se that contribute to the normal intestinal barrier against bacterial translocation include luminal factors, such as levels of bile acids, secretory immunoglobulin A (IgA), mucins, defensins, lysozyme and phopholipase A2. All of these mechanisms help to prevent attachment of bacteria to the epithelium — an event that has been proposed as an important first step in epithelial penetration.

Mucus also acts as a lubricant to reduce physical abrasion of the mucosa and participates in the protection of the mucosa from damage induced by acid and other luminal toxins. Mucosal secretions are rich in IgA antibodies that effectively bind and aggregate bacteria, preventing mucosal adherence and colonization (so-called immune exclusion). Bile acids exert a trophic effect on intestinal mucosa and inhibit intestinal bacterial overgrowth, especially that of Gram-positive species [8]. The absence of bile in the intestine has been shown to promote bacterial translocation; however, this could be attributed to malnutrition. In contrast, exposure to bile during bacterial growth decreases epithelial internalization of enteric bacteria. In cirrhosis there is a decrease in bile acid secretion [9].

12.1.5.4 Impaired Intestinal Immunity

Immune dysfunction is an important factor promoting bacterial translocation. The intestinal tract is an active immune organ, containing essentially all types of leukocyte involved in the immune response. The antigen-specific local immune system, termed gut-associated lymphatic tissue (GALT), is the largest immunological organ of the body, making up 25% of the mucosal cell mass. The GALT comprises more than half of the lymphoid cells in the body and hosts numerous plasma cells, macrophages, neutrophils, Paneth cells and specialized M-cells playing a key role in controling bacterial translocation. It appears that appropriate activation of intestinal T cells is critical in maintaining immunity against the translocation of enteric bacteria [9]. In liver disease, secretory IgA is decreased which suggests impaired intestinal immunity [10].

But until now little is known concerning the functional capacity of intestinal immunity in patients with cirrhosis and whether any disturbance of intestinal immune mechanisms contributes importantly to bacterial translocation, especially in the clinical setting.

12.2 Gut Microflora in the Pathogenesis of the Complications of Cirrhosis

The gut flora plays an important role in the pathogenesis of the complications of cirrhosis [11]. Major complications of cirrhosis include ascites infections, spontaneous bacterial peritonitis, hepatic encephalopathy, portal hypertension, variceal bleeding, and hepatorenal syndrome [12]. Cirrhotic patients are prone to develop bacterial infections, mainly the spontaneous infection of ascites or spontaneous bacterial peritonitis. Other complications of cirrhosis, such as variceal haemorrhage and ascites, occur mostly or solely as a consequence of portal hypertension [11].

The most important role of the gut flora in chronic liver disease relates to its contribution to the pathogenesis of the complications of cirrhosis. These complications are common to cirrhosis of all etiologies and account for the high morbidity, mortality and healthcare costs associated with cirrhosis [11].

The gut flora plays a role in the development of infections and also in the hyperdynamic circulatory state of cirrhosis and, although less prominently, it also plays a role in the pathogenesis of hepatic encephalopathy [11].

12.2.1 Bacterial Infections in Cirrhosis

Bacterial infections are a known complication of cirrhosis, with a reported incidence that ranges between 15% and 47% [18-20]. These figures contrast with a general hospital population rate of infection of 5% – 7%. Two factors are predictive of the development of bacterial infections in cirrhosis: the severity of the liver disease and gastrointestinal hemorrhage. Low serum albumin and admission for GI bleeding were also predictive of the development of a bacterial infection in cirrhosis.

Cirrhotic patients who develop an infection have a significantly higher mortality than uninfected patients. Patients with more severe liver disease are more susceptible to develop infections, while Child C and the occurrence of bacterial infection are independent predictors of mortality. Most infected cirrhotic patients die from sepsis, renal dysfunction and failure to control variceal hemorrhage [11].

12.2.2 Sources and Types of Bacterial Infection in Cirrhosis

Common infections in cirrhotic patients were mainly urinary tract infections, spontaneous peritonitis and pneumonia, 70% – 80% of which were caused by Gram-negative bacilli, especially *Escherichia coli*, suggesting that the gut was the main source of bacteria. However, the spectrum of bacteria causing infection in cirrhosis shows a significantly higher rate of Gram-positive cocci infections, probably due to an increase in the number of therapeutic invasive procedures and to the use of chronic antibiotic prophylaxis [11].

12.2.3 Bacteria Translocation in the Pathogenesis of Spontaneous Bacterial Peritonitis in Cirrhosis

Spontaneous bacterial peritonitis is a very common bacterial infection in patients with cirrhosis and ascites. When first discovered, its mortality exceeded 90% but it has been reduced to approximately 20% with early diagnosis and treatment [21-23].

The diagnosis of SBP is based on diagnostic paracentesis. All patients with cirrhosis and ascites are at risk of spontaneous bacterial peritonitis and the prevalence in outpatients is 1.5% – 3.5% and 10% in hospitalized patients. Half the episodes of SBP are present at the time of hospital admission while the rest are acquired during hospitalization [24, 25].

Patients with SBP may have one of the following symptoms: (i) Local symptoms and/or signs of peritonitis: abdominal pain, abdominal tenderness, vomiting, diarrhea, ileus; (ii) Signs of systemic inflammation: hyper or hypothermia, chills, altered white blood cell count, tachycardia, and/or tachypnea; (iii) Worsening of liver function; (iv) Hepatic encephalopathy; (v) Shock; (vi) Renal failure; (vii) Gastrointestinal bleeding. However, it is important to point out that SBP may be asymptomatic, particularly in outpatients [13].

The key pathogenic mechanism of spontaneous bacterial peritonitis is bacterial translocation. Blood dissemination and microbial growth in ascitic fluid resulting from spontaneous bacterial peritonitis are a consequence of damage to the immune system in cirrhosis. Hyperproduction of proinflammatory cytokines and other vasoactive substances contributes to the arterial vasodilation and renal failure that frequently complicate the course of spontaneous bacterial peritonitis [14].

12.2.4 Gut Flora and the Hyperdynamic Circulatory State in Cirrhosis

Vasodilatation and the subsequent development of the hyperdynamic circulatory state lead to a worsening of all complications of cirrhosis. The hyperdynamic splanchnic circulation leads to the developement of gastro-oesophageal varices, and their eventual rupture. Ascites also correlates with vasodilatation and activation of neurohumoral systems. In the renal circulation, the initial stages of the hyperdynamic circulatory state are marked by compensatory afferent arteriolar vasodilatation that preserves renal perfusion and glomerular filtration pressure. As vasodilatation worsens, however, renal perfusion pressure drops and arteriolar vasoconstriction can lead to the development of renal dysfunction (hepatorenal syndrome). In the pulmonary circulation, arteriolar vasodilatation is also observed, increasing pulmonary blood flow and the development of arteriovenous shunts, leading to abnormal gas exchange. In a small proportion of cases, this results in severe hypoxia that is resistant to oxygen supplementation (hepatopulmonary syndrome) [15]. Hepatic encephalopathy is a consequence, not only of shunting of toxins through porto-systemic collaterals, but also of brain oedema secondary to brain arterial vasodilatation [11].

The hyperdynamic syndrome should be better called "progressive vasodilatory syndrome", because vasodilatation is the factor that brings about all the vascular changes and finally leads to the multiorgan involvement observed as a consequence of this hemodynamic change [28]. It is characterized by low vascular resistance and mean arterial pressure and by increased heart rate, cardiac output and regional blood flow. Peripheral and predominantly splanchnic arterial vasodilatation induces arterial underfilling and hence relative hypovolaemia and reduction in arterial pressure, leading to activation of neurohumoral systems. Mediators from these systems lead to sodium and water retention by the kidneys, increasing plasma volume. The expansion in plasma volume and the corresponding increase in blood volume are essential for the full expression of HCS [11].

Many vasodilatory molecules, such as prostacyclin, carbon monoxide(CO), endothelium-derived hyperpolarizing factor, endocannabinoids, tumor necrosis factor alpha (TNF-α), hydrogen sulfide (H_2S), adrenomedullin [28], calcitonin-generelated peptide, substance P and glucagon, have been related to the pathogenesis of the hyperdynamic circulatory state. However, nitric oxide (NO) appears to be the key vasodilator responsible for the haemodynamic abnormalities of cirrhosis. This is most clearly shown in experiments, in which non-specific inhibition of NO synthesis leads to an almost complete normalization of splanchnic haemodynamics and to prevention of HCS. NO is synthesized by different isoforms of NO synthases (NOS); endothelial (e) NOS and neuronal (n) NOS are expressed constitutively, whereas inducible (i) NOS is expressed after induction by lipopolysaccharides, endotoxins and cytokines. The importance of cytokines, specifically TNF-α, in the development of HCS is evidenced by studies which show that TNF-α inhibitors ameliorate HCS in cirrhotic rats [11].

12.2.5 The Gut Flora and Hepatic Encephalopathy

Hepatic encephalopathy is a chronically debilitating complication of hepatic cirrhosis. It is a complication of cirrhosis considered a reversible metabolic encephalopathy. It occurs as a result of both liver insufficiency and increased portal-systemic shunting of gut-derived nitrogenous compounds and toxins. Bacterial infections are a common precipitant, not only of acute but also chronic hepatic encephalopathy.

In a randomized, double-blind, placebo-controlled trial, Bass NM and his colleagues randomly assigned 299 patients who were in remission from recurrent hepatic encephalopathy resulting from chronic liver disease to receive either rifaximin, at a dose of 550 mg twice daily (140 patients), or placebo (159 patients) for 6 months. Over a 6-month period, treatment with rifaximin, a minimally absorbed oral antimicrobial agent that is concentrated in the gastrointestinal tract, which has a broad-spectrum *in vitro* activity against Gram-positive and Gram-negative aerobic and anaerobic enteric bacteria, maintained remission from hepatic encephalopathy more effectively than did a placebo. Rifaximin treatment also significantly reduced the risk of hospitalization involving hepatic encephalopathy [29].

Ammonia is the main nitrogenous compound implicated in the pathogenesis of hepatic encephalopathy. Ammonia is generated in both the small bowel (from the effects of glutaminase on glutamine) and the large intestine (from urease activity of the colonic flora). Urease-producing bacteria in the colon include anaerobic bacteroides, aerobic coliforms, and aerobic and anaerobic streptococcal organisms. Once ammonia is absorbed it is metabolized by the liver. However, metabolism of ammonia is altered in cirrhosis because of porto-systemic shunting (portal blood containing ammonia is shunted away from the liver) and because of changes in the periportal (site of urea synthesis) and perivenous (glutamine synthesis) hepatocyte function. Other gut-derived toxins implicated in hepatic encephalopathy are gamma-amino-butyric acid (GABA) and benzodiazepine (BZD)-like substances, both of which may also be produced by specific colonic bacteria. Interestingly, hemoglobin increases the production of GABA and may contribute to the pathogenesis of hepatic encephalopathy after gastrointestinal bleeding [11].

12.3 Modulation of Intestinal Microbiota as a Therapeutic Strategy of Liver Disease

Healthy intestinal microbiota contributes to host resistance and infection through their involvement in the development of the host immune system and provision of colonization resistance. It is not surprising then that disruptions of the microbiota in liver disease can alter host susceptibility to infection. Additionally, the process of the infection itself results in a disturbance to intestinal microbiota, which would exacerbate the state of liver disease. More and more evidences show that manipulation of gut bacteria can restore the microbiologic and immunologic equilibrium in the intestinal wall in patients with liver disease and help in the treatment or prevention of infections.

12.3.1 Manipulation of Gut Flora and Its Effect on Infections in Cirrhosis

Complicated bacterial infection may have severe adverse clinical consequences in cirrhotic patients. The associated pro-inflammatory cytokine response exacerbates hepatic disfunction, encephalopathy and the haemodynamic disturbance that underlie the development of portal hypertension and hepatorenal syndrome. Most of the currently tested therapies geared at preventing infection in cirrhosis are aimed at decreasing bacterial overgrowth or changing the composition of gut flora.

12.3.1.1 Selective Intestinal Decontamination

Antibiotics that selectively eliminate intestinal Gram-negative bacteria, such as norfloxacin, have been shown to produce a marked reduction in fecal Gram-negative bacteria of patients with cirrhosis, without significant effects on Gram-positive cocci or anaerobic bacteria. Most studies use norfloxacin, a poorly absorbed quinolone with activity against aerobic Gram-negative bacilli. Selective intestinal decontamination has been shown to be effective in preventing bacterial infections in patients with gastrointestinal hemorrhage, and low ascites protein, as well as preventing SBP recurrence. There is evidence that antibiotic prophylaxis for SBP may be of benefit in three high risk groups with cirrhotic ascites, namely those who have survived a previous episode, those with low ascetic fluid total protein levels and those presenting gastrointestinal hemorrhage. A double-blind, placebo-controlled trial evaluating the long-term efficacy of norfloxacin (400 mg daily) in cirrhotic patients who had survived a previous episode of SBP found a significantly reduced rate of SBP recurrence in the treated group (12% *vs.* 35%) during a mean follow-up period of 6 months. The overall probability of SBP recurrence at one year from follow-up was 20% in the group receiving norfloxacin prophylaxis compared to 68% in the placebo group [30]. To investigate if oral, non-absorbable antibiotics prevent bacterial infections in cirrhotics with gastrointestinal hemorrhage, Rimola and his research group collected 140 consecutive patients who were randomly allocated into two groups: 68 patients were given oral, non-absorbable antibiotics (gentamicin + vancomycin + nystatin or neomycin + colistin + nystatin) from the inclusion into the trial up to 48 hours after cessation of the hemorrhage, or until emergency surgery or death in those cases with continued bleeding; and 72 patients did not receive oral, non-absorbable antibiotics. The incidence of infection was significantly lower in the antibiotic group. This difference was due to the fact that spontaneous bacteremia and peritonitis and urinary tract infection caused by enteric bacteria occurred almost exclusively in the control group. Their results indicate that prophylactic administration of oral, non-absorbable antibiotics markedly reduces the incidence of infections caused by enteric bacteria in cirrhotic patients with gastrointestinal hemorrhage [31]. Grange *et al.* found that primary prophylaxis with norfloxacin for 6 months is effective in the prevention of infections caused by Gram-negative bacilli in cirrhotic patients with low ascitic fluid total protein levels [32]. Preliminary studies also have shown that selective intestinal decontamination appears to ameliorate the hyperdynamic circulatory state of cirrhosis [16].

The efficacy of both short-term and long-term primary prophylaxis with norfloxacin has been assessed in cirrhotic patients. Long-term primary prophylaxis with norfloxacin may be preferable to prophylaxis only during periods of hospitalization, as long-term antibiotics prophylaxis has been associated with the development of antibiotic-resistant infections. According to the result of Novella's research, continuous long-term selective intestinal decontamination with norfloxacin is effective in preventing spontaneous bacterial peritonitis in cirrhotic patients at high risk. However, the overall incidence of infections caused by norfloxacin-resistant bacteria has a tendency to increase in the norfloxacin treatment group. The emergence of infections caused by norfloxacin-resistant bacteria must be weighed carefully against the benefits of continuous long-term prophylaxis [34].

12.3.1.2 Probiotics, Prebiotics and Synbiotics

Probiotics are defined as micro-organisms resistant to digestion, capable of adhering to the digestive tract mucosa and when ingested in sufficient amounts beneficial to the host's health. Prebiotics are non-digestible food ingredients that stimulate the growth and (or) activity of bacteria in the digestive system in ways claimed to be beneficial for health. Synbiotics refer to nutritional supplements combining probiotics and prebiotics in a form of synergism, hence synbiotics. Probiotics has been postulated as creating equilibrium in potentially pathogen bacterial populations and also reducing BT by increasing the immunologic capacity of the host at the gastrointestinal level. Probiotics have been proposed as a means of re-equilibrating gut flora in favor of protective anaerobic bacteria. Preliminary data are available in various in vitro and *in vivo* settings as to whether treatment with probiotics may be of benefit in reducing bacterial translocation and its infective complications. In order to determine the efficacy of *Lactobacillus casei* GG in reducing the rate of *Escherichia coli* C25 (*E. coli* C25) translocation, Mattar *et al.* used an *in vitro* enterocyte cell-culture model. The probiotic bacterium has been shown to inhibit translocation of *E. coli* in a dose-dependent manner in a cell culture model [17]. Administration of different *Lactobacilli* and *Bifidobacterium* strains have been reported to correct bacterial overgrowth, stabilize the mucosal barrier function, enhance hose defenses, and hence to decrease BT in rat models with acute liver injury and failure [18]. Patients with liver disease treated with a combination of probiotics (*Lactobacillus plantarum*) and prebiotics (fiber) had a lower rate of postoperative bacterial infections than those treated with selective intestinal decontamination [37]. *Lactobacillus johnsonii* La 1 in combination with antioxidants can be useful in preventing bacterial translocation in cirrhotic rats induced by CCL (4) [38]. However, controversial results also exist. Riordan *et al.* found that administration of a symbiotic Gram-positive gut flora regimen to patients with cirrhosis led to a further increase in peripheral mononuclear cell expression of TLR2 and circulating TNF levels, although the significance of this needs to be determined [39].

Bacterial DNA may be detected in the ascetic fluid of as many as one-third of cirrhotic patients with no obvious manifestation of infection. The presence of bacterial DNA can produce significantly higher amounts of TNF-α, IL-2, IL-6, and IL-12, inducible nitric oxide synthase, leading to increased vasodilatation of the mesenteric vasculature. There is evidence that synbiotics and fermentable fibre may improve the hepatic function and other cirrhotic complications in patients without overt infection. Treatment with synbiotics was found to be effective for management of minimal hepatic encephalopathy in patients with cirrhosis [40].

Safety of probiotics is an important problem we need to consider when selecting probiotics. *Bifidobacteria* and *Lactobacilli* are the main species to be considered within the category of probiotics. In general, they are species considered as non-pathogen to the human organism, which is implicit in the definition of probiotics. There have been reports of isolated cases (less than 2%) of endocarditis and hepatic abscess mainly in immunocompromised patients [19]. In such cases, the proposed action mechanism consisted of higher platelet aggregation collagen and fibrinogen binding, and production of glycosidases and proteases. However, upon collecting the data from these cases, it was confirmed that the *Lactobacillus* causing both episodes was not the product of any experiment, but it was secondary to intestinal native flora which was a product of the patient's diet. Based on the above data, it can be asserted that the use of probiotics in humans is safe and it has a very low risk of adverse reactions.

12.3.2 Manipulation of Gut Flora and Its Effect on Infections in Liver Transplants

Bacterial sepsis and wound complications after liver transplantation increase mortality, morbidity, hospital stay, and overall transplant costs. More and more evidence shows that manipulation of gut flora by probiotics can decrease the rate of bacterial infection in patients after liver transplant. Gurusamy *et al.* compared the effects of selective intestinal decontamination and active lactobacillus with fibres (probiotic with prebiotic), and inactivated Lactobacillus with fibres (prebiotic). Although there is no clear evidence of any intervention offering significant benefits in the reduction of bacterial infections and wound complications in liver transplantation, selective bowel decontamination increases the risk of infection and hospital stay compared to prebiotics and probiotics. The use of prebiotics and probiotics offers promise [20]. Further randomised clinical trials are necessary.

References

[1] Cesaro C, Tiso A, Del Prete A, *et al.* Gut microbiota and probiotics in chronic liver diseases. Dig Liver Dis, 2011, 43:431-438.
[2] Kan Z, Madoff D C. Liver anatomy: Microcirculation of the liver. Semin Intervent Radiol, 2008, 25:77-85.
[3] Roberts M S, Magnusson B M, Burczynski F J, *et al.* Enterohepatic circulation: Physiological, pharmacokinetic and clinical implications. Clin Pharmacokinet, 2002, 41:751-790.
[4] Abt M C, Artis D. The intestinal microbiota in health and disease: The influence of microbial products on immune cell homeostasis. Curr Opin Gastroenterol, 2009, 25:496-502.
[5] Wheeler M D. Endotoxin and Kupffer cell activation in alcoholic liver

disease. Alcohol Res Health, 2003, 27: 300-306.
[6] Szabo G, Bala S. Alcoholic liver disease and the gut-liver axis. World J Gastroenterol, 2010, 16: 1321-1329.
[7] Han DW. Intestinal endotoxemia as a pathogenetic mechanism in liver failure. World J Gastroenterol, 2002, 8: 961-965.
[8] Almeida J, Galhenage S, Yu J, *et al.* Gut flora and bacterial translocation in chronic liver disease. World J Gastroenterol, 2006, 12: 1493-1502.
[9] Bauer T M, Steinbruckner B, Brinkmann F E, *et al.* Small intestinal bacterial overgrowth in patients with cirrhosis: Prevalence and relation with spontaneous bacterial peritonitis. Am J Gastroenterol, 2001, 96: 2962-2967.
[10] Madrid AM, Hurtado C, Venegas M, *et al.* Long-term treatment with cisapride and antibiotics in liver cirrhosis: Effect on small intestinal motility, bacterial overgrowth, and liver function. Am J Gastroenterol, 2001, 96: 1251-1255.
[11] Morencos F C, de las Heras Castano G, Martin Ramos L, *et al.* Small bowel bacterial overgrowth in patients with alcoholic cirrhosis. Dig Dis Sci, 1995, 40: 1252-1256.
[12] Hashimoto N, Ohyanagi H. Effect of acute portal hypertension on gut mucosa. Hepatogastroenterology, 2002, 49: 1567-1570.
[13] Wiest R, Rath H C. Gastrointestinal disorders of the critically ill. Bacterial translocation in the gut. Best Pract Res Clin Gastroenterol, 2003, 17: 397-425.
[14] Norman K, Pirlich M. Gastrointestinal tract in liver disease: Which organ is sick? Curr Opin Clin Nutr Metab Care, 2008, 11: 613-619.
[15] Garcia-Tsao G, Wiest R. Gut microflora in the pathogenesis of the complications of cirrhosis. *Best Pract Res Clin Gastroenterol* 2004, 18: 353-372.
[16] Heidelbaugh J J, Sherbondy M. Cirrhosis and chronic liver failure: Part II. Complications and treatment. Am Fam Physician, 2006, 74: 767-776.
[17] Yoshida H, Hamada T, Inuzuka S, *et al.* Bacterial infection in cirrhosis, with and without hepatocellular carcinoma. Am J Gastroenterol, 1993, 88: 2067-2071.
[18] Deschenes M, Villeneuve J P. Risk factors for the development of bacterial infections in hospitalized patients with cirrhosis. Am J Gastroenterol, 1999, 94: 2193-2197.
[19] Caly W R, Strauss E. A prospective study of bacterial infections in patients with cirrhosis. J Hepatol, 1993, 18: 353-358.
[20] Garcia-Tsao G. Current management of the complications of cirrhosis and portal hypertension: Variceal hemorrhage, ascites, and spontaneous bacterial peritonitis. Gastroenterology, 2001, 120: 726-748.
[21] Tandon P, Garcia-Tsao G. Bacterial infections, sepsis, and multiorgan failure in cirrhosis. Semin Liver Dis, 2008, 28: 26-42.
[22] Gines P, Angeli P, Lenz K, *et al.* EASL clinical practice guidelines on the management of ascites, spontaneous bacterial peritonitis, and hepatorenal syndrome in cirrhosis. J Hepatol, 2010, 53: 397-417.
[23] Nousbaum J B, Cadranel J F, Nahon P, *et al.* Diagnostic accuracy of the

Multistix 8 SG reagent strip in diagnosis of spontaneous bacterial peritonitis. Hepatology, 2007, 45: 1275-1281.

[24] Evans L T, Kim W R, Poterucha J J, *et al.* Spontaneous bacterial peritonitis in asymptomatic outpatients with cirrhotic ascites. Hepatology, 2003, 37: 897-901.

[25] Gonzalez A R, Gonzalez G M, Albillos M A. Physiopathology of bacterial translocation and spontaneous bacterial peritonitis in cirrhosis. Gastroenterol Hepatol, 2007, 30: 78-84.

[26].Rasaratnam B, Connelly N, Chin-Dusting J. Nitric oxide and the hyperdynamic circulation in cirrhosis: Is there a role for selective intestinal decontamination? Clin Sci (Lond), 2004, 107: 425-434.

[27] Iwakiri Y, Groszmann R J. The hyperdynamic circulation of chronic liver diseases: from the patient to the molecule. Hepatology, 2006, 43(2 Suppl 1): S121-S131.

[28] Bass N M, Mullen K D, Sanyal A, *et al.* Rifaximin treatment in hepatic encephalopathy. N Engl J Med, 2010, 362: 1071-1081.

[29] Gines P, Rimola A, Planas R, *et al.* Norfloxacin prevents spontaneous bacterial peritonitis recurrence in cirrhosis: Results of a double-blind, placebo-controlled trial. Hepatology, 1990, 12(4 Pt 1): 716-724.

[30] Rimola A, Bory F, Teres J, *et al.* Oral, nonabsorbable antibiotics prevent infection in cirrhotics with gastrointestinal hemorrhage. Hepatology, 1985, 5: 463-467.

[31] Grange J D, Roulot D, Pelletier G, *et al.* Norfloxacin primary prophylaxis of bacterial infections in cirrhotic patients with ascites: A double-blind randomized trial. J Hepatol, 1998, 29: 430-436.

[32] Rasaratnam B, Kaye D, Jennings G, *et al.* The effect of selective intestinal decontamination on the hyperdynamic circulatory state in cirrhosis. A randomized trial. Ann Intern Med, 2003, 139: 186-193.

[33] Novella M, Sola R, Soriano G, *et al.* Continuous versus inpatient prophylaxis of the first episode of spontaneous bacterial peritonitis with norfloxacin. Hepatology, 1997, 25: 532-536.

[34] Mattar A F, Drongowski R A, Coran A G, *et al.* Effect of probiotics on enterocyte bacterial translocation *in vitro*. Pediatr Surg Int, 2001, 17: 265-268.

[35] Adawi D, Ahrne S, Molin G. Effects of different probiotic strains of Lactobacillus and Bifidobacterium on bacterial translocation and liver injury in an acute liver injury model. Int J Food Microbiol, 2001, 70: 213-220.

[36] Rayes N, Seehofer D, Muller AR, *et al.* Influence of probiotics and fibre on the incidence of bacterial infections following major abdominal surgery – results of a prospective trial. Z Gastroenterol, 2002, 40: 869-876.

[37] Chiva M, Soriano G, Rochat I, *et al.* Effect of Lactobacillus johnsonii La1 and antioxidants on intestinal flora and bacterial translocation in rats with experimental cirrhosis. J Hepatol, 2002, 37: 456-462.

[38] Riordan SM, Skinner N, Nagree A, *et al.* Peripheral blood mononuclear cell expression of toll-like receptors and relation to cytokine levels in cirrhosis. Hepatology, 2003, 37: 1154-1164.

[39] Liu Q, Duan Z P, Ha D K, *et al.* Synbiotic modulation of gut flora: Effect on minimal hepatic encephalopathy in patients with cirrhosis. Hepatology, 2004, 39: 1441-1449.

[40] Asahara T, Takahashi M, Nomoto K, *et al.* Assessment of safety of lactobacillus strains based on resistance to host innate defense mechanisms. Clin Diagn Lab Immunol, 2003, 10: 169-173.

[41] Gurusamy K S, Kumar Y, Davidson B R. Methods of preventing bacterial sepsis and wound complications for liver transplantation. Cochrane Database Syst Rev, 2008: CD006660.

13

Biliary Infection, Pancreatic Infection and Microecology

Jianwen Jiang, Zhigang Ren, Shusen Zheng *
Key Lab of Combined Multi-Organ Transplantation, Ministry of Public Health, Department of Hepatobiliary and Pancreatic Surgery, the First Affiliated Hospital, School of Medicine, Zhejiang University; Collaborative Innovation Center for Diagnosis and Treatment of Infectious Diseases, Hangzhou, 310003, China
* E-mail: shusenzheng@zju.edu.cn

The biliary tract system mainly transports bile secreted by hepatocytes and bile duct epithelial cells into the gut, and includes the intra-hepatic biliary tract and extra hepatic biliary tract. It starts from intra hepatic capillary biliary tract, and ends with pancreatic duct rendezvous with the Vater ampullary, opening into the duodenum nipple. The pancreas is the body's second largest digestive gland secondary to the liver, and has exocrine and endocrine functions. Normally, the pancreas stimulated by food, *etc.* secretes mounts of characteristic pancreatic juice into the gut. Thus, biliary tract and pancreas infections are closely associated with intestinal bacterial translocation and microecology imbalance.

13.1 Biliary Infection and Microecology

Biliary infection is mainly derived from intestinal bacterial translocation [1]. The inducements of intestinal bacterial translocation mainly include the dysbolism of cholic acid or bilirubin, the inhibition of intestinal movement, the disorder or overgrowth of intestinal bacteria, the dysfunction of the intestinal mucosa barrier, and the decline of the body's immune function [2-4]. To prevent the infection of the biliary tract and to keep the balance of intestinal microecology, many microecology modulators have been used to prevent intestinal bacterial translocation and disorders [5-7].

13.1.1 Microecology Foundation of Biliary Tract System

The structure and function of the biliary tract system constitutes the microecology foundation of the biliary tract system. Thus we will focus on explaining the structure and function of the biliary tract system.

13.1.1.1 The Structure of the Biliary Tract System

The biliary tract system refers to the tract that transports bile secreted by liver cells to the duodenum pipe, including the intrahepatic and extrahepatic biliary tract, which is up from the intrahepatic bile ducts of the capillary, and ends with the pancreatic duct rendezvous with the Vater ampullary, opening in the duodenum nipple. Intrahepatic biliary tract includes liver segmental bile ducts, lobe intrahepatic bile ducts and left-right hepatic duct. The extrahepatic biliary tract includes the extrahepatic left-right hepatic duct, gallbladder, cystic duct and common bile duct.

The gallbladder is pear-shaped, 8 – 12 cm in length, and has a width of 3 – 5 cm, a capacity of 30 – 60 mL. It can generally be divided into 3 parts: the end of the gallbladder, the gallbladder body and gallbladder neck, respectively. Under normal circumstances, the gallbladder has a volume of 15 mL, and the volume is 90 – 100 mL when it is full. Gallbladder pressure can be up to 9.8 kPa (100 mm H_2O), so gallbladder puncture would induce bile leakage. The arteries of the gallbladder come from the cystic artery. The common bile duct and hepatic duct have a rich blood supply, which forms a vein network in the serous plexus, mainly derived from the following sources: duodenal artery, posterior superior pancreaticoduodenal artery, pancreatic vascular arcades and the cystic artery. Lymphatic drainage of the gallbladder is similar to the liver, which flows through the liver, lymph nodes, gall bladder, neck lymph nodes, and omental foramen lymph node. Lymphatic drainage of the common bile duct would flow into the pancreatic head lymph nodes, hepatic artery and celiac artery lymph nodes and other lymph nodes.

13.1.1.2 The Function of the Biliary Tract System

The function of the biliary tract system mainly includes the storage, concentration and secretion function of the gallbladder, motor function of gallbladder, and the metabolism and secretion of bile.

The storage, concentration and secretion function of the gallbladder. In the inter-digestive period, the biliary sphincter in ampullary stays in a state of contraction. Thus the resistance at the end of the biliary tract increased, and the process of the flow of liver bile into the duodenum is blocked. Meanwhile, due to the diastolic gallbladder, the pressure at the end of the biliary tract is higher than gallbladder pressure and thus liver bile in turn flows into the gallbladder. Along with the bile continuously flowing into the gallbladder, the gallbladder expands and the intracapsular pressure increases. Thus the gallbladder transmural pressure and tension rises. Under normal circumstances, when the gallbladder pressure reaches 9.8 kPa (100 mm H_2O), equals or exceeds the tension of the biliary sphincter tone, the bile generated by the liver flows into the intestine.

The human gallbladder has a strong concentration function. According to the survey, the gallbladder can concentrate liver bile to 1/5 – 1/20. Through the strong concentration function, the gallbladder can store about half of 24 h of bile secreted by the liver, and the concentrated bile is isotonic to plasma. Bile water is rapidly absorbed, in which organic constituents such as cholesterol, bile pigment, bile salt and so on are retained and concentrated. Approximately 10% of lecithin in the gallbladder is absorbed, and a small amount of free bile acids is removed from the gallbladder by diffusion. Gallbladder mucosa easily absorb fat-soluble substances, but does not absorb water soluble highly molecular material.

The gallbladder also has a secretory function. Under normal circumstances, the gallbladder secretes about 20 mL of mucus every day to protect the gallbladder mucosa. The secreted bile is crystal, slightly milky, alkaline, and rich in mucin and bicarbonate. The process of gallbladder secretion is an active process, which is unaffected by the fluid mechanics and the effect of osmotic pressure. When the cystic duct is blocked, the bile pigment and bile in the gallbladder are absorbed gradually, and then are replaced with the mucus secreted by the gallbladder, which is clinically called "white bile".

The motor function of the gallbladder. The gallbladder has a motor function, *i.e.*, systolic function. The gallbladder contains smooth muscle layers composed with longitudinal muscle and circular muscle fiber. As it contracts, the gallbladder can provide the power of the bile flowing into the common bile duct. When it is diastolic, the bile within the biliary system can flow into the gallbladder for storage and concentration. Therefore, the contraction and diastole of the gallbladder have a function in regulating biliary duct pressure.

After eating, the gallbladder constriction is in a phase process. When the food emerges into the duodenum, gallbladder tensional contraction occurs immediately, and the gallbladder becomes hard and oval, appearing in the bile drain, which lasts several minutes, like the first phase. Subsequently from the gallbladder a suspended discharge of bile appears briefly, about 2 min, and then the main activities of the bile drain start, lasting about 30 min. About 2/3 of the gallbladder empties, and the bottom contraction and the body ring contraction can be visible in the gallbladder. The whole period of the bile drain can be from more than 10 min to 5 or 6 h, in different individuals or under different conditions.

In addition to the gallbladder, the biliary tract motor includes the biliary sphincter. Under normal circumstances, gallbladder contraction is accompanied by biliary sphincter relaxation, and gallbladder diastole is accompanied by sphincter contraction, both of which are in coordination.

The motor of the biliary sphincter has 4 functions [8, 9]: (i) Regulating the bile going into the intestine; (ii) Preventing intestinal contents going into the common bile; (iii) Controlling the filling of the gallbladder; (iv) Causing the erection of the nipple of the bile duct outlet, and producing the activity of squeezing-shooting the bile flow, to help the bile in the gallbladder and biliary tract to empty.

The metabolism and secretion of bile. Bile is a liquid of complex composition and function. It not only contains the necessary material for the process of food digestion and absorption in the intestines, but also is a metabolism drainage tract of the liver, dealing with a variety of macromolecular materials. Meanwhile, bile can convey the immune globulin secreted by the liver into the intestine, playing an important role in the intestinal local immune defense function. The components of bile are various, in which the components that are closely related are bile acid, cholesterol, phospholipids and bilirubin.

The metabolism of bile acid. Bile acids are the main component of bile, in a free state in bile. Bile acids can combine with the cations (mainly Na^+, K^+) in bile, and exist in the form of bile acid salts, also known as bile salts, accounting for about half of the solid component of bile.

Bile acids are the metabolite of cholesterol in the liver. The liver turns about 800 mg of cholesterol into bile acids every day. The generation of bile acids equals about 200 – 600 mg every day, and basically is equal to the loss of bile acids from feces. Therefore, under normal circumstances, the bile acid content of the body remains in a relatively stable state, about 3 – 4 g, known as the bile acid pool.

The conversion of cholesterol to bile acids is a complex process, including the decomposition of the cholesterol side chain and the core change. After blood cholesterol flows into liver cells, 80% – 90% of cholesterol first turns into 7 α-cholesterol in microsomes under the effect of 7 α-hydroxylase, and then turns into cholic and chenodeoxycholic acid through the two metabolic pathways.

The primary bile acid — cholic and chenodeoxycholic acids generated by liver cells, through the bile into the intestine, under the effect of terminal ileum bacteria (especially anaerobes), presents hydrolysis and releases the combined glycine and taurine, to turn the conjugation of bile acids into free bile acids. Under the effect of 7 α-off hydroxylase in the gut bacteria, the bile acids remove 7α-hydroxy into deoxycholic acids, and chenodeoxycholic acids turn into lithocholic acids, which are termed the secondary bile acids. And then in the brush border of the small intestinal mucosal epithelium, bile acids are removed from the intestinal lumen into the intestinal mucosal cells *via* the sodium gradient, and are again absorbed by intestinal mucosa. The bile acids return to the liver by the circulation of the portal system, and again flow into the gut *via* the biliary tract after the treatment of liver cells with the new synthesis that joins bile acids together. This process is called the enterohepatic circulation of bile acid.

The content of bile acids in the bile acid pool is about 3 – 4 g. After taking food, even if all bile acids pour into the small intestine, it is difficult to achieve the required critical concentration of lipids digestion. However, the enterohepatic circulation 2 – 4 times after every meal can make the limited bile acid pool maximize the role, thereby maintaining normal digestion and absorption for the lipid food.

The metabolism of cholesterol. Cholesterol in the body has two main sources: one is from the food intake; the other is synthesized in the human body.

The human body obtains about 400 – 600 mg cholesterol from the food intake every day. Not all cholesterol in food is completely absorbed, and the absorption rate of the cholesterol is about 40%. If the content of the cholesterol in food is very high, the absorption rate will decline, which may be a protective mechanism of the body.

The main organ of cholesterol synthesis is the liver, followed by the gut. The human body synthesizes about 900 mg of cholesterol every day. In the hepatic cytoplasm, 3 molecules of acetyl coenzyme A condense and form β-methyl-β-hydroxy e, two acyl coenzyme A (HMG-CoA), and then in the action of HMG-CoA reductase, form the mevalonate. This process is the key step in cholesterol synthesis. In the case of ATP supplying phosphate, the mevalonate gradually is phosphorylated and forms 5-pyrophosphate mevalonate, further decarboxylation to form 5 carbon pyrophosphoric acid isoamyl enolate, and further condensation to form an unsaturated long chain compound squalene, and then *via* cyclization to form lanosterol, and then finally converts to cholesterol.

Almost all the cholesterol in the bile is unesterified. When hepatic bile flows into the gallbladder, the cholesterol in the gallbladder is concentrated 10 times, released into the intestinal tract during the digestion period, and mixed with the cholesterol in the gut. Almost all the cholesterol in the gut is esterified, including that from food, small intestinal secretion and the shedding of the intestinal mucosal epithelial cell. Intestinal cholesterol is absorbed and mixed with the cholesterol in the epithelial cells of the small intestine, most of which again is esterified and returns into the liver in a chylous manner through the intestinal lymph. The unabsorbed cholesterol is drained in the form of the neutral cholesterol from the faeces.

The metabolism of phospholipids. The phospholipids in the body are mainly derived from the synthesis of liver cells, and also a little from food.

More than 90% of phospholipids in human bile are the phosphatidyl choline, mainly belonging to the phosphatidylcholine of the phosphorus acyl-dipalmitoyl phosphatidylcholine type. It is generated mainly by choline taking off the two glycerides in the smooth endoplasmic reticulum of liver cells, and the other part is generated by the glycerol phosphate ester formylation. The raw materials of phosphatidyl choline are glycerol, fatty acid and glucose. Every day about 3 – 6 g phosphatidylcholine is generated in the human body, and the majority is drained into the bile, which maintains a dynamic balance with the phosphatidyl choline content in the liver and plasma.

The endogenous phosphatidylcholine is drained into the intestine and mixed with the exogenous phosphatidylcholine in the food. The majority becomes the blood phosphatidyl choline in the hydrolysis of pancreatic lipase, is absorbed by the gut and secreted in the lymph, and then returns into the liver *via* the blood. The unabsorbed portion is drained by the faeces.

Bile acids are closely related to cholesterol and phosphatidylcholine in metabolism. If the gut lacks bile acids, the absorption of the endogenous and exogenous cholesterol will significantly weaken. Bile acids can also influence the synthesis, secretion and absorption of the phosphatidylcholine. When the gut lacks bile acids, the secretion of the phosphatidylcholine will be much lower than that of the cholesterol.

The metabolism of bilirubin. The normal human body generates about 300 mg of bilirubin daily. The known sources of the bilirubin include the following three aspects [10]. The first is derived from the senescent erythrocytes, accounting for approximately 80% – 85% of bilirubin. The senescent erythrocytes are destroyed by the cells of the reticulo-endothelial system (liver, spleen, bone marrow and so on), wherein the hemoglobin releases the heme, and is transformed into biliverdin in the hemoglobin oxygenation enzymes, and then biliverdin forms in the bilirubin under the catalysis of biliverdin reductase. The second is derived from the ineffective hematopoiesis, accounting for approximately 10% – 15%. In certain cases, the hemoglobin within the bone marrow decomposes into bilirubin before becoming erythrocytes. The third is derived from the non-erythrocytes, accounting for approximately 1% – 5%, mainly from myoglobin, cytochrome and peroxide enzyme *etc.*

The bilirubin metabolism in the liver can be divided into three steps, namely the uptake, esterification and excretion [11]. The bilirubin enters liver cells, and combines with the protein Y or Z in the intracellular fluids, to form bilirubin-Y protein or bilirubin-Z protein, which is transported to the smooth endoplasmic reticulum, where they are converted into bilirubin glucuronide. Protein Y has a very strong binding capacity with bilirubin, and is the main carrier protein of bilirubin. In general, protein Z will bind bilirubin only when bilirubin concentration is too high. After being transported to the smooth endoplasmic reticulum, in many enzymatic actions, the bilirubin becomes the carbohydrate binding bilirubin, of which 95% are bilirubin glucoside esters, 5% bilirubin glucosides and bilirubin glucoside, called bilirubin esterification. The non-ester bilirubin is almost insoluble in the aqueous solution with pH 7.4. Only when it is mixed with glucuronic acid is the hydrophilic capability significantly enhanced, is thereby easily soluble in water, and secreted in the bile. The liver cells exhaust the estered bilirubin from the billiary duct, which is mediated by the vector and is the rate-limiting step of hepatic clearance of bilirubin. Bile acid salt can increase the drainage of the ester type bilirubin in the liver.

13.1.2 Biliary Infection and Microecology

The causes, characteristics and clinical manifestation of biliary infection are closely associated with microecology. Thus, we will comprehensively discuss these contents of biliary infection.

13.1.2.1 The Causes and Characteristics of Biliary Infection

Normally, the bile is sterile. The bacteria found in the bile is taken from the portal vein or directly from intestinal reflux into the bile duct by the Oddi's sphincter [12]. Therefore, intestinal bacterial translocation is considered to be the main cause of biliary tract infection [6, 13-15]. Portal blood can often be cultured intestinal bacteria. Because of the liver's immune system and Kupffer cells phagocytosis as well as liver cell tight junctions, the bacteria is eliminated before entering the biliary system. Only when the small intestinal mucosa barrier Kupffer cells in the liver or tight junction structure of liver cells are destroyed, will intestinal bacteria and endotoxin in the portal circulation affect the general circulation, and the bacteria appears in the bile. In addition, the small amount of bacteria in the bile or bile duct through the Oddi's sphincter again flows into the intestine in the continuous erosion of the bile. In addition to the above two channels, the bacteria can enter the biliary tract through the lymphatic system.

The gut is an important bacterial repository in the human body, of which above 99% are anaerobic bacteria, while some other aerobic bacteria have an extremely disadvantaged distribution [16, 17]. Both are involved in the normal digestion and absorption of intestinal food, and are also the important components of the intestinal mucosal barrier [18]. Once intestinal floras are damaged, the bacteria can cross the intestinal mucosal barrier, and become an important source of systemic infection. There are many reasons leading to intestinal bacterial translocation and biliary infections [1, 19], in which the metabolism disorders of bile acids and bilirubin and the inhibition of intestinal movement induced by biliary tract diseases play an important role [19].

Some studies have shown that the stability of the biliary-intestinal microecological environment is associated with normal bile acids metabolism [19, 20]. During the enterohepatic circulation process of bile acids, in normal intestinal flora such as the anaerobic bacteria including *Bifidobacterium*, *Lactobacillus* and *Bacteroides fragilis*, and the aerobic bacteria including fecal true coli and *Staphylococcus* epidermidis, a part of the bile acids are oxidized (7α position dehydrogenation oxygen radicals) and generate chenodeoxycholic acid, and then chenodeoxycholic acid becomes lithocholic acid. But the gram-negative bacilli do not have this ability. Rudbach *et al.* thought that the deoxycholate was a surface active agent with the function of inhibiting gram negative bacilli, could decompose the endotoxin *in vivo*, and could be reversed to abolish its biological activity, thus becoming an important mediator in regulating intestinal flora balance. Intestinal anaerobic bacteria can inhibit the growth of the gram-negative bacilli through promoting the production of ursodeoxycholic acid salt, and maintain the advantageous distribution of anaerobic bacteria. In the condition of biliary obstruction, the enterohepatic circulation of bile acids is inhibited, and the deficiency of intestinal bile salt leads to the advantageous breeding of gram-negative bacilli, which inevitably produce large amounts of endotoxin. Thus, due to the endotoxin stimulus, liver cells, Kuffer cells, neutrophils, and macrophages are susceptible to producing large amounts of free radicals, thereby causing diseases [20, 21]. Moreover, in patients with long-term external biliary drainage, a lot of bile loss causes deficiencies in intestinal bile salts and bilirubin, while the bilirubin is an important antioxidant *in vivo* [22, 23], protecting the cells against damage from free radicals and maintaining the normal biliary tract function. Therefore, bile salts and bilirubin metabolism disorders are the important factors in intestinal bacterial translocation.

In addition, the inhibition of intestinal motility is also one reason for intestinal bacterial translocation induced by the obstructive jaundice and biliary duct post recanalization [24, 25]. Under normal circumstances, normal bowel peristalsis is an important mechanism of intestinal non-immune defense [25]. In patients with biliary obstruction and post recanalization, the direct effect of bile leakage or the irritation of the celiac nerve and other factors can inhibit intestinal movement to result in the stagnation of intestinal contents. Thus bacteria with overgrowth are prone to adhering on the surface of intestinal epithelial cells to penetrate the intestinal mucosa. Meanwhile, the accumulation of intestinal contents will increase intestinal intraluminal pressure to result in the decrease in the intestinal wall blood supply, which further induces mucosal damage and permeability changes, eventually inducing bacterial translocation.

13.1.2.2 The Clinical Manifestation of Biliary Infection

The clinical manifestation mainly includes gallstone disease, acute cholecystitis, and chronic cholecystitis. We will focus on discussing the clinical manifestation of these diseases.

Gallstone disease. Gallstone disease is the disease in which a stone occurs in any part of the biliary system (including the gallbladder and bile duct), and it has different types and composition. The clinical manifestations depend on whether the stone leads to biliary tract infection, biliary obstruction and the location and degree of the obstruction. The mechanism of gallstone formation is very complex, and there still is no conclusion so far. For example, as the most common disease, the cholesterol stone is considered to be associated with not only the lipid metabolism disorders, but also the excess slowness of gallbladder movement, the promotion of nuclear factors, cholesterol phospholipids bubble and other factors [26-28]. The clinical manifestation depends on the size, nature, dynamics, location and complication of the stones [29, 30].

The migrated gallstone in the biliary tract. When the gallstone migrates from the gallbladder to the cystic duct or common bile duct or from the dilated common bile duct to the ampulla, it generates incarceration. The smooth muscle of the gallbladder or common bile duct relaxes or spasms to attempt to discharge stones, resulting in biliary colic. Biliary colic usually happens within a few hours of eating too much or eating high-fat foods, or after an abdominal quake. The pain is more likely to present in the epigastric or right upper quadrant. At the beginning, it is persistently dull, and then gradually increases to an intolerable pain, when the patients are restless and bending, rolling and pressing the abdomen with their fists, or even crying. The pain often radiates to the right shoulder, and the patients often feel sweating, nausea and vomiting. It lasts a short time, rarely more than a few hours. If the gallstone goes back into the gallbladder or duodenum tube, the pain will completely disappear. Sometimes, because of the bile duct dilation, the pain can be relieved, but the pain will attack again when the gallstone moves to another place. In some cases it recurs a lot, but in other cases can lead to several months to several years of remission. A low-fat diet can reduce seizures.

The gallstone in the gallbladder. The stone usually does not induce signs of angina, known as the static stone. It can cause gastrointestinal and biliary dyskinesia and gastric and gallbladder emptying delay, indirectly influence the pancreatic digestive function, and generate excess stuffiness in the right upper quadrant or epigastric place, sometimes also along with heartburn, belching, acid attack and abdominal distention. If accompanied by infection, the body may have fever, right upper quadrant pain and other symptoms.

The gallstone within the cystic duct. When the gallstone blocks the cystic duct, it can produce biliary colic, and cause gallbladder swelling. In chronic cases, with the long-term incarceration of the gallstone or the scarring stenosis of the bile duct, the gallbladder often greatly inflates, and the capsule fills with mucus up to a maximum of several liters (gallbladder water). If accompanied by bacterial infection, it can cause cholecystitis and gallbladder empyema. The increase in intracavitary pressure, cystic duct inflammation or the blockage of the blood supply can cause some severe complications, such as gangrene and perforation of the capsule wall, a peri-gallbladder abscess, diffuse peritonitis and a gallbladder-GI fistula.

The gallstone within the common bile duct. The gallstone can come from the gallbladder or hepatic bile duct, and also from the common bile duct. When the gallstone first drops down to the common bile duct or migrates from the ectatic common bile duct to the ampulla, it can produce colic along with the signs of obstructive jaundice. Sometimes the gallstone can move up and down between the ampulla and the enlarged common bile duct, causing intermittent obstruction and cholangitis. When the common bile duct obstruction leads to acute obstructive suppurative cholangitis, it can present toxic shock signs, including jaundice, chills, fever, increased white blood cells, decreased blood pressure, delirium and even coma. When the gallstone moves away or is drained from the Vater ampulla, the symptoms of jaundice and inflammation can dissipate. Biliary tract infection can involve the liver, complicate with hepatitis, liver abscess, and hepatic vein thrombosis, causing hepatomegaly, hepatic tenderness, splenomegaly, ascites and gastrointestinal bleeding.

The gallstone embedded in Vater's ampulla. Persistent obstructive jaundice is a typical result of this condition. The colic often occurs before the jaundice, but the persistent jaundice exits along with the colic, often disappearing. The long-term incarceration of the gallstone can induce chronic cholestasis and biliary tract infection, and in a further way cause obstructive biliary cirrhosis. In addition, when the obstruction in Vater's ampullary leads to the bile refluxing in the pancreatic duct, it can also cause acute pancreatitis.

The gallstone in the intrahepatic bile duct. It means that the intrahepatic bile duct system exits gallstones, mostly with common bile duct stones. Most of these stones are of the yellow green type or the "sand-like" bile pigment stones, in the center of which ascaris eggs can often be found. Intrahepatic bile duct stones can originate primarily in the liver, and also be secondary to common bile duct stones or hepatic duct stenosis caused by other reasons. The stones can be found scattered around the different levels of biliary tracts in hepatic lobes, or confined to one or several places within the liver. These patients often have a history of abdominal pain, chills, fever, and recurrent jaundice since childhood. The liver function is damaged, but the gallbladder function may be normal. Many complications of this disease are serious, mainly including pyogenic intrahepatic cholangitis, liver abscess, biliary tract bleeding and so on.

Acute or Chronic Cholecystitis.

Acute Cholecystitis. Acute cholecystitis is an acute cholecystitis disease induced by chemical stimulation or bacterial infection, which in 96% of patients is accompanied by cystic duct obstruction [31-33]. The obstruction of the gallbladder function causes cholestasis and concentration, and the concentrated bile salt stimulates the mucosal epithelial of the capsule wall, resulting in chemical inflammation. The early stage of acute cholecystitis is often non-bacterial infection. However, because of gallbladder ischemia, damage and lowered immunity, 50% of patients can have a secondary bacterial infection a week after the disease. The causative agents mostly are intestinal parasitic flora, which can induce some severe complications, such as gallbladder empyema, gangrene and perforation.

The types of biliary bacterial infection are as follows [21, 34]. (i) The blood-borne type: For example, typhoid, paratyphoid, and *E. coli* can lead to whole body bacterial infections, and the pathogens can enter the gallbladder along with the blood flow. (ii) The intestinal and hepatic type: Intestinal bacteria can return to the liver *via* the portal vein. If these bacteria fail to be killed by the monocyte-macrophage, the intrahepatic bacteria can spread to the gallbladder *via* lymphatic vessels, or be drained out of the gallbladder with the bile, causing infections. And intestinal bacteria can also be carried into the bile duct with the entry of roundworm, and cause biliary obstruction. (iii) Other types: For example, in the condition of cystic trauma and surgery, the bacteria can directly violate the gallbladder from the wound.

The clinical manifestations of acute cholecystitis can be fever, right upper quadrant pain and tenderness, nausea, vomiting, mild jaundice, and increased white blood cells, *etc.*

Abdominal pain. More than 2/3 of patients will present right upper quadrant abdominal pain, but also some occurs in the upper abdomen. With the development of the inflammatory process, the visceral and parietal peritoneum of the gallbladder are stimulated by the inflammation. Thus the abdominal pain is often confined to the right rib area under the gallbladder, and the right subscapular area may present the radioactive pain. The pain often occurs at night, presenting a persistent and swelling pain. If the cystic duct is obstructed, there may be an intermittent biliary colic attack. The elderly have a much lower sensitivity to pain, thereby sometimes they have no severe abdominal pain or even no symptoms of abdominal pain.

Nausea and vomiting. 60% – 70% of patients have reflex nausea and vomiting, and severe patients can vomit bile, causing dehydration and electrolyte imbalance.

Systemic symptoms. 80% of patients have a moderate fever. When the patients suffer from suppurative cholecystitis, they may present chills, fever and irritability,

delirium and other symptoms, and even the severe patients may suffer from septic shock.
Signs. The patients often present acute tolerance, superficial and irregular breathing, and even the patients with severe vomiting may have signs of dehydration and exhaustion. 20% of patients have mild jaundice. Abdominal examination shows a little swelling in the right upper quadrant. The abdominal breathing is limited, and the gallbladder area under the right rib can have tense abdominal muscles, tenderness, rebounded pain, and a positive Murphy's sign. When abdominal tenderness and tense abdominal muscles extend to other regions or the whole abdomen, it indicates gallbladder perforation, or complications with acute peritonitis or hemorrhagic necrotizing pancreatitis.
Chronic cholecystitis. Chronic cholecystitis refers to chronic inflammatory disease in the gallbladder, most of which is chronic acalculous cholecystitis, and the minority is chronic non-calculous cholecystitis. For example, typhoid carriers always retain *Bacillus typhi* in the gallbladder resulting in chronic cholecystitis, without clinical symptoms. Most of the patients start with a chronic disease, but this can also be derived from the recurrent attacks of acute cholecystitis.

The main symptom of this disease is recurrent upper abdominal pain. The pain often occurs in the right upper quadrant abdominal or upper abdomen, radiating to the right subscapular area, and in some cases can occur in the substernal or left upper abdomen. Abdominal pain often occurs at night or after a full meal, and often is persistent. When the cystic duct or common bile duct suffers from gallstones incarceration, this can induce biliary colic. The pain usually relieves itself after lasting 1 – 6 h, and can be accompanied by reflex nausea, vomiting and other symptoms, but fever and jaundice is not common. During the intermittent period of the disease, the patients can present right upper abdominal distention and discomfort, or heartburn, belching, returning acid and other gastrointestinal symptoms. The above symptoms are not severe but are persistent, and are often aggravated after eating oily greasy food. When chronic cholecystitis presents acute exacerbation, or the concentrated mucus or gallstones in the gallbladder enter the cystic duct or common bile duct to cause obstruction, the patients can present the typical symptoms of acute cholecystitis or biliary colic.

Physical examination can find right upper quadrant tenderness and a positive Murphy's sign. When the gallbladder swells and expands, the cystic mass can be touched in the right upper abdomen.

13.1.3 Microecology Treatment of Biliary Tract Infection

Microecological research indicates that the gut is the body's largest bacterial library, with complex flora and a huge number of bacteria [16, 35], including a variety of aerobic, facultative anaerobic bacteria, anaerobic bacteria as well as some yeast, in which anaerobic bacteria accounts for the vast majority. Human intestinal normal cell formation has a long period of evolution, and has an important implication for human health, and is even indispensable for human life.

Intestinal microflora are very important for nutrient digestion and absorption, maintaining ecological balance of intestinal bacteria, preventing dysbacteriosis, resisting external bacterial invasion, and keeping normal bowel functions [36-42]. The intestinal microflora are relatively stable, but are prone to dysbacteriosis and functional disorders in the case of external factors, such as diseases (including abdominal infection, trauma, and surgical operation *etc.*), the overuse of antibiotics, and so on. Acute pancreatitis is a severe abdominal organ injury. Pancreatic location, structure and function are closely related to the gastrointestinal tract. The impaired gastrointestinal function is one of the common symptoms of the disorder, thus easily leading to intestinal Microecological damage.

There are a large number of microorganisms (about 400 – 500 different bacteria) in the human intestine belonging to normal flora [41, 42], when the body is in normal immune function mode. They are not harmful to the host. The bacteria inhibit and compete with each other, maintaining the balance of intestinal microflora. Obligate anaerobes like *Bifidobacterium*, *Bacteroides* and *Peptostreptococcus* account for about 99% of the total amount of bacteria in the intestine. Facultative anaerobes like *Enterobacteriaceae*, *Enterococcus* account for about 1% of the total amount of bacteria in the intestine. Microecological studies show that not all intestinal bacteria are involved in pathogenicity. In pathological conditions, the balance of gut microflora is broken, called Microecological imbalance, including dysbacteriosis and bacterial translocation. The former refers to the decline in original bacteria with the flourishing of potentially pathogenic bacteria while the latter means that the intestinal flora reaches local lymph nodes or more distant mesangial tissue through the intestinal mucosa.

Nowadays, antibiotic and microbial probiotics are the most common methods to prevent and control bacterial translocation in biliary tract infection

13.1.3.1 Antibiotic Therapy

For half a century, to solve the problem of the correct use of antibiotics in biliary tract infection, through clinical practice and experimental research, the experts could understand bacterial species in bile, antibiotics concentration in bile and the sensitivity of bacteria to antibiotics [43]. During the process of biliary tract infection therapy, they emphasize the choice of effective antibiotics for treating bacteria in the biliary tract, so as to achieve the most effective therapy for biliary tract infection.

In biliary tract infection, the main pathogens come from the gut, in which *E. coli* accounts for 50%. Moreover, the pathogens also include the gram-negative bacteria, such as *Pseudomonas aeruginosa*, *Klebsiella pneumoniae*, *Proteus*, *Alcaligenes*, *Klebsiella spp*, *Citric* acid bacteria, *Salmonella*, *Shigella*, *Enterobacter* and so on, and the gram-positive bacteria including *Enterococcus*, *Staphylococcus aureus*, hemolytic *Streptococcus*, *etc.* [44] In recent years, some reports have indicated that the positive rate of the anaerobic bacteria in bile accounts for 3.5% – 45% in patients with biliary tract infection, and *Bacillus spp.*

is more common, accounting for about 80% – 90%, particularily more in *Fragile bacilli spp*. In acute obstructive suppurative cholangitis, the positive rate of the anaerobic bacteria can reach 80%, mostly for aerobic and anaerobic mixed infection [44, 45]

Bile duct infection should require a high concentration of antibiotics in bile, and a preference for ampicillin and piperacillin. When the patients suffering from severe infections, gentamicin or amikacin are added; the combination of β-lactams and enzyme inhibitors, fluoroquinolones, and cephalosporins such as cefoperazone and cefotaxime, also can be used. In a combination with anaerobic bacteria infection, metronidazole or clindamycin should be used together. When the cystic duct or bile duct is obstructed, the drug concentration in the gallbladder and bile ducts will be significantly decreased, thus an early intravenous drug should be administered and early surgery should be performed to remove bile duct obstruction.

The successful prevention and treatment of antibiotics in biliary tract infection depends on how to rationally use antibiotics according to the different conditions of different patients. Only with a comprehensive measurement of the drug concentration in the bile, the influencing factors of drug concentration in the bile, the susceptibility of bacteria to the drug, the patients' status and other factors, can the right choice be made.

Antibiotics, whose concentration in the bile is much higher than in the blood, are as follow: Penicillin family including penicillin G, ampicillin, amoxicillin and mezlocillin; Cephalosporin family including cefazolin, cefadroxil, ceftazidime, pulling oxygen cephalosporins, cefoperazone, cefotaxime; Macrolides family including erythromycin and midecamycin *etc.*; Tetracycline family including lincomycin, rifampicin and quinolones *etc.*. Antibiotics, whose concentration in the bile is close to or a little higher than that in the blood, are as follow: Kanamycin, tobramycin, sulfur benzylpenicillin and metronidazole, *etc.* Antibiotics, whose concentration in the bile is lower than that in the blood, including carbenicillin, azlocillin, gentamicin, polymyxin B and E and so on.

The factors influencing antibiotics concentration in the bile include: (i) Liver function in patients: Antibiotics concentration in the bile in patients with abnormal liver function is generally lower than in patients with normal liver function. (ii) The cystic duct with or without obstruction: Antibiotics concentration in the gallbladder wall or bile in patients with cystic duct obstruction is significantly lower than that in patients without obstruction, and even could not be determined. (iii) The method of drug administration: Antibiotics concentration in the bile through injection is higher than that taken orally. For example, although ampicillin is easily absorbed orally, the concentrations in the gallbladder wall and bile are low. If intravenous administration is performed, its concentration in the gallbladder wall and bile is much higher than that in the blood. (iv) Medicinal dose: Antibiotics concentration in the bile is closely related to the dosage, and many antibiotics can be increased in concentration with the increased dosage, such as erythromycin, penicillin G and so on.

The incorrect use of antibiotics will cause some adverse effects. In the past, in most of the cases of biliary tract infection, the bacterial types in the bile and their sensitivity to antibiotics were not understood. Thus the treatment was only based on the doctor's experience, the common bacterial strains in the local bile and the efficacy of the previous antibiotic treatment. As a result, a considerable portion of the patient's medications were inconsistent with the bile bacterial culture and antibiotic sensitivity tests. Meanwhile, not only did the long-term, blind and repeated antibiotic treatment induce resistance to biliary strains in commonly used antibiotics,resulting in poor clinical efficacy, but the repeated attacks of biliary tract infection can also result in biliary cirrhosis, sepsis, and diffuse intrahepatic bile duct stenosis. Also, sometimes in order to inhibit bacterial growth in the bile, the blind application of antibiotics excreted by the liver in patients with bile duct obstruction can lead to a large number of antibiotics staying in the blood but not in the bile, resulting in potential drug toxicity.

13.1.3.2 Microecological Probiotics

Intestinal bacterial translocation is the main reason for biliary tract infection, thus regulating intestinal microflora is the key to preventing and treating biliary tract infection. When the biliary tract suffers from lesions, the physiological structure and function of the biliary tract are changed. Thus the normal flora settled in the gut can cross the intestinal mucosa into the blood, and then enter the hepatobiliary system through the portal vein, or directly enter the biliary tract through the Oddi's sphincter. This phenomenon of intestinal flora translocation is defined as the intestinal bacterial translocation. Intestinal bacterial translocation is also associated with the disturbance of intestinal flora, the overgrowth of intestinal bacteria, the damage to the intestinal mucosal barrier, and the decline of the immune defense function.

Normal human gut flora is very complex, including a variety of aerobic bacteria, facultative anaerobic bacteria and anaerobic bacteria [42, 46-48]. A total of bacteria in the human intestine can reach up to 100 trillion and the weight can reach 1 kg, forming a complex ecological system. Normal microecological flora is a balanced unit composed by microorganisms, host and environment. The imbalance in any link will lead to imbalance in the micro ecological system.

The mechanisms of intestinal microflora crossing intestinal mucosa and translocation are very complex, and have not yet been fully revealed. Generally, the different injuries of the host can cause different ways of intestinal bacteria translocation. A doctrine indicates that intestinal bacteria directly cross the intestinal mucosal barrier. For example, the *Salmonella typhi* cross the intestinal mucosa barrier mainly from the cell junction passage between the intestinal mucosa epithelial cells. Thus, the translocated bacteria must firstly adhere to the intestinal mucosa, and then migrate to extra-intestinal organs according to the following pathway. (i) The lymphatic pathway: Intestinal bacteria go into the intestinal submucosal lymphatic duct and mesenteric lymph nodes, and enter the thoracic duct through lymphatic circulation, and then enter the blood circulation, which is the main way. (ii) The portal system: Intestinal bacteria enter the submucosal capillary, and then go into the liver circulation through the portal venous system. (iii) In a few cases, intestinal bacteria directly cross the intestinal wall up to the abdominal cavity. Another theory suggests that intestinal bacteria are transported by the macrophages. When the bactericidal abilities of the macrophages are impaired, the macrophages can swallow intestinal bacteria and transport them to the outside of the intestinal tract, but cannot complete the intracellular bactericidal task, and release the bacteria.

Intestinal bacterial translocation is closely related to biliary infection, so the blockage of intestinal bacterial translocation is an important treatment measure. In recent years, a certain success has been gained in this respect. Currently, antibiotic application of bacterial translocation from the intestine has had some effects. However, due to the increasingly resistant bacteria and some certain difficulties in antibiotic choice, added to which antibiotics themselves can disrupt intestinal flora, reduce the biological antagonism of intestinal flora, and cause the imbalance in intestinal flora, the use of antibiotics has many limitations. In the light of the causes of intestinal bacterial translocation, the treatment of intestinal bacterial translocation of biliary tract infection should be established on the basis of restoring the normal anatomic relationships in the biliary system and the normal physiological function, regulating intestinal flora, protecting intestinal mucosa, and enhancing the immune function of the whole body.

It has been confirmed that currently some common intestinal microecological preparations have a very good effect in preventing and treating intestinal bacterial translocation and microflora imbalance [6, 13, 49-53]. Microecological agents are guided by the theory of micro ecology, regulate the Microecological imbalance, maintain the Microecological balance, and take advantage of some beneficial and harmless normal microbial members or their promoters, in order to enhance the health or health status of the host, including probiotics, prebiotics and synbiotics [54, 55].

The protective nature of certain microorganisms, in particular lactic acid bacteria, contained in fermented foods and drinks has a long history. Humans have been consuming live bacterial cultures for centuries in the form of fermented milk without any knowledge of the active ingredients or how they work. Probiotics are defined as "live microbial food supplements that beneficially affect the host by improving the intestinal microbial balance" [56, 57]. It is difficult to identify with certainty the first time the term probiotic was used, but it is believed that one of the earliest citations suggests that the intestinal microbial balance may be upset following antibiotic use and that it could be restored by a diet of probiotics, including fermented foods [56]. Probiotics are some microorganisms and their metabolic products that play a beneficial role by improving the Microecological balance to improve the health and health status of the host. The main criteria to be met by a microorganism to be characterized as probiotics are as follows [56, 58]. (i) Certain commercially available probiotics are not derived from humans, but it is a belief that if a probiotic is isolated from the human GI tract, it is safer for human consumption and may be more effective within the intestinal ecosystem. (ii) The safe status is generally granted by the FDA to food/food components that have been proven safe for human consumption by scientific procedures or by experience based on common use in food, resulting from a substantial history of consumption by a lot of individuals. For example, both *Bifidobacteria* and *lactobacilli* have a long history of safe consumption without a harmful role in human health. (iii) Probiotics must be resistant to gastric acidity and bile acid toxicity. A low gastric pH is one of the primary host defense mechanisms against ingested microorganisms, including probiotics. (iv) Probiotics must be capable of adhering to human intestinal mucosal epithelial cells and intestinal mucins. This characteristic may promote competitive exclusion of the potential pathogens from mucosal surfaces. (v) Probiotics must have the ability to produce antimicrobial substances against intestinal pathogens for the restoration of a healthy microflora composition. (vi) Probiotics must be safe in food and during clinical use, even in patients with an immune dysfunction. (vii) Probiotics must have their efficacy and safety proven in randomized, double-blind placebo-controlled human studies. Currently, applied probiotics in humans mainly include *Bifidobacterium*, *Lactobacillus*, *Enterococci*, *Escherichia coli*, *Bacillus subtilis*, wax-like bacillus, *Bacillus licheniformis*, and yeast.

The concept of prebiotics was first introduced in 1995 by Gibson and Roberfroid as an alternative approach that would overcome the survivability issues of probiotics during storage, distribution and in the GI tract. Prebiotics were defined as "nondigestible dietary ingredients that beneficially influence the host by selectively stimulating the growth and/or activity of one or a limited number of bacteria in the colon, thus improving host health" [56, 59]. According to the current definition, the criteria that must be satisfied for a dietary ingredient to be characterized as a prebiotic are as follows [56]. (i) Fermentability can be proved *in vitro* in fecal batch culture experiments simulating the pH and temperature conditions of selected regions of the human colon. Substrates that stimulate bacterial growth can be further evaluated in more complex in vitro continuous culture models, established to simulate the transit of luminal contents through the proximal, transverse, and distal parts of the colon as well as varying pH and temperature conditions therein. The promising substrates should be further evaluated in double-blind, placebo-controlled randomized human studies to confirm the observed effect *in vitro*. (ii) The main role of a prebiotic is as a selective substrate for one or more beneficial bacteria, which are enhanced to grow and/or are metabolically activated and consequently move the colonic microbiota of the host toward a healthier condition. Both criteria are critical for a dietary ingredient to be characterized as a prebiotic, but the selectivity is the most important and difficult to fulfill. In order to confirm the selectivity of a prebiotic, it is most important to accurately monitor the alterations of intestinal bacteria during the period of prebiotic administration. Prebiotics are capable of promoting probiotics growth. They are initially found to be a bifidus factor, and later some non-digestible food ingredients are added, such as various oligosaccharides, most commonly including lactulose, sucrose, raffinose oligosaccharides, and oligomeric maltose. These oligosaccharides cannot be decomposed and used for most of the intestinal bacteria, and can only be used by the probiotics to promote the growth of probiotics and suppress the harmful bacteria, to further adjust the intestinal flora.

According to the evolution of the probiotic and prebiotic terms, the symbiotic concept was first introduced, along with prebiotics, as the mixture of probiotics and prebiotics that beneficially affect the host by improving the survival and implantation of live microbial dietary supplements in the gastrointestinal tract, by selectively enhancing the growth and/or by activating the metabolism of one or a limited number of health-promoting bacteria, thus improving host "welfare" [59-62]. According to the current definition, there are two types of synbiotic approaches and they are are as follows [56]. (i) Complementary, whereby the probiotic is selected according to the specifically desired beneficial effects on the host, and the prebiotic is independently selected to selectively elevate concentrations of the beneficial bacterial components. The prebiotic may stimulate the growth and activity of the probiotic, but only indirectly as a part of its target range. (ii) Synergistic, whereby the probiotic is selected again according to the beneficial effects on the host, but the prebiotic is selected to specifically promote the growth and activity of the selected probiotic. Thus, the prebiotic is chosen to have a closer association with the probiotic and is selected to improve the survival and growth of probiotics in the host. It may also stimulate the growth of the intestinal beneficial bacteria in the host, but the primary target remains the ingested probiotics.

With the development and improvement of microecology andthrough bio-engineering methods, we can modify the bacterial physiological genes and develop more new microecosystem regulators. We should believe that we will make greater use of probiotics in the future to regulate intestinal bacterial flora, protect intestinal mucosa, and enhance the immune function of the whole body, so as to achieve the purpose of prevention and treatment of biliary tract diseases.

13.2 Pancreatic Infection and Microecology

Pancreas infection is closely associated with acute pancreatitis[6, 13, 49, 63-65]. On the one hand, pancreatic infection always appears secondary to acute pancreatitis (AP) and causes higher mortality in patients with AP [63]. It has been identified that intestinal bacterial translocation can induce pancreatic infection in AP patients. The injury of intestinal mucosa in AP patients is the major reason for intestinal bacterial translocation. Meanwhile, the increase in intestinal gram-negative aerobic bacteria, the disturbance of intestinal microecology, and the dysfunction of body immunity also contribute to intestinal bacterial translocation, and then causes pancreatic infection in AP patients. On the other hand, pancreatic infection sometimes can cause AP. Both bacteria and a virus can induce pancreatic infection. Pancreatic infection is considered to be a crucial factor in the prognosis of AP patients [63]. Thus the combined application of microecology modulators and antibiotics may be effective for pancreatic infection in clinical practice.

13.2.1 Microecology Foundation of the Pancreas

The pancreas is the body's second largest digestive gland after the liver, and has exocrine and endocrine functions. Normally, after the stimulation of food, *etc.*, the exocrine part of the pancreas composed of the pancreatic acini and duct system will secrete amounts of characteristic pancreatic juice in the gut. Pancreatic juice is a colorless, odorless, slightly viscous and transparent alkaline liquid, contains a variety of inorganic ions and digestive enzymes, and flows into the duodenum through the pancreatic duct, playing an important role in neutralizing gastric acid, promoting digestion and absorption. Pancreatic secretion is regulated by both the humoral and neural regulations, mainly depending on the humoral regulation. These regulations are prone to be influenced by many factors. Food is the most important stimuli of pancreatic secretion. In addition, intestinal vasoactive peptide, insulin, cholinergic drugs, histamine, ethanol, high-starch and high-protein diets can also enhance pancreatic secretion. On the other hand, sympathetic nervous stimulants, a cholinergic antagonist, pancreatic polypeptide, glucagon, somatostatin and carbonic anhydrase inhibitors can inhibit pancreatic secretion. Meanwhile, the pancreas is an important endocrine organ. Pancreatic islet cells are endocrine cells, and secrete insulin, glucagon and pancreatic polypeptide, *etc.*, playing a crucial role in regulating nutrient metabolism. Under normal conditions, these factors coordinate with each other, maintaining physiological balance. Once this balance is broken, diseases may occur. Many of the aforementioned regulatory factors are derived from the gut. Thus intestinal microecology balance and intestinal function stability have an important influence in maintaining the normal function of the pancreas.

Due to the anatomical location and the rich vascular and lymphatic vessels, the pancreas is tightly related to the adjacent vascular and abdominal organs, such as duodenum, common bile duct, stomach, spleen, transverse colon, portal vein, spleen artery, spleen vein and so on. Therefore, the generation and development of pancreatic diseases have a close association with the adjacent organs diseases, and they can involve each other. In anatomy, the pancreatic duct and common bile duct are closely related. The pancreatic segment of the common bile duct crosses the duodenum wall diagonally, and ends with the pancreatic duct rendezvous, forming Vater's ampulla. At the end of Vater's ampulla, about 15 mm of common bile duct and pancreatic duct transversely cross the duodenal parietal, and finally open into the duodenal papilla, 8 – 10 cm from the pyloric part. Oddi's sphincter embraces the end of the common bile duct and pancreatic duct, as well as Vater's ampulla, and controls the drainage of bile and pancreatic juice. Thus, biliary diseases have a direct influence on the pancreas.

The blood supply of the pancreas is mainly derived from the celiac artery and its branches, and partly from the superior mesenteric artery and its branches. These arterial branches march along the interlobular connective tissues of the pancreas, and some small branches enter pancreatic lobules, forming capillary networks and are distributed around the acini and islets. The veins go along with arteries, and finally directly or indirectly enter the portal vein through the spleen vein. Pancreatic lymph-flow starts from the capillaries around the acini. These capillaries merge into larger lymphatic vessels in the interlobular, converge into collected lymphatic vessels at the surface of the pancreas, and finally join the corresponding regional lymph nodes. Lymph of the common bile duct, duodenum and pancreatic head first join lymph nodes at the pancreatic head, flow into lymph nodes at the upper edge of the pancreas along the path of the pancreatic blood vessels bow, and then join lymph nodes around the celiac artery. Part of the lymph joins lymph nodes around the superior mesenteric artery along the path of the inferior pancreaticoduodenal artery. The lymph of the pancreatic body and tail merges into lymph nodes at the porta lienis and the surface of the pancreas, and then joins lymph nodes around the celiac artery along the spleen artery. Blood and lymph circulation of the pancreas not only provide a material basis for normal pancreatic function through passing various regulatory factors, but also become the pathological basis of the interaction between the pancreas and adjacent organs. The pancreas is a sterile organ, but pancreatic infection is always secondary. The occurrence, development and prognosis of pancreatic infection are closely associated with intestinal microecology.

13.2.2 Pancreatic Infection and Microecology

Pancreatic infection is closely associated with AP [13, 49, 63-66]. AP is a serious disease with an incidence that continues to increase worldwide [67]. It ranges from a mild, self-limiting illness to pancreatic necrosis and infected pancreatic necrosis with a mortality rate of up to 30% [68]. For one thing, pancreatic infection always appears secondary to acute pancreatitis and causes higher mortality in patients with AP [63]. For another thing, pancreatic infection sometimes can cause AP, and is considered to be a crucial factor in the prognosis of AP patients [63].

13.2.2.1 AP

AP is considered as an abdominal emergency and is related to high morbidity and mortality [69]. AP is pancreatic and surrounding tissue acute chemical inflammation response caused by its own digestive enzymes secreted by the pancreas. The disease begins with pancreatic acinar injury, and then the systemic inflammatory response evolves quickly, which may even become fatal [69]. Intra-acinar pathologic trypsinogen activation and subsequent autodigestive injury are the two major procedures of pancreatitis development, which has been widely recognized for over a century [70-73]. However, recent research in the mechanism of AP give us a novel explanation that intra-acinar trypsinogen activation contributes to early acinar injury, but local and systemic inflammation progress independently during pancreatitis [69, 73, 74]. Early intra-acinar nuclear factor κB (NF-κB) activation occurs parallel to, but independent of, trypsinogen activation, and may be crucial in acute pancreatitis [69, 73]. Moreover, some key pathogenic cellular events such as calcium signaling [75], mitochondrial dysfunction [76, 77], endoplasmic reticulum (ER) stress [78], autophagy [79] and impaired trafficking [80], and lysosomal and secretory responses [81] are also involved in the occurrence and development of pancreatitis [69].

According to histopathology and clinical manifestations, AP can be classified as mild acute pancreatitis (MAP) and severe acute pancreatitis (SAP). (i) MAP: The patient presents clinical manifestation and biochemical changes in AP, without organ dysfunction or local complications; the positivereaction to fluid supplement therapy; Ranson score$\leqslant$3, or APACHE-II score$\leqslant$8, or A, B or C in CT classification. (ii) SAP: The patient presents clinical manifestation and biochemical changes in AP, along with one of the following: local complications (pancreatic necrosis, pancreatic pseudocyst, pancreatic abscess), organ failure, Ranson score$\geqslant$3, APACHE-II score$\geqslant$8, D or E in CT classification.

Currently, the causes of acute pancreatitis mainly include [82]: (i) Common causes such as, cholelithiasis (including biliary micro stone), alcoholism [83], hyperlipidemia; (ii) Other causes: Oddi's sphincter dysfunction, drugs and toxins, post-operation of endoscopic retrograde cholangiopancreatography (ERCP), trauma, abdominal surgery, hypercalcemia, pancreas divisum, peri-ampullar carcinoma, pancreatic cancer, vasculitis, infectious diseases (Coxsackie virus, mumps virus, HIV, toxocara infestation), autoimmune diseases (systemic lupus erythematosus, Sjogren syndrome), α_1-antitrypsin deficiency, and so on. (iii) Idiopathic AP, the causes are uncertain after the imaging and biochemical examinations.

Infection can not only induce acute pancreatitis, but also become one of the common complications of acute pancreatitis. The following will describe pancreatic infection and microecology from "AP secondary infection" and "infection induced AP" respectively.

13.2.2.2 AP Secondary Infection and Microecology Imbalance

The majority of patients with MAP are self-limiting, and the patient recovers by himself/herself after a few days. The patient with SAP lies in a critical condition with a high mortality rate. From the clinical process of SAP in recent years, it is noted that the development of SAP presents bimodal pattern change. In the early phase, due to inflammation and tissue necrosis, pancreatic acini release many activated enzymes, tissue toxin, and cytokines, *etc.*, which will lead to the injury of remote organs such as liver, kidney and pulmonary [69, 84, 85]. Clinical manifestation mainly appears as a system inflammatory reactive syndrome (SIRS) and multiple organ dysfunction syndromes (MODS). In the middle or late phase, the infectious complications such as peri-pancreatic or pancreatic abscess and sepsis occupy the leading status in the clinical manifestation of SAP. From the data relating to death from SAP, the majority of previous cases of death appeared within 2 weeks of SAP occurrence. Currently, due to advanced care treatment, the mortality rate from multi-organ failure in the early phase has been significantly decreased, while the mortality rate from infectious complications has increased relatively. The incidence of SAP combined pancreatic infection has reached 40% –

70%, and death induced by secondary infection accounts for above 80% of deaths from SAP [86].

Clinical research has found that pancreatic necrosis and the degree of necrosis are not the only factors in determining the prognosis, while infection may play a more important role in the majority of patients [63, 87, 88]. In fact, pancreatic necrosis as secondary infection has a much poorer prognosis than aseptic necrosis [87]. The mortality rate in the former reaches 30% – 40%, but the prognosis of the latter is good with strict monitoring and supportive care, and the mortality rate is close to that of interstitial pancreatitis. Thus, effective control of pancreatitis as secondary infection will be the critical treatment to reduce the operation rate and improve the prognosis of patients. The pathologic essence of pancreatitis is the activation of pancreatic enzymes and the aseptic inflammation of pancreatic autodigestion. The sources of SAP secondary infection are divided into biliary origin and intestinal origin, but it has been confirmed that the bacteria inducing SAP secondary infection mainly comes from bacterial translocation of the gut, as intestinal origin. At present, in the pathological state, the process by which intestinal microorganisms (including bacteria, fungi, and viruses) and their products pass through the intestinal wall and enter the body's distant organs is known as bacterial translocation or microbial translocation.

The underlying etiologies of acute pancreatitis secondary to infection. The underlying etiologies of the secondary infection in AP mainly include circulatory insufficiency and shock, intestinal obstruction, biliary obstruction, fasting and parenteral nutrition (TPN), intestinal mucosal injury induced by inflammatory mediators, and so on.

Circulatory insufficiency and shock. During the process of AP, the body fluid loss induced by pancreatic edema and exudation, and the water loss caused by a lot of vomiting leads to circulatory insufficiency. At the same time, the excitability of the sympathetic-adrenal system and the redistribution of the systemic blood flow lead to the strong contraction of the small intestinal vascular mainly dominated by the α-receptor. Thus intestinal mucosal blood supply rapidly decreases, resulting in injury to the intestinal mucosa. Intestinal mucosal injury induced by circulatory decompensation in AP is an important cause of early bacterial translocation.

Intestinal obstruction. In AP, the strong stress response, the direct stimulation of lesions to the coeliac nervous plexus, the erosion of inflammatory exudate in the intestine, the large amount of toxin absorption and other reasons can cause intestinal motility inhibition, and even paralysis and obstruction. Some scholars attributed the causes of intestinal obstruction to systemic inflammatory response syndrome. They thought that intestinal obstruction was a GI manifestation of MODS. Regardless of its causes, the obstruction itself is the important condition of bacterial translocation. When a bowel movement is suppressed, the elimination capacity of harmful bacteria reduces, leading to excessive bacteria, flora imbalance, and the obvious increase in harmful bacteria. Moreover, the stasis of intestinal contents leads to bacterial adhesion on the surface of intestinal mucosa, and then bacteria penetrates the epithelium into the body. When intestinal obstruction becomes more aggravating, the concentration of the mass of intraluminal liquid and gas will increase the intraluminal pressure, leading to intestinal dilation and the occlusions of small vessels in the intestinal wall, as well as decreased mucosal blood supply, thus further causing intestinal mucosal ischemia or necrosis and complete dysfunction of the mucosal barrier. Currently, intestinal motility inhibition has been considered the most important inducement of bacterial translocation in acute pancreatitis.

Biliary obstruction. Biliary obstruction is not only the etiology of acute pancreatitis, as stones, ascaris lumbricoides, and inflammation cause biliary pancreatitis, but also can be a consequence of AP, as pancreatic tissue edema compresses the common bile duct causing bile duct dilatation. Whatever causes biliary obstruction, the influence of biliary obstruction on intestinal bacterial translocation is consistent. First of all, the bile is rich in cholic acid maintaining an intestinal mucosal barrier function, the prostaglandins, secretory IgA, various epithelial growth factors and so on. Biliary obstruction directly causes the lack of cholic acid in the gut, which impairs the mucosal barrier function and induces bacterial translocation. In addition, the jaundice induced by the complete obstruction of the biliary tract may be secondary to liver function damage, and impairs the defense ability of the liver Kupffer cells. Animal experiments also found that due to biliary obstruction, the bacteria in bile would reflux into the pancreas along with bile in AP, as a possible etiology of pancreas secondary infection. From a deeper perspective, intraluminal duodenal bile deficiency will responsively promote cholecystokinin (CCK) secretion dozens of times in the control group. CCK is the most important humoral factor in promoting pancreatic hypersecretion. Therefore, some scholars speculated that after biliary obstruction, the high CCK level induced by duodenal bile deficiency may be an important cause of biliary pancreatitis pathogenesis. In short, biliary obstruction influences the pathologic process of acute pancreatitis in many aspects.

Fasting and parenteral nutrition (TPN). In accordance with pancreatic rest as the basic principle of AP treatment, fasting and parenteral nutrition support has been one of the main treatment measures in AP. But a large number of experimental and clinical studies confirmed that fasting and long-term parenteral nutrition were important reasons in inducing bacterial translocation. The lack of GI intraluminal food stimulus will lead to the motility stasis of the digestive tract, the reduction in various digestive fluid secretions, the obvious changes in bile composition and secretion, and the decline in the GI endocrine hormone, thus causing GI mucosal atrophy. In addition, in recent years, it has been found that a substantial portion of intestinal nutrition supply is directly derived from the food products of intestinal lumen digestion. The nutritional needs of an estimated 50% of the small bowel and 80% of the large bowel rely on intraluminal food. So under the condition of long-term fasting, although the use of intravenous hyperalimentation can basically maintain systemic organ metabolism needs, the bowel remains in a state of chronic hunger. The new study also found that long-term TPN had a deep impact on the neuroendocrine metabolism regulation process of the body. Long-term TPN would extend the emergency response after trauma and infection and the high metabolism status, further resulting in weakened mucosal nutrition and immune function from the metabolic regulation angle. The overall consequences of the above-mentioned pathophysiological changes are a weakening intestinal barrier function and inducement to bacterial translocation. TPN are the most important causes of enterogenic infection in the middle or late phase of AP.

Intestinal mucosal injury induced by inflammatory mediators. The basic reasons for extra-pancreatic organ injury in AP are the excessive immune responses after pancreatic damage, including hypercytokinemia and SIRS induced by the second attack of white cells on autologous tissue. As a vital organ of the immune response, the intestine has apparently an immune overreaction and self-damages the main target organs. There have been an increasing number of experimental results indicating that some cytokines and inflammatory products play a crucial role in the intestinal mucosal barrier dysfunction. Currently, the cytokines of further studies are the platelet activating factor (PAF) and tumor necrosis factor (TNF). PAF is currently known as one of the most powerful mediums to promote intestinal ulcer formation, exerts a synergistic effect with other factors such as TNF *etc*., and becomes an important inducement causing intestinal mucosal damage in the early phase of AP.

The others. During the treatment of AP, some drugs and treatment modalities may also cause bacterial translocation. For example, as a potent painkiller, morphine can inhibit the intestinal motility, inducing bacterial translocation. Atropine is widely used to inhibit pancreatic exocrine secretion in the clinic, but it has a potential trend to promote bacterial translocation because the use of atropine always accompanies intestinal obstruction and exacerbations. Thus it has much less clinical application now. The improper application of broad-spectrum antibiotics leads to the damage of the normal bacterial membrane or the hyperplasia of the mould, which had also been proven to be the cause of bacterial translocation by corresponding experimental studies.

From the above discussion, we can conclude that the reasons for bacterial translocation in AP are very complex and almost include all currently known pathological factors that induce bacterial translocation. The main pathological factors are not the same in different stages of AP. The accumulation of a variety of factors significantly enhances the incidence of bacterial translocation in AP. Thus secondary endotoxemia and multiple organ dysfunctions, pancreatic and peri-pancreatic infection, and gut-derived sepsis constitute the primary cause of death in AP.

The mechanisms of secondary infection in AP.

Intestinal microecology imbalance causes bacterial growth. During the process of SAP, the alterations in the bowel function, such as weakened intestinal power, intestinal cavity effusion, intestinal gas, bowel dilatation, *etc*., create the conditions for intestinal bacterial overgrowth, which leads to the overgrowth of intestinal gram-negative aerobic bacteria, and the significant decrease in bifidobacterium and lactobacillus, thus destroying the balance of intestinal microecology [89].

The host immune defense function is weakened. Saidakhmedova *et al*. studied the immune function in patients with AP and healthy humans. They found that the function of the body's immune system significantly decreased, especially the function of T lymphocytes in patients with AP. The reduction in T lymphocytes can increase the body's susceptibility. Kylanpaa-Back *et al*. analyzed the immune function of macrophages in the blood of patients with AP by flow cytometry, and found that the decline in the immune function appeared in the early phase of the disease, along with the low phagocytic function of macrophages and the decreased capability of bacterial clearance.

The damage to the intestinal mucosal barrier. The gut is the largest bacterial and endotoxin pool of the body. Due to the existence of the integrated intestinal mucosal barrier function, these bacteria and endotoxins do not damage the body under normal circumstances. In patients with SAP, the intestinal mucosal barrier was damaged, and intestinal permeability was significantly increased. Thus these bacteria and endotoxins entered extra-gut organs mainly through the pathway of the mesenteric lymph nodes — the thoracic duct — systemic circulation. The bacterial translocation of the pancreas can lead to pancreatic necrosis secondary infection.

Any one of the above factors can increase the likelihood of bacterial translocation. Bacterial translocation can cause the formation of multiple organ secondary infection, or even systemic sepsis. The synergies of bacteria and endotoxins will also activate phagocytic cells, and release a series of inflammatory mediators with damaging effects, leading to systemic inflammatory response syndrome (SIRS).

The microorganism species of secondary infection in AP. The pathogenic bacteria inducing secondary infection in pancreatitis often present multiple bacterial floras. Because bacterial translocation is the major path to pancreatic infection, more than 70% of cultured bacteria in the infection foci belong to gut-derived bacteria. 75% of pathogenic bacteria are gram-negative bacillus, and 10% are anaerobic bacteria. The frequencies of occurrence of pathogenic bacteria respectively are *Escherichia coli* (35%), *Klebsiella pneumoniae* (25%), *Enterococcus* (24%), *Staphylococcus* (14%), *Pseudomonas* (11%), *Bacillus proteus* (8%), *Streptococcus* (7%), *Escherichia coli* (7%), *Bacteroides* (6%), and *anaerobic bacteria* (6%). In recent years, fungal infections such as *Candida albicans*, *Candida glabrata* and *Cryptococcus etc.* have increased a little, mainly in patients with long-term use of antibiotics. In addition, the infections of the Coxsackie virus and cytomegalo virus are uncommon.

During the invasive process, these bacteria of intestinal translocation may appear with enhanced virulence, such as increased adhesion ability, secreting a special enzyme to inhibit the bactericidal capability of macrophage lysosome, expression of heat shock protein to induce immune escape *etc.*, which may be some factors causing the expression gene activation of the bacterial virulence factor. Meanwhile, due to long-term fever, fasting, consumption and inadequate nutrient intake in patients with AP, the patients often present deficiencies in the immune function, together with the persistent application of amounts of broad-spectrum antibiotics, which easily lead to infection by multiple drug-resistant nosocomial infection bacteria, such as methicillin-resistant *Staphylococcus aureus* (MRSA), *Escherichia coli* and *Klebsiella pneumoniae*, extended-spectrum β-lactamases (ESBL), *Pseudomonas*, *Acinetobacter*, *Stenotrophomonas maltophilia*, onion Burke Hoddle bacteria, *etc.*

13.2.2.3 Infectious Pancreatitis and Microecology Imbalance

AP is usually considered as an aseptic chemical inflammation of the pancreas itself , but some cases in which the infection factor occupies the leading position are called infectious pancreatitis [90]. In fact, AP caused by severe infection is common. Infectious pancreatitis can be induced by bacteria (*Escherichia coli, Salmonella typhi, Streptococcus pyogenes, Mycobacterium tuberculosis* and so on) or a virus (such as mumps virus, EB virus, hepatitis virus, Sakirch virus, cytomegalovirus, *etc.*). The major mechanisms of infectious pancreatitis mainly include: (i) Sepsis caused by bacterial infection which can lead to acute suppurative pancreatitis. Now, because of the widespread use of high efficiency antibiotics, this etiology has been significantly reduced. (ii) Viral infections: For example, 15% of patients with AP have acute pancreatitis, and the patients with Coxsackie virus infection, infectious mononucleosis or viral hepatitis may also be complicated with pancreatitis. Virus infection can also occur after the surgical operation. (iii) Bacterial infections of the biliary tract: Bacterial bile refluxes into the pancreas to cause pancreatic infection. In terms of biliary ascariasis, the ascaris incarcerate in the Vater's ampulla to cause the obstruction and infection of the common bile duct, or the ascaris enter the biliary tract to induce cholangitis, and further the bile refluxes into the pancreatic duct to lead to pancreatic damage. Or even the ascaris enter the pancreatic duct to cause the obstruction and infection of the pancreatic duct.

A pancreatic abscess is a fatal complication of AP induced by infection, and often happens 1 – 3 d after symptoms of AP appear, with an incidence of about 2% – 6%. On the basis of biliary tract diseases, the incidence of acute bacterial pancreatitis secondary to a pancreatic abscess can be as high as 50%. Moreover, a pancreatic abscess can also be found in the secondary infection of pancreatic injury and pancreatic cyst. The abscesses can be confined to the pancreas, and also spread to the peri-pancreas to form multiple abdominal abscesses. Pathogenic bacteria include *Escherichia coli, Enterobacter, Proteus, Klebsiella, Pseudomonas aeruginosa, Enterococcus, Staphylococcus aureus, Streptococcus*, certain anaerobic bacteria and so on. However, most pancreatic abscesses are a mixed infection.

A pancreatic abscess can occur in the pancreatic head, body, tail and even in all parts of a whole pancreas, and also in the lesser sac around the pancreas, the mesenteric vessels, retroperitoneal space, paracolic space, subphrenic space, kidney and pelvis, *etc.* Some abscess pus can reach 1 L or more, and contain a lot of amylase. The characteristics of abscess pus vary due to the different pathogenic bacteria. Abscess erosion and penetration of the adjacent viscera can form a fistula, such as abscess gastric fistula, abscess duodenal fistula, abscess colon fistula *etc.* These abscesses can also encroach on adjacent vessels such as the splenic artery, splenic vein, superior mesenteric artery, mesenteric vein *etc.*, and lead to a septic hematoma or intra-abdominal hemorrhage. The pathophysiological changes are mainly associated with toxemia and septicemia, and can appear in disturbance in the water, electrolyte and acid-base balance, or even MODS in some severe cases.

Pancreatic infection often lacks a characteristic performance, but accompanies clinical manifestations of AP. The main clinical symptoms are infection-toxic manifestations and GI symptoms [91].

Fever. Fever often originates from acute inflammation, pancreatic necrosis secondary to the bacterial or fungal infections. The majority of moderate fevers and the minority of high fevers usually last 3 – 5 d. A fever persisting or increasing day by day suggests concurrent infection or a complicating pancreatic abscess. The fever often presents itself as a remittent fever or continuing fever with a temperature of about 39 – 40 ℃, accompanied by profuse sweating, restlessness or delirium *etc*.

Abdominal pain. Mostly, upper abdominal gas pains or dull pain with continuity may be associated with lower back pain. Physical examination often finds upper abdominal tenderness, rebound tenderness, rigid abdominal muscle, and mass tenderness. The mechanism of abdominal pain is as follows: (i) AP edema, inflammatory stimulation and traction on the nerve endings; (ii) Pancreatic inflammatory exudate and spilt pancreatic juice stimulate adjacent peritoneum and retroperitoneal tissue, resulting in localized peritonitis; (iii) Pancreatitis involving intestinal disorders and enteric dysbacteriosis cause intestinal inflammation, intestinal inflation, paralytic intestinal obstruction *etc*.; (iv) Pancreatic duct obstruction or associated cholecystitis, cholelithiasis cause pain.

Nausea and vomiting. Most of the patients have nausea and vomiting after eating, often vomiting stomach contents. In case of a stress ulcer, the patients can have haematemesis, melena and so have GI bleeding. Acute hemorrhagic necrotizing pancreatitis complicates GI bleeding, indicating a serious condition. Vomiting may be a reflective and defensive reflex to the stimulation of abdominal pain or pancreatitis disease. Vomiting may also be an intestinal flatulence, paralytic ileus or peritonitis. Vomiting with gallstone pancreatitis often occurs after abdominal pain, and vomiting with alcoholic pancreatitis often occurs with abdominal pain.

Abdominal mass. A pancreatic abscess caused by infection may present an abdominal mass of various sizes. A large abscess can be seen in the body surface, presenting a round or oval shape, obscure boundary, being relatively fixed, with obvious tenderness. The covered abdominal wall of a large abscess often appears as a sign of peritoneal irritation.

Hypotension and shock. Severe AP often presents hypotension or shock, patient irritability, pale and clammy skin, which is spent porphyritic, weak pulse. The mechanism of shock: (i) A lot of blood and plasma exudation cause hypovolemia and hypotension; (ii) Vomiting causes substantial loss of water and electrolytes; (iii) The complicated infection and endotoxemia lead to the activation of inflammatory cytokines, the increase in bradykinin, the increase in vessel permeability, and a fall in blood pressure, leading to DIC. Myocardial damage caused by toxins can lead to circulatory failure; (iv) GI bleeding.

Common complications. (i) Pulmonary atelectasis, pleuritis and pleural effusion and other lung diseases are often accompanied by pulmonary infection. Severe cases can lead to acute respiratory failure; (ii) Abscess gastric fistula, abscess duodenal fistula, and abscess colon abscess fistula *etc*.; (iii) Abscess erosion of adjacent vessels leads to abdominal bleeding; (iv) Pancreatic endocrine and exocrine dysfunction cause digestive dysfunction or diabetes; (v) Liver and renal insufficiency, urinary tract infections, bedsores, *etc.*

13.2.2.4 The Etiologic Diagnosis of Pancreatic Infection

The etiologic diagnosis of pancreatic infection primarily is bacterial culture. Under the guidance of β-type ultrasonography or CT, an abscess puncture can be done. Bacterial cultures of the puncture pus and a drug susceptibility test have an important significance in determining infectious pathogens and choosing effective antimicrobial treatment. Puncture sites need strict disinfection to prevent bacterial pollution growth. Meanwhile, an aerobic and anaerobic culture should be done. If necessary, we still need to add fungi cultivation, mycobacterium tuberculosis culture and so on. The pus should undergo a rapid smear and dyeing check, which will be helpful to determine early bacterial infections.

Blood cultures are useful in identifying bacteremia or sepsis which are caused by pancreas infection or a pancreatic abscess. An aerobic and anaerobic bacterial culture must also be done, and it is recommended to do a blood culture three times in 24 h.

When drainage fluid is cultured, we should identify the infected bacteria of the catheter and primary infection. If possible, the results should be compared with the blood culture results.

Many researchers concluded that the PCR method is more sensitive than a blood culture for detecting bacterial components in the blood of patients who might develop sepsis during or after major abdominal surgery [92, 93]. With limitations in certain clinical skills, it has not been widely used up till now.

13.2.3 Microecology Therapies for Pancreatic Infection

Most pancreatic infections are secondary to AP [94], thus anti-infection treatment actually is included in the whole treatment scheme and process of AP. Acute secondary infection in pancreatitis is closely related to bacterial translocation. On the one hand, AP causes intestinal disorder, intestinal peristalsis inhibition, paralytic ileus, and intestinal obstruction, *etc.* On the other hand, the increased intestinal permeability, intestinal bacterial translocation, increased invasion and intestinal dysbacteriosis cause the secondary infection. The two hands are reciprocal causation, eventually leading to severe organ damage and failure. Therefore, in the anti-infection treatment of pancreatitis, improving intestinal microflora and promoting intestinal function restoration have an important clinical significance.

Flora and bacterial translocation are interrelated [95, 96]. When intestinal flora imbalance occurs, the number of anaerobic bacteria, holding the original absolute advantage, will drastically reduce, while a large number of potentially pathogenic bacteria will breed, which will weaken the intestinal barrier function and cause intestinal bacterial translocation. Translocation of bacteria or endotoxin enters into the blood and abdominal organs, eventually leading to serious consequences like developed multiple organ failure, sepsis and even death. Now it is generally agreed that this process is the main source of endogenous infection. It is thought that increased gut permeability and bacterial translocation are important factors in the development of infectious complications in these patients with AP [97], so selective gut decontamination, whereby non-absorbable antibiotics are given enterally in addition to intravenous antibiotics, has also been tried [98].

Therefore, the Microecological treatments of AP broadly include internal medicine, surgery, nutritional support, and traditional Chinese medicine treatments *etc.*, such as inhibiting secretion, reducing exudation, anti-inflammation, reasonable use of antibiotics, early parenteral nutrition [99, 100], keeping the drainage unobstructed, and using herbal medicine and probiotics to improve the intestinal function and so on, to eliminate the various unfavorable factors influencing the microecological balance, thus to promote the recovery of the microecology balance and further to prevent and cure the diseases. This section focuses on the antibiotics, probiotics and Chinese medicine to discuss the Microecological prevention principles of pancreatic infection.

13.2.3.1 Antibiotics Therapy

Only from the basic pathophysiology of AP that AP is a process of the activation of pancreatic enzymes and the digestion of pancreatic self, the question of whether or not to use the antibiotic is still an argument [101-104]. Indeed, in some clinical studies of mild edematous pancreatitis, the use of antibiotics did not significantly reduce the patient's clinical symptoms and shorten the time of hospitalization [103]. But from the point of view of reducing bacterial translocation and preventing pancreatic infection, the early use of potent antibiotics is very necessary [94]. In general, antibiotics should be used after the onset within 1 week. Stopping a medicine could be decisive, and the abuse of antibiotics should be avoided.

The opportunity of antibiotics therapy. The use of antibiotics in AP should be an opportunity to grasp, mainly in these circumstances: (i) Biliary pancreatitis accompanied by obvious signs of infection, such as elevated white blood cell count, fever, or obvious biliary symptoms; (ii) With any need for surgical operation treatment; (iii) In AP, with a Ranson score $\geqslant$3 or more positive; (iv) In complicated pulmonary infection or urinary tract infection; (v) With a positive blood culture.

The principles of antibiotics therapy. Antibiotics reach the necrotic pancreatic tissue not by the blood pathway, but by the pancreatic duct and pancreatic juice diffusion. And most of the early pathogens are derived from the gut, such as *Escherichia coli*, *Enterobacter aerogenes*, *Klebsiella pneumoniae*, *enterococci* and the mixed infections. Thus, efficient broad-spectrum antibiotics should be selected. In patients with AP, antibiotics should be guided by the following principles. (i) Early use: The use of antibiotics should be early before the occurrence of the secondary infection, and its purpose is to prevent secondary infection and abscess formation; (ii) Enough use: The administration of medication should be at the level of bacteremia and sepsis; (iii) The purposed choice: The selected antibiotics should be effective on gut-derived gram negative bacilli, can pass the blood-pancreatic barrier, and keep high concentrations of antibiotics in the pancreatic tissue.

The blood-pancreatic barrier is defined by the capillary endothelial cell layer around the pancreatic acinar, basement membrane, pancreatic acinar cell layer, centroacinar cell layer and the intercalated ducts. The process whereby antibiotics enter the pancreatic juice through the blood pancreatic barrier can be affected by many factors, including: (i) Antibiotic polarity and soluble-fat. The small polar and high fat soluble antibiotics pass more easily through the barrier than the large polar and highly water soluble antibiotics; (ii) The binding rate of antibiotic serum protein, the molecular mass size of binding protein as a carrier, and antibiotic pH value can affect the penetration ability of antibiotics in the pancreatic juice.

Some studies determined the blood-pancreatic barrier penetration capability and the antibacterial spectrum of 12 antibiotics [105, 106], including cephalosporins, quinolones, aminoglycosides and metronidazole *etc.* The results showed that the penetrations of imipenem and ciprofloxacin are strong, with high tissue concentrations, so they are the preferred antibiotic in AP. The third generation cephalosporins can reach a high concentration in pancreatic tissue, but the bactericidal performance of the gram-positive bacterium and the anaerobic bacteria is poor, so should be used with a combination of other antibiotics. Aminoglycoside, penicillin and ampicillin *etc.* can not very well penetrate pancreatic tissue, so they should not be used. But this antibiotic can be used for extra-pancreatic infections (such as intestinal infections). The metronidazole can penetrate the blood-pancreatic barrier, and reach a high concentration in pancreatic tissue. Due to the narrow antimicrobial spectrum, it can be used in combination with other antibiotics.

Some other studies also showed that more than 30 antibiotics could enter the pancreas and reach 1/3 of the effective concentration. Some antibiotics such as penicillin G and some cephalosporin antibiotics cannot enter the pancreatic tissue; some antibiotics such as tetracycline, gentamycin and ampicillin cannot enter the pancreatic tissue enough to reach the effective concentration. Biichler studied 10 different antibiotics for their permeability in normal pancreatic tissue and studied the clinical effect of common bacterial infection of the pancreas. The antibiotics are divided into three groups: Group A, low concentration in pancreatic tissue, cannot reach the minimal inhibitory concentration (MIC), such as netilmicin, tobramycin; Group B, concentration in pancreatic tissue is enough to inhibit some bacteria but not cover all bacteria, such as piperacillin and cefotaxime; Group C, high concentration in pancreatic tissue, and is effective against most bacteria, such as ciprofloxacin, ofloxacin and imipenem. Studies later found that after intravenous administration, pefloxacin and metronidazole's concentration in necrotic pancreatic tissue could remain above MIC in most bacteria continuously when imipenem could not. The continuous administration of imipenem can improve the concentration in necrotic tissue. Other studies found that ciprofloxacin had a high concentration in both pancreatic tissue and pancreatic juice, but secondary pancreatic infection is often more a flora of mixed infection, thus often needing the combined use of antibiotics.

Most bacteria that infect pancreatic tissue are common bacteria that cause hospital infection with strong drug resistance, so the use of antibiotics should accord with drug sensitivity test results. *Escherichia coli* and Klebsiella pneumoniae are susceptible to producing extended spectrum β-lactamase (ESBL), resulting in resistance to all cephalosporins. The resistance of Pseudomonas aeruginosa and Acinetobacter to Tienam and ciprofloxacin is increasing year by year. We also found methicillin-resistant *Staphylococcus aureus* (MRSA) and *vancomycin-resistant* intestinal cocci (VRE), which made antibiotic treatment become very difficult. Therefore, for the prevention and treatment of secondary pancreatic infection, we are considering the choice of antibiotics, and we should take positive measures to improve the primary disease, such as controlling biliary tract infection, reducing acute pancreatitis, reducing leakage and necrosis, keeping a clear drainage, improving the bowel function, and inhibiting bacterial translocation *etc.*

Rational application of antibiotics. Here is a breakdown of the numerous antibiotics: (i) The carbapenems can cover the majority of gram-negative bacteria and anaerobic bacteria *etc.*, with the characteristics of potent, broad-spectrum and rarely drug-resistant bacteria, but they have an extremely low sensitivity to *Pseudomonas aeruginosa* and *Stenotrophomonas maltophilia*; (ii) The third generation cephalosporins are broad-spectrum antibiotics, but the long-term application can easily induce the multiplication of *Escherichia coli* producing extended-spectrum β-lactamase (ESBL); (iii) The antibiotics of quinoline re ketones can achieve a stable drug concentration in pancreatic necrosis tissues, and have a good antimicrobial activity against gram positive bacteria, but a poor effect against gram-negative bacteria; (iv) Metronidazole or tinidazole with a high fat-soluble activity can better pass through the blood-pancreatic barrier, and effectively inhibit the anaerobic bacteria. The combined application of metronidazole or tinidazole with (ii) and (iii) Are now recognized as adjuvant anti-inflammatory drugs; (v) The fluconazole is effective in the vast majority of candidiasis, can be taken orally or by intravenous injection, and can pass through the blood-brain barrier and blood-pancreatic barrier. Fluconazole is currently the preferred choice as an antifungal drug. In addition, when necessary, amphotericin B is also a good selection for anti-fungi. It can quickly remove the fungi, but does not improve prognosis.

Most of the bacteria inducing pancreatic tissue infections are nosocomial infection pathogens, and have a high resistance to antibiotics. The use of antibiotics should accord with the results of drug sensitivity tests. For example, *Escherichia coli* and *Klebsiella pneumoniae* readily produce ESBL, so they can resist almost all cephalosporin. The resistance of *Pseudomonas* and *Acinetobacter* to imipenem and ciprofloxacin is also increasing year by year. Moreover, MRSA and VRE can also be separated. All these resistances make antibiotic therapy become very difficult. Therefore, to prevent and treat pancreatic secondary infection, in addition to considering the choice of antibiotics, the most important task is to take positive measures to control the primary diseases, such as controlling biliary tract infection, reducing inflammatory condition, decreasing the exudation and necrosis, keeping the drainage unobstructed, improving the intestinal function and inhibiting bacterial translocation.

13.2.3.2 Intestinal Ecoimmunonutrition

Severely acute pancreatitis patients have a high catabolism rate, a long course of treatment, and a long-term negative nitrogen balance. Meanwhile, they can also show an intestinal motility disorder, intestinal flora imbalance, intestinal ischemia, excessive production of cytokines, growth factor deficiency, and excessive apoptosis of the intestinal mucosal epithelial cell, leading to the damage of the intestinal mucosal barrier function and intestinal failure, further causing intestinal bacterial translocation. These events can not only cause pancreatic necrotic infection, but also make the enterogenous endotoxin and bacteria enter the systemic circulation, inducing and aggravating MODS. Therefore, proper nutritional support plays a very important role in patients with severe AP [107-109]. Especially in recent years, the supplementation of immune nutrition and intestinal probiotics to intestinally normal flora has won more attention in the treatment of AP [107, 108, 110].

Glutamine, arginine, ω-3 polyunsaturated fatty acids, nucleotides, β-carotene and other micronutrients can regulate the immune function so as to reduce the process of SIRS, increase the blood flow of intestinal mucosa and liver, inhibit the increase in intestinal permeability, and maintain the integrity of the intestinal epithelial cell [111-115].

These nutrients are called immune nutrition. Ecoimmunonutrition refers to when the food fiber and other essential nutrients (including immune nutrition) are provided. Human resident bacteria (probiotics) are also provided, competing with enteric pathogenic bacteria, finally to recover intestinal normal flora, stimulate the immune system, enhance the specific and non-specific immune function, and increase intestinal immunity. Immune microecology includes probiotics, prebiotics, synbiotics and immune nutrition, *etc.*

Immunological nutrition. Glutamine plays an important role in the maintenance of the intestinal metabolism, structure and function[116-119]. In normal conditions, the small intestine can directly obtain glutamine from the intestinal lumen or the blood. Inside the mitochondria, glutamine generates ATP *via* three tricarboxylic acid cycles, provides intestinal epithelial cells with energy, and provides the synthesis of nucleic acids and proteins with a nitrogen source. Thus glutamine is an essential nutrient for intestinal cells. Under conditions of stress such as acute pancreatitis *etc.*, the intestine, kidney and various immune cells substantially absorb and utilize glutamine. Although the release of glutamine in the skeletal muscle and lung *etc.* is increased, the blood concentration of glutamine remains down to the normal 30%. Meanwhile, due to the fasting factor, the gut cannot get glutamine from enteral food. The above factors lead to a severe deficiency in body glutamine under stress response, a direct consequence of which is small intestinal malnutrition, manifesting itself as intestinal mucosal atrophy, increased permeability of the mucosa and damage to the barrier function. At the same time, glutamine also functions, maintaining the activity of immune cells, increasing the content of IgA in intestinal fluid, and reducing cell damage induced by oxygen free radicals by increasing glutathione content. It is because glutamine plays an important role in maintaining the intestinal barrier under stress response. Thus glutamine supplementation has been recognized as an effective measure for the prevention of bacterial translocation.

In recent years, arginine has been another amino acid preventing bacterial translocation [5, 120]. The mechanism of arginine preventing bacterial translocation may not be directly related to the nutrition of intestinal epithelial cells, but be related to the widely regulated roles of arginine on the immune function and metabolic hormone. For example, arginine not only promotes the T cells to secret IL-2, but also increases the number of IL-2 receptor on the membrane surface of lymphocytes, thereby promoting lymphocyte proliferation, differentiation and maturation. Arginine also directly increases the activity of monocytes and macrophages. Moreover, arginine promotes the secretion of the growth hormone in the pituitary and the insulin-like growth factor in liver cells. These two hormones can promote the growth of intestinal epithelial cells. Therefore, in patients with acute pancreatitis (AP), conventional arginine supplementation will be helpful to improve immunity and reduce enterogenic infection.

In addition, ω-3 polyunsaturated fatty acids, nucleotides and β-carotene can improve immune functions [121-125]. In short, the immunological nutrition can provide intestinal mucosa with nutrients to reduce intestinal mucosal atrophy and apoptosis, and can also improve the local and systemic immune function and reduce the infectious complications.

Microbial probiotics. Microbial probiotics, also known as a regulator, consist of the preparation based on Microecological principles to correct Microecological imbalance and maintain Microecological balance. It is made up of the beneficial and harmless members of normal microbial probiotics or their promotion materials and can improve body health [126]. Main types of probiotics include probiotics, prebiotics and synbiotics [127-132].

Probiotics, also known as eco-products or viable probiotics preparations, are able to promote the intestinal flora balance, and have beneficial effects on the host [133-136]. Probiotics are mainly composed of the normal intestinal flora members, including *Lactobacillus*, *Bifidobacterium spp*, *Streptococcus*, *Enterococcus*, *Bacillus* and some yeast. The action mechanisms of probiotics are mainly as follows: (i) Maintaining intestinal Microecological balance: Normally, the human intestinal Microecological system is balanced. On one hand, it is beneficial to the host, for it could help the body perform certain physiological processes; on the other hand, it is also beneficial to the microbes, so as to maintain a certain combination of microbial communities, to maintain their growth and reproduction. In the Microecological system, a small number of dominant species play a decisive role for the entire community. With the loss or destruction of a dominant species, the intestinal microflora balance would be lost, creating ecological and flora imbalance, causing a series of clinical symptoms, such as double infection and low immunity. As members of the dominant species of the host body's normal flora, microbial probiotics can adjust the flora imbalance to restore the ecological balance, to achieve the purpose of medical treatment. (ii) Biological antagonism: Normal intestinal flora is directly involved in the body structure of biological defense barriers, including chemical and biological barriers. The chemical barrier consists of the intestinal flora metabolites such as acetic acid, lactic acid, propionic acid, hydrogen peroxide and bacteriocins and other active substances that can stop pathogenic bacteria colonization in the body or kill them. The biological barrier is the biofilm-like structure formed of the normal flora planted in the mucosa or among the epithelial cells, which effects the colonization, occupation, growth and reproduction of the passing bacteria and foreign attacking bacteria through the plant protective effect. Probiotics are made of such members of the normal flora. They can be involved in the biological barrier structure, exerting the function of biological antagonism, to prevent the invasion of pathogenic bacteria and opportunistic pathogens. Probiotics in the intestine may also play a role in biological oxygen deprivation, so that the concentration of oxygen in the local environment is reduced and the redox potential decreases, resulting in an advantageous microenvironment for normal intestinal flora-anaerobe's growth. Finally, they will restore the ecological balance. (iii) Immunity: Probiotics can be used as non-specific immune regulatory factors, play a role in immune regulation, and promote macrophage activity and the capacity of B cells to produce antibodies. (iv) Antibacterial activity. Some probiotics such as *Lactobacillus* acidophilus probiotics can produce bacteriocin such as Acidophilus probiotics, lactobacillus and Spinosad that can inhibit Salmonella, *Shigella*, *Staphylococcus*, *Klebsiella*, *Pseudomonas*, *Bacillus* and other bacteria. *Lactobacillus* also produces some hydrogen peroxide that inhibits the growth of various bacteria. (v) Nutrition. Probiotics can synthesize many vitamins in the body, such as niacin, folic acid, nicotinic acid, *vitamin B*, *etc.*, to promote protein digestion and absorption in the body, promote the body's calcium, iron and vitamin D absorption, and improve digestion. Following the *Bifidobacterium*, probiotics have had a variety of therapeutic agents.

Prebiotics are a special class of colon food [137]. They improve the intestinal microbial environment by selectively stimulating the growth of intestinal dominant bacterium. Prebiotics are mainly functional oligosaccharides (bifidus factor), and also include soy oligosaccharides, FOS and lactulose. Synbiotics form a class of products combining probiotics and prebiotics. They can play a role in the probiotics' physiological activity in bacteria, but also selectively increase the number of such bacteria, to be more beneficial to lasting health effects [56, 61, 62, 138]. Many experimental studies and preliminary clinical applications proved that in patients with AP, the use of probiotics through oral or enema methods, combined with internal medicine, surgery, nutritional support and Chinese medicine treatment *etc.*, had very good application prospects in clinical practice.

Infectious complications and associated mortality are a major concern in AP [64, 139, 140]. There are high expectations for probiotics, as an adjunct to enteral nutrition, and they are currently winning worldwide popularity for their health-promoting effects. Certain strains of probiotic bacteria might prevent infectious complications by decreasing the overgrowth of intestinal bacteria, restoring the function of the intestinal barrier, and modulating the function of the systemic immune system. It has been reported that infectious complications have decreased in several clinical studies with probiotics, in patients with AP and in patients undergoing abdominal operations. However, in a multicentre randomised, double-blind, placebo-controlled trial, the application of probiotics did not reduce the risk of infectious complications and was associated with an increased risk of mortality in patients with predicted severely AP. Thus probiotics should not be used in this category of patients [64, 132, 139].

Meanwhile, probiotics cannot be considered as harmless adjuncts to enteral nutrition any more, especially in critically ill patients or patients at risk of non-occlusive mesenteric ischaemia [64, 139].

13.2.3.3 Chinese Medicine Therapy

Chinese medicine theory believes that the pathogenesis of AP is qizhi, shiji shiyun, rejie, fubi *etc.*, and that the key is the unfavorable air machine in mid-jiao, inducing movement disorders. According to Chinese medicine theory of blockage makes ache, fluency releases it, and six Organs communion with each other, to the heat aggregation evidence of AP, we can dredge of removing heat and blood circulation method, often access to good effect. Particularly, by means of diarrhea, early restoration of bowel movements and improving the bowel function have important clinical significance in controling symptoms and improving the prognosis. This shows that Chinese also recognize that AP is closely related to the intestinal function. Thus the emphasis of the treatment of AP is on the restoration of the intestinal function [141-143].

In response to secondary infection in AP, the use of purgation, removing heat and detoxication, and promoting blood circulation and removing blood stasis can often achieve a better curative effect, mainly reflected in the following: (i) The prevention and treatment of enterogenous infection and endotoxemia are helpful in reducing necrotic pancreatic tissue infection and abscess formation, thereby relieving the second MODS summit. (ii) The improvement of blood circulation within the abdominal cavity organs and the promotion of the inflammatory exudate absorption have a different degree of protection on the vital organs of the body. (iii) The degradation of Chinese medicine in endotoxin can inhibit inflammatory reaction induced by cytokines and other inflammatory mediators mediated by endotoxin. (iv) These methods can regulate the abnormal immune response in severe abdominal infection, and promote the recovery of the immune function.

Chinese medicine treatments also include Enema of Chinese herbs and acupuncture therapy [144, 145]. Acupuncture therapy can choose various points such as the Zusanli, xiajuxu, Neiguan, Zhongwan point *etc.* Acupuncture therapy can function in relieving spasms and pain, relieving spasms of the Oddi's sphincter, improving the bile and pancreatic juice drainage, resisting infection, reducing pancreatic secretion, and stopping vomiting *etc.*, by the mechanisms of facilitating and improving the immune function, strengthening the phagocytosis of immune cells to pathogenic microorganisms, improving blood circulation, reducing exudate, and removing toxins and so on.

Currently, in the prescription of Traditional Chinese Medicine used for the treatment of AP and infection, rhubarb and bupleurum roots are the main components. The heat-clearing and detoxifying pharmacological mechanism of rhubarb purgation is closely related to improving the intestinal microflora. In the treatment of AP, the effect of Rhubarb can be used alone or in combination, the main mechanism of which may be manifested in the following ways: (i) The purgative effects: Rhubarb is commonly used as purgative. Senna glycosides in rhubarb are the effective component in rhubarb to induce diarrhea. An experimental study of rhubarb confirmed that the major site of action in rhubarb lies in the large intestine, and is closely associated with intestinal bacteria. Further researches indicated that the mammalian GI tract contains a large number of anaerobic bacteria, a part of which have a β-glucosidase activity. This part of bacteria can break up its sennoside into sennoside elements by a series of hydrolysis, reduction and oxidation reactions. Sennoside elements are the active ingredients inducing diarrhea, and can directly stimulate the bowel or submucous plexus to strengthen the peristalsis. Meanwhile, these elements can also prevent Na^+-K^+-ATP enzymes inhibiting water absorption, thereby forming the purgative action of rhubarb. (ii) Promoting pancreatic secretion: Experiments have confirmed that rhubarb can effectively promote pancreatic secretion, but significantly inhibit the activities of trypsin, pancreatic lipase and pancreatic amylase. (iii) Choleretic effect: Rhubarb can promote animal bile secretion and increase bilirubin and bile acids content. (iv) The antibacterial effect: *In vitro* tests have showed that rhubarb has different degrees of inhibition on *Staphylococcus*, *Streptococcus*, intestinal bacilli and anaerobes. Rhubarb plays an anti-bacterial effect possibly through some of the ingredients inhibiting the function of the bacterial biological oxidation enzyme. (v) The anti-inflammatory effects: The anti-inflammatory effects of rhubarb may be related to the biosynthesis inhibition of TXA2 and 17-three acids of peanut four acid metabolites, and may also be associated with the effects of the immune regulation function. (vi) Improving microcirculation and hemostatic effect: Some reports have indicated that blood viscosity in patients significantly decreases after taking rhubarb. Moreover, rhubarb plays a hemostatic function through enhancing fibrinolytic activity and inhibiting antithrombin activity of its effective hemostasis components α-epigallocatechin gallate and gallic acid.

References

[1] Abdeldayem H, Ghoneim E, Refaei A A, *et al*. Obstructive jaundice promotes intestinal-barrier dysfunction and bacterial translocation: Experimental study. Hepatol Int, 2007, 1: 444-448.

[2] Pinzone M R, Celesia B M, Di Rosa M, *et al*. Microbial translocation in chronic liver diseases. Int J Microbiol, 2012: 629-694.

[3] Wang F, Jiang H, Shi K, *et al*. Gut Bacterial Translocation is associated with

Microinflammation in End Stage Renal Disease Patients. Nephrology (Carlton), 2012, 17: 733-738.
[4] Ilan Y. Leaky gut and the liver: A role for bacterial translocation in nonalcoholic steatohepatitis. World J Gastroenterol, 2012, 18: 2609-2618.
[5] Quirino I E, Cardoso V N, Santos R D, *et al*. The Role of L-arginine and inducible nitric oxide synthase in intestinal permeability and bacterial translocation. J Parenter Enteral Nutr, 2013, 37: 392-400.
[6] Lundell L. Use of probiotics in abdominal surgery. Dig Dis, 2011, 29: 570-573.
[7] Liu Z, Ma Y, Qin H. Potential prevention and treatment of intestinal barrier dysfunction using active components of *Lactobacillus*. Ann Surg, 2011, 254: 832-833; author reply 3.
[8] Sarna S K. Cyclic motor activity; migrating motor complex. Gastroenterology, 1985, 89: 894-913.
[9] Grivell M B, Woods C M, Grivell A R, *et al*. The possum sphincter of Oddi pumps or resists flow depending on common bile duct pressure: a multilumen manometry study. J Physiol, 2004, 558: 611-622.
[10] Zelenka J, Muchova L, Zelenkova M, *et al*. Intracellular accumulation of bilirubin as a defense mechanism against increased oxidative stress. Biochimie, 2012, 94: 1821-1827.
[11] Lamsa V, Levonen A L, Sormunen R, *et al*. Heme and heme biosynthesis intermediates induce Heme oxygenase-1 and cytochrome P450 2A5, enzymes with putative sequential roles in heme and bilirubin metabolism: Different requirement for transcription factor nuclear factor erythroid-derived 2-like 2. Toxicol Sci, 2012, 130: 132-144.
[12] Wi Y M, Peck K R. Biliary sepsis caused by *Ochrobactrum anthropi*. Jpn J Infect Dis, 2010, 63: 444-446.
[13] Correia M I, Liboredo J C, Consoli M L. The role of probiotics in gastrointestinal surgery. Nutrition, 2012, 28: 230-234.
[14] Chao C M, Lai C C, Tang H J, *et al*. Biliary tract infections caused by Aeromonas species. Eur J Clin Microbiol Infect Dis, 2013, 32: 245-251.
[15] Ortega M, Marco F, Soriano A, *et al*. Epidemiology and prognostic determinants of bacteraemic biliary tract infection. J Antimicrob Chemother, 2012, 67: 1508-1513.
[16] Shanahan F. Probiotics in perspective. Gastroenterology, 2010, 139: 1808-1812.
[17] Kunisawa J, Kiyono H. Peaceful mutualism in the gut: Revealing key commensal bacteria for the creation and maintenance of immunological homeostasis. Cell Host Microbe, 2011, 9: 83-84.
[18] Greenwood-Van Meerveld B. Intestinal barrier function in health and gastrointestinal disease. Neurogastroenterol Motil, 2012, 24: 889.
[19] Lorenzo-Zuniga V, Bartoli R, Planas R, *et al*. Oral bile acids reduce bacterial overgrowth, bacterial translocation, and endotoxemia in cirrhotic rats. Hepatology, 2003, 37: 551-557.
[20] Clements W D, Parks R, Erwin P, *et al*. Role of the gut in the pathophysiology of extrahepatic biliary obstruction. Gut, 1996, 39: 587-593.

[21] Ding J W, Andersson R, Soltesz V, *et al.* Obstructive jaundice impairs reticuloendothelial function and promotes bacterial translocation in the rat. J Surg Res, 1994, 57: 238-245.
[22] Clark J F, Loftspring M, Wurster W L, *et al.* Bilirubin oxidation products, oxidative stress, and intracerebral hemorrhage. Acta Neurochir Suppl, 2008, 105: 7-12.
[23] Liu Y, Li P, Lu J, *et al.* Bilirubin possesses powerful immunomodulatory activity and suppresses experimental autoimmune encephalomyelitis. J Immunol, 2008, 181:1887-1897.
[24] Kapan M, Tekin R, Onder A, *et al.* Thymoquinone ameliorates bacterial translocation and inflammatory response in rats with intestinal obstruction. Int J Surg, 2012, 10: 484-488.
[25] De Winter B Y, De Man J G. Interplay between inflammation, immune system and neuronal pathways: Effect on gastrointestinal motility. World J Gastroenterol, 2010, 16: 5523-5535.
[26] von Kampen O, Buch S, Nothnagel M, *et al.* Genetic and functional identification of the likely causative variant for cholesterol gallstone disease at the ABCG5/8 lithogenic locus. Hepatology, 2013, 57: 2407-2417.
[27] Xie M, Kotecha VR, Andrade JD, *et al.* Augmented cholesterol absorption and sarcolemmal sterol enrichment slow small intestinal transit in mice, contributing to cholesterol cholelithogenesis. J Physiol, 2012, 590: 1811-1824.
[28] Chai J, He Y, Cai S Y, *et al.* Elevated hepatic multidrug resistance-associated protein 3/ATP-binding cassette subfamily C 3 expression in human obstructive cholestasis is mediated through tumor necrosis factor alpha and c-Jun NH_2-terminal kinase/stress-activated protein kinase-signaling pathway. Hepatology, 2012, 55: 1485-1494.
[29] Ahmed M H, Hamad M A, Routh C, *et al.* Statins as potential treatment for cholesterol gallstones: An attempt to understand the underlying mechanism of actions. Expert Opin Pharmacother, 2011, 12: 2673-2681.
[30] Suo T, Peng P, Feng M, *et al.* Fixed-point and stratified analysis of the fine structure and composition of five gallstones with Fourier transform infrared (FT-IR) specular reflection spectroscopy. Microsc Res Tech, 2012, 75: 294-299.
[31] Kiriyama S, Takada T, Strasberg S M, *et al.* New diagnostic criteria and severity assessment of acute cholangitis in revised Tokyo guidelines. J Hepatobiliary Pancreat Sci, 2012, 19:548-556.
[32] Schmidt M, Dumot J A, Soreide O, *et al.* Diagnosis and management of calculous gallbladder disease. Scand J Gastroenterol, 2012, 47: 1257-1265.
[33] Lee S J, Cho Y H, Lee S Y, *et al.* A case of scrub typhus complicated by acute calculous cholecystitis. Korean J Fam Med, 2012, 33: 243-246.
[34] Cai D, Sorokin V, Lutwick L, *et al.* C. glycolicum as the sole cause of bacteremia in a patient with acute cholecystitis. Ann Clin Lab Sci, 2012, 42: 162-164.
[35] Lata J, Jurankova J, Kopacova M, *et al.* Probiotics in hepatology. World J Gastroenterol, 2011, 17: 2890-2896.

[36] Lee Y K, Mazmanian S K. Has the microbiota played a critical role in the evolution of the adaptive immune system? Science, 2010, 330: 1768-1773.

[37] Khanal T, Kim H G, Jin S W, *et al*. Protective role of metabolism by intestinal microflora in butyl paraben-induced toxicity in HepG2 cell cultures. Toxicol Lett, 2012, 213:174-183.

[38] Frick J S, Autenrieth I B. The gut microflora and its variety of roles in health and disease. Curr Top Microbiol Immunol, 2013, 358: 273-289.

[39] Guarino A, Wudy A, Basile F, *et al*. Composition and roles of intestinal microbiota in children. J Matern Fetal Neonatal Med, 2012, 1:63-66.

[40] Naik S, Bouladoux N, Wilhelm C, *et al.* Compartmentalized control of skin immunity by resident commensals. Science, 2012, 337:1115-1119.

[41] Macdonald T T, Monteleone G. Immunity, inflammation, and allergy in the gut. Science, 2005, 307:1920-1925.

[42] The Human Microbiome Project Consortium. Structure, function and diversity of the healthy human microbiome. Nature, 2012, 486: 207-214.

[43] Kaya M, Bestas R, Bacalan F, *et al.* Microbial profile and antibiotic sensitivity pattern in bile cultures from endoscopic retrograde cholangiography patients. World J Gastroenterol, 2012, 18: 3585-3589.

[44] Sung Y K, Lee J K, Lee K H, *et al*. The clinical epidemiology and outcomes of bacteremic biliary tract infections caused by antimicrobial-resistant pathogens. Am J Gastroenterol, 2012, 107: 473-483.

[45] Kager L M, Sjouke B, van den Brand M, *et al.* The role of antibiotic prophylaxis in endoscopic retrograde cholangiopancreatography; a retrospective single-center evaluation. Scand J Gastroenterol, 2012, 47: 245-250.

[46] Chow J, Lee S M, Shen Y, *et al.* Host-bacterial symbiosis in health and disease. Adv Immunol, 2010, 107: 243-274.

[47] Backhed F, Ley R E, Sonnenburg J L, *et al.* Host-bacterial mutualism in the human intestine. Science, 2005, 307:1915-1920.

[48] Methé B A, Nelson K E, Pop M. A framework for human microbiome research. Nature, 2012, 486:215-221.

[49] Bengmark S. Pro- and synbiotics to prevent sepsis in major surgery and severe emergencies. Nutrients, 2012, 4: 91-111.

[50] Resta-Lenert S, Barrett K E. Live probiotics protect intestinal epithelial cells from the effects of infection with enteroinvasive Escherichia coli (EIEC). Gut, 2003, 52: 988-997.

[51] Kinross J M, Markar S, Karthikesalingam A, *et al.* A meta-analysis of probiotic and synbiotic use in elective surgery: Does nutrition modulation of the gut microbiome improve clinical outcome? JPEN J Parenter Enteral Nutr, 2013, 37: 243-253.

[52] Holte K, Krag A, Gluud L L. Systematic review and meta-analysis of randomized trials on probiotics for hepatic encephalopathy. Hepatol Res, 2012, 42: 1008-1015.

[53] Furrie E, Macfarlane S, Kennedy A, *et al.* Synbiotic therapy (Bifidobacterium longum/Synergy initiates resolution of inflammation in patients with active ulcerative colitis: A randomised controlled pilot trial. Gut, 2005, 54: 242-249.

[54] Ichinohe T, Pang I K, Kumamoto Y, *et al.* Microbiota regulates immune defense against respiratory tract influenza A virus infection. Proc Natl Acad Sci USA, 2011, 108: 5354-5359.
[55] Lilly D M, Stillwell R H. Probiotics: Growth-promoting factors produced by microorganisms. Science, 1965, 147: 747-748.
[56] Kolida S, Gibson G R. Synbiotics in health and disease. Annu Rev Food Sci Technol, 2011, 2: 373-393.
[57] Fuller R. Probiotics in human medicine. Gut, 1991, 32:439-442.
[58] Dunne C, O'Mahony L, Murphy L, *et al. In vitro* selection criteria for probiotic bacteria of human origin: Correlation with *in vivo* findings. Am J Clin Nutr, 2001, 73: S386-S392.
[59] Gibson G R, Roberfroid M B. Dietary modulation of the human colonic microbiota: Introducing the concept of prebiotics. J Nutr, 1995, 125: 1401-1412.
[60] Eguchi S, Takatsuki M, Hidaka M, *et al.* Perioperative synbiotic treatment to prevent infectious complications in patients after elective living donor liver transplantation: A prospective randomized study. Am J Surg, 2011, 201: 498-502.
[61] Sugawara G, Nagino M, Nishio H, *et al.* Perioperative synbiotic treatment to prevent postoperative infectious complications in biliary cancer surgery: A randomized controlled trial. Ann Surg, 2006, 244: 706-714.
[62] Kinross J, Warren O, Silk D, *et al.* Perioperative synbiotic treatment to prevent postoperative infectious complications in biliary cancer surgery: A randomized control trial. Ann Surg, 2007, 245: 1000.
[63] Wang S Q, Feng Q X, Li S J, *et al.* The day when infection is confirmed is a better time point for mortality prediction in patients with severe acute pancreatitis. Pancreas, 2012, 41: 605-610.
[64] Besselink M G, van Santvoort H C, Buskens E, *et al.* Probiotic prophylaxis in predicted severe acute pancreatitis: A randomised, double-blind, placebo-controlled trial. Lancet, 2008, 371: 651-659.
[65] Olivieri C, Nanni L, Taddei A, *et al.* Acute pancreatitis associated with herpes simplex virus infection in a child. Pancreas, 2012, 41: 330-331.
[66] Jeppsson B, Mangell P, Thorlacius H. Use of probiotics as prophylaxis for postoperative infections. Nutrients, 2011, 3: 604-612.
[67] Yadav D, Lowenfels A B. Trends in the epidemiology of the first attack of acute pancreatitis: a systematic review. Pancreas, 2006, 33: 323-330.
[68] Gurusamy K S, Farouk M, Tweedie J H. UK guidelines for management of acute pancreatitis: Is it time to change? Gut, 2005, 54: 1344-1345.
[69] Sah R P, Garg P, Saluja A K. Pathogenic mechanisms of acute pancreatitis. Curr Opin Gastroenterol, 2012, 28: 507-515.
[70] Saluja A, Hashimoto S, Saluja M, *et al.* Subcellular redistribution of lysosomal enzymes during caerulein-induced pancreatitis. Am J Physiol, 1987, 253: G508-G516.
[71] Saluja A K, Bhagat L, Lee H S, *et al.* Secretagogue-induced digestive enzyme activation and cell injury in rat pancreatic acini. Am J Physiol, 1999, 276: G835-G842.

[72] Kolodecik T R, Shugrue C A, Thrower E C, *et al.* Activation of soluble adenylyl cyclase protects against secretagogue stimulated zymogen activation in rat pancreaic acin ar cells. PLoS One, 2012, 7: e41320.
[73] Dawra R, Sah R P, Dudeja V, *et al.* Intra-acinar trypsinogen activation mediates early stages of pancreatic injury but not inflammation in mice with acute pancreatitis. Gastroenterology, 2011, 141: 2210-2217 e2.
[74] Sah R P, Saluja A. Molecular mechanisms of pancreatic injury. Curr Opin Gastroenterol, 2011, 27: 444-451.
[75] Park C Y, Hoover P J, Mullins F M, *et al.* STIM1 clusters and activates CRAC channels via direct binding of a cytosolic domain to Orai1. Cell, 2009, 136: 876-890.
[76] Mukherjee R, Criddle D N, Gukovskaya A, *et al.* Mitochondrial injury in pancreatitis. Cell Calcium, 2008, 44: 14-23.
[77] Cardenas C, Miller RA, Smith I, *et al.* Essential regulation of cell bioenergetics by constitutive InsP3 receptor Ca^{2+} transfer to mitochondria. Cell, 2010, 142: 270-283.
[78] Lugea A, Waldron R T, French SW, *et al.* Drinking and driving pancreatitis: Links between endoplasmic reticulum stress and autophagy. Autophagy, 2011, 7: 783-785.
[79] Grasso D, Ropolo A, Lo Re A, *et al.* Zymophagy, a novel selective autophagy pathway mediated by VMP1-USP9x-p62, prevents pancreatic cell death. J Biol Chem, 2011, 286: 8308-8324.
[80] Ahmed F, Fogel E. Reply to Reiss G, Ramrakhiani S. Right upper-quadrant pain and a normal abdominal ultrasound. Clin Gastroenterol Hepatol, 2009, 7: 603. Clin Gastroenterol Hepatol, 2009, 7: 1256.
[81] Mashima H, Sato T, Horie Y, *et al.* Interferon regulatory factor-2 regulates exocytosis mechanisms mediated by SNAREs in pancreatic acinar cells. Gastroenterology, 2011, 141: 1102-1113, e1-8.
[82] Chen Y, Zak Y, Hernandez-Boussard T, *et al.* The epidemiology of idiopathic acute pancreatitis, analysis of the nationwide inpatient sample from 1998 to 2007. Pancreas, 2013, 42: 1-5.
[83] Dufour M C, Adamson M D. The epidemiology of alcohol-induced pancreatitis. Pancreas, 2003, 27: 286-290.
[84] Algul H, Tando Y, Schneider G, *et al.* Acute experimental pancreatitis and NF-κB/Rel activation. Pancreatology, 2002, 2: 503-509.
[85] Garcia M, Calvo J J. Cardiocirculatory pathophysiological mechanisms in severe acute pancreatitis. World J Gastrointest Pharmacol Ther, 2010, 1: 9-14.
[86] Andersen A M, Novovic S, Ersboll A K, *et al.* Mortality in alcohol and biliary acute pancreatitis. Pancreas, 2008, 36: 432-434.
[87] Hirota M, Satoh K, Kikuta K, *et al.* Early detection of low enhanced pancreatic parenchyma by contrast-enhanced computed tomography predicts poor prognosis of patients with acute pancreatitis. Pancreas, 2012, 41: 1099-1104.
[88] Bryner B S, Smith C, Cooley E, *et al.* Extracorporeal life support for pancreatitis-induced acute respiratory distress syndrome. Ann Surg, 2012, 256: 1073-1077.

[89] Besselink M G, van Santvoort H C, Renooij W, *et al.* Intestinal barrier dysfunction in a randomized trial of a specific probiotic composition in acute pancreatitis. Ann Surg, 2009, 250: 712-719.
[90] Runkel N S, Rodriguez L F, Moody F G. Mechanisms of sepsis in acute pancreatitis in opossums. Am J Surg, 1995, 169: 227-232.
[91] Penny S M. Clinical signs of pancreatitis. Radiol Technol, 2012, 83: 561-577.
[92] Ono S, Tsujimoto H, Yamauchi A, *et al.* Detection of microbial DNA in the blood of surgical patients for diagnosing bacterial translocation. World J Surg, 2005, 29: 535-539.
[93] Kane T D, Alexander J W, Johannigman J A. The detection of microbial DNA in the blood: a sensitive method for diagnosing bacteremia and/or bacterial translocation in surgical patients. Ann Surg, 1998, 227: 1-9.
[94] Carnovale A, Rabitti P G, Manes G, *et al.* Mortality in acute pancreatitis: Is it an early or a late event? JOP, 2005, 6: 438-444.
[95] Heimesaat M M, Boelke S, Fischer A, *et al.* Comprehensive postmortem analyses of intestinal microbiota changes and bacterial translocation in human flora associated mice. PLoS One, 2012, 7: e40758.
[96] Corradi F, Brusasco C, Fernandez J, *et al.* Effects of pentoxifylline on intestinal bacterial overgrowth, bacterial translocation and spontaneous bacterial peritonitis in cirrhotic rats with ascites. Dig Liver Dis, 2012, 44: 239-244.
[97] Deitch E A. The role of intestinal barrier failure and bacterial translocation in the development of systemic infection and multiple organ failure. Arch Surg, 1990, 125: 403-404.
[98] Luiten E J, Hop W C, Lange J F, *et al.* Controlled clinical trial of selective decontamination for the treatment of severe acute pancreatitis. Ann Surg, 1995, 222: 57-65.
[99] Maung A A, Davis K A. Perioperative nutritional support: immunonutrition, probiotics, and anabolic steroids. Surg Clin North Am, 2012, 92: 273-283.
[100]Anand N, Park J H, Wu B U. Modern management of acute pancreatitis. Gastroenterol Clin North Am, 2012, 41: 1-8.
[101]Su M S, Lin M H, Zhao Q H, *et al.* Clinical study of distribution and drug resistance of pathogens in patients with severe acute pancreatitis. Chin Med J (Engl), 2012, 125: 1772-1776.
[102]Qu R, Ji Y, Ling Y, *et al.* Procalcitonin is a good tool to guide duration of antibiotic therapy in patients with severe acute pancreatitis. A randomized prospective single-center controlled trial. Saudi Med J, 2012, 33: 382-387.
[103]Jiang K, Huang W, Yang X N, *et al.* Present and future of prophylactic antibiotics for severe acute pancreatitis. World J Gastroenterol, 2012, 18: 279-284.
[104]Whitcomb D C. Clinical practice: Acute pancreatitis. N Engl J Med, 2006, 354: 2142-2150.
[105]Burns G P, Stein T A, Kabnick L S. Blood-pancreatic juice barrier to antibiotic excretion. Am J Surg, 1986, 151: 205-208.
[106]Kang W, Zhao Y, Tao W, *et al.* Change of 5-fluorouracil penetration in blood-pancreatic barrier of rats after high-dose radiotherapy. Zhongguo Yi

Xue Ke Xue Yuan Xue Bao, 2000, 22: 457-459.

[107]Ong J P, Fock K M. Nutritional support in acute pancreatitis. J Dig Dis, 2012, 13: 445-452.

[108]Mirtallo J M, Forbes A, McClave S A, *et al.* International consensus guidelines for nutrition therapy in pancreatitis. JPEN J Parenter Enteral Nutr, 2012, 36: 284-291.

[109]Yi F, Ge L, Zhao J, *et al.* Meta-analysis: Total parenteral nutrition versus total enteral nutrition in predicted severe acute pancreatitis. Intern Med, 2012, 51: 523-530.

[110]Bordeje L L, Lorencio C C, Acosta E J. Guidelines for specialized nutritional and metabolic support in the critically-ill patient: Update. Consensus SEMICYUC-SENPE: Severe acute pancreatitis. Nutr Hosp, 2011, 26: S32-S36.

[111]Rangel-Huerta O D, Aguilera C M, Mesa M D, *et al.* Omega-3 long-chain polyunsaturated fatty acids supplementation on inflammatory biomakers: A systematic review of randomised clinical trials. Br J Nutr, 2012, 107: S159-S170.

[112]Liu D, Chen Z. The regulatory effects of glutamine on illness and health. Protein Pept Lett, 2011, 18: 658-662.

[113]Curi R, Newsholme P, Procopio J, *et al.* Glutamine, gene expression, and cell function. Front Biosci, 2007, 12: 344-357.

[114]Rossoni Junior J V, Araujo G R, Padua B C, *et al.* Annatto extract and beta-carotene enhances antioxidant status and regulate gene expression in neutrophils of diabetic rats. Free Radic Res, 2012, 46: 329-338.

[115]Katsuura S, Imamura T, Bando N, *et al.* β-Carotene and β-cryptoxanthin but not lutein evoke redox and immune changes in RAW264 murine macrophages. Mol Nutr Food Res, 2009, 53: 1396-1405.

[116]Bertrand J, Goichon A, Dechelotte P, *et al.* Regulation of intestinal protein metabolism by amino acids. Amino Acids, 2012, DOI 10. 1007/s00726-012-1325-8.

[117]Dai Z L, Li X L, Xi P B, *et al.* L-glutamine regulates amino acid utilization by intestinal bacteria. Amino Acids, 2012, DOI 10. 1007/s00726-012-1264-4.

[118]Lehmann C, Pavlovic D, Zhou J, *et al.* Intravenous free and dipeptide-bound glutamine maintains intestinal microcirculation in experimental endotoxemia. Nutrition, 2012, 28: 588-593.

[119]Feng Y, Ralls M W, Xiao W, *et al.* Loss of enteral nutrition in a mouse model results in intestinal epithelial barrier dysfunction. Ann NY Acad Sci, 2012, 1258: 71-77.

[120]Manzanares W, Heyland D K. Pharmaconutrition with arginine decreases bacterial translocation in an animal model of severe trauma. Crit Care Med, 2012, 40: 350-352.

[121]Braga M. Perioperative immunonutrition and gut function. Curr Opin Clin Nutr Metab Care, 2012, 15: 485-488.

[122]Han S C, Kang G J, Ko Y J, *et al.* Fermented fish oil suppresses T helper 1/2 cell response in a mouse model of atopic dermatitis via generation of $CD4^{+}CD25^{+}Foxp3^{+}$ T cells. BMC Immunol, 2012, 13: 44.

[123]Miles E A, Calder P C. Influence of marine N-3 polyunsaturated fatty acids on immune function and a systematic review of their effects on clinical outcomes in rheumatoid arthritis. Br J Nutr, 2012, 107: S171-S184.

[124]Bilku D K, Hall T C, Al-Leswas D, *et al.* Can enhanced recovery programmes be further improved by the addition of omega three fatty acids? Ir J Med Sci, 2012, 181: 453-457.

[125]Swanson D, Block R, Mousa S A. Omega-3 fatty acids EPA and DHA: Health benefits throughout life. Adv Nutr, 2012, 3: 1-7.

[126]Weichert S, Schroten H, Adam R. The role of prebiotics and probiotics in prevention and treatment of childhood infectious diseases. Pediatr Infect Dis J, 2012, 31: 856-862.

[127]Simren M, Barbara G, Flint H J, *et al.* Intestinal microbiota in functional bowel disorders: A Rome foundation report. Gut, 2012, 62: 159-176.

[128]D'Souza A, Cai C L, Kumar D, *et al.* Cytokines and toll-like receptor signaling pathways in the terminal ileum of hypoxic/hyperoxic neonatal rats: Benefits of probiotics supplementation. Am J Transl Res, 2012, 4: 187-197.

[129]Morrow L E, Gogineni V, Malesker M A. Probiotic, prebiotic, and synbiotic use in critically ill patients. Curr Opin Crit Care, 2012, 18: 186-191.

[130]Quigley E M. Therapies aimed at the gut microbiota and inflammation: Antibiotics, prebiotics, probiotics, synbiotics, anti-inflammatory therapies. Gastroenterol Clin North Am, 2011, 40: 207-222.

[131]Gourbeyre P, Denery S, Bodinier M. Probiotics, prebiotics, and synbiotics: impact on the gut immune system and allergic reactions. J Leukoc Biol, 2011, 89: 685-695.

[132]Zhang M M, Cheng J Q, Lu Y R, *et al.* Use of pre-, pro- and synbiotics in patients with acute pancreatitis: A meta-analysis. World J Gastroenterol, 2010, 16: 3970-3978.

[133]Ishikawa H, Matsumoto S, Ohashi Y, *et al.* Beneficial effects of probiotic bifidobacterium and galacto-oligosaccharide in patients with ulcerative colitis: A randomized controlled study. Digestion, 2011, 84: 128-133.

[134]Shimizu K, Ogura H, Asahara T, *et al.* Probiotic/synbiotic therapy for treating critically ill patients from a gut microbiota perspective. Dig Dis Sci, 2013, 58: 23-32.

[135]Tappenden K A. Probiotics are not a one-species-fits-all proposition. JPEN J Parenter Enteral Nutr, 2012, 36: 496.

[136]Kelly D, Mulder I E. Microbiome and immunological interactions. Nutr Rev, 2012, 70: S18-S30.

[137]Nagpal R, Kaur A. Synbiotic effect of various prebiotics on *in vitro* activities of probiotic lactobacilli. Ecol Food Nutr, 2011, 50: 63-68.

[138]van de Pol M A, Lutter R, Smids B S, *et al.* Synbiotics reduce allergen-induced T-helper 2 response and improve peak expiratory flow in allergic asthmatics. Allergy, 2011, 66: 39-47.

[139]Expression of concern − Probiotic prophylaxis in predicted severe acute pancreatitis: a randomised, double-blind, placebo-controlled trial. Lancet, 2010, 375: 875-876.

[140]Qin H L, Zheng J J, Tong D N, *et al.* Effect of Lactobacillus plantarum

enteral feeding on the gut permeability and septic complications in the patients with acute pancreatitis. Eur J Clin Nutr, 2008, 62: 923-930.
[141]Wang L, Li Y, Ma Q, *et al.* Chaiqin Chengqi Decoction decreases IL-6 levels in patients with acute pancreatitis. J Zhejiang Univ Sci B, 2011, 12: 1034-1340.
[142]Wan M H, Li J, Tang W F, *et al.* The influnence of dachengqi tang on acute lung injury and intra abdominal hypertension in rats with acute pancreatitis. Sichuan Da Xue Xue Bao Yi Xue Ban, 2011, 42: 707-711.
[143]Chen Y F, Sha J P, Wu Z M. Synergetic effect of yihuo qingyi decoction (see text) and recombinant staphylokinase in treatment of severe acute pancreatitis of rats. J Tradit Chin Med, 2011, 31: 103-106.
[144]Xue Q M, Ning L, Xue P, *et al.* Effect of electroacupuncture on serum proinflammatory cytokine levels and pancreatic nuclear factor κB expression in acute pancreatitis rats. Zhen Ci Yan Jiu, 2011, 36: 272-277.
[145]Xue Q M, Huang L, Li N. Effects of electroacupuncture at Tianshu (ST25) on pro- and anti-inflammatory cytokines in rats with severe acute pancreatitis. Zhong Xi Yi Jie He Xue Bao, 2011, 9: 658-664.

14

Infectious Microecology in Urinary Tract and Reproductive System

Zhoujun Shen *, Shan Zhong, Yu Zhu, Yuan Shao, Wei He, Chenjing Zhang, Xianjin Wang, Tao Li, Sakaliya, Hongchao He
Department of Urology, Ruijin Hospital, School of Medicine, Shanghai Jiao Tong University, Shanghai, 200025, China
* E-mail: shenzj6@sina.com

14.1 Introduction

Urinary tract infection (UTI) is an infection involving the kidneys, ureters, bladder, or urethra. These are the structures that urine passes through before being eliminated from the body. Symptoms include frequent feeling and/or need to urinate, pain during urination, and cloudy urine. The most common type of UTI is acute cystitis often referred to as bladder infection. An infection of the upper urinary tract or kidney is known as pyelonephritis, and is potentially more serious. Although they cause discomfort, UTI can usually be easily treated with a short course of antibiotics with no significant difference between the classes of antibiotics commonly used.

UTIs are more common in women than men, leading to approximately 8.3 million doctors visits per year. Although some infections go unnoticed, UTIs can cause problems that range from dysuria (pain and/or burning when urinating) to organ damage and even death. It may present as asymptomatic bacteriuria and complicated or uncomplicated infections in upper and lower urinary system. Ideal treatment results in symptomatic relief, prevents progressive renal damage and urosepsis with immediate bacterial eradication.

A simple UTI can be treated with a short course of oral antibiotics. A three-day course of antibiotics will usually treat most uncomplicated UTIs. However, some infections may need to be treated for several weeks. If the UTI is a complicated UTI, then a longer period of antibiotics is given and usually is started intravenously in the hospital. After a short period of intravenous antibiotics, then the antibiotics are given by mouth for a period up to several weeks. Kidney infections have usually been treated as a complicated UTI.

Many methods have been suggested to reduce or prevent UTIs. Some of these are home remedies and have been widely discussed. Incomplete bladder emptying and resisting the normal urge to urinate can allow pathogens to survive and replicate easier in a non-flowing system. Some clinicians recommend washing before and urinating soon after sex to reduce the chance of urethritis/cystitis.

In the future, scientists may develop a vaccine that can prevent UTIs from coming back. Researchers in different studies have found that children and women who tend to get UTIs repeatedly are likely to lack proteins called immunoglobulins, which fight against infection. Children and women who do not get UTIs are more likely to have normal levels of immunoglobulins in their genital and urinary tracts.

14.2 Infections of the Urinary Tract

UTI is an inflammatory response of the urothelium to bacterial invasion that is usually associated with bacteriuria and pyuria. UTI are usually monoinfections caused by the endogenous microflora including gram-negatives, such as *Escherichia coli*, or gram-positives, like enterococci. This allows for an empiric treatment of uncomplicated UTIs and a short duration of therapy to minimize the probability for the development of drug resistance.

14.2.1 Definitions

Bacteriuria indicates the presence of bacteria in the urine, which is normally free of bacteria, and implies that these bacteria are from the urinary tract and are not contaminants from the skin, vagina or prepuce.

Pyuria is the presence of white blood cells (WBCs) in the urine and is generally indicative of an inflammatory response of the urothelium to bacterial invasion. Bacteriuria without pyuria indicates bacterial colonization rather than infection. Pyuria without bacteriuria warrants evaluation for tuberculosis, stones, or cancer.

UTI refers to the presence of clinical signs and symptoms arising from the genitourinary tract plus the presence of one or more micro-organisms in the urine exceeding a threshold value for significance (ranges from 102 to 103 CFU/mL). Infections are localized to the bladder (cystitis), renal parenchyma (pyelonephritis) or prostate (acute or chronic bacterial prostatitis). Single UTI episodes are very common, especially in adult women where there is a 50-fold predominance compared with adult men. In addition, recurrent UTIs are also common, occurring in up to 1/3 of women after first-episode UTIs. Recurrences requiring intervention are usually defined as two or more episodes over 6 months or three or more episodes over 1 year (this definition applies only to young women with acute uncomplicated UTIs). A cornerstone of prevention of UTI recurrence has been the use of low-dose once-daily or post-coital antimicrobials; however, much interest has surrounded non-antimicrobial-based approaches undergoing investigation such as use of probiotics, vaccines, oligosaccharide inhibitors of bacterial adherence and colonization, and bacterial interference with immunoreactive extracts of *Escherichia coli*. Local (intravaginal) estrogen therapy has inconsistent results to date.

Acute pyelonephritis is a clinical syndrome of chills, fever, and flank pain that is accompanied by bacteriuria and pyuria, a combination that is reasonably specific for an acute bacterial infection of the kidney. The term should not be used if flank pain is absent; there may be serious difficulties in diagnosing spinal cord-injured and elderly patients who may be unable to localize the site of their discomfort.

Chronic pyelonephritis describes a shrunken, scarred kidney, diagnosed by morphologic, radiologic, or functional evidence of renal disease that may be postinfectious but is frequently not associated with UTI.

Cystitis is inflammation of the bladder, whether used as a histologic, bacteriologic, or cystoscopic description, or a clinical syndrome that is usually accompanied by an abrupt onset of dysuria, increased frequency, urgency, and suprapubic pain.

Urethritis, like cystitis, also refers to inflammation, but of the urethra rather than the bladder. It requires an adjective for modification, for example, nongonococcal urethritis.

14.2.2 Classification

UTI can be divided into four categories: (i) isolated infections, (ii) unresolved infections, (iii) recurrent UTIs that are reinfections, and (iv) recurrent infections resulting from bacterial persistence.

14.2.2.1 Isolated Infections

First infections or those isolated from previous infections by at least 6 months occur in 25% to 30% of women between the ages of 30 and 40 years, but these occur infrequently in men with a normal urinary tract. About 1/4 of such women experience a recurrence in the next few years. Isolated infections in domiciliary patients are usually susceptible to all antimicrobial agents.

14.2.2.2 Unresolved Bacteriuria during Therapy

The term unresolved indicates that the initial therapy has been inadequate. The absence of bacterial growth in the urine during therapy is a prerequisite for successful treatment and for characterization of the type of recurrence. The clinician often fails to recognize this problem because (i) Cultures of the urine are not obtained during treatment or (ii) If they are obtained, bacterial counts of less than 10^5/mL are misinterpreted as contaminants. Clearly, if any of the bacteria that caused the infection are present in the urine during therapy, regardless of how low the number, the bacteria have not been eradicated.

14.2.2.3 Recurrent Urinary Tract Infections

The term recurrent urinary tract infection applies either to re-infection from outside the urinary tract or to bacterial persistence in a focus within the urinary tract. It is also clear that until a UTI is resolved with proper antimicrobial therapy, the type of recurrence — bacterial re-infection or persistence — cannot be classified.

14.2.2.4 Reinfections

More than 95% of all recurrent infections in females are re-infections of the urinary tract. Re-infections in men are uncommon unless associated with an underlying abnormality of the urinary tract. Vesicointestinal and vesicovaginal fistulas are an unusual, but surgically correctable, cause of reinfections.

14.2.2.5 Bacterial Persistence

Once the bacteriuria has resolved (*i.e.*, the urine shows no growth for several days after the antimicrobial agent has been stopped), recurrence with the same organism can arise from a site within the urinary tract that was excluded from the high urine concentrations of the antimicrobial agent.

14.3 Diagnosis

Diagnosises of upper urinary tract infection and lower urinaru tract infection are as following.

14.3.1 Upper Urinary Tract Infection (UUTI)

There are clininal manifestations, laborafory findings, imaging studies, plain film of the abdomes, excretory urogram, *et al.*

14.3.1.1 Clinical Manifestations

Fever and flank pain usually indicates pyelonephritis, but in some cases patients with these symptoms do not have renal infection. In contrast, significant renal infection may be associated with an insidious onset of nonspecific local or systemic symptoms, or it may be entirely asymptomatic. The clinical spectrum ranges from gram-negative sepsis to mild flank pain. The classic clinical manifestation is an abrupt onset of chills, fever and flank or costovertebral angle pain and/or tenderness. These so-called upper tract signs are often accompanied by dysuria, urinary frequency, and urgency. In women who present 1 or more symptoms of UTI, the probability of infection is approximately 50%. Specific combinations of symptoms (*e.g.*, dysuria and frequency without vaginal discharge or irritation) raise the probability of UTI to more than 90%, effectively ruling in the diagnosis based on history alone [1].

On physical examination, there is often tenderness to deep palpation in the costovertebral angle. Variable clinical presentation have been recognized. Acute pyelonephritis may also simulate the gastrointestinal tract, which is accompanied by abdominal pain, nausea, vomiting, and diarrhea. Asymptomatic progression of acute pyelonephritis to chronic pyelonephritis, particularly in compromised hosts, may occur in the absence of overt symptoms. Renal failure may be present in the rare case [2] .

Pyelonephritis may transiently or permanently alter renal function, but non-obstructive pyelonephritis is no longer recognized as a major cause of renal failure [3]. However, pyelonephritis associated with urinary tract obstruction or granulomatous renal infection may lead to significant inflammatory complications, renal failure or even death.

14.3.1.2 Laboratory Findings

Presumptive diagnosis of UUTI is made by direct or indirect analysis of the urine and is confirmed by urine culture. Assessment of the urine provides clinical data about the status of the urinary tract. The urinary tract are normally free of bacteria and inflammation. False-negative urinalysis and culture can occur in the presence of UUTI, particularly early in an infection when the numbers of bacteria and WBCs are low or diluted by increased fluid intake and subsequent diuresis. Occasionally, the urine may be free of bacteria and WBCs despite bacterial colonization and inflammation of the uroepithelium [4]. False-positive urinalysis and culture are caused by contamination of the urine specimen with bacteria and

WBCs during collection. This is most likely to occur not only in voided specimens, but also during urethral catheterization. Suprapubic aspiration of bladder urine is least likely to cause contamination of the specimen; therefore, it provides the most accurate assessment of the status of bladder urine.

Urine Collection

Voided and Catheterized Specimens. Accuracy can be improved by reducing bacterial contamination when collecting the urine. In circumcised men, voided specimens require no preparation. For those who are not circumcised, the foreskin should be retracted and the glans penis needs to be washed and then rinsed with water before specimen collection. The first 10 mL of urine (representative of the urethra) and a midstream specimen (representative of the bladder) should be obtained. Prostatic fluid specimen is obtained by pressing prostatic gland with index and collecting the expressed prostatic fluid on a glass slide. In addition, collection of the first 10 mL of voided urine after massage reflects the prostatic fluid added to the urethral specimen. Usually, catheterization of a male patient for urine culture is not recommended unless the patient cannot urinate.

In women, contamination during the specimen collection is common, especially when the woman has difficulty in spreading and maintaining separation of the labia minor. Therefore, the female should be instructed to spread the labia minor, wash and rinse the periurethral area with a moist gauze, and then collect a midstream urine specimen. Cleansing with antiseptics is not recommended because they may contaminate the voided specimen and provide a false-negative urine culture. If the voided specimen shows evidence of contamination as indicated by vaginal epithelial cells and lactobacilli on urinalysis, catheterization should be performed and a mid-catheterized specimen collected.

The incidence of catheter-induced UTI is determined primarily by the population at risk, varying from 1% in non-hospitalized, healthy women [5] to 20% in women hospitalized on a medical ward [6]. The easiest way to prevent catheter-induced infections is to give a single dose of an oral antimicrobial agent, such as trimethoprim-sulfamethoxazole (TMP-SMX). However, because antibiotics can lead to bacterial resistance, prophylaxis should be limited to high-risk patients.

Suprapubic Aspiration. Suprapubic aspiration showed the lowest contamination rate and is highly accurate [7], but is limited because of its morbidity except on a patient who cannot urinate on command, such like patients with spinal cord injuries. It is highly useful in newborns [8] and in patients with paraplegia. A single aspirated specimen reveals the bacteriologic status of the bladder urine without introducing urethral bacteria, which can start a new infection.

Before a suprapubic aspiration is performed, the patient should force fluids into his system until the bladder is full. The site of the needle puncture is in the midline, between the symphysis pubis and the umbilicus and directly over the palpable bladder. The full bladder in the male is usually palpable due to its greater muscle tone; unfortunately, the full bladder in the female is frequently not palpable. In such patients, the physician performing the aspiration must rely on the observation that supra-pubic pressure directly over the bladder produces an unmistakable desire to urinate. After determining the approximate site for needle puncture, the local area is shaved and the skin is cleansed; a cutaneous wheal is raised with a 25-gauge needle and local anesthetic. A 3.5-inch spinal, 22-gauge needle is introduced through the anesthetized skin. The progress of the needle is arrested just below the skin within the anesthetized area, and with a quick plunging action, similar to that of any intramuscular injection, the needle is advanced into the bladder. Most patients experience more discomfort from the initial anesthetization of the skin than they feel during the second stage when the needle is advanced into the bladder. After the needle has been introduced, a 20-mL syringe is used to aspirate 5 mL of urine for culture and 15 mL of urine for centrifugation and urinalysis. The obturator is reintroduced into the needle, and both needle and obturator are withdrawn. A small dressing is placed over the needle site in the skin. If urine is not obtained with complete introduction of the needle, the patient's bladder is not full and is usually deep within the retropubic area. When no urine is obtained on the first attempt, it is probably wise to wait until the bladder is full.

Urinalysis

For patients with urinary symptoms, microscopic urinalysis for bacteriuria, pyuria, and hematuria should be performed. Urinalysis provides rapid identification of bacteria and WBCs and presumptive diagnosis of UTI. Urinalysis is not reliable for detecting urinary tract infections in febrile infants when compared with urine cultures [9]. Usually, the sediment from an approximately 5- to 10-mL specimen obtained by centrifugation for 5 minutes at 2,000 rpm is analyzed. Microscopic bacteriuria is found in more than 90% of infections with counts of 10^5 CFU/mL of urine or greater and is a highly specific finding[10]. However, bacteria are usually not detectable microscopically with lower colony count infections ($10^2 - 10^4$/mL). This important error (*i.e.*, a false-negative result) occurs because of the limitation imposed by the microscope on the volume of urine that can be observed. If the volume of urine that can easily rest beneath a standard 22-mm cover glass is carefully measured (0.01 mL) and the number of high dry fields (×570 magnification) present beneath the cover glass is estimated, it is disturbing to find that one high dry field represents a volume of approximately 1/30,000 mL. There are excellent studies showing that the bacterial count must be approximately 30,000/mL before bacteria can be found in the sediment, stained or unstained, spun or unspun [11]. For these reasons, a negative urinalysis for bacteria never excludes the presence of bacteria in numbers of 30,000/mL and less.

A false-positive result is the reverse of the first error: bacteria are seen in the microscopic sediment but the urine culture shows no growth. The voided urine from a female patient can contain thousands of lactobacilli and corynebacteria. These bacteria are readily seen under the microscope; and although they are gram-positive, they often appear gram-negative (gramvariable) if stained. Strict anaerobes, usually gram-negative bacilli, also make up a significant mass of the normal vaginal flora [12].

In practice, these problems can be minimized by using other information provided by urinalysis that can help the clinician to decide whether a patient has a UTI [13]. The validation of the midstream urine specimen can be questioned if numerous squamous epithelial cells (indicative of preputial, vaginal, or urethral contaminants) are present.

Pyuria and hematuria are good indicators of an inflammatory response. Although the number of WBCs per high-power field (HPF) in a centrifuged urine sample is useful, it is important to remember that other factors can influence the number of cells seen which includes the state of hydration; the intensity of tissue reaction; the method of urine collection; the volume, speed, and time of centrifugation; and the volume in which the sediment is re-suspended.

Significant pyuria can be determined simply and reliably with a microscope by accurately examining the centrifuged sediment or by using a hemocytometer to count the number of WBCs in the unspun urine. 1 – 2 WBCs per HPF in sediment from a centrifuged specimen represents about 10 WBCs/mL in an unspun specimen. More than 2 WBCs per HPF in a centrifuged specimen or 10 WBCs/mL of urine correlates well with the presence of bacteriuria and is rarely seen in non-bacteriuric patients [14]. In clinical studies, determination of pyuria in voided urine specimens has a reported sensitivity of 80% to 95% and a specificity of 50% – 76% for UTI (depending on the definition of infection, the patient population, and the method used to evaluate for pyuria) [15].

The absence of pyuria could cause the diagnosis of UTI to be questioned until urine culture data are available. In contrast, many diseases of the urinary tract produce significant pyuria in the absence of bacteriuria. Whereas tuberculosis is the well-recognized example of abacterial pyuria, staghorn calculi and stones of smaller size can produce significant pyuria with clumps of WBCs in the absence of UTI. Almost any injury to the urinary tract, from chlamydial urethritis to glomerulonephritis and interstitial cystitis, can elicit large numbers of fresh polymorphonuclear leukocytes (glitter cells). Depending on the stage of hydration, the intensity of the tissue reaction producing the cells, and the method of urine collection, any number of WBCs can be seen in the microscopic sediment in the presence of an uninfected urinary tract.

Microscopic hematuria is found in 40% – 60% of cases of cystitis and is uncommon in other dysuric syndromes [15]. Thus, microscopic bacteriuria and hematuria lack sensitivity but are highly specific for UTIs.

Wang analyzed the Sysmex UF-1000i (TOA Medical Electronics, Kobe, Japan) for accuracy in identifying RBCs and WBCs, casts, bacteria, and epithelia and evaluated its precision, linear estimation of results, carryover contamination rate, and anti-interference. UF-1000i agreement with manual counting was approximately 95% for RBCs and WBCs, epithelia, and casts. Its coefficiance of variation for bacteria ranged from 4.7% to 15.2%. UF-1000i screening for UTIs exhibited great sensitivity (97%), specificity (79%), positive predictive value (70%), negative predictive value (99%), and accuracy (85%) [16]. The Sysmex UF-1000i shows great promise in excluding more than 50% of true-negative samples, improving detection efficiency, and reducing laboratory costs.

Rapid Screen Methods

Biochemical and enzymatic tests have been devised to detect bacteriuria and pyuria [17]. The Griess test detects the presence of nitrite in urine that is formed when bacteria reduce the nitrate normally present in urine. Tests for detecting pyuria by determining leukocyte esterase activity have also been developed [18]. In a study comparing traditional urine culture with these indirect tests, the combination of nitrite and leukocyte esterase tests (either test positive) had a sensitivity of 71% and a specificity of 83% when compared with 10^3 CFU/mL or greater of urine cultures [19]. However, several investigators [20, 21] noted substantial variability in the sensitivity and specificity results, which could be markedly influenced by the types of patients and infections chosen to evaluate the tests. This concept of spectrum bias was illustrated by a study that reported differences in the sensitivity of reagent strip testing, ranging from 56% to 92%, by changing only the groups of patients included in the analysis. Although false-positive results are relatively uncommon, the borderline sensitivity of these tests, especially among patients with less characteristic symptoms of UTIs, does not allow these inexpensive tests to replace careful microscopic urinalysis in symptomatic patients [22]. Their main role is in screening asymptomatic patients.

Bacteriuria and pyuria, the hallmarks of UTI, are not predictive of renal infection. In contrast, patients with significant renal infection may have sterile urine if the ureter draining the kidney is obstructed or the infection is outside of the collecting system.

Urinary Pathogens

Urine Culture. Two techniques for urine culture are available. Direct surface plating of a known amount of urine on split-agar disposable plates is the traditional quantitative culture technique used by most microbiology laboratories. One half of the plate is blood agar, which grows both gram-positive and gram-negative bacteria, and the other is desoxycholate or eosin-methylene blue (EMB), which grows gram-negative bacteria (some of them, such as *E. coli*, in a very characteristic manner). Simple curved-tip eye droppers are sufficient to deliver about 0.1 mL of urine onto each half of the plate. After overnight incubation, the number of colonies is estimated, often identified (after some experience), and multiplied by 10 to report the number of cfu per milliliter of urine.

A simpler but somewhat less accurate technique is the use of dip slides. These inexpensive plastic slides are attached to screw-top caps; they have soy agar (a general nutrient agar to grow all bacteria) on one side and EMB or MacConkey's agar for gram-negative bacteria on the opposite side. A slide is dipped into urine, the excess is allowed to drain off, and the slide is replaced in its plastic bottle and incubated. The volume of urine that attaches to the slide is 1/100 – 1/200 mL. Hence, the colony count is 100 – 200 times the number of colonies that become visible with incubation. In actual practice, the growth is compared with a visual standard and reported as such. The species of bacteria is more difficult to recognize when this technique is used, but the technique is completely adequate.

It is emphasized that the urine must be refrigerated immediately on collection and should be cultured within 24 h of refrigeration. One advantage to the dip slide is the ease with which the urine can be immediately cultured without the necessity of refrigeration. Patients can culture their own urine at home, keep the slide at room temperature, and bring it to the office within 48 h.

Although most bacteria allowed to incubate for several hours in bladder urine reach cfu counts of 10^5/mL, this statistical number is fraught with two limitations.

The first is that 20% – 40% of women with symptomatic UTIs present with bacteria counts of 10^2 – 10^4 CFU/mL of urine [23-26], probably because of the slow doubling time of bacteria in urine (every 30 – 45 min) combined with frequent bladder emptying (every 15 – 30 min) from irritation. Thus, in dysuric patients, an appropriate threshold value for defining significant bacteriuria is 10^2 CFU/mL of a known pathogen [27]. Fortunately, most of these patients have symptoms of UTI and most have pyuria on urinalysis.

The second limitation of the 10^5 cutoff is over-diagnosis. Women susceptible to infection often carry large numbers of pathogenic bacteria on the perineum that contaminate otherwise sterile bladder urine. Uncircumcised men may harbor uropathogenic bacteria on their foreskins. A single culture of 10^5 cfu/mL or more had a 20% chance of representing contamination. There is no statistical way to avoid these two major limitations on the interpretation of the midstream voided culture in women and in uncircumcised men without careful preparation.

Common Organisms. *E. coli* is by far the most common cause of UTIs, accounting for 85% of community-acquired and 50% of hospital-acquired infections, wherase *Staphylococcus saprophyticus* is the second prevalent microorganism in UTI, accounting for 15% in acute community-acquired uncomplicated infections [28]. Other gram-negative *Enterobacter*iaceae, including Proteus and *Klebsiella*, and grampositive *E. faecalis* and *S. saprophyticus* are responsible for the remainder of most community-acquired infections. Prevalence of group b streptococcus is high among sexually active college students and sexual contact is associated with increased transmission [29]. Nosocomial infections are caused by *E. coli*, *Klebsiella*, *Enterobacter*, *Citrobacter*, *Serratia*, *Pseudomonas aeruginosa*, *Providencia*, *E. faecalis*, and *S. epidermidis* [30]. Less common organisms such as *Gardnerella vaginalis*, *Mycoplasma species*, and *Ureaplasma urealyticum* may infect patients with intermittent or indwelling catheters [31]. Women with a first UTI caused by *E. coli* are more likely than those with a non-*E. coli* first UTI to have a second UTI within 6 mon [32].

The prevalence of infecting organisms is influenced by the patient's age. For example, *S. saprophyticus* is now recognized as causing approximately 10% of symptomatic lower UTIs in young, sexually active females [33] whereas it rarely causes infection in males and elderly individuals. A seasonal variation with a late summer to fall peak has been reported [34]. Urinary tract infection caused by multi-resistant extended-spectrum beta-lactamase producing bacteria (ESBL) is a common complication among kidney transplant patients [35].

Fastidious Organisms. *Anaerobes in the Urinary Tract.* Most UTIs are caused by facultative anaerobes usually originating from the bowel flora. Uropathogens such as *S. taphylococcus epidermidis* and Candida albicans originate from the flora of the vagina or perineal skin.

Although symptomatic anaerobic infections of the urinary tract are documented, they are uncommon. However, the distal urethra, perineum, and vagina are normally colonized by anaerobes. Whereas 1% – 10% of voided urine specimens are positive for anaerobic organisms, anaerobic organisms found in supra-pubic aspirates are much more unusual [36]. Clinically symptomatic UTIs in which only anaerobic organisms are cultured are rare, but these organisms must be suspected when a patient with irritative bladder symptoms has cocci or gram-negative rods seen on microscopic examination of the centrifuged urine (catheterized, suprapubic aspirated, or voided midstream urine) and routine quantitative aerobic cultures fail to grow organisms [37].

Anaerobic organisms are frequently found in suppurative infections of the genitourinary tract. In one study of suppurative genitourinary infections in males, 88% of scrotal, prostatic, and perinephric abscesses included anaerobes among the infecting organisms [38]. The organisms found are usually Bacteroides species, including B. fragilis, Fusobacterium species, anaerobic cocci, and Clostridium perfringens [39]. The growth of clostridia may be associated with cystitis emphysematosa [40].

Mycobacterium tuberculosis and Other Non-Tuberculous Mycobacteria. Mycobacterium tuberculosis and other non-tuberculous mycobacteria may be found when cultures for acid-fast bacteria are requested; they do not grow under routine aerobic conditions and may be found during evaluation for sterile pyuria. It has been emphasized that the mere presence of mycobacteria may not indicate tissue infection. Therefore, other factors such as symptoms, endoscopic or radiologic evidence of infection, abnormal urine sediment, the absence of other pathogens, repeated demonstration of the organism, and the presence of granulomas should be considered before therapy is instituted [41, 42].

Chlamydia. Chlamydiae are not routinely grown in aerobic culture but have been implicated in genitourinary infections.

Tissue and Stone Cultures. It is clinically useful to culture stones removed from the urinary tract to document that bacteria reside within their interstices. Tissue cultures are primarily useful for research information.

Using sterile technique at the operating table, the surgeon places the stone or fragment of tissue into a sterile culture tube containing 5 mL of saline solution; the culture is packed in ice and sent to the bacteriologic laboratory, where, after agitation of the stone or tissue in the 5 mL of saline solution, 0.1 mL is surface-streaked on both blood agar and EMB agar. The saline solution is then poured off the specimen; and, with sterile forceps, the stone or tissue is transferred to a second 5 mL of sterile saline solution. After agitation to ensure a reasonable washing action, the saline solution is decanted again and the specimen is transferred to a third 5 mL of saline solution and finally to a fourth 5 mL of saline solution. This last saline solution wash is cultured quantitatively in the same manner as the first. The remainder of this fourth 5 mL of saline solution is poured with the stone into a sterile mortar and pestle dish.

After the stone is crushed (or the tissue is ground in a tissue blender) in the fourth saline solution wash, 0.1 mL is again cultured on both blood agar and EMB agar. The difference in colony counts between the first and the fourth saline solution washes represents the washing effect of the saline solution transfers on the surface bacteria of the stone or tissue. The difference between the fourth saline wash before and after crushing (or grinding, for tissue) represents the difference between surface bacteria and bacteria within the specimen.

14.3.1.3 Imaging Studies

Imaging studies are not required in most cases of UUTI because clinical and laboratory findings alone are sufficient for correct diagnosis and adequate management of most patients. However, infection in most men or a compromised host, febrile infections, signs or symptoms of urinary tract obstruction, failure to respond to appropriate therapy, and a pattern of recurrent infections suggesting bacterial persistence within the urinary tract warrant imaging for identification of underlying abnormalities that require modification of medical management or percutaneous or surgical intervention. In high-risk patients, including women with febrile infections and most men, radiologic studies may determine acute infectious processes that require further intervention or may ascertain the cause of complicated infections. A UUTI associated with possible urinary tract obstruction must be evaluated. These are patients with calculi, especially infection (struvite) stones; ureteral tumors; ureteral strictures; congenital obstructions; or previous genitourinary surgery, such as ureteral reimplantation or urinary diversion procedures, that may have caused obstruction. Patients with diabetes mellitus can develop special complications from UUTIs; they may acquire emphysematous pyelonephritis or papillary necrosis. Impacted necrotic papillae may cause acute ureteral obstruction. Patients with polycystic kidney disease who are on dialysis are particularly prone to developing perinephric abscesses. Urologic imaging is indicated in patients whose symptoms of acute clinical pyelonephritis persist after 5 – 6 days of appropriate antimicrobial therapy; they often have perinephric or renal abscesses. In addition, patients with unusual organisms, including urea-splitting organisms (*e.g.*, Proteus species), should be examined for abnormalities within the urinary tract, such as obstructing stones, strictures, or fungus balls. The second reason for radiologic evaluation is to diagnose a focus of bacterial persistence. In patients whose bacteriuria fails to resolve after appropriate antimicrobial therapy or who have rapid recurrence of infection, abnormalities that cause bacterial persistence should be sought. Although these patients are uncommon, it is important to identify them because they may have correctable urologic abnormalities that represent the only surgically curable causes of recurrent UUTIs.

14.3.1.4 Plain Film of the Abdomen

The plain film of the abdomen (kidney, ureter, and bladder) is useful for the rapid detection of radiopaque calculi and unusual gas patterns in emphysematous pyelonephritis. It may show abnormalities, such as an absent psoas or abnormal renal contour, that suggest perirenal or renal abscess, but these findings are nonspecific.

14.3.1.5 Excretory Urogram

The excretory urogram has been a roUUTIne examination to evaluate patients with complicated infection problems but is not required in uncomplicated infections. The radiologic features of acute clinical pyelonephritis are discussed in a later section on that subject. The excretory urography study is useful to determine the exact site and extent of urinary tract obstruction; however, it is not the best screening test for hydronephrosis, pyonephrosis, or renal abscess.

14.3.1.6 Voiding cystourethrography

Voiding cystourethrography (VCUG) is not recommended in most cases of UTI, but in infants, it is better to perform VCUG as soon as possible, provided the inflammation has subsided [43].In contrast, Muga concluded that it is unnecessary to carry out VCUG in an infant with an initial UTI [44].

14.3.1.7 Ultrasonography

The renal ultrasound study is an important renal imaging technique because it is noninvasive [45], easy to perform, and rapid, and offers no radiation or contrast agent risk to the patient. It is particularly useful in identifying calculi and hydronephrosis, pyonephrosis, and perirenal abscesses. A single radiograph for calculi should accompany ultrasonography. Ultrasonography is also useful for diagnosing post-void residual urine and intra renal vascular resistivity is significantly increased in recurrent UTI patients particularly in those sustaining renal scarring [46]. A disadvantage is that the study is dependent on the interpretative and performance skills of the examiner. Furthermore, the study may be technically poor in patients who are obese or who have dressings, drainage tubes, or open wounds overlying the area of interest.

14.3.1.8 Computed Tomography and Magnetic Resonance Imaging

The radiologic modalities that offer the best anatomic detail are Computed Tomography (CT) and Magnetic Resonance Imaging (MRI). They are more sensitive than excretory urography or ultrasonography in the diagnosis of acute focal bacterial nephritis, renal and perirenal abscesses, and radiolucent calculi [47-50]. When used to localize renal and perirenal abscesses, CT improves the approach to surgical drainage and permits percutaneous approaches. MRI has not superseded CT in the evaluation of renal inflammation, but it has provided some advantages in delineating extra-renal extension of inflammation.

14.3.1.9 Radionuclide Studies

Hippuran ^{131}I and technetium ^{99m}Tc glucoheptonate scans are used to detect focal parenchymal damage, renal function impairment, and decreased renal perfusion in acute renal infections [51]. Although gallium-67 scanning has been reported to be useful in the diagnosis of pyelonephritis and renal abscess, it is uncommonly required and may be positive in non-infectious entities. Indium-111-labeled WBC studies have limited efficacy in establishing the presence of an inflammatory focus, particularly when the patient's clinical presentation does not suggest an infectious process.

14.3.1.10 DMSA Renal Scintigram

In children younger than 2 years of age, DMSA renal scintigram should be recommended to localiz the site of urinary infection [52, 53].

14.3.2 Lower Urinaru Tract Infections (LUTI)

There are Uncomplicated Cystitis, asymptomatic bacteriuria, complicated cystitis, and unresolved UTIs.

14.3.2.1 Uncomplicated Cystitis

Most cases of uncomplicated cystitis occur in women. Each year, approximately 10% of women report having a UTI and more than 50% of all women have at least one such infection in their lifetime [54]. Uncomplicated cystitis occasionally occurs in pre-pubertal girls, but its incidence increases greatlyin late adolescence and during the second and fourth decades of life. 25% – 30% of women 20 – 40 years of age have a history of UTIs. Although it is much less common, young men may also experience acute cystitis without underlying structural or functional abnormalities of the urinary tract [55]. Risk factors include sexual intercourse and use of spermicides [56-58]. Sexual transmission of uropathogens has been suggested by demonstrating identical *E. coli* in the bowel and urinary flora of sex partners [59].

Clinical Presentation

The presenting symptoms of cystitis are variable and usually include dysuria, frequency or urgency, and supra-pubic pain. Hematuria or foul-smelling urine may develop. The probability of cystitis in a woman with these symptoms alone or in combination is 50% – 90%, respectively [1]. When a woman who previously has had cystitis has symptoms suggesting a recurrence, the probability that an infection is present is about 90% [60]. Because acute cystitis, by definition, is a superficial infection of bladder mucosa, fever, chills, and other signs of dissemination are not present. Some patients may experience supra-pubic tenderness, but most have no diagnostic physical findings. In women, physical examination should include the possibility of vaginitis, herpes, and urethral pathology, such as a diverticulum.

A remarkably narrow spectrum of etiologic agents with highly predictable profiles of antimicrobial susceptibility lead to infections in young women with acute uncomplicated cystitis. *E. coli* is the causative organism in 75% – 90% of cases of acute cystitis in young women [28, 33]. *S. saprophyticus,* a commensal organism of the skin, is the second most common cause of acute cystitis in young women, accounting for 10% – 20% of these infections [61]. Other organisms less commonly involved include *Klebsiella* and Proteus species and Enterococcus. In men, *E. coli* and other *Enterobacter*iaceae are the most commonly identified organisms.

Laboratory Diagnosis

Urine analysis. The presumptive laboratory diagnosis of acute cystitis is based on microscopic urinalysis, which indicates microscopic pyuria, bacteriuria, and hematuria. The presence of pyuria has a sensitivity of 95% and a specificity of 70%. The presence of bacteria is less sensitive but more specific (40% – 70% and 85% – 95%, respectively, depending on the number of bacteria observed) [62]. Indirect dipstick tests for bacteria (nitrite) or pyuria (leukocyte esterase) may also be informative and more convenient but are less sensitive than microscopic examination of the urine. Dipsticks are most accurate when the presence of either nitrite or leukocyte esterase is considered as a positive result (sensitivity of 75% and specificity of 82%, respectively) [21]. But Khasriya pointed out that the inadequacy of urinary dipstick and microscopy as surrogate markers of urinary tract infection in urological outpatients with lower urinary tract symptoms without acute frequency and dysuria and should be abandoned [63].

Urine culture. Urine culture remains the definitive test; and in symptomatic patients, the presence of 10^2 CFU/mL or more of urine usually indicates infection [13].

However, routine urine cultures are often unnecessary. It is generally more cost-effective to manage many patients who have symptoms and urinalysis findings characteristic of uncomplicated cystitis without an initial urine culture because treatment decisions are usually made and therapy is often completed before culture results are known [64]. This position was supported by a cost-effectiveness study [65] in which it was estimated that the routine use of pretherapeutic urine cultures for UTI increases costs by 40% but decreases the overall duration of symptoms by only 10%.

Thus, in women with recent onset of symptoms and signs suggesting acute cystitis and in whom factors associated with upper tract or complicated infection are absent, a urinalysis that is positive for pyuria, bacteriuria, or hematuria, or a combination should provide sufficient documentation of LUTI and a urine culture may be omitted [66]. A urine culture should be obtained for patients in whom symptoms and urine examination findings leave the diagnosis of cystitis in doubt. Pretherapeutic cultures and susceptibility tests are also essential in the management of patients with recent antimicrobial therapy or LUTI. In such situations, various pathogens may be present and antimicrobial therapy is less predictable and must be tailored to the individual organism [67].

14.3.2.2 Asymptomatic Bacteriuria

Asymptomatic bacteriuria is a microbiologic diagnosis based on the isolation of a specified quantitative count of bacteria in a properly collected specimen of urine from a patient who is without symptoms or signs referable to UTI. In healthy individuals, the absence of symptoms is clear cut, but in, for example, catheterized or neurologically comprised patients, it may be difficult to discern whether the UTI is truly asymptomatic. For asymptomatic women, two consecutive voided urine specimens with isolation of the same bacterial strain in quantitative counts of 10^5 CFU/mL is consistent with asymptomatic bacteriuria. In men, a single clean-catch voided specimen with similar counts is adequate. A single catheterized urine specimen with a solitary isolate with a quantitative count of 10^2 CFU/mL identifies bacteriuria in women or men [68]. The prevalence of pyuria with asymptomatic bacteriuria ranges from approximately 30% in young women [69] to 100% in catheterized patients. In addition, many coexisting factors, such as stones, can incite inflammation in these patients and therefore the presence or absence of pyuria is not sufficient to diagnose bacteriuria, nor does it differentiate symptomatic from asymptomatic patients or provide indication for antimicrobial treatment [68]. Asymptomatic bacteriuria in young women is common but rarely persists. It is a strong predictor of subsequent symptomatic urinary tract infection [69].

The prevalence of asymptomatic bacteriuria varies widely and depends on age, sex, and the presence of other genitourinary abnormalities. *E. coli* is the most common isolate among patients with bacteriuria, and it contains fewer virulence characteristics than isolates from patients with symptomatic infections [70]. Other *Enterobacter*iaceae (*e.g.*, *P. mirabilis*) and gram-positive uropathogens, including group B streptococci and coagulase-negative staphylococci, become more prevalent in concert with increased underlying abnormalities. For patients who are institutionalized and/or with indwelling urologic devices, *P. aeruginosa*, Proteus, and other highly resistant organisms are more prevalent.

14.3.2.3 Complicated Cystitis

Complicated UTIs are those that occur in a patient with a compromised urinary tract or that are caused by a very resistant pathogen (Table 14.1). These complicating factors may be readily apparent from the severity of the presenting illness or the past medical history. However, they may not be obvious at first and may only become evident from subsequent failure of the patient to respond to appropriate therapy.

Table 14.1 Complicating host factors

Functional/structural abnormalities of urinary tract
Recent urinary tract instrumentation
Recent antimicrobial agent use
Diabetes mellitus
Immunosuppression
Pregnancy
Hospital-acquired infection

Due to the wide range of host conditions and pathogens and a lack of adequate controlled trials, guidelines for empirical therapy are limited. For patients with mild to moderate illness who can be treated with oral therapy, the fluoroquinolones provide a broad spectrum of activity with excellent urine and tissue levels and safety. If the susceptibility pattern of the pathogen is known, TMP-SMX may be effective.

14.3.2.4 Unresolved UTIs

Clinical Presentation

Unresolved infection indicates that initial therapy has been inadequate in eliminating symptoms and/or bacterial growth in the urinary tract. If the symptoms of UTI do not resolve by the end of treatment or if symptoms recur shortly after therapy, urinalysis and urine culture with susceptibility testing should be obtained. If the patient's symptoms are significant, empirical therapy with a fluoroquinolone is appropriate, pending results of the culture and susceptibility testing.

The causes of unresolved bacteriuria during antimicrobial therapy are shown in Table 14.2. Most commonly, the bacteria are resistant to the antimicrobial agent selected to treat the infection. Typically, the patient has received the antimicrobial therapy in the recent past and developed bowel colonization with resistant bacteria. β-Lactams, tetracycline, and sulfonamides are notorious for causing plasmid-mediated R factors that simultaneously develop resistance to multiple antimicrobial agents. The second most common cause is development of resistance in a previously susceptible population of bacteria during the course of treatment of UTIs. This problem occurs in approximately 5% of the patients receiving antimicrobial therapy. It is easy to recognize clinically because the culture on therapy shows that the previous susceptible population has been replaced by resistant bacteria of the same species. It can be shown that resistant organisms were actually present before contact with the initial antimicrobial agent, but they were present in such low numbers that it was impossible to detect by *in vitro* susceptibility studies before therapy. When the antimicrobial concentration in the urine is insufficient to kill all the bacteria present, more resistant forms will emerge. This is seen characteristically in patients who are under-dosed or who are poorly compliant and hence have inadequate dose regimens. The third cause is the presence of an unsuspected, second pathogen that was present initially and is resistant to the antimicrobial therapy chosen. Treatment of the dominant organism unmasks the presence of the second strain. The fourth cause is rapid reintroduction of a new resistant species when the patient is undergoing initial therapy. Rapid re-infection that mimics unresolved bacteriuria should alert the clinician to the possibility of an enterovesical fistula.

Table 14.2 Causes of unresolved bacteriuria, in descending order of importance

Bacterial resistance to the drug selected for treatment
Development of resistance from initially susceptible bacteria
Bacteriuria caused by two different bacterial species with mutually exclusive susceptibilities
Rapid reinfection with a new, resistant species during initial therapy for the original susceptible organism
Azotemia
Papillary necrosis from analgesic abuse
Giant staghorn calculi in which the "critical mass" of susceptible bacteria is too great for antimicrobial inhibition
Self-inflicted infections or deception in taking antimicrobial drugs (a variant of Munchausen's syndrome)

If the culture obtained on therapy shows that the initial species is still present and susceptible to the antimicrobial chosen to treat the infection, the unresolved infection must be caused by either an inability to deliver an adequate concentration of antimicrobial agents into the urinary tract, or an excessive number of bacteria that 'override' the antimicrobial activity. In patients with azotemia, a determination of urinary antimicrobial concentrations usually shows that the level of the drug is below the minimal inhibitory concentration of the infecting organism.

In patients with papillary necrosis, severe defects in the medullary concentrating ability dilute the antimicrobial agent. A large mass of bacteria within the urinary tract is most commonly associated with a giant staghorn calculus. Even though adequate urinary levels of bactericidal drugs are present, the concentration is inadequate to sterilize the urine. It occurs because even susceptible bacteria cannot be inhibited once they reach a certain critical density, particularly if attached to a foreign body.

The last cause of unresolved bacteriuria occurs in those patients who have variants of Munchausen's syndrome. These patients secretly inoculate their bladders with uropathogens or omit their oral antimicrobial agents while steadfastly asserting that they never miss a dose. The patient with Munchausen's syndrome presents a horrendous clinical history and invariably a normal collecting system on excretory urography. Careful bacteriologic observations usually indicate the implausibility of the clinical picture.

Laboratory Diagnosis

Urinalysis and urine culture are mandatory to determine the cause of unresolved bacteriuria. The first four causes that are associated with resistant bacteria require no further evaluation. However, if reculture shows that the bacteria is sensitive to the antimicrobial agent the patient is taking, renal function and radiologic evaluation should be performed to identify renal or urinary tract abnormalities.

14.3.3 Factors Increasing Morbidity and/or Mortality

UTI is the most common infection experienced by humans after respiratory and GI infections. A variety of risk factors may be associated with the increasing morbidity or mortality of genitourinary infection. Associations have been established between UTI and age, pregnancy, sexual intercourse, use of diaphragm and a spermicide, delayed post-coital micturition, menopause and a history of recent UTI.

14.3.3.1 Elder

In 1982, the United Nations World Assembly on Ageing defined the "elderly" as persons aged 60 years and over. These people are especially prone to a variety of infectious diseases. Several studies have shown that UTI is more prevalent in older people. The study showed a 17% incidence of bacteriuria in 309 admissions of all ages, while over the age the incidence increased to 20% in males and 30% in females [71].

14.3.3.2 Catheterization

Urinary diversion or catheterization is necessary for some cases. It is evident that the main risk factor for catheter-related UTI is urinary catheter placement and its duration. The patients with catheterization for longer than 2 days were twice as likely to develop UTI than patients with catheterization of 2 days or less [72, 73].

14.3.3.3 Renal Transplantation

The incidence of UTI among kidney allograft recipients varies from 6% to 86%. UTIs are indeed common among renal transplant recipients with urinary reconstruction, with reported incidences of asymptomatic bacteriuria of 83%, of 5% – 65% symptomatic UTI, and of 43% pyelonephritis [74-77]. These events may potentially threaten the recipient and graft survival, even leading to a high mortality.

14.3.3.4 Pregnancy

UTI are the most common bacterial infections during pregnancy and these infections ifuntreated can be associated with serious obstetric complications. Females experience several alterations including emotional and physical as well as physiological, which make them more vulnerable to UTI. Besides the mentioned alterations, an increase in urine pH because of the reduction of the kidney's ability to concentrate urine and in glucose and amino acids excretion, increase the susceptibility of a pregnant women's urinary tract to infections, providing an appropriate bacterial growing media [78].

14.3.3.5 Sexual Intercourse

During sexual intercourse, normal bacterial flora may enter the urinary tract *via* ascent through the urethra from the bowel, vagina, or perineum. Sexual intercourse in the setting of periurethral colonization appears to be the most crucial trigger event for UTI among young women.

14.3.3.6 Urinary Tract Obstruction

Dysfunctional voiding causes functional urinary tract obstruction which increases intravesical pressure during storage and voiding, and contributes post-voiding residuals. When intravesical pressure exceed the ureteric opening pressure the vesicoureteral reflux emerges and UTI develops because of urinary stasis. Ureteropelvic junction obstruction and obstructive megaureter are common causes of upper urinary tract obstruction, which result in the hydronephrosis, increase in intrapelvic pressure, slow flow in renal circulation and decreasing resistance to pathogens.

Additionally, chronic comorbid conditions that alter immune function (chronic renal failure, diabetes mellitus, HIV) also contributed to the increasing genitourinary infection.

In intensive care unit, complicated nosocomial UTIs may lead to urosepsis, and increase patient morbidity and mortality [78]. The mortality rate due to nosocomial UTI was significantly higher in cases with bloodstream infection.

14.4 Prostatitis and Related Disorders

Prostatitis is a common syndrome that usually presents itself with voiding symptoms (irritative or obstructive) and pain (genitourinary, pelvic, or rectal) and is sometimes associated with sexual dysfunction (*e.g.*, ejaculatory discomfort and hematospermia). Characteristic features include a high prevalence, substantially impaired quality of life, and frequent recurrences [79]. Although some cases are clearly infectious, most men who receive a diagnosis of prostatitis have no evidence of a genitourinary bacterial infection and the cause is usually unknown [2, 80]. Disagreement persists over how to define prostatitis, including debates over the relative importance of various clinical, microbiological, and histopathological findings [3, 81]. Advances in the past decade, however, have spurred better-designed clinical trials and generated more robust evidence regarding treatment.

One major change was the development of a National Institutes of Health (NIH) consensus definition and classification system [4, 5, 82]. This scheme, although limited by the lack of a reliable comparison standard, clarified that a small minority of men with prostatitis have bacterial infection (*i.e.*, acute bacterial prostatitis) [ABP; category I] or chronic bacterial prostatitis [CBP; category II]) [6, 83]. The rest have nonbacterial prostatitis. If symptomatic, they have chronic prostatitis/chronic pelvic pain syndrome (CP/CPPS) — either IIIA, which is an inflammatory condition defined by leukocytes in the semen or postprostatic massage specimens, or IIIB, which is a non-inflammatory disorder. A new syndrome, asymptomatic inflammatory prostatitis (category IV), is defined by an abnormal semen analysis, elevated prostate-specific antigen (PSA), or incidental findings of prostatitis on examination of a biopsy specimen. The second advance was developing and validating an NIH-Chronic Prostatitis Symptom Index (NIH-CPSI) [7, 84]. This questionnaire scores disorders relating to pain, voiding, and quality of life. The maximum total score is 43, and a decrease of 4 – 6 points (or 25%) correlates with clinically significant improvement [8, 85]. The NIH-CPSI has proved to be useful for epidemiological studies and for assessment of patients over time [9, 86].

The greatest area of uncertainty in treating prostatitis concerns the approach to nonbacterial prostatitis. This review, however, will focus on treatment of bacterial prostatitis and will only briefly discuss nontreatment issues or nonbacterial disorders. Because of the familiarity of the prostatitis categories, we will generally refer to them by their classical (rather than NIH) designations. Our recommendations are derived from a comprehensive review of the literature and our combined clinical experience.

14.4.1 Epidemiology

Prostatitis is the most common urological diagnosis in men < 50 years of age and is the third most common diagnosis among those 50 years of age [10, 87]. Approximately 10% of men have chronic prostatitis-like symptoms; of these men, 60% have sought medical help [1, 11, 88]. The lifetime probability of a man receiving a diagnosis of prostatitis is > 25% [12, 13, 89], and prostatitis accounts for 25% of men's office visits for genitourinary complaints [14]. Reported rates of prostatitis are similar in North America, Europe, and Asia [15, 90]. In addition to discomfort, prostatitis syndromes are responsible for substantial physical and emotional distress [16, 17, 91] and financial costs [14, 92].

14.4.2 Pathophysiology

The prostate gland has several natural defenses against infection, including the production of antibacterial substances and mechanical flushing of the prostatic urethra by voiding and ejaculation [18, 93]. However, poor drainage of secretions from peripheral ducts or reflux of urine into prostatic tissue may lead to inflammation, fibrosis, or stones. Most bacterial prostatitis probably follows UTI, especially with uropathogens that demonstrate special virulence factors [19, 94]. Risk factors for developing prostate infection include urinary tract instrumentation, having a urethral stricture, or urethritis (usually due to sexually transmitted pathogens).

ABP, which accounts for < 1% of cases of prostatitis, is likely to be caused by infected urine ascending the urethra to intraprostatic ducts. The 10% of cases that follow genitourinary instrumentation generally occur in older patients, have a higher risk of recurrence or prostatic abscess, and are more often caused by non-*Escherichia coli* species [20, 95]. Despite antibiotic prophylaxis, 2% of men develop ABP after transrectal prostate biopsy, especially after repeat procedures [21, 96]. CBP complicates a minority of cases of ABP and often occurs without previous acute infection. The formation of either bacterial biofilm or prostatic calculi favors chronic, treatment-resistant infection [22, 97]. Histopathological findings in bacterial prostatitis are poorly defined, with infection primarily in the acinar rather than the interstitial spaces [22, 98] and primarily luminal rather than parenchymal.

14.4.3 Clinical Presentation and Diagnostic Evaluation

ABP typically presents itself abruptly with voiding symptoms and distressing but poorly localized pain and is often associated with systemic findings (*e.g.*, malaise and fever) [5]. Clinicians should enquire about urogenital disorders, recent genitourinary instrumentation, and new sexual contacts. Only 5% of men with ABP develop CBP, and 2% develop a prostatic abscess. CBP usually presents with more-prolonged (3 months) urogenital symptoms. The hallmark is relapsing UTI (*i.e.*, UTIs due to the same organism), but < 50% of patients with CBP have this history [23, 98]. Between symptomatic UTIs, patients may be asymptomatic, despite ongoing prostatic infection.

Physical examination should include obtaining vital signs and examining the lower abdomen (seeking a distended bladder), back (seeking costovertebral-angle tenderness), genitalia, and rectum. Digital prostate palpation in ABP can cause discomfort and can potentially induce bacteremia, but it is safe if done gently. In ABP, the gland is typically tender, swollen, and warm, whereas in CBP, there may be some tenderness, softening ('boggyness'), firm induration, or nodularity.

Few laboratory tests are diagnostically useful in evaluating possible prostatitis. Any patient at risk should be screened for sexually transmitted infections. All patients with possible prostatitis need a urinalysis and urine culture. Urine dipstick testing (for nitrites and leukocytes) in ABP has a positive predictive value of ~95%, but a negative predictive value of only ~70% [24, 99]. Blood cultures and a complete blood count are useful in ABP. For patients with possible CBP, the 4-glass test is considered as the diagnostic criterion standard. Diagnosis is based on finding substantially lower leukocyte and bacterial counts in voided bladder urine specimens from the urethra (VB_1) and bladder (VB_2), compared with counts in post-prostatic massage voided urine (VB_3) or expressed prostatic secretions (EPS). Adding a culture of ejaculated semen improves the diagnostic utility of the 4-glass test [25, 26, 100], but semen cultures are positive more often than are cultures of VB_3 or EPS in men with nonbacterial prostatitis [27, 101]. The 4-glass test is cumbersome, inadequately validated, and rarely performed, even by urologists [28, 29, 102]. It may be diagnostically helpful on first presentation, but its value is limited in previously treated patients with chronic symptoms. A simpler 2-glass test (comparing pre- with post-prostatic massage urine specimens) provides similar results [30, 103]. Leukocyte counts in expressed prostatic secretions do not correlate with the severity of symptoms in men with CP/CPPS [31, 104].

Evaluating patients with chronic prostatitis should always include administering the NIH-CPSI and perhaps measuring urinary flow rate and post-void residual urine; only selected patients need further urodynamic or imaging studies [32, 105]. Culturing prostatic tissue obtained by biopsy is neither sensitive (because infection is focal) nor specific (because ~25% of prostatectomy specimens are culture positive) [33, 106]. PSA levels are elevated in ~60% of men with ABP, 20% of men with CBP, and 10% of men with nonbacterial prostatitis [34, 107]; a decrease after antibiotic therapy (which occurs in ~40% of patients) correlates with clinical and microbiological improvement [35, 108]. Various imaging studies can detect a suspected prostatic abscess.

14.4.4 Causative Pathogens in Prostatitis

Aerobic gram-negative bacilli are the predominant pathogens in bacterial prostatitis. *E. coli* cause 50% – 80% of cases; other pathogens include *Enterobacter*iaceae (*e.g.*, *Klebsiella* and Proteus, which account for 10% – 30% of cases), Enterococcus species (5% – 10% of cases), and nonfermenting gram-negative bacilli (*e.g.*, Pseudomonas species; < 5% of cases). Some debate the role of gram-positive organisms other than enterococci [36, 37, 109], but most of them accept Staphylococcus and Streptococcus species as pathogens [37–39, 110]. The increasing prevalence of gram-positive pathogens may represent changing disease epidemiology (perhaps related to fluoroquinolone therapy) or acceptance of their pathogenicity by health care providers. Limited data suggested that obligate anaerobes may rarely result in chronic prostatitis [40, 111].

Some cases of prostatitis are caused by atypical pathogens [34]. A large prospective study of men with chronic prostatitis found that 74% had an infectious etiology; the most common isolates were Chlamydia trachomatis (37% of cases) and Trichomonas vaginalis (11%), whereas 5% of patients had infection due to

Ureaplasma urealyticum [41]. Classical bacterial uropathogens were found in 20% of patients, and more patients with these pathogens, compared with patients with nonbacterial pathogens, had prostatic specimens with leukocytes [41]. Other possible prostatitis pathogens include *Mycoplasma genitalium*, *Neisseria gonorrhoeae*, *Mycobacterium tuberculosis*, various fungi, and several viruses [34].

14.4.5 Treatment of Bacterial Prostatitis

The approach to treating bacterial infection of the prostate largely depends on appropriately selected antibotics. The best approach to treating nonbacterial prostatitis (NIH categories Ⅲ and Ⅳ) is less clear.

Overview of antibiotic therapy. Treatment of bacterial prostatitis is hampered by the lack of an active antibiotic transport mechanism and the relatively poor penetration of most antibiotics into infected prostate tissue and fluids. Most antibiotics are either weak acids or bases that ionize in biological fluids, which inhibits their crossing prostatic epithelium [23]. Only free, non-protein-bound antibiotic molecules enter tissues. Ordinarily, substances with molecular weights of < 1,000 pass through openings (fenestrae) between capillary endothelial cells, but prostate capillaries are nonporous. Passage of a drug through prostatic capillary endothelium and prostatic epithelium is enhanced by a high concentration gradient, high lipid solubility, low degree of ionization, high dissociation constant (pK_a; allowing diffusion of the unionized component into the prostate), low protein binding, and small molecular size [42]. A pH gradient allows electrically neutral molecules to pass through membranes, become ionized, and be trapped. Although ion trapping may increase prostatic drug concentration, the charged fraction has an unclear antimicrobial role. Fluoroquinolones are zwitterions that have a different pK_a in an acidic versus an alkaline milieu, allowing concentrations in the prostate to be 10% – 50% of concentrations in serum [43].

Normal human prostatic fluid has a pH of 7.3; in individuals with CBP, the prostatic fluid may become markedly alkaline (mean pH, 8.34) [44]. Many early studies of prostatic antibiotic penetration used dogs, which generally have acidic prostatic fluid. Human studies have mostly used adenoma tissue derived from prostate resection. These uninfected samples of mixed tissues and fluids with varied pH levels generally have antibiotic concentrations that exceed those in plasma. In humans, alkaline drugs (*e.g.*, trimethoprim and clindamycin) undergo ion trapping, which leads to high prostatic concentrations. Acidic drugs, such as β-lactams, achieve lower levels, but more drug is in the active unionized state.

Fluoroquinolones have emerged as the preferred antibiotics for treating bacterial prostatitis, and several have been approved by the US Food and Drug Administration (FDA) for this indication. Compared with concentrations in plasma, drug levels are generally higher in urine, similar in seminal fluid and prostatic tissue, and lower (albeit therapeutic) in prostatic fluid [43, 44]. One concern with these agents is the growing problem of fluoroquinolone resistance, which generally requires treatment with a third-generation cephalosporin (*e.g.*, ceftazidime or ceftriaxone) or a carbapenem (*e.g.*, imipenem or ertapenem) [45]. Table 14.2 provides information on other antibiotics that may be useful for treating bacterial prostatitis, based on pharmacodynamic data, case reports, or FDA approval for treating UTIs.

Although penicillin G achieves poor prostatic concentrations, piperacillin has good levels and has been successfully used to treat CBP. Cephalosporins, despite being weak acids with low lipid solubility, can attain therapeutic levels in prostatic fluid or tissue. Aztreonam, imipenem, and some aminoglycosides can attain levels in prostatic tissue that exceed the minimum inhibitory concentrations of most *Enterobacter*iaceae. Prostatic concentrations of minocycline and doxycycline are at least 40% of the corresponding serum concentrations. Erythromycin — and probably other macrolides, as well — can develop high prostate concentrations. Clindamycin and trimethoprim readily enter prostatic fluid, and levels of these drugs in prostatic fluid may exceed levels in plasma. The prostatic concentration of sulfamethoxazole is much lower, raising doubts that it synergizes with trimethoprim. Nitrofurantoin prostatic levels are likely non-therapeutic.

Antibiotic therapy for ABP. For systemically ill patients with ABP, parenteral antibiotic therapy is preferable, at least initially. Most antibiotic agents penetrate the acutely inflamed prostate, but experience favors empirical treatment with a broad-spectrum beta-lactam drug — either a penicillin (*e.g.*, piperacillin-tazobactam) or a cephalosporin (*e.g.*, cefotaxime or ceftazidime) — perhaps combined with an aminoglycoside for patients who are severely ill or who have recently received antibiotic therapy. Clinicians should consider local drug-resistance patterns in choosing antibiotics, especially with the emergence of extended-spectrum beta-lactamase-producing strains in complicated UTI [21], and should adjust therapy on the basis of culture results. Clinically stable patients may be treated with oral therapy (usually a fluoroquinolone). Duration of therapy for ABP is usually 2 weeks, although it can be continued for up to 4 weeks for severe illness or treatment of patients with concomitant bacteremia.

Two recent studies provided insights on treating ABP. A multicenter retrospective survey revealed that community-acquired infections were 3 times more common than nosocomial infections; *E. coli* remained the predominant pathogen, but nosocomial infections were more often caused by *Pseudomonas aeruginosa*, *enterococci*, or *Staphylococcus aureus*, and these organisms were associated with higher microbiological and clinical failure rates [46]. A similar study found a high rate of ciprofloxacin-resistant pathogens and that nosocomial acquisition or prior instrumentation were associated with increased antibiotic resistance and higher rates of clinical failure [47]. Ancillary measures for ABP include ensuring adequate fluid intake and urinary drainage.

Antibiotic therapy for CBP (category II) or inflammatory nonbacterial (category IIIA) prostatitis. CBP should be treated with 4 – 6 weeks of antibiotics. When persistent infection is caused by infected prostate stones or other types of genitourinary pathology, patients who have shown some response may benefit from more-prolonged antibiotic therapy [48]. In contrast with treatment of ABP, treatment of CBP can usually be delayed until culture and susceptibility results are available. Fluoroquinolones are the preferred drugs, except when resistance to these agents is confirmed or strongly suspected. The overall rates of clinical and microbiological response for CBP treated with fluoroquinolones are 70% – 90% at the end of therapy, but only ~60% after 6 mon [38]. Clinical and microbiological response rates are similar in those whose prostatic specimens grow either well-accepted uropathogens or coagulase-negative *Staphylococcus* or *Streptococcus* species [39]. Giving repeated courses of antibiotics is generally unwise. Surgically removing infected prostatic stones may help when other stategies fail. Some case reports suggest apparent benefit from direct injection of antimicrobials into the prostate, but the evidence is insufficient to recommend this approach. Long-term suppressive therapy with low doses of oral antibiotics (*e.g.*, trimethoprim-sufamethoxazole) may reduce symptomatic recurrences, but evidence is lacking.

Although < 10% of men who receive a diagnosis of prostatitis have a proven bacterial infection, approximately one-half are treated with antibiotic therapy [49]. Clinicians often treat nonbacterial prostatitis because of concern over missing infections that are due to pathogens that are difficult to culture, and because many apparently uninfected patients appear to respond to treatment. Most treatment studies have been poorly designed, but several, including randomized controlled trials, note improved symptoms in 50% of patients with CP/CPPS managed with a fluoroquinolone [50]. In one study, however, patients with CP/ CPPS who had received multiple prior treatments (including treatment with antimicrobials) had similar symptom response rates (20% – 30%) after 6 weeks of therapy with either fluoroquinolones or placebo [23]. In the subset of patients who had been symptomatic for a shorter duration and had not recently received antibiotics, the response rate was as high as 75% [23]. One prospective study involving men with CP/CPPS found that the percentage of patients who responded to antibiotic therapy was similar for those with and those without bacterial prostatitis [3]. This may be at least partly related to the fact that some antibiotics (*e.g.*, macrolides and tetracyclines) have direct anti-inflammatory effects.

There is no validated test of cure for bacterial prostatitis. If the patient's symptoms resolve after therapy, we would generally not treat asymptomatic bacteriuria, if it is present. If symptoms that are thought to be related to prostatitis persist, culture-directed antibiotic therapy with a more prolonged course, higher dosage, or different agent should be considered.

To interrogate the literature on the possible value of antibiotic therapy for chronic prostatitis (bacterial or presumed nonbacterial), we identified studies published in the previous decade that reported rates of either symptom improvement or microbiological eradication. All but 1 of the studies used an oral fluoroquinolone for treatment of at least some of the patients, and the duration of therapy was typically ~4 weeks; the comparator arms varied. In all 8 trials involving patients with CBP, the patients experienced significant symptomatic and microbiological improvement (usually defined by improved prostate symptom scores and infection eradication) with antibiotic therapy. Of the 5 trials that involved patients with CP/CPPS treated with antibiotics, 2 showed no advantage for fluoroquinolone therapy over placebo. Thus, these studies show clear benefit from fluoroquinolone therapy for CBP but not for CP/CPPS.

Older studies have shown that longer (6 weeks) duration of therapy with trimethoprim-sulfamethoxazole for probable CBP is more effective than a shorter duration of therapy. Outcomes in treating CBP with trimethoprim-sulfamethoxazole, however, are not as good as those with fluoroquinolones [51]. A single, limited (< 6 weeks) course of antibiotic therapy may be appropriate for some patients with CP/CPPS patients but repeated courses are not necessary.

Because antibiotics are unhelpful for most cases of nonbacterial prostatitis, many non-antibiotic agents and procedures have been recommended, most of which are inadequately studied. Recently published expert recommendations, based on data from prospectively designed, randomized, placebo-controlled trials that enrolled a well-defined population of men with CP/CPPS and employed the NIH-CPSI, offer some guidance [50]. Adding an alpha blocker to antibiotic therapy appears to improve symptomatic outcomes, especially for patients with newly diagnosed disease and patients who are alpha blocker naive [52], but there is no support for 5-α reductase inhibitor therapy. Anti-inflammatory drugs are rarely effective alone but may help some patients as part of multi-modal therapy. There is no definitive evidence of efficacy for most other conventional or alternative medications [52]. Few controlled trials support various non-pharmacological treatments, such as repetitive prostatic massage, physical therapy, acupuncture, biofeedback, or local heat [53]. In a well-designed systematic study, less than one-third of patients with CP/CPPS had even modest improvement during 1 year of follow-up [54]. Finally, no surgical procedure, whether minimally invasive or more extensive, has proven to be effective for treating prostatitis [53].

14.4.6 Conclusions

Considering the high prevalence of symptoms attributed to prostatitis and the many studies conducted during the past 50 years that have attempted to define its causes and optimal treatments, it is surprising how little we know about this syndrome. Although bacterial prostatitis constitutes a small minority of cases, we now have enough data on the causative pathogens and a better understanding of the most appropriate antimicrobial treatment regimens. Fluoroquinolones are currently the major weapon in our therapeutic arsenal, but growing resistance to these agents will require that we find other drugs that adequately penetrate the prostate (and are perhaps active in the presence of biofilm) to effectively treat CBP. Moving this 'stuck' field forward will require developing accurate diagnostic tests to differentiate bacterial prostatitis from nonbacterial syndromes and new antimicrobials that demonstrate efficacy in properly designed clinical trials.

References

[1] Liang C Z, Li H J, Wang Z P, *et al.* Treatment of chronic prostatitis in Chinese men. Asian J Androl, 2009, 11: 153-156.

[2] Weidner W, Wagenlehner F M, Marconi M, *et al.* Acute bacterial prostatitis and chronic prostatitis/chronic pelvic pain syndrome: Andrological implications. Andrologia, 2008, 40: 105-112.

[3] Nickel JC, Downey J, Johnston B, *et al.* Predictors of patient response to antibiotic therapy for the chronic prostatitis/chronic pelvic pain syndrome: A prospective multicenter clinical trial. J Urol, 2001, 165: 1539-1544.

[4] Doble A. Chronic prostatitis. Br J Urol, 1994, 74: 537-541.

[5] Krieger J N, Nyberg L Jr, Nickel J C. NIH consensus definition and classification of prostatitis. JAMA, 1999, 282: 236-237.

[6] Weidner W, Anderson R U. Evaluation of acute and chronic bacterial prostatitis and diagnostic management of chronic prostatitis/chronic pelvic pain syndrome with special reference to infection/inflammation. Int J Antimicrob Agents, 2008, 31: S91-S95.

[7] Turner J A, Ciol M A, Von Korff M, *et al.* Validity and responsiveness of the national institutes of health chronic prostatitis symptom index. J Urol, 2003, 169: 580-583.

[8] Propert K J, Litwin M S, Wang Y, *et al.* Responsiveness of the National Institutes of Health Chronic Prostatitis Symptom Index (NIH-CPSI). Qual Life Res, 2006, 15: 299-305.

[9] Clemens J Q, Calhoun E A, Litwin M S, *et al.* Validation of a modified National Institutes of Health chronic prostatitis symptom index to assess genitourinary pain in both men and women. Urology, 2009, 74: 983-987.

[10] Collins M M, Stafford R S, O'Leary M P, *et al.* How common is prostatitis? A national survey of physician visits. J Urol, 1998, 159: 1224-1228.

[11] Nickel J C, Downey J, Hunter D, *et al.* Prevalence of prostatitis-like symptoms in a population based study using the National Institutes of Health chronic prostatitis symptom index. J Urol, 2001, 165: 842-845.

[12] Roberts R O, Lieber M M, Rhodes T, *et al.* Prevalence of a physician-assigned diagnosis of prostatitis: The olmsted county study of urinary symptoms and health status among men. Urology, 1998, 51: 578-584.

[13] Calhoun E A, McNaughton Collins M, Pontari M A, *et al.* The economic impact of chronic prostatitis. Arch Intern Med, 2004, 164: 1231-1236.

[14] Krieger J N. Classification, epidemiology and implications of chronic prostatitis in North America, Europe and Asia. Minerva Urol Nefrol, 2004, 56: 99-107.

[15] McNaughton C M, Pontari M A, O'Leary M P, *et al.* Quality of life is impaired in men with chronic prostatitis: The chronic prostatitis collaborative research network. j gen intern med, 2001, 16: 656-662.

[16] Turner J A, Ciol M A, Von Korff M, *et al.* Health concerns of patients with nonbacterial prostatitis/pelvic pain. Arch Intern Med, 2005, 165: 1054-1059.

[17] Fair W R, Parrish R F. Antibacterial substances in prostatic fluid. Prog Clin Biol Res, 1981, 75A: 247-264.

[18] Johnson J R, Kuskowski M A, Gajewski A, *et al.* Extended virulence genotypes and phylogenetic background of *Escherichia coli* isolates from patients with cystitis, pyelonephritis, or prostatitis. J Infect Dis, 2005, 191: 46-50.

[19] Millan-Rodriguez F, Palou J, Bujons-Tur A, *et al.* Acute bacterial prostatitis: Two different sub-categories according to a previous manipulation of the lower urinary tract. World J Urol, 2006, 24: 45-50.
[20] Ozden E, Bostanci Y, Yakupoglu K Y, *et al.* Incidence of acute prostatitis caused by extended-spectrum beta-lactamase-producing *Escherichia coli* after transrectal prostate biopsy. Urology, 2009, 74: 119-123.
[21] Nickel J C, Olson M E, Barabas A, *et al.* Pathogenesis of chronic bacterial prostatitis in an animal model. Br J Urol, 1990, 66: 47-54.
[22] Nickel J C, Moon T. Chronic bacterial prostatitis: An evolving clinical enigma. Urology, 2005, 66: 2-8.
[23] Etienne M, Pestel-Caron M, Chavanet P, *et al.* Performance of the urine leukocyte esterase and nitrite dipstick test for the diagnosis of acute prostatitis. Clin Infect Dis, 2008, 46: 951-953.
[24] Zegarra Montes L Z, Sanchez Mejia A A, Loza Munarriz C A, *et al.* Semen and urine culture in the diagnosis of chronic bacterial prostatitis. Int Braz J Urol, 2008, 34: 30-37; discussion, 38-40.
[25] Magri V, Wagenlehner F M, Montanari E, *et al.* Semen analysis in chronic bacterial prostatitis: Diagnostic and therapeutic implications. Asian J Androl, 2009, 11: 461-477.
[26] Budia A, Luis Palmero J, Broseta E, *et al.* Value of semen culture in the diagnosis of chronic bacterial prostatitis: A simplified method. Scand J Urol Nephrol, 2006, 40: 326-331.
[27] Rizzo M, Marchetti F, Travaglini F, *et al.* Prevalence, diagnosis and treatment of prostatitis in Italy: A prospective urology outpatient practice study. BJU Int, 2003, 92: 955-959.
[28] McNaughton C M, Fowler F J Jr, Elliott DB, *et al.* Diagnosing and treating chronic prostatitis: Do urologists use the four-glass test? Urology, 2000, 55: 403-407.
[29] Nickel J C, Shoskes D, Wang Y, *et al.* How does the pre-massage and post-massage 2-glass test compare to the Meares-Stamey 4-glass test in men with chronic prostatitis/chronic pelvic pain syndrome? J Urol, 2006, 176: 119-124.
[30] Schaeffer A J, Knauss J S, Landis J R, *et al.* Leukocyte and bacterial counts do not correlate with severity of symptoms in men with chronic prostatitis: The National Institutes of Health Chronic Prostatitis Cohort Study. J Urol, 2002, 168: 1048-1053.
[31] Nickel J C. Clinical evaluation of the patient presenting with prostatitis. Eur Urol, 2003, 2: 11-14.
[32] Lee J C, Muller C H, Rothman I, *et al.* Prostate biopsy culture findings of men with chronic pelvic pain syndrome do not differ from those of healthy controls. J Urol, 2003, 169: 584-587; discussion 587-588.
[33] Wise G J, Shteynshlyuger A. Atypical infections of the prostate. Curr Prostate Rep, 2008, 6: 86-93.
[34] Schaeffer A J, Wu S C, Tennenberg A M, *et al.* Treatment of chronic bacterial prostatitis with levofloxacin and ciprofloxacin lowers serum prostate specific antigen. J Urol, 2005, 174: 161-164.

[35] Krieger J N, Ross S O, Limaye A P, *et al.* Inconsistent localization of gram-positive bacteria to prostate-specific specimens from patients with chronic prostatitis. Urology, 2005, 66: 721-725.
[36] Naber K G. Management of bacterial prostatitis: What's new? BJU Int, 2008, 101: S7-S10.
[37] Naber K G, Roscher K, Botto H, *et al.* Oral levofloxacin 500 mg once daily in the treatment of chronic bacterial prostatitis. Int J Antimicrob Agents, 2008, 32: 145-153.
[38] Nickel J C, Xiang J. Clinical significance of nontraditional bacterial uropathogens in the management of chronic prostatitis. J Urol, 2008, 179: 1391-1395.
[39] Szoke I, Torok L, Dosa E, *et al.* The possible role of anaerobic bacteria in chronic prostatitis. Int J Androl, 1998, 21: 163-168.
[40] Skerk V, Krhen I, Schonwald S, *et al.* The role of unusual pathogens in prostatitis syndrome. Int J Antimicrob Agents, 2004, 24: S53-S56.
[41] Charalabopoulos K, Karachalios G, Baltogiannis D, *et al.* Penetration of antimicrobial agents into the prostate. Chemotherapy, 2003, 49: 269-279.
[42] Naber K G, Sorgel F. Antibiotic therapy-rationale and evidence for optimal drug concentrations in prostatic and seminal fluid and in prostatic tissue. Andrologia, 2003, 35: 331-335.
[43] Wagenlehner F M, Weidner W, Sorgel F, *et al.* The role of antibiotics in chronic bacterial prostatitis. Int J Antimicrob Agents, 2005, 26: 1-7.
[44] Shigehara K, Miyagi T, Nakashima T, *et al.* Acute bacterial prostatitis after transrectal prostate needle biopsy: clinical analysis. J Infect Chemother, 2008, 14: 40-43.
[45] Etienne M, Chavanet P, Sibert L, *et al.* Acute bacterial prostatitis: Heterogeneity in diagnostic criteria and management. Retrospective multicentric analysis of 371 patients diagnosed with acute prostatitis. BMC Infect Dis, 2008, 8: 12.
[46] Ha U S, Kim M E, Kim C S, *et al.* Acute bacterial prostatitis in Korea: Clinical outcome, including symptoms, management, microbiology and course of disease. Int J Antimicrob Agents, 2008, 31: S96-S101.
[47] Magri V, Trinchieri A, Pozzi G, *et al.* Efficacy of repeated cycles of combination therapy for the eradication of infecting organisms in chronic bacterial prostatitis. Int J Antimicrob Agents, 2007, 29: 549-556.
[48] de la Rosette J J, Hubregtse M R, Meuleman E J, *et al.* Diagnosis and treatment of 409 patients with prostatitis syndromes. Urology, 1993, 41: 301-307.
[49] Nickel J C. Treatment of chronic prostatitis/chronic pelvic pain syndrome. Int J Antimicrob Agents, 2008, 31: S112-S116.
[50] Kurzer E, Kaplan S. Cost effectiveness model comparing trimethoprim sulfamethoxazole and ciprofloxacin for the treatment of chronic bacterial prostatitis. Eur Urol, 2002, 42: 163-166.
[51] Murphy A B, Macejko A, Taylor A, Nadler R B. Chronic prostatitis: management strategies. Drugs, 2009, 69: 71-84.
[52] El-Hakim A, Shah D K, Smith A D. Advanced therapy for prostatitis:

Minimally invasive and invasive therapies. Curr Urol Rep, 2003, 1: 44-50.
[53] Nickel J C, Downey J, Ardern D, *et al.* Failure of a monotherapy strategy for difficult chronic prostatitis/chronic pelvic pain syndrome. J Urol, 2004, 172: 551-554.
[54] Juricic C, F. K K, Sietzen W, *et al.* Concentration of amoxicillin and clavulanate in the prostate tissue and in serum: A pharmacokinetic study. Proceedings of the 15th International Congress of Chemotherapy, 1987.
[55] Fraschini F, Scaglione F, Falchi M, *et al.* Pharmacokinetics and tissue distribution of amoxicillin plus clavulanic acid after oral administration in man. J Chemother, 1990, 2: 171-177.
[56] Klotz T, Braun M, Bin Saleh A, *et al.* Penetration of a single infusion of ampicillin and sulbactam into prostatic tissue during transurethral prostatectomy. Int Urol Nephrol, 1999, 31: 203-209.
[57] Goto T, Makinose S, Ohi Y, *et al.* Diffusion of piperacillin, cefotiam, minocycline, amikacin and ofloxacin into the prostate. Int J Urol, 1998, 5: 243-246.
[58] Symes J M, Jarvis J D, Tresidder G C. An appraisal of cephalexin monohydrate levels in semen and prostatic tissue. Chemotherapy, 1974, 20: 257-262.
[59] Litvak A S, Franks C D, Vaught S K, *et al.* Cefazolin and cephalexin levels in prostatic tissue and sera. Urology, 1976, 7: 497-499.
[60] Smith R P, Schmid G P, Baltch A L, *et al.* Concentration of cefaclor in human prostatic tissue. Am J Med Sci, 1981, 281: 19-24.
[61] Vree T B, Hekster Y A. Pharmacokinetics and tissue concentrations of cefuroxime. Pharm Weekbl Sci, 1990, 12: 262-266; discussion 267.
[62] Becopoulos T, Georgoulias D, Constantinides C, *et al.* Acute prostatitis: which antibiotic to use first. J Chemother, 1990, 2: 244-246.
[63] Adam D, Schalkhauser K, Boettger F. Diffusion of cefuroxime into the prostatic and other tissues of the urogenital region [in German]. Med Klin, 1979, 74: 1867-1870.
[64] Fraschini F, Scaglione F, Mezzetti M, *et al.* Pharmacokinetic profile of cefotetan in different clinical conditions. Drugs Under Experimental and Clinical Research, 1988, 14: 547-553.
[65] Takeuchi N, Kinukawa T, Matsuura O, *et al.* A study of prostatic tissue levels of latamoxef, cefoperazone and cefotaxime [in Japanese]. Hinyokika Kiyo, 1986, 32: 1831-1841.
[66] Fujita K, Fujita H M, Fujii K, *et al.* Cefotaxime concentration in the prostatic tissue [in Japanese]. Jpn J Antibiot, 1983, 36: 1465-1468.
[67] Novick W J Jr. Levels of cefotaxime in body fluids and tissues: A review. Rev Infect Dis, 1982, 4: S346-S353.
[68] Martin C, Viviand X, Cottin A, *et al.* Concentrations of ceftriaxone (1,000 milligrams intravenously) in abdominal tissues during open prostatectomy. Antimicrob Agents Chemother, 1996, 40:1311-1313.
[69] Morita M, Hatakeyama T, Suzuki K. Ceftazidime concentration in human prostatic tissue and serum following intravenous injection]. Hinyokika Kiyo, 1991, 37: 659-662.

[70] Morita M, Nakagawa H, Suzuki K. Cefixime concentration in human prostatic tissue following 3-days of administration [in Japanese]. Hinyokika Kiyo, 1991, 37: 1581-1584.
[71] Naber K G, Kinzig M, Adam D, *et al.* Concentrations of cefpodoxime in plasma, ejaculate and in prostatic fluid and adenoma tissue. Infection, 1991, 19: 30-35.
[72] Madsen P O, Dhruv R, Friedhoff L T. Aztreonam concentrations in human prostatic tissue. Antimicrob Agents Chemother, 1984, 26: 20-21.
[73] Whitby M, Hempenstall J, Gilpin C, *et al.* Penetration of monobactam antibiotics (aztreonam, carumonam) into human prostatic tissue. Chemotherapy, 1989, 35: 7-11.
[74] Cannon G M Jr., Smaldone M C, Paterson D L. Extended-spectrum betalactamase gram-negative sepsis following prostate biopsy: Implications for use of fluoroquinolone prophylaxis. Can J Urol, 2007, 14: 3653-3655.
[75] Baker S D, Horger D C, Keane T E. Community-acquired methicillin-resistant Staphylococcus aureus prostatic abscess. Urology, 2004, 64: 808-810.
[76] Pierce JR Jr., Saeed Q, Davis WR. Prostatic abscess due to community acquired methicillin-resistant Staphylococcus aureus. Am J Med Sci, 2008, 335: 154-156.
[77] Bergmann M, Lederer B, Takacs F. Tissue concentrations of sulfametroltrimethoprim in the human prostate [in German]. Urologe A, 1979, 18: 335-337.
[78] Dan M, Golomb J, Gorea A, *et al.* Concentration of ciprofloxacin in human prostatic tissue after oral administration. Antimicrob Agents Chemother, 1986, 30: 88-89.
[79] Naber C K, Steghafner M, Kinzig-Schippers M, *et al.* Concentrations of gatifloxacin in plasma and urine and penetration into prostatic and seminal fluid, ejaculate, and sperm cells after single oral administrations of 400 milligrams to volunteers. Antimicrob Agents Chemother, 2001, 45: 293-297.
[80] Drusano G L, Preston S L, Van Guilder M, *et al.* A population pharmacokinetic analysis of the penetration of the prostate by levofloxacin. Antimicrob Agents Chemother, 2000, 44: 2046-2051.
[81] Wagenlehner F M, Kees F, Weidner W, *et al.* Concentrations of moxifloxacin in plasma and urine, and penetration into prostatic fluid and ejaculate, following single oral administration of 400 mg to healthy volunteers. Int J Antimicrob Agents, 2008, 31: 21-26.
[82] Wagenlehner F M, Lunz J C, Kees F, *et al.* Serum and prostatic tissue concentrations of moxifloxacin in patients undergoing transurethral resection of the prostate. J Chemother, 2006, 18: 485-489.
[83] Naber K G, Adam D, Kees F. *In vitro* activity and concentrations in serum, urine, prostatic secretion and adenoma tissue of ofloxacin in urological patients. Drugs, 1987, 34: S44-S50.
[84] Giberti C, Gallo F, Rosignoli M T, *et al.* Penetration of orally administered prulifloxacin into human prostate tissue. Clin Drug Investig, 2009, 29: 27-34.
[85] Foulds G, Madsen P, Cox C, *et al.* Concentration of azithromycin in human

prostatic tissue. Eur J Clin Microbiol Infect Dis, 1991, 10: 868-871.

[86] Giannopoulos A, Koratzanis G, Giamarellos-Bourboulis E J, *et al.* Pharmacokinetics of clarithromycin in the prostate: Implications for the treatment of chronic abacterial prostatitis. J Urol, 2001, 165: 97-99.

[87] Grabe M, Bishop M C, Bjerklund-Johansen L, *et al.* Management of urinary and male genital tract infections. European Association of Urology, 2008: 84-88.

[88] Cai T, Mazzoli S, Bechi A, *et al.* Serenoa repens associated with Urticadioica (ProstaMEV) and curcumin and quercitin (FlogMEV) extracts are able to improve the efficacy of prulifloxacin in bacterial prostatitis patients: Results from a prospective randomised study. Int J Antimicrob Agents, 2009, 33: 549-553.

[89] Jeong C W, Lim D J, Son H, *et al.* Treatment for chronic prostatitis/chronic pelvic pain syndrome: levofloxacin, doxazosin and their combination. Urol Int, 2008, 80: 157-161.

[90] Giannarini G, Mogorovich A, Valent F, *et al.* Prulifloxacin versus levofloxacin in the treatment of chronic bacterial prostatitis: A prospective, randomized, double-blind trial. J Chemother, 2007, 19: 304-308.

[91] Magri V, Trinchieri A, Ceriani I, *et al.* Eradication of unusual pathogens by combination pharmacological therapy is paralleled by improvement of signs and symptoms of chronic prostatitis syndrome. Arch Ital Urol Androl, 2007, 79: 93-98.

[92] Chen W M, Yang C R, Ou Y C, *et al.* Combination regimen in the treatment of chronic prostatitis. Arch Androl, 2006, 52: 117-121.

[93] Ziaee A M, Akhavizadegan H, Karbakhsh M. Effect of allopurinol in chronic nonbacterial prostatitis: A double blind randomized clinical trial. Int Braz J Urol, 2006, 32: 181-186.

[94] Alexander R B, Propert K J, Schaeffer A J, *et al.* Ciprofloxacin or tamsulosin in men with chronic prostatitis/chronic pelvic pain syndrome: A randomized, double-blind trial. Ann Intern Med, 2004, 141: 581-589.

[95] Nickel J C, Downey J, Clark J, *et al.* Levofloxacin for chronic prostatitis/chronic pelvic pain syndrome in men: A randomized placebo-controlled multicenter trial. Urology, 2003, 62: 614-617.

[96] Bundrick W, Heron S P, Ray P, *et al.* Levofloxacin versus ciprofloxacin in the treatment of chronic bacterial prostatitis: A randomized doubleblind multicenter study. Urology, 2003, 62: 537-541.

[97] Naber K G. Lomefloxacin versus ciprofloxacin in the treatment of chronic bacterial prostatitis. Int J Antimicrob Agents, 2002, 20: 18-27.

[98] Hu W L, Zhong S Z, He H X. Treatment of chronic bacterial prostatitis with amikacin through anal submucosal injection. Asian J Androl, 2002, 4: 163-167.

15

Infectious Microecology in the Diseases of the Respiratory System

Jianying Zhou *, Zhou Hua
The First Afflilated Hospital, School of Medicine, Zhejiang University, Hangzhou, 310003, China
* E-mail: drzjy@163.com

The respiratory system is an organ system with the largest surface area through which the body keeps in continuous contact with the external environment. At rest, an adult breathes about 10,000 L of air into and out of the respiratory tract every day. Since the respiratory tract communicates with the external environment, respiratory organs are most vulnerable to invasion of microorganisms, allergens and harmful gases in the environment, resulting in diseases [1]. The respiratory system's microecological space has the following functions.

15.1 Ecological Space and Microecological Characteristics of the Respiratory System

In this section, it includes physical defense, humoral defense and cellular defense.

15.1.1 Non-Specific Defense Mechanism

Non-specific Defense Mechanism includes physical defense, humoral defense and cellular defense.

15.1.1.1 Physical Defense

The upper respiratory tract has many anatomical features that can prevent inhaling particles from the air into the lung. The inner wall of the whole upper respiratory tract is covered by mucosa, on whose surface cilia keep moving, pushing up the mucosa layer and attached particles, as well as aged and dead cells of the lower respiratory tract and finally delivering them to the pharynx. At that point they are then swallowed into the stomach. If there are microorganisms, they will be killed by gastric acid or expelled out of the respiratory tract as foreign bodies through reflex action such as coughing. Cilia in the respiratory tract are an important defense structure, which plays an important role in eliminating respiratory tract secretion and foreign particles and preventing respiratory tract infection.

15.1.1.2 Humoral Defense

Lysozyme: There are high concentrations of lysozymes in specific azurophil granules of polymorphonuclear leucocytes, which are released outside cells and can lyse the bacterial wall or interact with other components on the bacterial surface, thus killing bacteria and also enhancing the complement function.
Interferon: Interferon is a group of species specific glycoproteins with wide biological activities. It mainly inhibits viral replication in cells. The most important activity of type I interferon is an anti-viral effect. Under the action of a virus, inducer and interferon itself, leucocytes generate antiviral proteins and thus inhibit viral replication. Type II interferon is an important activator of macrophages. It may be important for the oxygen-dependent killing mechanism of intracellular pathogens.
Complement: Complement and its activation products have specific and non-specific anti-infective effects, which may be important in early defense against infection.

15.1.1.3 Cellular Defense

Polymorphonuclear leucocytes provide effective defense against bacterial and fungal infection. The germicidal mechanism is their early binding to the bacterial wall which increases wall permeability. Alveolar macrophage is an important inherent defense line of the lower respiratory tract against pathogens and other invaders. Besides the non-specific bactericidal defense function, it also has important effects in processing and presenting antigen, participating in cell-mediated specific immune reaction and acting as an effector cell.

15.1.2 Specific Defense Mechanism

Immune organs, immune competent cells, and immune globulins and lymphokines have different defense mechanisms.

15.1.2.1 Immune Organs

The thymus, lymph nodes, lymph nodules and spleen are collectively called immune organs. One function of the thymus is to cultivate and deliver large amounts of lymphocytes. The other function is to produce thymosin in order to assist T lymphocyte proliferation and differentiation. Lymph nodes are peripheral immune organs mainly distributed at non-mucosal sites. Lymph nodules are lymphatic tissues with incomplete structure, mostly distributed at mucosal sites. Spleen and lymphatic tissues construct a reticuloendothelial system, which is the largest lymphatic tissue in the body. It belongs to peripheral immune organs and is the residence of various immune cells as well as the base of immune cell proliferation and immune response generation.

15.1.2.2 Immune Competent Cells

T lymphocytes are involved in a series of processes including regulation of specific cellular immunity, recognition of antigen, cell activation, transfer of activation information, receipt of cytokine and subsequent cell proliferation and differentiation. The main function of B lymphocytes is to generate antibodies. NK cells originate in the bone marrow. They have various immune functions, such as inhibiting and killing tumor cells, preventing tumor growth, metastasis and antiviral effect.

15.1.2.3 Immune Globulins and Lymphokines

In humoral immunity, the substances closely related to anti-respiratory tract infection are IgA, IgG and IgM. The main component in the secretion of the upper respiratory tract is IgA, while the main component in serum is IgG. The main effect of immune globulin-mediated humoral immunity in anti-infective immunity is to act as opsonin and phagocytosis through activating complement-mediated phagocytes, as well as neutralizing toxin by preventing exotoxin from binding to the target cell receptor. Lymphokines are low molecular mass polypeptides activating lymphocytic secretion. They act in immune response and regulation. General pathogens such as Bacillus tuberculosis, mold and some mold spores causing allergic alveolitis are hard to digest even after phagocytosis. However, in a body with immunity, under the action of lymphokines released by sensitized lymphocytes, not only does the number of macrophage increase but also the synthesis of lysozymes and various germicidal substances increases, thus greatly enhancing phagocytosis and the killing power against pathogens. In a word, cytokines are involved in the whole process of immune response. The final manifestation pattern, intensity and site largely depend on function expression and interaction of the cytokine network.

15.2 Microecology and Microecology Changes in Respiratory System Infection

A healthy respiratory tract is colonized by certain kinds of microorganisms, *i.e.*, normal flora of the respiratory tract. They maintain a dynamic balance in respect of flora species and number under the effect of the host and external environment [2]. Once the normal flora ecological structure is destroyed, it may cause flora imbalance, leading to respiratory tract infection or even death.

Usually 13 – 17 bacterial species in 6 – 8 genera can be isolated from a healthy pharynx posterior wall. In the order of detection rate, they are *Streptococcus*, *neisseria*, *corynebacteria*, *Staphylococcus*, *Peptostreptococcus*, fusiform bacteria and actinomycete. Population abundance and density vary with age and season. Common bacteria colonizing the human nasal pharynx include *Streptococcus viridans*, *neisseria*, *Streptococcus pneumoniae*, *Staphylococcus aureus*, *Klebsiella oxytoca*, *Staphylococcus epidermidis*, *Hemophilus influenza*, *Hemophilus parainfluenzae*, *Moraxelle catarrhalis*, Bacteroides, bacteroid, *Bacterium lacticum*, *Fusobacterium*, *Veillonella*, peptostreptococcus and other anaerobic streptococci and anaerobic gram-negative bacteria; the most common one is *Streptococcus viridans*. They maintain the balance of "normal microorganisms" and usually do not cause diseases. Physiological *Streptococcus viridans* flora has developed biological antagonism and colonization resistance to allied bacteria and has effects in balancing mucosal flora and relieving inflammatory reaction. However, this balance is disturbed due to many factors, such as low resistance, irrational antibiotic use and excessive bacterial proliferation [3], in which case bacteria occupy receptors of susceptible host cells, take nutrition and cause disease.

From a microecological perspective, the normal flora of the upper respiratory tract is a natural defense barrier of the body. When this barrier is destroyed by various harmful factors, such as body injury, decreased resistance, *etc.*, flora loses balance in number and species, the defense barrier is broken, exogenous pathogens invade (virus, bacteria from other sites) or there is mass proliferation of certain endogenous bacterium, causing infection.

Normal flora is important in maintaining the body's ecological balance and the stability of the internal environment. A healthy human's upper respiratory tract is colonized by many aerobic, microaerobic and anaerobic bacteria, including more than 200 species in 21 genera. Anaerobic bacteria have the highest concentration. In the gum, the secretion can achieve 10^8/mL. The upper respiratory tract colonizing bacteria begin to appear soon after birth. They are in ecological balance and never cause disease, and are also known as inherent bacteria or symbiotic bacteria. They are mainly from human or inanimate objects that are in contact. Their long-term existence in the upper respiratory tract may be a result of natural selection. These bacteria may vary in terms of type or concentration with season and environment but maintain relative stability, composing normal flora of the upper respiratory tract. In oropharynx, there is *Streptococcus viridans*, staphylococci, *Streptococcus pyrogenes*, *Moraxelle catarrhalis*, neisseria, *Lactobacillus lactis*, non-fragile bacteroides, *Candida albicans* and occasionally gram-negative bacilli and protozoa; in the nasopharynx, there are also staphylococci (including *Staphylococcus aureus*), streptococci (including *Streptococcus pneumoniae*), Moraxelle catarrhalis, neisseria and *Hemophilus influenza*, which is especially common. Normal flora has a certain meaning to the host respiratory tract defense mechanism. It can be involved in substance metabolism, nutrition conversion and synthesis, bile metabolism, cholesterol metabolism and hormone conversion; can act as an antigen to stimulate the host to generate antibodies, activate macrophages and other elimination mechanisms and enhance interferon activity; it can constitute a biological barrier, *e.g.* bacterin, bactericin and toxic final metabolites released by normal flora, against exogenous bacterial invasion. Due to the existence of normal flora, its metabolites such as fatty acid and bacteriocin can resist invasion of exogenous bacteria. This ability is called "colonization resistance". It is important to the host metabolism, nutrition and substance digestion and absorption. It can convert some carcinogens into non-carcinogenic substances. Existence of normal flora is an important factor in maintaining the health of the human body. It reflects a state where under normal circumstances, the host, normal flora and external environment co-adapt and keep in equilibrium. Besides the beneficial effects mentioned above, normal flora in the oropharynx can also have unfavorable effects, such as weakening host resistance and causing lower respiratory tract infection as a secondary or opportunistic pathogen in the host with viral respiratory tract infection or compromised immunity. Therefore, many factors can lead to the above ecological imbalance. If the host is ill, colonization of gram-negative bacilli in the upper respiratory tract increases. Medical measures include therapeutic intervention, especially respiratory therapy and application of H_2-receptor blocker. In patients with tracheal intubation, the colonization rate of gram-negative bacilli in the lower respiratory tract is as high as 45% – 100% with Pseudomonas as the most common one. Moreover, in case of exogenous bacterial invasion and use of an antimicrobial agent, ecological imbalance will cause flora disturbance, manifested by diarrhea or colonitis. In severe cases, this may cause various dual infections. Normal flora can also be transferred from the original site to other sites, causing infection, known as colonization transfer. For example, *Escherichia coli* can be transferred from the colon to the respiratory tract, causing pneumonia, or to the urinary tract, causing nephropyelitis and cystitis. When there is a sudden increase in bacterial number in the upper small intestine, deficiency of fat and vitamin B may be induced, causing nosocomial infection, also known as "opportunistic infection", *i.e.* due to effects of various factors, normal flora that originally exists in the bowel, mouth or pharynx cause infection in other susceptible sites. This is especially common in immune deficiency. Excessive use of antibiotics can inhibit normal flora in the oropharynx, destroy ecological balance and create an ecological vacuum, which is soon filled with gram-negative bacteria. Researchers inoculated large amounts of gram-negative bacilli in a healthy oropharynx but failed to make them proliferate and the bacteria were soon eliminated. However, in patients with severe diseases, 50% developed colonies. Why do these patients tend to develop gram-negative bacilli? *In vitro* studies have confirmed that Streptococcus viridans inhibits growth of *Streptococcus pyogenes*, *Streptococcus pneumoniae*, *Mycobacterium tuberculosis*, legionella, *Neisseria meningitis* and other gram-negative bacilli. In the mouth, the most common non-fragile bacteroides can inhibit growth of *Klebsiella pneumoniae*, *E. coli*, mycoplasma and serratia [4]. Excessive use of antibiotics can change the ecological balance among body bacteria, leading to rapid growth of gram-negative bacilli. In some circumstances, such as hypoxia, lung edema, acid poisoning, viral respiratory tract infection, alcoholic poisoning and azotemia, the net removal of bacteria in the respiratory tract decreases, thus increasing susceptibility.

15.3 Microecology of Respiratory System Fungal Infection

Fungus is one of the important pathogens causing pulmonary infection in an immunosuppressed population. At present, due to wide use of broad-spectrum antibiotics and increase in organ transplantation and hematopoietic stem cell transplantation, the incidence and mortality of invasive mycosis have significantly increased. Invasive pulmonary fungal infection refers to acute/chronic infection with corresponding tissue pathological injury caused by direct fungal invasion in the lung and the bronchi [5].

15.3.1 Pulmonary Candidiasis

This is an acute, subacute or chronic respiratory infection caused by candida. It is relatively common. The main pathogen is *Candida albicans*. Others include *Candida tropicalis* and *Candida krusei*. These bacteria widely exist in nature as well as in excretion of birds and mammals. They also parasitize human skin, the mouth and gastrointestinal tract, usually acting as an opportunistic pathogen. After invading tissue, candida will be transformed to mycelial form and proliferate in great numbers.

Therapy requires elimination of inducing factors, such as discontinuation or change of narrow-spectrum antibiotics, hormones and immunosuppressants. In addition to enhancement of supportive therapy and improving patient's immunity, a broad-spectrum antifungal drug can be used for patients with severe infection.

Amphotericin B and its liposome, triazole antifungal agents and echinocandins can be selected.

15.3.2 Pulmonary Aspergillosis

Aspergillus is widely distributed in nature and most are bacterial parasites. It is a common pathogen in the human body. Aspergillus fumigatus is a main pathogen causing pulmonary infection. There are aspergillus spores in the air, which also exist in moldy grain, feed and brewing products. Aspergillus spores are 2.5 – 3 μm in size. They float in the air and are easily inhaled. Inhalation of aspergillus spores does not necessarily cause disease. Only major inhalation will cause acute pulmonary aspergillosis. Aspergillus may temporarily reside in the upper respiratory tract. In the case of underlying disease, such as pulmonary tuberculosis, bronchiectasis and pulmonary carcinoma, or long-term use of glucocorticosteroid, the body has poor disease resistance and may be infected. In severe cases, the pathogen can invade the bloodstream and cause systemic dissemination.

Therapy: Remove precipitating factors, stop hormones, reduce use of antibiotics and immunosuppressants, enhance supportive therapy and improve body immunity. An intravenous drip of amphotericin B is effective, which can also be administered through aerosol inhalation. Voriconazole and itraconazole are new broad-spectrum triazole antifungal drugs. They are effective in the treatment of pulmonary aspergillosis. Echinocandin antifungal drugs such as caspofungin can also be used for salvage therapy of invasive pulmonary aspergillosis. Delayed treatment may lead to a severe condition for which antifungal therapy may be ineffective. Therefore, early diagnosis and treatment are very important.

15.3.3 Pulmonary Cryptococcosis

Pulmonary cryptococcosis is caused by *Cryptococcus neofomans*, which is widely distributed in the environment, such as in bird manure, mice droppings, soil, air, fruit and vegetables. Onset is mainly due to inhalation of spores of *Cryptococcus neofomans*, which can also be spread through the bloodstream and invade the brain, the meninges and other organs. Poor immunity is an important inducing factor for cryptococcosis. In a healthy body, infection can have a spontaneous cure and the focus is only limited to the lung.

The therapeutic regimen of choice is a combination of amphotericin B and fluctyosine. In combined treatment, the doses can be slightly reduced. In recent years, fluconazol among triazoles has been used in the treatment of cryptococcal meningitis. Generally, it is considered that in the treatment of meningitis in patients with immune deficiency or other high risk factors, a combination of amphotericin B and fluctyosine is appropriate for initial therapy, followed by long-term fluconazol therapy after the condition is stablized to reduce the risk of recurrence.

15.3.4 Pulmonary Coccidioidomycosis

Pneumocystis, formerly called *Pneumocystis carinii*, usually causes opportunistic pulmonary infection in patients with immune deficiency or immune suppression, characterized by diffuse interstial pneumonia. With increased concern about acquired immune deficiency syndrome (AIDS) and an increase in population with immune suppression due to the wide use of organ transplantation, more and more attention is paid to pneumocystosis.

Pneumocystosis is highly prevalent in the global population. Studies have demonstrated that in America, 75% of people have acquired immunity against this disease by the age of 4. Therefore, in healthy populations, most infections are asymptomatic. Pneumocystosis is the most common concomitant disease in HIV patients. Pneumocystis may be inhaled through the respiratory tract and reside in the human body with low virulence and slow proliferation. Therefore, only when the host immunity has serious and persistent injury, will there be a clinical onset.

A definite diagnosis of pneumocystosis depends on confirmed existence of *Pneumocystis carinii* in the respiratory tract secretion and tissue slice.

In the treatment of pneumocystosis, the regimen of choice is trimethoprim/ sulfamethoxazole (TMP/SMZ, SMZco). When SMZco therapy fails or there are side effects, pentamidine or other drugs can be selected. The therapeutic course of all drugs is 21 d.

Pulmonary fungal infection increasingly threatens the health and lives of patients with a tumor, corticosteroid use or immunosuppressant use. Besides, the diagnostic rate of fungal infection is lower than the actual infection rate. In most cases, it is hard to distinguish between fungal colonization and fungal infection. Therefore, a strategy to prevent fungal infection is to improve body immunity. When the body has poor immunity, it is susceptible to fungal infection. Drugs used to improve immunity like adjunctive therapy for fungal infection will bring a new breakthrough in the treatment of clinical fungal infection.

15.4 Microecology of Respiratory System Viral Infection

Viral pneumonia is a pulmonary inflammation caused by various viruses invading pulmonary parenchyma. It usually results from a downward spread of upper respiratory tract viral infection. This disease can occur in any season but mostly in spring and winter. In addition to viral virulence, infection route and infective dose, the host's age, respiratory tract local and systemic immune status are also important influencing factors of pneumonia, *e.g.*, human cytomegalovirus infection exists widely among the population. In people with normal immunity, after initial infection, there is usually no clinical manifestation or mild symptoms and the disease tends to be a latent infection, while in hosts with impaired immunity, relapse of primary infection or latent infection can induce active infection where viruses abundantly proliferate and spread, involving multiple organs, resulting in many clinical diseases. The most common one is cytomegalovirus pneumonia. In severe cases, this can cause respiratory failure and consequently death. If diagnosis of cytomegalovirus is early,rapid and accurate and antiviral therapy is provided in time, clinical symptoms are sure to be effectively improved and mortality will be reduced. In recent years, as organ transplantation techniques have become increasingly mature and there are more and more transplantation cases, severe cytomegalovirus infection occurs as a common postoperative complication, which usually causes viremia and serious pulmonary infection.

Respiratory tract viral infection injures the mucosal epithelium of the respiratory tract and stimulates the bronchial sensor, which is located beneath the tight junction of the airway mucosal epithelium. In case of infection, it tends to be exposed and the sensitivity increases when stimulated by antigen or excessive mucous secretion, causing airway hyper-reactivity and induces asthmatic attack, which may be even in a persistent state. In circumstances of low temperature and sudden weather change, especially in autumn and winter, the population is susceptible to RSV infection, inducing asthma. If asthma patients' resistance to RSV is increased to avoid RSV infection, this is a protection against asthmatic attack.

At present, there is no specific antiviral drug and the main option is symptomatic therapy.

General treatment: Keep warm, keep the respiratory tract clear, provide oxygen, correct the water, electrolyte and acid-base imbalance in time.

Antiviral drug therapy: Amantadine, ribavirin, acyclovir, ganciclovir; some herbal drugs have a certain efficacy against viral infection, such as *Radix isatidis*, *Radix astragalus*, *Lonicera japonica*, isatis leaf, *Forsythia suspense*, *Cyrtomium fortunei* and *Chrysanthemum morifolium*. These drugs can have certain therapeutic effects.

Biological agent therapy: With continuous development of various biological agents recently, interferon and the transfer factor can prove effective in the treatment of severe viral pneumonia.

In case of secondary bacterial infection, corresponding sensitive antibiotics should be provided to control and eliminate bacterial infection.

15.5 Microecology of Respiratory System Mycobacterial Infection

The mocroecology of respiratory system mycobacterial infection will be elaborated on fromaspects of pulmonary tuberculosis and non-tuberculous mycobacterial disease.

15.5.1 Pulmonary Tuberculosis

Mycobacterium tuberculosis has no exotoxin, endotoxin or invasive enzymes as the material base of its pathogenicity. However, the disease is related to some components in bacterial cells and the bacterial wall, such as lipoid, phospholipid, polysaccharide, bacterial wall and wax D and bacterial protein [6]. Its pathogenicity results from interaction of various factors in the bacteria-host immune system [7].

Tuberculosis has two types of onset, primary infection and secondary infection. Primary infection means the body is first infected with *Mycobacterium tuberculosis*, mostly in children and young people. Primary tuberculosis tends to be spontaneously cured. Most infections are limited to bacterial and pathological onset, without developing into clinical tuberculosis. Only when the body's resistance decreases will the disease continue to develop caseous pneumonia, systemic miliary tuberculosis or dissemination along lymphatic vessels and bronchi, leading to obvious clinical symptoms. In a few cases, *Mycobacterium tuberculosis* becomes latent and serves as the pathogens of secondary infection. Secondary tuberculosis can occur at any time after primary infection, mostly in adults. At present, it is considered that most secondary tuberculosis is due to residual *Mycobacterium tuberculosis* in the primary infection foci or early dissemination foci after primary infection. When the host's resistance decreases, *Mycobacterium tuberculosis* proliferates again and causes disease [8]. Only a few patients are pronee to repeated exogenous infection.

One microecological feature of the infection is dynamicity, *i.e.* dynamic balance between the etiological factor (microorganism), host and immunity, which is destroyed by the occurrence of infection. In turn, the occurrence, development and outcome of infection is exactly the process from imbalance to balance. As mentioned above, *Mycobacterium tuberculosis* does not produce toxin or invasive enzyme. Therefore, the host response plays a more important role in the pathogenesis of tuberculosis than in other infectious diseases. The core of the response is the host's specific immune response.

The feature of cellular immunity in tuberculosis is that T cell-mediated cellular immunity with macrophages as effector cells is predominant in the body's acquired anti-tuberculosis immunity, which is decisive in the pathogenesis of tuberculosis. This is a series of interrelated processes including macrophages engulfing and processing *Mycobacterium tuberculosis*, antigen presentation, specific recognition of antigens by T lymphocytes, binding, proliferation, differentiation, releasing various cytokines, activating macrophages and killing *Mycobacterium tuberculosis*.

Though T cell-mediated cellular immunity is the main immune response of tuberculosis, B lymphocyte-mediated humoral immunity is not only different, but also closely linked to it. The two types of immune response coordinate with each other. Now it is thought that humoral immunity can regulate cellular response and participate in the regulation of the anti-tuberculosis protective immune response.

Main clinical manifestations include coughing, expectoration, hemoptysis, chest pain, short breath, fever, night sweat and other systemic toxic symptoms. Other manifestations are follicular kerato-conjunctivitis, skin erythema nodosum and multiple arthralgia or arthritis caused by tuberculous allergic reaction. Tuberculous rheumatism is also known as Poncet's disease, characterized by multiple arthritis and erythema nodosum with rheumatoid process. It is also caused by tuberculous allergic reaction. The efficacy of anti-rheumatic therapy is poor but anti-tuberculosis therapy is effective.

A tuberculin test has been used to determine whether there has been *Mycobacterium tuberculosis* infection for a long time. Though it has many shortcomings, as there is a lack of a better detection method, it is still an important test that clinical practice relies on. At present, a purified protein derivative of the tuberculin test (PPD test) is widely used in clinical practice. Detection of *Mycobacterium tuberculosis* in the sputum is the most reliable evidence for definite diagnosis of pulmonary tuberculosis. Taking an X-ray is necessary procedure for pulmonary tuberculosis in order to know the lesion site, range, nature, development process and selection of therapy. For bronchial tuberculosis, bronchoscopy is the most sensitive and specific method. Using the immunological method to detect serum changes related to specific antigen of *Mycobacterium tuberculosis* has an encouraging clinical application perspective. However, so far no tuberculosis antigen with both strong antigenicity and high species specificity has been discovered. Therefore, tuberculosis serological examination can only serve as an auxiliary diagnostic technique with reference value. With continuous maturity and development of molecular biological techniques, especially the mapping of the *Mycobacterium tuberculosis* genome, the molecular biological technique for diagnosis of tuberculosis shows us an attractive application perspective. At present, DNA probe, polymerase chain reaction and DNA sequencing are used for diagnosis of tuberculosis and investigation and detection of the drug resistance mechanism of *Mycobacterium tuberculosis*, which have achieved encouraging findings.

Antituberculosis drugs are the basis of chemical therapy for tuberculosis. Since the development of streptomycin in the 1940s, especially wide application of armazide and rifampicin in the late 1970s, as well as re-recognition of pyrazinamide, tuberculosis control has obtained remarkable achievements. During the early part of the 21st century, research on antituberculosis drugs has made further progress. Armazide (INH, H) has a bactericidal effect inside the outside cells. It is not influenced by environmental pH value and is a 'full-valent bactericidal drug'. Streptomycin (SM, S) has a bactericidal effect only on extracellular *Mycobacterium tuberculosis* with vigorous growth and metabolism in a basic environment and thus is a 'semivalent bactericidal drug'. Rifampicin (RFP, R) is a 'fullvalent bactericidal drug', effective on *Mycobacterium tuberculosis* inside and outside cells. Ethambutol (EMB, E) in a routine dose has a bacteriostatic effect. At present, it has been widely used in combined therapy as a main adjuvant to powerful antituberculosis drugs. Pyrazinamide (PZA, Z) is one of the most effective bactericidal drugs in short-course chemotherapy. It has the strongest action in an acid environment and has a specific bactericidal effect against *Mycobacterium tuberculosis* inside macrophages and caseous foci. Current research has shown that PZA also has an effect on bacteria outside cells in an acid environment. Sodium paraminosalicylate (PAS, P) only has weak bacteriostatic action and is usually used as an adjuvant in combined regimen. It can enhance the bacteriostatic effect of other antituberculosis drugs and delay resistance to main drugs. Ethyl mercaptan armazide (1314Th) and propyl mercaptan armazide (1321Th) have a similar chemical structure to INH, but their bactericidal actions are weaker, but with higher toxicity. They can be used for drug resistant bacterial strains. Thiacetazone (TB1) is a bacteriostatis drug. It is seldom used in short-course chemotherapy [9]. Besides, due to no delay in the growth period, it cannot be used in intermittent chemotherapy.

With increased incidence of HIV infection with concurrent tuberculosis and the emergence of multi-drug-resistant bacterial strains, as well as the anticipation that the frequency of rifampicin resistant strains will increase in the future, development of new high-performance, low-toxicity antituberculosis drugs with long half lives, which are suitable for wide-interval observed treatment and are well accepted by patients, is urgent [10]. So far, certain achievements have been obtained in this respect. The most remarkable drugs are rifamycins and quinolones. Investigation of new derivatives of rifampicin is always a hot topic. So far, several derivatives of rifamycins with antituberculosis activity have been developed successively, including rifabutin, benzoxazine rifamycin and rifapentine. However, their bactericidal effect is inferior to rifampicin. Among the third generation of fluoroquinolones successively developed in the 1980s, many drugs have strong anti-*Mycobacterium tuberculosis* activity and are also effective on non-tuberculous mycobacteria (except *Mycobacterium avium*/intracellulare complex). Moreover, as *Mycobacterium tuberculosis* has a low spontaneous mutation rate under the effect of fluoroquinolones, about $1/10^6 - 1/10^7$, and these drugs have no cross resistance with other antituberculosis drugs, the development of fluoroquinolones opens a wider perspective in clinical treatment of tuberculosis. Now they have become the main drugs for drug resistant tuberculosis.

As described above, infection is a result of imbalance between microorganism, host and immunity and its occurrence, development and outcome is the process from imbalance to balance of this triangle of dynamicity. Such a dynamic feature of infection suggests that in antituberculosis therapy, we should not only pay attention to killing microorganisms, but also recognize that how to effectively regulate the immune response bridging host and microorganism in the dynamic triangle balance through appropriate methods is also a key point in successful antituberculosis therapy. (i) *Mycobacterium vaccae*: It is the only immunotherapeutic agent recommended by the World Health Organization for immunotherapy. *Mycobacterium vaccae* vaccine is a two-way immunomodulator. In people with normal or impaired immunity, it has a stimulatory function and enhances immunity. In people with hyperactive immunity and in a hypersensitive state, it can inhibit or lower immunity. This two-way immunoregulation effect helps the body adapt to external changes and achieve a self-stabilizing function. Studies show that, compared with simple chemotherapy, it can accelerate elimination of sputum bacteria and closure of cavities, increase the cure rate of chemotherapy, improve a patient's general condition and strengthen the physique. (ii) *Mycobacterium phlei* vaccine: It has a specific stimulatory effect on the T lymphocyte system, increases release of lymphokines to promote proliferation and activation of macrophages and enhance the body's ability to eliminate *Mycobacterium tuberculosis*. Currently, *Mycobacterium phlei* is considered to have a definite immune enhancement effect in the treatment of tuberculosis. (iii) Interferon: It is a defensive substance generated by the body itself to resist invasion of exogenous viruses and maintain self-stability of the body and cell functions. It has an important regulating effect in the body's immune response. (iv) Interleukin-2: It can promote proliferation of activated T and B cells, increase the activity of NK cells and interferon level, elevate $CD4^+$ T cell level and restore a normal $CD4^+/CD8^+$ ratio. (v) Thymosin: Through inducing each stage of T cell differentiation and development, it enhances mature T cell response to antigen and thus enhances the body's cellular immunity.

Emergence of multi-drug-resistant tuberculosis (MDR-TB) raises a critical challenge to today's tuberculosis prevention and treatment work. Regarding the definition of MDR-TB, there is still controversy. One opinion is that MDR-TB refers to concurrent resistance to INH and RFP, regardless of sensitivity to other drugs; the other opinion is that MDR-TB refers to resistance to 3 or more of the 5 main antituberculosis drugs, *i.e.*, H, R, Z, E and S. Which opinion is more appropriate still requires discussion. At present, most authors accept that MDR-TB refers to concurrent resistance to INH and RFP.

According to the time of occurrence, MDR-TB can be divided into primary and acquired. Primary MDR-TB refers to patients infected with MDR-TB who have no history of tuberculosis and never receive antituberculosis therapy. It is mostly caused by infection with the MDR-TB strain, common in HIV patients. Acquired MDR-TB refers to patients infected with a sensitive strain who developed drug resistance during antituberculosis therapy (at least exceeding 1 mon) due to inappropriate treatment. It is most common in cavitary or bilateral pulmonary tuberculosis patients. At present, the relatively agreeable opinion is that occurrence of acquired MDR-TB is secondary to inappropriate chemotherapy, including monotherapy, inadequate drug dose and patients who are incompliant with the planned regimen. Subsequently, the spontaneous resistant mutant strain is selected to become the predominant strain, causing therapeutic failure.

Definite diagnosis of MDR-TB depends on the results of a laboratory bacteriological test. Some authors have proposed that patients testing positive for sputum bacteria who meet any of the following three criteria can be suspected of having MDR-TB and can have preliminary screening as follows: (i) Regularly complete antituberculosis retreatment chemotherapy regimen; (ii) Receive at least 2 courses of antituberculosis drug therapy as required; (iii) Receive irregular antituberculosis therapy $\geqslant$ 2 years. Suspected patients will receive further bacteriological examination. If bacteriological examination reports *Mycobacterium tuberculosis* is resistant to INH and RFP or resistant to 3 or more drugs among S, H, R, Z and E, MDR-TB is diagnosed.

Currently, there is no ideal drug that can replace INH or RFP, and treatment of MDR-TB is very difficult. It is reported that the cure rate of MDR-TB is only 56%. The therapeutic principle of MDR-TB is to consult previous medical history on the basis of a drug susceptibility test, establish an individualized chemotherapy regimen by selecting at least 3 sensitive drugs, supervise the whole process, administer drugs every day and prolong the therapeutic course to 21 mon. Some authors emphasize that for patients resistant to most or all first-line antituberculosis drugs, drugs should continue to be administered for 24 mon after sputum bacteria become negative. WHO recommends first-line and second-line antituberculosis drugs can be combined to treat MDR-TB. Among first-line drugs, except INH and RFP to which it is resistant, SM, PAZ or EMB can be selected according to susceptibility. Second-line drugs are the main drugs for MDR-TB, including amikacin (AMK), capreomycin (CPM), thioamides (1314TH, protionamide), ofloxacin (OFLX) among fluoroquinolones, levofloxacin, cycloserine, sodium para-aminosalicylate, rifabutin and isoniazid aminosalicylate (pasiniazid). If there are no susceptible test results but MDR-TB is clinically considered, the chemotherapy regimen that can be selected is AMK or (CPM) + TH + PZA + OFLX in the intensification period, TH + OFLX in consolidation period. The intensification period should last for at least 3 mon and the consolidation period at least 18 mon. The overall therapeutic course is more than 21 mon. If sputum bacteria is converted to negative within 4 mon of therapy, it suggests therapeutic success; otherwise, it suggests resistance to the selected regimen. For patients with limited lesion range, who have no sputum bacteria negative conversion after chemotherapy for 4 mon, or are only sensitive to 2 – 3 drugs whose efficacy is poor and are resistant to all the other anti-tuberculosis drugs, surgical operation should be considered of the essence.

15.5.2 Non-Tuberculous Mycobacterial Disease

Non-tuberculous mycobacteria (NTM) refers to mycobacteria other than the *Mycobacterium tuberculosis* complex (including *Mycobacterium tuberculosis*, *Mycobacterium bovis*, *Mycobacterium africanum* and *Mycobacterium microti*) and *Mycobacterium leprae*. They are widely distributed in soil, dust, water, plants, foods and environment. Infection with NTM which causes pathological changes to relevant tissues and organs is called NTM disease. So far, more than 100 NTMs have been identified. Most have no pathogenicity in the human body. However, reports on diseases caused by certain strains are increasing. Moreover, with the prevalence of HIV, the incidence of NTM disease in foreign countries has rapidly increased. Therefore, we have to pay close attention to NTM in our daily work and master relevant knowledge.

NTM is a kind of environmental mycobacteria, mainly originating from sewage, soil and aerosols. The environmental factor is the reason for human NTM infection, while human-to-human transmission is extremely rare. On a global scale, tuberculosis tends to decrease while NTM disease tends to gradually increase. NTM is the main opportunistic pathogen in AIDS cases. HIV positive patients with low $CD4^+$ are usually susceptible to disseminated *Mycobacterium avium* infection, which mostly occurs in late AIDS. It is reported *Mycobacterium* avium-intracellulare can be isolated in 20% – 40% of AIDS patients in autopsy.

NTM can cause pathological changes in multiple organs and systems including the skin, bone, lymph node, genitourinary system, eye, GI tract and lung. In people with normal immunity, the lung is the main site of NTM infection. Symptoms of pulmonary NTM disease are usually unapparent, manifested by coughing, expectoration, dyspnea and hemoptysis.

The American Thoracic Society (ATS) recommends the following diagnostic criteria of pulmonary NTM disease: (i) Consistent with clinical manifestations; (ii) At least two positive results in acid fast stain test of sputum (or bronchial lavage material) smear and culture confirms mass growth of NTM; (iii) Exclude infection caused by other pathogens. Therefore, to improve diagnostic accuracy for pulmonary NTM disease in clinical practice, firstly, medical workers have to elevate their awareness and alertness to NTM; secondly, NTM culture and identification have to be considered as the key points. NTM routine culture is characterized by slow growth and complex operation. Besides, as conventional classification and identification systems are mainly based on phenotypic features such as morphological, physiological and biochemical properties, they are less accurate. Therefore, establishment of a rapid, accurate identification method is extremely important.

NTM is resistant to most conventional anti-tuberculosis drugs. At present, there is no specific drug. Therefore, NTM disease is hard to treat with poor prognosis. "Non-Tuberculous Mycobacterial Disease Diagnosis and Treatment Guidelines" issued in 2000 in China states that there are no consistent criteria for a rational chemotherapy regimen and therapeutic course for NTM disease. Usually it is recommended to use combined therapy of 4 – 5 drugs. After acid-fast bacilli negative conversion, the therapy should be continued for 18 – 24 mon, and for 12 mon at a minimum. Monotherapy should be avoided in treatment and attention should be paid to adverse drug reactions.

Mycobacterium kansasii is usually sensitive to rifampicin, mildly resistant to armazide, ethambutol and streptomycin, sensitive to macrolides, sulfonamides, amikacin and rifabutin, and resistant to pyrazinamide.

Both *in vitro* trials and small animal trials have shown that mycobacterium avium-intracellular complex (MAC) is resistant to most antituberculosis drugs. In people with normal immunity, treatment of MAC infection requires long-course therapy of multiple drugs (18 – 24 mon).

Mycobacteria with fast growth, including *Mycobacterium fortuitum*, *Mycobacterium chelonae* and *Mycobacterium abscessus*, are highly resistant to conventional anti-tuberculosis drugs, but sensitive to conventional antibiotics (amikacin, cefoxitin, doxycycline, fluoroquinolones, imipenem/cilastatin belonging to carbopenems, new macrolides and sulfonamides).

With continuous increase in NTM incidence, it is necessary to spread knowledge of NTM and continuously improve the diagnosis and treatment level. In addition, development of rapid and accurate bacterial identification, a reliable susceptibility test and high-efficacy, low-toxicity chemotherapy drugs will be the main direction of NTM disease study in the future [11].

15.6 Main Measures of Microecological Prevention/Treatment and Respiratory System Ecological Prevention/Treatment in Anti-Infective Therapy

An anti-infective drug is a double-edged sword. Rational use of anti-infective drugs is critically important to improve clinical efficacy of anti-infective therapy, shorten duration of therapy and reduce medical costs. Considering the effect of anti-infective drugs in inhibiting and killing pathogens, there is currently no alternative preparation. Therefore, rational use of anti-infective drugs in microecological prevention/treatment is extremely necessary. By contrast, inappropriate use of anti-infective drugs not only does no favor to timely control infection, but destroys the microecological balance in patients, increasing the incidence of double infection and other adverse reactions, and also causes rapid increase in drug resistant bacteria and bacterial resistance, further enhancing the threat of infectious disease to human health. Anti-infective drugs, as a precious health resource, must be carefully used. This is common sense to the domestic and foreign medical profession. In application of antibacterial drugs, clinicians should follow several basic principles: strictly master indications of anti-infective drugs; select antibacterial drugs accordingly; establish a therapy regimen according to pharmacological properties of the anti-infective drugs and specific infection condition; enhance laboratory monitoring during anti-infective drug therapy. Moreover, it should be sufficiently recognized that anti-infective therapy is not all about antibacterial drugs. Auxiliary measures, such as enhancing local drainage and improving systemic symptoms, help to improve the efficacy of antibacterial drugs.

According to whether the selection of antibacterial drugs is based on etiological diagnosis and an *in vitro* drug susceptibility test, anti-infective therapy can be divided into empirical therapy and specific etiological therapy. Empirical therapy occupies a certain position in anti-infective therapy, 'perform anti-infective therapy mainly according to previously accumulated etiological diagnosis and therapeutic experience as well as evidence based medicine'. However, so-called empirical therapy is not simple and blind use of broad-spectrum antibacterial drugs is wrong, but should be based on a full understanding of the prevalence of pathogens and drug resistance in a specific region and specific population. Even though, during therapy, a therapeutic regimen should still be corrected in time according to results of bacterial culture and a drug susceptibility test, in hematological patients or patients with a malignant tumor or after organ transplantation, whose self defense mechanism are impaired, empirical therapy often seems to be more important. To avoid rapid dissemination of infection, once these patients develop fever or other signs of infection, venous administration of antibacterial drugs should immediately be offered. However, selection of antibacterial drugs should not follow the orderless shift of high-price drugs, but should be made according to the specific conditions of the hospital, department and patient to choose relatively targeted antibacterial drugs.

The object of antibiotic application is the patient, so in addition to the severity of infection, therapy should also be provided based on the health state of the patient. In the case of elderly or severe pneumonia patients with high risk of several diseases, adequate antibacterial drugs should be selected for initial empirical therapy. Some foreign clinical observations also confirmed that in severe pneumonia, the mortality of the initial empirical therapy group was significantly lower than that of the inadequate therapy group.

To maintain ecological balance, broad-spectrum antibiotics should be used at the very beginning for severe pneumonia to cover all possible pathogens. Subsequently (48 – 72 h), antibiotics are adjusted according to the results of the microbiological test, *i.e.* replaced by narrow-spectrum antibiotics to reduce coverage. This is known as step-down therapy.

The main measures of respiratory system ecological prevention/treatment include the following aspects:

(i) Follow natural rules, support normal flora, protect natural microecological environment and promote microecological balance during therapy. Air is an important factor for humans and other living creatures. Stimulation of a few microbiological antigens from the air is favorable in improving body immunity. For example, hygiene hypothesis deems that due to improvement in hygenic conditions, the surrounding environment of a human residence is too clean. The opportunity for people, especially children, to contact pathogenic microorganisms is significantly reduced. Natural bacterial and viral infection can stimulate the body's immune system, inducing Th1 response. A decrease in such infections weakens the body's Th1 response. Meanwhile, Th2 response is relatively enhanced and the body is susceptible to allergic diseases. Therefore, an atmospheric environment with normal physical properties, chemical composition and certain microorganism content is necessary to maintain a human ecological balance. Air is an important factor in which humans and other microorganisms live.

(ii) Improve the microecological environmentThe micro-environment has a direct and very important influence on the normal biological population. The microecological environment includes the biological environment, referring to the host's biological environment and micro-environment, mainly consisting of the local physical and chemical environment. Any pathological change in the host can cause an imbalance in the microecological environment factors. For example, according to surveys of smokers, smoking can make bronchial epithelial cilia shorter and irregular and inhibit cilia movement, cause hyperplasia of bronchial goblet cells, increasing mucous secretion and weakening bronchial purifying capacity, cause congestion, edema and mucous accumulation of bronchial mucosa and weaken the function of alveolar phagocytes; it can also cause bronchial spasm. All of these favor bacterial colonization in bronchi. Therefore, chronic pulmonary inflammation leads to an increase in protease released by leucocytes and macrophages, further aggravating pulmonary tissue and alveolar wall injury. Giving up smoking can relieve congestion and edema of respiratory tract mucosa and prevent respiratory tract infection.

Congenital abnormality of the respiratory system anatomical structure and tumor compression can become micro-environmental factors causing ecological imbalance. If comprehensive therapy including surgical correction of abnormal anatomical structure and tumors is available, the micro-environment can be improved to promote microecological balance.

For any patient with compromised resistance to colonization, such as patients having organ transplantation and chemotherapy, improvement of the micro-environment is particularly important to avoid invasion of exogenous microorganisms. Currently, the use of an isolation chamber, air disinfection and filtration, water and food disinfection and avoidance of contamination by improving the artificial micro-environment to protect the host against invasion of exogenous microorganisms are important.

(iii) Improve the host's immunity and enhance the host's adaptability.

Artificial active and passive immunity is applied to improve the host's immunity. The surface of *Streptococcus pneumoniae* is a polysaccharide capsule, which can avoid the host's phagocytes. A passive or active immunity measure is adopted to make the body generate anti-capsule antibodies and thus improve phagocytosis, enhancing the body's defense against *Streptococcus pneumoniae* invasion. The currently used 23-valent *Streptococcus pneumoniae* vaccine can cover more than 90% of invasive *Streptococcus pneumoniae*.

Though there is still controversy over the effectiveness of vaccine, in adults with normal immunity, the overall effective rate of vaccine prevention is 75%. In the majority of healthy adults, capsule antibodies increase by more than 2 times within 2 – 3 weeks of inoculation. These specific antibodies can be maintained for several years.

Home oxygen therapy and respiratory muscle function exercises, including Qigong, Tai-Chi, breathing exercises, quantitative walking or climbing stairs, can enhance the host's adaptability.

The development of science and technology not only allows us to enjoy the achievements of modern civilization, but also brings us the various negative consequences of modern civilization. One consequence is pollution of the human environment, especially air pollution which is closely related to respiratory tract diseases. Therefore, besides taking good care of our home, the Earth, how to identify and treat respiratory tract diseases secondary to air pollution as early as possible, protect respiratory tract mucosa from the impact of air pollution through effective measures, maintain stability of the respiratory tract microecological environment and thus reduce respiratory tract infections, are all sure to attract more and more research interest in the future.

References

[1] Lowry P W, Tompkins L S. Nosocomial Legionellosis: A review of pulmonary and extrapulmonary syndromes. Am J Infect Control, 1993, 21: 21-27.

[2] Victor L, Emanuel Y, Vergis N L. In: Comprehensive Respiratory Medicine. Albert R, Spiro S, Jett J (eds.). York: Harcourt Brace and Company Limited

UK, 1999, 2235-2246.
[3] Mandell L A. Sequential antibiotic therapy. Nether J Med, 1997, 93-96.
[4] Cunba B A. Community acquired pneumonia. Postgrad Med, 1996, 99: 113-119.
[5] Richardson M D, Wamock D W. Fungal Infection, Diagnosis and Management. 2nd Ed. Oxford: Blackwell Science, 1997, 20-58.
[6] Collins H L, Kaufmann S H. Prospects for better tuberculosis vaccines. Lancet Infect Dis, 2001, 1: 21-28.
[7] Caws M, Drobniewski F A. Molecular techniques in the diagnosis of *Mycobacterium tuberculosis* and the detection of drug resistance. Ann N Y Acad Sci, 2001, 953: 138-145.
[8] Somoskovi A, Parsons L M, Salfinger M. The molecular basis of resistance to isoniazid, rifampin, and pyrazinamide in *Mycobacterium tuberculosis*. Respir Res, 2001, 2: 164-168.
[9] Kobayashi K, Kaneda K, Kasama T. Immunopathogenesis of delayed-type hypersensitivity. Microsc Res Tech, 2001, 53: 241-245.
[10] Soini H, Musser J M. Molecular diagnosis of mycobacteria. Clin Chem, 2001, 47: 809-814.
[11] Barrow W W. Treatment of mycobacterial infections. Rev Sci Tech, 2001, 20: 55-70.

16

Infectious Microecology of Skin

Yina Wang, Hong Fang *
Department of Dermatology, the First Affiliated Hospital, School of Medicine, Zhejiang University, Hangzhou, 310003, China
* E-mail: fanghongzy@sina.com.cn

As a covering on the surface of the human body, skin serves as an active border between internal and external environment. As a great ecosystem, skin consists of several ecological compartments. Like different countries and nations on the earth, different microorganism groups also colonize in different parts of the skin [1-2]. Similar to the importance of geological structure to global geographic environmental research, a better understanding of cutaneous structures will also lead to a better understanding of cutanous microecology.

16.1 Histological Structures of Skin

The skin is composed of three layers: epidermis, dermis and subcutaneous tissue, with the distribution of plenty of blood vessels, lymph vessels, nerves, muscles and adnexa derived from epidermis. The epidermal adnexa include hair, sebaceous glands, and nails of fingers or toes. Skin is the largest organ of the human body in weight and in area, which accounts for 16% of weight and 1.2 – 2.0 m^2 in surface area. The surface area of skin in a newborn is about 0.21 m^2.

16.1.1 Epidermis

The epidermis is the outermost layer of the skin. There is considerable regional variation in its relative thickness. The epidermis is the thickest on the palms and soles, measuring approximately 1.5 mm. It is very thin on the eyelid, where it measures less than 0.1 mm. The epidermis is mainly composed of two basic cell types: keratinocytes and dendritic cells.

16.1.1.1 Keratinocytes

Kerationocyts are the principal cells of the epidermis, which account for more than 80% of the epidermis. It has the specialized function of the production of keratin during differentiation process. Keratinocytes have direct connections with adjacent kerationcytes by desmosomes. The epidermis may be divided into the following layers, beginning with the innermost layer: basal layer, prickle layer, granular layer, clear layer and stratum corneum. The basal layer is connected with the dermis by basement membrane zone.

Basal layer. It lies in the innermost zone of epidermis, and consists of a single layer of columnar cells, which are aligned in the shape of a bar vertical to the basal membrane zone. Onofilament can be observed in the cytoplasm of basal cells under an electron microscope. Basal cells proliferate and eventually differentiate into new keratinocytes, which serve as the origin of epidermic cells, and are of great importance in self-recovery as well as in the traumic recovery process. Usually, about 10% of basal cells undergo karyokinesis everyday and move in a stepwise fashion to the stratum corneum to fall off at last. A whole renewal cycle for keratinization takes about 41 – 57 d.

Prickle layer. It lies upon the basal layer and consists of 4 – 10 layers of cornified cells. As the upper layer moves, the better is the cell differentiation, and keratinocytes tend to flatten out in shape. A tonofilament can also be observed in the cytoplasm under an electron microscope, which is fashioned into bundles and attached to desmosomes. Orbicular membrane-coating keratinosome in the cytoplasm is called the Odland body.

Granular layer. It consists of 2 – 4 layers of flat cells of fusiform shape. Amorphous particulate material can be observed in the cells. In this layer, the Odland body moves closer to the cellular membrane and enters intercellular space, which contains bipolar phosphonolipid and converts to a barrier shield.

Clear layer. It can only be observed on palms and soles, which consists of 2 – 3 layers of flat eosinophilic cells without nucleus. There's plenty of hydrophobic phospholipid in the clear layer, which together with tonofilaments, establishes a barrier to water and electrolyte.

Stratum corneum. It functions as the safeguard of skin. It consists of 5 – 20 layers of dead flat cells without nucleus. The outmost layer of the stratum corneum tends to fall off easily, while the inner layers are tightly-connected. Derived from amorphous particulate materials, electron-dense interfilamentous protein matrix and tonofilaments fill in the cytoplasm, leading to a powerful barrier function. Due to the barrier function, water and electrolyte cannot pass through the epidermis freely. Furthermore, the tight and ordered alignment of keratin and lipids strengthens the barrier function of the epidermis, helps to inhibit the proliferation of microorganisms and protects the skin from invasion of microorganisms. The metabolic disturbance of epidermal lipids for example, and shortage of fatty acid, will damage the barrier function of skin, and result in pathologic symptoms such as dryness, desquamation, inflammation and infections by microorganisms.

16.1.1.2 Dendritic Cells

Dendritic cells constitute only a small part of epidermic cells, which are characterized by extending dendrites. They are composed of three types of cells: melanocytes, Langerhans' cells and Merkel cells.

Melanocytes. Melanocytes account for about 10% of basal cells. The main functions of melanocytes are production of melanin and transportation of melanin to adjacent keratinocytes. Melanin can block and reflect the sunlight, therefore protect the deeper tissues against radiation damage. A melanin granule in keratinocytes lies on the top of the nucleus which, like an umbrella in shape, helps to alleviate direct impact on the nucleus by photons, and therefore protects the nuclear DNA.

Langerhans' cells. Langerhans' cells are normally found scattered at the middle part of the epidermis. They originate in activated lymphocytes of bone marrow, with intracytoplasmic organelles called Birbeck granules, which are of great importance in antigen presentation. Functionally, Langerhans' cells are similar to phagocytes, and play a role in uptaking, processing and presenting antigens. The exogenous proteins can be digested to immunogenic peptide by Langhans' cells and then be presented to effector T cells.

Merkel cells. Merkel cells are distributed among basal cells, with a lot of neurosecretory-like granules in the cytoplasm. At the bottom of Merkel cells, there are synapse organelles, which connect with the demyelinated nerve endings, can act as touch receptors, and take part in the afferent pathway of the slow adapting type of mechanical receptors in skin.

16.1.2 The Dermis

The dermis is derived from mesoderm, and there is considerable regional variation in its relative thickness. The dermis is thinnest on the eyelid, where it measures only 0.3 mm; and thickest on the back, where it is 30 – 40 times as thick as the overlying epidermis. The components of the dermis are collagen, a small amount of elastic fibers and ground substance. Blood vessels, lymph vessels, nerves, cutaneous adnexal structures and cell components, such as mast cells, macrophages, Langhans' cells and melanophages, are scattered among them. The ground substance is rich in proteoglycan and aminopolysaccharide. The aminopolysaccharide in the dermis is mainly composed of hyaluronic acid and dermatan sulfate, with some chondroitin sulfate and heparan sulfate, and plays an important role in maintaining the water content of skin. The molecules in the ground substance form a molecular sieve of three-dimensional configuration with many micropores, and enable the transportation of water, electrolyte and nutrient substance to promote the material exchange with the epidermis. The ground substance is the place for the exchange of metabolic materials, and has a barrier function against the invasion of bacteria.

16.1.3 Subcutaneous Tissue

The principal components of subcutaneous tissue are loose connective tissue and sebaceous lobules. There is considerable variation in its relative thickness according to the regional difference and nutritional status. The subcutaneous tissue is very thick on the abdomen and buttocks, while it's much thinner on the nasal tip and breast bone. The eccrine sweat glands, hair follicles, blood vessels, lymph vessels and nerves can also be found in subcutaneous tissue.

16.1.4 Cutaneous Appendages: The Adnexa

Skin appendages, which include hair, nails and sweat glands, are composed of two distinct components: an epidermal portion, which produces the differentiated product, and the dermal component, which regulates differentiation of the appendage.

16.1.4.1 Hair and Hair Follicle

The hair is formed from cornified keratinocytes, and is divided into three types of hair: long hair, short hair and woolly hair. Hair is distributed throughout all skin sites except for fingers, the extensor aspect of paratelum toes, palms and soles, nipples, vermilion of the lip, glans penis and clitoris. The growth of hair can be modulated by various factors such as genes, health, nutrition, drugs and hormones. Hair follicles grow down into the dermis and subcutaneous tissue. Normally, sebaceous glands and apocrine sweat glands form an outgrowth from the upper portion of the hair follicle. The hair follicle consists of three parts: the infundibulum, from the ostium at the surface of the skin to the open part of the sebaceous gland; the isthmus, from the base of the infundibulum above to the place where arrector muscles of hair adhere; and the foot, which stretches from the base of the isthmus to the bottom of the hair follicle.

16.1.4.2 The Eccrine Sweat Unit

As one type of simple coiled tubular gland, the eccrine sweat unit is composed of small sweat glands, and the secretory portion of the unit is found within the dermis and subcutaneous fat. Eccrine sweat units are found at virtually all skin sites except for vermilion of the lip, glans penis, prepuce wall, labium minus and clitoris. The total amount of eccrine sweat units reaches about 2 – 4 million, with an average amount of 143 – 339/cm^2. They are most abundant on the palms, soles, forehead, and axillae, with an amount of 620/cm^2; while much rarer on the back, with an amount of 64/cm^2. The eccrine sweat unit functions as a secretion of sweat and regulator of body temperature, and the secretion is mainly mediated by cholinergic innervations.

16.1.4.3 The Apocrine Unit

As one type of tubular gland, the apocrine sweat unit is composed of large sweat glands, and the secretory portion of the unit is found within the subcutaneous fat. The tubular structure of the apocrine unit is similar to that of the eccrine sweat unit. Mainly, the excretory portion of the duct opens into the infundibular portion of the hair follicle, which is regarded as the follicle apocrine unit. A small part of the duct, however, opens directly into the epidermis, and it is regarded as the non-follicle apocrine unit. Apocrien units of the human body are generally confined to the following sites: axillae, umbilicus, areola of breast, anus and cunnus. Apocrine secretion is mediated by neuroendocrine innervation and the secretion peaks in the juvenescent phase.

16.1.4.4 The Sebaceous Gland

The sebaceous gland is a type of exocrine gland of honeycomb appearance. It has no glandular cavity, and glandular cells gradually enlarge from outside to inside, with abundant lipid in their cytoplasm. The lipid is continuously being extruded through the short sebaceous duct. Most ducts of the sebaceous gland open at the infundibular portion of the hair follicle and constitute a hair follicle sebaceous gland unit. Some ducts can open directly at the skin surface. Sebaceous glands are found in greatest abundance on the face, scalp, breast, back and axillae, though they are distributed throughout all skin sites except the palms, soles, flexor aspect of fingers and toes. The components of sebum vary depending on different ages, diets and race. Cutaneous sebum is rich in lipid and essential amino acids. Regulated by endocrine secretion, the sebaceous gland changes with increase in age. It is primarily regulated by the androgen level, and also can be affected by progesterone and ACTH.

16.1.4.5 The Nails

The nails include fingernails and toenails. Nails are at the extensor aspect of the paratelum on fingers or toes, which consists of compact and solid keratinocytes. Nails act to assist in grasping small objects and in protecting the fingertip from trauma. Nails may provide a hiding place for microorganisms and is the most important tool for scratching.

16.2 Functions of Skin

Skin is the surface of the human body, and serves as the first-line protection against damage. It plays an important role in maintaining the healthy state of the human body. Functionally, the skin protects the organs and tissues against mechanical, physical and chemical injuries, and prevents the loss of various kinds of nutrition, electrolyte and water. In addition, skin affords a unique barrier function to defend the invasion of microorganisms. Recently, with the development of molecular biology, there's a growing understanding of the biochemical processes of skin, and much more attention has been paid to the functions of skin and the relationship between skin and microecology [3].

16.2.1 Biological Barrier Function of Skin

Primarily, microorganisms colonize the superficial sites of skin, such as hair follicle, infundibulum of sebaceous glands, opening of sweat units and epidermic lipid layers. Usually, due to the defensive function of skin, most microorganisms do no harm to the human body. The outermost 2 – 3 layers of stratum corneum easily fall off, while the inner layers are tightly aligned. Keratinocytes connect with each other *via* desmosomes which, together with the hydrophobic protein-bind phospholipid as well as tonofilament, scattered in the granular layer, clear layer and stratum corneum, function as a barrier against the invasion of microorganisms. Generally, bacteria with a diameter of 200 nm, and a virus with a half diameter cannot enter the skin. Moreover, the outmost layer of keratinocyte continues to shed, and results in elimination of colonized microorganisms. The dry environment and acid pH of the skin surface can limit the growth of microorganisms. Furthermore, due to the catalysis of lipoidase (mainly produced by *Propionibacterium acnes* and *pityrosporum*), long chain free saturated fatty acid and oleinic acid decomposed from the epidermic lipid also have an inhibitory effect on *Staphylococci*, *Streptococcus* and *Candida albicans*. After adolescence, certain kinds of unsaturated fatty acids, such as propanoic acid, octanoic acid and undecylenic acid, can inhibit dermatophyte such as *T. mentagrophytes*. That's why a spontaneous cure occurs in post-adolescence patients with tinea Alba. As to the microorganisms invading the dermis, the molecular sieve structure of the dermis helps to limit the extension of microorganisms and the phagocytosis of leucocytes.

16.2.2 Immune Function of Skin

Several immune-related types of cells in skin, including keratinocytes, Langerhans' cells, lymphocytes, endothelial cells and macrophagus, constitute the immunological system of the skin, and play a key role in the cutaneous defense process against microorganisms.

16.2.2.1 Keratinocytes

With the expression of IgG-Fc receptor and MHC-II, keratinocyte participates in the immune response mediated by T lymphocytes. More than 10 cytokines are produced by keratinocytes, including IL-1, IL-6, IL-8, IL-10, TGF-α, TNF-α, CSF and MCAF. Therefore, stimulation of keratinocytes may result in the activation of keratinocytes and a cascade of secretion of cytokines, which lead to a network of epidermal cytokines and finally have an important role in immunological homeostasis and immune response.

16.2.2.2 Langerhans' Cells

There are plenty of receptors on the surface of Langerhans' cells, including leucocyte common antigen, IgG receptor (FcγRII), C3biR and CDIa antigen. Similar to macrophage, Langerhans' cells can recognize, uptake and process the haptens, and present them to helper T cells. With the secretion of IL-1, Langerhans' cells may promote Th cells to produce IL-2, and lead to the activation and proliferation of T cells.

16.2.2.3 Lymphocytes

A small quantity of lymphocytes, primarily $CD8^+$ T lymphocytes, is found in the basal layer of the epidermis. In the dermis, however, $CD4^+$ T lymphocytes are the primary type, and $CD8^+$ T lymphocyte are secondary. T lymphocytes are distributed around blood vessels. Their differentiation and maturation are promoted by IL-1, which is produced by keratinocytes. Mature T lymphocytes function in mediating the immune response.

16.2.2.4 Endothelial Cells

Several adhesion molecules are expressed on endothelial cells, which serve as lymphocyte homing receptors. The lymphocyte homing receptor is regarded as one of the factors that lead to the exfiltration of inflammatory cells during the process of infective skin disorders.

16.2.2.5 Macrophagus

Macrophagus scatters among the superfacial layer of dermis and also participates in the immune response. It can decompose IL-1, IFN and some enzymes, and therefore mediate the specific immune response and non-specific immune response against external microorganisms.

16.2.2.6 Mast Cells

Mast cells locate around the vessels in the papillary layer of dermis, with Fc receptor of IgE on the surface, and can combine with IgE. They are intimately associated with type I hypersensitivity reaction, and take part in the occurrence of type IV hypersensitivity reaction. The activated mast cells triggered by both immune and non-immune mechanisms can produce and release various biologically active mediums, such as the chemotactic factor, active enzymes and vascular active compounds.

16.2.2.7 Dermal Fibroblasts

Dermal fibroblasts locate nearby the collagen fibers, and primarily function as the synthesis tool of collagens, elastic proteins and ground substance. Also, fibroblasts produce enzymes that can decompose the components mentioned above and thereby maintain the balance of the metabolism. Also, fibroblasts produce the keratinocyte growth factor, and have interactions with cytokines secreted by keratinocytes, which is very important to the self-balance of the cutaneous immune system.

Up to now, little is known about humoral immunity of skin. IgA antibody, IgG antibody and IgE antibody have been found in sweat. If skin is depleted of IgA, the ability to defend pyogenic infection is decreased.

Stimulated by external factors, the nerve endings of skin can release neuropeptide, participate in the local chemotactic response and inflammatory response mediated by immune cells.

16.2.3 Functions of the Sweat Gland and Sebaceous Gland

A number of functions have been attributed to the sweat glands, including odoriferous roles as sexual attractants, territorial markers, and warning signals, a role in increasing frictional resistance and tactile sensibility, as well as a role in increasing evaporative heat loss in some species. The sebaceous gland produces sebum that mixes with moisture such as sweat and is emulsified on the skin surface to form fatty acids that coat the skin. Sebum and sebaceous glands prevent invasion and infection by pathogens and toxic substances. Additionally, the sebaceous glands control water loss from the skin and maintain moisture in the horny cell layers.

16.2.3.1 Eccrine Sweat Gland

Mainly there are two kinds of activity during secretion process of the eccrine sweat gland. One is the secretion of ultrafiltrate, which, similar to blood plasma, is secreted by glandular clear cells under the innervation of acetylcholine. The other is the reabsorption of natrium by the duct. Sweat is one type of hypotensive liquid, with a specific gravity of 1.001 – 1.006 and a pH value of 5.5 ± 0.5. Water accounts for 99% – 99.5% in sweat, while the solid component accounts for only 0.5% – 1%. The solid component in sweat is composed of inorganic substance and organic substance. The former contains primarily sodium chloride, which may increase in concentration with the increase in sweat secretion, and reach a highest concentration of 120 mmol/L. The other ingredients in the inorganic substance include kalium ion, bicarbonate ion, calcium, magnesium, phosphonium and iron. The concentration of kalium ion is constant, with a variation range of 5 – 10 mmol/L. The organic substance in sweat is mainly composed of lactic acid and urea. Lactic acid is produced by clear cells *via* a glycolysis process. The concentration of lactic acid depends on the secretion of sweat. A higher concentration of lactic acid (30 – 40 mmol/L) is observed with a little secretion of sweat. However, the concentration of lactic acid may decrease to 10 – 15 mmol/L when a considerable quantity of sweat is secreted. The concentration of urea in sweat is higher than that in blood by 1.5 – 2.0 times. Some free amino acids can also be found in sweat, including glycine, histidine, threonine, asparagine, serine, aspartate, proline, alanine, carbamyl ornithine, valine, methionine, isoleucine, leucine, tyrosine, phenylalanine, ornithine and oysine. The nitrogen content of sweat varies from 1.5 – 4.76 mg%, and is only one quarter of the content in urine. Most proteins in sweat are small molecules with a molecular weight less than 10,000 Dalton, and the concentration of protein in sweat is about 20 mg/dL. Also, there's a small quantity of immunoglobulin such as IgG, IgA and proteolytic enzyme in sweat. It is reported that IgE content in sweat of patients with atopic dermatitis is higher than that in healthy controls.

The secretion of the eccrine sweat gland is associated with the environmental temperature. At room temperature, few eccrine sweat glands secrete. While the environmental temperature increases to 32 ℃, the eccrine sweat gland can be activated and the secretion of sweat is also increased, resulting in general perspiration. The secretion of sweat also relates to other factors such as psychological state, drugs and foods.

Physiologically, the eccrine sweat gland functions as: (i) Thermolysis: A high environmental temperature leads to rapid perspiration of 12 L/d. The continuous evaporation of water takes heat away. (ii) Excretion: Like the kidney, the eccrine sweat gland helps to excrete the metabolic production. (iii) Softens the horny layer: The sweat can maintain the water content of the stratum corneum and make the skin soft, smooth and moist. (iv) Acidifies the skin: The acid sweat helps to maintain the acid environment of the skin, and protects the skin from invasion of microorganisms. (v) Emulsifies the lipid: There's an emulsification effect between sweat and skin lipid, which helps to form an emulsifier on the skin surface, and supplies nutrition for the growth of normal microorganisms on the skin.

16.2.3.2 Apocrine Sweat Gland

The secretion of the apocrine sweat gland is sticky and milky, and the components are still unclear. The majority is water, and only a small part is solid, including iron, fluorescent material, colorful material, odorous material and lipid such as fatty acid, neutral fat and cholesterol. The fluorescent material is solvable in acetone, and gives fluorescence upon UV radiation. The colorful material can make the secretion yellow, green, red or black, and stain the skin and clothes. The odorous material is different according to different race, gender and age, and can produce short-chain fatty acid, amonia and a special smell in decomposition by some bacteria (for example, *Diphtheroid bacilli*). The exact function of apocrine secretion in humans is not known, although it serves as a sexual attraction and an activity range mark in horses.

16.2.3.3 Sebaceous Gland

The production of the sebaceous gland is sebum. The secretion of sebum varies in different people as well as on different skin sites. It is most productive on the scalp, with a secretion of 150 – 300 $\mu g/cm^2$, while on limbs only 5 – 10 $\mu g/cm^2$. With a complex mixture of lipids, sebum mainly contains triglyceride, wax ester and squalene, and a small quantity of cholesterol. At the site of the gland duct, the triglyceride can be hydrolized to monoglyceride by microorganisms such as *Propionibacterium* acnes and *Malassezia furfur*. The secretion of sebum depends on many factors such as race, age, gender, food, nutrition, and climate and skin site. The sebaceous gland endures two significant changes throughout the whole life of humans. One is at infancy shortly after birth, which is affected by the mother's sex hormone level (mainly androgen) and becomes active in sebum production, leading to similar components of sebum in infants to that in adults. During 2 – 8 years of age, the secretion of the sebaceous gland decreases gradually with decreased components of wax ester and squalene, and increased cholesterol and cholesteryl ester. Another change is at puberty. In response to endogenous androgens, sebaceous glands enlarge and become increasingly productive once again and, at the age of 10 – 15 years, the sebum components of an adolescent are close to those of an adult. At menopause, the sebaceous gland in women begins to wane, while in men it has residual secretion. Throughout all the age phases, the sebaceous gland is more productive in the male than in the female.

The sebum mixes with sweat and water, and is emulsified on the skin surface to form fatty acids that coat and moisturize the skin and hair. As a type of holocrine gland, sebum contains a mixture of cells, proteins, sugars and enzymes, which supply nutrition for normal microorganisms in skin. The free fatty acids have an inhibitory effect on some pathogenic fungi and bacteria.

16.2.4 Nutritional Metabolism of Skin

The skin participates in the metabolisms of carbohydrates, lipids, protein, water, electrolyte, vitamin and microelements.

Carbohydrate in the skin mainly composed of glycogen, glucose and mucopolysaccharide. The content of glucose in skin is 600 – 810 mg/L, which accounts for about 2/3 of the blood sugar level. The content of glucose is highest in the epidermis. The cutaneous glucose level is even higher in patients with diabetes mellitus, and it facilitates the proliferation of fungi as well as bacteria. The glycogen can be synthesized in skin by the monosaccharide condensation pathway (mainly) and glycolysis pathway and the enzyme lies in smooth endoplasmic reticulum of keratinocyte. Carbohydrate in skin mainly serves as the energy provider; also it can form the substrates for biological synthesis of mucopolysaccharide, lipid, glycogen, nuclear acid and protein. Two pathways are involved in the decomposition of carbohydrate: The aerobic oxidation pathway and anaerobic glycolysis pathway. The anaerobic glycolysis process in skin is much quicker than in other tissues, which might be attributed to the higher production of lactic acid resulting from the shortage of blood vessels and low oxygen content in the epidermis. The anaerobic glycolysis process also plays a role in the acid reaction of epidermis.

Two proteins, fiber-associated protein and non-fiber-associated protein, constitute the epidermic proteins. The former contains keratin, collagen and elastin. The keratin is the main structural content of keratinocyte, hair and nail, and tonofilament is the most important material that maintains the intracellular and extracellular tension. Non-fiber-associated proteins include nuclear protein and extracellular enzymes. Non-fiber-associated protein, in combination with mucopolysaccharide, can form mucoprotein, and participate in all other cellular functions except for keratinization. Multiple amino acids constitute the cutaneous protein. The content of tyrosine, cystine, histidine and tryptophane in epidermis is 3 – 4 times of that found in dermis. A higher content of proline, hydroxyproline, alanine and phenylalanine is found in dermis. The proteins are decomposed by proteinases, which participate in the release of chemotaxis polypeptides during the inflammatory process.

The cutaneous lipids include fat and lipoid. Fat locates at the subcutaneous tissue and supplies energy *via* β-oxidative degradation. The lipoid includes phospholipid, glycolipid, and cholesterol and cholesteryl ester. The contents of choleterol and phospholipid in epidermis are higher than that in dermis. The cholesterol exists in a free way. 7-dehydrogenation can synthesize vitamin D upon UV radiation. An important function of keratinocyte during its differentiation is to synthesize the lipids. From the terminal differentiation to the death of keratinocytes, the eipidermic lipids are synthesized on top of the granular layer, and lamellated granules (keratinosome) are then formed and fill in among keratinocytes. The lipid is the main content of epidermic sebum. The oxidation and decomposition processes of epidermic lipids are similar to those in other tissues. They are hydrolyzed to glycerol and fatty acid in cytoplasm. Some resident bacteria on skin, such as *Corynebacterium acnes* and *Pityrosporum furfur*, also release esterase and can decompose the triglycerid into free fatty acid. The most abundant essential fatty acids in epidermis are linoleic acid and arachidonic acid. The linoleic acid esterizes with phospholipid in the epidermic cell membrane and maintains the barrier function of the skin. Arachidonic acid is the precursor of prostaglandin and other arachidonic acid metabolic products.

The majority of water in skin is stored in the dermis. The metabolism of water is affected by the water metabolism all over the body. There are abundant electrolytes in skin, including trace elements such as natrium, potassium, magnesium, chlorine, calcium, phosphonium, zincum and cuprum, which account for 0.6% of the weight of skin. The electrolytes play an important role in maintaining the functions of epidermic cells.

16.2.5 Thermo-Regulation of Skin

The body temperature refers to the average temperature of interior tissues and organs of the human body, which ranges from 36.5 – 37.5 ℃. The skin temperature refers to the temperature on the surface of the skin, and it varies with the difference in environmental temperature and anatomical region. When the environmental temperature is 23 ℃, the temperature in the frontal region is 33.0 – 34.0 ℃, on the trunk is 32 ℃, on the hands is 30 ℃, on the feet is 27 ℃, and a little higher on reductus regions such as axil and inguina. When the environmental temperature is higher than 32 ℃, the difference between different skin sites decreases. In a cold environment, the temperature on hands and feet decreases significantly. The variation in temperature in the head is relatively mild. Skin temperature correlates with local blood flow closely. Factors that have an influence on the contraction of cutaneous blood vessels can also affect the skin temperature. Agitation may increase the level of angiotensin and contract the blood vessels, and result in decreased skin temperature. The cold environment also contracts the blood vessels and decreases the blood flow, and leads to decreased skin temperature. However, in a hot environment, it is just the contrary. The alteration in skin temperature has a relationship with the development of cutaneous microorganisms to some degree. In reductus regions, the skin temperature is relatively higher which, together with the humidity, facilitates the proliferation of microorganisms. The best temperature for the growth of *Dermatophyte* is 25 ℃, and the temperature in the sites of interdigital and inguinal regions during spring and summer offers optimized circumstances for them.

16.3 Characteristics of Cutaneous Microecology

Skin is sterile at birth. However, due to the environmental microorganisms and the surface location of skin, it becomes the first-line protection from microorganisms that are in constant contact with skin, and offers a large region for colonization of exterior microorganisms. Soon after birth, various kinds of fungi and bacteria begin to colonize the surface of skin. All these microorganisms, with a long-term co-adaptation on the skin surface, can inhabit the skin for a very long time. The category and quantity of cutaneous microorganisms are in accordance with the microecological rule, from the newborn, infants, adolescents, adults to the elderly. Although there is a variance among different individuals or among different sites, the cutaneous microorganisms can maintain a relative balance with the skin. These microorganisms are regarded as the normal microbial community of the skin. Commonly, they are non-pathogenic and constitute an indispensable part of life. They play a key role in maintaining the ecological balance of skin, offering the outmost biological barrier, defending the invasion of external pathogens and participating in the physiological functions. The changes in the internal or external environment on the human body will result in damage to the microecological balance as well as interactions among the normal microbial communities. Such conditions facilitate the invasion of pathogens, and turn the normal microbial community into a pathogenic one, therefore leading to the production of toxins and the infectious disorders of the skin and organs [4].

16.3.1 Normal Microbial Community of the Skin

Two groups of microorganisms, resident flora and transient flora constitute the normal microbial community of the skin. The category of the cutaneous normal microbial community only accounts for a very small part of the total variety of external microorganisms. There are estimated to be 60,000 – 80,000 types of microorganisms per square centimeter in skin. Generally, the normal microbial community of the skin is located on the outermost layer of epidermis (between the loosest stratum corneum and epidermic sebum membrance) and the openings of hair follicles, with most microbials existing in the form of a minute colony.

There's great variance in the cutaneous normal microbial community among different individuals as well as different biological sites. The main categories of the cutaneous normal microbial community are listed below.

16.3.1.1 The Resident Flora of Skin

Mainly, the resident flora of skin include coagulase-negative staphylococci (CNS), micrococci, Corynebacterium, mycoflora, protozoan and virus.

The resident flora refers to the microorganisms that can proliferate and inhabit the skin permanently, and do not usually comprise pathogenous microorganisms.

Coagulase-negative staphylococci (CNS). Eighteen types of CNS have been isolated from the skin. The most common *CNS* is *S. epidermidis*. Other *CNS* include *S. haminis*, *S. capitis*, *S. auricularis*, *S. saccharolyticus*, *S. warneri*, *S. hemolyticus*, *S. saprophyticus*, *S. cohnii*, *S. xylosus* and *S. simulans*, with the first two types being the next most abudant. As the most dominant microorganism in human skin, *S. epidermidis* is one of the main symbiotic bacteria of skin, and serves as one of the most important members of the cutaneous normal microbial community. *S. epidermidis* is most abundant on the superior part of the trunk, where it accounts for more than 50% of resident *staphylococci.* It is of great importance in maintaining the cutaneous microecological balance, while upon the disturbance of cutaneous microecology it may lead to the infection of organs as one of the opportunistic pathogens. Another important symbiotic bacterium of skin is *S. haminis.* It tends to inhabit sites with a prosperous secretion of glands, such as axil, buttocks, pubic symphysis, perineum, inguina and legs. *S. capitis* inhabits primarily the sites of scalp, frontal region, eyebrows, face, neck, external acoustic meatus and secretory opening of sebaceous glands, while on other sites of the skin it lives in the form of transient flora. *S. auricularis* inhabits primarily the sites of external acoustic meatus, together with *S. capitis*, constituting the dominant symbiotic bacteria around the external acoustic meatus. As one kind of anaerobe in *Staphylococci*, *S. saccharolyticus* accounts for 20% of the normal microbial community, and lives primarily in the sites of the frontal region and cubital fossa, and in other sites such as fingers, buttocks and waist, or even hair follicles distributed in the sites mentioned above. Although it grows slowly and is scanty in quantity, *S. saccharolyticus* is thought to be one kind of resident flora, for it meets the definition of resident flora. It was overlooked in the study of cutaneous microecology. However, now it is realized that *S. saccharolyticus* might be an important category of cutaneous microbials. Other *staphylococci*, such as *S. warneri*, *S. hemolyticus*, *S. saprophyticus*, *S. cohnii*, *S. xylosus* and *S. simulans*, are grouped as transient flora.

Micrococci. Although it is not so common in skin as *staphylococci*, eight types of *Micrococci* have been isolated in skin, including *M. luteus*, *M. varians*, *M.lylae*, *M. nishinomiyacnsis*, *M. kristinae*, *M. sedentarius*, *M. agieis* and *M. roseus*, with the most common type being *M. luteus* which, together with *M. varians*, constitutes dominant symbiotic bacteria. *M. lylae* and *M. kristinae* are much more common in children's skin, while *M. lylae* is more common in cold seasons.

Corynebacterium. It is one of the gram-positive microbials with pleomorphism. The most common *Corynebacterium* is *Diphtheroid*, which accounts for the majority of cutaneous resident flora, and is divided into two groups: Aerobic group and anaerobic group. *Bacillus brevis* is the next most common in *Corynebacterium*, *Aerobic Diphtheroid.* It is distributed on reductus and humid sites such as axil, inguina, buttocks internatal groove, interdigital regions, nose, pharynx, conjunctiva and external acoustic meatus, and is even more common in sweaty individuals. It can be divided into two types, lipophilic type and unlipophilic type. The fomer is the predominant type, and can be encouraged in growth by oleic acid of the cutaneous sebum. As one common type of lipophilic *Corynebacterium*, *C. minutissimum* is capable of producing porphyrin, resulting in superficial erythrasma on the sites of axil and inguina. It was regarded as a single microbial. However, now it's clear that *C. minutissimum* is one kind of compound bacteria that is composed of 8 types of microbial. *C. tenuis* is the pathogen of trichomycosis axillaris, inhabiting the hair cuticles of axillary hairs and pubic hairs in an intracellular or an intercellular way. Nevertheless, it does no damage to hair roots or skin.

Anaerobic Diphtheroid. It is one of the resident floras in hair follicles and in sebaceous glands, and is one of the dominant categories of cutaneous normal microbials. The classification of this microbial is still a matter of debate. However, according to the type of colony and susceptivity to the decomposition of bacteriophage, it is commonly divided into three types. The most common one is *C. acanes* which, as one of the most predominant populations on human skin, is the main member of the cutaneous microecological system. It plays an important role in maintaining the stability of the cutaneous microecological system and in the metabolism of cutaneous lipids. Disturbance to cutaneous microecology may result in the over-proliferation of *C. acanes*, which is the one of the most important pathogentic factors for acne. The quantity of the microbial is parallel to the production of sebaceous glands on sebum-rich sites such as the scalp, frontal region, and upper breast and back. The amount of *C. acanes* reaches a peak in adolescence, and then it's stable in adults, and doesn't decrease until old age. *C. granulosum*, with an amount in quantity only less than *C. acanes* (accounts for about 20% in *Corynebacterium*), is distributed everywhere sporadically, although it can be more easily isolated in sebum-rich sites. It is rather common in comedones, and it has been regarded as one of the pathogens of acne. Another category of *Corynebacterium*, *C. avidum*, tends to inhabit humid reductus sites such as axil, perineum and nasal cavity.

Bacillus brevis. It accounts for a certain percentage in *Corynebacterium*. *Bacillus brevis* may produce protease and grow very quickly. It can be observed in interdigital regions in patients with tinea pedis, and is capable of producing an awful smell in feet.

Mycoflora. Fungi, especially some yeast fungus, also account for a certain percentage in normal cutaneous microorganisms. It is also recognized that *Mycoflora* is one of the predominant microorganisms in normal cutaneous microorganisms.

Pityrosporum. As one type of lipophilic yeast, *Pityrosporum* requires an environment with a high content of fat for growth, and olive oil is essential in the *in-vitro* culture. There are two types of *Pityrosporum* according to the shape, *P. ovale* and *P. orbiculare*, although sometimes these two types are regarded as the same one. *Pityrosporum* exists in blastospore form among normal cutaneous microflora. In conjunction with the production of cutaneous lipids, it's most abundant on the back, and *Pityrosporum* turns into a hyphal form after it enters the deep layers of stratum corneum. It is now generally accepted that *Malassezia furfur*, the pathogen of pityriasis versicolor, and *Pityrosporum*, are exactly the same fungus.

Candida. Normally, the isolation rate of *Candida* in the oral mucous membrane is as high as 40%. The isolation rate of *C. albicans* in normal skin is 15%. Pythogenetic *C. parapsilosis* and *C. tropicalis* are non-lipophilic yeasts, with a higher isolation rate in interdigital regions. Broad-spectrum antibiotics, immunosuppressive agents and a high dose of glucocorticoid may result in the mass proliferation of *Candida* and then lead to diseases. Dermatophyte in interdigital regions related to tinea pedis is also regarded as one part of cutaneous normal microflora.

Protozoan. *Protozoan* (for example, *Demodex folliculorum*) inhabits hair follicles and sebaceous glands, with a higher isolation rate in sebum-rich sites such as face and scalp.

Virus. It is still a matter of debate as to whether a virus should be included in the cutaneous normal microflora. Nevertheless, *Herpes simplex virus* and *Varicella-zoster virus* can stay in certain parts of the skin for a long time or even for a lifetime. The former primarily lies on the border between the skin and mucous membrane, such as oral lips and genital organs, and the virus might be in incubation, which colonizes basal cells of a rare type. Generally, the virus has difficulty proliferating due to the local protective system; however, as soon as there's an opportunity, such as decreased resistance of the host and local immune deficiency, the virus will result in disorders.

16.3.1.2 Transient Flora of Skin

Mainly, the transient flora of skin include Staphylococcus aureus, Streptococcus, Sarcina, Neisseria and gram-negative rods.

The transient flora refers to the microorganisms that inhabits the skin temporarily, and which may then disappear after a period of time.

Staphylococcus aureus. Due to the natural resistance of human skin to the coagulase positive *Staphylococcus aureus*, it is very difficult for *Staphylococcus aureus* to inhabit healthy skin. However, if qulitation is considered, it's easier to trace *Staphylococcus aureus* in the whole cutaneous ecological system. The positive rate of *Staphylococcus aureus* in reductus sites is very high, for example, in the perineal region it is 20%, while on the nose it is higher. The persistent carriage rate in the population is estimated to be 20% – 40%. The carriage rate is even higher in hospitals, as well as in patients with diabetes mellitus, vein addicts and dialysis. *Staphylococcus aureus* can be found all through the skin of patients with psoriasis and atopic dermatitis. It is the most common bacterium that leads to pyogenic infection in skin and the mucous membrane.

Streptococcus. A type of gram-positive bacteria, *Streptococcus* is spherical or orbicular in shape, with an alignment by way of germination or in a chain. It can be divided into three types according to the haemolytic character: α, β and γ *Streptococcus*. It has various types and a wide range of distribution. Generally, α *haemolytic Streptococcus* or γ *non-haemolytic Streptococcus* can be found in the nose and laryngea pharyngis of healthy individuals, but is hard to detect on skin. Nevertheless, at the anaphase of newly born, the cutaneous isolation rate of α *haemolytic Streptococcus* or γ *non-haemolytic Streptococcus* is rather high. β *haemolytic Streptococcus* has a robust virulence and leads to pyogenic infection readily.

Sarcina. It has a high isolation rate in healthy skin of infants.

Neisseria. As one type of gram-negative bacteria, it has a special requirement for oxygen and produces oxidase and catalase. The isolation rate of *Neisseria* is high in the nasopharynx in healthy individuals, but on skin it's rare. Only *Neisseria intracellularis* or *Neisseria gonorrhoeae* is dangerous in humans.

Gram-negative rods. These are not common among cutaneous resident flora due to dryness. However, as one kind of cutaneous transient flora, it usually results from contamination of the discharge inthe GI tract. It can be detected in healthy individuals on sites of humid reductus such as perineum, axil, interdigital region and nasal mucosa. It is mainly composed of the types listed below.

Acinetobacter. As one kind of anaerobe, *Acinetobacter* can be found extensively in nature. It can be detected on the skin in more than 25% of healthy individuals. The isolation rate in males is higher than that in females. The quantity of *Acinetobacter* is especially higher in summer due to the increased secretion of sweat and high humidity.

Esherichia. As a group of dynamic gram-negative rods, *Esherichia.* is regarded as one of the normal microflora in the human intestinal tract. The most common type, *E. coli*, can be detected in the normal skin of infants and children. It's not dangerous in humans; on the contrary, it helps to synthesize vitamin B and K in the intestinal tract and is good for the human body. Nevertheless, a certain type of *E. coli* may cause cutaneous infection in a condition of immune-depression.

Proteus. As another kind of normal microflora in the human intestinal tract, *Proteus* can facilitate the phagocytosis *via* its pilus and then result in a decreased virulence. It exists in human skin as a type of transient flora. Normally, it is non-morbid. However, in unusual conditions, it may proliferate quickly, which leads to an increased quantity and makes *Proteus* a conditional pathogenic bacteria.

Pseudomonas. It can be found extensively in nature and consists of many types, with the most important type being *P. aruginosa.* Although it's another kind of resident flora in the human intestinal tract, *P. aruginosa* is a type of transient flora in skin. The amount of *P. aruginosa* may increase significantly in immunodeficient patients or hospitalized patients. The virulence of *P. aruginosa* comes from the structural components, toxins and enzymes, which can result in infections from operative incisions, burn wounds and severe drug eruption or, even worse, result in septicemia.

Alkaligenes faecalis. It's one kind of resident flora in the human intestinal tract. Yet, it also can be isolated in skin in a small part of healthy individuals.

16.3.2 Influential Factors of Cutaneous Normal Microflora

There's a relatively standard system of normal microflora in skin. However, the quantity as well as the construction of flora may change due to many influential factors. These influential factors include the interior factors, the environmental factors, and the interaction among different bacteria [4, 5].

16.3.2.1 Climate (Temperature and Humidity)

Usually, resident flora can be affected by an external 'big circumstance' and local ecologic circumstances. The alteration in the environmental climate, namely in temperature and humidity, may have an influence on the ecological circumstances. A higher temperature and increased humidity will lead to increased hydration of the stratum corneum. Generally speaking, a humid environment promotes the proliferation of bacteria, while a dry environment inhibits it. It has been indicated that an eligible temperature and humidity is essential for the proliferation of bacteria. A study reported that when inoculated into skin, the bacteria lived longer on wet skin than on dry skin. Bacteria on the forearm skin increased 10,000 times when watered for 24 h, and with gram-negative rods, gram-negative *Corynebacterium* and *Candida* increased much quickly than *coccobacteria.* Also, the increased temperature and humidity caused by the wafer may result in a change in the microorganisms, such as yeasts (*Candida* and *Pityrosporum*) and *Dermatophyte*, from non-pathogenic types to pathogenic types. Other factors, *i.e.* the increased CO_2 level in the local region, join forces in the influence on the proliferation of bacteria. However, the quantity of bacteria may decrease to a normal range once the wafer is removed. Interestingly, some bacteria, such as *nicrococcus*, favor a dry and cold environment.

16.3.2.2 Age

Age is another important influential factor. The carriage rate of *Micrococcus*, *Corynebacterium* and gram-negative bacteria in infants is much higher than that in children or adults. Furthermore, the carriage rate of pathogenic bacteria in infants is higher. Before adolescence, cutaneous resident flora mainly consists of *Staphylococcus epidermidis* and *sarcina*, and *E. coli* can be isolated in the inguina and perineum, while the isolation rate of *P. ovale* and *Corynebacterium acnes* is low. It's very hard to find *round Pityrosporum* in children less than 5 years old. However, the amount of bacteria will increase over the following 10 years, and will be close to the adult level by the age of 15. The amount of *Corynebacterium acnes* also increases in adolescence, due to the increase in sebaceous secretion and free fatty acid.

16.3.2.3 Position

The constitution of the normal cutaneous microflora varies according to the difference in the skin site. At the UV-exposed areas such as face, neck and hands, the transient flora take advantage; while at lipid-rich areas such as face and upper trunk, lipophilic bacteria become the predominant type. As a special ecological region, the scalp has a high density of *Staphylococci*, *Propionibacterium acnes* and *Pityrosporum*. Relatively closed areas, such as axillary fossa, the perineal region and interdigital region, have a higher temperature and humidity, and supply a special ecological environment for the cutaneous colonized bacteria. The normal flora in the axillary fossa includes *Staphylococci* and *Corynebacterium*, while the perineal region includes *Corynebacterium minutissimum*, which is the pathogen for erythrasma. Many other bacteria, with the most part being gram-negative bacteria, *Dermatophyte* and some conditional pathogenic bacteria have also been isolated in the interdigital region. The amount of bacteria in the upper arms and upper legs is much less due to the dryness. Nevertheless, there are some bacteria that can proliferate anywhere, such as *Staphylococcus epidermidis*.

16.3.2.4 Sebum and Sweat

The sebum, secreted by the sebaceous gland, is a key factor that affects the regional distribution of normal microflora in skin. Before adolescence, epidemic microflora is relatively low due to the low secretion of sebum. During adolescence, with the gradual increase in the sebum secretion, a relatively standard lipophilic microflora will come into shape at the lipid-rich areas, such as the face and upper trunk. Among them, the anaerobic *Propionibacterium acnes* takes predominance, which produce extracellular lipase and decompose the triglyceride into glycerol. Due to the anoxic structure of the close crypt, anaerobic bacteria, for example *Propionibacterium acnes*, favor the deep part of the sebaceous gland, while lipophilic aerobic bacteria, such as *Malassezia*, favor the upper site of the infundibulum in the hair follicle. At sebum-scanty areas, the major influential factor is water. For example, the axil, inguinus and interdigital regions are full of sweat glands, with less sebaceous glands as well as less air circulation, which leads to the proliferation of hydrophilic bacteria (mainly gram-negative rods and *Staphylococcus aureus*). Many hydrophilic bacteria may proliferate in dry skin (such as the forearm) after being wafered. However, in sebum-rich areas such as the scalp, no significant change in cutaneous flora is observed after wafer treatment.

16.3.2.5 PH Value of Skin

The pH for the growth of cutaneous resident flora (mainly *Staphylococcus epidermis*) ranges from 6.5 to 8.5, with the best range from 7.5 to 8. Although pH in normal adult skin is about 4.5 – 6, cutaneous normal microflora is tolerant to pH and can grow well. The pH value in the skin of the newborn and infants is higher than that in adults, which reaches 6.0 – 7.0, and is more suitable for the growth of the resident flora than in adults' skin. The low pH in adults' skin should be mainly attributed to the increased production of fatty acid in the sebum layer. The total amount of flora on adults' skin is not less than that in infants, which results from the proliferation of lipophilic *Pityrosporum* and *Corynebacterium diphtheroides*. Marples *et al.* reported that a 5-d-wafer on the forehead resulted in the increase in pH (from 4.38 to 7.05) and average cfu (colony forming unit), from 1.8×10^6 CFU/mL to 4.5×10^6 CFU/mL). However, a different opinion pointed out that cutaneous pH should not be regarded as an independent influential factor, and it might have a combined action with temperature and humidity to affect the growth of the flora.

16.3.2.6 Oxygen and Carbon Dioxide

There are anaerobic bacteria (such as *Corynebacterium acnes*), specific aerobic bacteria (such as *Brevibacteriaceae*) and facultative anaerobic bacteria (such as *staphylococci* and *Corynebacterium*) in cutaneous normal microflora. Therefore, the concentration of cutaneous oxygen and carbon dioxide is very important for the habitat of microorganisms. Although the epidermis is in contact with the outside directly, the intracellular oxygen is supplied by small dermal vessels. Epidermic PO_2 is lower than that in arteries, while PCO_2 is comparable with that in arteries, indicating that oxygen is essential for the metabolism of epidermic keratinocytes and microorganisms.

How to explain the phenomenon that some aerobic bacteria (such as *Corynebacterium acnes*) can proliferate in the skin? The concept of biofilm was raised recently. Biofilm is a self-closed system resulting from the epidermal aggregation of microflora as well as their metabolic products. The nutrition and air are dispersed into the system, while PO_2 exists there at escalated levels. The balance alteration in PO_2 and PCO_2 may lead to changes in microbial category and quantity. For example, the wafer may result in a decrease in PO_2 and an increase in PCO_2, and in the long run lead to a remarkable increase in gram-negative rods and *Corynebacterium*.

16.3.2.7 Ultraviolet

Ultraviolet (UV) is capable of inhibiting or even killing some normal cutaneous microfloras. *In vitro* study has revealed that UVA with a dose of 50 mJ/cm^2 can kill *Pityrosporum*, while UVB with a dose of 250 – 900 mJ/cm^2 can also kill

Pityrosporum and *Candida albicans*. However, *Staphylococci* are not sensitive to UV radiation. *Staphylococcus epidermidis* will not be killed until a dose of 900 mJ/cm^2 UVB radiation is used, and *Staphylococcus aureus* can only be inhibited by the dose. This might explain why sunlight is effective in the treatment of seborrheic dermatitis. Both UV light and sunlight have a beneficial effect on psoriasis; this might be attributed to the production of vitamin D in skin induced by radiation, or to the direct or indirect effects of radiation on the cutaneous microorganisms. However, whether UV radiation can cause the alteration in cutaneous microorganisms requires further investigation. Previous studies have observed the effect of PUVA on psoriasis, and failed to find a significant difference in cutaneous normal microflora on a local radiated area.

16.3.2.8 Adhesive Power of Bacteria

The first step for microbial habitation is adhesion. As to microorganisms, there's a positive association between habitation ability and adhesive power. The surface molecule responsible for adhesion is called adhesin. Adhesin takes effect *via* a special receptor on the surface of the host molecule. These special receptors, with the main content being glycose or glycoconjugate, are regarded as adhesive receptors. Adhesin is the bridge between microorganisms and host molecules. The epidermic cells in different areas of skin have different adhesive receptors, which can explain why normal cutaneous microflora varies among different sites in skin.

Teichoic acids, one component of the cell wall in *Staphylococci* and *Streptococci*, serve as one kind of adhesin, can combine with its corresponding adhesive receptor (fibronectin, Fn) in epidermic cells. Fn is a type of glycoprotein receptor. It is revealed that adhesin may combine with Fn from at least two parts of epidermal cells, and dissolvable Fn can inhibit the combination between *Streptococci* and epidermal cells. Multiple receptors can combine with one single adhesin, while one single receptor can also be competitively combined with multiple adhesins.

The possible explanation as to why *Staphylococcus aureus* and some *Streptococci* can't colonize in skin is that under normal conditions the corresponding adhesive receptors in the epidermal keratinocytes are not exposed. However, in patients with atopic dermatitis, the cutaneous flora damage caused by scratching leads to the exposure of the Fn receptor, and increases the adhesive power of *Staphylococcus aureus*. Yet there's a different opinion, which believes that interior factors in AD patients increase the adhesive power of bacteria, and finally favor the cutaneous colonization of *Staphylococcus aureus*.

As to gram-negative bacteria, the most important adhesin is pilus, which is constituted of pilin. For example, the adhesins of *E. coli* include common pilus, P pilus and S pilus, and most *E. coli* express common pilus under appropriate conditions, which enables *E. coli* to adhere to almost all epithelial cells in humans. Also, it has been proved that pilus of *Neisseria gonorrhoeae* can adhere to the epithelial cells at the cervix or vagina. Increasing evidence indicates that the fibril may promote the adhesion of *Streptococcus pyogenes* to epithelial cells, and the adhesive effect is related to the hydrophobicity.

Recently, non-pilus adhesin has been gradually recognized. For example, it has been revealed that all pathogenic *E. coli* contain a plasmid with 55 – 57 Mbp, which encodes adhesive factors, and is proved *in vitro* to have an association with the adhesion of *E. coli* to HEP-2 cells or Hela cells.

As to *Candida albicans*, the adhesin is one kind of mannitol-protein compound, which can combine with glycoprotein protein on the surface of the host and therefore cause adhesion. Furthermore, the adhesive power is related to plasmids. It has been observed that pathogenic *E. coli* has a special protein adhesin, which is linked to the R factor.

Not only microbial factors but also host factors can have an influence on the adhesive power. Some patients are more sensitive to certain kind of pathogens, which might be attributed to the higher adhesive power of their keratinocytes to the pathogens. It is observed that due to the increased adhesive receptors, keratinocytes in AD patients are more adhesive to *Staphylococcus aureus* than healthy controls. Also, among susceptable individuals carrying *Staphylococcus aureus* on the nasal mucosa, the expression of HLA is intimately associated with bacterial colonization.

16.3.2.9 Interactions between Microflora

The normal cutaneous microflora serves as the barrier against the invasion of bacteria, which together with the human body and the environment, constitutes a harmonious system[7-8]. The mechanisms of the microbial barrier are complex, including the interaction between the microflora and host, and the interaction between microorganisms. Both antagonism and enhancement effect are of great importance to the integrity of this barrier and the balance of normal cutaneous microecology.

Reciprocal antagonism among microflora. The mechanisms of reciprocal antagonism between the microflora include: Competitive consumption of nutrition among microflora; production of pH value or oxidoreduction of electric potential to inhibit the growth of other bacteria; competitive combination of the adhesive receptor to interfere with the colonization of other bacteria; and production of inhibitory materials to restrain the growth of other bacteria. Up to date, the inhibitory materials mainly include:

Decomposed products of lipids. Sebum and its metabolic products may inhibit the growth of bacteria. Many resident floras have the activity of lipidase, for example *Corynebacterium acnes* are capable of decomposing triglyeride to free fatty acid. Long-chain free saturated fatty acid and oleic acid have an inhibitory effect on streptococcus pyogenes and gram-negative bacteria, but they don't have an inhibitory effect on *Staphylococci*. *Pityrosporum* has the activity of lipoxygenase and can turn oleic acid to azelaic acid, which can inhibit both *Corynebacterium*, *Staphylococci* and a certain type of fungi. Propionic acid produced by *Corynebacterium* has an inhibitory effect on *Trichophyton gypsum*. Other products decomposed from resident flora, such as short-chain fatty acid, can also have an inhibitory effect in a high local concentration.

Bacterial hydrolases. Corynebacterium acnes may produce bacterial hydrolases and inhibit the growth of *Staphylococci* and other *Corynebacterium. Staphylococci* can produce lysozyme and inhibit the colonization of other microorganisms. *Spore-forming* can also produce materials similar to hydrolases and have an inhibitory effect on other microorganisms.

Antibiotics. There're a lot of normal cutaneous microflora that can synthesize antibiotics. For example, a certain kind of fungi is capable of producing antibiotics such as streptomycin, penicillin and actinomycin, which leads to the detection of resistant bacteria around the border of the dermatomycosis lesion, and the isolation of penicillin in the lesion. Some dermatophyte may produce peptide and inhibit the growth of *Bacillus brevis* or even a virus, and therefore inhibit the awful smell in the feet. *Coagulase-negative staphylococcus* and a small quantity of *Corynebacterium* may produce round polypeptin, and inhibit or even kill the bacteria that are close in taxology (such as *Staphylococcus aureus*). The possible mechanism involves the adsorption of special receptors on the envelope of bacteria.

Other inhibitors. Candida albicans can produce CO_2 and therefore inhibit the growth of other fungi. This might explain why it's difficult to detect other fungi in lesions with infection of *Candida albicans. Streptococci* and aerobic *coccobacteria* can produce H_2O_2, and inhibit the growth of *Staphylococcus aureus*.

Reciprocal enhancement among the microflora. There are reciprocal enhancements among different microorganisms in the cutaneous microflora, which is called symbiosis. For example, there is a symbiosis between *Corynebacterium acnes* and *Staphylococcus epidermidis*, which can both live in hair follicles and sebaceous glands. Due to the consumption of oxygen and the decrease in local pH value caused by *Staphylococcus epidermidis*, there are good conditions for the proliferation of *Corynebacterium acnes;* while the decomposition of cutaneous keratin and the secretion of probiotics by *Corynebacterium* and *Bacillus* can also stimulate the growth of *Staphylococcus epidermidis.* Also, the decomposition of sebum by *Corynebacterium* and *Staphylococcus epidermidis* can enhance the growth of *bacillus.* Compared with the complicated inhibitors mentioned above, there are less enhancing factors, which are produced by the cutaneous microflora and include some lipids, amino acids and coenzymes. It is confirmed that *Nicrococcus* is capable of synthesizing some nutritious factors, and then enhances the proliferation of fungi that can produce antibiotics. The growth of one kind of anaerobe may result in the decreased sensitivity to oxygen of other anaerobes. Lloyd and Noble reported that in the mice model, some *Staphylococci* can enhance the cutaneous infection caused by *Dermatophilus congolensis*, which might be attributed to the chemotaxis effect caused by *Staphylococci. Staphylococci* can produce CO_2 and some enhancing factors, and therefore lead to the migration of *Dermatophilus congolensis* to the infected lesion. Clinically, it is very common that in inflammatory lesions of skin there's reciprocal enhancement. It is also believed that *Staphylococci* can produce hyaluronidase or other unknown materials, and finally enhance the growth and adhesive power of anaerobes.

16.3.3 Physiological Function of Normal Cutaneous Microfloras

Normal cutaneous microfloras inhabit the surface of skin, having a correlative dependence on the host and the environment, and constitute an ecological system which accompanies the whole process of occurrence, development and elimination of the host. Like the microfloras in intestinal tracts, the normal cutaneous microfloras participate in the important physiological functions of the human body and have a close relationship with human health [9, 10].

16.3.3.1 Defence Functions

The primary function of normal cutaneous microfloras is protection and defence. *Corynebacterium acnes* and *Staphylococcus epidermidis* are capable of decomposing the sebum and producing free fatty acid and an acid emulsified lipid membrane, which lead to the acidity of the skin surface and inhibit the growth and proliferation of pathogens such as *Staphylococcus aureus*, *Streptococci* and fungi. These normal microfloras colonize the skin and form a biological barrier, functioning as a protection of naked epidermis from the colonization of external pathogens, therefore making the host avoid infection of the skin or even internal organs.

16.3.3.2 Immune Function

Skin is an important immune organ. Normal cutaneous microfloras can serve as natural non-specific antigens to constantly stimulate the immune system, and enhance the immunity of the human body. *Staphylococcus epidermidis* may enhance defense ability by inducing the immune response of the host. Evidence indicates that the inflammation clearance ability of the host may decrease without the trigger effect of *Staphylococcus epidermidis*. *Staphylococcus epidermidis* can enable cutaneous innate immunity *via* a toll like receptor (TLR) signal transductor system and produce effective response against pathogens in keratinocytes, therefore avoiding the skin or internal organs from being infected by pathogens.

Another study reported that normal cutaneous microflora is capable of secreting antibacterial peptides, which can help the host to eliminate the pathogens. For example, *Staphylococcus epidermidis* produces lantibiotics which, as one type of antibacterial peptides containing lantibiotics (Lans), doesn't hurt epidermic cells but is toxic to other microorganisms such as *Staphylococcus aureus* and *group A Streptococcus*. *Pseudomonas aeruginosa* is also protective for the human body, because it produces PsVP-10 which, as another type of antibacterial peptide, possesses a defensive ability against *Streptococci*. Also, *Pseudomonas aeruginosa* can produce chemical compounds such as pyocyanin, pyrrolnitrin and oxyhydroxide phenazine, and therefore can kill or inhibit the growth of fungi, and establish an unfavorable environment for fungi infection, as well as prevent the transformation of fungi from yeast form to a pathogenic hyphal form.

16.3.3.3 Nutrition Function

During the differentiation of epidermis, keratinocytes gradually migrate from the basal layer to stratum corneum, and finally lose organelles, keratinize and fall off. The residues of the keratinization process, such as phospholipids and amino acids, can be used by both cells and bacteria. Intracellular glucose, water and electrolytes (such as potassium, natrium and calcium) supply nutrition for the growth of cutaneous microflora. Phospholipids, sterin and keratin produced by cutaneous microflora can also be absorbed by cutaneous cells, which can enhance the growth of cells, prevent the skin aging progress and reduce wrinkles.

16.4 Microecological Disturbance and Cutaneous Disorders

The relationship between normal cutaneous microflora and human health has aroused attention recently. During the long history of biological evolution, normal cutaneous microfloras form a correlative dependent relationship with the host, and they constitute a harmonious union, which plays an important role in maintaining the physiological functions of skin. The harmony can only be maintained under the condition that there's a dynamic ecological balance among microorganisms, the host and environment. The disturbance of this ecological balance is called a microecological disturbance. Namely, the physiological balance turns to a pathological imbalance due to the internal or external factors. Once there is a condition of microecological disturbance, external pathogens easily invade the host, and the normal cutaneous microfloras may turn to pathogenic bacteria and therefore lead to infectious disorders [8, 9].

16.4.1 Bacteria and Cutaneous Diseases

The most common cutaneous disease caused by bacteria is pustular dermatitis, which is mainly caused by *pyogenic coccus*, such as *Staphylococci* and *Streptococci*. Other infectious diseases caused by bacillus (for example, *Bacillus leprae*, *Bacillus tuberculosis* and *gram-negative* bacteria, such as *Bacillus proteus*, *Pseudomonas* and *E. coli*) are relatively less common. The following focuses on pustular dermatitis [10, 11].

16.4.1.1 Etiology

Primarily, the pathogens leading to pustular dermatitis include *S. aureus*, *S. epidermidis*, *S. saprophyticus* and *β-hemolytic streptococcus*. Due to the protection of normal skin, the bacteria don't cause infection under common conditions although they can habitat or colonize in the Skin. However, the pathogens may invade in the circumstances listed below: skin damage, including external injury, surgical incision, burning, bites from arthropods and repetitive scratching due to pruritus, which lead to the destruction of skin integrity and the invasion of bacteria; The robust virulence of one strain or magnanimous proliferation of one kind of bacteria; Chronic diseases of the host (*e.g.*, diabetes mellitus, pulmonary tuberculosis, dystrophy and metabolic disturbance); Thin and immature condition of children's skin, which may cause a decreased barrier function of skin; Thin, exfoliating and dry condition of the elderly's skin, which also causes a decreased physiological barrier; Immuno-compromise or immune-deficiency, which may result in a decreased immunological barrier function and the invasion of pyogenic bacteria.

16.4.1.2 Pathogenesis

Staphylococci belong to *gram-positive* bacteria, and *S. aureus* has the most powerful pathogenicity in *Staphylococci*. *S. aureus* can produce toxins and enzymes including coagulase, staphytolysin, leukocidin, enterotoxin, exfoliatin and toxic shock syndrome toxin 1 (TSST-1). Leukocidin can kill neutrophils and macrophages, and may protect *S. aureus* against phagocytosis. *S. aureus* are divided into 4 groups, among which group II mainly produce exfoliation and is the most common strain that leads to pustular dermatitis. The highest rate of resistance to penicillin G is found in group I *S. aureus*.

β-*hemolytic Streptococcus (group A)* is also known as *Streptococcus pyogenes*, which, accounting for about 90% in the infection caused by *Streptococci*, is the most pathogenic type in *Streptococci*. The pathogenicity is associated with its surface structure, and the extracellular enzymes or toxins produced by *Streptococci*. The pathogenic materials mainly consist of adhesin, streptolysin, pyrogenic exotoxin, hyaluronidase, streptokinase and streptodornase. The pyrogenic exotoxin is also known as erythrogenic toxin, which may cause erythema and heat on the skin. Hyaluronidase is one kind of spreading factor, which may decompose the hyaluronic acid in tissues and increase the permeability.

Normally, there's a well-being immune function in the human body. A small quantity of bacteria may pass through epidermis and reach the dermis, but it will be killed and phagocytized by leucocytes, macrophages, specific factors and non-specific factors from serum. Even if a higher quantity of bacteria invades the deep tissue, it can be restricted to a local area by leucocytes and serum factors. Due to the effect of chemotactic factors, there will be infiltration of leucocytes, tissue edema, vessels damage and pyogenic necrosis at a local lesion. However, if the host is suffering from an immune-compromise, the pathogen will diffuse and invade the blood circulation, and finally lead to septicemia.

16.4.1.3 Clinical Presentation

Bacteria infection-caused clinical presentation can be of great diversity. Three superficial bacterial "infections" occur in the stratum corneum and hair follicles, associated with overgrowth of normal flora at sites of occlusion and high surface humidity.

In general, a skin infection can follow three different events.

Primary infection. This refers to a cutaneous lesion directly caused by a pathogen. The clinical manifestation changes as the pathological area changes. If the pathogen only invades the epidermis and causes impetigo, the symptoms will appear as erythema, vesicles and pustules. Bullous impetigo in the newborn is caused by coagulase positive *Staphylococcus aureus* (phage group II type 71) , which may present symptoms of scalded slack bullous and the exfoliation of epidermis. The infection of hair follicles or tissue around hair follicles is called folliculitis, which may present symptoms of papulopustule. If the lesion is deep and the root of the hair follicle and tissue around it are involved, a furuncle will occur and the lesion will be present in the form of an inflammatory nodule. A carbuncle is a further aggravated furuncle whose inflammation spreads to multiple peripheral hair follicles. It is accompanied by swelling plaque, necrosis of suppurated tissue, pustular plugs at the top or even ulcers. Usually, primary infection is caused by infection of one single pyogenic pathogen, yet sometimes it could be mix-infected by two species of bacteria. Antibiotics are effective in the treatment.

Secondary infection. A secondary infection is an inflammation that occurs secondarily after a primary or other skin lesion, such as dermatitis, eczema and insect-bites. The primary lesion will be aggravated (*e.g.*, redding, erosion, effusion and papulopustule) and the disease course elongate without characteristic clinical manifestation. It is usually a mix-infection, and the single therapy of an antibiotic is not enough. It is generally acknowledged that secondary infection only occurs if there are more than $1 \times 10^6/cm^2$ bacteria in the lesion.

Cutaneous presentation of systemic infection. Bacteremia or septicemia caused by *Staphylococci* and *Streptococci* may result in both internal organ damage and skin lesions. It is very common in a clinic that crushing a furuncle may destroy the local barrier of the skin and facilitate the blood invasion of bacteria. Scarlet fever is erythematous exanthema and general toxic symptoms caused by erythrogenic toxin from *group-A β-hemolytic streptococci* (GAS).

16.4.2 Fungi and Cutaneous Diseases

As a large class of microorganisms, fungi exist in nature extensively and have a huge amount of categories (more than 1.5 million). However, very few (no more than 100) categories are pathogenic. Fungi can be divided into dermatophyte, yeast and mould. Fungal diseases commonly composed of superficial mycosis and deep mycosis, and the former is much more common in clinics [12-15].

16.4.2.1 Dermatophytoses

Dermatophytoses is abbreviated as tinea, which is a superficial dermatophyte infection of the epidermis, hair, finger nails and toe nails. It may be generally classified as tinea capitis, tinea corporis, tinea cruris, tinea manuus, tinea pedis and tinea unguium by skin location.

Etiology. There are three genera of dermatophytes: *Microsporum*, with the most common types being *M. canis*, *M. gypseum*, *M. ferrugineum* and *M. audouinii*; *Epidermophyton*, with the most common one *E. floccosum*; and *Trichophyton* with the most common ons *T. rubrum*, *T. mentagrophytes*, *T. tonsurans*, *T. violaceum*, *T. schoenleinii* and *T. verrucosum*. The dermatophytes feed on keratin. They usually infect the epidermal horny cell layer, nails and hair follicles. Generally, due to the dry skin and the continuous shedding of stratum corneum, only a few fungi can live in the skin as non-pathogens. However, the fungi may invade the skin under special conditions as below: Warm and humid surroundings, which facilitate the invasion and spread of pathogenic fungi; Skin wounds, including external injury and impregnation, which also facilitate the invasion of bacteria; A thick stratum corneum due to skin disorders, which favors the growth and proliferation of fungi; Low content of fatty acid in sebum due to non-complete development of the sebaceous gland in children, which is favorable to the growth of pathogenic fungi (after adolescence, with the development of the sebaceous gland, certain kinds of unsaturated fatty acid, especially fatty acids that contain carbon chain 7, 9, 11 and 13, can inhibit the growth of dermatophyte). That's why a spontaneous cure occurs in adolescent patients with tinea capitis. Broad-spectrum antibiotic, glucocorticosteroid, immunosuppressive agents, antineoplastics, organ transplantation, burning, pipes, catheters and intravenous nutrition increase the infection rate of conditional pathogenic bacteria significantly; Chronic diseases, senescence and immunocompromise, may result in an increased rate of dermatophytoses and refractory disease.

Pathogenesis. Pathogenic fungi produce mycotoxin and enzymes that digest keratin and sustain the existence of fungi in host cells. For example, *Microsporum gypseum* can produce elastase to digest keratin, elastin and collagen. Some dermatophytes synthesize keratinases and other enzymes such as S-sulfo-cysteine protease and sulfo-alanine protease, which can separate the disulfide bonds in keratin and decompose the keratin, so that the keratin may be digested by non-proteases. The component of mannitol in cell walls of *Trichophyton rubrum* is capable of inhibiting the responsive proliferation of lymphocyte aimed at the antigen; Therefore this inhibits the immune response of the host.

Immunological factors and non-immunological factors are involved in the protective mechanisms of the host against dermatophytes. The former is mainly mediated by T lymphocytes. Dermatophytes initiate the antigen presentation of Langhans' cells to sensitized T lymphocytes, and activate cell-mediated immunity and inhibit the growth and proliferation of fungi. It has been observed that patients with dermatophytoses have alterations in the quantity and distribution of epidermic Langhans' cells. Due to a decreased amount or abnormal distribution of Langhans' cells, the antigen presentation to T lymphoctyes is arrested. Then T lymphocytes cannot be activated and fail to secrete lymphokines, and therefore dermatophytes cannot be eliminated. During the acute period of infection, LCs decrease primarily due to the abnormal distribution of LC (most LC are found in dermis), while, during the chronic period, the main reason for LCs decrease is that dermatophytes have long-term and direct damaging effect on LCs. Furthermore, evidence indicates that patients suffering from refractory and extensive fungi infection usually have a negative response to a trichophytin test, and the possible explanation is that the antigen determinant of *Trichophyton rubrum* may block the cellular immune induction, or the persistent infection may activate a specific immune response of inhibitory T lymphocytes. The latter, namely the non-immunlogical reactions, are mainly mediated by monocytes and neutrophils, which may initiate phagocytosis and respiration, and produce oxidative metabolics such as superoxide anions, hydrogen peroxides and myeloperoxidases, and therefore complete the intracelluar and extracelluar disinfection. Patients with chronic dermatophytoses are usually found to have decreased or disfunctional neutrophils. The peroxidase produced by dermatophytes can antagonize the myeloperoxidase system, which is capable of killing dermatophytes.

Clinical presentation. Tinea capitis includes favus, "gray patch" tinea capitis, "black dot" tinea capitis and kerion. Clinical presentation of favus produces thick yellow adherent crusts (scutula) composed of skin debris and hyphae. Hair is not glossy and is easy to shed, with cutaneous atrophy, scar formation, and scarring alopecia. "Gray patch" tinea capitis is clinically present in the form of a gray scaly patch, showing numerous broken-off hairs slightly above the scalp for 3 – 4 mm and dull gray from their coating of arthrospores. It can be self-healed without scarring alopecia after adolescence. "Black dot" tinea capitis presents dots of scaly patches occurring in the scalp, with broken-off hairs near the openings of hair follicles, giving the appearance of "black dots". Kerion is clinically characterized by a dark red abscess on the scalp, which drains pus from multiple openings of the hair follicles, like honeycomb. Frequently a scar formation is present after healing.

Tinea corporis clinically presents single or multiple annular lesions of dark red scaly plaques, with papules and papulovesicles at margins, showing annular puffiness at margins and central scale. Peripheral enlargement and central pigmentation or clearing produce centrifuged development of the lesions. Tinea cruris refers to dermatophytosis of the groin, pubic regions, and thighs.

Clinical presentation of tinea manum and tinea pedis changes with the different infection type of dermatophytes, and the most common one is the interdigital macerated type, presenting scaly impregnation and erosion at the 3, 4 and 5 intertriginous sites. A moccasin type may present scaling patches; Hyperkeratosis and scaling confined to palms and soles, and have fissures in winter. A bullous type may present sporadic or widespread vesicles, being scaly after drying. Onychomycosis may present thickened and brittle fingernails or toenails without gloss, yellow or brown in color.

16.4.2.2 Candidiasis

Candidiasis is caused by *Candida*, with infections involved in skin, the mucous membrane and internal organs.

Etiology. Sometimes *Candida* is commensal in humans. Frequently, *C. albicans* may be detected at the gastrointestinal tract, oropharyngus and vagina. It can also be isolated on the skin surface in a very small amount.

More than 270 species of the genus have been identified, and *C. albicans* is the most common one in pathogenic species. Other pathogenic species include: *C. tropicalis*, *C. glabrata*, *C. parapsilosis*, *C. krusei*, *C. guilliermondii* and *C. pseudotropicalis*.

Generally, *Candida* is neither commensal, nor pathogenic in humans. Candidiasis occurs in the setting of local or general immune-compromise. It could be an endogenous infection or an exogenous one, and clinically is associated with broad-spectrum antibiotics, corticosteroids and immunosuppressive agents.

Pathogenesis. Like other infectious diseases, the occurrence of candidiasis depends on the interactions among pathogen, host and environment. The development of candidiasis is associated with pathogenic virulence, pathogenic quantity, the invasion path of the pathogen and the host response to the pathogen. The factors below are involved in the pathogenicity of *Candida*.

Adhesiveness. The adhesiveness of *Candida* to epithelial cells is the first step in the habitation and invasion, which indicates the pathogenicity of *Candida*. *Candida albicans* possesses the highest adhesiviness to mucosa and skin, and therefore has the strongest pathogenicity. The mannan glycoprotein in the cell wall is the molecular basic for adhesiveness. The influential factors of adhesiveness include the hydro-phobicity of the cell surface, the local pH value and the restraints of other normal skin flora.

Extracellular protease. Candida is capable of producing protease and phosphatidase to enhance its adhesivness and invasiveness. Aspartic proteinase (Sap) is a key proteinase related to the virulence. Mannan glycoprotein in the cell wall of *Candida albicans* may enhance the production of inhibitory T lymphoctyes, inhibit the macrophagus secretion of IFN-γ and TNF, and inhibit the immune response. It has been observed that the hypha of *Candida albicans* may inhibit the chemotaxis, adsorption and phagocytosis of PMN by secreting some materials, and the possible mechanism involves the selective inhibition of a respiratory burst of PMN.

The form of Candida. Candida is displayed in yeast forms and hypha forms. Generally, yeast forms are non-pathogenic for humans, and existing in a commensal state. Yeast forms may turn to hypha forms under certain conditions and become pathogenic. This is because that budding spore of the yeast form is easily digested or killed by macrophages, while hypha is difficult to kill. Also, *Candida* in a hypha form has an increased adhesiveness. Isolation of hypha forms of *Candida* in clinical samples usually indicates a high risk of infection.

The protective mechanisms of humans against *Candida* include cutaneous barrier function, inflammatory reaction caused by complements, phagocytosis and T lymphocytes-mediated immunological response. Infections of *Candida* usually arise after defence functions decrease or the immune function is in disturbance. For example, skin trauma, occlusion, maceration and humidity may damage the barrier function of the stratum corneum and of the dermis, and facilitate the invasion and infection of *Candida*. Chemotactic factors mediated by complements are regarded as an early defensive reaction to *Candida* invasion; while complement deficiency decreases the protective effect against the invasion of *Candida*. Phagocytes play a key role in defending the invasion of *Candida*, and candidiasis is usually associated with the abnormal phagocytes. A consistent decrease in leucocytes is usually observed in patients with systemic candidiasis. Cellular immunity is more important than humoral immunity in the protection against *Candida* infection. Factors such as long-term usage of corticosteroids, immune-suppressive agents, abnormal cellular immunity or immune-compromise, abnormal T lymphocytes and lymphokines' product, may cause acute or chronic candidiasis of the skin and mucosa.

Clinical presentation.

Cutaneous candidiasis. Candidiasis intertrigo occurs in moist, occluded cutaneous sites, such as interdigital, inguina, axillae and sub-mammary regions. Initial patches on the erythematous base become eroded, with small pustular lesions at the periphery and scales on the surface. Patients may feel itching. Candidiasis of the nail apparatus (*Candida* onychia) can cause dark red swellings of the tissues around the nails with pus. Nail folds are inflamed and thickened, with turbidness and disfiguration. General candidiasis is common in infants, showing scaly red patches with clear margins and small papules and vesicles at the periphery. It tends to occur in the area covered by a diaper and inter-digital areas.

Mucous candidiasis. Candidiasis of the oropharyngeal mucosa (OPC) is also called oral candidiasis (Thrush). It presents exanthema, maceration and erosion at the angle of the mouth, with white plaques on the surface. Monilial vaginitis and candidal balanitis can cause flush and erosion in mucosal areas with white membrane on the surface.

Chronic mucocutaneous candidiasis. It usually occurs in infancy or early childhood with underlying immune-compromise.

Systemic candidiasis. It may involve the GI tract, bronchus and lung, urinary system, endocardium and meninges, and cause septicemia. It's usually associated with diabetes mellitus, malignant tumors such as leukemia and lymphoma.

16.4.2.3 Pityrosporum and Cutaneous Diseases

Pityriasis (tinea) versicolor (PV) is a chronic asymptomatic scaling epidermomycosis associated with the superficial overgrowth of the hyphal form of Pityrosporum, characterized by well-demarcated scaling patches with variable pigmentation, occurring most commonly on the trunk.

Yeasts of the genus pityrosporum belong to the normal microflora of the human skin. They are now divided into eleven species. Different species initiate or aggravate different skin diseases, the most frequent of which is pityriasis versicolor. Pityrosporum yeasts are also thought to be associated with seborrheic dermatitis, dandruff and pityrosporum folliculitis.

Etiology. *Pityrosporum*, also known as *Malassezia furfur*, is lipophilic yeast that normally decomposes sebum for its nutritional source. Lipids enhance the growth of *Pityrosporum*. It is a normal skin flora and can be isolated on the skin surface of almost all adults, and is most common in sebum-rich areas such as the face and upper trunk. *Pityrosporum* may cause pityriasis versicolor and *Malassezia* folliculitis and has been recently implicated in the pathogenesis of seborrheic dermatitis.

Pityrosporum is an opportunistic organism. Normally it's non-pathogenic and turns to a pathogen (from yeast form to hyphe form) under certain favorable conditions. 50% of people in tropical zones suffer from pityriasis versicolor, which may be related to the warmth, humidity and the slower renewal cycle of epidermis. Corticosteroids also cause pityriasis versicolor, and it might be due to the elongated renewal cycle of epidermic cells. Patients with organ transplantation are easily infected by *Pityrosporum* due to the long-term usage of immunosuppressive agents. It has been reported by virgili that 20 out of 73 cases of renal transplantation suffered from pityriasis versicolor. It's the most common type of cutaneous fungal infection in organ transplantation patients, with a significantly increased incidence rate. Family susceptibility to *Pityrosporum* is observed and it is thought to associate with oily and sweaty skin. Also, dystrophy, extracellular glycogen deposition and ill health may lead to repeated infections of *Pityrosporum*.

Pathogenesis. It has been revealed recently that due to the lipid component in the cell wall, *Pityrosporum* can induce immune escape and avoid the protection of the host immune system. Kesavan *et al.* carried out a 48-h co-incubation of live *Pityrosporum* (three serotypes A, B and C), *Pityrosporum* stored in formaldehyde and PBMC (peripheral blood mononuclear cells). It was observed that both strains of *Pityrosporum* significantly inhibited the ability of PBMC to produce IL-1β, IL-6 and TNF-α when the ratio of yeast cells to PBMC was 20：1. Also, Kesavan made a comparison between lipid-free *Pityrosporum* and normal *Pityrosporum*, and found the former can induce PBMC to produce much more IL-1β, IL-6 and TNF-α than control ($P < 0.05$), therefore indicating that the lipid layer of *Pityrosporum* is capable of inhibiting the yeast from inducing inflammatory reactions. Moreover, an abnormal humoral immunity has been observed in patients with pityriasis versicolor. Silva *et al.* reported that the level of the IgG antibody to *Oval Malassezia furfur* in patients with pityriasis versicolor is significantly higher than that in controls.

Clinical presentation. Initially, there are macules around the openings of hair follicles, with fine scales on the surface, varying in color such as light brown, dark brown, light red and skin-color. Individual lesions may enlarge and merge, forming extensive geographic areas. It's much more common in adults and male patients. Occasionally, it can be found in children on the face. The lesions tend to occur at the neck and upper trunk, with aggravation in summer and improvement in winter. Globose spores and filamentous hyphae can be detected under direct microscopic examination.

Pityrosporum folliculitis is much more common in youths, and adult males. It mainly involves sebum-rich areas such as the upper trunk, and is less common on upper arms, legs, face and abdomen. It presents a small red follicular papule, sometimes accompanied by a small pustule, with erythema at the periphery. Long-term application of corticosteroids or broad-spectrum antibiotics may cause disturbance of normal skin flora, and *Malassezia furfur* overgrows under such conditions, resulting in sebum storage in hair follicles, aggregation of cellular fragments and obstruction of ducts. Therefore, stimulation from free fatty acids may induce the expansion and breakage of hair follicles, leading to the release of components to tissues and local inflammatory reactions.

16.4.3 Virus and Cutaneous Diseases

A virus is a tiny and simple microorganism, which normally resides in the cells and proliferates by way of replication. Some viruses may incubate in the body for a very long time in an inactivated way, such as the herpes simplex virus and varicella-herpes zoster virus. Generally they are non-pathogenic to the host. However, they can be activated again and become pathogenic under the stimulation of favorable factors. Viral skin diseases refer to the skin and mucous lesions caused by virus infection, which may be due to the direct damage of the virus in skin and mucosa, or due to the allergic reactions of skin and mucosa to the viral antigen [11, 16].

16.4.3.1 Virus Infection and the Defensive Mechanisms of Human

A virus is a center of nucleic acid enclosed by structural proteins, which is called a neucleocapsid. If a neucleocapsid is being enclosed by a lipoprotein envelope, it is defined as an enveloped virus, while a naked neucleocapsid without the lipoprotein envelope is regarded as a non-enveloped virus. Nucleid acid (DNA or RNA) contains the genome, and is the material basic for replication, pathogenicity and variation of the virus.

The infection of a virus involves the interaction between virus and host. The virus combines with a virus receptor in infected cells *via* contactin. After invasion, it synthesizes nucleid acid and the protein envelope in host cells, assembles another integrated virus and releases it outside of the cells. A virus may cause host damage *via* direct pathological damage, immunologic pathological damage and indirect inflammatory reaction. Two mechanisms, specific and non-specific, are involved in the defensive mechanisms of the host.

Non-specific mechanism. Integrated skin has a natural barrier function against a virus. The tight connection of the horny layer, the free fatty acid in the sebum layer, the lactic acid in sweat and the lysozyme in mucous secretion, may join hands in preventing the invasion of a virus. The activated macrophages produce interferon to inhibit the proliferation of a virus. Besides, macrophages are capable of handling antigen and helping B lymphocytes to produce antibodies. NK cells can kill the target cells infected by a virus, and can be activated by interferon. Damage to the non-specific mechanism may favor the invasion of a virus.

Specific mechanism. Cellular immunity is of great importance in the anti-virus effect. The virus antigen stimulates T lymphocytes, makes them differentiate and proliferate and finally be activated. A second stimulation on the lymphocytes enhances the secretion of cytokines, including interferon, the macrophage migration inhibition factor and lymphotoxin, therefore causing damage to the virus. It is observed in clinics that early application of interferon in the treatment of herpes zoster helps to alleviate the symptoms of vesicles and neuralgia, and to decrease the incidence rate of dissemination. Patients with cellular immunity defects tend to suffer from recurrent herpes simplex virus infection.

The humoral immunity is also helpful in the anti-virus effect. Viral antigens and the antigenic determinants may stimulate lymphocytes to produce antibodies against the virus, with the first antibody being the IgM type. The specific IgM antibody can be detected four days after clinical symptoms of primary infection of the herpes simplex virus, and lasts for 8 weeks. A later IgG type of antibody is produced, which combines with the virus and fixes complements, having the effect of a neutralizing virus. The IgG type of antibody is capable of passing through placenta. The IgA type antibody is produced by mucosa. Furthermore, there is a small quantities of IgD and IgE type antibodies in the host, which are also helpful in defence against the virus.

16.4.3.2 Herpes Simplex

Herpes simplex virus (HSV) infection, whether first-symptomatic or recurrent, may ‘typically’ present clinically with grouped vesicles arising on an erythematous base on keratinized skin or mucous membrane.

HSV is a DNA virus that cause acute skin infections and are present as grouped vesicles on an erythematous base. Herpes labialis is the most common infection caused by HSV type 1 (HSV-1), whereas genital herpes is usually caused by HSV type 2 (HSV-2). Most infections are recurrent and tend to reappear at or near the same location.

Etiology. Herpes simplex is caused by HSV, which belong to human herpes virus group A, with neutropism and double-stranded DNA. It encodes eight specific glycoproteins, *e.g.*, gB, gL, gD, gE, gC, gH, gI and gJ. It contains two subtypes, HSV-1 and HSV-2, according to the biochemical feature and antigenicity. HSV-1 mainly affects skin and mucosa in the mouth, nose and eyes, while HSV-1 affects genital skin and mucosa. HSV enters the skin by nasal, pharyngeal, oral and oculas mucosa as well as through damaged skin, then replicates and proliferates at the epidermis and dermis, and spreads *via* the bloodstream or nerves. Only 10% of cases with primary HSV infection present clinical manifestations. After regression of the primary infection, the virus ascends the peripheral sensory nerves and enters the lumbosacral sensory, where latency is established. Approximately 70% – 90% of adults can be infected by HSV. There's no lifetime immunity to HSV in humans, and recurrences usually occur under conditions such as fever, fatigue, menstruation, pregnancy, infection, GI tract disorders and emotional stress, when the cellular immunity is temporarily impaired and the latent HSV is reactivated, traveling along sensory neurons to skin and mucosa.

Clinical presentation.

Primary herpes simplex. Primary HSV infection occurs mainly in infants with 0.5 – 5 year-old or in a dystrophic state. 90% of cases are clinically asymptomatic and may lead to a person shedding the virus for life. Clinical manifestations include herpetic gingivostomatitis and widespread vesicles, occasionally with high fever and intumescent local lymph nodes.

Recurrent herpes simplex. It may occur on any site of the body, particularly at the border of skin and mucosa such as the lips and their periphery, nose and genital mucosa. It begins with scorching heat, itching and tightness in local areas, and then erythema appears and small blisters with thin walls and flush at the periphery. The blisters soon form pustules, erosion and crusts. Intumescent local lymph nodes and concomitant low-grade fever are common. Eye herpes simplex presents dendritic keratitis, leaving pannus at the cornea and damaged visual acuity. Herpes labialis is easy to break and cause pain when eating. Herpes genitalis is often transmitted through sexual activity and can be regarded as a sexually transmitted disease (STD).

16.4.3.3 Varicella-Herpes Zoster

Varicella-zoster virus (VZV) is a human herpes virus that infects 98% of adult populations. Primary VZV infection (varicella or chickenpox) is nearly always symptomatic and characterized by disseminated pruritic vesicles.

Zoster is a common, predominantly dermal, and neurological disorder caused by VZV, a virus morphologically and antigenically identical to the virus causing varicella (chickenpox). The difference in clinical manifestations between varicella and zoster apparently depends on the immune status of individual patients; those with no prior immunological exposure to varicella virus, most commonly children, develop the clinical syndrome of varicella, while those with circulating varicella antibodies develop a localized recrudescence, zoster.

Etiology. It is caused by VZV, which belongs to human herpesvirus group B, with double-stranded DNA and neutropism. It encodes three glycoproteins, *e.g.* gPI, gPII and gPIII. VZV is thought to enter through mucosa of the upper respiratory tract, then disseminates to skin *via* the bloodstream, and blisters and unapparent infection occur. VZV can pass from the skin lesions to the sensory nerves, travel to the sensory ganglia, and establish long-term latent infection in the root of spinal nerves or cerebral ganglion. Approximately 70%

of children under 15 year-old are infected by VZV.

The reactivation of VZV is closely associated with host immune-compromise, including fever, malignant tumors, trauma, fatigue, connective tissue disease, long-term usage of corticosteroids and immunosuppressive agents. The reactivated VZV will travel down the sensory nerve to the skin, and cause herpes zoster at the corresponding nerve-dominated area.

Clinical presentation. Varicella is the primary infection caused by VZV. First lesions begin on the trunk, spreading gradually to the face and extremities. Lesions are intensive on the face and trunk, while sparse on extremities. It is characterized by successive crops of vesicles that evolve to pustules, crusts, and at times scars, with erythema at the periphery and thin vesicle walls. Because the eruptions continue to appear, preexisting eruptions are found together with newly formed ones.

Herpes zoster presents multiple herpetic vesicles in band-like patterns over certain innervated regions with an edematous base. The unilateral blisters soon rupture, become erosions, and heal after crust formation in several days to weeks. The disease is characterized by neuralgic pain, which may occur several days before, or concomitant with, the onset of eruptions. The severity of pain rises with increasing age. VZV may be widespread in patients who are immune-compromised as a result of aging, with malignant tumors and connective tissue diseases. Small widespread blisters may spread on the whole body, showing manifestation of generalized herpes zoster, and even being accompanied by hyperpyrexia, pneumonia and encephalitis. Usually, there is lifetime immunity after recovery of herpes zoster. However, a few patients may endure a second infection.

16.4.4 Warts

Warts are benign proliferations of skin and mucosa caused by the human papilloma virus (HPV). Warts are transmitted by direct or indirect contact, and predisposing factors include disruption to the normal epithelial barrier. Treatment can be difficult, with frequent failures and recurrences. Some warts, however, resolve spontaneously within a few years.

16.4.4.1 Etiology

Papovaviruses are small DNA viruses, with a globular appearance and a diameter of 45 – 55 nm, presented as an icosahedron with 72 nucleocapsids and without envelope. HPV inhabits the cellular nucleus and has more than 200 subtypes depending on their nucleotide sequence of double-stranded DNA. Certain human HPV types commonly infect skin, which enter keratinocytes *via* minor trauma with breaks in the skin or mucosa, inhabit the cellular nucleus of keratinoyctes, and incubate in the cellular nucleus of replicated basal cells. The infection may be latent and symptomless, or result in replication of HPV in the nucleus and proliferation of cells, showing pathological changes such as hyperkeratosis, parakeratosis, acanthosis and papillomatous hyperplasia. Sole friction and sweat may enhance the occurrence of plantar warts. HPV infection is transmitted by direct skin-to-skin contact, or by indirect contact of contaminated items. The incubation period may last for 4 – 6 mon. Approximately 2/3 of cases experience a natural regression in 2 years. The immune function is closely related to the occurrence of warts. Immuno-compromise, such as renal transplantation, lymphoma and leukemia, is associated with an increased incidence of warts.

16.4.4.2 Clinical Presentation

Certain HPV types tend to occur at particular anatomic sites. The primary clinical manifestations of HPV infection include common warts, genital warts, flat warts, and deep plantar warts. Less common manifestations of HPV infection include focal epithelial hyperplasia, epidermodysplasia verruciformis.

Verruca vulgaris (common warts). Verruca vulgaris (common warts), mainly caused by HPV types 2, 4 and 7, occur at sites of the dorsum of hands, dorsum of feet, fingers, toes, edge of feet and clinically present firm hyperkeratotic papules, single or multiple, with cleft surface and skin-color (or dust color), being 1 – 10 mm or larger. The surface may be rough and have a cauliflower-like appearance. Evident subjective symptoms are rare.

Verruca plantaris (plantar warts). Verruca plantaris (plantar warts), mainly caused by HPV types 1, 2 and 4, occur on soles. Are related to local trauma and frictions, and clinically present yellowish, keratotic papules, with black dots in the center. The lesion could be single or multiple. The patient may have some degree of pain when crushed.

Verruca Plana (flat Warts). Verruca Plana (flat Warts), mainly caused by HPV types 3, 5, 8, 9, 10 and 11, occur at dorsum of hands and face, and clinically present flat papules (1 – 5 mm); with flat and smooth surface, round or oval in shape, skin-colored or light brown. Lesions that arise after trauma may have a linear arrangement. Evident subjective symptoms are rare. Occasionally, there's an itching feeling before regression.

Genital warts. Genital warts is mainly caused by HPV types 6, 11, 16, 18, 31 and 33, and HPV-16 and HPV-18 have high malignant potential, and occur at the cunnus and crissum. Are also known as condyloma acuminata, which is the most prevalent sexually transmitted disease.

AIDS. Acquired immunodeficiency syndrome (AIDS) is caused by the human immunodeficiency virus (HIV). HIV attacks T helper lymphocytes and damages the cellular immune response, resulting in infections of conditional pathogenic bacteria and malignant tumors. The relationship between HIV infection and microecology will be described in Chapter 21.

16.5 Ecological Prevention and Treatment of Cutaneous Diseases

Disturbance of cutaneous microecology refers to the pathological relations between the cutaneous microflora and the host, which may turn from a physiological state and be affected by both intrinsic and extrinsic environments. According to the microecological point of view, skin infections are actually due to the interactions of invaded microorganisms and the host, namely the ultimate imbalance between the invasion of the microorganism and the cutaneous response to the invasion. Skin infections are divided into 4 types depending on the pathogens: Endogenous infection, exogenous infection, mixed infection and contamination. Contamination is the ecological colonization of exogenous bacteria in the area of normal skin flora.

The objective of ecological prevention and treatment of cutaneous diseases is to regulate the disturbance of cutaneous microecology, as well as to regain and enhance the microecological balance. Ways to carry out the prevention and treatment include: Protecting the macroecological environment; Improving the microecological environment; Using antibiotics appropriately; And applying microecological reagents to support normal flora, enhancing the microecological balance, increasing the immunity of the host, and reinforcing the adaptability of the host.

16.5.1 Protect the Macroecological Environment

The macroecological environment of humans, which is constituted of air, water, sunlight, soil and vegetation, is indispensable for humans and other creatures. To enhance the ecological balance, the natural rules should be obeyed and the natural environment protected. Reckless reconstitution and destruction of nature will lead to atmospheric pollution, water pollution and soil pollution, and do harm to the health of human beings [9, 17].

16.5.2 Improve the Microecological Environment

Under the guidance of these principles of dermal microecology, the prevention or treatment of skin diseases should be aimed at restoring or promoting the microecological balance of skin normal flora. Therefore, ecological measurement should include the protection of the macroecological and microecological environment, rational usage of antibiotics, positive control of skin diseases, application of probiotics that support normal flora, and improvement of host immunity.

16.5.2.1 Keep Skin Dry and Clean

Regular exfoliation of keratinocytes, secretion and excretion of sweat glands and cutaneous glands, direct contact of the skin with the external environment, adhesion of external dusts, microorganisms, sensitizers and stimulators for the skin, together with sweat and sebum, form deposits and have an impact on the skin metabolism. The barrier function of stratum corneum may be impaired by sweat. Then it will be colonized and infected by exogenous pathogens, or be susceptible to allergic skin diseases by allergens. Therefore, in the routine health maintenance prevention of cutaneous diseases, it's especially important to keep the skin dry and clean. To get rid of the dirt, the skin should be regularly cleaned by appropriate skin cleaning. Alkaline soap or contract solution is suitable for oily skin; Soft soap or fat-free skin cleaning solution is better for neutral skin; While fat-rich soap or skin cleaning cream is better for dry skin.

16.5.2.2 Keep Food Reasonable

Glucose, fat, protein, vitamins, water, electrolytes and microelements are indispensable for both human life and cutaneous metabolism. Diversification and a reasonable match of food are necessary for keeping the skin healthy and maintaining normal cutaneous functions. Grains should be the main food. Vegetables and fruits, milk, beans, fish, meat, eggs and poultry are essential for the balance of cutaneous nutrition. Furthermore, meals should be arranged regularly, with no redundancy or diet in one meal.

16.5.2.3 Keep an Optimistic Mind

The human body is an organized integral whole. The optimistic mind is favorable for regulating the neuroendocrine and immune functions, maintaining the balance of hormones, and keeping the skin healthy. However, tension, anxiety and melancholy may inhibit the neuroendocrine and immune systems, and finally do harm to the microecological balance of the skin.

16.5.2.4 Control Skin Diseases Positively

Pruritic skin diseases, such as ezcema, dermatitis and psoriasis, lead to scratches and friction on skin, which may cause the defect of epidermis and inflammatory exudation, and therefore impair the barrier functions of the skin and encourage the infection of exogeneous pathogens (*e.g.*, *Staphylococcus aureus*). It has been demonstrated that the infection rate of *Staphylococcus aureus* at a cutaneous lesion is significantly higher than that in healthy controls. Accordingly, the positive treatment of skin diseases is important in the regulation of the microecological balance of skin.

16.5.3 Use Antibiotics Appropriately

The antibiotic aims to inhibit or kill pathogens and cure the diseases. It's indispensable in the treatment of cutaneous infectious diseases. Antibiotics have saved millions of lives and made contributions in the long-term defence against infectious diseases. However, antibiotics abuse inhibits both the pathogens and cutaneous normal microflora, which may disturb the interactions between microorganisms and cause microecological imbalance. Under such conditions, commensal bacteria and colonized bacteria may overgrow and turn to opportunistic pathogens, therefore producing toxins and lead to endogenous infection or dual infection. At the same time, the application of antibiotics enhances the emergences of resistance-plasmid (R-plasmid), which may be transferred to sensitive strains of bacteria, and clinically lead to wide-range resistance to antibiotics. Alarmingly, the development of new antibiotics could not keep up with the emergence of resistant strains of bacteria and the transfer speed of R-plasmid in recent years. The widespread application of broad-spectrum antibiotics or associated application of antibiotics may lead to co-infection of pathogens and resistance-strains, and consequently make illness even more complicated. For antibiotics used for cutaneous infectious diseases, the following advice is recommended: Choose sensitive, narrow-spectrum antibiotics; Choose antibiotics having less impact on host colonization ability; Choose antibiotics having less of a side-effect; Use cheap antibiotics; Choose external application. Microecological reagents should be supplied after antibiotics to regain the physiological relationship between microflora and host, which was in a pathological state due to the disturbance caused by the antibiotic [5].

16.5.4 Apply the Microecological Reagents

In 1958, a study on cutaneous normal flora was reported in the 1st Cambridge Dermatology Symposium by Gleeso White, which focused on the relationship between cutaneous flora and skin diseases. Recently, microecology has aroused attention, and the initial success of microecological reagents has been achieved for the treatment of skin diseases [18-20].

Cutaneous microecological reagents are microecological products originating from cutaneous normal flora and their enzymes with biological activity which, being effective in inhibiting or killing pathogens, may regulate the disturbance of cutaneous microecology, and lead to successful prevention and treatment of skin diseases. The theoretical foundation for cutaneous microecological reagents is listed below.

16.5.4.1 Ecological Balance Theory

The host, the cutaneous normal microflora and external environment constitute a microecological system, which is in a dynamic state of balance. On one hand, the microecological system is beneficial to the host, because it's indispensable to the host by participating in the physiological functions of skin; On the other hand, it's also beneficial to microorganisms, for it's helpful to keep the combination of microflora and maintain their growth and proliferation. There are interactions among cutaneous microflora, with a few predominant flora playing a deterministic role. The damage to the predominant flora will result in the disintegration or alternation of the whole microflora. For example, due to the sudden change in external conditions, *e.g.* long-term application of broad-spectrum antibiotics, systematic diseases, operation and trauma, the disturbance of cutaneous normal microflora (imbalance of microecology) will emerge, and exogenous bacteria will invade, and normal flora will turn to pathogens which, in the long run, will lead to cutaneous infection. The therapy of probiotic bacteria is aiming to supply normal microflora and regulate the imbalance of flora, therefore regaining the microecological balance and curing the disease.

16.5.4.2 Ecological Antagonism Theory

Cutaneous normal microflora serves as both biological barrier and chemical barrier, and directly participates in the barrier structure against the invasion of microorganisms. The biological barrier refers to the biological membrane-like structure formed by normal microflora, which affects the colonization, occupation, growth and proliferation of passenger bacteria or external bacteria. The chemical barrier is made of free fatty acid and lactic acid, with the former originating from sebum, which is decomposed by lipidase (produced by normal flora such as *Pityrosporum* and *Propionibacterium acnes*), and the latter originating from sweat. The chemical barrier can inhibit the proliferation of passenger bacteria or external bacteria. Besides, certain type of normal flora may produce antibiotic-like materials, for example *Propionibacterium acnes* produce hydrolase to inhibit the activity of *Staphylococcus aureus* and *Streptococcus pyogenes*; while aerobes produce H_2O_2 to inhibit the growth of *Staphylococcus aureus.*

16.5.4.3 Immune Theory

Cutaneous microecological reagents can function as non-specific immunolo-regulatory factors to activate the immune cells of the host, and lead to the production of cytokines and antibodies. Cutaneous microecological reagents also promote the phagocytosis activity of phagocytes. Besides, they may serve as adjuvant, therefore preventing the habitat of pathogens and inhibit the proliferation of external bacteria.

16.5.4.4 Anti-Aging Theory

Normal microflora participate in not only the prevention and treatment of skin diseases, but also the health care process of skin (anti-aging effect). Senescence microecology believes that the increase in age is responsible for the skin aging process, that the quantity and quality of cutaneous normal microflora change with age, and accordingly the functions of the host change, which lead to the decay of the physiological functions of the skin. Therefore, cutaneous probiotics can enhance the skin functions by supplying additional predominant flora. Also, they may activate the intracellular superoxide dismutase, catalase and glutathione peroxidase, promote the production of anti-oxidant, decrease the damage of free radicals, thus performing an anti-aging effect and promoting longevity.

Increasing evidence has demonstrated that cutaneous probiotics are beneficial in treating acne, chloasma, chapped skin and perianal eczema. Furthermore, they can prevent the infections caused by trauma and enhance the healing process of the wound.

In China, one cutaneous probiotic has been developed by Prof. Xiong De xi, containing two species of facial beneficial flora, vitamin E, proteins, moisturizing components and biological enzymes. It is proven that this probiotic is rather effective in treating sebaceous dermatitis when combined with other routine therapies. The total effective rate in the combination group is 92.5%, while in the routine control group it's only 72.2%. The routine therapies include anti-bacteria, anti-inflammation and exfoliating, which may result in side-effects such as dry skin. On the contrary, the cutaneous probiotic aims to promote the growth of beneficial bacteria, decompose the over-secreted sebum, clear out the hair follicles and enhance the ventilation, thus inhibiting the anaerobe and regulating the balance of the microecology. Therefore, combined therapy may eliminate the side-effects of routine therapy. Besides, ingredients of the cutaneous probiotic (including proteins, amino acids, vitamin E and moisturizing components) supply material nutrition for epidermic cells, enhance the skin's defensive ability against pathogens, thus treating both the symptoms and the cause.

You *et al.* [19] reported that cutaneous probiotic cream is effective in the treatment of acne, chloasma and cosmetic dermatitis. Among 142 cases with acne, 81 cases were cured, 45 treatments was classified as effective, while 16 cases were classified as ineffective. Among 43 cases with chloasma, 11 cases were cured, in 23 cases treatment was deemed effective, while in 9 cases ineffective. Among 64 cases with cosmetic dermatitis, 54 cases were cured, in9 cases the treatment was deemed effective, while in 1 case ineffective. The investigators pointed out that as a natural ecological cosmetic, cutaneous probiotic is promising in the treatment and care of skin disorders.

Jianfusheng, a complex microecological reagent containing *Propionibacterium acnes*, *Staphylococcus epidermidis* and physiological fermentation products of bacteria, has been approved by the Chinese Department of Health as a biotic additive. It's the only microecological reagent being approved for the prevention and treatment of cutaneous diseases in China. The positive indications include treatment of acne, chloasma, chapped skin, sebaceous dermatitis, cosmetic dermatitis, tinea cruris, *etc.*. Jianfusheng regulates cutaneous flora, supplies additional cutaneous resident flora, increases the skin anti-colonization ability, enhances the skin immune function *via* non-specific stimulation to the immune system, and achieves the ecological balance of cutaneous normal microflora, finally leading to the inhibition or elimination of pathogens.

In particular, the overall view should be emphasized in ecological prevention and treatment. An ecological reagent does not equal ecological therapy. Ecological therapy includes the adjustment from the macroscopic to the microcosmic environment, the full-scale regulation of nutrition, the reasonable application of antibiotics, the application of immune modulation agents, traditional therapy and the use of a microecological reagent. A cutaneous microecological reagent is one type of biological agent as well as cosmetics, mainly functioning as the regulator of the cutaneous microecological balance. It's not a panacea and should be properly used in clinics. Antibiotics, antifungal agents and antiviral drugs are leading therapies in the treatment of infectious cutaneous diseases, while a microecological reagent should be considered as the adjuvant treatment. Conditional pathogens are supposed to be treated by decolonization therapy or immune modulation agents first. A cutaneous probiotic can only be considered in conditions where a disturbance of cutaneous predominant bacteria occurs.

16.6 Prospects

As a branch of life science, microecology is a new field and has developed quickly in recent years. Skin microecology, being one sub-branch of microecology, is about the complicated interactions between host, environment and microflora, which inhabit the skin as well as its appendages. This subject involves the structure of cutaneous microecology, the theory of skin microecological balance and disturbance, the ecological prevention and treatment of skin diseases, and the theory of skin health care and anti-aging.

Cutaneous microecological reagents are developed based on the application study of cutaneous microecology. They differ from other drugs in that they may avoid the side-effects of dysbacteriosis or drug-resistance caused by antibiotics, with the objective of curing disease, preventing disease and improving health care. Cutaneous microecological reagents have few side effects and have displayed a beneficial effect on skin diseases as well as skin health care and in the anti-aging field. The effect of cutaneous microecological reagents on skin diseases have been confirmed in dermatological clinics. Thus, cutaneous microecological reagents show promise for further development and application.

The skin lies on the surface of the human body, with a huge quantity of microflora and species. The study of cutaneous microecology is still in the initial stage. Up to now, the study mainly has focused on the investigation of species and amounts of cutaneous microflora, and many further questions remain unclear. For example, the cutaneous microflora in different ecological systems varies in distribution, evolution and interaction with the difference in age, area of the anatomy and physiological function. Is there any rule to obey? How does the enormous microflora interact with host epidermic cells to keep a dynamic balance? What's the relationship between the evolution of the microflora and the cutaneous pathophysiological process? How to make the cutaneous microflora evolve in an advantageous direction for the human body? All these topics, together with other questions, deserve further investigation.

With the development of molecular biology, the species and strains of cutaneous microflora will be identified gradually, and it is believed that their functions, as well as their commensal relationship with the host, will be further elucidated. Study of these fields undoubtedly will lead to the development of cutaneous microecology, and bring a whole new viewpoint and methods which, in the long run, will advance medical science.

References

[1] Fan Z M. Senescent Microecology. Beijing: International Culture Press, 1993.
[2] Qian L S. Medicine Microecoogy. Shanghai: Shanghai Medical College, 2000.
[3] Zhao B. Chinese Clinical Dermatology. Nanjing: Jiangsu Science and Technology Press, 2010.
[4] Zhang X J, Liu W D, He C D. Basic for Modern Dermatology. Beijing: People's Sanitation Press, 2010.
[5] Webster G F. Skin microecology: The old and the new. Arch Dermatol, 2007, 143: 105-106.
[6] Xiong D X. Modern Microecology. Beijing: Chinese Science and Technology Press, 2000.
[7] Xiong D X. Microecological balance between functional textiles and human skin. Knitwear industry, 2006: 13-18.
[8] Foulongne V, Sauvage V, Hebert C, *et al.* Human skin microbiota: High diversity of DNA viruses identified on the human skin by high throughput sequencing. PLoS One, 2012, 7: e38499.
[9] Grice E A, Segre J A. The skin microbiome. Nat Rev Microbiol, 2011, 9: 244-253.
[10] Cogen A L, Nizet V, Gallo R L. Skin microbiota: A source of disease or defence? Br J Dermatol, 2008, 158: 442-455.
[11] Ruocco E, Donnarumma G, Baroni A, *et al.* Bacterial and viral skin diseases. Dermatol Clin, 2007, 25: 663-676.
[12] Chen S P. Fungal Infections. Shenyang: Liaoning Science and Technology Press, 2000.
[13] Liao W Q, Wu S X. Development on Fungal Diseases. Shanghai: The Second Military Medical College, 1998.
[14] Schmidt A. Malassezia furfur: A fungus belonging to the physiological skin flora and its relevance in skin disorders. Cutis, 1997, 59: 21-24.
[15] Treat J, James W D, Nachamkin I, *et al.* Growth inhibition of trichophyton species by pseudomonas aeruginosa. Arch Dermatol, 2007, 143: 61-64.

[16] Zhang X J. Dermatology and Venerealogy. 7th Version. Beijing: People's Sanitation Press, 2008.
[17] Gfatter R, Hackl P, Braun F. Effects of soap and detergents on skin surface PH, stratum corneum hydration and fat content in infants. Dermatology, 1997, 195: 258-262.
[18] You B E, Chen X P, Wang L L. Application of cutaneous probiotic cream in facial cosmetology. Chin J Microecology, 2000, 12: 45-46.
[19] You B E, Tan Z J, Li J Q. Effect of combined therapy of cutaneous probiotic cream and piyanxiao on facial sebaceous dermatitis. Chin J Microecology, 2001, 3: 98.
[20] Huang K, Liu W H, Yu X F. The effect of ecological cream on cutaneous flora during treatment of acne vulgaris. J Mudanjiang Med College, 2006, 27: 15-17.

17

Infectious Microecology of the Hematological System

Jie Jin [1], **Jian Huang** [2]
Department of Hematology, the First Affiliated Hospital, School Medicine, Zhejiang University, Hangzhou, 310003, China
[1] E-mail: jiej@hzcnc.com;
[2] E-mail: househuang@hotmail.com

Microecology infections are important for the hematological system. We will explain in five Sections: Defensive Function of Blood, Molecular Ecology and Hematological Disease, Microecological Changes and Hematologic Diseases, Treatment of Hematologic Diseases and Infective Microecology, and Molecular Ecological Treatment.

17.1 Defensive Function of Blood

Blood is an important component of microecology, representing ecology at the cellular and molecular level. Blood has both cellular and non-cellular components. The cellular components include red blood cells, white blood cells and platelets. The non-cellular component is plasma, which contains a variety of proteins, electrolytes and water [1]. Under normal circumstances, both cellular and non-cellular components have defensive functions. Besides, its main function of supplying oxygen, erythrocytes also plays a role in the immune response of the human body. White blood cells are the main cellular elements which have immune and anti-inflammatory properties. In recent years, our understanding of the defensive function of non-cellular components, especially the important roles of cytokines, has become more explicit [2].

17.1.1 Cellular Components of Blood

Granulocytes including neutophils, eosinophils and basophils, are the body's first line of defense, and its main function is to engulf and kill invading bacteria and other pathogenic micro-organisms. The function of neutrohpils is regulated not only by various cytokines such as IL-1, IL-6, TNF-α, IL-4 and IL-10, but also by active blood lipids such as arachidonic acid, leukotrienes and neuroendocrinal hormones.

Monocytes and macrophages have the function of chemotaxis, phagocytosis, immune regulation, cytotoxicity and production of various cytokines. Under normal circumstances, invading micro-organisms, such as bacteria, can activate a complement and release chemokine which induces monocyte-macrophage cells to migrate into the inflammatory sites and engulf invading micro-organisms. At the same time, monocyte-macrophage cells process antigens and present them to lymphocytes to initiate cell immunity and humoral immunity. Monocyte-macrophage cells also secrete soluble factors and regulate lymphocyte responses. Many cytokines secreted by monocyte-macrophage cells have an effect on the immune system and hematopoiesis. Also, monocytes have the capabilities of lysis and killing of target cells through Fc receptors which can recognize the antibody on the target cell surface.

Lymphocytes particapte in both humoral and cellular immunity, and can be divided into B cells, T cells, NK cells and K cells [3].

B cells account for 10% – 15% of lymphocytes. Various B cell clones can produce thousands of immnunoglobins in order to recognize different antigens and play different roles in immune functions.

T cells are composed of a group of lymphocytes with different functions. According to the difference in antigen recognition receptors, T cells can be divided into TCRαβ T cells and Tγδ T cells. In the peripheral blood, TCRαβ T cells account for 95% of T lymphocytes, while Tγδ T cells only account for 1% – 10%. Both types of T cells are CD2 and CD3 positive, while Tγδ T cells are CD4 and CD8 double-negative; Tcrαβ T cells are CD4 or CD8 single-postive. According to the difference in phenotype, Tcrαβ T cells can be divided into $CD4^+$ cells and $CD8^+$ cells. And T cells can also be divided into regulatory T cells and effector T cells by the difference in function. The former includes T helper cells (TH) and suppressor T cells (TS). The latter includes cytotoxic T cells (TC) and delayed type hyersensitivity T cells (Tdth).

NK cells and K cells are a group of lymphocytes with neither T cell nor B cell surface markers. These cells are derived from stem cells and mature in the bone marrow. The NK cells are also called natural killer cells, which can directly kill tumor cells, virus-infected cells and transplanted cells without being activated by antigens, and have no killing effect on normal tissues. Morphologilly, NK cells are large granular lymphocytes, which account for about 3% of peripheal blood lymphocytes. NK cells are able to dissociate from the target cell after killing it, and again kill other target cells. The killing mechanism may be through the release of cytotoxic factors, including tumor necrosis factor, lymphotoxin, natural killer factors and perforin. Interferon and IL-2 can activate NK cells' killing activity, while estrogen, glucocorticoids can inhibit its activity. K cells, namely killer cells, are antibody-dependent lymphocytes. The origin of K cells is not clear, and some evidence shows that it derives from bone marrow pleuripotent hematopoietic stem cells which develop into K cells by the stimulation of GM-CSF. The surface of K cells has no T or B cell markers, has Fc and C3 receptors but without sIg. Therefore, they do not have the function of recognization of specific antigens. K cells are able to kill target cells though Fc receptors.

17.1.2 Non-Cellular Components of Blood

Blood contains water, electrolytes, other inorganic and organic materials and proteins, which include a variety of enzymes, complements, immune-globulins and cytokines, and have anti-inflammatory and anti-infection properties.

There are lots of cytokines in blood, which are produced by lymphocytes, monocyte-macrophage cells, endothelial cells and fibroblast cells. Each cytokine has its own function to maintain the normal ecological balance of blood.

Stem cell factor (SCF) Serves to promote growth of hematopoietic stem cells and differentiation of hematopoietic stem cells into hematopoietic progenitor cells.

Granulocyte - colony stimulating factor (GM-CSF) is produced by endothelial cells and fibroblasts. The role of GM-CSF is to promote development and differentiation of pluripotent hematopoietic stem cells into granulocytes and monocyte-macrophage cells. And together with K cells and TNF-α, GM-CSF can also promote differentiation of stem cells or monocyte-macrophage cells into dendritic cells that present the antigen to T lymphocytes to kill tumor cells or infected cells.

Granulocyte colony stimulating factor (G-CSF) is also produced by the endothelial cells and fibroblast cells. The role of G-CSF is to promote proliferation and differentiation of pluripotent hematopoietic stem cells into granulocytes that play a part in infection resistance.

Vascular endothelial growth factor (VEGF) is produced by bone marrow stromal cells and lymphocytes. VEGF is a multifunctional cytokine involved in angiogenesis, inflammation and wound healing. The VEGF production increases in various human malignancies and is associated with tumor growth and metastasis.

Tumor necrosis factors α/β (TNF-α/β) are produced by lymphocytes. They are associated with bone marrow hematopoietic regulation, and the growth of hematopoietic cells will be inhibited by a high expression of TNF-α and TNF-β. The TNF-α is triggered by endotoxin and involved in cytokines activation. Also, when bacteria and lipopolysaccharide (LPS) from the Gram-negative bacteria cell wall invade the human body, TNF-α rapidly increases in circulation and induces the production of IL-1 and IL-6. The former aggregates neutophils to initiate an acute inflammatory response and the latter has a synergistic effect with the former one.

Interferon (IFN) is produced by monocytes and lymphocytes. IFN has an anti-virus and anti-tumor effect.

Interleukin (IL) is produced by fibroblast cells and endothelial cells, and there are many Interleukin ILs *in vivo*. Different types of cells may produce the same IL. Many ILs are now well-known, including IL-1, IL-2 … IL-17 and more and more new ILs have been found in recent years. Different ILs have different functions, for example, IL-1, IL-2, IL-4 and IFN-γ are able to develop K cells into NK cells; IL-4, GM-CSF and TNF-α are able to promote transformation from hematopoietic stem cells or mononuclear cells to dendritic cells. IL-4 is an important macrophage inhibitory factor, which reduces the production of IL-8 by monocytes. IL-3 can function as a hematopoietic stem cell factor and it can also regulate the function of macrophages. When LPS exists, IL-3 can stimulate the production of IL-1, IL-6 and TNF. IL-6 is a myeloma cell growth factor. The production of IL-6 increases at the time of serious infection or LPS invasion, and the level of IL-6 is associated with prognosis of the patients. Although polymyxin can reduce the level of plasma endotoxin, it has no effect on IL-6 response. IL-8 is related to neutophil inflammatory aggregation and adhesion, and is able to regulate the permeability of vascular endothelial cells, reduce the aggregation of multinucleated giant cells with acute inflammation and alleviate the inflammatory effect. TNF-α and IL-1 can stimulate the production of IL-8. IL-11 can promote cell growth and differentiation, especially promoting the differentiation of macrophages, which play an important part in inflammatory response. IL-12, an important member of inflammatory media, is produced immediately by macrophage after infection, and this process is considered an effective immune response to vascular endothelail cells associated disease. Members of the Cytokine network have a very complex ecological relationship, with some cytokines promoting inflammatory response and some cytokines inhibiting inflammatory response. They interact and are dependent on one another, composing a complex inflammatory reaction process.

The immunoglobulins in blood include IgG, IgA, IgM, IgD and IgE. They play an important role in humoral immunity.

17.2 Molecular Ecology and Hematological Disease

The pathogenesis of many hematological diseases is closely associated with changes in molecular ecology. External factors such as infection, including viruses, bacteria, fungi and mycoplasma infection can cause cellular or moleculular ecological changes. Internal factors such as gene mutation can also lead to molecular ecological changes and cause hematological disease. This section will focus on the relationship between molecular ecological changes and hematological disease.

17.2.1 Apoptosis and Hematological Disease

Cell apoptosis is an orderly death caused by a specific intracelluair protease cascade reaction that is induced by intracellular and extracellular factors. Morphologically, characteristic apoptotic features include cell shrinkage, chromatin condensation and appearance of apoptotic bodies. An apoptotic body has an intact membrane structure, which envelops part of the cytoplasm, organelles and fragmentation of the nuclear components. The apoptotic body is always removed by the adjacent normal cells or phagocytic cells. Therefore, its content will not leak and cause an inflammatory response.

Cell apoptosis can occur in exceptional cases, for instance, when the temperature, pH and osmotic pressure change or the cell encounters a variety of hits, like DNA damage, mitochondrial damage, infection and malignancy; all of these conditions can induce cell apoptosis and remove unhealthy cells. Apoptosis can also occur in normal physiological conditions, for example, normal tissue renewal and clearance of senescent cells are regulated by apoptosis.

Apoptosis-related factors include protease and calcium ion.

Caspase enzymes (Caspases, cystein aspases) are the basic structure causing cellular apoptosis. They exist in all kinds of cells as inactive plasminogens. The activation of caspase enzymes is involved in all processes of apoptosis. The activation of correlated proteases is necessary and shared, though the mechanisms of apoptotic processes induced by different factors are not the same. According to its homology, Caspase can be divided into 3 subfamilies, Caspase-1, Caspase-2 and Caspase-3. The Caspase-1 subfamily includes Caspase 1, 4, 5, 11, 12, 13 and 14. The Caspase-2 subfamily includes Caspase 2, 8, 9 and 10, while the Caspase-3 subfamily includes Caspase 3, 6 and 7. The Caspases are activated through cascade reaction triggered by apoptotic signal: One of the Caspases serving as promoters (*e.g.* Caspase-8 and 9) will be activated firstly, and then Caspases serving as effectors (*e.g.* Caspase-3, 6 and 7) will be activated. Different promoters transduce different apoptotic signals, for example, Caspase-8 is correlated with death receptors associated apoptosis, and Caspase-9 is correlated with apoptosis induced by cytotoxic factors.

Caspase can induce cell apoptosis in two different ways, through the death receptor pathway and the mitochondrial pathway.

The death receptor pathway: Cytokines can regulate cell proliferation and differentiation by binding with their specific receptors on the target cells. The tumor necrotizing factor (TNF) subfamily consists of Fas Ligand (FasL), TNF, lymphotoxin, CD30 ligand, 4-BB ligand, CD40 ligand, CD27 ligand and TNF-related apoptosis inducing ligand (TRAIL), in which TNF, FasL and TRAIL are termed as death factors. Membrane-associated TNF and TRAIL can be lysed into soluble cytokines by membrane-type metalloproteinase. Since membrane-associated TNF has stronger activation ability to type II TNF receptors than soluble TNF, it's generally considered that under physiological conditions FasL and TNF can function locally through cell-cell interaction, and this function would be attenuated once they were lysed into soluble TNF or FasL. The soluble TNF and FasL both exist in the form of trimers, and although the membrane -associated type has not yet been proven to be in the form of trimers, it's generally believed that they have the trimer-forming potential. The death factor receptors belong to the TNF receptor (TNF-R) family, which includes more than ten receptor members. TNF receptors are type I membrane protein, about 25% of the extra-cellular sequence is similar among the members, but their intra-cellular parst are rarely similar except for Fas, TNF-R1, and two other members, which have a homologous functional domain composed of about 80 amino acid residues. Mutation analysis revealed that this domain is responsible for death signal transduction, so it's called death domain (DD). TNF can induce apoptosis, activate NF-κB and also promote thymocyte proliferation. Both TNF-R1 and TNF-R2 can transduce signals for apoptosis and NF-κB activation, but in most cases TNF-R1 is responsible for the transduction of these signals and TNF-R2 for the proliferation of the TNF-induced thymocytes. Fas can bind with FasL or conjugate with an agonistic antibody and induce apoptosis in cells that have Fasl. Most of the other TNF receptor family members can transduce active or excitatory signals, but some can also induce cell apoptosis.

It's known that the binding of the growth factor receptor and its ligand can result in the dimerization of the receptor, and the X-ray diffraction analysis of the complex formed by the death factor and their receptors, such as TNFβ-TNF receptor complex, revealed that TNFβ trimer can bind with the extracellular domains of three TNF-R molecules and form a complex, suggesting that the binding of the death factor and their receptors may induce the trimerization of the receptors, and the trimerized intracellular domain can then transduce the corresponding signals.

Since Fas and TNF-R1 mediated apoptosis can still occur in the presence of RNA or protein-synthesis atagonists, and in de-nucleated cells apoptosis can still be induced once Fas is activated, which suggests that the necessary ingredients required for the corresponding apoptotic signal transduction have long been in the cells and the activation of Fas is just a triggering event. Two related signal transduction proteins have been identified: FADD (Fas-associating protein with death domain, also termed MORT1 protein) and TRADD (TNF-R1 associated death domain). The former is associated with signal transduction after Fas activation, and the latter with signal transduction after TNF-R1 activation. The C terminal of FADD contains a death domain (DD), by which it can be anchored to activated Fas, and its N-terminal is associated with signal transduction; this domain is called DED (death effector domain). TRADD also has DD, through which it binds with TNF-R1, but TRADD does not have DED; it can bind with FADD through DD interaction and pass down signals by FADD, meaning Fas and TNF-R1 can share the same FADD signal-passing-down structure. Yet TNF-R1 has another apoptosis-inducing pathway, the RIP (receptor interacting protein) pathway. RIP was originally called Fas-binding protein; it has the priority to bind TRADD. It's known that RIP is a serine/threonine kinase, and it also has a death domain, by which it binds with TRADD. The N-terminal of Caspase-8 has two DEDs, and its C-terminal is similar to that of ICE family members especially Caspase 3. In this way, signals can be passed from FADD to Caspase-8 through DED to subsequently cause the lysis of caspase-3 and caspase-1, and evoke proteinase cascade reaction.

The mitochondrial pathway: Recent studies have found that a mitochondrion is unexceptably involved in, and participating in, the process of apoptosis induced by other signal pathways, so a mitochondrion is sometimes called a death signal integrator. Many intrinsic and extrinsic factors can lead to the disclosure of the permeability transition pores in mitochondria, thus opening the outer membrane pores and leading to phagocytosis and breakage of the cell organelles, resulting in the release of mitochondrial cytochrome C (Cyt C). The Cyt c released to the cytoplasm can activate Apaf-1 (apoptosis protease activating factor-1) and caspase-9 zymogen, and form a complex with them termed the apoptotic body. Caspase-9 can further activate the downstream Caspase, the most important of which are Caspase-3 and Caspase-8, and finally cause DNA breakage and apoptosis. The Cyt c mediated Apaf-1 oligomerization and caspase-9 activation can be negatively regulated by heat shock protein 90 (Hsp 90), which can form a cytoplasmic complex with Apaf-1, and so to inhibit the formation of Apaf-1-caspase-9 complex. The immunodepletion of Hsp90 can lead to Apaf-1 depletion, and thus inhibit the Cyt c mediated caspase-9 activation. Adding purified Apaf-1 to Hsp90 depleted cytoplasmic extracts can reconstruct Cyt-c's ability to activate caspase-9. Various kinds of DNA damaging agents can dissociate the Hsp90-Apaf-1 complex, and relieve its inhibition to caspase-9 activation.

Mitochondria are death signal integrators. In cell apoptosis triggered by different factors, mitochondria are inevitably involved in the death signal integration and the development of cell apoptosis, in which the Bcl2 family proteins, which are located on the mitochondrial membrane, prove to be very important.

At least 15 bcl2 family members have been identified in mammals, according to their function; they can be classified into apoptosis-inhibitory or anti-apoptotic genes and apoptosis-promotive or pro-apoptotic genes.

Apoptosis inhibitory genes include *Bcl-2*, bcl-xl, bcl-w and Mcl-1, *etc. Bcl-2* is an oncogene that can inhibit cell apoptosis, over-expression of *Bcl-2* can block cell apoptosis caused by irradiation, anti-tumor medicine and c-myc but not apoptosis induced by antigens. The C-terminal of *Bcl-2* has a hydrophobic transmembrane anchoring motif composed of 19 amino acids; it's located on the outer membrane of mitochondria, the endoplasmic reticulum, and the nuclear membrane, and is closely related to *Bcl-2*'s function of regulating endoplasmic reticulum Ca^{2+} concentration, participating in trans-nuclear transportation and mitochondrial permeability transition. Membranous location of *Bcl-2* is critical for its apoptosis inhibitory function. *Bcl-2* has four homologous domains, BH1, BH2, BH3 and BH4. All apoptosis-inhibitory genes have BH1 and BH2, the most similar ones to *Bcl-2* genes have all 4 domains, and all apoptosis-promoting genes have a BH3 domain.

The exact mechanism thatas to how *Bcl-2* or Bcl-xl inhibit apoptosis is not well known; the tertiary structure of Bcl-xl has two hydrophobic alpha-helix, which is similar to the pore-forming domain of diphtherotoxin and colicins, suggesting that Bcl-xl may form pores on the membrane, and *Bcl-2* family members may act as ion channel and docking protein in cell apoptosis regulation. It has been proved that when mitochondrial injury causes transition of its permeability, nuclear apoptosis can be induced, and this mitochondrial permeability transition and nuclear apoptosis can be blocked by *Bcl-2*. Also, *Bcl-2* and Bcl-xl can inhibit Fas-mediated apoptosis *in vivo* and *in vitro*. Activation of Fas can impair the mitochondrial function, but this impairment could be inhibited by caspase inhibitors; these suggest that during Fas-induced apoptosis, mitochondrial injury occurs downstream of caspase cascade reaction, and thus may be a secondary effect.

Apoptosis promoting members include Bac, Bak, Bad, Bik, Bid, Bim, Bok, Bcl-xs and the newly discovered gene Noxa. Bax is another main member of the *Bcl-2* family which is highly homologous to *Bcl-2*. Bax, Bak and Bok also have highly conservative BH1, BH2 and BH3 domains whereas other members only have the BH3 central short domain composed of 9 – 16 residules. Bax has three forms, namely α, β and γ. Baxα belongs to membranous proteins while Baxβ and Baxγ are both cytoplasmic proteins. Bax can antogonize the apoptosis-inhibiting role of *Bcl-2*, and its overexpression can induce cell apoptosis.

Under pro-apoptotic stimuli, Bax and Bid can specifically translocate to mitochondria which suggests that on the mitochondrial surface there might be special proteins working as their molecular targets. Several mitochondrial surface proteins have already been identified to interact with Bax protein directly or indirectly. These proteins can form a tight complex, which exits on the junction of the inner and outer mitochondrial membrane. During pathological conditions, the complex can "transform" and become non-selective pores on the inner membrane which is called permeability transition pores, and Bax can promote the disclosure of these pores.

Regulation of the *Bcl-2* family: The expression of Bax can be induced by P53; Bad protein in IL-3 stimulated haemopoietic stem cells can undergo phosphorylation and dissociate from Bcl-xL and its signal is transmitted from PI-3K to Bad through Akt; *Bcl-2* can be activated by serine 70 phosphorylation, but deactivated by its loop area phosphorylation which may be mediated by JNK.

Mechanism of the influence of *Bcl-2* on cell apoptosis: Researchers found that *Bcl-2* members can form dimmers autologously or with each other. Within the cells, the ratio of the inhibitory and excitatory *Bcl-2* members can determine whether cell apoptosis would happen by influencing dimmer formation.

Contrary to earlier conclusions, it's now known that the heterogenous dimmer formation is not necessary for the anti-apoptotic function, but necessary for promoting apoptosis. Bcl-xS and Bcl-xL have opposite functions. Bcl-xL can inhibit cell apoptosis while Bcl-xS can promote apoptosis. Mechanism research of Bcl-xL revealed its molecular mechanism in regulating apoptosis. Bcl-xL and Apaf-1 can interfere with each other and block the combination of caspase-9 to CARD (caspase recruitment domain) on its N terminal, so as to block the activation of caspase-9 while the apoptosis-promoting factor Bik would release Apaf-1 from Bcl-xL.

Bcl-2 can block the release of Cyt c from mitochondria directly or indirectly. Cyt c and ATP could together change the conformation of Apaf-1 and activate caspase-9. The specific pore-forming property of Bax and Bax-like proteins is related to their function of inducing non-caspase dependent cell death.

Mechanism of mitochondria inducing the release of apoptosis factors: During many apoptosis processes, the inner transmembrane potential (Δφm) of mitochondria is low which indicates that the mitochondrial permeability transition pore (PT pore) is open. But this can only explain the release of Cyt c whose molecular weight is small (134 kD). Actually, AIF and other proteins whose molecular weight is larger than Cyt c could also be released, even adenylate kinase and sulfide oxidase, whose molecular weight is as large as 100 kD can also be released from the intra-membranous gap together with the pro-apoptotic proteins. This unselective loss of the proteins in the membrane gap indicates that the outer membrane forms a big gap instead of simply forming a channel protein, meaning the cracking of the mitochondrial outer membrane, which has been confirmed by electron microscopy. The mechanism of the cracking of the mitochondrial outer membrane is still unclear. One view is that the disclosure of the PT pores can lead to the equilibrium of ions between the mitochondrial matrix and intra-membranous gap, resulting in the disappearance of the H^+ concentration gradient on both sides of the endomembrane, and also leading to the disorders of mitochondrial volume caused by matrix hypertonicity. Finally, the outer membrane is damaged and the caspase-activated protein existing in the inner and outer membranes is released. However, this viewpoint has not been verified. Although it has been confirmed that inhibiting the opening of PT pores can block some apoptotic processes, *Bcl-2* could block PT and Bax could induce apoptosis as well as the occurrence of PT. But there still exists some evidence to show that the release of Cyt c and the activation of caspase can occur before the disappearance of the Δφm. Other studies suggested that caspase enzymes could induce the opening of PT pores, and the released cytochrome c and AIF from the pores in turn induced the activation of caspase enzymes which formed a feedforward amplification loop.

Calcium ion is another apoptosis-related factor. The normal concentration of Ca^{2+} existing in the plasma is an indispensable factor in cellular physiology and various enzymatic reactions. Under pathological conditions, cell calcium excess caused by various reasons may induce the apoptosis of related cells.The increase in Ca^{2+} within the cells can damage the energy homeostasis of the cell by inducing its downstream events, such as the generation of active oxygen and nitrogen ions, and finally lead to the death of the cell. The Ca^{2+} in cytoplasm will be pumped out of the cell or be pumped into endoplasmic reticulum by ion-motive ATPase. Mitochondria could also buff a part of cytoplasmic calcium. However, excessive accumulation of Ca^{2+} in mitochondria would damage the reaction of oxidative phosphorylation, and promote the generation of active oxygen ions such as superoxide (O^{2-}), hydrogen peroxide (H_2O_2) through the electron transport chain. Ca^{2+} accumulated in mitochondria can also change the permeability of the mitochondrial membrane, inhibiting the formation of ATP and promoting necrosis. Recent studies have found that calcineurin, a phosphatase activated by Ca^{2+}, could induce apoptosis. It caused dephosphorylation of Bad which is one pro-apoptotic member of the *Bcl-2* family, leading its translocation to the mitochondria and promoting the formation of dimers of Bad and Bcl-xL which finally promoted cell apoptosis.

In addition, Ca^{2+} can directly activate some cellular enzymes, such as Ca^{2+}/Mg^{2+} activated endonuclease, Ca^{2+} sensitised phospholipase protease and so on, to switch on the cytotoxic cascade reactions. Some Ca^{2+} activated enzymes are related to the formation of free radicals. For example, calpain can transform xanthine dehydrogenase into xanthine oxidase and the latter is a kind of enzyme which can promote the generation of superoxide. Hydrogen peroxide can be generated from superoxide, and it can convert into highly reactive hydroxyl ions (OH^-) by iron enzymatic reaction.These active oxygen ions can damage lipids, proteins and nucleic acids. Besides, Ca^{2+} can also activate the enzyme regulated by calmodulin and nitric oxide synthase (NOS) to generate a large amount of nitric oxide whose combination with superoxide can form a more lively peroxynitrite union (OONO—) OONO— can further damage plasma membrane and lead to the oxidation and nitrification of aromatic amino acids such as tyrosine. It can also be the source of the generation of active hydroxyl ions. The cell's oxidative stress will induce apoptosis.

Hematopathy is related to the changes in apoptosis molecular ecology.

Aplastic anemia and myelodysplastic syndrome are related to the increase in apoptosis.

A large number of studies about aplastic anemia suggested that bone marrow biopsy could prove that mononuclear cells from patients with aplastic anemia had a significantly higher incidence of apoptosis than in normal individuals. *In vitro* experiments showed that interferon (IFN) and tumor necrosis factor (TNF) could inhibit the haemopoiesis, including the colony formation of early and late progenitor cells and the production of the long-term culture initiating cell (LTCIC). These cytokines could induce the expression of Fas antigen of $CD34^+$ cells and cause apoptosis of hematopoietic cells. And the reaction of the cytokines which were locally secreted had significantly stronger activity than those added to the medium. In the bone marrow of patients with aplastic anemia, there are activated cytotoxic lymphocytes, high levels of TNF-γ and TNF-α and elevated expression of Fas in $CD34^+$ bone marrow cells. Therefore, the apoptosis of hematopoietic cells induced by the death factor and its receptor play an important role in the pathogenesis of aplastic anemia.

Myelodysplastic syndromes (MDS) are a group of malignant diseases that are derived from myeloid progenitor cells or pluripotent hematopoietic stem cells, characterized by a high tendency to transform to acute leukemia. Flow cytometry detected that the content of the sub-diploid of MDS bone marrow cells was significantly higher than the normal ones. The ISEL assay of an MDS bone marrow biopsy also proved a higher apoptosis incidence in MDS cells than in normal cells. The mechanism studies of increased apoptosis in MDS patients have shown that the increased endonuclease activity in cells is one of the reasons why MDS bone marrow cells are prone to apoptosis. During the mechanism study of the transformation of MDS to acute leukemia, we found MDS bone marrow cells had a low expression rate of C-*myc* and *Bcl-2* in the low-risk group, and this was significantly increased in the high-risk group. High expression of *Bcl-2* can inhibit apoptosis, and then enable the long-term survival of cells. With the cooperation of C-*myc*, it further promotes malignant transformation of cells which at last evolves into acute myeloid leukemia [4].

Further study also found that the apoptosis rate of cells and protein expression levels of *Bcl-2* and C-*myc* showed no difference between MDS $CD34^+$ bone marrow cells and normal ones. But the apoptosis rate of CD34-cells was significantly higher than in the control group and it had a positive correlation with the reduced expression of *Bcl-2* and elevation of C-*myc*/*Bcl-2* ratio. During the cell kinetics study of MDS bone marrow cells, we found bone marrow cells of MDS were characterized by high proliferation. But the high production rates were offset by high mortality rates caused by apoptosis. So we considered that the primary injury of MDS bone marrow cells may due to some abnormal expression of death regulatory genes of the CD34-cell, which made the process of apoptosis easy [5].

Chronic myelogenous leukemia is related to a reduction in apoptosis. The genetic characterization of chronic myelogenous leukemia (CML) is the existence of ph chromosome [t(9; 22) (q34; q11)], which is actually the BCR-ABL fusion gene formed by the reciprocal translocation of the oncogene abl, located on chromosome 9, and gene split concentrated zonebcr, located on chromosome 22. The BCR-ABL fusion gene encoded the p210 protein which has a very strong activation of tyrosine kinase. Studies showed that the proliferation rate of bone marrow stem cells of CML patients had no difference compared with the normal ones, and p210 protein alone did not induce proliferation of hematopoietic stem cells, which indicated that the effect of the BCR-ABL fusion gene was not to promote proliferation [6]. Studies also suggested that the BCR-ABL fusion gene could inhibit cell death and prolong survival time of CML cells, which can be reversed by the use of antisense oligonucleotides. Thus, it suggested that to inhibit apoptosis by the BCR-ABL fusion gene may be one of the main pathogenesises of CML.

17.2.2 Oncogenes, Tumor Suppressor Genes and Signal Conducting Molecules

The *P53* gene is an important tumor suppressor gene which is located on the short arm of chromosome 17(17p13.3). Its biological function is to monitor the integrity of genomic DNA in the G_1 phase. If DNA is damaged, *P53* protein stops the cell at the G_1 phase, and won't let it enter the M phase until it is repaired. If the damage cannot be repaired, apoptosis will be induced and acid in carcinogenesis. *P53* will lose this function when mutations happen.

Yonish and others first reported in 1991 that the wild-type *P53* gene could induce apoptosis in leukemia cells. After that, many studies have further confirmed that one of the *P53* anti-tumor mechanisms was triggering apoptosis of tumor cells. *P53*-mediated apoptosis may be blocked by numbers of physical and chemical signals or the activation of a special signal transduction pathway. For example, myeloid leukemia cells or mice erythroleukemia cells will undergo apoptosis when *P53* is over-expressed and activated, but the apoptosis mediated by *P53* can be blocked by treatment with IL-6 on myeloid cells or with erythropoietin (EPO) on erythroid cells. So the communication between *P53* and a certain signal transduction pathway can reverse the apoptosis mediated by *P53*. P53 is also involved in apoptosis induced by abnormal expression of a virus or cell oncogenes or by lack of some tumor suppressor gene products. Take the adenovirus E1A protein as an example; its expression in rat embryonic fibroblast cells stabilizes and activates the *P53* protein, and then results in apoptosis. Just like the adenovirus E1A protein, human papilloma virus E7 protein can also induce apoptosis mediated by *P53*, while its E6 protein can bind to *P53* and degrade it. Also, E1A and E7 proteins can bind to the RB protein, making it inactive in the regulation of E2F-DP-1 activity. E2F-1's out of control can also trigger the course of apoptosis by activating *P53* and so can the C-*myc*'s over-expression.

As reported, under special conditions such as DNA damage, shortage of essential survival factors, activation of oncogene or forcing the cells into the replication cycle, *P53* mediated-apoptosis will be given priority and the cells will be removed. In the signals or upstream events triggering *P53*-mediated apoptosis, DNA damage was studied the most. DNA damage causes phosphorylation of *P53* protein and extends its half-life.

In the mechanism of *P53* mediated-apoptosis, the transcriptional activity of P53 protein and the non-transcriptional effect may both be involved, but there are variations in different cells. In some cells, DNA damage or activation of c-myc does not depend on new protein synthesis. On the contrary, in young rat kidney cells or mice cells, the mutations which affect *P53*-specific DNA binding or transcription activation make *P53* protein lose the function of inducing apoptosis. Therefore, *P53* protein may trigger apoptosis by its transcriptional activation and / or direct protein signaling (protein - protein interactions or other activity).

As is known, genes transcriptionally activated by *P53* protein are WAF1, mdm2, gadd45, KAI1, cyclinG, Bax, Noxa, Fas, PIGs, KILLER/DR5,PERP and insulin-like growth factor binding protein-3(IGF-BP3)genes. Among them, the activation of gene expression of Bax, Fas, PIGs, KILLER/DR5,PERP,and IGF-BP3 may affect the cells going into the decision-making process of apoptosis. Bax protein encoded by the Bax gene may antagonize its activity to block apoptosis by binding to *Bcl-2* protein. And the proportion of *Bcl-2* and Bax determines whether cells will go into apoptosis. With transcriptional activation of P53, the proportion of Bax protein increases and forms the homodimer of Bax, which leads to the apoptosis of cells [7].

Receptor tyrosine kinases (RTKs) are a class of membrane-bound enzymes, with typically an extracellular ligand binding region, transmembrane region, highly conserved intracellular kinase domain, and a C terminal tail [8]. RTKs consists of 20 sub-family members, subtype I including epidermal growth factor receptor I , subtype II including insulin-like growth factor-1 (IGF-1), subtype III including the PDGF receptor (PDGFR), c-FMS, c-KIT, FLT3, and subtype IV including fibroblast growth factor receptor (FGFR). Among them, subtype III contains five subtypes of immunoglobulin-like extracellular regions and two types of intracellular kinase domains which are inserted and separated with kinases. Mutations in these areas often happen which have some relationship with the occurrence and development of hematologic malignancies [9]. The mechanisms of the persistent activation of RTKs are mainly chromosomal translocation and various mutations involved in its regulatory regions. Chromosomal translocation can always produce fusion proteins that are formed by the dimerization or polymerization of the RTKs' cytoplasmic part and its fusion motif, resulting in the dimerization and activation of RTKs. On the other hand, a missense mutation, insertion and deletion of regulatory regions can activate the RTKs via inhibiting their negative feedback pathway. Activating mutations of RTKcan activate downstream signaling molecules such as Ras/MAPK, PI3K, phospholipase C-γ, JNK, STATs and NF-κB pathway by the phosphorylation of tyrosine or serine/threonine, which changes these downstream signaling molecules in terms of nature and quantity, leading to disease. And Ras/MAPK, PI3K, and STATs are the major signal transduction pathways [10].

Ras/MAPK signal transduction pathway: The Ras family proteins belong to a major subfamily of GTP enzyme families, located in the cytoplasmic membrane surface. Ras proteins play key roles in many RTKs-mediated signaling pathways. Ligand activation and RTKs' autophosphorylation into the receptor proteins, such as Shc, Grb2 and so on, can affect guanine exchange factors (GEFs), which activate Ras proteins and produce phosphate binding sites. Once Ras proteins are activated, they can further activate Raf serine/threonine kinase and induce the activation of MAPK kinase by phosphorylation. After MAPKs or extracellular signal-regulated kinases (ERKs) are activated, they enter the nucleus to activate nuclear transcription factors such as Elk-1. ERKs can also activate other kinases, such as RSKs (also known as MAPK-activated protein kinase), to regulate the cell cycle and cell apoptosis. ERK-activated RSK kinases can affect Bad, a pro-apoptotic protein of the *Bcl-2* family, thereby inhibiting Bad-mediated apoptosis. And the Ras-Raf-MEK-ERK cascade can regulate the expression of many proteins to influence cell proliferation, including cell cycle regulators (such as cyclin D1, p21, p27, cdc25A) and transcription factors (such as C-*fos*, C-*jun*, C-*myc*). The activity of MEK and ERK is abnormal in both AML and CML.

PI3K signal transduction pathway: PI3K is another important signaling pathway to control the serine/threonine phosphorylation. PI3K consists of two subunits, p85 regulatory subunit and p110 catalytic subunit. P85 subunit can bind ligand- activated phosphorylated RTKs. PI3K/Akt pathway can activate many downstream target proteins, including p70RSK, forkhead transcription factor (FOXOs) and NF-κB. Serine/threonine kinase Akt is an important regulatory factor in the cell survival mechanism. PI3K-activated Akt can induce the activation of many downstream signaling molecules. For instance, Akt can phosphorylate NF-κB inhibitor-IκB. IκB can be degradated by the 26S proteasome to release NF-κB, entering the nucleus and inducing transcription of many target genes which can affect cell survival, such as Bcl-XL, IAPs. AKT can also phosphorylate pro-apoptotic protein Bad, and increase free type Bcl-XL to inhibit caspase9. In addition, the tumor-suppressor gene PTEN, a phosphatase, allows

PIP3, 4, 5 to release phosphate radicals at the 3' position of the inositol ring. PTEN is a negative regulator of Akt. And mTOR is another important downstream signaling molecule of PI3K/Akt, which can also mediate cell survival and proliferation as serine/threonine kinases. PI3K/Akt signaling pathway is not necessary for haemopoiesis under normal physiological regulation. It is tumor-specific, which is unique to hematological malignancies.

STATs protein family: STATs are ten different proteins encoded by six kinds of known mammalian genes, including a variety of isomers of STAT1, 3, 4, and 5. Similar to other transcription factors, the structure of STATs comprise a DNA-binding domain, a conserved NH_2-terminal domain, a COOH-terminal transactivation domain, an SH_2-like domain and an SH_3-like domain. After RTKs are activated to induce tyrosine phosphorylation, STATs are activated and induce dimerizations, entering into the nucleus to activate specific target genes. STATs-mediated cytokine-dependent cell growth comes into play mainly through regulating the expression of its target genes such as *cyclins*, C-*myc* and *Bcl-XL*. In addition to the normal ligand-activated cytokine receptor pathway, TKD mutations of c-Kit can induce non-ligand-dependent activation of STAT3 and STAT1 [11]. FLT3 receptor is a tyrosine kinase of a membrane-bound receptor, which plays a key role in the maintenance, proliferation and differentiation of hematopoietic tissues [12]. FLT3 mutations are found in many patients with AML and MDS [13]. The mutations consist of two types: point mutations in the juxtamembrane domain of internal tandem duplication (FLT3-ITD) and in the highly conserved tyrosine kinase domain (FLT3-TKD) [14, 15]. Both these two kinds of mutations lead to persistent activation of FLT3, and further activate its downstream signaling pathway. FLT3-ITD can continue to activate STAT3 and STAT5 and this effect is more effective than ligand-activated wild-type FLT3 [16]. Recent studies have shown that these two mutations have different signal transduction properties. MDS patients with FLT3-ITDs mutation are easier to transform into AML with a poor prognosis and short survival time [17, 18]. Recent studies have discovered that the FLT3/ITD mutant also exists in $CD34^+$ $CD38^-$ leukemic stem cells [19]. STATs have played a key role in the development of hematological malignancies caused by mutant-activated RTKs [20, 21].

The ABL gene of BCR-ABL gene is a highly conserved sequence, homologous to Abelson murine leukemia viral oncogene v-abl, encoding non-receptor tyrosine kinase. Breakpoints in the ABL gene can occur in any part of more than 300 kb in the 5' region, in the upstream of Ib, or in the downstream of Ia, but the most common position is between the two. But breakpoints in the BCR gene are concentrated in three regions. In most CML and one-third of ph (+) acute lymphoblastic leukemia (ALL) patients, the breakpoints are located inside the 5.8 kb within the span of BCR exon e12-e16, known as the major breakpoint cluster region (M-bcr). Due to different splicing, b2a2 and b3a2 are two forms of the transcripts, encoding P210 fusion protein. To the remaining patientscharacterized by clinically significant mononucleosis and very few CML patients, their breakpoints are located in the 54.4 kb region between exons e2' and e2, known as the minor breakpoint cluster region (m-bcr), producing e1a2 transcripts and translated into P190 fusion protein.

Studies have shown that part of ABL is almost constant, while part of BCR changes greatly. Therefore, ABL sequence of fusion gene BCR-ABL is considered to carry the principles of gene transcription, while the BCR sequence determines the different disease phenotypes.

Bcr-Abl gene encodes P210 fusion protein, with a strong tyrosine kinase activity, leading to chronic myeloid leukemia. Studies show that there are three BCR-ABL-mediated malignant transformation mechanisms: Changing adhesion properties of cells to stromal cells and the extracellular matrix; activating mitogenic signaling continually; inhibiting apoptosis. Another mechanism may be that *Abl* inhibits the protease-mediated protein degradation.

Changes in cell adhesion properties: The adhesion of hematopoietic cells to bone marrow stromal negatively regulates cell proliferation. When the adhesion of CML progenitor cells to bone marrow stromal cells and the extracellular matrix weakens, the cells will proliferate unlimitedly. α-Interferon seems to reverse the adhesion defect [22]. Studies have shown that β integrin plays an important role in adhesion of bone marrow stromal to the progenitor cell. CML cells express β1 integrin-dependent adhesion variant molecules and these variant molecules do not exist in normal progenitor cells. The combination of integrin molecules and receptors can trigger normal signal transduction from extracellular to intracellular space, thereby inhibiting cell proliferation. But as for CML cells, normal signals to inhibit proliferation are blocked. *Abl* has the ability of signal transduction in cells. Since there are a lot of Bcr-Abl fusion proteins in CML cell cytoplasm, the normal signaling function has been greatly weakened.

Activation of mitogenic signals: Bcr-Abl is known to activate Ras and MAP kinase pathways, Jak-Stat pathway, PI3 kinase pathway and Myc pathway. The activations of these kinases allow cells to proliferate indefinitely.

Inhibition of apoptosis: Apoptosis is inhibited in BCR-ABL-positive cells, which depends on the tyrosine kinase activity and is associated with Ras activation. *Bcr-Abl* may also prevent cytochrome c release from mitochondria and activate cytochrome c-mediated caspase enzymes. The upstream activation of caspase enzymes may be mediated by the *Bcl-2* family. The third possible mechanism is that *Bcr-Abl* inhibits apoptosis by reducing interferon consensus sequence-binding protein (ICSBP). Existing data shows ICSBP knockout mice had myeloid proliferative syndrome, while hematopoietic progenitor cells' response to cytokine in ICSBP-negative mice also had been changed.

17.2.3 Telomere-Telomerase

Telomere, with a length of about 2 – 15 kb, is a special DNA structure in the intracellular chromosome terminal. Under normal circumstances, telomeres lose 50 – 200 bp at each cell division. With the increasing number of cell divisions, telomeres progressively shorten. Shortened to a certain extent, cells lose the ability to proliferate, thereby entering aging and death. If cells are infected by a virus, or some tumor-suppressor genes such as *Rb*, *P53 mutation*, gene molecular ecology can be changed. Thus, telomerase can be activated and telomere length will be recovered, which allow cells to escape death, into an immortalized state, thus leading to malignancy.

Telomere is a special structure in the chromosome terminal composed of GTP-rich hexamer repeating sequences. The repeating unit is (5'-TTAGGG-3') n with a length of about 2 – 15 kb. Telomere is very important to chromosome stability and genome integrity, providing a protective 'cap' for the chromosome terminal. In this way, it can prevent chromosomal DNA degradation, terminal fusion, informal restructuring and deletion in order to maintain the stability of chromosomes. Studies demonstrate that due to an 'end-replication problem', telomeres progressively shorten with constant DNA replications. To a certain degree, cells enter senescence and death. Thus the number of cell divisions can be estimated to reflect the age of cells by calculating the telomere length, known as the "mitotic clock".

The elomere-telomerase hypothesis is put forward by Harley. It believes that telomeres will shorten with the loss of some sequences of telomerase in the process of mitosis. When the telomere shortens to a certain extent, a certain signal may be triggered which makes cells enter a crisis point of the M1 phase. Cells stop dividing at this moment, exit the cell cycle and die. If cells are transfected by the virus or some tumor-suppressor genes such as *P53*, *Rb* mutation, cells can divide over the M1 and continue on, and the length of the telomere continues shortening. A critical threshold is eventually reached and cells enter another crisis point of the M2 period when the length of the telomere is insufficient to maintain its features so as to induce death of most cells. Only in a very few cells is the telomerase activity activated and telomere lengths have been recovered to maintain homeostasis. These cells become immortal cells.

Telomerase consists of three components: human telomerase RNA component (hTR), human telomerase reverse transcriptase (hTRET) and telomerase-associated protein (TP1) [23]. The RNA component of human telomerase has been cloned. The RNA template region encompasses 11 bp (5'-CUAACCCUAA-3'), complementary to the telomeric repeat sequences (TTAGGG) n. As the templete in synthesising telomeric G-rich DNA chain, the complementary RNA sequences can resolve the end replication problem. Studies have shown telomerase reverse transcriptase was most closely associated with telomerase activity. In 1997, Nerkanara and Meyerson reported their successful cloning of the *hEST2* and *hTRT* genes (later named as unified *hTERT* in the gene bank) sequentially, which is in fact the telomerase reverse transcriptase gene. Studies have found *hTERT* was highly expressed in primary telomerase-positive tumors, while it was not expressed in telomerase-negative cell lines. In the process of tumor development, inducing the expression of *hTERT* mRNA is accompanied by the activation of telomerase. In contrast, research of the protein component of telomerase progresses slowly. Researchers obtained two telomerase proteins with a weight of 80 kD, 95 kD when they purified the tetrahymena telomerase. Gene encoding mammalian telomerase associated protein (TP1) has yet been cloned. It is reported that phosphorylation and dephosphorylation of telomerase protein component or telomerase regulatory protein can regulate the telomerase activity reversibly.

A high level of telomerase activity can be detected in most human cancer cells. And the level of telomerase activity is found to be closely related to the prognosis of certain tumors which could be indicators of tumor diagnosis and prognosis.

Viruses, bacteria and other infections can damage DNA or change cell biological characteristics partially, shorten the chromosome telomere and activate telomerase. Infection can also lead to gene mutations directly and activate telomerase, which results in cell immortalization and cancer development.

17.3 Microecological Changes and Hematologic Diseases

There is a close correlation between microecology and hematologic diseases. On the one hand, some microecological changes can lead to hematologic diseases. On the other hand, certain hematologic diseases can alter the microecology. Thus, to discuss their relationship is of important clinical significance.

17.3.1 Helicobacter Pylori and Primary Gastric Lymphoma

Although primary gastric lymphoma accounts for only 3% of gastric malignant tumors, it is the most common type of extranodal non-Hodgkin's lymphoma. In recent years, the close association between helicobacter pylori (HP) infection and primary gastric lymphoma has been extensively researched. HP is the most common bacterial infection in the stomach, but to date [24], whether it belongs to colonization or infection remains unclear. Mucosa associated lymphoid tissue (MALT) lymphoma is the most frequent HP-induced malignant tumor and its treatment by eradicating HP has aroused great interest.

The relation between HP infection and primary gastric lymphoma was demonstrated by previous studies in the following aspects: (i) HP infection was found in stomachs of patients with primary gastric lymphoma; (ii) The rate of HP infection was positively correlated with primary gastric lymphoma; (iii) Greater risk of primary gastric lymphoma was found in patients with HP infection than in those without HP infection; (iv) HP infection had been confirmed before primary gastric lymphoma was diagnosed; (v) It was demonstrated that erasing HP could reduce the risk of primary gastric lymphoma.

The first clue of confirmation that HP can cause chronic gastritis and primary gastric lymphoma is that the characteristics of HP infection were observed in almost all gastric lymphoid tissues of patients with primary gastric lymphoma. Subsequently, Wotherspoon *et al.* studied 110 gastric lymphoma samples by general and immunohistochemical staining, showing that the HP infection rate was 92%. Further epidemiological evidence from the study by Doglioni *et al.* was that the incidence of gastric lymphoma in Feltre, a region of northeast Italy, was 13 times as great as in three central regions of the United Kingdom, and the prevalence of gastric lymphoma in Feltre was 87% compared to 67% in the UK. A recent report by Parsonnet *et al.* revealed that the incidence of HP infection was 85% in the gastric lymphoma group and 55% in the control group and a significant difference was observed in both. Moreover, the study also verified that existence of HP infection was prior to gastric lymphoma.

It is found that not only do lymphoid follicles form but also B lymphocytes infiltrate the epithelial layer, the formation of the histological features of MALT, in HP-related gastritis. Data indicate that immune responses induced by HP colonization in the gastric mucosa result in the formation of MALT, which provides a histopathological basis for the occurrence of gastric lymphoma. Much research shows that HP can stimulate the proliferation of gastric MALT lymphoma cells, which depends on T cells activated by HP and its secretory products including IL-2. Hussel *et al.* co-cultured HP and tumor cells taken from surgical specimens with low-grade gastric/MALT lymphoma. The results showed that these cells were activated and expressed IL-2 receptor (CD25), which turned back and promoted the proliferation of tumor cells and the synthesis of tumor immune globulin. This effect was HP strain-specific and dependent on non-tumor T cells because removal of them eliminated the effect. These findings offered some evidence on the level of cytobiology for the close association of primary gastric MALT lymphoma with HP infection, explained why its distant metastasis rarely occurred in the long term and provided a theoretical basis for the phenomenon that HP eradication therapy can result in its clinical remission.

Many histopathological features indicate primary gastric MALT lymphoma has immune response characteristics, in which HP may be the key antigen and HP eradication, *i.e.* HP antigen removal, might inhibit its growth. It remains unclear whether HP eradication therapy is effective for patients at all stages of the disease or only for those at the early stage. Some studies showed that anti-HP treatment could be only suitable for early gastric MALT lymphoma and led to complete or partial regression in all 48 cases with early-stage gastric MALT lymphoma while surgery, chemotherapy or radiation therapy were more preferable for cases at an advanced stage, indicating that HP eradication might be suitable for only early-stage patients. Horstman *et al.* reported that one patient with gastric MALT lymphoma showed improved symptoms and their relapse was along with HP reinfection, suggesting that the long-term effect of simple HP eradication needs to be further and carefully researched.

Although many studies have confirmed that primary gastric lymphoma, especially MALT lymphoma, is closely correlated with HP infection, it occurs in only a very small part of cases infected with HP, which account for nearly half of the world population. Thus, it is deduced that there must be many other factors playing an important role, so that the body is more sensitive to other risk factors. However, further research is needed to demonstrate whether the regression of gastric MALT lymphoma tumor after elimination of HP is due to gene regulation or simply the death of tumor cells.

17.3.2 Microecological Changes and Erythrocyte Disorders

Aplastic anemia can be caused by viral infection. These viruses include hepatitis, Cytomegalovirus, Epstein-Barr virus (EBV), Human pavovirus B19 (HPVB19).

In 1955, Lorenz proposed the concepts of hepatitis associated aplastic anemia (HAAA) and hepatitis-aplastic anemia syndrome. Clinically, the latent period of HAAA, which occurs most in the recovery of acute hepatitis, is about 6 – 12 weeks after hepatitis virus infection, and the longest one is up to more than 1.5 years after diagnosis of hepatitis. HAAA is more commonly seen in hepatitis patients aged 18 – 22 years and the male to female ratio is 2 – 4 : 1. Although clinical manifestations of HAAA are not very serious, its therapeutic effect and prognosis are poor compared with aplastic anemia caused by other factors. Can any types of hepatitis virus cause aplastic anemia? According to a large number of clinical observations in recent years, hepatitis A virus (HAV) and hepatitis B virus (HBV) infections have no relation to HAAA [25]. The study by Jerome *et al.* showed 80% of patients with HAAA had hepatitis C virus (HCV) infection, indicating that HAAA is closely associated with hepatitis C virus (HCV) infection [26]. Continuous reports by Zaidi *et al.* and Byrneset *et al.* reveled that no serum markers of hepatitis virus but anti-HGV or HGV-RNA were detected in blood samples of two patients diagnosed with HAAA. Recent reports, however, denied the possibility of HGV as a cause of aplastic anemia. Therefore, though the name of HAAA has been used for decades and many cases with HAAA, possibly due to several types of hepatitis virus, were reported, epidemiological investigations of large samples of the association between any type of hepatitis virus and aplastic anemia would be still needed.

Aplastic anemia has been listed in CMV associated diseases. To date some animal models of AA-related cytomegalovirus (CMV-AA) have already been established. Thus, it is proven that CMV can lead to bone marrow suppression. However, up to now, CMV-AA has been observed in more patients who received organ transplants, including renal, liver, heart and bone marrow transplants, but not reported in healthy individuals after acute CMV infection. CMV viral particles could be found in blood samples of these transplant patients. CMV-DNA could be detected by PCR in their bone marrow, and even CMV strains can be isolated from inoculated animals. In view of organ transplant patients having different levels of immune suppression, it is speculated that the occurrence of CMV-AA could be associated with individual immune status, in which case CMV infection might be regarded as an opportunistic infection after transplantation.

A large number of reports described that aplastic anemia occurred after EBV infection, often accompanied by clinical manifestations of other systems, such as pneumonia, choroiditis and central nervous system symptoms, the seriousness of which sometimes made doctors ignore the existence of aplastic anemia.

HPVB19 is currently the only known human pathogenic parvovirus and the relationship between it and aplastic anemia is considered important. Some studies have demonstrated that HPVB19 had a high degree of affinity for bone marrow and pure red cell aplasia often occurring after HPVB19 infection in those patients who had potential dangers of anemia of haemolysis, such as sickle cell anemia, thalassemia, etc. HPVB19 infection could cause severe pancytopenia or severe aplastic anemia in patients with immune dysfunction. Like CMV, HPVB19 infection often occurs in organ transplant patients. For example, the incidence of aplastic anemia is high in liver transplant patients who had a fulminant hepatic failure caused by hepatitis virus, which had been considered as a special case of HAAA for a long time. However, anti-HPVB19-IgM and HPVB19 DNA in bone marrow or blood samples of these patients were detected by ELISA and PCR, indicating a high rate of infection of HPVB19 in them. According to statistical analyses, the correlation of hepatitis viruses with aplastic anemia was much lower than that of HPVB19 in HAAA patients. Therefore, it was suggested that HPVB19 might be another hepatotropic virus which remains to be recognized and its infection might result in similar clinical manifestations of hepatitis virus and thus cause some serological tests for the hepatitis virus to show false positive. HBVB19 may be the real pathogen of HAAA and some people even suggested that HPVB19 inspection should be necessary for all patients with chronic anemia.

The possible pathogenesis of aplastic anemia includes mainly the damage to the hematopoietic stem cell, the dysfunction of the immune system and the destruction of the bone marrow microenvironment. It is currently considered that the occurrence of aplastic anemia caused by viral infection could be associated with all three points mentioned above. It was found that in peripheral blood or bone marrow of patients with aplastic anemia caused by viral infection, $CD8^+$ T cells significantly increased and $CD4^+$ T cells decreased, which leads to an imbalance in the CD4/CD8 ratio. Meanwhile, these activated T cells released aberrant INF-γ, tumor necrosis factor β (TNF-β) and lymphotoxin α (LT-α) which induce $CD4^+$ cells in bone marrow to express huge amounts of Fas antigen (Fas-Ag). Fas-Ag, a messenger of apoptosis, makes cytotoxic T cells kill hematopoietic stem cells and sensitise bone marrow cells to another lymphokine-clonal inhibitory factor (CIF), which results in hematopoietic failure. Interleukin -6 (IL-6) is an important positive regulatory factor in the process of bone marrow hematopoiesis. However, its expression decreases because the storage of mRNA used to synthesis IL-6 in bone marrow stromal fibroblasts is reduced after virus infection, together with the function of other stromal cells affected by the virus, so that the bone marrow microenvironment is finally destroyed. As early as 1955 it was suggestedthat the bone marrow and the liver should be homologous in embryogenesis. Owing to its biological characteristics, the hepatitis virus causes both liver and hematopoietic stem cells to be damaged (chromosome aberrations), the latter of which leads to HAAA. HLA-DRB1 • 0405- restricted T-lymphocytes, a kind of cytotoxic T cells, appear in bone morrow due to EBV infection and attack hematopoietic stem cells [27]. CMV can cause the erythropoietin (EPO) gene to be in loss of regulation or dysfunction of positive regulation so that EPO cannot be produced normally or it is out of action, finally resulting in occurrence of aplastic anemia. HPVB19 reproduces only in the erythroid progenitor cell, which is determined by the characteristics of receptors on it and glucoside esters in the red blood cell. Thus, hematopoietic depression caused by HPVB19 is mainly red cell aplastic anemia. However, it is not the only mechanism for bone marrow suppression. HPVB19 infection also stimulates the bone marrow to produce a large amount of platelet associated IgG (PA-IgG) which causes platelets in peripheral blood to decrease.

Principles and measurements for treating the aplastic anemia induced by virus infection are basically the same as those for treating idiopathic aplastic anemia, and here we will just focus on the antiviral treatment. Acyclovir was once the commonly used drug; however, ganciclovir (GCV) is now more widely used abroad, which can remove viruses, but early and continuous administration are still recommended. Generally speaking, the patient should be medicated in the preliminary stage of the viremia, and patients who have potential infection risk from the above viruses (such as those patients who suffered an organ transplant) can even obtain prophylactic treatment with a course lasting 3 – 4 weeks.

Hemolytic anemia has been related to microbial infection, such as protozoan infections that may cause malaria and Kala-azar, while bacterial infection can cause bartonella bartonellosis and Clostridium perfringens sepsis [28].

Malaria is a mosquito-borne infectious disease of humans and other animals caused by protists (a type of microorganism) of the genus Plasmodium.

There are four types of Plasmodium which can infect and be transmitted by humans: P. falciparum, P. malariae, P. vivax and P. ovale. Plasmodium merozoites can asexually proliferate inside erythrocytes to form schizonts which crack to release merozoites, resulting in the rupture of erythrocytes, resulting in hemolytic anemia. Therefore, the more erythrocytes that are infected by Plasmodium, the more severe the hemolytic anemia will be. Within erythrocytes, the proliferation of Plasmodium utilizes and consumes the cellular components such as hemoglobin, enzymes and glucose, which can impair the metabolism of the erythrocyte greatly, modify its permeability and increase its osmotic fragility. In addition, when merozoites enter into erythrocytes, part of the erythrocyte membrane will follow merozoites to enter the cytoplasm as well, which can cause abnormal morphology as well as reduced deformability of the erythrocyte. Furthermore, the invaded erythrocyte can produce new membrane antigens, inducing an immune response to present a positive result in the Coombs test, and such an immune response may also play a role in hemolysis induced by Plasmodium. As a result, the erythrocytes without parasitic Plasmodium may also be destroyed. Hemolysis will still last four or five weeks after Plasmodium *in vivo* is completely removed. The abnormal erythrocytes with parasitic Plasmodium are mainly removed in the spleen. Splenomegaly and hypersplenism can aggravate a malaria patient's anemia. The hematopoietic function of the bone marrow of malaria patients may be influenced, and worsen the patient's anemia, and it is hypothesized to be mediated by TNF.

Clinical manifestations and laboratory examinations: In addition to regular shivering and high fever, symptoms of malaria also include anemia. The severity of anemia is related to the type of Plasmodium, the patient's immune status, nutritional status, with or without other complications and so on. For falciparum malaria patients, anemia is the most common and serious symptom. A small number of falciparum malaria patients may suddenly develop acute intravascular hemolysis, namely blackwater fever with manifestations of soy-colored urine, chills, high fever, vomiting and organ failure. In more severe cases, some patients may develop acute renal failure.

Lab tests show different degrees of anemia. The number of leukocytes decreases or stays normal. The platelet of two-third falciparum malaria patients decreases; this may be associated with platelet aggregation or DIC caused by the release of adenosine diphosphate due to the destruction of erythrocytes.

After a prompt and effective radical treatment of malaria, hemolytic anemia will gradually disappear; no special treatment is needed except in the case of the severe anemia. Blackwater fever can be treated according to the principles of treating acute hemolysis. Splenectomy is feasible for patients with splenomegaly and hypersplenism.

Kala-azar, also known as visceral leishmaniasis, is caused by Leishmania donovani transmitted by female sandflies. Leishmania donovani grows and reproduces in reticuloendothelial cells, resulting in hepatomegaly, splenomegaly, lymphadenectasis, anemia, leukopenia and thrombocytopenia. The anemia is normochromic and normocytic, and the survival period of the erythrocyte is shortened. Destruction of the erythrocytes in spleen is the main cause of anemia. An antimonial agent is effective in these cases. When the theraputic effects emerge, hematological abnormalities can be recovered. Since these patients are usually in poor nutritional condition, it is necessary to pay attention to improving their nutrition level and avoiding nutritional anemia.

Bacterial infection can cause bartonella bartonellosis and Clostridium perfringens sepsis.

The pathogen of bartonella bartonellosis has a rod-shaped [(1 – 2) μm×(0.2 – 0.5) μm] or circular (diameter of 0.3 – 1.0 μm) form, and can attach to the surface of erythrocytes, presenting itself in single form or in pairs, or even taking the shape of "V" and "Y" by end connections. Bartonella bacilliformis can be seen after microscopic examination of peripheral blood smears with normal staining. Such erythrocytes adhered to by a pathogen are removed by phagocytes of the liver and spleen, resulting in extravascular hemolysis.

The disease prevails in South America, especially in the Andes area of Peru. There are 2 – 3 weeks of latent period before the onset and the acute phase is called Oroya fever with manifestations of chills, fever, joint and muscle pain, enlargement of lymph node and failure followed by rapid hemolytic anemia. The patient's condition in this period is very severe and can be fatal.

Toxin-α produced by Clostridium perfringens is a kind of phosphatidylcholine enzyme and can affect the cell-membranous phosphatidylcholine and thereby hydrolyze to lysophosphatidylcholine which can cause intravascular hemolysis. Sepsis caused by this bacterium often occurs in septic abortion and also can be found in uterine infection after dystocia.

Some Gram-positive and Gram-negative bacterial infections can also cause hemolytic anemia. The reported pathogens include *Staphylococcus aureus*, streptococcus, pneumococcus, *Escherichia coli*, *Vibrio cholerae* and so on.

17.3.3 Microecological Changes and Leukocyte Diseases

Infections and viruses are highly related to neutropenia, malignant lymphoma, adult T cell leukemia (ATLL) and human herpesvirus 8 (human herpesviruses-8, HHV8).

Neutropenia could appear in many infectious diseases; the main mechanism may be due to the shortened life cycle of neutrophil in peripheral blood, and some infection could also inhibit the generation of neutrophil in the bone marrow.

Bacterial infection: Thyphoid fever and paratyphoid fever always accompany leucopenia, and mild leukocytosis could be seen in the early stage of the disease (the first week), but granulocyte decreased when bacteremia occurred, yet the granulocyte count is hardly lower than 0.6×10^9/L.

Viral infection: Leukopenia (granulopenia) could be found in many viral infections. The cell count begins to decrease on the second day of the fever in virus hepatitis, yellow fever and sandfly fever, and touches the bottom on the days 4 – 6, when granulocyte and lymphocyte decrease simultaneously.

Rickettsia infection: Leukopenia (granulopenia) is usually seen in the first week of most rickettsia infection, such as in the epidemic typhus, Ginger worm disease and so on.

Parasitic infection: In the case of a malaria attack, mild leukocytosis could be found, followed by leukopenia (granulopenia). The leukocyte count could be up to 15×10^9/L (15,000/mm^3) in the febrile phase of the relapsing fever, but in the interval of febrile stage leukopenia (granulopenia) could happen.

In the pathogenic mechanism of the malignant lymphoma, virus theory has attracted most of our attention, especially the EV virus (EBV) theory.

EBV infection is related to the incidence of Hodgkin's disease (HD). Liu *et al.* found 62.9% of HD cases to be infected with EB after the examination of 70 samples [29]. Among these the lymphocyte depletion type (LD) has the greatest infection rate, followed by the mixed cellularity (MC) type, the nodular sclerosis type (NS), and finally the lymphocyte predominance type (LP). According to a study of the pathogenesis and the origin of HD, some researchers think that HD should be divided into two types: classical type HD, including NS, MC and LD, and LP type HD (LPHD). EBV infection has been shown to be closely related to classical type HD, but not to LPHD. A molecular biology study of a single classical HD cell by Sein *et al.* showed that in most cases R-S cells appear as the over proliferation of the late germinal center B cells, which have lost the ability to express Ig. In physiological conditions, B cells that do not express immunoglobins are erased by apoptosis but, in many classical HD derived from the B cell, although the R-S cells have lost the ability of expressing Ig, they can escape apoptosis. This might be the main mechanism underlying the pathogenesis of classical HD, and EBV might be involved in this process.

Non-Hodgkin's lymphoma (NHL) can be divided into B cell lymphoma and T cell lymphoma, the pathogenesis of which is now considered partly associated with EBV infection, described as follows.

Burkitt's lymphoma: EBV is present as a latent infection in B cells, but the exact role that EBV plays in the pathogenesis of Burkitt's lymphoma is unknown.

The commonly seen chromosomal translocation and activation of myc oncogene in Burkitt's lymphoma may have some relationship with EBV. Some studies showed that EBV might be involved in the regulation of cell apoptosis. The study by Gutierrez *et al.* showed that apoptosis mediated by Fas was inhibited in many B lymphocytes. This inhibition was generated by many factors, of which the pre-apoptotic gene Bax might play an important role in the apoptosis mediated by Fas in Burkitt's lymphoma.

Immunodeficiency-related B cell lymphoma: Compared with normal people, patients with AIDS and those who are administered immunosuppressive drugs after organ transplantation are more susceptible to malignant lymphoma, and within this cohort, B rather than T cell lymphoma is more commonly seen. Studies show that this kind of lymphoma is related to EBV infection. In normal cell immunity, the T cell can recognize the heterogenous protein EBNA2 and LMP-1, and attack the transformed B cells. But in immune-deficient conditions, these B cells will just proliferate and develop into lymphoma.

T cell lymphoma: EBV is related to T cell lymphoma. The study by Huh *et al.* showed that the incidence of EBV infection in peripheral T-cell lymphoma of South Korean people is quite high, especially in the angiocentric lymphomas and primary extranodal subtype. The positive EBV rate tested in T cell NHL was 63.6% and in the nose pharynx ministry, T cell lymphoma was as high as 76.1%.

The mechanism by which EBV enters into the T cell is still unknown. Some researchers pointed out that the T cell might express EBV/c3d receptor (CD21) in some stage of cell development, so EBV might directly infect the T cell through the c3d receptor. It is thought that the T cell is the host of EBV, in which the DNA structure of EBV can keep its linear molecule structure instead of cyclizing itself. And this low DNA cyclization rate can also explain the low incidence of T cell transformation when induced by EBV. This is the first time that a mechanism by which EBV can lead to T cell proliferation and T lymphoma is proposed.

So far, in the latest research, researchers separated the peripheral blood T lymphocytes from a normal person, infected them with EBV, then successfully induced them into lymphoblastoid cell lines (LCL) under the synergistic effect of IL-2, and these LCL were proved to have been derived from CD4 positive T cells. These results suggested that EBV is closely associated with T cell lymphoma.

Now it has been unequivocally established that the T-cell leukemia virus I (*HTLV-1*) can cause adult T cell leukemia (ATLL).

In 1980, Miyoshi *et al.* in Japan isolated *HTLV-1* from ATLL. ATLL incidence is higher in southern Japan, the Caribbean and Central Africa, consistent with infection of this virus in these areas. *HTLV-1* is classified as one of the C-type RNA viruses, its genome is 9 kb, and it can be vertically transmitted, or horizontally spread via breast feeding, sexual intercourse, blood transfusion and sharing of needles. Epidemiological studies have shown that *HTLV-1* positive or *HTLV-1* antigen positive was significantly correlated with ATLL incidence. In the leukemic cells of ATLL, the copy of the *HTLV-1* provirus can be detected. *HTLV-1 in vitro* infection of T cells causes $CD4^+$ T cell immortalization and expresses IL-2 receptor and HLA-DR antigens.

The study of its pathogenesis revealed that *HTLV-1* is different from other retroviruses; in its 3' end there is an X zone of approximately 1.6 kb; it is related to T-cell immortalization. This X zone can encode at least two transacting regulatory proteins; one is the tax product 40 kD protein, a transcription regulator; another is rex product 27 kD phosphoprotein, a post-transcriptional regulator. The X-zone encoding products can reversely activate IL-2 and IL-2 receptor genes, abundant IL-2 and IL-2 receptor expression can activate T cells, forming the T-cell autocrine loop, so the X-zone is also known as the tat gene. Recently it was reported that the tat gene product can also activate the fos gene and IL-3 gene. Although the location of the integrated *HTLV-1* gene in host cells is random, integration in certain regions can lead to malignancy. In ATLL cells chromosome 7 trisomy and chromosome 14 translocation have been found .

Multiple myeloma (MM) is considered highly related to human herpesvirus 8 (human herpesviruses-8, HHV8) [30]. Rettig *et al.* and Said detected this in long-term cultured dendritic cells (DC) and bone marrow biopsy specimens from MM patients by PCR and in situ hybridization (ISH). They found that the HHV8 DNA-positive rate is as high as 100%. In monoclonal gammopathy of undetermined significance (MGUS) the positive rate is up to 25%. HHV8 infection is only detected in DC but not myeloma cells. It was also confirmed that there is virus interleukin 6 (VIL-6) transcription in DC. MM patients with viral infection tend to relapse or are resistant to chemotherapy. Virus-free patients could much more easily reach complete remission with a VAD regimen or autologous peripheral blood stem cell transplantation (APBSCT) treatment. This evidence showed us that the clinical manifestations of MM are related to viral infections.

It is not completely certain that MM is caused by HHV8 infection, or HHV8 infection must occur before the onset of MM rather than after the onset. But the HHV8 gene is extraordinarily important in the pathogenesis of MM; it may play a causative role via IL-6. It has been demonstrated that: (i) IL-6 can induce *in vitro* growth of freshly isolated myeloma cells; (ii) Myeloma cells can spontaneously produce IL-6 and express IL-6 receptor; (iii) Anti-IL-6 antibody inhibited growth of MM cells or cell lines; (iv) IL-6 monoclonal antibody treatment of MM is effective; (v) IL-6 can also inhibit apoptosis induced by a variety of drugs such as dexamethasone. These facts indicate that IL-6 plays an important or even a key role in MM pathogenesis. But the source of IL-6 is still controversial; some thought it comes from tumor cell autocrine, yet more research supports the paracrine secreted by bone marrow stromal cells (including DC, fibroblasts and macrophagescells), osteoclasts, osteoblasts. It has been found that in HHV8 there is a homologous sequence of high similarity with human IL-6, and later it was found that its encoded protein has similar functions to IL-6. Rettig *et al.* confirmed that the DC from MM patients does have VIL-6 expression. Burger *et al.* have proved that VIL-6 from HHV8 may send signals through the IL-6 receptor and gp130 to stimulate the human myeloma cell line INA-6 proliferation. But to achieve the human IL-6 DNA synthesis rate, the amount of required VIL-6 protein will be much larger (approximately 4,000 times the amount of the human IL-6). Therefore, though it is speculated that VIL-6 may play a role in the pathogenesis of MM, because of the lack of *in vivo* quantitative data, how important its role is in MM onset is still unclear.

Overexpression of *Bcl-2* in tumor cells can cause resistance to chemotherapy and prevent cells from apoptosis. In B-cell malignancies *Bcl-2* plays an important role. However, the majority of MM cells and cell lines also have *Bcl-2* overexpression. In addition, the potentially undiscovered chromosomal changes that may lead to *Bcl-2* upregulation might also be associated with HHV8 infection. Whether there is expression of *Bcl-2* homologue by HHV8 in MM and whether it plays a role in MM onset are yet to be determined.

HHV8-encoded viral interferon regulatory factor (VIRF) and human interferon regulatory factor (hIRF) have low but very important homology. The main function of hIRF is to regulate IFN signal transduction between cells; vIRF is mainly expressed in HHV8-infected B cells, which is an active oncogene in B-cell malignancies and can provide a unique immune escape mechanism for the infected cells. The role of vIRF in MM is yet to be confirmed.

G–protein-coupled-receptor (GPCR): ORF74 of HHV8 encodes a G protein coupled receptor, named HHV8-GPCR. Constitutive activation of the GPCRs will lead to the transformation of the cells. HHV8-GPCR is one of the oncogenes. The secondary messengers of HHV8-GPCR-induced signaling pathway are JNK/SAPK and P38MAPK. JNK/SAPK and P38MAPK can induce angiogenesis, as well as the mitosis of KS cells and B cells by the similar mechanisms as some inflammatory factors. Thus, HHV8-GPCR driven-signaling induces the transformation of the cells and plays an important role in tumorigenesis. HHV8- GPCR an also transform the cells to the phenotype of angiogenesis by a growth factor VEGF (vascular endothelial growth factor), which can promote angiogenesis and the proliferation of KS cambiform cells.

Up to now, most of the topics concerning the relationship between HHV8 and MM are restricted to hypothesis or presumption. We anticipate that further investigations will be concentrated on the existence of HHV8-infection in MM, as well as the biological role of HHV8 cellular homologenes in the transformation of plasmacytes into myeloma cells.

17.3.4 Infection and Bleeding Disorders

Idiopathic thrombocytopenic purpura (ITP) is one of the most common hemorrhagic diseases encountered in medical practice. The cause of hemorrhage is thrombocytopenia. Because the anti-platelets antibodies can be detected in peripheral blood of ITP patients, idiopathic thrombocytopenic purpura is also named immune thrombocytopenic purpura (ITP) [31]. According to the patients' pathogenetic condition, ITP was classified as acute ITP or chronic ITP. Acute ITP often occurs 2 – 21 days after infection. It is generally believed that there are relationships between the pathogenesy of acute ITP and viral infection, but not chromic ITP [32]. Up to now, several viruses have been found to be the potential cause of ITP, such as the herpes virus, human parvovirus B19, measles virus, epidemic parotitis virus, rubella virus and hepacivirus.

Human herpesvirus(HHV) is a kind of double-stranded DNA virus. It includes herpes simplex virus (HSV)(type I and type II) (that's also named HHV-1 and HHV-2 , respectively), varicella zoster virus (VZV, HHV-3), Epstein-Barr virus (EBV, HHV-4), human cytomegalovirus (HCMV, HHV-5), as well as the newly-discovered HHV-6, 7 and 8. Infection from these viruses is the potential cause of autoimmune diseases. After analyzing the amino acid sequence of Epstein-Barr viral capsid antigen and the IE2 protein of the human cytomegalovirus, scientists found that they were homologous to the major human histocompatibility complex class II antigen (HLA-DR). These viral proteins may react in an antigen-antibody way with histocompatible antigens or other proteins which have similar antigenic structure as histocompatible antigens. Thus, the mechanism of molecular mimicry may play an important role in the development of autoimmune diseases.

After infection, our body produces a kind of IgM auto-antibody, specific for the Gly-Ala repetitive sequence of EBV nuclear antigen. Then, cross reaction will occur between these autoimmune antibodies and the proteins in the normal tissues whose construction is similar to that of the EBV nuclear antigen. The autoimmune reaction happens. ITP caused by EBV infection is often accompanied by monocytosis, autoimmune anemia or neutropenia. One month after infection, there will be mild to mid-range thrombocytopenia. Megakaryocytes in bone marrow will be normal or increased in count. Most of the patients can be treated effectively and have a eusemia. The platelets will recover in two months. Patients with severe ITP caused by EBV infection often have a high level of platelet-associated antibody IgG (PAIgG). And its antigenic determinants are located in the platelet membrane glucoprotein GPIIb. Direct binding of the antibodies to platelets accelerates the clearance of platelets; that's the pathogenesis of EBV causing ITP. However, little is known about the relationship between EBV infection and the generation of autoantibodies.

The clinical manifestation of VZV infection was varicella in children, while herpes zoster in adults. 1 – 2 weeks after varicella, some patients will be ill with ITP, but only a few cases have been reported after herpes zoster. In the bone marrow of VZV-related ITP patients, the megakaryocyte count is increased. But there are viral inclusion bodies in the cells, which indicate that VZV directly affects the megakaryocytes and decreases the generation of platelets. At the same time, VZV can increase the destruction of platelets. They enzymolyze the platelets directly or indirectly by releasing neuraminidase. Wright *et al.* detected the VZV specific antibodies in the blood serum of VZV infected ITP patients, and found the protein structure of the antigenic determinants on the surface of viruses is similar to that of platelet membrane glucoprotein. So virus specific antibodies react with platelet membrane glucoprotein by the mechanism of molecular mimicry, and destroy the platelets. Also, the antibodies change the surface construction of platelets and activate the platelet surface complement C1q. Then, activating the cascade reaction of the complementary system, C3b, the split product of C3, attaches to the surface of platelets and binds to the C3b receptors of phagocytes, which leads to increased destruction of platelets.

Parvovirus B19 (B19) is a kind of DNA parvovirus. Infection of B19 has various effects on the hematopoietic system. It can induce not only hemolytic anemia, aplastic anemia and chronic marrow failure of immunosuppression patients, but also ITP. There are two ways that B19 causes ITP: (i) By autoimmune response. The anti-virus antibodies' level of IgG and IgM increases in the blood serum after B19 infection. Both IgG and IgM positive indicate a recent infection, while IgG positive and IgM negative indicate that it has been at least 3 months since infection. (ii) Bone marrow depression. Though DNAs of B19 Genome cannot replicate in megakaryocytes, they can completely transcript to RNAs, which may impare the megakaryocytes and result in decreased generation of platelets. Recently, both *in vivo* and *in vitro* research has found that this virus can greatly inhibit the megakaryocytes, and leads to the low generation of platelets. Most of the scientists now believe that ITP in the prophase of B19 infection results from the suppression of the bone marrow, while humoral immunity mediated the pathogenesy of ITP in the advanced stage of infection.

Measles virus and mumps virus. Measles virus and mumps virus both belong to the paramyxovirus genus. There is a kind of non-glycosylated membrane protein inside the capsule of such viruses. While viruses aggregate and pullulate their proteins interacting with actins to change their configuration, this result in the exposure of potential antigen sites in the hosts' immune system and the formation of new antigen determinants. These antigen determinants will eventually induce the production of autoimmune antibodies; this may be one of the ways these kinds of viruses contribute to autoimmune diseases. In addition, the cytotoxic effect of antibodies against normal lymphocyte cells is observed after infection. Thrombocytopenia and over destruction of platelets in the bone marrow are usually the only symptoms demonstrated in patients diagnosed with ITP and measles. The increase in cells expressing T10 activation antigen and IL-2 receptor in peripheral blood of patients with acute measles infection suggests that a large number of peripheral blood lymphocytes are activated. It needs further research to see if this is related to measles infection in the complication of ITP. As to ITP caused by mumps virus, it may not be due to the effect of anti-platelet antibodies, but the direct impairment from viruses and (or) the deposition of immune complexes on platelets.

Hepatitis virus. Recent studies found that the infection of the hepatitis virus, especially hepatitis A (HAV) and hepatitis C (HCV) could cause ITP. It's reported that the presence of antiplatelet autoantibodies or the deposition of non-specific circulating immune complexes on platelets may account for the destruction of platelets among ITP patients infected with HAV [33]. Kosugi's research suggests that ITP patients who are diagnosed with chronic HCV appear to show a significant elevation on serum PAIgM titer, sometimes accompanied by PAIgG titer, though antiplatelet autoantibodies in normal platelets are not found. Moreover, Rochi reports that antiplatelet autoantibodies of IgG, IgM are detected in those patients who have an incidental onset of ITP after injection with recombinant hepatitis B virus vaccine. This recommends that such type of ITP is also mediated by immune mechanisms, though the pathogenesis remains to be further studied.

In summary, the pathogenesis of ITP induced by a virus is complex. It may include immune complexes combined with virus antigen and antibody deposit on platelets and megakaryocytes to enlarge the destruction. Viruses modify the structure of platelet membrane glycoprotein to change its antigen, which induces auto-antibodies to destroy platelets. Antiviral antibodies *in vivo* induce cross reaction of glycoproteins on the surface of platelets via molecular simulation mechanisms, or activate the complement system to injure platelets. Magakarycytes are transformed into intranuclear viral inclusion by viruses directly, which decreases the production of platelets. The autoimmune epitopes induced by virus infection are usually located on the platelet membrane glycoprotein GPII b/ IIIa, GPV and GPI b/ IX.

Infections play a role in coagulation disorders such as disseminated intravascular coagulation (DIC), a syndrome caused by multiple microbial infections [34].

The etiology and pathogenesis conjunction of DIC include bacterial infection, viral infection and rickettsial infection.

Gram-negative bacterial infection accounts for about 65% of bacterial infections which lead to DIC. Intestinal bacteria such as *E. coli*, Salmonella, proteus, *Shigella flexneri* and meningococcus are commonly seen in those cases. The activation and destruction of vascular endothelial cells caused by bacterial endotoxin play a major role in the pathogenesis of DIC. There are direct and indirect activation mechanisms. Direct activation requires the presence of serum LBP and soluble CD14 [35]. LPS in conjunction with LBP combines with soluble CD14, or LPS binds soluble CD14 directly. Then they bind another receptor on the surface of vascular endothelial cells and deliver signals from endotoxin to endothelial cells. Thus endothelial cells are activated and damaged. It's cell-dependent in indirect activation. Endotoxin activates monocytes and neutrophils in blood vessels as well as macrophages in tissues to induce the release of TNF-α, IL-1 β, IL-8 and other cytokines. These proinflammatory cytokines contribute to the activation and injury of vascular endothelial cells.

DIC induced by the activation and injury of vascular endothelial cells via bacterial endotoxin.

The main viral infections are epidemic hemorrhagic fever, dengue fever, flu, hepatitis B, rubella, and AIDS virus (HIV). Serious viral infections inducing DIC by directly destroying cells is one of the pathological characteristics. Viruses injure endothelial cells to activate the coagulation process. Some viruses such as cytomegalovirus (HIV, CMV) assemble thromboplastin on their surface to amplify the coagulation process. Furthermore, viral infections may aggravate the bleeding symptom of DIC through direct thrombocytopenia.

The interaction between viral hepatitis and DIC is considerable at present. Viral hepatitis is common in throughout China. It indicates that 7% of viral hepatitis is complicated by DIC. The mechanism is as follows: Among patients with severe liver disease, the clearance of the activated coagulation factor is depressed because of the dysfunction of the reticuloendothelial system, and the endogenous coagulation pathway activated by the injury of vascular endothelial contributes to the production of abnormal thrombosis, which finally results in ITP;

the synthesis of coagulation factors (except for calcium and tissue factor) and anticoagulant protein (antithrombin III, liver cofactor II, protein C, protein S) is reduced and the excessive consumption of coagulation and anticoagulation protein, the increase of circulating anticoagulant and blood FDP, thrombocytopenia and dysfunction of the platelet. As the liver function is severely impaired in severe viral hepatitis patients, the content and activity of antithrombin III (AT III) are significantly reduced. As a result, the treatment for liver diseases complicated with DIC with heparin is ineffective and the prognosis is poor.

Rickettsial infection leads to diseases such as typhus, ginger worm disease and so on. Rickettsia are strict cytozoic prokaryotic microbial cells. They primarily invade vascular endothelial cells and reproduce after infecting the human body. The excessive proliferation leads to the rupture of cells and the exposure of pathogens, which result in diffuse vasculitis, intravascular thrombosis and exudative reaction. There are other infections such as fungal infections, spirochetal infection, mycoplasma infection etc.

The clinical manifestations of DIC are as follows:

Hemorrhage. It may appear as large cutaneous ecchymosis, hemorrhage from wounds or injection sites and visceral hemorrhage in the gastrointestinal tract, respiratory tract, the urinary tract and so on.

Shock. It is characterised by low blood pressure and decreased urinary volume.

Multiple organ failure. Microvessel embolism may occur in every organ of the body, commonly in renal, lung, adrenal glands and skin, followed by the stomach, intestine, liver, brain, pancreas and heart.

Infection complicated by DIC usually presents an acute or fulminant course. Patients with severe infection should be closely observed, diagnosed and treated early.

Treatment for infection complicated by DIC. For treatment for primary infection it's fundamental to search for the primary foci of infection and choose a different therapy according to the pathogen. Gram-negative bacterial infection is common clinically. Potent broad-spectrum antibiotics should be applied before the exact pathogenic bacteria are found.

Anticoagulation therapy. Heparin is the primary choice to prevent and treat DIC as an effective anticoagulant.

It's reported that the application of a monoclonal antibody to specifically block cytokines such as IL-1, IL-6, IL-8, TNF effectively reduces the incidence of shock induced by endotoxin, but the clinical application and efficacy remain to be further evaluated.

Chronic infection may lead to anemia manifested as anemia of chronic disease (ACD). Recent research suggests that ACD is related to hepcidin *in vivo*. Hepcidin is a newly found polypeptide containing 25 amino acids produced by the liver. Its major function is to regulate iron homeostasis. Human hepcidin gene HAMP (hepcidin antimicrobial peptide) is located on chromosome 19q13.1, which contains 3 exons and 2 introns. The upstream USF2 (upstream stimulatory factor 2) gene is associated with the regulation of hepcidin expression. Fpn 1 is the target where hepcidin modulates irons. Hepcidin binds with Fpn 1 on the cell membrane to induce tyrosine phosphorylation and reduce Fpn 1 on the cell surface by internalization and degradation in lysosomes. When the concentration of plasma hepcidin increases, it reduces the Fpn 1 on the basal membrane surface of intestinal mucosal epithelial cells and locks irons inside the cells, which makes irons lose the exfoliative of intestinal mucosa cells and thus reduces the absorption of irons. On the other hand, the low concentration of plasma hepcidin will increase Fpn 1 in intestinal mucosal cells and thus elevate the absorption of irons. The level of plasma hepcidin is regulated by inflammatory cytokines, plasma iron, anemia, hypoxia and so on. The upregulation factor of hepcidin is plasma irons and inflammatory cytokines, while the down regulation factor is hypoxia.

Inflammatory cytokines, especially IL 6, may induce the secretion of hepcidin when infection occurs. Consequently, intestinal iron absorption decreases and microcytic hypochromic anemia ensues.

17.4 Treatment of Hematologic Diseases and Infective Microecology

For patients who suffer from hematological malignancies, the disease itself and chemotherapy and (or) radiotherapy and other factors make them immunosuppressed. They may suffer from a decrease in and dysfunction of the neutrophil, T lymphocytes and B lymphocytes deficiency, complement deficiency and secondary hypergammaglobulinemia. All these contribute to serious infection in the patients.

17.4.1 Predisposing Factors and Pathogens

Predisposing factors include deficiency of cellular and humoral immunity, damage to the cellular defense function, destruction of skin and mucosal barrier, influence of the tumor itself, and change in the normal physiological environment.

An apparent dysfunction of humoral immunity occurs in patients with chronic lymphocytic leukemia or multiple myeloma, so does cellular immunity in patients with lymphoma. Both chemotherapy and radiotherapy do damage to T lymphocytes and B lymphocytes, resulting in deficiency of the cellular and humoral immunity, which makes patients easily infected by bacteria, viruses and fungi from the environment.

After chemotherapy, leukocyte and neutrophil decreased, which caused the decrease in anti-infection ability. The extent and duration of neutropenia is closely related to the severity of bacterial or fungal infection. Once the number of neutrophils is less than $0.5 - 1.0\times10^9$/L, the chance of infection will increase by 14%, while when the number is less than 0.1×10^9/L, the incidence rate of infection is as high as 24% – 60%. If neutropenia lasts more than 5 weeks, the incidence rate will be 100% and if neutrophils $<0.5 - 1.0\times10^9$/L last more than 10 days, the risk is high enough to cause a serious infection.

Because of the damage of chemotherapy drugs to mucosa, including mucosa of mouth, esophagus, stomach and intestines, the defensive ability of mucosa will be weakened or disappear, resulting in the invasion of bacteria. For example, in the intestinal tract, under normal circumstances resident bacteria and the crossing bacteria interact and keep in balance with each other. But when the intestinal mucosa is damaged, this balance will be broken, resulting in a change in dominant flora and imbalance in microecology, which is one of the basic factors of endogenous infection. In addition, chemotherapy drugs cause nausea and vomiting, change the pH of the stomach and esophagus; and acid refluxing to the esophagus will lead to the ulceration of the esophageal mucosa. Repeated venipuncture, and a venous catheter can also damage the skin barrier, resulting in microbial invasion to cause infection.

Destruction of skin and mucosal barrier:

The tumor tissue itself may be complicated by the occurrence of edema, erosion, ulceration, necrosis, oppression and obstruction, which is conducive to infection. In cancer patients, weakened immunity from the use of chemotherapy and immunosuppressive agents, especially the invasion of endotoxins, increase mucosal and vascular permeability, and destroy cell metabolism, which is a mechanism for endogenous infection in hosts.

Change in the normal physiological environment: Chemotherapy-induced vomiting can create bile in the stomach and change the acidic environment of the stomach, resulting in a decrease in the defense capability of the stomach to resist bacteria. Then micro-organisms are much easier to clone and cause infection.

The majority of the pathogens in cancer patients are Gram-negative bacteria, especially *Escherichia coli*, *Klebsiella*, and *Pseudomonas aeruginosa*. Since 1987, infection of *Staphylococcus* aureus and *Staphylococcus* epidermidis significantly increased, which may be a result of the general application of third-generation cephalosporins and quinolones. *Streptococcus viridans* and *Corynebacterium* can cause severe infections in long-term low granulocyte patients. Fungi are also important pathogens, especially in patients with immune dysfunction who have neutropenia for a long time and receive antibiotics as preventive treatment. *Candida genus*, *Aspergillus*, *Mucor*, and *Cryptococcus neoformans* are the major pathogenic fungi. In addition, protozoan and viral infections are also important pathogenic microorganisms such as pneumonia caused by *Pneumocystis carinii*, and herpes simplex, herpes zoster and cytomegalovirus infection; the incidence of viral hepatitis in cancer patients is also higher.

The main pathogens that cause death are fungi (45%), bacteria (24%), cytomegalovirus (CMV) (12%) and *Pneumocystis carinii* (6%).

For patients with neoplastic hematologic disorder and immune dysfunction, a series of changes appear in the characteristics of their infection: Pathogens turn to the endogenous from the exogenous. Pathogens that are less virulent normal flora become a cause of infection. In recent years, among the pathogens of hospital-acquired infections, the infection of viruses, Gram-negative bacilli, anaerobes, and deep fungal infections have increased. The infection often involves multiple-organs. The pathogens may be different in the same sites of infection. The pathogens are extensively drug-resistant and difficult to treat, and can lead to a high mortality.

For patients with immune dysfunction, once infected by bacteria, viruses, fungi, parasites, the clinical symptoms and signs are not typical. They usually are present as a cross-hybrid of the underlying disease and with new infectious lesions and seem minor but actually produce important physical weakness and slow response, and it's easy for them to deteriorate. Fever is the most common clinical manifestation in infected patients. Although in patients with neoplastic hematologic disorder the disease itself can lead to a fever, in most of the patient's fever is due to secondary infection.

For the immunocompromised, due to mucosal damage after chemotherapy, the bacteria which are originally cloned in the respiratory tract or the intestinal tract invade the bloodstream, causing sepsis or infection of other parts of the body. In addition, non-pathogenic bacteria under normal circumstances can lead to severe infections and high fever when the patient is immune-compromised, and antibiotics are not easily effective in these situations.

The types of infection in patients with neoplastic hematologic disorder include pneumonia, septicemia, skin infections, urinary tract infections, abdominal infections and central nervous system infections. Routes of infection involve the skin, mucous membranes, the soft tissue, the respiratory tract, the gastrointestinal and urinary system.

17.4.2 The Principles of Treatment

The principles of treatment for patients with malignant hematologic diseases are the prevention of endogenous infection, maintenance of environmental hygiene and early application of empirical antibiotics.

Prevention of endogenous infection: for patients with neoplastic hematologic disorder, whose mucosal barrier is damaged due to immune dysfunction and a variety of radiotherapies and chemotherapies, with intestinal flora then imbalanced, it's easier to be infected by Gram-negative bacteria and fungal opportunistic pathogen, secondarily or endogenously. So it's routine to give patients 200 mL of fermented milk containing *Bifidobacterium longum* and *Lactobacillus acidophilus* in a concentration of 107 CFU/mL, respectively, which can reduce the intestinal facultative Gram-negative bacilli and fungi.

Maintenance of environmental hygiene: It's better for patients with granulocytopenia to stay in a sterile layer flow chamber and eat food sterilized by microwaves, and then go back to the general ward after the granulocyte increases. To reduce cross-infection risks, the general patient should also be sent to a single room and have reduced visits.

For patients with hematologic malignancies, these are often complicated by immune dysfunction so that, once infection occurs, it is often ferocious and rapidly progresses, especially in Gram-negative bacilli infections. Early application of empiric antibiotics greatly reduces the mortality of patients infected with granulocytopenia. Therefore, proper handling in a timely manner is essential to infection.

For patients with hematologic malignancies complicated by infections, sufficient and carefully historical and physical examinations are required. Samples of blood, sputum, urine, cerebrospinal fluid, should be prepared for tests in time. Fast and appropriate empirical therapy has to be given. As soon as pathogens are identified, sensitive antibiotic infusion will be applied. Antibiotic treatment of tumor infection should follow the following principles: Intravenous administration; application of broad-spectrum antibiotics, which can kill the common Gram-negative and positive pathogens; bactericidal antibiotics; use of definitely effective antibiotics; adequate duration of treatment.

17.5 Molecular Ecological Treatment

Molecular ecological treatment has played a more and more important role in recent times. We will explain into two parts below the genic ecological treatment and immune ecological treatment.

17.5.1 Genic Ecological Treatment

Tyrosine kinase inhibitor STI571 is the first effective drug developed according to molecular-biology principles to treat a malignant tumor. BCR-ABL fusion protein of Chronic myelogenous leukemia (CML) has a tyrosine kinase activity which could inhibit cell apoptosis and lead to persistent cell proliferation. The most effective method to inhibit its signal conducting is to inhibit tyrosine kinase directly. The signal conducting inhibitor-STI571, which is designed with molecular biology principles, can prevent the protein from acting in the proto-oncogene pathway. It has been widely used in clinics and has achieved a significant therapeutic effect. STI571 has been found more effective in suppressing PDGFR and c-Kit. And it has been proved that STI571 is quite effective in treating the GIST carrying the c-Kit or PDGFR mutation; the phase III clinical trials are underway. Also, STI571 have a therapeutic effect for CML patients with TEL-PDGFRβ fusion gene produced by t(5; 12) (Q33; P12) and HES/CEL patients with FIP1L1-PDGFR α fusion gene produced by 4q12 deletion.

FLT3 inhibitors include CEP-701, CEP-5214, SU5416, SU5614, SU11248, MLN518, PKC412 [36], L-000021649 and AG1295, in which SU5416 and PKC412 have been used for the treatment of MDS patients, and achieved very good clinical effects [37]. PKC412 (75 mg, tid) has been used for the treatment of high-risk MDS with FLT3 mutations and relapsed AML; the response rate was 85%, of which 30% of patients' blast count decreased by more than 50% in bone marrow, 55% of patients' blast count decreased by 5% – 50%. And it was found that in these

PKC412-effective patients, their FLT3 autophosphorylation had been suppressed, thereby confirming the targeted theraputic effect of PKC412 treatment [38-40].

RAS mutations are quite common in patients with hematologic malignancies. The key regulatory sites of RAS, such as codon 12, 13 and 61 can have point mutations and result in the activation of RAS proteins. The activation of RAS protein can extend the half-life of RAS-triphosphare guanosine (RAS-GTP) by eliminating the endogenous or GAP-stimulated enzymatic activity of GTP. Then, RAS-GTP can bind to, and activate, its downstream effector molecules, including RAFs, MEK kinase, PI3K/AKT, to regulate the proliferation of tumor cells. The appropriate location of RAS proteins in the cell is also essential to conduct mitogenic/exogenous stimulus signals properly. In order to combine with the plasma membrane proteins effectively and obtain full biological activity, RAS protein needs post-translational modification, including prenylation, protein hydrolysis, methylolated and sixteen acylation. Every step of the post-translational modification is regulated by specific rate-limiting enzymes, including farnesyl transferase (FTase), human RAS converting enzyme 1 (hRCE1), isoprene cysteine hydroxymethyl transferase (ICMT) and sixteen acyltransferase (PAT). Specific inhibitors of these enzymes have been synthesized, among which the research of farnesyl transferase inhibitors (FTIs) is the most sophisticated. There are 5 types of FTIs currently undergoing clinical trials, including R115777 [41,42], SCH66336, L-778123, BMS-214662 and CP-609754. Although FTIs were originally designed to target RAS- prenylation, some other proteins containing the CAAX motif have also become novel targets for FTIs, such as small G protein, Rho, Rap2, Rheb, centromere binding protein, tyrosine phosphatase, visual signal transduction proteins and nuclear envelope structure protein and inositol signaling protein[43]. FTIs can initiate a series of cellular effects, including changes in the cellular proliferation cycle, inducing cell apoptosis and morphological changes to the cells.

In clinical trials, the total effective rate of R115777 in MDS patients was 20% – 30%, in which the complete remission rate was 5% – 10%; the main side effects are myelosuppression and neuropathy [41, 42]. SCH66336 is also used in MDS treatment; the total effective rate was 20% – 30%, especially in patients with dependence on platelet transfusion;, the main side effects are neutrocytopenia, neuropathy, diarrhea, anorexia, weight loss, weakness and bradycardia[44]. BMS-214662 is a thiol, nonpeptide competitive FTI, which is highly selective to FTase. Treating 30 patients with AML or MDS (42 – 157 mgm, once a week, with 4 weeks as a course of treatment, 2 weeks as intermission) resulted in a total efficiency of 17%, including 6.7% complete remission, 6.7% nearly complete remission. The main side effects were the elevation of transaminase, nausea, vomiting, diarrhea, fatigue, anorexia and leucocytopenia.

DNA methylation is known as one of the most common epigenetic modifications involved in malignant hematologic disease. A large number of genes are highly methylated in the process of AML and MDS, including P15INK4B, P21CIP1/WAF1, estrogen receptor gene, CALC1, E-cadherin, HIC-1, RASSFIA and so on [45, 46]. Up-regulation of DNA methyltransferases DNMT1, 3A, and 3B in myelodysplastic syndrome has been demonstrated in a recent experiment [47]. Cytidine derivatives including 5-aza-cytidine(Aza C), -aza-2, -deoxycitidine

(Dacitabine, DAC) and Zebularine, as DNA methylation inhibitors, have been widely used in treatment.

5-Aza-cytidine (Aza C) is the first agent approved for treatment of myelodysplastic syndrome. According to data submitted to the U.S. Food and Drug Administration for marketing approval of azacitidine as injectable suspension for treatment of patients with myelodysplastic syndrome, the overall response rate (including complete remission and partial remission) was 15.7% in the azacitidine treatment group and there were no responders in the observation group. An additional 19% of azacitidine-treated patients had less than partial responses, most becoming transfusion independent. The most common adverse events attributed to azacitidine are gastrointestinal reaction and myelosuppression [48].

While sharing the similar mechanism of action with Aza C, Dacitabine (DAC) can be phosphorylated by a series of enzymes, among which the form of Dacitabine triphosphate is becoming incorporated into genomic DNA more easily and this irreversible binding results in enzyme depletion and DNA demethylation. The cellular toxicity of 5-aza-dC is mainly mediated by the involvement of Dnmt3a and Dnmt3b [49]. In patients with high-risk myelodysplastic syndromes (MDS), Dacitabine can induce a 50% blood response and a 38% cytogenetic response to the treatment [50-53].

Zebularine, an inhibitor of cytidine deaminase, has been recently identified as a general inhibitor of DNA methylation. This inhibition of DNA methyltransferase (DNMT) is hypothesized to be mechanism-based and results from formation of a covalent complex between the enzyme and zebularine-substituted DNA. Metabolic activation of Zebularine thus requires that it be phosphorylated and incorporated into DNA [54, 55]. Zebularine aqueous solution , not like Aza C and DAC, is quite stable. Zebularine has been proved to have little toxic effect in animal experiments. The low bioavailability limits its potential, because of its oral administration. This drug has not been put into clinical practice [56].

Gene transcription is modified by the reversible acetylation of histone proteins on lysine residues in the N-terminal tail. Typically, these reactions are catalyzed by enzymes with histone acetyltransferase (HAT) or histone deacetylase (HDAC) activity. They play an important role in the process of the regulation of gene expression, assembly of histone proteins, replication of cells. The imbalance between acetylation and deacetylation leads to inhibition of the expression of some genes, which regulate the proliferation and differentiation. This situation can be reversed by HDAC inhibitors, which promote the acetylation of histone and enlongation of DNA, resulting in exposure of some key promotors and the re-expression of the genes regulating proliferation, differentiation and apoptosis. Also, HDAC inhibitors can enhance the acetylation of histone. HATs and HDACs can also modify the acetylation status of non-histone proteins as well, leading to the upregulation of proapoptotic protein (such as Fas/FasL, P53, Bak, Bax, Bim and Caspase 3) or cell cycleregulatory protein P21WAF1/CiP1, resulting in cell-cycle arrest and apoptosis. Recently, HDAC inhibitors proved to have a function in anti-tumor angiogenesis [57].

Several structurally distinct classes of HDAC inhibitors have been developed, including small molecule carboxylates (VPA, for example), hydroxamic acids

(NVP-LAK974, for example)[58], benzamides (MS275, for example), cyclic peptides, epoxy ketones, and hybrid molecular classes[59]. Some HDAC inhibitors have shown clinical promise when combined with other therapeutic agents. In patients with MDS and recurrent AML, VPA combined with low-dose DAC, achieved an overall response rate of 75%. 25% of patients achieved complete remission, while 25% had a bone marrow response to the treatment. The most common side effects are hemocytopenia and neurotoxicity.

17.5.2 Immune Ecological Treatment

Immunotherapy refers to the use of the body's immune system to treat the disease, which is not a direct treatment for disease but is influenced by the host's immune response or immune activity. The main purpose of immunotherapy is to induce and increase the effective immune response to cancer cells, achieving the goal of treatment of malignant tumors. Immunotherapy strategies make use of T cells, including LAK, TIL, to remove the abnormal cells. The key to cellular immune-therapy is the responding cells required for induction *in vitro* which will be transfused to the patients afterwards.

The immune response begins with antigen presenting cells (APC) that capture the antigens, process them and then present the antigens to T cells to activate the specific immunity response. There are various kinds of APC *in vivo*, such as monocyte-macrophage cells, B cells, Langerhans cells and dendritic cells (DCs). DC, also known as a part of the mononuclear phagocytes system, is a kind of full-time APC. DCs differentiate and express surface molecules with the action of certain cytokines and growth factors, accelerating the activation of T cells. For the key role DCs play in the immune response, it is natural that research of immune-therapy should be focused on DCs. Recently, the application of DC-based immunotherapy has become a hot topic all over the world.

Dendritic cells: DCs were first described by Langerhans in 1868 and their function was identified in 1970. As a full-time APC, DCs have the ability to induce the primary and secondary T cell responses. DCs are derived from bone marrow progenitor cells, umbilical cord blood hematopoiefic stem cells and peripheral blood mononuclear cells, which are normally less than 0.1% and have no characteristics of DCs. With the stimulation of GM-CSF, IL-4, TNF, SCF, c-Kit ligand and Flt-3 ligand, these precursor cells produce DCs. There are different kinds of DCs: interstital DCs, interdigitating DCs, blood or lymphoid DCs, Langerhans cells, follicular dendritic cells and so on. Their functions are to capture antigens, process them and present them to T lymphocytes, activating a series of immune responses. Different DCs have different functions, for example interstitial dendritic cells can directly stimulate B cells to produce antibodies, while lymphoid DCs dependent on IL-3 may be involved in the induction of immune tolerance.

Because of the key role DCs playing in the immune response, people began to explore the clinical application of DCs in immunotherapy. Recently, immunotherapies based on DCs developed very fast. They are not only used for treatment of malignant tumors, but also for infection, autoimmune diseases, graft rejection, GVHD and so on. DCs have become the hottest topic in cancer immunotherapy and there are a number of international research centers working in this area. Banchereau used DCs, after sensitization by peptides derived from four melanoma antigens, to transfuse to the patient twice a week ($0.7-7\times10^7$ cells every time and every two weeks for a course), and found that all the patients had some reactions. Doctor Gilba from Duke University Medical Center transfected DCs with tumor cRNA and then transfused them to cancer patients. No side effects were found in phase I clinical studies. In Lotze's study, 60 mL heparinized peripheral blood was drawn from 24 patients once a week. Mononuclear cells were isolated and the adherent cells were collected as DCs. After culturing the DCs with cytokine GM-CSF and IL-4 at 37 degrees, Lotze transfused the cells back to the patients, 1×10^6 cells each one. The results showed that two cases showed a complete response, one of them lasted one year and half of the remaining cases had a partial response after 3 courses of treatment. Currently, DC has been used for treatment of breast cancer, prostate cancer, melanoma, lymphoma, diabetes and arthritis.

Choudhury and other researchers have found that the leukemia cells isolated from peripheral blood of patients with myelogenous leukemia could differentiate into DCs *in vitro*. And specific CTL could be induced to kill leukemia cells. As DCs differentiated from Leukemia cells still preserved tumor antigens, they are likely to be applied in leukemia combination therapies, to remove residual leukemia cells by activating specific immunity inside the body.

Currently some problems still exist. Firstly, DC immunotherapy may cause autoimmune diseases. Lotze reported one case that was treated by DCs where the patient developed rheumatoid arthritis, which finally developed into systemic arthritis. But his tumor did not relapse after steroid treatment.

The second problem is how to amplify sufficient DCs for clinical use. When cancer occurs, the normal function of DC is to migrate to the tumor, identifying the antigens on the tumor cells and processing them. Then DCs migrate to the lymph nodes, stimulating natural T cells and specific T cell clones. Finally, the expanded T cells migrate to the site of the tumor to destroy tumor cells. In addition, DCs also allow the tumor-infiltrating lymphocytes to survive. Therefore, if there is not a sufficient number of DCs around the tumor, or the DCs are dead, all the tumor immune pathways will be blocked and this is another new mechanism explaining how the tumor cells escape immune recognition.

Finally, whether or not DCs entering a patient can live long enough, a number of clinical cases and animal model studies have shown the abnormality of DCs in tumor patients. Their function was reduced and production was inhibited. Also, the co-stimulatory molecules expressed on the surface of DCs reduced in the tumor microenvironment which is very important for the proliferation of T cells. Studies found that there exist tumor release factors inside cancer patients, such as Fas ligand which could induce apoptosis of T cells as well as NK cells, eosinophils and DC cells. And this suggests that many immune effector cells in the tumor microenvironment will die when a tumor occurs. It is known that IL-12, IL-15,

TNF α and CD40L could protect DCs to escape the tumor induced apoptosis, but when they will be used in the clinic is still unknown.

To date, the most effective monoclonal antibody which has been ripe for clinical use is the CD20 monoclonal antibody. It is mainly used for treatment of lymphoma. The monoclonal antibody can induce apoptosis of tumor cells and increase their chemotherapy sensitivity. As a result, a significant effect has been achieved in the treatment of B-cell lymphoid tumors, especially in highly malignant diffuse large B-cell lymphoma and mantle cell lymphoma. Combination with chemotherapy could significantly improve the efficacy.

In addition, IL-2 exerts an anti-tumor effect by stimulating proliferation of NK cells, and it has been used in the treatment of minimal residual diseases in myeloid leukemia.

References

[1] Erika Isolauri, Yelda Sütas, Pasi Kankaanpää, *et al.* Probiotics: Effects on immunity. Am J Clinl Nutr, 2001, 73: S444- S450.

[2] Donowitz R D, Maki D G, Crnich C J, *et al.* Infection in the neutropenic patient-new view of an old problem. Hematology-ASH Education Program book, 2001: 113-133.

[3] Long ZZ. Lymphocyte, Clinical Hemotology. Shanghai: Shanghai Scientific and Technical Publishers, 2001.

[4] Morgan M A, Ganser A, Reuter C W. Therapeutic efficacy of prenylation inhibitors in the treatment of myeloid leukemia. Leukemia, 2003, 17: 1482-1498.

[5] Shih L Y, Lin T L, Wang P N, *et al.* Internal tandem duplication of fms-like tyrosine kinase 3 is associated with poor outcome in patients with myelodysplastic syndrome. Cancer, 2004, 101: 989-998.

[6] O'Dwyer M E and Druker B L. STI571: An inhibitor of the BCR-ABL tyrosine kinase for the treatment of chronic myelogenous leukaemia. Lancet Oncol, 2000: 207-211.

[7] Gutierrer M I, Cherney B, Hussain A, *et al.* Bax is frequently compromised in bukitt lymphoma with irreversible resistence to Fas-induced apotosis. Cancer Res, 1999, 59: 696-703.

[8] Krause D S, Van Etten R A. Tyrosine kinases as targets for cancer therapy. N Engl J Med, 2005, 353: 172-187.

[9] Mizuki M, Ueda S, Matsumura I. *et al.* Oncogenic receptor tyrosine kinase in leukemia. Cell Mol Biol, 2003, 49: 907-922.

[10] Chalandon Y, Schwaller J. Targeting mutated proteinbtyrosine kinases and their signaling pathways in hematologic malignancies. Haematologica, 2005, 90: 949-968.

[11] Chian R, Young S, Danilkovitch-Miagkova A, *et al.* Phosphatidy-linositol 3 kinase contributes to the transformation of hematopoietic cells by the D816V c-Kit mutant. Blood, 2001, 98: 1365-1373.

[12] Hayakawa F, Towatari M, Kiyoi H, *et al.* Tandem-duplicated Flt3 constitutively activates STAT5 and MAP kinase and introduces autonomous cell growth in IL-3-dependent cell lines. Oncogene, 2000, 19: 624-631.
[13] Choudhary C, Schwable J, Brandts C, *et al.* AML-associated Flt3 kinase domain mutations show signal transduction differences in comparison to Flt3 ITD mutations. Blood, 2005, 106: 265-273.
[14] Blume-Jensen P, Jiang G, Hyman R *et al.* Kit/stem cell factor receptorinduced activation of phosphatidylinositol 3'-kinase is essential for male fertility. Nat Genet, 2000, 24: 157-162.
[15] Levis M, Murphy K M, Pham R, *et al.* Internal tandem duplications of the FLT3 gene are present in leukemia stem cells. Blood, 2005, 106: 673-680.
[16] MizukiM, Fenski R, Halfter H *et al.* Flt3 mutations from patients with acute myeloid leukemia induce transformation of 32D cells mediated by the Ras and STAT5 pathways. Blood, 2000, 96: 3907-3914.
[17] Kissel H, Timokhina I, Hardy M P *et al.* Point mutation in kit receptor tyrosine kinase reveals essential roles for kit signaling in spermatogenesis and ogenesis without affecting other kit responses. EMBO J, 2000, 19: 1312-1326.
[18] Ning Z Q, Li J, Arceci R J. Signal transducer and activator of transcription 3 activation is required for Asp (816) mutant c-Kit-mediated cytokine independent survival and proliferation in human leukemia cells. Blood, 2001, 97: 3559-3567.
[19] Sternberg D W, Tomasson M H, Carroll M, *et al.* The TEL/PDGFbetaR fusion in chronic myelomonocytic leukemia signals through STAT5-dependent and STAT5-independent pathways. Blood, 2001, 98: 3390-3397.
[20] Rane SG, Reddy E P. JAKs, STATs and Src kinases in hematopoiesis. Oncogene, 2002, 21: 3334-3358.
[21] Darnell JE Jr. STATs and gene regulation. Science, 1997, 277: 1630-1635.
[22] Bonifazi F, de Vivo A, Rosti G, *et al.* Chronic myeloid leukemia and interferon-alpha: A study of complete cytogenetic responders. Blood, 2001, 98: 3074-3081.
[23] Ducrest A L, Szutorisz H, Lingner J, *et al.* Regulation of the human telomerase reverse transcriptase gene. Oncogene, 2002, 21: 541-552
[24] Xie GJ. Helicobacter pylori and primary gastric lymphoma. International Journal of digestive system, 1996, 16: 5.
[25] Rochi F, Cecchi P, Marsciani A, *et al.* Thrombocytopenic purpura as adverse reaction to recombinate hepatitis B vacine. Arch Dis Child, 1998, 78: 273-274.
[26] Chen T, Lee J. Viral infection related aplatic anemia. Foreign Medical Sciences(Section of Virology), 1997, 4: 119.
[27] Kanegene H, Bhatia K, Cutierrez M, *et al.* Asyndrome of peripheral blood t-cell infection with Epstein-Barr virus followed. Blood, 1998, 91: 2085-2091.
[28] Qi J Y, Zhang Z N. Microbial Infection Caused Hemolytic Anemia Clinical Hemotology, Shanghai, Shanghai Scientific and Technical Publishers, 2001.
[29] Liu S M, Chow K C, Chin C F, *et al.* Expression of Epstein-Barr virus in patients with Hodgkin disease in Taiwan. Cancer Sur, 1998, 83: 367-371.

[30] Dong Y J. The association of Kaposi's sarcoma-associated herpesvirus and multiple myeloma. Foreign Medical Sciences (Section of Blood Transfusion and Heanatology), 1999: 22.

[31] Karpatkin S. Autoimmune thrombocytopenic purpura. Lancet, 1997, 349: 1531.

[32] Wright J F, Blanchette V S, Wang H, *et al.* Characterization of platelet-reactive antibodies in children with variclla-associated acute immune thrombocytopenic purpura. Br J Haematol, 1996, 95: 145-152.

[33] Kosugi S, Mai Y, Kurata Y, *et al.* Platelet-associated IgM elevated in patients with chronic hepatitis C contains in anti-platelet auto antibodies. Liver, 1997, 17: 230-237.

[34] Chen F P. Infectious and disseminated inravascular coagulation. chinese Journal of Practical Internal Medicine, 2000, 20: 329.

[35] Tapping R I, Tobias P S. Cellular binding of soluble CD14 requires lipopoly saccharide and lps-binding protein. J Biol Chem, 1997, 272: 23157-23164.

[36] Growney J D, Clark J J, Adelsperger J, *et al.* Activation mutations of human c-KIT resistant to imatinib are sensitive to the tyrosine kinase inhibitor PKC412. Blood, 2005, 106: 721-724.

[37] Giles F J, Stopeck A T, Silverman L R, *et al.* SU5416, a small molecule tyrosine kinase receptor inhibitor, has biologic activity in patients with refractory acute myeloid leukemia or myelodysplastic syndromes. Blood, 2003, 102: 795-801.

[38] Chen J, Lee B H, Williams I R, *et al.* FGFR3 as a therapeutic target of the small molecule inhibitor PKC412 in hematopoietic malignancies. Oncogene, 2005, 24: 8259-8267.

[39] Gotlib J, Berube C, Growney J D, *et al.* Activity of the tyrosine kinase inhibitor PKC412 in a patient with mast cell leukemia with the D816V KIT mutation. Blood, 2005, 106: 2865-2870.

[40] Stone R M, DeAngelo D J, Klimek V, *et al.* Patients with acute myeloid leukemia and an activating mutation in FLT3 respond to a small-molecule FLT3 tyrosine kinase inhibitor, PKC412. Blood, 2005, 105: 54-60.

[41] Kurzrock R, Albitar M, Cortes J E, *et al.* Phase II study of R115777, a farnesyl transferase inhibitor, in myelodysplastic syndrome. J Clin Oncol, 2004, 22: 1287-1292.

[42] Kurzrock R, Fenaux P, Raza A, *et al.* High-risk myelodysplastic syndrome (MDS): First results of international phase 2 study with oral Farnesyl transferase inhibitor R115777 (ZARNESTRA). Blood, 2004, 104: 23a.

[43] Lancet J E, Karp J E. Farnesyltransferase inhibitors in hematologic malignancies: New horizons in therapy. Blood, 2003, 102: 3880-3889.

[44] Ravoet C, Mineur P, Robin V, *et al.* Phase I – II study of a Farnesyl Transferase Inhibitor (FTI), SCH66336, in patients with Myelodysplastic Syndrome (MDS) or Secondary Acute Myeloid Leukemia (sAML). Blood, 2002, 100: 794a.

[45] Hasegawa D, Manabe A, Kubota T, *et al.* Methylation status of the p15 and p16 genes in paediatric myelodysplastic syndrome and juvenile myelomonocytic leukaemia. Br J Haematol, 2005, 128: 805-812.

[46] Johan M F, Bowen D T, Frew M E, *et al.* Aberrant methylation of the negative regulators RASSFIA, SHP-1 and SOCS-1 in myelodysplastic syndromes and acute myeloid leukaemia. Br J Haematol, 2005, 129: 60-65.
[47] Langer F, Dingemann J, Kreipe H, *et al.* Upregulation of DNA methyltransferases DNMT1, 3A, and 3B in myelodysplastic syndrome. Leuk Res, 2005, 29: 325-329.
[48] Kaminskas E, Farrell A, Abraham S, *et al.* FDA Approval summary: Azacitidine for treatment of myelodysplastic syndrome subtypes. Clin Cancer Res, 2011, 2005, 11: 3604-3608.
[49] Oka M, Meacham A M, Hamazaki T, *et al.* De novo DNA methyltransferases Dnmt3a and Dnmt3b primarily mediate the cytotoxic effect of 5-aza-2'- deoxycytidine. Oncogene, 2005, 24: 3091-3099.
[50] van den Bosch J, Lubbert M, Verhoef G, *et al.* The effects of 5-aza-2'-deoxycytidine (Decitabine) on the platelet count in patients with intermediate and high-risk myelodysplastic syndromes. Leuk Res, 2004, 28: 785-790.
[51] Issa J P, Garcia-Manero G, Giles FJ, *et al.* Phase 1 study of low-dose prolonged exposure schedules of the hypomethylating agent 5-aza-2'-deoxycytidine (decitabine) in hematopoietic malignancies. Blood, 2004, 103: 1635-1640.
[52] Saba H, Rosenfeld C, Issa J P, *et al.* First report of the phase III north american trial of decitabine in advanced myelodysplastic syndrome (MDS). Blood, 2004, 104: 23a.
[53] Lübbert M, Wijermans, P W, Rüter, B H. Re-treatment with low-dose 5-aza-2'-deoxycytidine (decitabine) results in second remissions of previously responsive MDS patients. Blood, 2004, 104: 405a.
[54] Ben-Kasus T, Ben-Zvi Z, Marquez VE, *et al.* Metabolic activation of zebularine, a novel DNA methylation inhibitor, in human bladder carcinoma cells. Biochem Pharmacol, 2005, 70: 121-133.
[55] Guo H, Rao N, Xu Q, *et al.* Origin of tight binding of a near-perfect transition-state analogue by cytidine deaminase: Implications for enzyme catalysis. J Am Chem Soc, 2005, 127: 3191-3197.
[56] Holleran J L, Parise R A, Joseph E, *et al.* Plasma pharmacokinetics, oral bioavailability, and interspecies scaling of the DNA methyltransferase inhibitor, zebularine. Clin Cancer Res, 2005, 11: 3862-3868.
[57] Drummond D C, Noble C O, Kirpotin D B, *et al.* Clinical development of histone deacetylase inhibitors as anticancer agents. Annu Rev Pharmacol Toxicol, 2005, 45:495-528.
[58] Speranza A, Pellizzaro C, Coradini D. Hyaluronic acid butyric esters in cancer therapy. Anticancer Drugs, 2005, 16: 373-379.
[59] Kuendgen A, Strupp C, Aivado M, *et al.* Treatment of myelodysplastic syndromes with valproic acid alone or in combination with all-trans retinoic acid. Blood, 2004, 104: 1266-1269.

18

Infectious Microecology in Solid-Organ Transplantation

Shusen Zheng *, Jian Wu
Hepatobiliary Pancreatic Surgery, the First Affiliated Hospital, Zhejiang University; Collaborative Innovation Center for Diagnosis and Treatment of Infectious Diseases, Hangzhou, 310003, China
* E-mail: shusenzheng@zju.edu.cn

Solid-organ transplantation is often the last alternative in many patients with end-stage organ disease. Liver, kidney, heart, and lung transplantation have became standard therapy for selected end-stage diseases, although increasingly potent immunosuppressive agents, surgical techniques, organ preservation, and overall management of transplant recipients have dramatically reduced the incidence of rejection of transplanted organs and improved graft and patient survival. However, infection and malignancy have become major causes of morbidity and mortality following solid organ transplantation. The main reason for infection is related to immuno-suppression, which depresses cell-mediated immunity. The resultant depressed cell-mediated immunity leads to increased susceptibility to intracellular pathogens, such as bacterial, fungal, and herpes viruses, similar to the situation in human immunodeficiency virus (HIV)-infected patients. Otherwise, the presence of metabolic abnormalities, such as protein malnutrition, uremia and hyperglycemia, and the presence of damage to mucocutaneous barriers and foreign bodies that interrupt these barriers, such as intravenous lines, endotracheal tubes, urinary catheters, and chest and biliary tubes, are important factors.

These infections, including bacterial, fungal, viral, and parasitic depend on numerous factors, such as immunosuppressive strategy, type of organ transplant, and the period after transplantation. The successful prevention of infection in the solid organ transplant recipient requires an understanding of these factors for developing a preventive treatment adapted for each individual. Recognition of infection as a serious complication following transplantation supports the need for pre-transplant screening, early diagnosis of active infection, and modalities for prevention and treatment of various infections [1].

In this chapter, we present a review of infections in the solid-organ transplant recipient including pre-transplantation evaluation, pathophysiology, incidence, prevention and treatment strategies of common post-transplant infections.

18.1 Screening of Donor and Recipient Prior to Solid-Organ Transplantation

It is difficult to measure the patient's risk of infection after transplantation. The condition of allograft rejection, the intensity of immuno-suppression, and other factors may contribute to his or her susceptibility to infection. Prophylactic strategies are based on the patient's known or likely exposure to infection according to the results of serologic testing and epidemiologic history before transplantation.

Four objectives of pre-transplant infectious disease screening we need to mention are: (i) To identify conditions which may disqualify either donor or recipient; (ii) To identify and treat active infection before transplantation; (iii) To define the risk of infection and determine strategies for preventing and treating post-transplant infection; (iv) To implement preventative interventions during operation, such as vaccination [2]. Here we will briefly demonstrate the following categories of epidemiologic exposure including donor-derived infections, and recipient- derived infections.

18.1.1 Donor-Derived Infections

The transplanted organs will transmit the infected organism from donor to recipient. These infections, including cytomegalovirus (CMV) infection [3], tuberculosis [4], and *Trypanosoma cruzi* infection are latent in transplanted tissues. Transmission may also be due to active donor infection such as viremia or bacteremia that was undiscovered at the time of organ procurement. Organ donors also may become infected with nosocomial organisms that are resistant to routine surgical antimicrobial prophylaxis, and they may transmit these organisms to recipients [5].

Because the immunosuppressive is given after transplantation, the organ-transplant recipients with these infections can suffer rapid progression, permanent neurological damage, and death is common. Up until now, the screening methods for transplant donors have not been satisfactory in the short period during which the organ is harvested from deceased donors. At present, the routine evaluation of donors for infectious disease generally depends on antibody detection using the serum from donors. Since seroconversion may not occur during acute infections and the sensitivity of these tests is not 100%, some active infections remain undetected. Meanwhile, unidentified pathogens will inevitably be implanted if there is not any information about this pathogen. Improved donor screening will require the use of more sensitive (*e.g.*, molecular) and rapid assays by organ-procurement organizations. Some documented infections, such as sepsis and HIV infection, preclude organ donation. Organs from donors with specified known infections may be considered for specific recipients provided there is appropriate informed consent based on the urgency of the need for transplantation and the availability of effective antimicrobial therapies. For example, some livers from donors who were seropositive for Chagas' disease have been used successfully with benznidazole prophylaxis in regions where the disease is endemic.

Similarly, although organs from donors infected with the hepatitis B virus (HBV) and who had test results that were positive for antibodies against hepatitis B core antigen and negative for antibodies against hepatitis B surface antigen were rejected in the past, they are currently used for some recipients who have been vaccinated or who were previously infected, provided there is treatment with specific antiserum and anti-HBV antiviral agents [6]. The use of organs infected with the hepatitis C virus (HCV) remains controversial and is generally reserved for HCV-infected recipients [7].

Transplantation of organs from deceased donors who had fever or viral syndromes is controversial, and the uncertainty highlights the need for improved microbiologic screening tools. In cases in which the need for transplantation is relatively less urgent, it is reasonable to avoid the use of organs from donors with unexplained fever, rash, encephalitis, or untreated infectious syndromes.

18.1.2 Recipient-Derived Infections

Transplant recipients are at risk from infections related to complications of organ failure. So, active infection in transplant recipients should be eradicated before transplantation, since immune-suppression will exacerbate the infection process. Patients awaiting renal transplants may have some type of infection because of hemodialysis or peritoneal dialysis with catheters, or complicated with urinary tract infections. Candidates awaiting liver transplants are at risk of aspiration pneumonia, spontaneous bacterial peritonitis, urinary tract infection and infections associated with intravenous catheters. Candidates for heart transplants may have infections related either to indwelling intravenous catheters, or to ventricular assist devices (VADs) utilized as a bridge to transplantation [8]. In addition, heart candidates are also at risk of pneumonia in the setting of congestive heart failure and debilitation.

Common recipient-derived pathogens include *Mycobacterium tuberculosis*, viruses (*e.g.*, CMV, EBV, HBV, HCV, HIV, and herpes simplex virus), certain parasites (*e.g.*, *Strongyloides stercoralis* and *T. cruzi*), and endemic fungi (*e.g.*, *Histoplasma capsulatum, Coccidioides immitis,* and *Paracoccidioides brasiliensis*). The importance of recipient-derived screening for solid organ transplantation is to find the suspected infection after transplantation. Active bacterial infections should be emphasized before the patient receives organ transplantation. Infections of the respiratory tract, the urinary tract or other focal sites should be thoroughly treated. The potential kidney candidate with urinary tract infection should be investigated to rule out upper tract involvement.

For the *Mycobacterium tuberculosis* infection, all patients should receive a PPD (tuberculin skin test) performed prior to transplant, and those who have a positive skin test, or a history of active *tuberculosis*, should undergo additional screening to rule out active disease. The patients with a history of positive PPD or radiographic evidence of prior TB with no previous treatment should be considered for isoniazid prophylaxis. Prophylaxis can be started while the patient is on the transplant waiting list and completed after transplantation if a donor organ becomes available and at least 1 – 2 months of isoniazid has been administered. The prophylaxis course (9 – 12 months) can be completed after transplantation [9].

Aspergillus is common in lung transplant recipients, particularly in cystic fibrosis patients. Such colonization should prompt a rigorous evaluation to exclude active infection. Although post-transplant *aspergillosis* is a feared complication, transplant clinicians have generally relied more on post-transplant pre-emptive and prophylactic strategies rather than pre-transplant antifungal therapy for colonized patients. A pre-transplant candidate with invasive fungal infection (rather than colonization) should be treated at least until there is radiographic, clinical and microbiologic resolution in order to minimize the risk of this high-mortality infection post-transplant [10].

Active primary infection with viruses such as CMV, EBV or HBV at the time of transplant is uncommon. Nonetheless, if an active viral infection is detected in a potential recipient, transplantation should likely be delayed until the infection resolves in order to allow for development of natural immunity prior to transplant immune-suppression. This recommendation also extends to candidates who present themselves for transplantation with clinical symptoms suggestive of an acute community-acquired viral infection. If there is any chance of exposure to HIV pre-transplant, the potential recipient should have an HIV molecular detection test as well as HIV antibody testing [11, 12]. Successful transplantation has been achieved in HIV-infected patients treated with highly active antiretroviral therapy. In such recipients, the toxic effects of drugs and interactions between calcineurin inhibitors and antiretroviral agents require careful monitoring.

18.2 Timeline of Infection Post-Transplantation

The timeline is used to establish a differential diagnosis for infectious syndromes at various stages after transplantation. The timeline of post-transplant infections reflects the post-transplantation relationship between the recipient's epidemiologic exposure and the immunosuppressive strategy employed. Changes in immunosuppressive regimens, routine prophylaxis and improved graft survival have altered this timeline. Learning about the timeline of infection after transplantation is important in selecting immuno-suppression regimens to prevent complications [1].

18.2.1 Early Period (1 – 4 weeks)

During the early period after transplantation, the opportunistic infections are absent since the full effect of immune-suppression needs more time to develop. Infections such as viremia and candidemia in this period are generally donor-derived or recipient-derived, or they are associated with technical complications of surgery. Therapy must be guided by antimicrobial-susceptibility data, making microbiologic analysis of aspirates or biopsy specimens essential. *Clostridium difficile* colitis is common in this setting. Early graft injuries (*e.g.*, ischemia of bile ducts or pulmonary reperfusion injury) may later become foci for liver or lung abscesses. Unexplained early signs of infection, such as hepatitis, pneumonitis, encephalitis, rash and leukopenia, may be donor-derived.

18.2.2 Intermediate Period (1 – 6 months)

Viral pathogens and allograft rejection are responsible for the majority of febrile episodes that occur during the period from 1 – 6 months after transplantation. Trimethoprim-sulfamethoxazole prophylaxis generally prevents most urinary tract infections and opportunistic infections such as pneumocystis pneumonia, *Listeria monocytogenes* infection, *T. gondii* infection, and infection with sulfasusceptible *Nocardia* species. Some infections (cholangitis, pneumonia, *C. difficile* colitis) persist from the perioperative period. Infection due to endemic fungi, *Aspergillus*, *Cryptococcus*, *T. cruzi,* or *Strongyloides* may occur. Herpesvirus infections are uncommon with antiviral prophylaxis. However, other viral pathogens, including polyomavirus BK, adenovirus, and recurrent HBV or HCV, have emerged. Viral infections may cause "direct" (tissue invasive) disease or "indirect effects" which may be manifested in several virus-associated phenomena. These viral effects include systemic (CMV) or enhancing other opportunistic infections or PTLD and an increased risk of acute and chronic graft injury or rejection. Given the array of potential pathogens, in the future multiplex quantitative assays will be used to monitor acute infections.

18.2.3 Late Period (After 6 months)

The risk of infection diminishes 6 months after transplantation, since immunosuppressive drugs usually have a low concentration in recipients who have a satisfactory allograft function. However, transplant recipients have a persistently increased risk of infection due to community-acquired pathogens, with a limited risk for most opportunistic infections without other factors. In some patients, chronic viral infections may cause allograft injury (*e.g.* cirrhosis from HCV infection in liver-transplant recipients, bronchiolitis obliterans in lung-transplant recipients, accelerated vasculopathy in heart-transplant recipients with cytomegalovirus infection) or a malignant condition such as post-transplantation lymphoproliferative disorder (PTLD) or skin cancers. Recurrent infection may develop in some patients despite minimization of their immunosuppression. These patients are at increased risk of opportunistic infection with *Listeria* or *Nocardia* species, invasive fungal pathogens such as *Zygomycetes* and *dematiaceous moulds*, and unusual organisms (*e.g.*, *Rhodococcus species*). Minimal signs of infection merit careful evaluation in such high-risk patients; they may benefit from long-term trimethoprim-sulfamethoxazole or antifungal prophylaxis. Such prophylaxis may have some risk of microbial resistance to the prophylactic agents or future drug interactions.

18.3 Prevention of Infection in Solid-Organ Transplantation

Solid-organ transplant recipients are considered to be at "high risk" of developing infection because of immune-suppression therapy. Individual risk is determined by a relationship between the epidemiologic exposure of the individual and the patient's "net state of immune-suppression" [1]. The successful prevention of infection in the solid organ transplant recipient requires finding out these factors in order to develop a preventive strategy adapted for each individual. Moreover, epidemiologic exposure may happen many years before transplantation or at any time following the transplant operation. Thus, clinicians must obtain a detailed history of encounters with potential pathogens, even if the exposure was relatively remote.

18.3.1 Viral Infections

Viral infection in solid organ transplant recipients may be caused by several viruses, such as cytomegalovirus (CMV), hepatitis B (HBV) and hepatitis C virus (HCV), herpes simplex virus (HSV), varicella zoster virus (VZV), Epstein Barr virus (EBV), adenovirus, *et al.*

18.3.1.1 CMV infection

Despite remarkable development in the diagnostic and therapeutic choice of its management, cytomegalovirus (CMV) remains one of the most important factors to impact the outcome of solid organ transplantation. CMV could influence the direct morbidity and many short-term or long-term indirect effects, thus resulting in reducing recipient survival. Prevention of CMV infection should be the key issue in ensuring the outcome of solid organ transplantation [13].

CMV infection may cause both invasive disease and "direct effects" in transplant recipients. Most invasive disease occurs in the first year after completion of prophylaxis. The most common manifestation is fever and neutropenia. Other invasive diseases include lymphadenopathy, pneumonitis, gastrointestinal invasion, pancreatitis, or meningoencephalitis. CMV infection is also associated with increasing the risk of additional infections, such as EBV-associated PTLD and hepatitis. In addition, CMV infection may contribute to vasculopathy in heart-allograft recipients.

Universal antiviral prophylaxis could not only help to prevent CMV infection, but also other viral infections such as herpes simplex virus, EBV, and HSV 6 infections. In addition, prevention of CMV infection may reduce episodes of both early and late acute rejection in renal-transplant recipients, cardiac vasculopathy in heart-transplant recipients. Although optimal regimens remain undefined, most centers provide anticytomegalovirus prophylaxis for the first 3 – 6 months after transplantation, using valacyclovir, high-dose acyclovir, ganciclovir, valganciclovir or, less commonly, cytomegalovirus hyperimmune globulins [14, 15]. Although recipients with solid organ transplants who are seropositive or who receive transplants from seropositive donors generally receive prophylaxis for at least 6 – 12 months, longer courses of antiviral prophylaxis were suggested if they lack evidence of protective immunity, or if they require much more immunosuppression. However, longer courses of universal antiviral prophylaxis using ganciclovir or valganciclovir may result in marrow suppression.

Otherwise, ganciclovir resistance is due to mutations in the cytomegalovirus UL97 gene (a viral protein kinase that phosphorylates the drug) or the UL54 gene (cytomegalovirus DNA polymerase). Such resistance happens not so often and may present a slowly responsive or relapsing infection in patients who receive inadequate or prolonged doses of oral ganciclovir or valganciclovir, especially during active infection, or in patients with intensified immunosuppression [16].

Quantitative diagnostic assays for CMV infection include molecular assays (polymerase-chain reaction [PCR] and other amplification assays) and antigen-detection (pp65 antigenemia) assays. In some patients, blood-based cytomegalovirus assays may be negative, while invasive procedures such as colonoscopy with a biopsy may be necessary. Invasive disease and the CMV syndrome (which is manifested as fever and leukopenia) need anti-virus therapy, generally with intravenous ganciclovir. Otherwise, oral ganciclovir therapy for CMV disease is also acceptable [17]. Intravenous ganciclovir is currently preferred for the initiation of therapy for CMV associated invasive gastrointestinal disease.

18.3.1.2 Epstein-Barr Virus and Post-Transplantation Lymphoproliferative Disorder

Post-transplant lymphoproliferative disorder (PTLD) is recognized as potentially one of the most devastating complications of organ transplantation. Most of the PTLD cases are associated with Epstein-Barr virus (EBV) infection and account for more than half of post-transplantation malignant conditions in pediatric solid organ transplant recipients. These syndromes range from infectious mononucleosis to true malignancies [18]. Risk factors for PTLD include primary EBV infection after transplantation in seronegative recipients from seropositive donors, allograft rejection, exposure to antilymphocyte antiserum, and cytomegalovirus co-infection.

The EBV virus is known to play a major role in the development of PTLD occurring early (within the first year) after solid organ transplantation. The pathogenesis of these disorders is related to EBV's ability to transform and immortalize B lymphocytes. The high virus peak results in massive infection of the B-cell pool and perhaps other cells not normally infected (T cells, NK cells, memory B cells), thereby setting the stage for secondary events that lead to malignancy.

The clinical presentation of EBV-associated PTLD includes infectious mononucleosis (fever, malaise, lymphadenopathy, hepatosplenomegaly and atypical lymphocytosis), specific organ diseases such as hepatitis, pneumonitis, gastrointestinal symptoms and hematological manifestations such as leucopenia, thrombocytopenia, hemolytic anemia and hemophagocytosis. Some of these manifestations may be identical to the features of PTLD. Both B-cell and T-cell PTLD may infiltrate allografts and may be confused with allograft rejection or other viral processes.

Quantitative EBV viral-load testing, flow cytometry, analysis of immunoglobulin gene rearrangements, and histologic analysis with staining for EBV-derived RNA are helpful in guiding the diagnosis of PTLD [19]. Pathology remains the gold standard for PTLD diagnosis. Fine needle biopsy is acceptable when larger biopsies are impractical, as in the case of allograft organ biopsy.

Given the absence of reliably effective therapy for PTLD, it would be ideal to have an effective preventive strategy. It is important to identify the patient at high-risk of PTLD development prior to transplantation. EBV serostatus should be determined for all transplant recipients. The patients at risk of primary CMV infection or severe CMV disease or receiving antithymocyte globulin for induction or rejection should be monitored carefully for clinical symptoms (fever, diarrhea, lymphadenopathy, allograft dysfunction, *etc.*) and investigated aggressively for PTLD. Allograft biopsies from these patients should be performed to find the evidence of early PTLD. Then immunosuppression should be minimized and aggressive immunosuppression should only be prescribed in acute rejection proved by biopsy.

The treatment of PTLD remains a challenge. There is no specific treatment approach that could be followed for patients with PTLD. The general approach to therapy involves a stepwise strategy that starts with reduced immunosuppression, but poses the risk of allograft rejection. The progression of disease requires alternative approaches that may include the administration of chemotherapy, irradiation (for the central nervous system disease), and treatment with anti-CD20 antibodies. Adoptive immunotherapy (T-cell transfer) is under investigation as a treatment strategy for PTLD.

18.3.1.3 Herpes Simplex Virus and Varicella Zoster Virus

Herpes simplex virus (HSV), usually reactivated from latent infections, can be presented as a primary infection from the donor within the first few weeks after transplantation. Incidence of HSV is highest in kidney transplant recipients, and antibodies are found in 75% of adult kidney transplant recipients [20]. The two most common viral strains, human herpes virus 1 (HHV-1) and HHV-2, are responsible for mucocutaneous infections. Complications associated with high mortality rates include HSV-related pneumonia, hepatitis, and meningoencephalitis [21]. Laboratory tests include HSV tissue culture, enzyme immunoassays, and polymerase chain reaction (PCR), commonly used if a cerebrospinal fluid infection is suspected, will aid HSV diagnosis.

Approximately 90% of kidney transplant recipients are *Varicella zoster* virus (VZV) seropositive at the time of transplantation. Immunocompetency is an important factor in the process of reactivation. The diagnosis is established according to clinical symptoms, or common laboratory tests include VZV DNA in bronchial washings and liver biopsy specimens. Moreover, PCR will give a rapid result in the cerebrospinal fluid antibody when VZV meningoencephalitis is suspected.

The goal of HSV prophylaxis is to suppress the viral reactivation in patients with latent infection or to prevent primary infection if the donor is seropositive for HSV. Acyclovir has been the gold standard for many years, while valganciclovir and ganciclovir have a suppressive effect against HSV. The duration of prophylaxis is usually 30 – 90 days. Other antiviral agents that may be used include valacyclovir and famcyclovir. The same antiviral regimen applies to patients with VZV prophylaxis.

To treat localized, mucocutaneous HSV infection includes oral acyclovir, famciclovir, and valacyclovir. For patients with severe, noncutaneous HSV, high-dose intravenous acyclovir is suggested. Because VZV is less sensitive to acyclovir than HSV, primary VZV or VZV associated with organ involvement should be treated with high-dose intravenous acyclovir [21]. Other agents that can be used include ganciclovir, cidofovir and foscarnet, but these agents are rarely used because of their toxicities.

18.3.2 Bacterial Infections

Bacterial Infections is including *clostridium difficile* and Mycobacterium

Tuberculosis.

18.3.2.1 *Clostridium Difficile*

Clostridium difficile is known to produce protein endotoxins that cause colonic mucosal inflammation and injury. The clinical manifestations may include fever, abdominal pain, mild diarrhea, toxic megacolon, or even perforation because of swelling. Antimicrobial therapy and hospitalization are well-known risk factors for *C. difficile* infections, but it can occur without previous antibiotic treatment [22].

The preferred treatment choice for *C. difficile* infections is metronidazole, of which 250 – 500 mg are to be taken orally, 3 – 4 times daily, for 10 – 14 days. If patients fail to respond to metronidazole, vancomycin 125 mg orally 4 times daily for 10 – 14 days may be prescribed. Concern about increased vancomycin resistance in other pathogens, such as *Enterococci*, further discourages use of oral vancomycin as first-line therapy for *C. difficile* infection [23].

18.3.2.2 Mycobacterium Tuberculosis

Latent tuberculosis (TB) infection can be reactivated because of immunosuppression after solid organ transplantation. The infection may also be acquired and transmitted from donors. Most transplant centers screen transplant candidates for latent TB infection preoperatively and treat with isoniazid postoperatively [24].

Medications used to treat TB infection are not benign, particularly when combined with immunosuppressive medications. The immunosuppressant drugs will increase the risk of hepatotoxicity and sometimes the side effect is unacceptable [25]. Drug interactions challenge the treatment of TB when using immunosuppressive medications. Cyclosporine, sirolimus and tacrolimus are all substrates of cytochrome P450 (CYP) 3A isoenzymes, and a significant dose increase of these immunosuppressant agents may be necessary to maintain therapeutic drug concentration in the presence of rifampin, a CYP enzyme inducer [26].

18.3.3 Fungal Infections

18.3.3.1 Candida Species

The incidence of candidal infection ranges from 5% to 50% in transplant recipients, depending on the type of organ transplanted. Several factors will be involved in candidal infections. Pretransplant endogenous gastrointestinal colonization, pancreas and liver transplant recipients who have a complicated course of surgery, Roux-en-Y anastomosis or retransplantation, all will increase risk factors. Moreover, the use of the muromonab-CD3 monoclonal antibody and immunomodulatory anti-viral treatment, including CMV and human herpes virus 6, are also factors that increase the risk of invasive fungal infection [27].

Amphotericin B is the first choice for candidal infections (0.5 – 0.7 mg/kg per day), even the nephrotoxic effect. Lipid formulations with less nephrotoxic may be considered for patients who cannot tolerant the conventional drug, or for recipients receiving calcineurin inhibitors. In addition, fluconazole may be a reasonable substitution for infections caused by *Candida albicans* [28]. Some centers are looking at antifungal susceptibilities to determine appropriate treatment because interpretive breakpoints are available for *Candida spp* against fluconazole, itraconazole and 5-flucytosine [29]. *Candida glabrata*, for example, has reduced susceptibilities to fluconazole and requires either higher doses of fluconazole or the use of nonazole agents. *Candida krusei* is generally deemed resistant to fluconazole, requiring maximal doses of amphotericin B.

Combination antifungal therapy has strong potential because limited agents are available for the treatment of systemic candidal infections, particularly in patients with persistent conditions. A randomized and blind multicenter trial compared high-dose fluconazole plus placebo with fluconazole plus amphotericin B in the treatment of candidemia in nonneutropenic patients.

18.3.3.2 Aspergillus Species

Invasive *aspergillosis* has been most commonly reported in lung and heart-lung transplant recipients, mostly occurring within 6 months after transplantation [27]. These transplant recipients may be at higher risk because the respiratory tract is a portal of entry for filamentous fungi and it may also be colonized with spores. In addition, denervation of the airways during the transplant procedure damages the defense mechanisms, such as cough reflexes or mucociliary reflexes, thus heightening the risks of infection [30].

Amphotericin B is considered the first-line treatment for *aspergillosis*, but may cause nephrotoxicity, particularly when the patient is taking calcineurin inhibitors. Lipid formulations of amphotericin B are less likely to cause nephrotoxicity and can serve as substitutes, especially when chronic treatment is required for invasive aspergillosis [27, 30].

Itraconazole is active against *Aspergillus spp*, but as monotherapy it has been associated with higher relapse rates than amphotericin B. However, itraconazole is an option as a step-down oral therapy [31]. Voriconazole and caspofungin have demonstrated activity against invasive *aspergillosis* and may be used to avoid amphotericin B nephrotoxicity. Itraconazole and voriconazole, as inhibitors of P-glycoprotein and CYP 3A isoenzymes, carry potential for significant interactions with cyclosporine, tacrolimus and sirolimus. The use of voriconazole in combination with sirolimus is contradicated, and close monitoring of cyclosporine and tacrolimus levels is warranted when these azole-calcineurin inhibitor combinations are used [32].

18.3.3.3 Pneumocystis carinii

Pneumocystis carinii pneumonitis (PCP) is a common opportunistic infection in the transplant recipients receiving immunosuppression treatment. Although PCP prophylaxis is a routine practice, the drug regimen and the duration of therapy vary depending on transplant centers and the type of organ transplanted. Trimethoprim-sulfamethoxazole (single or double strength orally once daily to thrice weekly) provides excellent prophylaxis against PCP. Pentamidine (300 mg inhalation monthly) or dapsone (50 – 100 mg orally once daily) may also be used. The risk of dose-dependent hemolytic anemia, particularly in individuals with glucose-6-phosphate dehydrogenase deficiency, requires careful evaluation before the initiation of dapsone. 1500 mg of Atovaquone taken once daily with food is another option for both primary and secondary PCP prophylaxis.

18.3.4 Parasitic Infections

Toxoplasmosis is usually a rare opportunistic infection following most solid-organ transplantations, with the exception of heart transplantation; heart transplant recipients carry a high risk of being infected with *Toxoplasma gondii.* Without proper prophylaxis, patients who are seronegative and receive a transplant from a seropositive donor have a 50% – 75% risk of disseminated toxoplasmosis [33]. Patients become infected when *Toxoplasma cysts*, often located in the allograft muscle, become reactivated as a result of post-transplant immunosuppression. Infection usually occurs 2 – 6 months after transplantation, but may be present as early as day 1 after transplantation. Diagnosis is usually made by biopsy or detected by bronchoalveolar lavage. Clinical complications of *T gondii* infection include pneumonitis, myocarditis, pericarditis, brain abscess, meningoencephalilis and disseminated disease [33].

Toxoplasmosis is prevented with trimethoprim-sulfamethoxazole (TMP/SMX) for 6 months or longer, after transplantation. For patients with a sulfa allergy, pyrimethamine is recommended. There is current debate as to whether TMP/SMX is efficacious enough to prevent *toxoplasmosis* in high-risk patients. Many centers are using the combination of pyrimethamine, folinic acid and sulfadiazine as primary prophylaxis, arguing that this may be a more potent prophylactic regimen. However, this combination does not cover certain organisms that TMP/SMX does, such as *Nocardia* and *Listeria*.

In patients with *toxoplasmosis*, treatment consists of the pyrimethamine, folinic acid, and sulfadiazine combination. In patients with a sulfa allergy, dapsone may be used in place of sulfadiazine. Another alternative is clindamycin. The duration of treatment is controversial in transplant recipients.

References

[1] Fishman J A. Infection in solid-organ transplant recipients. N Engl J Med, 2007, 357: 2601-2614.

[2] Avery R K. Prophylactic strategies before solid-organ transplantation. Curr Opin Infect Dis, 2004, 17: 353-356.

[3] Kalil A C, Levitsky J, Lyden E, *et al.* Meta-analysis: The efficacy of strategies to prevent organ disease by cytomegalovirus in solid organ transplant recipients. Ann Intern Med, 2005, 143: 870-880.

[4] Hernandez-Hernandez E, Alberu J, Gonzalez-Michaca L, *et al.* Screening for tuberculosis in the study of the living renal donor in a developing country. Transplantation, 2006, 81: 290-292.

[5] Delmonico F L. Cadaver donor screening for infectious agents in solid organ transplantation. Clin Infect Dis, 2000, 31: 781-786.

[6] Madayag R M, Johnson L B, Bartlett S T, *et al.* Use of renal allografts from donors positive for hepatitis B core antibody confers minimal risk for subsequent development of clinical hepatitis B virus disease. Transplantation, 1997, 64: 1781-1786.

[7] Abbott K C, Lentine K L, Bucci J R, *et al.* The impact of transplantation with deceased donor hepatitis c-positive kidneys on survival in wait-listed long-term dialysis patients. Am J Transplant, 2004, 4: 2032-2037.

[8] McCarthy P M, Schmitt S K, Vargo R L, *et al.* Implantable LVAD infections: implications for permanent use of the device. Ann Thorac Surg, 1996, 61: 359-365; discussion 372-353.

[9] Manuel O, Humar A, Preiksaitis J, *et al.* Comparison of quantiferon-TB gold with tuberculin skin test for detecting latent tuberculosis infection prior to liver transplantation. Am J Transplant, 2007, 7: 2797-2801.

[10] Ruiz I, Gavalda J, Monforte V, *et al.* Donor-to-host transmission of bacterial and fungal infections in lung transplantation. Am J Transplant, 2006, 6: 178-182.

[11] Halpern S D, Shaked A, Hasz R D, *et al.* Informing candidates for solid-organ transplantation about donor risk factors. N Engl J Med, 2008, 358: 2832-2837.

[12] Ahn J, Cohen S M. Transmission of human immunodeficiency virus and hepatitis C virus through liver transplantation. Liver Transpl, 2008, 14: 1603-1608.

[13] Eid A J, Razonable R R. New developments in the management of cytomegalovirus infection after solid organ transplantation. Drugs, 70: 965-981.

[14] Slifkin M, Doron S, Snydman D R. Viral prophylaxis in organ transplant patients. Drugs, 2004, 64: 2763-2792.

[15] Efferth T, Romero M R, Wolf D G, *et al.* The antiviral activities of artemisinin and artesunate. Clin Infect Dis, 2008, 47: 804-811.

[16] Shapira M Y, Resnick I B, Chou S, *et al.* Artesunate as a potent antiviral agent in a patient with late drug-resistant cytomegalovirus infection after hematopoietic stem cell transplantation. Clin Infect Dis, 2008, 46: 1455-1457.

[17] Razonable R R. Immune-based therapies for cytomegalovirus infection.

Immunotherapy, 2: 117-130.
[18] Preiksaitis J K. New developments in the diagnosis and management of posttransplantation lymphoproliferative disorders in solid organ transplant recipients. Clin Infect Dis, 2004, 39: 1016-1023.
[19] Green M, Webber S A. EBV viral load monitoring: unanswered questions. Am J Transplant, 2002, 2: 894-895.
[20] Patel R. Infections in recipients of kidney transplants. Infect Dis Clin North Am, 2001, 15: 901-952, xi.
[21] Smith S R, Butterly D W, Alexander B D, *et al.* Viral infections after renal transplantation. Am J Kidney Dis, 2001, 37: 659-676.
[22] Johnson S, Gerding D N. Clostridium difficile-associated diarrhea. Clin Infect Dis, 1998, 26: 1027-1034; quiz 1035-1026.
[23] Chavers L S, Moser S A, Benjamin W H, *et al.* Vancomycin-resistant enterococci: 15 years and counting. J Hosp Infect, 2003, 53: 159-171.
[24] Blumberg H M, Burman W J, Chaisson R E, *et al.* American Thoracic Society/Centers for Disease Control and Prevention/Infectious Diseases Society of America: treatment of tuberculosis. Am J Respir Crit Care Med, 2003, 167: 603-662.
[25] el-Agroudy A E, Refaie A F, Moussa O M, *et al.* Tuberculosis in Egyptian kidney transplant recipients: study of clinical course and outcome. J Nephrol, 2003, 16: 404-411.
[26] Hebert M F, Fisher R M, Marsh C L, *et al.* Effects of rifampin on tacrolimus pharmacokinetics in healthy volunteers. J Clin Pharmacol, 1999, 39: 91-96.
[27] Singh N. Fungal infections in the recipients of solid organ transplantation. Infect Dis Clin North Am, 2003, 17: 113-134, viii.
[28] Rex J H, Bennett J E, Sugar A M, *et al.* A randomized trial comparing fluconazole with amphotericin B for the treatment of candidemia in patients without neutropenia. Candidemia Study Group and the National Institute. N Engl J Med, 1994, 331: 1325-1330.
[29] Rex J H, Pfaller M A, Galgiani J N, *et al.* Development of interpretive breakpoints for antifungal susceptibility testing: Conceptual framework and analysis of *in vitro-in vivo* correlation data for fluconazole, itraconazole, and candida infections. Subcommittee on Antifungal Susceptibility Testing of the National Committee for Clinical Laboratory Standards. Clin Infect Dis, 1997, 24: 235-247.
[30] Patterson J E. Epidemiology of fungal infections in solid organ transplant patients. Transpl Infect Dis, 1999, 1: 229-236.
[31] Stevens D A, Kan V L, Judson M A, *et al.* Practice guidelines for diseases caused by Aspergillus. Infectious Diseases Society of America. Clin Infect Dis, 2000, 30: 696-709.
[32] Romero A J, Le Pogamp P, Nilsson L G, *et al.* Effect of voriconazole on the pharmacokinetics of cyclosporine in renal transplant patients. Clin Pharmacol Ther, 2002, 71: 226-234.
[33] Baden L R, Katz J T, Franck L, *et al.* Successful toxoplasmosis prophylaxis after orthotopic cardiac transplantation with trimethoprim-sulfamethoxazole. Transplantation, 2003, 75: 339-343.

19

Microecology of Infections Associated with Surgery and Trauma

Hongqi Chen [1], **Huanlong Qin** [2*]
[1] The Sixth People's Hospital Affiliated to Shanghai Jiao Tong University, 600 Yishan Road, Shanghai, 200233, China
[2*] Shanghai Tenth People's Hospital of Tongji University, 301 Yanchangzhong Road, Shanghai, 200072, China
[1] E-mail: hqchen08@hotmail.com
[2*] E-mail: huanlong_qin@live.cn

Trauma is one of the leading worldwide causes of death at present and fatal trauma cases are the fourth highest cause of death in youths. As a consequence of improved emergency medical treatment, early death rates (48 h post injury) have been significantly reduced. However, death rates resulting from trauma-related complications have not diminished. The most common and dangerous complication that can develop post surgery/trauma is infection caused by opportunistic pathogens, which pose a significant challenge to the healing process [1]. In addition to interference with the healing process and direct damage to infected tissues caused by the infecting organism, systemic complications, including acute respiratory distress syndrome (ARDS), multiple organ dysfunction syndrome (MODS), systemic inflammatory response syndrome (SIRS) and acute renal failure (ARF) can develop. These complications significantly exacerbate a patient's already deteriorating health and increase recovery times that correlate directly with increased mortality. Therefore, these infections are a major threat to trauma patients, early detection and control of respective infectious agents is essential in decreasing the rates of post trauma-associated morbidity and mortality.

19.1 Main Pathogenic Bacteria Associated with Surgical and Trauma-Related Infections

Damage to the epidermis or to physical barriers that maintain homeostasis between the body and the external environment provides an opportunity for opportunistic organisms to invade and establish infections. This process is facilitated by decreased immunity associated with various states of trauma and/or potential damage to internal tissues. Damage to internal physical barriers, including damage to the intestine can result in dysbacteria-related infections. Although strains and their respective distributions vary between countries, regions and even medical units, these types of infection have a similar presentation and can be treated in similar fashion. For example, it has been reported that the prevention and treatment of surgical and trauma-associated infections are similar between USA and UK, suggesting that differences in the causative agents of infections between these countries do not affect treatment modalities.

19.1.1 Changes to the Spectrum of the Pathogenic Bacteria

Due to the continuous improvement of anti-infection therapies (*e.g.* surgical procedures, post surgical support and critical care strategies) in recent years, a reduction in post-surgical and post-trauma-related infections has been observed. However, concomitant with the reduced rates of infection associated with these improvements has been the rise of increasingly drug-resistant bacteria that are also more virulent. The changes in the microbial landscape have required modifications in some cases to existing treatment protocols. Historically, Gram-positive bacteria were the primary pathogens associated with traumatic infections. However, the effective use of penicillin and related antibiotics during the developing years of the antibiotic era increased the number of infections caused by Gram-negative bacteria. Presently, due to the abuse and overuse of antibiotics, both Gram-positive and Gram-negative bacteria (in addition to fungal infections) have become associated with serious post trauma/surgical infections. There has also been a parallel increase in the number of nosocomial infections that pose an increasing threat to the success of clinical therapies and the prevention of post-traumatic and post-surgical infections.

19.1.1.1 The Rise of Drug-Resistant Bacteria and Changes in Pathogenicity

Antibiotics were developed to control bacterial infections and penicillin, the first antibiotic discovered and the first prepared for commercial use, had a significant effect on the control of both Gram-positive and Gram-negative cocci and its use spread worldwide. However, excessive and undisciplined use of antibiotics, combined with patients not completing their respective treatment doses has resulted in the selection of bacterial strains resistant to multiple classes of antibiotics (sometimes accompanied with increased virulence). To date, bacterial infections represent a serious hazard to human health, and these types of infections are becoming more common and refractory in nature. The clinical spectrum of infections currently is focused on infections caused by methicillin-resistant *Staphylococcus aureus* (MRSA), penicillin resistant *Streptococcus pneumoniae* (PRSP), Vancomycin-resistant Enterococcus (VRE), Super-spectrum-Lactamase (SSBL) bacteria, axiomatic multi-platform C bacteria (AmpC), metallo-β-lactamase bacteria, *M. tuberculosis*, coagulase-negative staphylococci, *Escherichia coli*, pneumonia crayresearch bacteria, *Pseudomonas aeruginosa*, *Acinetobacter baumanii* and *Enterobacter cloacae*. Because of the development of resistant strains and changes in the pathogenic spectrum, post-operative and post-traumatic infections have become more and more difficult to control.

19.1.1.2 Changes in the Pathogenic Spectrum of Community Acquired Infections

Community acquired infections often occur at the injury site and, despite the use of preventative measures (including antibiotics), the significant incidence and severity of post-traumatic infections needs to be addressed. Improvements to support therapies facilitating the survival of severe trauma patients are threatened by emerging antibiotic-resistant pathogens, making treatment of post-traumatic infections challenging, especially since even the best treatment options are not guaranteed to prevent infections.

Wound infections. Most primary wound suture infections are caused by *S. aureus*. Farming accidents and other outdoor traumas often include inoculation of wounds with Gram-negative bacteria, including *P. aeruginosa*. The identification of the causative agent of industrial accident or mechanized farming trauma-related wounds is very important since atypical bacteria may be present on the equipment associated with these accidents. Suppurative infections caused by insect stings are primarily caused by *S. aureus*, therefore, if a Gram-positive cocci is identified, the likely cause of the infection can be deduced without additional microbiological testing.

Bacterial infections. Bacterial infections can develop into necrotizing fasciitis that can be caused by group A *Streptococcus*, aerobic or anaerobic bacteria. Necroinflammation may develop relatively slowly in infections caused by multiple organisms. Various gram-negative enteropathogens that are typically obligate anaerobes are associated with infections resulting from intestinal puncture wounds, and *S. aureus* infections can also result from these types of trauma. Group A Streptococcus and *S. aureus* are also associated with toxic shock syndrome resulting from hemodynamic instability and soft tissue necrosis. In addition, deep wound infections can be caused by gangrenous anaerobic spore-forming bacteria, and drainage of respective wounds is typically associated with short gram-positive bacteria.

Open fractures. Soft tissue infections (STI) are frequently associated with open fractures, and infections are typically caused by *S. aureus* infections. However, nosocomial infections occur more frequently than infections caused by environmental wound infections.

Intra-abdominal infections. Bacteria associated with peritonitis resulting from intestinal trauma are typically of intestinal origin. However, celiac cultures may not be an effective means of identifying microbial species during emergency laparotomies, and the nature of the respective pathogens is consistent with the region of the intestine damaged. Therefore, the accuracy of hypothesizing, as to the nature of the causative agent responsible for respective infections resulting from bacteria of intestinal origin, is limited to the experience of the physician. The types of bacteria, which can cause infections presenting an abdominal abscess, are highly variable, further supporting the observations that damage to different regions of the intestine can result in vastly different types of infections. For example, patients suffering from a ruptured colon can become infected with intestinal gram-negative bacteria and/or anaerobic bacteria. Abscesses caused by *S. aureus*, *Pseudomonas* species and other nosocomial, drug-resistant bacteria can often occur in patients with abdominal drainage tubes. The most common pathogen-associated pleural cavity trauma-related infections are *S. aureus* related and these infections typically result from bacterial colonization of abrasions present on the patient's skin or contamination of indwelling devices. Contamination of indwelling devices with Gram-negative organisms should be considered, if pyothorax infections are secondary to sub-diaphragm infections.

19.1.1.3 Nosocomial Infections

Most patients suffering from traumatic injuries receive timely treatment in hospitals, including antibiotic prophylaxis and wound debridement. Therefore, any resulting infections are nosocomial, typically occur superficially and are generally easy to manage. Most infections of this nature develop within 48 h in hospital and are the most common type of infection observed in burn and trauma patients. Unfortunately, the rates of nosocomial infections continue to rise, in part due to the development of more invasive (and frequent) therapies, the misuse of antibiotics and iatrogenic contamination. Currently, nosocomial infections have seriously affected the quality of medical treatment, resulting in significant economic losses and wasted resources [2]. Although most burn and trauma patients are relatively young, the incidence of nosocomial infections is highest in this population since these types of injuries predispose patients to infections for the following reasons: loss of skin due to burns leaves an unprotected area vulnerable to infection by numerous pathogens, damage to the intestine can expose the body to flora not typically present in extra-intestinal tissues; treatment involves the use of indwelling devices that also provide a conduit for infectious pathogens; the use of conventional antibiotic prophylaxis may not prevent infections with virulent antibiotic resistant antibiotics and massive bleeding and serious damage may affect the host immunity. Respiratory and urinary tract infections are the most common sites of post-traumatic nosocomial infections, compared to wound (incision infections), blood, chest and abdominal infections, which are rarer.

Nosocomial infections of burn and trauma patients can be caused by various pathogens, including gram-negative organisms such as *P. aeruginosa*, *E. coli*, enterbacteriaceae, crayresearch bacteria and gram-positive bacteria, including *S. aureus* and coagulase-negative *Staphylococci*. In addition, fungal infections caused by *Candida albicans* and other yeasts are also common. The rate of drug-resistant nosocomial infections is 91.6% and the rate of MSRA infections in burn and trauma patients was higher than in other in-hospital patient populations. Although nosocomial infections from different geographic locations vary in the context of the organisms most frequently identified, the group of pathogens associated with these infections remains constant worldwide.

The anatomic sites of trauma-related nosocomial infections vary with the nature of the injury. Minor injuries with an injury severity score (ISS) of 1 – 15 are typically UTIs (76.8%). Moderate injuries (ISS 16 – 25) are associated with higher incidence rates of respiratory infections (from 6.7% to 8%) and blood infections (from 1.5% to 41.7%). UTI infections in this group occurred at a reduced frequency (from 76.8% to 35%). Severe injuries (ISS 19) presented with even higher incidence rates of respiratory infections ranging from 51.9% to 41.7% and blood and UTI infections remained at 8% and 2.6%, respectively. In summary, mild injuries are associated with UTIs and moderate and severe injuries with respiratory and blood infections. It has been reported that blood infection rates in patients with serious wounds reached as high as 19%. Blood infections commonly were reported to occur 1-week post injury. In addition, since trauma patients often present multiple wounds, it is common for more than one site to become infected. Studies have shown that 58% of post-traumatic nosocomial infections may involve two or more sites, while single site infections accounted for only 42% of infections.

The nature of pathogens associated with nosocomial infections has similar distribution patterns in hospitals. For example, pathogens associated with respiratory infections in intensive care units (ICUs) are often the result of iatrogenic infections caused by respiratory pathogens. In addition, due to the ICU environment and the systemic use of antibiotics, many infections are caused by *Pseudomonas* species and *S. aureus*. Previous research has demonstrated that antibiotics can change a patient's normal intestinal flora by eliminating commensal populations and selecting antibiotic resistant organisms that are also typically pathogenic, including strains of *Pseudomonas* and other Gram-negative bacteria, including pathogenic *E. coli*, crayresearch bacilli and fungi.

Bacteria commonly associated with blood infections include *P. aeruginosa*, *E. coli* and coagulase-negative *S. aureus*. Bacteria isolated in the context of contaminated indwelling devices, such as catheters, include *S. aureus*, coagulase-negative *S. aureus*, *C. albicans* and intestinal bacteria.

Common wound infection-associated bacteria include coagulase-negative *S. aureus* and *Pseudomonas* species. In addition, the causative agents of wound infections are related to the type of injury sustained, that is, wound infections following severe trauma are more likely to be caused by intestinal pathogens such as *Acinetobacter* species, *Proteus* species, crayresearch bacteria and *Pseudomonas* species.

19.1.2 Predominant Pathogenic Bacteria

Predominant pathogenic bacteria associated with infections post-surgery or associated with trauma include Gram-positive cocci, Gram-negative bacilli, anaerobic bacteria and fungi. In addition, dysbacteriosis resulting from antibiotic use or the translocation of normal gut flora into extra-intestinal sites caused by biological barrier damage (resulting from either surgery or trauma) can lead to severe infections.

19.1.2.1 Gram-Positive Cocci

S. aureus and *Streptococcus* species are the predominant pathogenic bacteria associated with surgical procedures or trauma.

Staphyloccocus. A significant percentage of the natural flora of humans is comprised of *Staphylococcus* species. These organisms can colonize the upper respiratory tract and skin surfaces and are also the most common pathogens associated with post-surgical and trauma-related infections (primarily *S. aureus* and *S. epidermidis*). Many *S. aureus* isolates can produce hemolysins, leukotoxins, tissue destructive enzymes (such as hyaluronidase), T cell super antigens (such as toxic shock syndrome toxin I), B cell super antigens (such as protein A), immunomodulators (such as the MHC analog protein [Map or Eap] and the extracellular fibrinogen-binding protein [Efb]) and numerous antibiotic resistant determinants. These organisms are the most common pathogens associated with systemic blood-borne infections.

S. epidermidis is also a predominant organism associated with the natural human skin flora and is responsible for numerous opportunistic infections associated with post-surgical procedures or trauma-related damage. *S. epidermidus* isolates secrete extracellular slime substance (ESS) which adheres to bacterial surfaces and interferes with cellular immune responses and prevents antibiotics from reaching the bacteria. Both *S. aureus* and *S. epidermidis* have acquired multiple antibiotic-resistance genes including methicillin resistance *i.e.* methicillin-resistant *S. aureus* (MRSA) and methicillin-resistant *S. epidermidis* (MRSE). In addition, *S. aureus* has acquired the ACME element from *S. epidermidis* that facilitates the intracellular survival of *S. aureus*. This panel of virulence factors makes prevention and treatment of infections caused by these pathogens a significant health-risk priority.

Enterococcus. *Enterococcus* species are part of the normal intestinal flora of both humans and other mammals, and the prevalence of infections from these organisms are only second to *S. aureus* infections. In addition, Enterococcal infections are common causes of UTI and sepsis cases post surgery. Treatment complications are associated primarily with enterococcal strains that are resistant to a broad range of antibiotics.

Streptococcus. *Streptococcus* genera include 40 subtypes and these organisms are a part of the normal oral cavity and gastrointestinal flora. Not surprisingly, infections with these organisms are common agents of post-surgical or post-trauma infections. Group A *streptococci* are some of the more virulent *streptococci*, in part due to their ability to produce numerous enzymes and exotoxins, including the M protein, hyaluronidase, streptokinase, streptodornase, streptococcal hemolysin and pyrogenic exotoxin. Streptococcal infections can be present as allergic diseases, infections resulting in heart valve and kidney damage, bacteremia, sepsis, septicopyemia or purulent infections.

19.1.2.2 Gram-Negative Bacilli

Gram-negative bacilli associated with post surgical infections typically are caused by *Pseudomonas sp.*, *Escherichia sp.*, *Klebsiella sp.*, *Serratia sp.*, *Proteus sp.*, *Acinetobacter sp.*, *Citrobacter sp.* or enterobacteriaceae.

Pseudomonas. *Pseudomonas sp.* is widely distributed in nature and comprises more than 200 species, some of which comprise members of normal human flora. Depressed immune responses associated with post surgical procedures and trauma, in addition to invasive treatment protocols (*i.e.* tracheotomies and the use of indwelling devices) lead to *Pseudomonas sp.* infections, making infections with this pathogen important to consider in the context of nosocomial infections. *P. aeruginosa* infections are the fourth leading cause of Gram-negative sepsis, and the leading cause of sepsis-related deaths. In addition, exposure to *Pseudomonas sp.* can easily progress into respiratory infections, infective endocarditis, UTI, central nervous system infections, digestive tract infections, joint infections and skin and soft tissue infections.

Escherichia. *Escherichia sp.* infections are primarily caused by different *E. coli* strains even though many non-pathogenic *E. coli* strains comprise a significant percentage of the healthy intestinal flora. Pathogenic *E. coli* have been divided into five primary categories: enteropathogenic *E. coli* (EPEC), enterotoxigenic *E. coli* (ETEC), enteroinvasive *E. coli* (EIEC), enterohemorrhagic *E. coli* (EHEC) and enteroaggregative *E. coli* (EAEC). The most common infection route associated with these *E. coli* strains results from parenteral infections that are present as UTI and acute diarrhea. These types of *E. coli* infections are common post surgery and trauma.

Klebsiella. *Klebsiella sp.* is a Gram-negative bacilli comprising 7 strains that include *K. pneumonia*, the primary *Klebsiella sp.* associated with iatrogenic infections. *K. pneumonia* isolates have been demonstrated to produce extracellular toxic compounds (ETC, the primary [63%] capsular polysaccharide), lipid polysaccharide (30%) and small amounts of protein (7%). Some strains produce thermolabile (LT) and thermostable (ST) toxins and capsule formation is also related to pathogenicity.

Serratia. *Serratia sp.* is small, gram-negative bacteria which can produce water-insoluble yellow, purple and red pigments. *Serratia sp.* has been discovered in soil, water, plants and in animal and human intestinal tract and urinary tract samples. *S. marcescens* is the main nosocomial-associated *Serratia sp.*, and infections with this pathogen are associated with pneumonia, sepsis, blood transfusion infections and surgical, postoperative infections and UTI.

***Proteus*.** *Proteus sp.* is distributed widely in nature, including soil, water, and waste and in the intestines of humans and other mammals. *Proteus sp.* can be responsible for a variety of infections, including respiratory infections, diarrhea, UTI, peritonitis, otitis media, mastoiditis, infective endocarditis, meningitis, sepsis and food poisoning. Finally, Proteus infections are a primary cause of post-surgical and wound infections.

***Acinetobacter*.** This genus can be divided into 6 species that are also widely distributed in external environments such as water, soil, and other damp environments. *Acinetobacter sp.* comprises 25% of the normal human skin flora of humans in addition to 7% of bacteria recoverable from pharynx (7%) cultures. In addition, it has been found to colonize the conjunctiva, gastrointestinal tract and has been identified in both saliva and vaginal secretions. Due to their adherence properties, these organisms are difficult to remove from medical instruments and are a primary cause of post-surgical iatrogenic infections and trauma. Their prevalence in clinical settings is only second to that of *P. aeruginosa*.

***Enterobacter*.** *Enterbacteriaceae* in general comprise a group of pathogens that are widely distributed in nature and are members of the intestinal flora of both humans and other animals. *E. aerogenes* and *E. cloacea* are common causative agents of UTI and respiratory infections occurring post-surgery or trauma. Infections with these pathogens can progress to sepsis or meningitis.

***Citrobacter*.** *Citrobacter sp.* exists widely in nature and is present in the intestines of both humans and animals under normal conditions. However, opportunistic infections caused by members of this genus typically manifest themselves as hemorrhagic enteritis and infections are common post-surgery and trauma.

19.1.2.3 Anaerobes

Anaerobes comprise a significant portion of the normal human intestinal flora and can also colonize human skin and the deep mucosa of the sinus tract. A feature that is characteristic of patients' post-surgery and trauma is tissue ischemia and local necrosis associated with low oxygen concentrations that make patients vulnerable to anaerobic bacterial infections. Most anaerobic bacteria are aerogenic, such as the tetanus bacillus and other *Clostridium sp.* Most sporeless anaerobic bacterial infections are often the result of endogenous strains of bacteria (*i.e.* autoinoculation) except for *C. tetani* infections. Infections caused by anaerobic bacteria are the result of mixed infections that include bacteroid (*Bacteroides sp.*) and digestive streptococci, often in combination with other bacteria such as *E. coli*. Furthermore, *B. fragilis* can produce a beta amide enzyme which significantly reduces the efficacy (and can even inactivate) penicillin-based antibiotics at the site of infection making the choice of antibiotic important. Anaerobic bacterial infections are often accompanied by a purulent smell. Furthermore, the growth of sporeless anaerobic bacteria is slow, therefore infections with these pathogens may not be present until later. For example, wound infections post surgery and trauma often appear several days after stitches are removed.

19.1.2.4 Fungi

Fungi are a type of eukaryotic microbial cell widely distributed in nature and can be considered opportunistic human pathogens in most cases. Pathogenic fungi are divided into shallow and deep fungi. The former mainly exists on skin, hair and nails and treatments to eradicate fungi from these surfaces are difficult, although these fungi do not generally pose a severe threat to human health. Infections with the latter, however, may result in systemic, splanchnic infections that may result in death. Fungi cause diseases in animals, plants and humans, including various types of infections and allergic diseases. Infections typically only occur following major surgery, severe trauma, in burn victims, patients with intestinal disorders or in immune-compromised individuals. Common fungal infections are caused by *C. albicans*, *Aspergillus sp.* or mucor fungi. The most common fungal infections are caused by *C. albicans* infections that are usually present as upper respiratory tract, oral cavity, vaginal or bowel commensal infections. Infections caused by *Aspergillus sp.* and mucor fungi are also responsible for serious post-surgery or trauma-related infections.

19.1.2.5 Viruses

Viral infection rates of post-surgery or trauma are lower than observed for either bacteria or fungi. Cytomegalovirus and herpes simplex virus infections have been reported post-surgery (mainly in children) and cytomegalovirus antibodies have been detected after surgery and trauma [3]. Cytomegalovirus inclusion bodies have also been seen in adult burn victims. However, blood transfusions post-surgery and trauma cannot be ruled out as the cause of the infections. Serious injury requiring significant amounts of transfused blood has also been associated with serum hepatitis and cytomegalovirus infections.

19.1.3 Primary Pathogenic Factors

Infection and the nature of disease presentation depends on various host-pathogen interactions. The arsenal of virulence factors associated with respective pathogens will greatly influence disease presentation. Factors that can influence colonization efficacy are adhesins that facilitate microbial adherence to host extracellular matrix structures (infections cannot begin without microbial attachment). In addition, receptors that bind to host receptors typically mediate pathogen internalization into host cells, facilitating immune evasion and causing damage to host tissues. Some bacteria produce capsules that are resistant to phagocytosis, while toxins and invasive enzymes can damage various tissues, including immune system components.

19.1.3.1 Exotoxins and Invasive Enzymes

Exotoxins and invasive enzymes secreted by some bacteria during the initial infection stages allow bacteria to create a protective niche facilitating bacterial reproduction. For example, *P. aeruginosa* exotoxin A can interfere with (mammalian) protein synthesis, resulting in toxic effects that can facilitate tissue destruction leading to gangrene, the formation of deep abscesses, hypotension, edema, hepatic steatosis, kidney hemorrhagic necrosis and death.

The *S. aureus* toxic shock syndrome toxin-1 (TSST-1) is a T cell super-antigen which ligates T cells to major histocompatability antigen II (MHC-II), expressing antigen presenting cells in a non-antigen specific manner resulting in polyclonal activation and deletion of up to 20% of T cells. As a consequence of this process, T cells release vast quantities of IL-1 and TNF, which results in fever, desquamation, rashes, osmotic imbalances, hypotension, multiple organ damage and shock. It also can increase the sensitivity of the host to endotoxins and alter capillary permeability resulting in cardiovascular disorders. In addition, *S. aureus* coagulase mediates fibrin clot formation, protecting *S. aureus* from immune clearance and access by antibiotics.

The streptococcal thermophilus toxin (SPE) is responsible for causing scarlet fever in humans and is associated with local or systemic roseola, fever, pain, nausea, vomiting and general malaise. *Streptococci* also produce hyaluronidase that hydrolyzes hyaluronic acid, resulting in tissue destruction associated with bacterial dissemination.

19.1.3.2 Endotoxins

Endotoxins, such as the gram-negative cell wall component lipid polysaccharide (LPS), can be released from the bacterial surface under normal physiological conditions. The chemical constituents of LPS include lipid A, polysaccharides and amino acids. Lipid A from different bacteria is similar; therefore different bacterial LPS molecules will have similar effects on human physiology. LPS is heat stabile, weakly antigenic and a potent immune-stimulator. LPS is the primary factor driving septic shock as a consequence of non-specific stimulation of macrophages, endothelial cells and granulocytes, stimulating the release of inflammatory mediators such as IL-1, TNF-α and IL-6. In addition, endotoxin can activate the alternative complement pathway and promote B lymphocyte proliferation. Endotoxemia post-surgery results in the systemic inflammatory response syndrome (SIRS) and multiple organ dysfunction syndrome (MODS).

19.2 Post-Surgery- and Post-Trauma-Related Wound Infections

Infections have been the leading cause of death in both surgery and trauma patients, and surgical wound infections are one of the main complications associated with these patients since these infection types severely affect wound healing, extend hospital stays, raise treatment costs and can result in severe and even fatal infections [4]. Due to the constant change in microbial virulence patterns and increased resistance to antibiotics, the incidence and severity of wound infections remain a problem that cannot be ignored in the treatment of surgery and trauma patients.

19.2.1 Etiology and Pathogenesis of Wound Infections

Surgery and trauma create significant damage to protective skin barriers, exposing subcutaneous tissues to environmental pathogens. Commensal skin flora and environmental opportunistic pathogens can invade surgical- or trauma-related wounds resulting in life-threatening infections. Although pathogens can be identified from all wound infections, not all organisms will necessarily cause disease in all individuals since many host (immune status, health) and pathogen-related factors (virulence factors) come into play during the infection process. However, the right combination of virulence factors in the context of an environment conducive to infection (surgery or trauma), combined with diminished host defenses, greatly increase the risk of developing a wound infection. Risk factors also associated with developing a wound-related infection are extensive antibiotic use and hormone therapies that will change the natural flora of both the skin and intestine, thereby increasing the risk of developing an endogenous infection. The formula for the risk of developing a wound infection described by Altemier *et al.* is as follows: the risk of wound infection = the quantity × toxicity of pathogens / host resistance.

19.2.2 Prevention and Treatment of Wound Infections

Wound infections are the most common complications associated with surgical and trauma patients. In China, surgical wound infections accounted for 13% – 18% of inpatient cases. In mild cases, postoperative infections increased pain and hospital stays by about 10 days, resulting in significant increases in health care costs. In serious cases, postoperative infections led to surgical failure and in some cases death.

Since the causes and anatomical location of wound infections vary, so does clinical presentation. If the wound results from a surgical incision, the wound should be relatively clean. However, wounds caused by non-sterile objects may lead to infected necrotic tissues, hematomas and foreign body infections. Open fracture wounds are often associated with different degrees of tissue injury and are more prone to infections. Once the infection develops, treatment becomes more difficult, especially in cases of traumatic osteomyelitis which may affect limb recovery time and function, leading to physical disabilities and potential life-threatening infections [5].

19.2.2.1 Wound Infection Prevention

Expedient surgical debridement is important in preventing wound infections and is the most effective measure in reducing wound infections. Debridement not only prevents infections, but also creates an environment conducive to the repair and healing of deep tissues and in the restoration of function and structural integrity of the injured area. The process of debridement includes the removal of necrotic tissue, hematomas and/or foreign bodies while retaining viable tissues. The beneficial effects of debridement are more effectively observed if this process is carried out no later than 6 – 8 h post injury, since bacteria during this time frame have not yet penetrated into deeper tissues of the contaminated wound. For smaller wounds resulting in less serious injuries contamination is less likely and less severe, particularly in cooler climates. Smaller wounds can be sutured 24 h post injury if they are correctly cleaned. For severely contaminated wounds, debridement is more difficult 24 h after injury and in warmer climates debridement must be carried out quickly to prevent pathogen dissemination. Studies have demonstrated that repeated debridement every 48 – 72 h is important in the removal of necrotic tissues not removed following the initial debridement procedure. Full thickness wound resection, followed by a second resection, are the most effective debridement procedures. For some wounds, which may not have been completely resected all at once, should be repeatedly washed to remove all contaminants prior to continuing with the resection process. Open fracture wound closures should be carried out as soon as possible to reduce infection rates since avoiding dehydration of deep tissues will accelerate fracture healing.

Wound debridement should be considered, based on the degree of wound contamination, size and depth. In general, primary sutures should only be put on clean wounds no more than 12 h post debridement. Large and deep wounds require drainage after primary suture. For wounds in tissues with a rich blood supply (head, neck and face), it is important that sutures are not made until thorough debridement is carried out, even if it means that the injury will take longer to heal. Primary sutures put in after debridement are important in protecting wounds from infection and environmental contaminants. However, for seriously contaminated wounds of soft tissues, or tissues where efficient debridement cannot be performed, primary sutures are not recommended and should not be put in until the wound can be better stabilized. In addition, high impact-induced soft tissue contusions or lacerations generally do not require primary sutures due to the inflammation at the injury site. In summary, sutures should be utilized to stabilize wounds following debridement procedures.

In dealing with wound treatment, the systemic application of antibiotics, especially when used prophylactically, remains controversial and the utilization of antibiotics in surgical procedures has become more widely spread. In order to prevent the development of postoperative wound infections, prophylactic utilization of antibiotics has become routine even though postoperative infections have not been reduced. However, antibiotic use has had some beneficial effect on preventing wound infections not associated with surgical procedures. However, the overuse of antibiotics and the utilization of the incorrect antibiotic type does not only fail to achieve the desired effects but results in increased side effects and in the selection of resistant strains.

Prophylactic antibiotic use for the prevention of wound and/or surgical infections does have merit but must be carried out appropriately, that is the respective antibiotic used must be applied in the correct dosage and for an amount of time conducive to bacterial clearance or, in the case of surgical prophylaxis, at levels shown to prevent bacterial contamination [6]. Antibiotics should be administered intravenously at a rate that ensures that the correct antibiotic concentration is achieved in blood and tissues. If necessary, antibiotics can be administered during surgical procedures and the antibiotics selected should be used short-term and not utilized post-surgery. Some researchers believe that a single dose of antibiotics can be used to treat purulent *S. aureus* infections when combined with debridement, irrigation and primary sutures. However, systemic antibiotic use is unnecessary following wound closure, particularly since antibiotic use at this juncture will result in microflora imbalances and will select antibiotic resistant organisms without reducing wound infection rates. Early use of antibiotics for the prevention of infections associated with open fractures, however, is well accepted. Studies have indicated that bacterial contamination of open fracture wounds likely occurs within the hospital environment and early use (3 – 5 days) is recommended prior to treatment. Lee *et al.* reported that positive rates of bacterial cultures from fracture wounds were not high before debridement, are not predictive of infection nor did they reflect the species of bacteria that may be the causative agent of an infection (if it develops) [7]. In summary, repeated, thorough debridement is key to preventing wound infections and even prophylactic antibiotic use cannot replace debridement as the primary anti-infective strategy prior to suture use.

19.2.2.2 Treatment of Wound Infections

Effective clinical diagnosis and treatment of wound infections is needed to prevent systemic disease and permanent tissue damage. Wound infections are generally characterized by inflammation, redness, heat, pain, and when pus is present it is indicative of an infection. Cellulitis around the wound is more common. In addition, persistent tissue necrosis around the wound is another sign of wound infection. A key first step to wound infection treatment is thorough pus drainage. Mechanical removal of fibrin deposition and bacterial debris is critical to the healing process. Typically, these procedures are sufficient and do not require the systemic use of antibiotics, assuming that wound treatment is initiated in a timely manner. However, more serious cellulitis or persistent necrosis of the surrounding tissues will necessitate the use of antibiotics, fluid replacement therapy and supportive therapy (following debrediment) [8].

Two severe soft tissue infections must be considered separately due to their severity.

Acute necrotizing fasciitis. Acute necrotizing fasciitis is a form of severe subcutaneous tissue and fascia necrosis accompanied by severe systemic symptoms [9]. The most common cause of this form of infection is trauma, especially deep penetrating wounds that physically localize microorganisms to the fascia surface where there are reduced numbers of blood vessels surrounding the muscle bundles of the fascial spaces. Therefore, proliferating microorganisms can spread rapidly at the fascia surface, sometimes resulting in fascia necrosis. Surgical contamination of hollow organs as a result of inappropriate antibiotic use and serious primary suture infections can often result in necrotizing fasciitis. Infections resulting in necrotizing faciitis are the result (in most cases) of mixed infections comprised of both aerobic and anaerobic bacteria resulting in the production of fetid pus production. Rea and Wyrick confirmed the presence of gram-positive bacteria (including *Streptococcus* and *S. aureus*), gram-negative bacteria and anaerobic bacteria in necrotizing faciitis tissues. The most prominent clinical manifestations of disease are fascia necrosis, cellulitis, edema, skin discoloration and gas gangrene. Symptoms associated with disease onset include (at the site of infection) reddening, inflammation, heat, pain and rapid dissemination of the infection. Severe cases can result in sepsis, toxic shock and death.

Necrotizing faciitis cases are not common but are progressing rapidly. An additional complication is that early symptoms are not necessarily indicative of the severity of the infection that can result in a delayed diagnosis that in turn leads to higher mortality rates. The most important parameter in recognizing the symptoms of early necrotizing faciitis is physician awareness. Extensive necrosis of subcutaneous fat and fascia and subcutaneous, tunnel-like changes in the absence of muscle involvement (early) are important diagnostic signs. Cellulitis with hemorrhagic spots on the skin are mostly a consequence of group A streptococcal infections. The main manifestation is severe pain that can be caused by gentle palpation of the wound site, with pain extending to the edge of trauma areas. Cases may present little or no purulent secretions but systemic symptoms are more serious than those at the wound site. Bacteriological examination of pus smears is important in diagnosis.

Early debridement and drainage are key to the treatment of this infection type, including complete removal of necrotic tissues followed by repeated debridement and drainage as needed. The use of antibiotics is based on bacterial diagnosis from smear results initially, followed by modifications following more sensitive bacterial culture and antibiotic sensitivity testing. Fluid replacement therapy and nutritional support are provided as necessary.

Clostridium muscular necrosis (gas gangrene): Clostridium muscular necrosis (gas gangrene) is a rapid developing, severe, acute infection caused by infections with *Clostridium perfringens*; a gram-positive, short and thick spore-forming anaerobic bacteria. This type of infection is common following severe trauma, especially in patients with open fractures, significantly contaminated muscle wounds and infections in tissues with poor blood circulation. In addition, abdominal surgery may lead to Clostridium muscular necrosis, due to cross-infections or endogenous infections typical of intestinal surgery. *C. perfringens* may disseminate to deeper tissues following stab wounds or other injuries to deeper tissues resulting in gas gangrene. The incubation period of this infection is shorter, *i.e.* 3 or more days after injury is common. Wound pain is one of the first symptoms to present itself and is severe and difficult to control. Edema develops around the wound site and the skin becomes pale, shiny, rapidly turns purple and finally turns a dark gray, accompanied by the emergence of large and small blisters containing foul-smelling serous or blood liquid. Crepetation can be heard following palpation of most soft tissue infection cases. Due to deep compartment infections, limbs can be lost in a short time period. In addition, severe systemic manifestations can present themselves as a consequence of exotoxin production.

Gas gangrene diagnosis is usually not difficult, but physicians need to remain vigilant to diagnose this type of infection. Early diagnosis and treatment are critical not only to rescue affected limbs but also to the patient's life. Palpable crepitation around the wound, gas discernable in muscle X-ray and thick gram-positive staining of secretion smears are the main diagnostic criteria. Once disease is diagnosed, it must be treated actively and rapidly to control the infection.

Due to the rapid progression of necrotizing fasciitis, timely and extensive debridement is needed to remove necrotic tissues and foreign bodies. Systemic antibiotic therapy is also recommended and hyperbaric oxygen therapies may be used as surgical adjuvant therapy.

19.3 Intra-Abdominal Infections Associated with Surgery and Trauma

Intra-abdominal infections result from significant damage to the host's abdominal cavity caused by the invading microorganism. Intra-abdominal infections resulting from surgical trauma, also known as surgical intra-abdominal infections, are infectious diseases of the abdominal cavity developing after abdominal surgery or trauma and requiring surgical interventions as part of the treatment process. Intra-abdominal infections are caused by gut flora (most commonly *E. coli* and anaerobic *Bacteroides sp.*) often present with suppuration and necrosis resulting in the destruction of the intestinal architecture and needing surgical treatment such as incision and drainage or puncture and drainage as part of the treatment process.

19.3.1 Acute Peritonitis Resulting from Surgery and Trauma

Peritoneal cavity infections represent the most important complication associated with intra-abdominal infections and are also the most difficult to diagnose and treat. Trauma, combined with intra-abdominal infections is commonly observed in the gastrointestinal tract, biliary tract, bladder and other hollow organs. Abdominal infections post-surgery develop as a consequence of colon content leakage, leakage of GI tract anastomosis, biliary or pancreatic leakage or perforations to the colon due to endoscopic procedures. Because there are a small number of bacteria within the stomach and duodenum, the incidence of severe intra-abdominal infections is low when injuries occur at these sites. However, the incidence of intra-abdominal infections is relatively high, due to the large amounts of intestinal bacteria in the distal gut ($>10^{10}$/g of feces in the colon and rectum).

Pathological changes associated with peritonitis developing post-surgery and trauma depend on the bacterial species present, its concentration, virulence, host immune status and how rapidly treatment is administered. For patients with intestinal rupture, the longer the operation time the greater the likelihood of bacterial dissemination within the abdominal cavity.

Intra-abdominal infections are multi-bacterial in nature, comprised of aerobic and anaerobic bacteria. Most strains belong to bacteria common to the human intestinal tract, and skin. *E. coli* is the dominant aerobic bacteria, and *B. fragilis* the dominant anaerobe associated with these types of infections. Infections caused by gram-negative bacilli are more common following surgery of the GI tract and pathogenic bacteria associated with infections of the spleen or pancreas are mainly caused by *staphylococci*. Bacterial virulence and pathogenicity resulting from mixed infections can be significantly increased as a consequence of bacterial synergy. Aerobic bacteria utilize oxygen from the surrounding environment, making the environment conducive to anaerobic bacterial growth. In turn, anaerobic bacteria release enzymes, growth factors and host response inhibitors that facilitate survival and propagation of aerobic bacteria. In addition, aerobic bacteria also provide vitamin K used for growth and reproduction of anaerobic bacteria. Furthermore, mixed infections reduce the efficacy of antibiotic therapies.

When non-virulent bacteria are introduced into the abdominal cavity of an immunocompetent host, post-surgical and post-traumatic peritonitis is transient. Intra-abdominal abscess formation is a sign of localized inflammation associated with bacterial clearance or control of the infection and will only form when the bacterial concentration present in the abdominal cavity is too large to be cleared by the body's defenses.

Fever, with or without chills, may be the only manifestation of abdominal infections. Some gram-negative bacterial infections may only be present with a low-grade fever, with maximum temperatures in the afternoon or evening. Pain may be helpful in determining the site of infection. However, pathogen characterization will require bacterial culture of drained samples or pus, using methods conducive to the growth of both aerobic and anaerobic bacteria.

Prompt repair of the wound and closure of the damaged intestinal tract and the immediate use of broad-spectrum antibiotics capable of clearing both aerobic and anaerobic bacteria may be effective in preventing or reducing post-traumatic acute peritonitis.

Correct surgical procedures are critical to the prevention of infections resulting from colon injury. Adopted surgical methods that can affect the outcome include fecal bypass procedures that depend on the type of wound, severity, patient health, how long the wound has been open, the extent of the intra-abdominal infection, combined injury to other organs, shock and the technical level of the surgeon. Clinical treatment should be individualized. For example, if a patient is generally in good health, primary surgical resection should be adopted. In addition, regardless of the type of surgery, immediate closure of the rupture is imperative, followed by thorough debridement, early full washing and postsurgical drainage. Close attention must be paid to anastomosed regions, especially to pressure and blood supply. The importance of emptying the bowel cannot be ignored. Because preoperative bowel preparation cannot always be carried out prior to emergency surgery, it is important to carry out a complete GI decompression and bowel irrigation. Additionally, omentum or fat can be used to cover respective anastomosis to reduce the incidence of anastomotic leakage. In order to reduce blood clots, thorough and repeated washing of the abdominal cavity with the right temperature saline solution during the operation is needed to remove foreign bodies and reduce the number of bacteria present.

19.3.2 Intra-Abdominal Abscesses after Surgery and Trauma

Intra-abdominal abscess formation, including subphrenic abscesses, pelvic abscesses and interbowel abscesses are comprised of a localized accumulation of liquor puris in the abdominal cavity due to liquefaction necrosis of surrounding tissues such as the bowels, viscera, the abdominal wall, omentum or mesenterium. Intra-abdominal abscesses often occur after abdominal surgery and trauma. X-ray and radionuclide scanning are helpful in their diagnosis and BUS and CT scans can accurately locate the majority of abscesses and guide the puncture and catheter drainage procedures. The pathogenic bacteria associated with abdominal abscess are different and typically abscesses are comprised of intestinal gram-negative bacilli and obligate anaerobes that are sensitive to conventional antibiotic therapies. Some intra-abdominal abscesses in patients with intra-abdominal drainage tubes can be caused by staphylococci. *Pseudomonas sp.* and other drug-resistant bacterial strains associated with other nosocomial infections can also lead to intra-abdominal infections following long-term use of antibiotics. The identification of pathogenic bacteria is dependent on bacterial culture.

In order to prevent postoperative intra-abdominal infections and abscess formation, preventive measures need to be in place and include careful surgical procedures, protecting the surgical field, preventing and reducing endogenous bacterial contamination. Bacteria can be cultured from the abdominal cavity if surgical procedures are too long and the abdominal cavity is exposed to environmental contaminants during this time. Antibiotics should be administered before surgery to ensure adequate antibiotic concentrations in tissues.

Treatment of intra-abdominal abscesses includes nutritional support, antibiotic therapy and adequate drainage. Intra-abdominal abscesses may be absorbed by the body if early antibiotics and supportive therapy are administered. BUS superpositioning puncture and drainage are simple and effective treatments for a subphrenic abscess, deep abscess and small abscess filled with thin pus. However, surgical drainage is applicable for an abscess with large cavities, thick-walls and/or multilocular.

19.4 Enterogenic Infections Associated with Post-Surgery and Trauma

Large numbers of clinical studies have shown that enterogenic infections are the most common endogenous infections associated with surgery and trauma. Enterogenic infections are not only associated with the development of refractory shock caused by severe trauma and refractory sepsis, but also play an important role in the pathogenesis of MODS.

Humans are colonized on their skin, genitourinary tract, GI tract, oral cavity and lung epithelial surfaces by a myriad of different types of bacteria that constitute the normal flora. The nature of this flora has evolved with humans and serves various beneficial purposes at respective anatomic sites. The quantity of microorganisms comprising the human intestinal flora is larger than that in any other organ in the body and the number of bacteria (10^{14}) is 10 – 20 times the number of human cells comprising the human body. Intestinal microflora form a biological intestinal barrier that protects the intestinal epithelium from colonization with pathogenic bacteria, participates in host metabolic processes and contributes nutritional value that promotes (host) growth and health. In 2001, Hooper *et al.* found that the major human intestinal bacteria formed part of the host biological and immune barriers that contribute to intestinal growth, symbiotic bacterial metabolism, nutrient absorption and angiogenesis. Therefore, the balance and health of the intestinal microflora is a sign of health. The functional integrity of the normal intestinal flora (and the intestinal barrier they provide) has an important role not only in preventing bacterial infections, but also in diminishing inflammation, ulcer formation and preventing allergic and neoplastic diseases.

When the intestinal barrier function is impaired, intestinal bacterial translocation can occur. The currently accepted view is that bacterial translocation is the process by which microorganisms and their respective toxins cross the intestinal wall into otherwise sterile tissues, including the mesenteric lymph nodes, portal vein and other distant organs or systems, providing primary bacterial foci that contribute to various infections. Following surgery and trauma, the intestinal barrier function can be impaired due to shock, hypoxemia, malnutrition and reduced immune function which facilitate intestinal flora translocation leading to intestinal infections. Therefore, understanding the structure and function of the intestinal barrier will help with the control of intestinal infections.

19.4.1 Intestinal Barrier Function

Intestinal barrier function indudes structural basis of the intestinal barrier and intestinal barrier dysfunction.

19.4.1.1 Structural Basis of the Intestinal Barrier

The intestinal barrier is comprised of mechanical, chemical, biological and immunological barriers. This complex barrier system effectively limits invasion of intestinal flora into extra-intestinal sites and constitutes an extremely important part of the body's defense system. Intestinal barrier integrity is an important factor in preventing bacterial or toxin translocation out of the intestine, although some degree of bacterial translocation constitutes the normal physiological process and plays an important role in the formation and maintenance of intestinal and systemic immune systems.

Mechanical barrier. This barrier consists of intestinal epithelial cells, tight junctions and plaque. The intestinal mechanical barrier is effective in preventing deep penetration of bacteria into mucosal tissues and establishes the physical architecture of the intestinal barrier. The intestinal epithelium consists of cells associated with gut absorption, goblet cells, Paneth cells and some endocrine cells. Gut absorption takes place on luminal surfaces and adjacent cells are connected by junctional complexes including tight junctions, gap junctions, adhesion junctions and desmosomes. Tight junctions are the most important type of cell-to-cell connection and can be of two types: (i) Villus epithelial tight junctions which have a smaller pore size and complex structure allowing only water molecules and other small molecules to selectively pass between cells; (ii) Duct cell tight junctions which have larger pore sizes and are simple in structure, allowing larger molecules to pass between cells. Tight junctions form a narrow band structure located outside of the top of the epithelial cell membrane. Adjacent cells form a fusion or match point, connected to form a continuous fishing-net-like structure. There are many proteins involved in the formation of tight junctions, including structural proteins (*e.g.* occludin, claudin, and JAM) and regulatory proteins (*e.g.* E-cadherin, actin, myosin, and cingulin). Tight junctions not only maintain cell polarity, but also selectively modulate the passive transfer of ions and macromolecules, forming a physical permeability barrier that limits bacterial and bacterial product transfer, making tight junctions important intestinal mechanical barrier components.

Chemical barrier. This barrier is comprised of intestinal mucus secreted by intestinal epithelial cells and a variety of digestive and intestinal digestive enzymes mediating various chemical reactions. Mucus lubricates the intestinal mucosa and protects it against mechanical and chemical injury. Glycoproteins and glycolipids secreted by intestinal goblet cells can be combined with bacteria, so that bacteria are discharged with intestinal fluid under the force of fluid flow. Lysozyme cleaves gram-negative cell-wall peptidoglycan bonds resulting in bacterial lysis. Bile can combine with endotoxin to form detergent-like complexes that prevent endotoxin absorption through the intestine and cholic acid can degrade endotoxin molecules. Secretory IgA (SIgA) present in bile can opsonize bacteria preventing their attachment to epithelial surfaces. These broad-ranging components are important members of the intestinal barrier system.

Biological barriers. The GI tract is sterile in fetuses and following exposure to the external environment after birth the neonatal intestine becomes colonized *via* the oropharynx which then leads to GI tract colonization which is also exposed to microbes present in food. The small intestine and the colon are the last intestinal components to become colonized. Bacterial growth and reproduction in the colon establish an important intestinal biological barrier. Intestinal flora under physiological conditions remain relatively stable and the intimate interactions of the various bacterial populations between themselves and the intestinal epithelium create a physiological bacterial barrier consisting mainly of intestinal anaerobic bacteria combined with intestinal mucosa that prevents colonization with potential pathogens. However, the use of broad-spectrum antibiotics can cause bacterial imbalances, tipping the scale in favor of antibiotic-resistant pathogens not part of the normal flora. In summary, the normal flora maintains a dynamic balance in the intestine by preventing colonization with pathogens (contact inhibition), competing with pathogens for nutrients and producing bacteriocin and other anti-microbial factors.

Immunological barrier. This barrier is comprised of the intestinal immune system that neutralizes pathogenic bacteria and their products *via* cellular and humoral responses. Lymphocytes comprise one of the primary effector cell types of the immune system and the body's largest lymphoid tissue, the intestinal mucosa-associated lymphoid tissue (MALT), is situated in the gut. There are two different intestinal lymphocyte phenotypes: intestinal intraepithelial lymphocytes (iIEL) scattered throughout the intestinal epithelium and submucosal tissues rich in lymphocytes. Peyer's patch lymph nodes are another important lymphoid tissue. The cavity surface of Peyer's patch lymph nodes is covered with follicular epithelium which constitutes a single-cell complex comprised of columnar epithelial cells, goblet cells and microvilli. This complex forms lymphoid follicles under the epithelium. SIgA is the major intestinal immunoglobulin. The main function of SIgA is antigen binding that prevents antigens from passing through the intestinal mucosa. In addition, intestinal mucosal cells may also secrete IgE, IgM and IgG that are also important in mediating humoral immunity in the intestine.

Intestinal barriers made up of mechanical, chemical, biological and immune barriers that are key to host defenses against intestinal bacteria and endotoxin translocation. Any structural changes to these barriers are likely to lead to intestinal barrier dysfunction resulting in bacterial translocation.

19.4.1.2 Intestinal Barrier Dysfunction

Understanding the complexities associated with intestinal dysfunction can be extrapolated from understanding intestinal failure, a concept proposed by Irving in 1956. Intestinal failure was defined as 'the overall reduction of functional gut, unable to digest and absorb food' as was observed in patients with short bowel syndrome (SBS). In 1981, Fleming and Remingtor stated that 'it was hard to maintain intestinal minimum needs of digestion and absorption of nutrients as the intestinal function fell'. In 1986, Carrico stated that 'the gut is the start organ of multiple organ failure'. In 1992, in the diagnosis criteria of multiple organ failure, Deitch established standards of intestinal injury, namely, bloating and food intolerance for more than 5 d and that the standard of failure should be stress ulcers and acute cholecystitis. In 1996, Wilmore proposed that the intestine is one of the primary organs affected in patients under stress and, in 2001, Nightingale stated that 'due to the reduction of intestinal absorption, supplementation of nutrition and water, electrolytes are needed to maintain health and (or) growth'. Although nearly 50 years have passed, our understanding of the intestinal function is still limited to the processes of digestion and absorption. In the 1980s, the basic understanding of the intestinal function was its role in food digestion, nutrient absorption and GI hormone secretion. When the body is under stress, the gut is in 'hibernation'. During shock, intestinal blood is redistributed to the liver, lungs, kidneys and other organs. In the 1970s, when our understanding of 'multiple organ failure' began to be better defined, there were no established criteria for defining intestinal failure. During the 1980s, it was noted that burn victims not suffering from wound infections could have bacteria recoverable from the blood. Furthermore, these bacteria were derived from the intestine and described as an 'enterogenic infection'. It has been confirmed in animal experiments that the intestinal mucosa has barrier function properties and that hypoxia and ischemia can impair the intestinal mucosal barrier function, defined by bacterial and endotoxin translocation into the lymph or blood circulation [10]. This phenomenon can be observed in animal models but in the human body it is still difficult to obtain direct proof. However, many recent reports indirectly confirmed the existence of intestinal bacterial and endotoxin translocation in the human body. Surgical procedures or illnesses contributing to the development of intestinal hypoxia and ischemia are associated with gut mucosal barrier dysfunction resulting in bacterial translocation associated with SIRS and sepsis that can progress to MODS.

Attention must be paid to complications associated with endotoxin and bacterial translocation that can result in MODS or facilitate endogenous infections (bacterial, fungal infections). Intestinal function definitions are no longer limited to descriptions of digestion and nutrient absorption, but also include descriptions of the intestinal barrier function. It has therefore been suggested that the definition of intestinal dysfunction should be 'the damage of intestinal substance and (or) intestinal function, resulting in the serious disorder of digestion and absorption of nutrients and (or) barrier'. It is involved in the patho-physiological changes of the body under stress and considered 'the central organ of body stress', 'the motor of multiple organ dysfunction'.

It is generally understood that the intestinal barrier function describes the intestine as preventing bacterial translocation into the body from the intestinal lumen. In recent years, a large number of studies on the epithelial barrier function were performed. It has been widely believed that ischemia and hypoxia-induced bowel dysfunction are associated with intestinal necrosis. Under conditions of bowel dysfunction resulting from malnutrition, atrophy of the intestinal mucosa can occur and the villi height decreased. Intestinal permeability can be defined using the L/M (mannitol/lactulose) test that allows physicians to determine whether damage to the intestinal epithelial barrier has occurred. Although the clinical importance of maintaining an adequate intestinal barrier function is better appreciated today, further study is needed regarding mechanisms mediating this process that will allow for earlier diagnosis of gut barrier dysfunction.

To date, clinical definitions and methods of monitoring the intestinal barrier function are not well established. Indirect methods of monitoring the intestinal barrier function are widely used however, and include determination of intestinal permeability to macromolecules to indirectly confirm bacterial translocation. Clinical measurements of intestinal permeability to macromolecules can be carried out using a variety of water-soluble probe molecules that cannot be metabolized, can partly pass through the mucosal barrier and are quickly recoverable in urine. Commonly used molecular probes include PEG400 (5.3Å), mannitol (6.7Å), rhamnose (8.3Å), lactulose (9.5Å), cellobiose (10.5Å), ^{51}Cr-EDTA (11Å), 99mTc-DPTA (11Å). PEG400, mannitol and other small molecules can directly pass through villus epithelial tight junctions. Data derived from using these probes were used to estimate villus epithelial tight junction permeability. Lactulose, 99mTc-DPTA and other large molecular probes are mainly absorbed through duct cell tight junctions and the recovery of these probes is a reflection of duct cell tight junction permeability. Combined use of small and large molecular probes, such as the dual-sugar method M/L ratio method can be used to assess the permeability of two types of tight junctions. In recent years, "Pulsed Electrochemical Detection in High Performance Liquid Chromatography (HPLC-PED)" represents the most advanced method of measuring M/L permeability. Studies have shown that the M/L ratio correlated with the degree of trauma. In recent years, reports have also demonstrated that serum D-lactate and diamine oxidase (DAO) concentrations reflected intestinal barrier permeability. The using Chamber is another instrument used for studies assessing intestinal drug absorption and intestinal barrier function and is considered the most effective and direct advanced equipment to monitor the electrophysiological changes occurring in intestinal mucosal cells.

19.4.2 Intestinal Barrier Function and Bacterial Translocation

Intestinal bacterial translocation does not always result in a state of pathology. Data show that bacteria can be found in the mesenteric lymph nodes of healthy individuals with incidence rates of 5% – 10%. Limited bacterial translocation is likely to be a normal physiological process associated with luminal antigen processing by the MALT as a means of regulating immune responses against luminal antigens. Under pathological conditions, significant bacterial translocation may likely cause a series of patho-physiological changes and sepsis.

It has been hypothesized that bacterial translocation is most likely to take place across ileal and colonic epithelium. However, it was recently demonstrated that bacterial translocation rates were equal in the jejunum, ileum and colon using ^{14}C-labeled *E. coli*. This study also demonstrated that bacteria translocated from the jejunum were more likely to survive.

Intestinal flora that colonized epithelial mucosal surfaces makes it difficult for pathogenic bacteria to colonize and translocate. Once the micro-ecological balance is disrupted as a consequence of antibiotic use or intestinal obstruction, however, the dominant bacterial population (usually gram-negative bacteria) may breach the mucosal barrier and translocation is then likely to occur. The most common bacteria associated with translocation are aerobic bacteria such as *E. coli*, *Klebsiella sp.* and *Proteus sp.*

Typically, bacterial translocation occurs after damage to the mucosal mechanical barrier takes place, in which case bacteria move from the mucosal layer into the tissue, adhere, colonize and translocate. If mucosal integrity is maintained, bacteria entering the peritoneum do so *via* endocytosis, followed by transfer by subepithelial macrophages to mesenteric lymph nodes where they are exocytosed. Bacteria not killed in macrophages by this process are free to elicit peritoneal infections. The functional status of macrophages, as carriers of bacteria, is considered to exert influence on bacterial translocation. However, in cases of trauma, shock or stress, the situation is different. Due to ischemia (reperfusion injury), the mucosa can suffer varying degrees of damage, including micro-structure changes and cell tight junction damage which are exploited by bacteria in the translocation process leading to infection. Bacterial endotoxins can more easily pass through the mucosal barriers compared to whole bacteria, so bacterial translocation as a result of stress is often accompanied by endotoxemia which in turn causes further damage to mucosal barriers and further increases mucosal permeability facilitating bacterial translocation.

However, endogenous bacterial translocation does not mean infection will develop since this process can be still impeded by the intestinal immune barrier. The induction sites of the immune barrier include the GALT (gut associated lymphoid tissue), MALT, isolated lymph nodes and the appendix. Immune effector sites include the intestinal lamina propria and intraepithelial lymphocytes. Intestinal tract-derived antigens can reach resting B cells present in the lymph nodes which in turn can migrate (via the lymphatic circulation) to the mesenteric lymph nodes and thoracic duct where they can in turn take up residence in the intestinal lamina propria. This process results in the production of IgA-positive B cells that continue to expand and differentiate into mature IgA-positive plasma cells that produce polymeric IgA that can in turn be translocated to luminal surfaces in the form of SIgA. SIgA can also be released into the intestine in bile. Antigen-specific SIgA can then opsonize bacterial targets forming antigen-antibody complexes that prevent adhesion and invasion of intestinal epithelial cells.

T lymphocytes can also differentiate and mature in response to intestinal antigens. Mature $CD4^+$ T cells make up a considerable percentage of lamina propria lymphocytes and $CD8^+$ T cells make up 5% – 15% of the intestinal epithelial cell barrier, producing a series of cytokines that include IL-2, IL-6, TGF-α, TNF-α, and IFN-γ which help prevent bacterial translocation. However, under conditions of severe traumatic stress, chemotherapy, radiation therapies or local or systemic immune suppression, large numbers of bacteria will break through the intestinal barrier, translocate into blood and distant organs causing intestinal infections resulting in MODS. MODS in turn increases damage to the intestinal mucosa, further increasing bacterial translocation thereby feeding a vicious cycle.

Intestinal infections have their own characteristics in that they are primarily caused by bacteria normally colonizing the human body, and infections are caused primarily by anaerobic and aerobic mixed infections (with a higher proportion of anaerobic bacteria).

19.4.3 Clinical Manifestations of Enterogenic Infections

Intestinal bacterial translocation into mesenteric lymph nodes, portal circulation and other distant organs or systems is a gradual process that occurs as follows.

19.4.3.1 The Sub-Clinical Infection Stage

Bacterial translocation limited to the mesenteric lymph nodes results is in a stalemate. Infection does not typically develop or clinical manifestations are not obvious although fever may occur.

19.4.3.2 Septic Stage

As the bacteria break through the local defense barrier of mesenteric lymph nodes, pathogens continuously secrete toxins or endotoxins that are released into the circulation, stimulating systemic cytokine responses. If untreated, the resulting cytokine storm results in septic shock accompanied by respiratory distress. At this time point bacteria have not yet entered into the circulation and blood cultures are negative for bacterial growth.

19.4.3.3 Bacteremic Stage

In this stage, bacterial containment fails and bacteria break through all barriers into the blood, resulting in culture-positive blood cultures. This comprises the late stage of enterogenic infections.

19.4.3.4 SIRS or MODS Stage

Systemic inflammatory responses result as a consequence of severe infections characterized by body temperatures > 38 ℃ or < 36 ℃, heart rate > 90 beats/min, respiration rates of > 20 times/min or $PaCO_2$ < 4.3 kPa (32 mmHg), white blood cell counts > 12×10^9/L or < 4×10^9/L or immature neutrophils > 10%. MODS can be simultaneous or sequential, *i.e.* dysfunction of two or more organs at 24 h after serious infection or trauma. MODS is the most serious pathological process and the final outcome.

19.4.4 Prevention and Treatment of Enterogenic Infections

Clinically, it is important to maintain and improve intestinal mucosal permeability and barrier function and to prevent intestinal infections resulting from bacterial translocation. Intestinal integrity can be maintained by improving oxygen supply and blood flow to affected areas to meet the metabolic needs of respective tissues, prevent intestinal bacterial overgrowth using selective intestinal decontamination methods, maintain intestinal integrity by means of maintaining enteral nutrition and maintaining the gut microflora balance *via* probiotic use [11].

The intestinal mucosa is particularly sensitive to hypo-perfusion due to its high metabolism and capillary structure. After low blood volumes are corrected, decreased GI blood flow and visceral vascular contractions will continue for some time. The effect of adrenalin or other vasoactive drugs on intestinal perfusion is not clear; visceral vascular responses in individuals also vary significantly. During sepsis and SIRS, cardiac output is not representative of visceral blood flow. Visceral tissue perfusion can be re-distributed by administering vasoactive drugs. Despite improvements to the blood flow, there may still be a shortage of local perfusion. In general, increased cardiac output can increase visceral blood flow in patients with no impaired vascular regulatory mechanisms. Therefore, small doses of dopamine can then be used to improve visceral blood flow. In addition, improving intestinal mucosal perfusion also improves metabolism and further reduces intestinal mucosal damage. Therefore, improving the efficacy of intestinal mucosal perfusion therapies is key to protecting the intestinal barrier function.

In critically ill patients, another feature associated with intestinal barrier dysfunction is uncontrolled intestinal bacterial overgrowth which can be treated by antimicrobial drugs such as neomycin, gentamicin, metronidazole or by application of mechanical cleaning methods to remove bacteria which results in a reduction in bacterial translocation. These methods are still recommended in the clinic, although there have been no strict clinical control reports to date, and the benefits from these treatments remain inconsistent.

The nutritional status of the intestine can be affected by hunger, excess nutrients, weight loss of more than 10% of normal body weight, and these respective changes can alter intestinal morphology and function.

Nutrients within the intestinal lumen play an important role not only in affecting the growth and function of intestinal epithelial cells, but also in stimulating the secretion of GI hormones that also promote intestinal cell growth. Therefore, in clinical situations where the GI function remains intact, early enteral nutritional interventions should be implemented in postoperative patients or critically ill patients after recovery. This can be carried out using a nasal feeding tube or *via* a jejunostomy tube when patients are incapable of orally taking in food. To promote cell growth and to regulate the intestinal immune function, arginine, glutamine, fish oil and nucleotides can be incorporated into food sources [12]. Animal experiments suggested that epidermal growth factor and insulin-like growth factors could stimulate protein synthesis and intestinal cell proliferation. It has now been confirmed that early enteral nutrition plays an important role in the maintenance of the normal mucosal structure and barrier function, preventing intestinal flora disorders, enhancing host resistance to infection and preventing hypermetabolism after surgery and trauma. Early enteral nutrition is an important countermeasure in the control of enterogenic infections.

Bifidobacteria comprise the dominant probiotic used in humans, can attach to the intestinal epithelium through teichoic acids, and can form a natural biological barrier with anaerobes such as *Lactobacilli*. *Bifidobacterium* and *Lactobacillus* produce acid as a metabolic by-product that lowers intestinal pH that negatively impacts the survival and proliferation of gram-negative bacteria and reduces the production and absorption of endotoxin (therefore reducing plasma endotoxin levels). In addition, oral administration of *Bifidobacteria* can increase IgA secretion, enhance non-specific phagocytosis and enhance local intestinal immunity. Furthermore, *Bifidobacteria* produce amino acids, cell proteins and vitamins that can be used by humans. Therefore, probiotic organisms like *Bifidobacteria* can play an important role in maintaining the ecological balance of the intestine, generating an environment that is not conducive to pathogen growth, protecting the intestinal mucosal barrier and improving the immune function [13]. It has been confirmed that oral administration of *Bifidobacterium* can be effective in preventing post-traumatic enterogenic infections.

19.5 Prevention and Prognosis of Surgical and Traumatic Infections

A combination of factors today has compounded the difficulties associated with treating surgery and trauma-related infections. Some of these compounding factors include a growing population of elderly patients, patients undergoing chemo- and radio-therapies, increased incidence of diabetes and cancer and a growing immune-suppressed population. These risk factors, combined with the weakened resistance of surgical patients following the use of anesthesia and surgical trauma, traumas caused by a variety of accidental injuries, the increased number of major surgeries and increased frequency of prosthetic surgical procedures, have significantly increased the chances for patients to develop intra-abdominal infections. Despite the use of multiple new antibiotics and novel treatment methods, sepsis and other complications caused by traumatic infections are still difficult to prevent and treat. In the United States, severe wound infections leading to sepsis and shock (especially multiple organ failure) are the 10th leading cause of death in surgical patients. About 30% of surgical ICU patients suffer from sepsis leading to multiple organ dysfunction or failure that has become one of the major causes of death in surgical patients. It is clear from these observations that surgeons need to focus on establishing novel prevention strategies designed to curb the frequency of surgery and trauma-related infections.

19.5.1 Prevention of Post-Surgical and Post-Traumatic Infections

We will introduce the following measures to prevent post-surgical and post-traumatic infections in this section.

19.5.1.1 Improvement of Pre-Surgery Prophylaxis

It has been proven that strengthening preoperative treatments can significantly reduce the incidence of surgery and trauma-related infections. Awareness of surgical asepsis must be emphasized and operating room management must be strengthened [14]. In particular, emergency surgeries, preoperative hospital stays, operation times and the patient's health must be considered in the context of bacterial infections. The risk of surgical site infections can be significantly reduced if the following risk factors are addressed [15].

Maintaining normal temperature. Independent of the use of general, spinal or epidural anesthesia, patient body temperature can decrease by 1 – 3 ℃, resulting in significant reductions in the immune function and reduced surgical site blood flow, resulting in an increased risk of SSIs. Therefore, the use of thermal equipment and measures to maintain body temperatures at or above 36 ℃ during surgical procedures is recommended.

Optimizing oxygen tension. Post surgical administration of 80% oxygen has been shown to reduce post-surgical wound infections and postoperative nausea and vomiting rates [16]. Therefore, it is recommended that intubated surgical patients receive 80% oxygen (>12 L/min) *via* inhalation using a non full-face breathing mask for up to 2 hours post surgery.

Blood glucose control. Maintaining blood glucose levels at a stable level is especially important for patients undergoing coronary artery bypass graft surgery. Furthermore, control of pre-surgical blood sugar levels is particularly critical since elevated blood glucose increases the risk of postoperative infections (POI). Hyperglycemia has been linked to leukocyte function abnormalities, including granulocyte adherence, impaired phagocytosis and delayed chemotaxis. These leukocyte deficiencies are associated with increased infections and infection risks have been shown to be reduced if glycemic control is maintained. Using a closed-looped artificial pancreas system to maintain constant glycemic levels without increasing the risk of hypoglycemia might be a safe and useful blood glucose control system in critically ill surgical patients [17]. At present, however, the preoperative optimal target range for blood glucose to reduce POI remains undefined.

Depilation of the surgical site. Traditional shaving one day prior to surgery has proven ineffective in preventing infections and in some cases promotes colonization of the shaved sited since some microorganisms colonize shaving-associated micro abrasions. Surgical areas that have limited amounts of hair do not necessitate shaving and areas with more hair can be shaved using clippers or an electric razor and this can be carried out immediately before surgery in the operating room (particularly if razors are to be used, *e.g.*, craniotomy).

Surgical environment. It is important to strictly abide by the principles of aseptic surgery. Infections of the incision site are related to loss of live tissues, residual foreign bodies, blood clots or dead space. To date, wound irrigation with antibiotic-containing solutions is not common practice.

Hospital stay. A simple, cost-effective means of reducing infections is to remove patients from hospital environments as quickly as possible so as to remove them from an environment containing an array of pathogens capable of infecting the surgical site.

19.5.1.2 Prophylactic Antibiotic Therapies

Antimicrobial prophylaxis for the preventions of SSI is one of the most widely accepted and used surgical practices. Effective prophylaxis requires adhering to safe surgical practices in combination with the appropriate antimicrobials administered in correct dosages for finite periods of time.

There is no question that antibiotics serve as an effective prophylactic strategy for the prevention of SSI but they are not needed for all operations. Generally, Class I incisions (clean incisions) are carried out under strict aseptic conditions and should not require prophylactic antibiotic use. Prophylactic antibiotics are used for Class II incisions (clean incision) and the less contaminated Class III incisions. Most serious contamination is associated with Class III and IV type incisions. These include open wounds and gastrointestinal perforations that should be treated with antibiotics before and after surgery.

Specific indications of prophylactic antibiotic use. (i) Class II incision cleaning involves wounds with contamination at the wound site, typically involving surgery of the GI tract, oropharynx, respiratory tract and the female reproductive tract. (ii) The use of artificial materials or manually operated devices, such as artificial heart valve replacement surgery, artificial vascular grafts or artificial joint replacement surgery. (iii) Major surgery or surgery lasting for many hours, trauma patients or in the case of infections, serious, consequences such as craniotomy and heart surgery, establishing portosystemic shunts or carrying out splenectomies. (iv) Patients with risk factors for infection, such as old age, diabetes, immune dysfunction or malnutrition.

Parameters associated with prophylactic antibiotic usage. (i) The first dose should be administered before incision. The timing of administration is critical and should start 20 – 30 min before surgery (during induction of anesthesia) to ensure delivery prior to any potential contamination and so that appropriate antibiotic tissue concentrations are reached (MIC > 90). Antibiotics should be administered in the operating room. In order to avoid infusion reactions, fluoroquinolones and vancomycin should be administered 2 h before surgery. The use of vancomycin as a prophylactic is limited to use when there exists a potential risk of MRSA infections. Post surgery antibiotic treatments should be stopped within 24 h since it has been demonstrated that extending prophylactic antibiotic usage past this time does not diminish infection risks. (ii) The entire dosage of prescribed antibiotics should be administered for maximum efficacy. The upper range of the dose should be considered for larger patients or those undergoing longer operations. Specifically, antibiotic dosing should be based on a patient's weight or body surface area. Cephalosporins that have a serum half-life of 1 – 2 h are commonly used to ensure that effective drug concentrations are maintained during the entire surgical procedure, particularly if the operation exceeds 3 – 4 h. In these cases, up to three dosages may need to be administered. Antibiotics should be administered intravenously and administration completed within 20 – 30 min. Delivery should take place drop-wise and slowly to achieve effective concentrations. (iii) Re-dosing should be considered for longer surgery, for example, patients undergoing surgery that extends beyond two antibiotic half-lives should be re-dosed during surgery. Generally, the antibiotics selected for this process should be of short-range use and should not be used during elective surgeries. If the patient presents significant risk factors for infection, particularly during surgery involving prosthesis or implants, antibiotic use can be extended for several days or even until the stitches are removed to reduce the risk of developing SSIs. (iv) Choosing an appropriate antibiotic is also critical to ensuring success. Antibiotics should be chosen on the basis of their effectiveness against pathogens most likely to be encountered during the course of a particular surgery rather than selecting broad-spectrum antibiotics. Common microbial targets considered during antibiotic selection include microbes comprising the normal skin flora (*e.g.* staphylococci). In these cases, first-generation cephalosporins are frequently selected. Although intravenous administration is the most common route, combinations of oral and intravenous routes can also be used. If anaerobic infections are expected, metronidazole and other anti-anaerobic-selective drugs can be used. From an economic, safety and antibiotic-resistance perspective, it is recommended that relatively narrow-spectrum antibiotics be selected and to avoid the use of new, broad-spectrum antimicrobial agents. The preventive treatments for fungal infections remain controversial.

19.5.2 Conventional Treatments Associated with Surgery and Wound Infections

Conventional treatments associated with surgery and wound infections include antibiotic therapies controlling the infection source.

19.5.2.1 Antibiotic Therapies

At present, treatment of surgical and traumatic infections primarily employs the use of antibiotics. However, antibiotics are not capable of effectively controlling some infectious agents due to the development of widespread drug-resistance. This problem should prompt us to modify our use of antibiotics to minimize the emergence of resistant strains by first determining the antibiotic resistance profile of respective clinical isolates. This will allow for the selection of the most effective antibiotic for the treatment of specific infections.

A better understanding of how antibiotics function (including their mechanism of action and its impact on microorganisms and the human body) will also help to select the most effective antibiotics for the treatment of different infections. This knowledge is important since some antibiotics increase bacterial toxin production (*e.g.* lipopolysaccharide and peptidoglycan). The release of antibiotic-induced endotoxin is both dose- and drug-dependent. Experiments measuring *E. coli* endotoxin release in the presence of antibiotics at doses 50-fold, the MIC showed that endotoxin release was unchanged in the presence of meropenem or imine, but differences were observed in the presence of ceftazidime. Selecting antibiotics that confer their antimicrobial effects *via* different mechanisms may circumvent the problems observed with antibiotics that increase toxin production, *e.g.* rifampicin led to minimal release of lipoproteins and teichoic acids; half as much as were released in the presence of dalfopristin and a quarter of the amount released in the presence of ceftriaxone. Clinically, we want not only to clear the pathogen but also to minimize toxin production associated with antibiotic usage. This will require experiments that define toxin-production profiles for different antibiotics.

When patients are being treated with antibiotics post-surgery, it is important to pay close attention to the clinical response. If infections develop, the original treatment strategy should be modified to include an increased dose or frequency of administration, consider combination therapies to widen the antibacterial spectrum, carefully select the antimicrobial drugs used and consider anti-fungal treatments when fungal infections are suspected. In addition, surgical interventions (including drainage and debridement) should be considered while actively searching for the focus of infection. Once the patient's vital signs stabilize, that is, infection symptoms and signs have disappeared, body temperature and white blood cell counts reach normal levels for 3 d, antibiotic treatment can be stopped.

19.5.2.2 Controlling the Infection Source

Dealing with the source of infection is important in the treatment of surgical and wound infections. Treatments include abscess drainage, removal of necrotic tissues and potentially infected devices and controlling/limiting bacterial contamination. Once a focus of infection is identified, it should be dealt with promptly. Draining abscesses encased in a cellulose capsule will help drain necrotic tissues. If located percutaneously, abscess drainage can be carried out under the guidance of ultrasound or CT which significantly increases the success rate.

Debridement is the process of removing necrotic, solid tissues while leaving healthy tissues intact as a means of facilitating the healing process. Identifying healthy tissues, however, can sometimes be challenging, especially under conditions of low perfusion or bowel necrosis caused by thrombus. These complications in the gut make it difficult to determine the scope of bowel resection required. Regarding necrotic tissue surrounding the pancreas, it is important to delay necrotic tissue removal and debridement should be conducted at multiple times since necrosis develops in stages in this tissue. In addition, avoiding uncontrolled bleeding is facilitated following CT examination that can be used to assess tissue health before and after surgery.

In the context of heart surgery, clinicians should give priority to the removal and replacement of vascular access devices after it has been established that other indwelling devices are not the source of infection. Treatments are more difficult when left ventricle assisting devices are in place and patients have artificial heart valves. Mortality rates are high from these types of postoperative complications.

Once the infection foci are identified, selecting the most effective removal strategy is the key to minimizing post-operative morbidity and mortality. In general, measures used to control the infection source ought to minimize damage to the body so that patients can recover more rapidly. Removal of foci should be rapid and, if possible, cause little or no damage to the patient. Patients suffering from necrotic cholecystitis or secondary sepsis should be treated by percutaneous cholecystostomy under radiation. Percutaneous drainage is superior to surgical drainage for intra-abdominal abscesses presenting formed capsules. Utilizing minimally invasive surgical procedures such as laparoscopy should be considered in cases of severe surgical infection. The advantage of laparoscopy is that it has few complications and patients have a quick recovery time even in cases of complicated intra-abdominal abscesses, bile peritonitis or delayed hemoperitoneum. The choice of the surgical procedure selected and when the procedure is carried out must be determined in the context of the patient's health. If the patient's condition does not improve, the treatment approach will have to be modified. Failure to control the source of infection will result in a series of complications that include nutritional and metabolic imbalances accompanied, potentially, by multiple organ failure.

19.5.3 Complications of Severe Surgical and Wound Infections Prevention and Treatment of Sepsis and MOF

Sepsis and MOF are serious complications associated with surgery and trauma. Their treatment, in addition to the above-mentioned conventional methods, requires intensive care and adoption of early goal directed therapy (EGDT) during the early stages of sepsis or shock. Patients suspected of sepsis/shock should receive anti-shock therapy, timely glucocorticoid treatment, have blood glucose levels stabilized and treated with anticoagulants in order to prevent MOF when septic shock occurs. It is also important to provide effective nutritional support, particularly in septic patients also affected by malnutrition due to the energy consumed by the high metabolic processes associated with immune function decline (resulting from hyper-stimulation and the resulting cytokine storm) resulting in a sub-clinical MOF state. Treatment measures must be carried out according to the organs affected during the course of MOF, since the mortality rates of septic patients with acute renal dysfunction was significantly higher and about 60% required renal replacement therapy.

It is still undetermined if it is beneficial to reduce inflammatory mediators *via* continuous renal replacement therapy (CRRT), but some studies have shown that early and continuous veno-venous hemodiafiltration (CVVHDF) can improve the prognosis of patients with sepsis and organ failure. In addition, the efficacy of specific antibodies and antagonists used to limit immune dysfunction during the course of MOF remains inconclusive. The application of Chinese medicine also needs to be further studied in the context of MOF.

The goal-directed therapy commonly used in ICU refers to a titration-style treatment under the guidance of dynamic and quantitative targets while at the same time broadening the treatment according to different levels of monitoring indicators and combined with appropriate therapeutic tools needed to complete the treatment.

If the goal-directed therapy specifically targets disease at a certain stage, the timing of the therapy becomes critical to success, for example EGDT that targets severe infections and septic shock. At the early stage of sepsis and septic shock, appropriate treatment strategies used to treat or prevent local or systemic hypoperfusion and hypoxemia resulting from infections, hypotension, mental confusion, oliguria, lactic acid build-up, decreased oxygen levels in the superior vena cava, can all lead to an improved health status.

One study demonstrated that using EGDT, combined with blood transfusions and vasoactive and inotropic drug therapies to improve myocardial contractility of septic patients, enhanced tissue perfusion and reduced hypoxemia. These results showed that EGDT significantly reduced the incidence of both MOF and mortality. Reuben *et al.* recommended that patients that developed severe sepsis and septic shock should receive EGDT as soon as possible. EGDT administration, however, requires that clinicians act in a timely manner and accurately diagnose the patient's condition.

At present, it is believed that nutritional support provided to septic and MOF patients serves to maintain the structure and function of tissues and organs and the improved metabolism reduces deterioration of the intestines and promotes healing, resulting in improved survival rates in septic patients. However, nutritional support should be administered only at appropriate time. A patient's electrolyte imbalance during the early stages of shock must first be normalized prior to the administration of nutritional support.

Typically, nutritional support is implemented when respiratory, circulatory function, liver and kidney functions are stable and blood sugar levels are under control. Nutritional support is administered to septic patients parenterally (parenteral nutrition, PN) and/or enterally (enteral nutrition, EN). The advantages of these methods are that they reduce complications associated with single-use and the cost of the treatment. At the same time, PN can supplement nutrients that cannot be delivered by EN. It is important in the context of disease prevention that the structure and function of the small intestine should be maintained in septic patients. PN is conducive to maintaining intestinal integrity and function and in stabilizing the growth of normal intestinal flora, while at the same time reducing endotoxin release and bacterial translocation. PN should be administered as long as the patient's intestinal anatomy and function remain intact. PN can be first administered when sepsis patients with different degrees of intestinal dysfunction cannot receive EN. As the disease stabilizes, both PN and EN can be administered prior to transitioning to EN or oral diets. The main principles of gastrointestinal dysfunction treatment in the context of MODS are to prevent occult shock, improve GI ischemia, restore the intestinal architecture and GI motility. Once GI function is restored, early enteral nutrition can be started. In addition, EN will effectively stimulate the gastrointestinal mucosal barrier, reduce inflammation and prevent gut-pathogen associated sepsis. In addition, EN has been shown to improve the immune function.

Nutritional support is a part of comprehensive sepsis treatments. In order to stabilize patients, rational nutritional support combined with effective control of the infection should be considered.

Use of traditional Chinese medicines to treat sepsis has a long history and this field has uncovered new information relating to novel treatment strategies. Domestic scholars have verified using animal models and clinical studies show that rhubarb and other Chinese medicines possess therapeutic properties capable of restoring GI motility, protect the GI mucosa and prevent microecological disturbances. At present, traditional Chinese medicine used for experimental and clinical treatment of MODS can also be used to treat other maladies. Traditional Chinese medicine treatment modalities can be divided into either detoxification, Huayu, Tongfu or Fuzheng. The main medicinal components used for detoxification and Huayu are reduqing (honeysuckle, dandelion, Daqingye, heartleaf houttuynia herb), 912 (astragalus, Chinese angelica, red peony, Danshen) and Xuebijing (safflower, red peony, Chuanxiong, Danshen, Chinese angelica). The main medicinal components used for Tongfu are Cudong particles (rhubarb, Magnolia officinalis). The Shuangqing particles (Bupleurum root, astragalus, gypsum, Zhimu) not only have a role in detoxification and Huayu, but also in Fuzhengyangyin and Qixueliaqing. The above-mentioned traditional Chinese medicines have been proved effective in animal studies and small-scale clinical applications.

At present, it is imperative for these drugs to be tested in prospective studies carried out at multiple sites as a means of enrolling significant numbers of patients to study the effects of these treatments on sepsis and MODS and then based on these data, develop integrated programs combining Chinese and Western medicines to improve the prevention and treatment of sepsis and MOF.

19.5.4 Future Directions

In recent years, a reduction in the number of novel antibiotics available for the treatment of gram-negative and gram-positive bacterial infections (in addition to fungal infections) due to the rise in drug-resistant organisms has resulted in significant challenges in treating pre- and post-surgical/trauma-related infections. Novel antimicrobial peptides (as an alternative to antibiotics) have been identified from a variety of organisms (including bacteria, fungi, plants, insects, amphibians, fish, birds, mammals and humans) that represent a potentially new method of treating bacterial infections in the future. In addition, new treatment methods are constantly evolving. For example, one study found that a number of factors that contribute to cellular apoptosis (including lymphocyte-, neutrophil-, dendritic cell- and GI epithelial cell-related factors) have an important role in controlling the progression of sepsis, shock and MODS. Results in mice demonstrated that animals fed oral protease inhibitors not only presented a decrease in apoptosis of lymphocytes, but also presented an increase in sepsis survival rates, suggesting that inhibiting apoptosis in lymphocytes may represent a new strategy for the treatment of sepsis. We believe that problems relating to surgery and trauma-related infections will be solved by a combination of therapies resulting from the discovery of novel treatment modalities.

References

[1] Fry D E. A systems approach to the prevention of surgical infections. Surg Clin North Am, 2009, 89: 521-537.

[2] Kamp-Hopmans T E, Blok H E, Troelstra A, *et al.* Surveillance for Hospital-acquired infections on surgical wards in a Dutch university hospital. Infect Control Hosp Epidemiol, 2003, 24: 584-590.

[3] Martin L W, Altemier W A, Reyes P M Jr. Infections in pediatric surgery. Pediatr Clin North Am, 1996, 16: 735-766.

[4] Lee J A, Kim M S, Kim D H, *et al.* Postoperative infection and survival in osteosarcoma patients. Ann Surg Oncol, 2009, 16: 147-151.

[5] Gross T, Kaim A H, Regazzoni P, *et al.* Current concepts in posttraumatic osteomyelitis: A diagnostic challenge with new imaging options. J Trauma, 2002, 52: 1210-1219.

[6] Steven M, Gordon, M D. Antibiotic prophylaxis against postoperative wound infections. Clevel and Clinic Journal of Medicine, 2006, 73: 42-45.

[7] Lee CK, Hansen SL. Management of acute wounds. Clin Plast Surg, 2007, 34: 685-696.

[8] Sarkar B, Napolitano LM. Necrotizing soft tissue infections. Minerva Chir, 2010, 65: 347-362.

[9] Rea W J, Wyrick W J Jr. Necrotizing fasciitis. Ann Surg, 1970, 172: 957-964.

[10] Hang C H, Shi J X, Li J S, *et al.* Alterations of intestinal mucosa structure and

barrier function following traumatic brain injury in rats. World J Gastroenterol, 2003, 9: 2776-2781.

[11] Shen T Y, Qin H L, Gao Z G, *et al.* Influences of enteral nutrition combined with probiotics on gut microflora and barrier function of rats with abdominal infection. World J Gastroenterol, 2006, 12: 4352-4358.

[12] Quan Z F, Yang C, Li N, *et al.* Effect of glutamine on change in early postoperative intestinal permeability and its relation to systemic inflammatory response. World J Gastroenterol, 2004, 10: 1992-1994.

[13] Xia Y, Yang Z, Chen H Q, *et al.* Effect of bowel preparation with probiotics on intestinal barrier after surgery for colorectal cancer. Zhonghua Wei Chang Wai Ke Za Zhi, 2010, 13: 528-531.

[14] Daniel I, Sessler M D. Non-pharmacologic prevention of surgical wound infection. Anesthesiol Clin, 2006, 24: 279-297.

[15] Mangram A J, Horan T C, Pearson M L, *et al.* Guideline for prevention of surgical site infection. Infect Control Hosp Epidemiol, 1999, 20: 250-278.

[16] Belda F J, Aguilera L, García de la Asunción J, *et al.* Supplemental perioperative oxygen and the risk of surgical wound infection: a randomized controlled trial. JAMA, 2005, 294: 2035-2042.

[17] Hanazaki K, Maeda H, Okabayashi T. Relationship between perioperative glycemic control and postoperative infections. World J Gastroenterol, 2009, 15: 4122-4125.

20

Infective Microecology of Chemotherapy and Radiotherapy

Nong Xu *, Chenyu Mao
Department of Medical Oncology, the First Affiliated Hospital, School of Medicine, Zhejiang University, Hangzhou, 310003, China
* E-mail: xunong@medmail.com.cn

This chapter mainly discusses the Mechanism of Chemotherapy and Radiotherapy, Chemotherapy and Radiotherapy Effect on Microecology, and Microecology of Infection Caused by Radiotherapy and Chemotherapy.

20.1 Mechanism of Chemotherapy and Radiotherapy

This section provides an overview of the mechanism of chemotherapy and radiotherapy. The topics to be discussed include the classification and mechanism of anticancer agents and molecular targeted agents and their infection-related toxicity, as well as the mechanism of radiotherapy and infection-related toxicity of radiotherapy.

20.1.1 Chemotherapy

Most chemotherapeutic agents currently in use appear to exert their effect primarily on cell multiplication and tumor growth.

20.1.1.1 Classification and Mechanism of Anticancer Agents

Presently there are more than 50 kinds of anticancer agents used in clinics [1]. Traditionally, the anticancer agents are classified into alkylating agents, antimetabolism agents, antibiotics, medicinal plants, platinum and others (including methyl hydrazine bian, L-asparaginase and hormones, *etc.*). Another classification is based on the targets for anticancer drugs: (i) Drugs that target DNA structure (including alkylating agents, anthracyclines and platinum); (ii) Drugs that interfere with the synthesis of nucleic acid (antimetabolism agents); (iii) Drugs that interfere with DNA transcription by acting on the DNA template or inhibit RNA synthesis by inhibition of DNA-dependent RNA polymerase; (iv) Drugs that interfere with protein synthesis (such as homoharringtonine, taxanes, vinblastine and podophyllotoxin); (v) Other types (such as hormones).

20.1.1.2 Classification and Mechanism of Molecular Targeted Agents

With the progress in knowledge of molecular cancer biology, the mechanism of the molecular signal pathway on proliferation and growth of tumor cell was clarified. Then the treatment target cell receptor, key gene and regulatory molecules began to be developed. The new generation of molecular targeted agents can block tumor growth and induce apoptosis by intervening in tumor stroma, tumor blood and lymphatic vessels and the cell signaling pathway. Presently, they are mainly classified into drugs that block the epidermal growth factor receptor (EGFR), monoclonal antibodies aiming at proliferation-related receptors, drugs against certain cancer genes and genetic marks of a cancer cell, tumor angiogenesis drugs, tumor vaccines, antisense oligonucleotides and small interfering RNA *etc.* [2].

Distinguished from traditional chemotherapeutic drugs, molecular targeted agents have the following characteristics: Are noncytotoxic and targeted; have a moderating effect and cell stabilization; do not necessarily achieving the dose-limiting toxicity and maximum tolerated dose; show a great difference from cytotoxic drugs regarding range and clinical manifestation of adverse reactions; have better effects when combined with conventional treatment (chemotherapy, radiotherapy) and so on. Currently, several molecular targeted agents have already entered clinical use, monoclonal antibodies including rituximab, trastuzumab, cetuximab, bevacizumab, *etc.*, signal transduction inhibitors including imatinib gefitinib and erlotinib *etc.*, and angiogenesis inhibitors including endostatin and so on.

20.1.1.3 Infection-Related Toxicity of Anticancer Drugs

Most of the existing chemotherapy drugs will be toxic to normal cells, especially rapidly proliferating cells when inhibiting or killing tumor cells, including bone marrow hematopoietic cells and gastrointestinal epithelial cells. Therefore, they can both attack the immune system and damage the barrier function of skin, mucosa and mouth [3, 4].

Skin and Mucosal Lesion. Some chemotherapy drugs can induce chemical stimulation in the tissue, causing chemical inflammation. Some drugs can make the tissue form a water blister (blistering agent such as nitrogen mustard). Intravenous injection compartment and can cause chemical phlebitis, leakage or leakage into the extravascular can develop redness, swelling and pain, even necrosis and ulcers of local subcutaneous or deep tissue. The blistering agents include nitrogen mustard, actinomycin D, mitomycin, anthracyclines and vinblastine *etc.*

The rapidly proliferating mucosal tissue is one of the most vulnerable to damage from chemotherapy drugs. It clinically manifests itself in stomatitis, glossitis, esophagitis, mucosa and gastrointestinal ulcers, causing eating pain, even severe bloody diarrhea. The damage to the mucosal barrier can lead to bacterial invasion and occurrence of infection. The drugs which cause mucositis include MTX, 5-Fu, actinomycin D and mitoguazone *etc.* The severity of mucositis is positively correlated to the dose and continuous administration time of drugs.

Myelosuppression. Myelosuppression is the most common major limiting toxicity of chemotherapy. The minimum half life of granulocyte is 6 – 8 h, so leukopenia is always performed first. The half life of a platelet is 5 – 7 d, so the decline in the platelet appears later and not so severe. The half life of erythrocytes is 120 d, therefore the influence of chemotherapy is minor and the decline is usually not obvious.

The extent, occurrence and duration of myelosuppression and time of marrow function recovery are all different among different types of chemotherapy drugs. The alkylating agent of nitrogen mustards, podophyllotoxins, anthracycline antitumor antibiotics, MTX, Ara-C, nitrosoureas, carboplatin, thiotepa and other substances develop severe bone marrow suppression. Myelosuppression induced by vincristine, pingyangmycin, asparaginase, plicamycin and cisplatin is minor. Myelosuppression induced by CTX, HN_2, anthracyclines, MTX, Ara-C, podophyllotoxins, hydroxyurea, vincristine and cisplatin occurs immediately and recovers fastly. The minimum value of leucopenia appears 1 – 2 weeks after chemotherapy and recovery in 2 – 3 weeks. The minimum value of leucopenia induced by nitrosoureas, MMC, procarbazine, Busulfan appears later between 3 – 8 weeks and recovery time is 1 – 2 mon.

Patients can have up to 90% more chance of serious bacterial, fungal or viral infection when leucopenia $< 1.0\times10^9$ / L, in particular granulocytes $< 0.5\times10^9$ /L for 5 d.

Main Infection-Related Toxicity of Molecular Targeted Agents. Adverse reaction incidence and severity vary in different molecular targeting drugs. The specific adverse reactions of a monoclonal antibody are allergic reactions, bronchospasm, stridor, hoarseness, speech difficulties, rubella and hypotension. Both the monoclonal antibodies and small molecule kinase inhibitors targeting EGFR have acneiform rash, xerosis, eczema and other dermal toxicity [5]. Acneiform rash is the common adverse reaction in anti-EGFR targeted drugs and happens dose-dependently. Diarrhea caused by anti-EGFR therapy is usually mild or moderate, only 6% of patients are 3 or 4 grades, usually temporarily, and only 1% of patients required reduction because of diarrhea. Inhibition of the

EGFR-mediated signaling pathway affects keratinocytes by inhibiting growth, promoting apoptosis, inhibiting cell migration, increasing cell adhesion and differentiation, stimulating inflammation, which will result in characteristic skin manifestations of acneiform rash, postinflammatory effects, dry skin, rhagadia, paronychia. Toxicity in inhibiting the VEGF signaling pathway leads to hypertension, proteinuria, bleeding, perforation, and poor wound healing [6]. Multi-target kinase inhibitors can induce hand-foot syndrome, manifested as disesthesia, paresthesia, erythema, edema, hyperkeratosis, dry or chapped skin, sclerosis-like blisters, decortication and so on [7]. It usually occurs at the onset of treatment, hits the hardest two weeks after medication and then reduces gradually, or even disappears after 6 – 7 weeks of treatment. The incidence of hand-foot syndrome will decrease over time with treatment. Most of the small molecule kinase inhibitors have gastrointestinal side-effects, including diarrhea, nausea, vomiting, abdominal distension and pain. Diarrhea is common and often appears as loose stools (not watery) with increased defection frequency. Molecular targeted agents mostly have slight blood toxicity.

20.1.2 Radiotherapy

Radiotherapy in use appears to exert an ionizing radiation effect on cell multiplication and tumor growth. Ionizing radiation is energy sufficiently strong to remove an orbital electron from an atom. This radiation can have an electromagnetic form, such as a high energy photon, or a particulate form, an electron, proton, neutron, or α particle.

20.1.2.1 Mechanism of Radiotherapy

Radioactive sources used in radiotherapy include three types: (i) α, β, γ-ray released by radionuclides; (ii) X-ray and electron beam generated from energy electron accelerator; (iii) Proton beam, neutron beam, negative π-meson beam, and other heavy particle beams produced by a radiation therapy device. These sources can produce damage and kill cancer tissue, which exerts an anti-tumor effect.

Cancer has "unlimited cell reproduction capacity", so a tumor can be controlled as long as the X-ray can inhibit the reproduction of tumor cells to induce tumor cell proliferation and death occurring (the loss of unlimited reproductive capacity), and those tumor cells which cannot reproduce without limit will eventually die and be excluded from the body. The sensitivity of cells to radiation is the highest in the division period, the lowest in the DNA synthesis phase. Radiotherapy can only destroy most abnormal proliferating cancer cells and reduce them, but with less damage to normal surrounding tissue [8].

20.1.2.2 Infection-Related Toxicity of Radiotherapy

The main toxicity associated with infection produced by radiotherapy is the damaged barrier function of the skin and mucosa, a suppressed immune function and an attack on the phagocytosis of the neutrophil [9].

Acute radiation reactions and damages to skin. Local skin of radiation field flushing is erythema response to radiation, also known as the grade I skin reaction to radiation. Hair follicles or papules, depilation or alopecia, local dry peeling and so on appear in the skin of the radiation field. Patients may have a local skin burning sensation, which isknown as a grade II skin reaction to radiation. Edema and inflammatory infiltration cause further aggravation in the skin of the radiation field, cells of the folliculosebaceous and sweat glands will degenerate. Water blisters, skin exudate and skin exfoliation will appear on the skin in the radiation field. This is a grade III skin reaction to radiation, also known as wet radiation skin reaction. If radiation therapy continues at this time, the skin precursor cells within the basal layer cannot produce new cells, mature skin cells suffer necrosis, the endothelial cells of small subcutaneous vascular swell or suffer thrombosis. Clinically acute radiation skin ulcers can occur, which is a grade IV skin reaction to radiation.

Chronic radiation reactions and damages to the skin. Chronic radiation skin reactions can be divided into three categories: (i) Local skin presents atrophy and thining irregularly, lighter texture, pigmentation and hyper-pigmentation mixed at the Grannophyric phase, hair follicle atrophy, telangiectasia. This category is called chronic atrophic radiation skin damage. (ii) Repeated radiation causes excessive keratinization and proliferation of skin, and then pigmented patches appear. Skin becomes rough and chapped. This category is called chronic proliferative radiation skin damage. (iii) The persistent ulceration caused by infection and parts of acute radiation skin ulcers are called chronic radiation skin ulcers.

The radiation response and damage to the mouth. The radiation response and damage to the mouth are mainly related to oral mucositis, xerostomia, radiation caries and necrosis of the jaws.

Radiation response and damage to the larynx and trachea. The radiation response and damage to the larynx and trachea are mainly manifested as sore throat. Laryngeal cancer patients with a large tumor or infection will become dyspnea. Severe cases can lead to asphyxia. The laryngeal and tracheal cartilage are generally more tolerant of radiation. However, when the cartilage is infected and invaded by a tumor, necrosis of the cartilage and aspiration pneumonia happen easily.

Radiation response and damage of lung. The radiation response and damage to the lung can generally be divided into three phases: (i) Exudation period (the period during radiotherapy and 1 – 2 mon after radiotherapy). (ii) Proliferation phase (2 – 6 mon after radiotherapy). (iii) Fibrosis phase (6 months after radiotherapy).

Radiation injury to bone marrow. Bone marrow is one of the organs that are most sensitive to radiation. Due to the decline of the original mother cell after radiation, the components of blood and bone marrow are damaged by radiation. Poor bone marrow regeneration can exist if a large volume of marrow is irradiated. Sometimes a plastic anemia will happen.

20.2 Chemotherapy and Radiotherapy Effect on Microecology

Microecological balance depends on the environment, host and microorganism which is direct, primary and mutual when patients with cancer receive chemotherapy and radiotherapy.

20.2.1 Cancer Patients and Microecology

Microecological balance is mainly determined by the environment, the host and microorganism. These factors are comprehensive and interconnected, forming a dynamic ecological balance. In the balance, the normal microorganism is an important factor in transformation of the disease and health. The main factor causing microecological imbalance is the environment, including the environmental change in the host, such as immunodeficiency or using an immunosuppressant. Besides, many factors will affect the dynamic microeological balance, such as the destruction of the barrier of skin, mucosa, phagocytes, declined colonization resistance in the intestinal mucosa, the destruction of the immunosystem and malnutrition when cancer patients whose cancer is caused by self-immunological dysfunction receive anti-cancer therapy.

20.2.1.1 Destruction of the Host Defense Function in Cancer Patients

The cancer or anti-cancer therapy in cancer patients will cause immunodeficiency. Hodgkin lymphoma or non-Hodgkin lymphoma patients have an abnormal cell immune system, which will increase the risk of a virus (Herpes simplex and Herpes zoster) and fungi (*Cryptococcus*) infection. On the other hand, acute leukemia patients are susceptible to severe Gram-negative bacteria infection as a result of neutropenia or granulocyte dysfunction. Chronic lymphocytic leukemia and multiple myeloma patients are prone to *staphylococcus*, *streptococcus*, particularly *streptococcus pneumoniae* infection. Various treatments, like chemotherapy, bone marrow transplantation, corticosteroid treatment and regional or wide range radiotherapy, will cause immunodeficiency in the patients. The most significant risk factor of severe bacteria infection in cancer patients is neutropenia caused by chemotherapy. An allogeneic bone marrow transplantation patient is prone to virus infection due to T cell defense deficiency. Other changes in the bacterial community in the host, such as damage to the natural barrier by skin and mucosa, malnutrition caused by eating disorders, are all risk factors for infection [10].

Skin and mucosa barrier. Skin and mucosa are the first protective barrier against invasion of endogenous and exogenous microorganisms. Regional cancer invasion, surgery, radiotherapy, chemotherapy and molecular targeted therapy will destroy the skin barrier. The toxicity of gastrointestinal mucosa in radiotherapy and chemotherapy leads to mucositis, which makes patients prone to microorganisms in the gastrointestinal tract. And much diagnosis technology (including fingertip needling, venipuncture, marrow suction, and vein indwelling catheter) will also destroy the integrity of the skin, which will be the primary site of pathogen colony formation and final spread of cancer.

Phagocytes barrier. Neutrophils are the main cellular barrier against most bacterial and fungal infections. Tumor and/or chemotherapy or radiotherapy may inhibit the formation of bone marrow granulocytes, thus causing neutropenia, which will increase the chances of bacterial and fungal infections. Bodey *et al.* have found the extent and duration of neutrophil that were significantly correlated with incidence of serious infections by bacterial and fungal [11]. It is generally believed that patients who have cell equalling to or less than 0.5×10^9/L (500/μL) have a high risk of severe infection.

For hematological malignancies, neutropenia not only has volume deficiency, but also qualitative abnormalities. The latter includes chemotaxis, phagocytosis and low ability of anti-bacteria and the respiratory outbreaks. Although the phagocytic activity of white blood cells of untreated patients with lymphoma and sarcoma remains normal, the spontaneous migration and chemotaxis decrease, suggests that there may exist a cyclical inhibitor in the blood serum of such patients. Chemotherapy and radiotherapy can cause defects in the neutrophil function. For example, corticosteroids can weaken the phagocytic ability and migration capacity of neutrophil. Joint application of prednisone, vincristine, asparaginase or prednisone, 6-mercaptopurine and methotrexate can significantly decrease the capacity of phagocytic and death.

Cellular immunity and humoral immunity. Abnormal lymphocytes caused by a malignant tumor or chemotherapy and radiotherapy can affect the humoral immunity (mainly mediated by the B cell) and cellular immunity (mainly mediated by the T cell) in the immune response. Some malignant tumors, such as chronic lymphocytic leukemia, Hodgkin lymphoma, multiple myeloma, significantly affect the humoral immune response. The ability to produce pathogens, specifically neutralizing antibodies, can change significantly. The production of antibodies is reduced in untreated patients. Chemotherapy leads to a weakened modulation, insufficient effects of agglutination and dissolved bacteria, deficiency in neutralization of bacterial toxins by affecting the B and T cell function. Therefore, patients with defects in humoral immunity and antibody production are susceptible to pyogenic bacteria and susceptible to encapsulated bacteria (such as *Streptococcus* pneumoniae, *Haemophilus* influenzae, and *Meningococcus*) even without neutropenia.

T cell immune response in malignant tumor patients (such as Hodgkin lymphoma and non-Hodgkin lymphoma) is damaged. Those patients are susceptible to some fungi (disseminated histoplasmosis, cryptococcosis), viruses (especially herpes-like virus, cytomegalovirus) and bacteria which are intracellular replicated (such as mononucleosis of *Listeria monocytogenes*, *Salmonella spp*). The incidence of pneumocystis carinii pneumonia is also increased in patients with T cell functional defects. Corticosteroids and radiotherapy can aggravate the dysfunction of the T cell. Patients who receive T cell bone marrow transplants are susceptible to a virus, particularly cytomegalovirus (CMV), which further suppresses the host defense function.

Chemotherapy can reduce the amount of helper T cell ($CD4^+$). Mackall *et al.* indicated that after several cycles of chemotherapy, the number of neutrophils, monocytes and platelets were restored to 50% of baseline levels and above, but the number of lymphocytes could not quickly be restored [12]. In fact, lymphopenia can last several months after chemotherapy. The regeneration capacity of $CD4^+$ T cells seems to decline with age.

Splenectomy. The spleen plays a role as mechanical filter and early conditioner. Splenectomy patients will develop reduced antibodies production after specific antigen stimulation, lack of promotion of peptide of phagocytic cells, a decreased level of immunoglobulin M (IgM) and properdins. Therefore, splenectomy patients are susceptible to slime-forming bacteria (in particular *Streptococcus* pneumoniae, meningococcus, *Haemophilus* influenzae) septicemia and septic Babesia [13].

Nutrition. Malnutrition is a common complication in patients with malignant tumors. Anti-cancer treatment can lead to loss of the integrity of the skin and mucosal barrier, impaired phagocytosis, decreased mobilization of macrophages and inhibition of lymphocyte function [14]. Total parenteral nutrition aims to reduce drug-induced myelosuppression and reduce mucosal cell injury. Nutrition serves multi purposes in the maintenance of ability to resist infection, including regulating intestinal microbial flora andimpacting various resistance factors, thereby affecting the nutritional status and resistance of various body organs.

20.2.1.2 Pathogenic Microorganisms in Cancer Patients

In the past 50 years, the microbial spectrum of infections has changed. In the mid-twentieth century, the isolated bacterium in immune-suppressed patients was *Staphylococcus* aureus . *Staphylococcus* aureus infections can be treated by a new type of penicilin which can effectively resist *Staphylococcus* aureus β-lactamase. Then Gram-negative bacilli had become the major bacterial pathogens (especially *E. coli*, *Klebsiella* and *Pseudomonas aeruginosa*). In the 1980s, *Pseudomonas aeruginosa* infection was in decline. The main isolated Gram-negative bacilli were members of Enterobacteriaceae (Such as *E. coli*, *Klebsiella* and *Enterobacter spp*). Nosocomial infection or antibiotic resistance with non-pseudomonas aeruginosa (such as *Pseudomonas maltophilia*, and *Pseudomonas cepacia*, *Pseudomonas Amur*) increased in cancer patients. Multidrug resistance to gram-negative bacterial infections in inpatients increased. The most common bacteria include *E. coli*, *Klebsiella* and *Salmonella*, which result from antibiotics abuse. Infection caused by resistant Gram-negative bacteria led to a high mortality because of ineffective antibiotic treatment.

Recently, several Gram-positive bacteria have been an important cause of refractory infection in cancer patients. α-Hemolytic *Streptococcus*, which is the cause of bacteremia, is usually sensitive to vancomycin and inactivated in blood culture by vancomycin. The sepsis syndrome is hard to cure by antibiotics and the mortality rate is around 10% – 20% [15].

Enterococcus faecalis (*E. faecalis*) and *Fecal Streptococci* (*S. faecium*) in *Enterococcus* have become common pathogens because of the acquisition of resistant plasmid. Outbursts of nosocomial infection, caused by strains resistant to vancomycin and aminoglycosid in severe patients, have increased gradually. Infections caused by multi-drug resistant enterococcal are resistant to current antibiotics and have high mortalities. Because of the misuse of antibiotics, especially vancomycin, resistant *Enterococci* have become important nosocomial pathogens as resistant Gram-negative bacteria [16].

Anaerobic bacteria doesn't play an important role in early infection of cancer patients. Compared with aerobic bacteria, about 5% of the bacteremia and mixed infections of the mouth and anus are caused by anaerobic bacteria. There are reports about isolation of clostridium perfringens (clostridia perfringens) and *Septicemia Clostridia* (*C. septicum*). The third spindle bacillus (*C. tertium*), considered as contaminated bacteria in the past, is considered related to serious infection now. Only 50% of the isolated third spindle bacillus are sensitive to the standard anti-anaerobic drugs (such as clindamycin, metronidazole), but most of them are sensitive to vancomycin. Bacillus is related to infections in cancer patients. The bacteria are difficult to remove until the indwelling silicone tube can is unplugged.

The infection of mycobacterium is uncommon in cancer patients. In the patients with an indwelling catheter, some rapid-growing mycobacteria occur, such as M chelonei and M fortuitum. They can lead to chronic ulcer lesions at the exit of the catheter. Drug-fast mycobacterium tuberculosis is gradually prevailing in AIDS patients. It's estimated that it will also prevail in cancer patients.

Fungus is one of the main pathogens, particularly in persistent neutropenia and immune-suppressed patients. The main fungi include candida, aspergillus, *Cryptococcus neoformans* and phycomycete. Aspergillus infection is increasing. The prevalence of Mucor (mucor, RSA and Rhizopus species) is low, but it can induce lung disease or invade the nasal sinus causing nose-brain syndrome. Lymphoma patients are easily infected by *Cryptococcus*. The infections of *Fusarium*, *Dreshlera*, *Pseudallescheria boydii* and *Malassezia* in cancer patients have been reported [17].

Except for bacteria and fungi, parasitic and viral infections are also important. *Pneumocystis carinii* is an important cause of pneumonia, particularly in cancer patients with the application of corticosteroid. The incidence of virus infection is high in immune-suppressed patients, and Herpes simplex virus (HSV), herpes zoster virus (VZV) and cytomegalovirus (CMV) are the most common ones. HSV can cause unexpected infection although the preventive treatment of acyclovir significantly reduces the HSV infection, particularly in patients with leukemia and bone marrow transplants. VZV is easily activated in lymphoma patients and results in herpes zoster. CMV infection can lead to immune suppression. Some other viruses are typical benign viruses in the normal host, such as adenovirus, respiratory syncytial virus (RSV) and human herpes virus 6 (HHV-6). However, they can cause severe lower respiratory tract infection in bone marrow transplantation.

20.2.1.3 Factors of Microdysbiosis

There are many factors affecting microdysbiosis, mainly concerning the anatomical structure and immune function of the host, the effect of antibiotics on microorganisms.

Chemotherapy can injure the defense function in cancer patients, such as destroying the skin and mucosal barrier, damaging the phagocytes barrier for neutropenia, affecting cellular and humoral immunity. To cancer patients, the immune function is suppressed because of protein-energy malnutrition, thus increasing the chance of infection.

Radiotherapy can destroy the microecological balance between normal microbiota and the host. Microorganisms can invade tissue and blood, causing several kinds of inflammation. The invasive process can increase the sensitivity of the host to microorganisms, thus increasing host morbidity and mortality.

A natural defense mechanism will be destroyed and the tolerance content of microbes in the respiratory tract and intestinal tract will increase when cancer patients undergo chemotherapy and radiotherapy. The function and quantity of the phagocytic cell will decrease, the barrier function of the lymph will be weakened, the nonspecific germicidal effect of serum will disappear or reduce, and the immune response will be significantly damaged. Molecular targeted therapy is a new field of cancer treatment. Although the adverse reactions have different manifestations, with a lower incidence and lighter severity when compared to chemotherapy, there are still some severe adverse reactions that are hard to recover from. Some special adverse reactions of targeted agents have craised wide concern in clinics, especially skin reactions, interstitial lung disease (ILD) and adverse cardiovascular events. Rash is a common toxicity of molecular targeted agents, particularly severe dermal toxicity such as acneiform rash which may damage the barrier of the skin and mucous membrane. Whether it will cause microecological imbalance or not has still not been concluded.

20.2.2 Microdysbiosis in Cancer Patients

Microdysbiosis is defined as when the physically combined state between the microorganism and the host change into a pathological state because of the effect of the environment. Environment is the critical factor causing microdysbiosis. It includes the biological environment (the host) and abiotic environment. In recent years, the microecological opinion has been to suggest that infection is the interaction between a microorganism and the host caused by a microorganism anomaly infesting the host. Infection is a manifestation of the microecology phenomenon and is controlled by the balance and imbalance mechanism of the cause and the host. With the advance in reinforcement chemotherapy and radiotherapy with higher efficacy, the number of cancer patients with weakened immunity is increasing. The relationships among malignant tumors, weakened immunity and infection-induced morbidity and mortality are clear. The fatal infectious complications limit the efficacy of anti-tumor therapy. Therefore, it is necessary to be familiar with the risk factors of infection, apprehend infectious syndromes and learn how to treat them effectively.

20.2.2.1 Clinical Manifestation

Clinical manifestations of microdysbiosis in cancer patients include fever, other non-specific performance and some other symptoms associated with specific organs.

Fever. Fever is a common symptom for cancer patients. Although it is usually caused by infection, non-infection factors are also significant. Some drugs (particularly cytotoxic drugs, such as bleomycin, cytarabine, and blood products), allergic reactions and cancer itself may cause fever. It should also be clear whether the patient is using drugs that may mask fever reaction (such as steroids, and antipyretic analgesics).

Fever may be the first and only one symptom in infection with neutropenic patients. The criteria for fever include an oral temperature over 38.3 ℃ once or exceeding 38 ℃ in 1 h, which can be diagnosed as fever in cancer patients with neutropenia [18]. Clinical practice shows that the absolute value (including segmented and staff leukocytes)$\leqslant 0.5\times10^9$/L (500/μL) should be considered as neutropenia. Actually, even if the absolute value of neutrophils is $(0.5-1.0)\times10^9$/L (500 – 1,000/μL) it is also considered as neutropenia, because there is always a rapid decline due to the recent anti-tumor treatment. Physical examination must be repeated frequently in patients when the fever lasts a long time and the infectious primary sites are indefinite.

Other Symptoms. Except for general symptoms such as fatigue, weakness, and other non-specific performance, there are still some other symptoms associated with specific organs, such as cough and expectoration in pulmonary infection, bellyache and diarrhea in abdominal infection, frequent micturition and urgency in urinary tract infection.

Physical Signs. (i) Some special infections which are often caused by gram-negative bacteria will manifest themselves as hypotension and shock. (ii) Tachycardia infection should be suspected in emerging or unexplainable tachycardia. (iii) The emergence of inflammatory reaction indicates infection. However, patients with neutropenia do not show typical inflammation performance during infection. There are usually no infiltrates on the chest radiograph or even a negative result in sputum culture when the patient has bacterial pneumonia.

20.2.2.2 Examination Method

Examination methods include physical examination, cultivation, imaging examination and other examinations.

Physical examination. Oropharynx, perianal and skin need to be carefully monitored.

Cultivation. (i) General samples: Samples from any location with the possibility of infection should be cultured. All the collected samples should be tested for bacterial counts. (ii) Blood: At least two Petri dishes are needed (one anaerobic dish and one aerobic dish). Catheters needs to be scissored from the indwelling central venous catheter for another sample, except when taking a blood sample. (iii) Urine: It is better to be collected from a clean urethral catheter or under cystoscopy. A fine needle for bladder puncture can be used if necessary. (iv) Sputum: Cough up deep sputum naturally after gargling. Three percent saline can be used for an aerosol if the patient has difficulty coughing up sputum. (v) Cerebrospinal fluid: A lumbar puncture needs to be taken if there are neurological abnormalities. Gram stain, bacterial and fungal culture, cell count, carbohydrate (simultaneous determinate serum glucose) and quantity of protein and cytology must be tested in the cerebrospinal fluid. (vi) Defecation: Clostridium is the most common toxigenic anaerobic bacteria of diarrhea. Salmonella and listeria monocytogenes are common causes of diarrhea and septicemia caused by hospital infection. (vii) Virus culture: The incidence of and mortality from virus infection are progressing in tumor patients. Herpes simplex virus, varicella zoster virus and cytomegalovirus are the main sources of infection.

Imaging examination. (i) X-ray: All patients with suspected infection should take a routine chest examination. Paranasal sinuses radiography is necessary to search for infection foci. Dental radiography needs to be taken in neutropenia patients with fever to detect cusp abscess. (ii) CT: It is helpful to scan the chest, abdomen, brain, sinus, head, neck, spine and other regions for the diagnosis of infection. (iii) Ultrasound: echocardiography can be taken if infective endocarditis is suspected. Transesophageal echocardiography is more sensitive compared with conventional echocardiography, but it is invasive. Conventional ultrasound is valuable for the diagnosis of ascites, cholecystitis, liver and pancreatic disease.

Other examinations. (i) Invasive examination: Bronchial lavage, skin biopsy, lung biopsy, bone marrow biopsy, percutaneous liver biopsy and so on can be taken according to the infection site if patients are suspected of having infection. (ii) Laboratory: examinations. Blood cell count, liver function, urinalysis and erythrocyte sedimentation rate should be taken for patients with infection or suspected infection.

20.3 Microecology of Infection Caused by Radiotherapy and Chemotherapy

Except for the prevention of infection-related radiotherapy and chemotherapy toxicity and reasonable treatment for chemotherapy-related infections, more attention should be paid to microecology following the principles of ecology. Therefore, controlling of infection caused by radiotherapy and chemotherapy mainly consists of the following three aspects: prevention and treatment of infection-related radiotherapy and chemotherapy toxicity [19]; reasonable control of infection caused by radiotherapy and chemotherapy; ecological control.

20.3.1 Prevention and Treatment of Infection-Related Chemotherapy Toxicity

Prevention and treatment of infection-related chemotherapy toxicity include prevention and treatment of injury to skin and mucosa, myelosuppression and control of molecular targeted drug-related toxicity.

20.3.1.1 Prevention and Treatment of the Injury on Skin and Mucosa

In this section, prevention and treatment of injury to skin and mucosa will be included.

Prevention. Intravenous injection should be into the head tube of the proximal vein and avoid the hands and the area near the joints. Chemotherapeutics can be administered by intravenous drip or injected only after venipuncture success is confirmed and the transfusion is done. Deep vein catheterization chemotherapy is more helpful for preventing and reducing chemotherapy-induced phlebitis. It reduces the pain of repeated vein puncture in long-term chemotherapy. In addition, medical staff should read drug instructions before medication.

Treatment. If chemotherapeutic drugs extravasation occur, normal saline should be subcutaneously injected immediately to dilute the drugs. The application of antidote: 10% sodium thiosulfate 4 mL with 6 mL water can be injected into the extravasation site if nitrogen mustard extravasated; 1 – 2 mL 50% – 100% DMSO can be smeared on the extravasation site if mitomycin C and anthracyclines extravasated; hyaluronidase 300 U plus 1 – 2 mL normal saline can be injected into the extravasation if vinblastines, VP-16, VM-26 extravasated. Surgical treatment should be considered for severe, prolonged unhealed necrosis and ulcer lesions. A gargle of 20% lidocaine solution 15 mL can be used for the treatment of mucositis.

20.3.1.2 Prevention and Treatment of Myelosuppression

Chemotherapeutic drugs with myelosuppression are unsuitable when w leukocytes are $< 3.5\times10^9$/L, platelets $< 80.0\times10^9$/L (except acute leukemia). The condition of hematopoiesis should be considered (leukocytes and platelets count and myelogram) to adjust the dose of drugs to avoid serious inhibitory effect on the marrow function.

When leukocytes are $< 1.0\times10^9$/L and neutrophils $< 0.5\times10^9$/L, antibiotics should be applied appropriately to prevent infection. Blood culture and drug susceptibility should immediately be taken into account if fever occurred. Antibiotics with broad spectrum efficiency should be applied. Granulocyte colony stimulating factor (G-CSF), granulocyte-monocyte colony stimulating factor (GM-CSF) or granulocyte infusion should be appropriately provided if necessary.

When platelets are $< 50.0\times10^9$/L, prednisone or hemostatics like ethamsylate can be applied to prevent hemorrhage. It is critical when platelets are $< 20.0\times10^9$/L. Platelet transfusion, a larger dose of ethamsylate and prednisone should be applied.

20.3.1.3 Control of Molecular Targeted Drug-Related Toxicity

Toxicity of targeted agents mainly involves dermal toxicity and diarrhea.

Dermal toxicity. If patients who are receiving anti-EGFR treatment have a rash (inflammatory papules, pus herpes) and/or other skin reaction, do not immediately stop the treatment. Adjust treatment and/or dose with reference to rash grading. Patients with skin rash of grade 1 or 2 can continue with the original dosage, simultaneously adopting some supportive care, including strengthening the skin care, avoiding secondary infection, avoiding stress or friction, local administration of emulsion or lubrication agent containing urea and corticosteroids, providing antifungal or antibiotic agents as the occasion requires [20]. Treatment of hand-foot syndrome caused by multi-targeted multi-kinase inhibitor is similar to skin rash. Anti-acne cream can be a popular application for acne-like rash. Pulsed dyed laser and emollient cream can be used in the reaction after inflammation and dry skin. Hydrocolloid and propylene glycol can be used when the skin cracks.

Diarrhea. Treatment should begin when the first signs of diarrhea appear. Loperamide and diphenoxylate can be applied. The starting dose of loperamide is 4 mg and then 2 mg every 4 h, 2 mg every 2 h for severe patients. Water should be supplied appropriately. The dose should be reduced in patients with severe diarrhea (if loperamide treatment fails).

20.3.2 Prevention and Treatment of Infection-Related Radiotherapy Toxicity

Prevention and treatment of infection-related radiotherapy toxicity primarily consists of control of skin injury, oral damage, throat and trachea damage, lung injury and myelosuppression.

20.3.2.1 Control of Skin Injury

Control of skin injury involves prevention and treatment of acute injury and chronic injury.

Prevention and Treatment of Acute Injury. (i) Take multi-field irradiation technique and minimize the radiation dose at the same sites while doing conventional radiation. (ii) Keep enough distance between the reference point and the skin to avoid a high dosage radiation of the skin during interstitial brachytherapy. (iii) It's forbidden to use the afterloading source brachytherapy on the subcutaneous metastatic carcinoma or superficial lymph node. (iv) When doing conventional long-distance irradiation at the location where the skin is wrinkled such as the postaurale, armpit, inguinal region, perineum and gluteal fold, moist skin radiation reaction is likely to occur. Therefore, keep the local area clean and dry when using radiotherapy. (v) It's forbidden to use a smear with iodine and mercury if radiation skin reaction exists. (vi) Sea buckthorn oil, vitamin E drops, quick kouyangxiao, Hirudoid, Visk can be used for radiation skin reactions of grade 1 – 3.

Prevention and Treatment of Chronic Injury. (i) Protect the local skin in the radiation field and prevent damage and infection. (ii) Avoid small amounts of radiation and chronic radiation. Wear lead gloves to prevent chronic exposure to the hands. (iii) When medicine, hyperbaric oxygen and other conservative treatments are ineffective for radiation-induced skin ulcers, surgical treatment is required.

20.3.2.2 Control of Oral Damage

By blocking the mouth with a diving weight and pressing the tongue with a small bottle, unnecessary radiation can be reduced in radiotherapy. In the period of radiotherapy, the patient should keep the mouth clean with chlorhexidine rinse or other methods. After radiotherapy, brush the teeth with fluorinated tooth paste to protect from caries. Mucosal reaction can be treated with sea buckthorn oil, vitamin E, smecta powder and WISK spray, which all have good effects.

20.3.2.3 Control of Throat and Trachea Damage

Laryngeal edema is unavoidable in throat radiotherapy, so the patient should pay close attention to that. Anti-infection and hormone prophylaxis can be applied. A tracheotomy can be taken in advance if necessary. Overexposure to the trachea may lead to a severe cough, which is beyond the control of medicine. Therefore, it should be avoided.

20.3.2.4 Control of Lung Injury

The pulmonary fibrosis of normal tissue in the radiation field of lung cancer after conventional radiation is inevitable. Acute radiation pneumonitis is related to the radiation area, dose, speed and infection. Infection is the key inducing element. Active anti-inflammatory therapy should be administered to patients with respiratory infection. Sensitive antibiotics should be adopted according to sputum culture, together with a large amount of dexamethasone, bronchodilator, oxygen inhalation and other symptomatic supportive treatments. A high dose of adrenocortical hormone can be applied if necessary.

20.3.2.5 Control of Myelosuppression

Leukocyte, platelet, fresh blood and medicines like rhG-CSF can be infused. Bone marrow transplantation can also be considered if necessary. Bone marrow radiation should be reduced as much as possible. The dose should be carefully considered when radiation is inevitable. Bone marrow transplantation should be considered before using whole or subtotal bone marrow radiation with a high dose.

20.3.3 Prevention and Treatment of Cancer Patients Infection

Prevention and treatment of cancer patients' infection include primary treatment and a recommended regimen.

20.3.3.1 Primary Treatment

Infection exists if the oral temperature is over 38.3 ℃ once or exceeds 38 ℃ in 1 h. Initial risk evaluation and appropriate treatment protocols should be applied to patients with reducing febrile neutropenia [10, 19].

Recommended regimen. (i) Usage of intravenous antibiotics (choose one only): Cefepime, ceftazidime, imipenem/cilastatin, meropenem, piperacillin/tazobactam. (ii) Combined treatment of intravenous antibiotics: aminoglycosides + anti-*P. aeruginosa* penicillin ± extended spectrum β-Lactamases inhibitor; aminoglycosides + broad-spectrum cephalosporins (cefepime + ceftazidime); ciprofloxacin + anti-*P. aeruginosa* penicillin; vancomycin, linezolid, daptomycin or dalfopristin/quinupristin are normally used. (iii) Combined usage of oral antibiotics: Ciprofloxacin, amoxicillin and clavulanate potassium are applicable for low risk patients; ciprofloxacin and clindamycin for penicillin allergic patients; if fluoroquinolone is used in prevention, oral antibiotics is recommended.

Course of Treatment. (i) Primary antibiotics treatment should go on until the neutrophil amount ⩾ 0.5×10^9/L. The individual treatment in the antibacterial period should be based on neutrophil recovery, fast antipyretic, special infection site, potential disease of patients, *etc.* (ii) Skin/soft tissue infection: 7 – 14 d. (iii) Blood infection: Gram-negative, 10 – 14 d; Gram-positive, 7 – 14 d. *Staphylococcus aureus*, blood culture become negative and echocardiography should be normal in no less than 2 weeks. Yeast, blood culture should become negative in no less than 2 weeks. (iv) Sinusitis: 10 – 21 d. V. Bacterial pneumonia: 10 – 21 d. (v) Fungi (yeast or mould) infection: Candida, blood culture become negative after no less than 2 weeks. (vi) Viral infection: Herpes simplex virus/Varicella-zoster virus, aciclovir, Valaciclovir, famciclovir 7 – 10 d (only for skin). Influenza: oseltamivir 5 d.

Anti-Fungus Therapy. Anti-fungus therapy should be applied to neutropenia patients being given broad spectrum antibiotics after more than 3 – 5 d but still with fever. Amphotericin B is the preference drug in anti-fungus therapy with wide coverage like candida, aspergillus (except terrein), conjugated vaccine, rare fungi, cryptococcus and histoplasma capsulatu. Fluconazole is effective for candida, histoplasma capsulatu and cryptococcus. Itraconazole is useful for candida, aspergillus niger, histoplasma capsulatu, cryptococcus and some rare fungi. Voriconazole is useful for candida, aspergillus niger, histoplasma capsulatu, cryptococcus and some rare fungi as a standard treatment against invasive fungi.

20.3.3.2 Therapeutic Strategy

Therapeutic strategy should be adjusted according to the drug sensitivity test result, which includes special site evaluation and treatment, antiviral therapy and application of a biological response modifier.

General principle. Therapeutic strategy should be adjusted according to the drug sensitivity test result. Two types of sensitive antibiotics should be used on neutropenia patients with gram negative infection, but one type of antibiotics for Gram-positive infection.

Special site evaluation and treatment. (i) Mouth/mucous membrane: ①Necrosis and ulcer: Anti-herpes simplex virus treatment and general anti-epiphyte therapy should be adopted for herpes simplex virus, fungal infection or suspected fungal infection in biopsy, which are confirmed by culture and gram stain. ②Fungal stomatitis: Anti-epiphyte therapy should be carried out by using fluconazole first, voriconazole, posaconazole and itraconazole can also be applied if it's resistant. (ii) Esophagus: Anti-inflammatory therapy can be taken after culture confirmation if symptoms like substernal burning and dysphagia/swallowing aching appear. If the treatment is invalid, an endoscope should be taken after absolute neutrophil number recovery. (iii) Orbit/paranasal sinuses: High resolution CT/orbit MRI are available for diagnosis of paranasal sinuses haphalgesia, orbit cellulitis, rhinelcos, unilateral eye tear. Culturing and staining/biopsy can also be applied if necessary. Vancomycin can be added to treat orbit cellulitis. Liposomal amphotericin B can be applied if CT/MRI lead to a suspicion of aspergillosis and mucormycosis infection. (iv) Abdominal pain: Abdominal CT (prior consideration) or B-ultrasound. Metronidazole can be used to insure enough anaerobic treatment if anaerobic bacteria infection is suspected. (v) Rectum peripheral pain: Rectal examination; abdominal/pelvic CT should be taken. Enough anaerobic treatment should be given. Local treatment (sitz bath) can be applied. (vi) Diarrhea: Spindle bacillus is hard to detect. Rotavirus and norovirus can be arranged in winter or when there is an outbreak through faecal bacteria culture and/or parasite detection. Oral metronidazole can be strengthened when suspecting anaerobic bacteria and when waiting for laboratory report. (vii) Vascular path equipment: Insert swab into outlet for drainage culture. Vancomycin can be used at first or added after 48 h of failed empiric therapy. If the catheter//bag hole is infected or there is sepsis phlebitis, the indwelling catheter should be removed. (viii) Lung infiltration: Blood and sputum culture can be operated on. Bronchoalveolar lavage can be applied for failed primary treatment or diffused infiltration. Azithromycin or fluoroquinolone can be strengthened and aimed at untypical bacteria; oseltamivir can be used in an influenza outbreak; vancomycin or linezolid, G-CSF or GM-CSF can be used if MRSA is suspected. Blood serum galactomannan or β-gluca can detect suspected fungal infection.

Antiviral therapy. (i) Acyclovir: Serious cutaneous herpes simplex virus infection, single dermatomere varicella zoster virus, disseminated herpes simplex virus or varicella zoster virus infection can be treated with acyclovir. (ii) Valaciclovir: Herpes simplex virus infection or varicella zoster virus can be treated with valaciclovir. (iii) Famciclovir: Herpes simplex virus or varicella zoster virus can be treated with famciclovir. (iv) Ganciclovir: Treat cytomegalovirus infection. Intravenous immunoglobulin can be added to treat cytomegalovirus pneumonia and infection in other sites.

Application of Biological Response Modifier. Clinical trials have proven the significant effect of G-CSF/GM-CSF in shortening myelocytic duration and raising the myelocytic low point due to chemotherapy. G-CSF/GM-CSF can reduce fever and associated antibiotics by 50% [21].

20.3.3.3 Infection Prevention

Infection prevention involves environmental factors, microbiological cultivation and preventive medication.

Environmental factors. (i) Wash hands and wear respirator. (ii) Protective isolation. Avoid public contact and avoid contact with children who have just been injected with vaccine and other patients. (iii) Eat cooked food.

Microbiological cultivation. Cultivate physical secretion even if there are no signs of infection in order to accurately analyze a patient's condition.

Preventive medication. (i) Antibiotics: Preventive medication of antibiotics must be based on the risk degree of tumor patients infections. ①Low-risk infections mean that neutropenia lasts less than 7 d when assessing standard chemotherapy for a solid tumor. Prevention is unnecessary. ②Medium-risk infections mean neutropenia lasts 7 – 10 d, such as autologous stem cell transplantation, lymphoma, multiple myeloma, leukemia, purine analog therapy (for example, fludarabine, 2-CdA). The risk of infection is high. To use quinolones prophylaxis or not is based on the patient's condition. ③High-risk infections mean neutropenia lasts more than 10 d, including allogeneic hematopoietic stem cell transplantation, acute leukemia (induction, consolidation), alemtuzumab therapy, a high-dose hormone treating graft-versus-host disease(GVHD). Concrete preventive medication is the application of trimethoprim and cotrimoxazole in alemtuzumab therapy for at least 2 mon and CD4 $\geqslant$ 200. Use preventive medication with penicillin and TMP or SMX in high-dose hormone therapy for GVHD. (ii) Antifungal agents: Antifungal prophylaxis should be considered in medium and high risk infections. Available drugs are fluconazole, itraconazole, Micafungin, voriconazole, posaconazole, amphotericin B product, until neutropenia recovers or 75 d after transplantation. Posaconazole, voriconazole, hydatid interleukin and amphotericin B products can be used as preventive medication for exceptional GVHD until recovery. (iii) Antiviral drugs: For low-risk infection there is no need to prevent the herpes simplex virus. For medium-risk infections, acyclovir, famciclovir and valaciclovir are needed to be applied during neutropenia and for at least 30 d in autologous stem cell transplantation to prevent herpes simplex virus and herpes zoster virus. In a high-risk infection condition like acute leukemia (induction, consolidation), alemtuzumab therapy and allogeneic hematopoietic stem cell transplantation, acyclovir, famciclovir and valaciclovir can be used to prevent herpes simplex virus, herpes zoster virus and cytomegalovirus. (iv) Immunity: Vaccinate the patients and apply biological response modifiers, such as intravenous immunoglobulin, monoclonal antibodies, varicella-zoster virus immune globulin and others. (v) Others: Replace ductus venosus every 72 h in a neutropenia patient. Take a survey temperature via the rectum, use a rectal suppository and unnecessary rectal examination should be avoided; in addition, patients should be encouraged to exercise.

20.3.4 Microecological Control of Cancer Patients

Ecological control of microecology means the eco-network structure and its regularity adapts to the microecological environment, adroitly guiding action according to circumstances, improving the microecological environment, establishing microecological balance with higher biomass, not arbitrarily imposing any interference that is beyond its control capability or damaging any link, any factor of the microecosystem, or interfering with the normal process of substances, energy and gene flow. Comprehensive ecological control measures in principle include the improvement of the microecological environment, enhancement of the adaptability of the host, reasonable nutrition adjustments, effective and short-term use of antibiotics, increasing the colonization resistance, applying mixed physiological vaccine (also known as ecological agents), offering Chinese herbal medicine for strengthening the body's resistance to eliminate pathogenic factors, and so on.

Conditioned pathogens are the normal part of the microecology of the host, which only cause disease after the destruction of the normal homeostasis. So it is difficult to choose a particular type of antibiotic to kill these bacteria without affecting the other physiological bacteria. Conditioned pathogens that have penetrated into the mucous membrane can be cleared only by enhancing the body's resistance to maintain the body's normal physiological process. The immunity is deficient in cancer patients and the body's defense barrier is damaged because of radiotherapy and chemotherapy. It is easy to create favorable conditions for conditioned pathogens to cause opportunistic infections. It's necessary to consider controlling this type of infection by microecology, which might be a new treatment strategy by combining an ecosystem cure on the basis of the conventional control of radiotherapy and chemotherapy toxicity and infection. An ecosystem cure includes using antibiotics, probiotics and comprehensive ecological control measures.

Choosing antibiotics with a microecological perspective: (i) When choosing antibiotics, if the effect (sensitivity), toxicity and price are similar, the first choice is the drug with noninterference colonization resistance, especially antibiotics that can both control the infection and be a selective removal colonization. (ii) If the sensitive antibiotic interferes with colonization resistance, the dose, course and route of administration should also be controlled. In principle, narrow-spectrum antibiotics cause less interference than broad-spectrum ones. If clinically sensitive drugs with interference colonization resistance have to be adopted, a parenteral route of administration would be better, thus avoiding direct damage to bacteria of intestinal origin and ecosystems as much as possible. (iii) When choosing antibiotics in the clinic, a drug combination that causes damage to the anaerobe should be avoided. (iv) Selective removal colonization treatment of the pharynx and GI tract are applied to the control of the endogenous infection of all patients whose resistance is significantly reduced. In short, when choosing an antibiotic from the microecological perspective, we should not only consider the sensitivity, toxicity and cost, but also whether it interferes with colonization resistance or not.

The cancer patient's immunodeficiency and the toxicity of radiotherapy and chemotherapy greatly increase the possibility of opportunistic infections. The application of a new antimicrobial method to control conditional pathogenic bacteria has great significance. Application of microecological preparation can control both microflora and immunity. Pre-biotics can directly interact with pathogenic bacteria or conditional pathogenic bacteria, interact with physiological processes and immune response, nutrient competition and 'steric hinderance'. The latter can directly affect the colonization of other bacteria, enhance the immune mechanism, involve both humoral and cellular immunity. Metabolites such as short chain fatty acids and hydrogen peroxide can directly sterilize and stimulate intestinal peristalsis, and bacteriocin substances, enzymes and others, can improve the body's colonization resistance.

Combined ecological control is particularly important. The patient's microenvironment can be improved for cancer patients after radiotherapy and chemotherapy, such as air disinfection and filtration. We should improve the immunity of the host, enhance nutrition, apply a combination of antibiotics and microecological preparation, apply a combination of immune-potentiators and probiotics. Immuno-potentiators, including some biological agents, currently play a considerable role in auxiliary therapy of clinical diseases such as malignant tumors, autoimmune disease and exclusive reaction. For example, when biological agents relieve the condition of cancer patients, adding the ecological agents can prevent the side effects of chemotherapy such as alopecia and anorexia. They can also enhance colonization resistance and prevent infection.

Microecological prevention has been widely used in the prevention of many clinical diseases. With the continual expansion of its applications, it has gradually become one of the major biological weapons struggling with disease [22-25]. Microecology can resist the damage caused by carcinogens, adjuvantly reduce the side effects of radiotherapy and chemotherapy. A large number of experimental studies have confirmed that bifidobacterium and surface molecules, such as surface molecules lipoteichoic acid (LTA), cell wall whole peptidoglycan (WPG) *etc.*, can resist the damage from mutagens, carcinogens on the DNA. Therefore, long-term use of ecological agents like bifidobacterium can prevent tumorigenesis. In addition, GI side effects are common in radiotherapy and chemotherapy. Microecology also has adjuvant treatment effects in alleviating the side effects caused by radiotherapy and chemotherapy and improves the intestinal function.

20.4 Prospects

Infection epidemiology, diagnosis, therapy and prophylaxis of cancer patients who receive antitumor therapy will continue to be a great challenge. Accurate classification according to different degrees of hazard of cancer patients is helpful for making a proper choice of anti-infection measures. Injections of recombined cytokines and stem cells can shorten the duration of neutropenia and make it realistic to receive chemotherapy in high-doses. However, intensive chemotherapy will destroy the surface of mucosa and make it more susceptible to unknown pathogenic infection. The shortening of neutropenia duration can promote the application of new methods, such as oral administration of antibiotics to control fever and infection in milder oral mucositis cases. However, with the increasing application of stronger cytotoxic drugs, neutropenia duration and immunosuppression aggravation we are constantly facing the challenge of new pathogens. These pathogens are mostly multi-drug resistant bacteria and pathogenic fungi. Invasive mycosis, especially resistant yeast and filamentous fungi like aspergillus infection are difficult to diagnose and treat. Virus infection because of immune-deficiency, especially cell immunodeficiency, is also a challenge. Though infections make the anti-tumor therapy complicated, the improvement in diagnosis, the development of antibiotics, the application of immune-potentiators and the introduction of the concept of microbial control will provide great help to doctors in the diagnosis, therapy and prevention of infection.

Microecological therapy is a new strategy in the therapy of specified infections, such as radiotherapy, chemotherapy and molecular agented therapy. Microecological therapy includes using adapted antibiotics, vaccines, immune preparations and Chinese traditional medicine. Generally, antibiotics or vaccine preparation are firstly used for the pathogens from natural foci. For conditioned pathogens, we can choose an immune-potentiator. For a fungus with strong symbiosis, the only choice is probiotics, namely microecological preparation. Ecological control is a unified plan, simple use of ecological agents is not ecological control therapy. Microecological adjustments, especially the improvement of the microbial eco-system, the increase in the adaptability of the host, more attention to nutritional adjustment, using adapted antibiotics, targeted and flexible use of probiotics and a combined ecological therapy of different patients in various conditions, can cure patients by promoting the microecological balance and improving colonization resistance of the patients. Microecological therapy is one of the medical measures in clinical application now. It is worth further experimental study and clinical research. Currently, clinical research materials are still lacking for microecological control of cancer patients after radiotherapy, chemotherapy and molecular targeted therapy. Exploratory development remains to be continued in order to develop the application of probiotics in the prevention and therapy of infection in cancer patients after radiotherapy, chemotherapy and molecular targeted therapy.

References

[1] Chabner B A , Longo D L. Cancer Chemotherapy and Biotherapy: Principles and Practice. 4th Ed. Philadelphia: Lippincott-Raven Inc, 2006.
[2] Ma W W, Adjei A A. Novel agents on the horizon for cancer therapy. CA Cancer J Clin, 2009, 59: 111-137.
[3] Segal B H, Walsh T J, Gea-Banacloche JC, *et al.* Infections in the Cancer Patient. In: Cancer Principles & Practice of Ononlogy. 7th Ed. Philadelphia: Linppncott Williams & Wilking Inc, 2005.
[4] Perry M C. Management of Drug Toxicy. 4th Ed. Philadelphia: Lippincott Williams & Wilkins Inc, 2008: 116-294.
[5] Li T H, Perez-Soler R. Skin toxicities associated with epidermal growth factor receptor inhibitors. Targ Oncol, 2009, 4: 107-119.
[6] Higa G M, Abraham J. Biological mechanisms of bevacizumab-associated adverse events. Exper Review of Anticancer Ther, 2009, 9: 999-1007.
[7] Anderson R, Jatoi A H, Robert C, *et al.* Search for evidence-based approaches for the prevention and palliation of hand-foot skin reaction (HFSR) caused by the multikinase inhibitors (MKIs). Oncologist, 2009, 14: 291-302.
[8] Mundt A J, Roeske J C, Weichselbaum R R. Physical and biologic basis of radiation oncology. In: Cancer Medicine. 5th Ed. Philadelphia: Harcout Inc, 2000: 465-478.
[9] Perez C A, Brady L W, Halperin E C, *et al.* Principles & Practice of Radiation Oncology. 4th Ed. Philadelphia: Lippincott Williams & Wilkins Inc, 2005.
[10] Maschmeyer G, Haas A. The epidemiology and treatment of infections in cancer patients. Int J Antimicrob Agents, 2008, 31: 193-197.

[11] Bodey G P, Buckley M, Sathe Y S, *et al*. Quantitative relationship between circulating leukocytes and infection in patients with acute leukemia. Ann Inteen Med, 1966, 61: 328-340.

[12] Mackall C l, Fleisher T A, Brown M R, *et al*. Age, thymopoiesis and $CD4^+$ T-lymphocyte regeneration after intensive chemotherapy. N Engl J Med, 1995, 332: 143-149.

[13] Sun T, Tenenbaum M J, Greenspan J, *et al*. Morphologic and clinical observations in human infection with Babesia microti. J Infect Dis, 1983, 148: 239-248.

[14] Santos J I. Nutrition, infection and immunocompetence. Infec Dis Clin North Am, 1994, 8: 243-267.

[15] Bochud P, Calandra T, Francioli P. Becteremia due to viridans streptococci in neutropenic patients: A review. Am J Med, 1994, 97: 256-264.

[16] Wells VD, Wong ES, Murray BE, *et al*. Infectios due to β-lactamaes producing, high-level gentamicin-resistant *Enterococcus faecalis*. Ann Intern Med, 1992, 116: 285-292.

[17] Walsh T J, Pizzo P A. Nosocomial fungal infections: a classification for hospital-acquired fungal infections and mycoses arising from endogenous flora or reactivation. Annu Rev Microbiol, 1988, 42: 517-545.

[18] Hughes W T, Armstrong D, Bodey G P, *et al*. From the infectious diseases society of America. Guidelines for the use of antimicrobial agents in neutropenic patients with unexplained fever. J Infect Dis, 1990, 161: 381-396.

[19] NCCN practice guidelines for prevention and treatment of cancer related infections. 2009 v2. http://www.nccn.org.

[20] Gridell C, Maione P, Amorosob D, *et al*. Clinical significance and treatment of skin rash from erlotinib in non-small cell lung cancer patients: Results of an experts panel meeting. Crit Rev Oncol Hematol, 2008, 66: 155-162.

[21] NCCN practice guidelines for myeloid growth factors. 2011 v1. www.nccn.org.

[22] Park J, Floch M H. Prebiotics, probiotics, and dietary fiber in gastrointestinal disease. Gastroenterol Clin North Am, 2007, 36: 47-61.

[23] Lenoir-Wijnkoop I, Sanders ME, Cabana MD, *et al*. Probiotic and prebiotic influence beyond the intestinal tract. Nutr Rev, 2007, 65: 469-489.

[24] Khani S, Hosseini H M, Taheri M, *et al*. Probiotics as an alternative strategy for prevention and treatment of human diseases: A review. Inflamm Allergy Drug Targets, 2012, 11: 79-89.

[25] Yan F, Polk D B. Probiotics and immune health. Curr Opin Gastroenterol, 2011, 27: 496-501.

21

Infectious Microecology in Immunodeficiency Diseases

Jin Yang, Nanping Wu *
State Key Laboratory for Diagnosis and Treatment of Infectious Diseases, Institute of Infectious Diseases, the First Affiliated Hospital, School of Medicine, Zhejiang University; Collaborative Innovation Center for Diagnosis and Treatment of Infectious Diseases, Hangzhou, 310003, China
* E-mail: flwnp@yahoo.com.cn

Extensive research during the past two decades has provided new insight into the pathogenesis of the human immune deficiency virus-1 (HIV-1). With the emergence of the new theory of chronic immune activation, it has now become clear how the continuous T cell depletion occurred in HIV-1 progression. The improved understanding of mucosal immunity, especially in gut-associated lymphoid tissue (GALT), has led to a more rationale analysis of HIV-1 pathogenesis. With the role of microbiota looming on the horizon, new clues linking the gap between mucosal immunity dysfunction and immune activation depict an intricate continuum of the battle between the virus and the host. Altogether, the overview of the HIV-1 pathogenesis, in the view of the microecology presented here, further highlights the complexity of the mechanism of HIV-1, and sheds light on the novel strategy to control this notorious agent.

Residing in the host, dynamic microbial ecosystems establish and maintain the microbiota in humans [1], shaping a multitude of vital functions both for the microorganism and the host. In this regard, Lederberg has emphasized the importance of having a broad ecological view of our relationships with microbes [2], that could affect the physiology and pathology of the host. As the largest reservoir of human flora, the gut and GALT accounts for 70% of the body's immune system [3], while worldwide, gastrointestinal (GI) disease continues to account for a high proportion of the presentations of symptoms of HIV-1 infection [4].

21.1 HIV, Immune Deficiency, Old View

Discovered in 1981 and identified in 1983, HIV, linked to immune deficiency in general as characterized by an insidious deterioration of the cellular immune system, invariably leads to the deadly syndrome that is known as acquired immunodeficiency syndrome (AIDS). AIDS remains one of the most devastating pandemics in human history based on its global dissemination and lethality [3]. Though currently highly active antiretroviral therapy (HARRT) represents a halfway triumph, there is no cure for HIV-1 infection, and the possibility of developing a vaccine in the near future appears unlikely [5].

The high mutation rates derived from the reverse transcription step in the life cycle, and inherent propensity for recombination of HIV-1 contribute to the remarkable plasticity of its quasispecies [6], at least in part, incarnating the ecological behavior. The diversity of the HIV-1 population was shown among infected individuals (interhost variability), and also within each infected individual (intrahost variation) over time and among different compartments [4]. The micro-ecosystem of HIV-1 provides the viral generations with the capacity to adjust to immunologic, pharmacologic or other micro-environmental selection pressure. Thus, biological properties of HIV-1 strains differ from the chemokine receptor usage in cell tropism, phenotype and virulence.

Due to the high affinity of the primary receptor CD4 for the viral gp120, $CD4^+$ T cells constitute the main target cells for HIV-1. HIV-1 preferentially infects $CD4^+$ T helper lymphocytes and leads to elimination of the host immune cells, which is called immune deficiency. Many lymphocytes expressing CD4 and such co-receptors as CCR5 or CXCR4 may potentially be infected with HIV, and serve as vehicles or reservoirs that disseminate or cover up the viral infection. The degree of immunodeficiency, as reflected by the onset of opportunistic diseases, correlates closely with plasma $CD4^+$ T cell counts. At the advanced stage of the infection, the rate at which immune-suppression develops also closely reflects the levels of HIV-1 RNA in plasma, such that the higher the HIV-1 viral load, the greater the loss of circulating $CD4^+$ T cells [7]. Overall, HIV-1 is responsible for subverting the host immune system mainly attributable to rapid production and mutation of the virus, integration into the host genome, down-regulation of MHC-I, impairing Th1 response of CD4 helper T lymphocyte, infecting cells in the different compartments, *etc.* What is true is that individuals in whom HIV infection has been established cannot eliminate the virus from their bodies [8]. The characteristic feature of HIV-1 has long focused on the gradual depletion of $CD4^+$ T cells, as indicated in the 'tap-and-drain' model [9], but the classical view of HIV pathogenesis could not reasonably explain immunodeficiency ($CD4^+$ T cell depletion) and immunosuppression (immune activation) in HIV natural history [10].

21.2 Immune Activation in HIV Infection

In fact, the degree of productive HIV-1 infection of peripheral blood $CD4^+$ T cells is very low (0.01% – 1%) [11]. HIV-1 preferentially infects activated $CD4^+$ T cells, most of which are destined to die despite of their infection, because of the activation process itself [12]. High death rates of the $CD4^+$ T cell do not change immediately with the initiation of HARRT [13]. The evidence is mounting that HIV does not directly cause the effect of T cell continuous depletion.

First observed in mice, chronic immune stimulation may lead to immunodeficiency. Following the kinetics of HIV-1 infection, altered immune states such as increased T cell turnover, polyclonal B cell activation, increased frequencies of T and B cells with activated phenotypes and increased serum concentrations of proinflammatory cytokines and chemokines, could be observed during all the stages of HIV-1 infection and termed "immune activation" [14]. In the simian immunodeficiency virus (SIV) infection model, immune activation was a critical factor that distinguishes pathogenic from nonpathogenic SIV infection [15]. The clinical studies suggested that chronic immune activation is associated with impaired immune reconstitution in patients on HAART. It was soon supposed to be one of the strongest predictors of disease progression [16].

An "immune-activation model" was proposed to explain HIV-1 pathogenesis. The Brenchley group suggested that immune activation could affect T cell depletion by imposing an unremitting homeostatic strain, such that persistent rounds of activation and death would slowly drain the memory and naive T cell pools [17]. Due to the finiteness of the source of naive $CD4^+$ T cells, persistent rounds of activation and death would lead to the $CD4^+$ T cell pools being singled out for depletion while leaving the $CD8^+$ T cell pools relatively intact until the advanced stages of the disease.

Apart from the positive feedback of promoting T cell proliferation and the partial restoration of memory $CD4^+$ T cells to limit virus replication, immune activation has an overwhelmingly negative effect on immune dysfunction. A line of evidence disclosed the mechanism of T cell depletion induced by immune activation: High turnover of $CD4^+$ and $CD8^+$ T cells imposes a strain on their homeostatic mechanisms, resulting in a decrease in the overall half-life of T cells [18]; Clonal exhaustion of T cells may ultimately drain memory T cell pools [11]; Inflammatory damage to lymphoid tissues may trigger thymic dysfunction and TGF-β mediated fibrosis of lymph nodes which are, in turn, associated with abnormal retention of effector type T cells and poor immune reconstitution with HAART [14]; Immune activation results in the generation of activated T cell targets for the virus itself, further driving viral replication [14].

Based on this theory, through the induction of immune activation, HIV-1 generates its own substrate for replication. Moreover, as a "self-renewing" source for replenishing tissue memory CD4 T pools, central memory CD4 T cells, undergoing the persistent rounds of activation, infection and depletion, may correlate most closely with progression to AIDS [19]. Thus, for a quasi-perpetuating relationship between virus and immune activation, constant damage to the cellular sources of the $CD4^+$ T cell compartments, further exacerbates the progressive net loss in $CD4^+$ T pools and inevitably leads to AIDS [14].

Immune activation theory seems perfect to meet the observations found in HIV infection. Some questions remain regarding this hypothesis, however. In the chronic phase of infection, HIV-1 infects only a small minority (<1.0%) of $CD4^+$ T cells [20], yet a much higher percentage of many different cell types possess the activated phenotype [18]. How does increased activation or apoptosis of a small percentage of infected $CD4^+$ T cells lead to activation of large populations of uninfected cells [7]?

When revisiting the natural history of HIV-1 infection in the acute phase, a rapid and striking loss of $CD4^+$ T cells from the intestinal lamina propria was found in the SIV macaque model within days of infection, regardless of the inoculation route [21]. At the peak of viremia, 60% of mucosal memory $CD4^+$ T cells are infected following the 80% of infected cells that are depleted within only 4 days. Similarly in human HIV-1 infection, a preferential and profound depletion of $CD4^+$ T cells occurs within the GI tract [22], at a time when little or no $CD4^+$ T cell depletion was evident in the peripheral blood. This depletion is chiefly a direct consequence of interaction of the virus with its $CD4^+$ T cells targets, corresponding to a direct link between immune activation and the bulk of $CD4^+$ T cells depletion in chronic HIV infection.

The critical observation of the T cell depletion occurring in gut mucosal is that the mucosal $CD4^+$ T cells compartment accounts for most total-body $CD4^+$ T cells. Intestinal $CD4^+$ T-cells depletion might be an early feature of HIV-1 infection and intestinal $CD4^+$ T-cells depletion is more significant than depletion of $CD4^+$ T cells in the peripheral blood [22]. In fact, significant GI tract $CD4^+$ T-cell depletion was characteristic of all stages of HIV-1 infection. Thus, the cause of immune activation needs further investigation based on its seedbed.

21.3 Breakdown of Mucosal Immunity

In 1984, Kotler and colleagues documented that "The histological finding suggests that a specific pathologic process occurs in the lamina propria of the small intestine and colon in some patients with the syndrome" [23]. Indeed, occurring from the acute phase of the infection through the advanced stage, HIV-1 enteropathy, involving diarrhea, increased GI inflammation, increased intestinal permeability, and malabsorption [24], is the most common sign of the disease.

The GI tract has attracted ever-increasing attention in HIV-1 infection in recent years, as the mucosal surface of the GI tract forms a unique anatomical and physiological niche, serving as a structural and immunological barrier against the microorganisms of the outside world [14]. It has been well established that the GI tract harbors the majority of the body's complement of immune cells [25]. When compared with the peripheral blood lymphocytes, more than 90% of intestinal lymphocytes exhibit a memory phenotype [26]. GI tract lymphocytes are significantly more activated than peripheral blood lymphocytes because of constant exposure to a myriad of food and microbial antigens. Furthermore, up to

70% of GI tract lymphocytes express CCR5 whereas only approximately 20% of peripheral blood lymphocytes express CCR5 [27]. Thus the vast population of activated memory $CD4^+$ T cells with abundant expression of co-receptors provides HIV-1 with an ideal environment to establish infection [28]. Gene expression profiles of GI tract biopsies reveal that genes associated with cell cycle regulation, lipid metabolism, epithelial cell barrier and digestive functions are down-regulated in HIV-1 infected individuals [29].

Considering that most HIV infections are established through mucosal transmission, defects of mucosal immunity are consistently present in all stages of HIV-1 disease [29]. Studies of SIV-infected rhesus macaques show that although most infected monkeys are rapidly depleted of mucosal $CD4^+$ T cells in the acute phase, monkeys that can partially reconstitute mucosal $CD4^+$ T cells with T cell proliferation after the acute phase progress to AIDS at a slower rate than those that are unable to mount a reconstitutive response [30]. Thus, the integrity of the mucosal immune system, particularly the ability to maintain $CD4^+$ T cell numbers above some critical threshold, seems to be a key determinant of the rate of progression to AIDS.

Further studies comparing T-cell and virus dynamics in different anatomical sites including the GI tract, bronchoalveolar lavage (BAL) and blood indicated the following: Depletion of GI $CD4^+$ T cells is associated with high frequencies of infected $CD4^+$ T cells, while BAL $CD4^+$ T cells are not massively depleted during the chronic phase; HIV-specific T cells, are present at low frequencies in the GI tract compared to blood. On the contrary, higher frequencies and increased functionality of HIV-specific T cells were observed in BAL compared to blood; Infection frequencies of BAL $CD4^+$ T cells are similar to those in blood; A rapid, dramatic, and largely irreversible depletion of mucosa-associated lymphoid tissue-based memory $CD4^+$ $CCR5^+$ T-cells is a key determinant of disease progression in HIV-1 infected individuals [29, 31]. Taken together, these data suggest the global depletion of mucosal $CD4^+$ T cells, accompanying the different effect in different anatomical sites.

How to explain this differentiation at the mucosal surface level? According to the available data, it has become apparent that the GI mucosal barrier suffers a serious structural and immunological attack immediately in HIV-1 infection and that this damage may be linked to the successive immune activation in the advanced stage of the disease [14]. HIV-1 enteropathy involves inflammatory infiltration of lymphocytes and damage to the GI epithelial layer, including villous atrophy, crypt hyperplasia and villous blunting but, importantly, these pathological changes could occur in the absence of a detectable pathogen [32]. These phenomena raise a new issue if there are other key mediators linking the mucosal immunity and chronic immune activation. Or in another word, are there any factors that classify these phenomena as a whole?

21.4 Solving the Problem from the Microecological Viewpoint

A line of evidence for understanding the causes of immune activation and mucosal immunity in HIV could provide mechanistic insights into the biological mechanism of HIV-1 pathogenesis [17]. Immune activation theory, based on immunological and virological sense, could adequately account for the overall course of the disease. Even though immune activation may not be the primary cause of memory $CD4^+$ T cell depletion, it determines the rate of progression to AIDS.

The human endogenous intestinal microflora is an essential "organ" in providing nourishment, regulating epithelial development, and instructing innate immunity, though as yet basic features remain poorly described [33]. A combination of the physical barrier generated by mucosal cells, innate immune factors such as phagocytic activity, secreted immunoglobulin A, intraepithelial lymphocytes and antigen-specific T cells either prevent or restrict systemic translocation of potentially pathogenic commensal and noncommensal flora. The mucosal barrier limits local inflammation and blocks pathogens accessing the immune system, but remains mostly "ignorant" of the potential antigens derived from the flora [34], which would have unexpected responses if the integrity is broken.

Damage to the barrier function of the GI tract, as occurs in inflammatory bowel disease and immune-suppressive therapy for hematopoietic cell transplantation, results in the microbial translocation of lipopolysaccharide (LPS), which correlates with systemic immune activation. Brenchley *et al.* first reported that chronically HIV-1 infected individuals have significantly increased levels of plasma LPS when compared to uninfected individuals [35].

LPS is an indicator of the translocation of microbial products that stimulate the immune system. As early as 1979, in a gnotobiotic mouse model, the translocation of certain indigenous bacteria from the GI tract to the mesenteric lymph nodes and other organs was found [36]. *In vivo*, the magnitude of exposure to LPS is commensurate with a graded dose response of an acute-phase inflammatory response [37]. Increased LPS is correlated with up-regulated levels of soluble CD27, CD14, tumor necrosis factor (TNF), and LPS binding protein, as well as down-regulated levels of antibodies directed against LPS core antigen. In addition, LPS levels are associated with both the frequency of activated memory $CD8^+$ T cells and plasma levels of the proinflammatory cytokine IFNα. Elevated plasma LPS concentration was found to be associated with defective innate and mitogen responsiveness [38].

However in those elite controllers, the degree of $CD4^+$ T cell depletion is closely associated with the level of T cell activation, which is in turn associated with significantly increased levels of plasma LPS [14]. In pediatric HIV-1 infection, microbial translocation and LPS production are associated with persistent monocyte/macrophage activation independent of viral replication or T-cell activation [39]. Plasma LPS levels remained elevated in patients receiving effective HAART [33]. These findings implicate microbial translocation as a major force driving chronic inflammation in HIV-1 infected individuals, even receiving HAART.

It is tempting to speculate that after dysregulation of mucosal integrity initiated by HIV-1 invasion, local mucosal immunity in GI reduces control of commensal microorganisms and leads to consequent increased systemic translocation. In addition to an impaired mucosal immune function, damage to the integrity of the epithelial barrier itself would exacerbate matters further. Mediators from translocation, such as LPS, were secreted into the systemic circulation and result in chronic immune activation.

The finding that immune activation decreases with the inhibition of viral replication is consistent with an integral function for HIV-1 in compromising mucosal pathogen control and perpetuating immune activation [17]. The relationship between immune activation and HIV-1 may not always be direct, but it is clear that with no virus there is no immune activation. It should not be concluded, however, that the virus or immune system itself plays a major role in disease progression in pathogenic infection. On the contrary, in an infectious microecological manner, the factors such as mucosal integrity, microbe translocation, HIV-1 invasion and local systemic immune responses play important roles at all stages of the disease (Fig. 21.1).

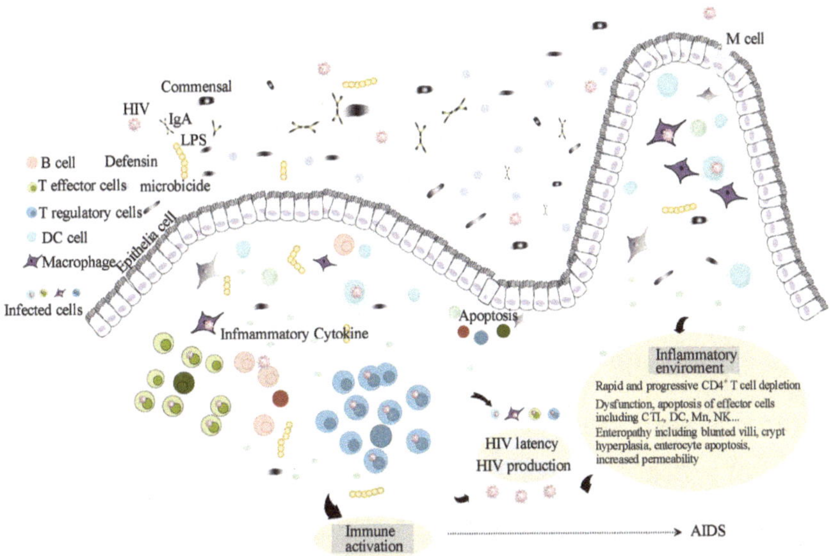

Fig. 21.1. Mucosal immunity in HIV infection: The original physiological balance between the environmental factors and mucosal immunity is disrupted by HIV-1 infection. Damage to the mucosal barrier results in the translocation of commensal bacteria and the associated component, like LPS, which stimulates the immune system to activation. Chronic immune activation, with persistent excessive inflammation, enteropathy, further magnifies the continual and accelerated production of the virus, along with the immune system exhaustion. Ultimately, AIDS occurs

Controversy remains as to whether most lymphocytes in humans really reside in the gut [40]. Given that there is greater understanding of the biology of HIV-1 based on the mucosal immunity and microorganism translocation, it is logical to extend the pathogenesis of HIV-1 disease to other mucosal space and microbiota dysregulation.

21.5 HIV-1 Infection in Mucosa Scope

It is well established that the intestinal mucosa serves as a viral reservoir, regardless of levels of plasma viral load or HAART. HIV-1 RNA was detected in distal duodenal mucosa in HAART-naive patients with high plasma viral loads, as well as in patients on HAART with plasma viral loads below the limit of detection and in patients on HAART with virological failure [41]. Jejunal DC phenotype cells isolated from jejunal lamina propria rapidly took up cell-free HIV-1 through the mucosa and transmitted in transit to blood or intestinal lymphocytes, and subsequently to T cells [42].

Moreover, at different anatomic sites, columnar, polarized epithelial cells in intestinal, rectal, colonic and endocervical mucosa form tight junctions to divide the cells into apical and basolateral domains [32], whereas pluristratified epithelial cells in ectocervical and vaginal epithelium lack a polarized plasma membrane and tight junctions, allowing a rich network of intraepithelial dendritic cells and langerhans cells (LCs) to integrate into the epithelium [43]. HIV-1 uses different strategies, including columnar epithelial cell transcytosis and dendritic/langerhans cell transport [42], to cross the epithelium barrier. These successive studies demonstrate an effective infection potential of HIV-1 in almost all the mucosal types.

Studies of the foreskin model showed that HIV-1 infected cells form virological synapses with apical foreskin keratinocytes, and lead to a polarized budding of HIV-1, which is rapidly internalized by LCs in the inner foreskin. In turn, LCs migrate toward the epidermis-dermis interface to form conjugates with T cells to transfer HIV-1 [44].

The predominant mode of HIV-1 transmission globally is from sexual practices. Study of the early events of HIV-1 infection of human cervico-vaginal tissue showed that virtually all T cells are of the effector memory phenotype with CCR5 expression in cervico-vaginal tissue, preferentially supporting the productive infection of R5 HIV-1. HIV-1 occurred preferentially in activated $CD38^{+}CD4^{+}$ T cells and was followed by a similar activation of bystander $CD4^{+}$ T cells that may amplify viral infection [45]. In different *ex vivo* models, HIV-1 rapidly penetrated both intraepithelial vaginal LCs and $CD4^{+}$ T cells. By contrast, HIV-1 entered $CD1a^{+}$ Langerhans cells primarily by endocytosis, by means of multiple receptors and virions persisted intact within the cytoplasm for several days [46]. These findings shed light on the very earliest steps of mucosal HIV infection *in vivo*.

Importantly, the vagina is an original biotype with its own ecosystem, with a dynamic but very unstable equilibrium. Many studies showed that HIV-1 is detected more frequently and at higher levels in the lower genital tract of HIV-seropositive women with bacterial vaginosis (BV). Increasingly abnormal flora or severe BV is associated with increasing risk of HIV. The prevalence of HIV-1 infection increased linearly with an increasing Nugent's score (BV score of Gram stain), while the prevalence of *T. vaginalis* increased suddenly from the patients with a Nugent's score of $<$ or $=$ 3 to 4 [47]. In this regard, *T. vaginalis* might be responsible for the change in normal vaginal flora and be one of the ecological causes of HIV-1 infection. Similar studies showed a significantly reduced quantity of lactobacilli in HIV-infected women. The prevalence of H_2O_2-producing lactobacilli was lower in HIV-1 positive as compared to HIV-1 negative women [48]. Recent *in vitro* studies showed that genital tract secretions from women with BV or flora associated with BV induce HIV expression in infected cells. Thus, alteration in vaginal microbial communities is associated with an increased risk of HIV-1 infection and transmission [49].

21.6 Through Microbiota and Beyond

It is not surprising to witness that there has been an explosion of interest in identifying microbial inhabitants of humans and understanding their role in health and disease [50], due to the emerging links between microbial community structure, function, infection and disease. Microbiota (microbial genetic repertoire) is approximately 100-times greater than the human host, and co-evolution has shaped this human-microorganism interaction into a homeostasis relationship in which gut bacteria extends well beyond the metabolic function, but to the unified immune adaptations [51]. An ever-increasing body of evidence implicates the homeostasis of microbiota in defining the state of health [52], while perturbation of this homeostatic coexistence has been strongly associated with disease.

From the viewpoint of microbiota interacting with the immune system, homeostasis is maintained by a hierarchy of immunological barriers. Immune mediators limiting direct contact between the intestinal bacteria and the epithelial cell surface, detecting and killing bacteria that manage to penetrate intestinal tissues, decrease the likelihood of pathogen invasion. By distinctive anatomical adaptations, it is the systemic immune "ignorance" towards the microbiota [51].

The gut-barrier acts by protecting against pathogens by elaborating and releasing protective peptides, cytokines, chemokines, and phagocytic cells. The mucosal surface is constantly sampling luminal contents and making molecular adjustments at its frontier [53]. A large epidemic study evaluating plasma levels of bacterial LPS and 16S rRNA (16S rDNA) in chronic HIV-infected individuals showed that plasma levels of 16S rDNA were significantly higher in infected subjects than in uninfected subjects, and correlated well with LPS levels. Higher levels of 16S rDNA were associated with higher levels of T cell activation and with lower levels of CD4 T cell restoration during antiretroviral therapy. HARRT therapy reduces but does not fully normalize plasma levels of bacterial 16S rDNA. During therapy , high levels of 16S rDNA are strongly correlated with reduced increases in the $CD4^+$ T cell count, irrespective of plasma HIV-1 RNA levels. These findings are consistent with the importance of microbial translocation in immunodeficiency and T cell homeostasis in chronic HIV-1 infection [53]. Though flora constitution is largely unknown due to technological limits, evidence highly suggested the conditional pathogen is co-functioning during HIV-1 infection.

Apart from the proinflammation mediators induced by microbe secretion [54], a recent study by the Kantor group identified the metabolites processed by flora assisting the HIV-1 infection in the gut [55]. The majority of reverse-transcribed, nuclear-imported viral genomes remain episomal in the viral life cycle, either as linear or circular DNA. Gene expression and replication of non-integrating HIV-1 was significantly stimulated upon exposure to histone deacetylase (HDAC) inhibitors in the form of various short-chain fatty acids (SCFAs) ,which were known to be endogenously produced by normal microbial-gut flora. Furthermore, genetic and functional crosstalk between episomal and integrated viral genomes, resulted in recombination between integrated and non-integrated HIV-1, as well as mobilization of episomal vector genomes by productive viral particles encoded by integrated viral genomes [56].

$CD4^+$ regulatory T (Treg) cells are an essential component of host-microbiota mutualism. Recent study showed a prominent human commensal, Bacteroides fragilis, mediate the conversion of $CD4^+$ T cells into $Foxp3^+$ Treg cells that produce IL-10 during commensal colonization, to direct the development of $Foxp3^+$ Treg cells with a unique "inducible" genetic signature [57]. Functional $Foxp3^+$ Treg cells are also produced by polysaccharide A (PSA) during intestinal inflammation through Toll-like receptor 2 signaling. The significance of this study lies in the demonstration of *B. fragilis* co-opting the Treg lineage differentiation pathway in the gut to actively induce mucosal tolerance [57]. In HIV-1 infection, increased Treg turnover status, indicated by higher expression of proliferation marker Ki-67 and apoptosis marker caspase-3 and Annexin-V, was observed. The turnover level of Treg was positively associated with disease progression and immune hyperactivation. The HAART treatment decreased the turnover and activation levels of Treg cells in complete responders [58]. Since regulatory responses mediated by Treg cells subsets are necessary for homeostasis (regulation of effector T cells), especially in the presence of more potentially pathogenic microbial constituents in the microbiota [51], a possible axis linking opportunistic infection, Treg, HIV repeat priming, immune activation would be anticipated.

Commensal microbiota control Th17 differentiation in the gut [59]. Previously observed in monkeys, Th17 cells are only depleted in the pathogenic simian immunodeficiency virus (SIV) infection of rhesus macaques, which correlates with the progression to AIDS in these primates, whereas they remain intact in the nonpathogenic SIV infection of African green monkeys or sooty mangabeys. Recent studies on patients with recurrent mucocutaneous candidiasis have led to the discovery of mutations in two genes, CARD9 and DECTIN-1, which are important for the production of the Th17-driving cytokines IL-1β, IL-6, and IL-23 [60]. Studies of the peripheral blood of HIV-1 positive patients have shown a decreased Th17:Th1 ratio [60].

A pronounced loss of mucosal Th17 $CD4^+$ T cells in the SIV-infected rhesus macaque model of AIDS is linked to impaired immune responses in the gut mucosa to an enteric pathogen, *Salmonella typhimurium*, which leads to the lack of local control of the pathogen and its translocation [56]. Recent studies suggested that the replenishment of Th17 $CD4^+$ T cells in the gut mucosa during HAART, or during nonpathogenic SIV infections in the nonhuman primate models, correlates with better restoration and function of the gut mucosal immune system [61]. Failure to restore Th17 cells in GALT during HAART might impair both the recovery of the gut mucosal barrier and the clearance of microbial products [62].

Therefore, HIV-1 mediated loss of Th17 cells from the GALT impairs mucosal integrity and innate defense mechanisms against gut microbes. Translocation of microbial products from the gut, in turn, correlates with increased immune activation in chronic HIV-1 infection and may further damage the immune system by increasing viral and activation-induced T cell death, by reducing T cells reconstitution due to tissue scarring, or by impairing the function of other cell types, such as γδT cells and epithelial cells [63].

Occurring at the level of the mucosal surface, we would predict that the refractory HIV pathology lies in the dysfunction of the homeostasis between microbiota and host immunity (Fig. 21.2). Through the microbiota system, and beyond, the virus further impedes local immune reconstitution *via* continued immune activation and depletion of target cells during all phases of infection. Similar turmoil may occur at all mucosal surfaces [17], though this needs further investigation.

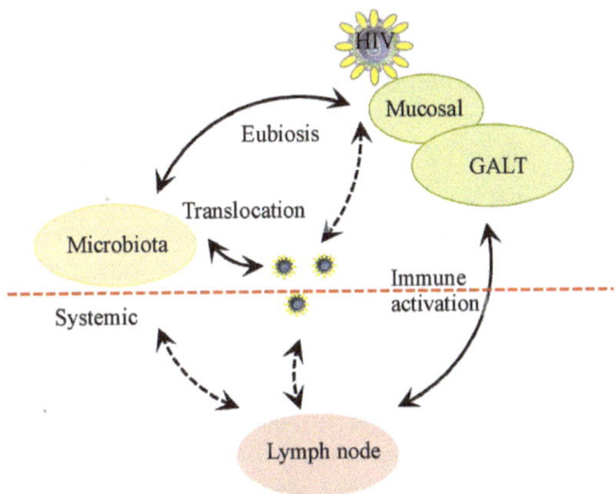

Fig. 21.2. Central role of microbiota in HIV immunopathogenesis

Mucosal barrier invaded and disrupted by HIV-1, leads to the imbalance between microbiota and local immunity, and finally results in microbial translocation. Ectopia bacteria and related components stimulate immune cells systemically, creating the proinflammatory milieu associated with chronic HIV infection. Chronic immune activation leads to the generation of more target cells for the virus, *via* the homing pathway of lymph tissue, thus perpetuating viral replication and mucosal damage, and consequently resulting in lymph node fibrosis and immune deficiency. Microbiota related immune status is disordered, initiated by HIV-1 infection at the mucosal level, and this turbulence is further reinforced at the systemic level following progress of HIV-1 disease.

21.7 Probiotics and HIV

Ifinfectious microecology hypotheses plays an important role in HIV-1 pathogenesis, then the therapy targeting the microbiota regulation may have an advantageous effect [64]. Probiotics are live microbial feed supplements that beneficially affect the host by improving eubiosis and promoting health benefits. In contrast to the colitogenic effects of enteric bacteria, clinical and experimental studies showed that specific probiotic strains prevent pathogen adherence and invasion of the epithelium. Short chain fatty acid, produced by the metabolic effects of probiotics, affect epithelial cell metabolism, turnover and apoptosis. Probiotics alter expression and redistribution of tight junction proteins and reduce intestinal permeability. Through secreted molecules, probiotics influence the innate inflammatory response of epithelial cells to stimuli from the gut lumen, and reduce mucosal inflammation. Through effects on dendritic cells , they influence naive T cells in the lamina propria of the gut and thus influence adaptive immunity [65]. Although the molecular understanding of probiotics is improving, the overall mechanisms induced by probiotics are just starting to be unraveled. Derived from the microecology view of HIV pathogenesis, along with HIV infections having episodes of diarrhea and frequently experiencing malabsorption associated with possible bacterial overgrowth, it is logical to speculate that probiotics may be therapeutic in preventing HIV infection. The earlier study showed that oral administration of the probiotic *Lactobacillus plantarum* 299v could improve nutrient status and promote growth in children congenitally exposed to HIV. After oral supplementation, *Lactobacillus plantarum* 299v was colonized and elicited a specific systemic immune response [66].

A recent investigation was conducted on 77 HIV-1 infected children receiving a probiotics formula containing *Bifidobacterium bifidum* and *Streptococcus thermophilus*. There was an increase in the mean CD4 count in the probiotics group as compared with a small decrease in the control group. A similar reduction in liquid stool consistency was also observed [67]. Another study was investigated in sub-Saharan Africa. Conventional yogurt fermented with *Lactobacillus delbruekii* var bulgaricus and *Streptococcus thermophilus* was supplemented with probiotic *Lactobacillus rhamnosus* GR-1 and *L. reuteri* RC-14. The mean CD4 cell count remained the same or increased in almost all the probiotic-treated subjects. Diarrhea, flatulence, and nausea were resolved in all the probiotic-treated subjects within 2 days [67].

Although significant progress has been made in this field, more fundamental research is required to better understand vaginal ecology to maximize probiotic formulations, since the significant diversity of microorganisms add complexity to the microbiota regulation. Probiotic *Lactobacillus rhamnosus* GG (LGG) formula was evaluated in ameliorating gastrointestinal symptoms in HIV-infected patients on antiretroviral therapy. No significant differences in noninfectious diarrhea or gastrointestinal symptoms compared to a placebo could be observed in a crossover study [68]. But indeed, exogenously supplied substances used as a probiotic may prove a cost-effective, natural-initiated method to prevent or treat HIV-1 infection.

A probiotic may act indirectly through treating and preventing recurrent bacterial inflammation or directly by secreting endogenous and exogenous substances that block HIV-1 transmission. Several new strategies have been proposed.

A highly colonizing probiotic strain of *E. coli*, Nissle 1917, secreting HIV-gp41-hemolysin A hybrid peptides that block HIV fusion and entry into target cells was engineered. By using an appropriate combination of cis- and transacting secretory and regulatory signals, micromolar secretion levels of the anti-HIV peptides were achieved. The Nissle 1917 were capable of colonizing mice for periods of weeks to months, predominantly in the colon and cecum, with lower levels present in the rectum, vagina, and small intestine. Histological and immunocytochemical examination of the colon revealed bacterial growth and peptide secretion throughout the luminal mucosa and in association with epithelial surfaces [69].

The probiotic organism *Lactobacillus reuteri* RC-14 colonizing in the human vagina, has been genetically modified to produce anti-HIV proteins which were capable of blocking the main steps of HIV-1 entry into human peripheral blood mononuclear cells. The HIV-1 entry or fusion inhibitors were fused to the native expression and secretion signals of BspA, Mlp or Sep in *L. reuteri* RC-14 and the expression cassettes were stably inserted into the chromosome. *L. reuteri* RC-14 expressing fusion proteins were able to bind a single or dual tropic co-receptor-using HIV-1 primary isolates [70]. This is the first study to show that a well-documented and proven human vaginal probiotic strain can express potent functional viral inhibitors, which may potentially lower the sexual transmission of HIV. Preclinical and clinical studies to test probiotic bacteria for these purposes are underway [70].

The critical role of the "forgotten organ", the microbiota, in generating a variety of functions to sustain health has been suggested in recent years [71]. Central to this beneficial interaction between the microbiota and host is the manner in which the bacteria contained within the gut "talk" to the immune system [72]. Into this landscape comes the novel understanding of HIV-1 pathogenesis in an infectious microecological manner. It is particularly important in the dilemma of curing or preventing HIV-1 infection today.

References

[1] Claesson M J, Jeffery I B, Conde S, *et al.* Gut microbiota composition correlates with diet and health in the elderly. Nature, 2012, 488: 178-184.

[2] Lederberg J. Infectious history. Science, 2000, 288: 287-293.

[3] Wilcox C M, Saag M S. Gastrointestinal complications of HIV infection: Changing priorities in the HAART era. Gut, 2008, 57: 861-870.

[4] Anastassopoulou C G, Kostrikis L G. Viral correlates of HIV-1 disease. Curr HIV Res, 2005, 3: 113-132.

[5] Baden L R, Dolin R. The road to an effective HIV vaccine. N Engl J Med, 2012, 366: 1343-1344.

[6] Eberle J, Gurtler L. HIV types, groups, subtypes and recombinant forms: errors in replication, selection pressure and quasispecies. *Intervirology*, 2012,

55: 79-83.
[7] Smith S M. The pathogenesis of HIV infection: Stupid may not be so dumb after all. Retrovirology, 2006, 3: 60.
[8] Durand CM, Blankson JN, Siliciano RF. Developing strategies for HIV-1 eradication. Trends Immunol, 2012, 33:554-562.
[9] Amadori A, Zamarchi R, Chieco-Bianchi L. CD4: CD8 ratio and HIV infection: The "tap-and-drain" hypothesis. Immunology today, 1996, 17(9): 414-417.
[10] Lane H C. Pathogenesis of HIV infection: Total $CD4^+$ T-cell pool, immune activation, and inflammation. Top HIV Med, 2010, 18(1): 2-6.
[11] Grossman Z, Meier-Schellersheim M, Sousa AE, *et al.* $CD4^+$ T-cell depletion in HIV infection: Are we closer to understanding the cause? Nat Med, 2002, 8(4): 319-323.
[12] Brenchley J M, Hill B J, Ambrozak D R, *et al.* T-cell subsets that harbor human immunodeficiency virus (HIV) *in vivo*: Implications for HIV pathogenesis. J Virol, 2004, 78: 1160-1168.
[13] Davey R T Jr, Bhat N, Yoder C, *et al.* HIV-1 and T cell dynamics after interruption of highly active antiretroviral therapy (HAART) in patients with a history of sustained viral suppression. Proc Natl Acad Sci USA, 1999, 96: 15109-15114.
[14] Douek D C, Roederer M, Koup R A. Emerging concepts in the immunopathogenesis of AIDS. Annu Rev Med, 2009, 60: 471-484.
[15] Silvestri G, Paiardini M, Pandrea I, *et al.* Understanding the benign nature of SIV infection in natural hosts. J Clin Invest, 2007, 117: 3148-3154.
[16] Stein J H, Hsue P Y. Inflammation, immune activation, and CVD risk in individuals with HIV infection. JAMA, 2012, 308: 405-406.
[17] Brenchley J M, Price D A, Douek D C. HIV disease: Fallout from a mucosal catastrophe? Nat Immunol, 2006, 7: 235-239.
[18] Deeks S G, Hoh R, Grant R M, *et al.* $CD4^+$ T cell kinetics and activation in human immunodeficiency virus-infected patients who remain viremic despite long-term treatment with protease inhibitor-based therapy. J Infect Dis, 2002, 185: 315-323.
[19] Kovacs J A, Lempicki R A, Sidorov I A, *et al.* Identification of dynamically distinct subpopulations of T lymphocytes that are differentially affected by HIV. J Exp Med, 2001, 194: 1731-1741.
[20] Brenchley J M, Schacker T W, Ruff L E, *et al.* $CD4^+$ T cell depletion during all stages of HIV disease occurs predominantly in the gastrointestinal tract. J Exp Med, 2004, 200: 749-759.
[21] Veazey R S, DeMaria M, Chalifoux L V, *et al.* Gastrointestinal tract as a major site of $CD4^+$ T cell depletion and viral replication in SIV infection. Science, 1998, 280: 427-431.
[22] Mehandru S, Tenner-Racz K, Racz P, *et al.* The gastrointestinal tract is critical to the pathogenesis of acute HIV-1 infection. J Allergy Clin Immunol, 2005, 116: 419-422.
[23] Kotler D P, Gaetz H P, Lange M, *et al.* Enteropathy associated with the acquired immunodeficiency syndrome. Ann Intern Med, 1984, 101: 421-428.

[24] Smale S, Tibble J, Bjarnason I. Small intestinal permeability. Curr Opin Gastroenterol , 2000, 16: 134-139.
[25] Mowat A M. Anatomical basis of tolerance and immunity to intestinal antigens. Nat Rev Immunol, 2003, 3: 331-341.
[26] Schieferdecker H L, Ullrich R, Hirseland H, *et al.* T cell differentiation antigens on lymphocytes in the human intestinal lamina propria. *J Immunol*, 1992, 149: 2816-2822.
[27] Anton P A, Elliott J, Poles M A, *et al.* Enhanced levels of functional HIV-1 co-receptors on human mucosal T cells demonstrated using intestinal biopsy tissue. AIDS, 2000, 14: 1761-1765.
[28] Li Q, Duan L, Estes J D, *et al.* Peak SIV replication in resting memory $CD4^+$ T cells depletes gut lamina propria $CD4^+$ T cells. Nature, 2005, 434: 1148-1152.
[29] Paiardini M, Frank I, Pandrea I, *et al.* Mucosal immune dysfunction in AIDS pathogenesis. AIDS Rev, 2008, 10: 36-46.
[30] Picker L J, Hagen S I, Lum R, *et al.* Insufficient production and tissue delivery of $CD4^+$ memory T cells in rapidly progressive simian immunodeficiency virus infection. J Exp Med, 2004, 200: 1299-1314.
[31] Brenchley J M, Paiardini M, Knox K S, *et al.* Differential Th17 CD4 T-cell depletion in pathogenic and nonpathogenic lentiviral infections. Blood, 2008, 112: 2826-2835.
[32] Bomsel M, Alfsen A. Entry of viruses through the epithelial barrier: Pathogenic trickery. Nature reviews Molecular cell biology, 2003, 4: 57-68.
[33] Cassol E, Malfeld S, Mahasha P, *et al.* Persistent microbial translocation and immune activation in HIV-1-infected South Africans receiving combination antiretroviral therapy. J Infect Dis, 2010, 202: 723-733.
[34] Macpherson A J, Harris N L. Interactions between commensal intestinal bacteria and the immune system. Nat Rev Immunol, 2004, 4: 478-485.
[35] Brenchley J M, Price D A, Schacker T W, *et al.* Microbial translocation is a cause of systemic immune activation in chronic HIV infection. Nat Med, 2006, 12: 1365-1371.
[36] Berg R D, Garlington A W. Translocation of certain indigenous bacteria from the gastrointestinal tract to the mesenteric lymph nodes and other organs in a gnotobiotic mouse model. Infect Immun, 1979, 23: 403-411.
[37] Suffredini A F, Hochstein H D, McMahon FG. Dose-related inflammatory effects of intravenous endotoxin in humans: evaluation of a new clinical lot of *Escherichia coli* O:113 endotoxin. J Infect Dis, 1999, 179: 1278-1282.
[38] Nowroozalizadeh S, Mansson F, da Silva Z, *et al.* Microbial translocation correlates with the severity of both HIV-1 and HIV-2 infections. J Infect Dis, 2010, 201: 1150-1154.
[39] Wallet M A, Rodriguez C A, Yin L, *et al.* Microbial translocation induces persistent macrophage activation unrelated to HIV-1 levels or T-cell activation following therapy. AIDS, 2010, 24: 1281-1290.
[40] Ganusov V V, De Boer R J. Do most lymphocytes in humans really reside in the gut? Trends Immunol, 2007, 28: 514-518.
[41] Belmonte L, Olmos M, Fanin A, *et al.* The intestinal mucosa as a reservoir of

HIV-1 infection after successful HAART. AIDS, 2007, 21: 2106-2108.
[42] Shen R, Smythies L E, Clements R H, *et al*. Dendritic cells transmit HIV-1 through human small intestinal mucosa. J Leukoc Biol, 2010, 87: 663-670.
[43] Shattock R J, Moore J P. Inhibiting sexual transmission of HIV-1 infection. Nature reviews Microbiology, 2003, 1: 25-34.
[44] Ganor Y, Zhou Z, Tudor D, *et al*. Within 1 h, HIV-1 uses viral synapses to enter efficiently the inner, but not outer, foreskin mucosa and engages Langerhans-T cell conjugates. Mucosal Immunol, 2010, 3: 506-522.
[45] Saba E, Grivel J C, Vanpouille C, *et al*. HIV-1 sexual transmission: Early events of HIV-1 infection of human cervico-vaginal tissue in an optimized *ex vivo* model. Mucosal Immunol, 2010, 3: 280-290.
[46] Hladik F, Sakchalathorn P, Ballweber L, *et al*. Initial events in establishing vaginal entry and infection by human immunodeficiency virus type-1. Immunity, 2007, 26: 257-270.
[47] Moodley P, Connolly C, Sturm AW. Interrelationships among human immunodeficiency virus type 1 infection, bacterial vaginosis, trichomoniasis, and the presence of yeasts. J Infect Dis, 2002, 185: 69-73.
[48] Knezevic A, Stepanovic S, Cupic M, *et al*. Reduced quantity and hydrogen-peroxide production of vaginal lactobacilli in HIV positive women. Biomed Pharmacother, 2005, 59: 521-523.
[49] Frank D N, Manigart O, Leroy V, *et al*. Altered vaginal microbiota are associated with perinatal mother-to-child transmission of HIV in African women from Burkina Faso. J Acquir Immune Defic Syndr, 2012, 60: 299-306.
[50] Chung H, Kasper D L. Microbiota-stimulated immune mechanisms to maintain gut homeostasis. Current opinion in immunology, 2010, 22: 455-460.
[51] Hooper L V, Macpherson A J. Immune adaptations that maintain homeostasis with the intestinal microbiota. Nat Rev Immunol, 2010, 10: 159-169.
[52] Fujimura K E, Slusher N A, Cabana M D, *et al*. Role of the gut microbiota in defining human health. Expert review of anti-infective therapy, 2010, 8: 435-454.
[53] Sharma R, Young C, Neu J. Molecular modulation of intestinal epithelial barrier: Contribution of microbiota. Journal of biomedicine & biotechnology, 2010, 2010: aticle ID 305879.
[54] Gori A, Tincati C, Rizzardini G, *et al*. Early impairment of gut function and gut flora supporting a role for alteration of gastrointestinal mucosa in human immunodeficiency virus pathogenesis. J Clin Microbiol, 2008, 46: 757-758.
[55] Kantor B, Ma H, Webster-Cyriaque J, *et al*. Epigenetic activation of unintegrated HIV-1 genomes by gut-associated short chain fatty acids and its implications for HIV infection. Proc Natl Acad Sci USA, 2009, 106: 18786-18791.
[56] Dandekar S, George M D, Baumler A J. Th17 cells, HIV and the gut mucosal barrier. Curr Opin HIV AIDS, 2010, 5: 173-178.
[57] Round J L, Mazmanian S K. Inducible Foxp3^{+} regulatory T-cell development by a commensal bacterium of the intestinal microbiota. Proc Natl Acad Sci

USA, 2010, 107: 12204-12209.
[58] Xing S, Fu J, Zhang Z, *et al.* Increased turnover of Foxp3 high regulatory T cells is associated with hyperactivation and disease progression of chronic HIV-1 infection. J Acquir Immune Defic Syndr, 2010, 54: 455-462.
[59] Ancuta P, Monteiro P, Sekaly RP. Th17 lineage commitment and HIV-1 pathogenesis. Curr Opin HIV AIDS, 2010, 5: 158-165.
[60] Milner J D, Sandler N G, Douek D C. Th17 cells, Job's syndrome and HIV: opportunities for bacterial and fungal infections. Curr Opin HIV AIDS, 2010, 5: 179-183.
[61] Macal M, Sankaran S, Chun T W, *et al.* Effective $CD4^{+}$ T-cell restoration in gut-associated lymphoid tissue of HIV-infected patients is associated with enhanced Th17 cells and polyfunctional HIV-specific T-cell responses. Mucosal Immunol, 2008, 1: 475-488.
[62] Hunt P W. Th17, gut, and HIV: Therapeutic implications. Curr Opin HIV AIDS, 2010, 5: 189-193.
[63] Hofer U, Speck R F. Disturbance of the gut-associated lymphoid tissue is associated with disease progression in chronic HIV infection. Semin Immunopathol, 2009, 31: 257-266.
[64]Schellenberg J J, Plummer F A. The microbiological context of hiv resistance: vaginal microbiota and mucosal inflammation at the viral point of entry. International journal of inflammation, 2012, 2012: article ID 131243.
[65] Ramakrishna BS. Probiotic-induced changes in the intestinal epithelium: Implications in gastrointestinal disease. Trop Gastroenterol, 2009, 30: 76-85.
[66] Cunningham-Rundles S, Ahrne S, Bengmark S, *et al.* Probiotics and immune response. Am J Gastroenterol, 2000, 95: S22-S25.
[67] Trois L, Cardoso E M, Miura E. Use of probiotics in HIV-infected children: A randomized double-blind controlled study. J Trop Pediatr, 2008, 54:19-24.
[68] Salminen M K, Tynkkynen S, Rautelin H, *et al.* The efficacy and safety of probiotic *Lactobacillus rhamnosus* GG on prolonged, noninfectious diarrhea in HIV Patients on antiretroviral therapy: A randomized, placebo-controlled, crossover study. HIV Clin Trials, 2004, 5:183-191.
[69] Rao S, Hu S, McHugh L, *et al.* Toward a live microbial microbicide for HIV: Commensal bacteria secreting an HIV fusion inhibitor peptide. Proc Natl Acad Sci USA, 2005, 102: 11993-11998.
[70] Bolton M, van der Straten A, Cohen CR. Probiotics: Potential to prevent HIV and sexually transmitted infections in women. Sex Transm Dis, 2008, 35: 214-225.
[71]Quigley EM. Prebiotics and probiotics; modifying and mining the microbiota. Pharmacol Res, 2010, 61: 213-218.
[72] van Loveren H, Sanz Y, Salminen S. Health claims in Europe: Probiotics and prebiotics as case examples. Annual review of food science and technology, 2012, 3: 247-261.

22

Microecology Intervention in Prevention and Treatment of Infectious Diseases

Jieli Yuan *, Ao Xie
Department of Microecology, School of Basic Medical Sciences, Dalian Medical University, Dalian, 116044, China
* E-mail: yuanjieli@163.com

Several characteristics of current infectious diseases show that most patients of infectious disease are immune tolerant hosts. Most infection-induced microorganisms come from patients themselves and people around them with normal indigenous microflora. The effectiveness of antibiotics are weakened with the emergence of multidrug resistant strains. The model of 'infection→antibiotics→reinfection→reuse of antibiotics' must not be followed in order to more effectively fight against infection from various critical points such as outbreaks and spreads. Infectious microecology provides theoretical and practical support to avoid and control infections. The introduction of mechanisms and applications of probiotics in prevention and control of infectious diseases will be introduced and discussed below in details.

22.1 Theoretical Basis of Microecological Prevention and Treatment

In recent years, worldwide attention has been paid to the study and application of micoecological preparation in the field of medical science. The concept of biological therapeutical preparation proposed by Elmer *et al.* is a particular case. In 1993, the International Study Group on New Antimicrobial Strategies (ISGNAS) was established by 14 of the most famous experts in German. They clearly pointed out: the emergence of antibiotic-resistant microorganisms had become a great challenge to human beings, even more severe than pandemic in World War I, while during the same period, the WHO recommended microbial interference treatment (MIT) to control and prevent infections. The research and application of microecologial preparation had been commonly recognized as a weapon fighting against microorganism with its advantages focusing on it being non-toxic and side-effect-free, the product of microorganism and enzyme which can contribute to regulation of microbiological balance or enhancement of immunity.

22.1.1 The Principle of Microecology Balance

From the point of view of microecology, large amount of normal indigenous microflora colonize inner and outer surfaces of animals and plants. A microecological system is composed of host, indigenous microbiota and external circumstances; the system is dynamically balanced under normal conditions. The system either favours the host by assisting in some of its physiological processes, or benefits the microbes that can maintain the composition of microbiota and help with its reproduction. A minority of microbiota play a dominant role in the microecological system composed of the host, external and internal circumstances, instead of that, the individual microflora are dominant in the micro-community. Once the dominant microflora is absent, the micro-community will disintegrate. Once the dominant individuals are absent, the microecology disturbance may occur. For instance, the following factors may change the microbial balance: antibiotics, radiation therapy, chemical therapy, surgery and an allergy induced normal microflora change [1]. Microecological balance is disturbed which leads to an array of clinical symptoms such as superinfection and immunocompromised conditions. The probiotics, a member of normal dominant microflora in the host, can adjust and restore the balance of microflora, which treats and prevents disease.

22.1.2 Principles of Biological Antagonism

Normal microflora colonized on the intestinal tract is a part of barrier structures for biological defense, which includes mechanical barriers, chemical barriers, biological barriers and immunological barriers. Mechanical barriers include epithelial cell loss, ciliary movement, peristalsis and mucus secretion; chemical barriers are enzymes, peptides and metabolites produced by enteric microflora, such as acetic acid, lactic acid, propionic acid, peroxide and bacteriocin, which prevent or remove the pathogens from colonization; biological barrier are the normal microflora colonized on mucus or epithelial cells which form biofilm-like structure protecting indigenous bacteria; SIgA (mucosal immunity), IgA, IgD (humoral immunity) and types of immune cells and cytokines (cellular immunity) are preformed to construct the immune barrier. Probiotics, a member of the normal microflora, directly participates in the constructing of biological barrier structure so as to exert the effect of antagonism [2].

22.1.3 Biological Oxygen Consumption Hypothesis

The hypothesis of biological oxygen consumption was first proposed by Wei Xi and Kang Bai (1980) on the basis of normal flora colonization. Microbes colonized after the human beings and animals were given birth. The order of colonization goes on from aerobe through facultative aerobe to anaerobe. There is too much oxygen in the body that anaerobe cannot colonize first, unless the aerobe or facultative aerobe consume a large amount of oxygen. Although it cannot be colonized first, the anaerobe takes up the dominant place in the whole microecological system and keeps the balance. The oxygen concentration declines and oxidation reduction potential decreases by utilizing the non-toxicity, harmless and non-pathogenic microorganisms (like *Bacillus cereus*) exhibited in the intestinal tract temporarily, which constructs the optimal microecological conditions for dominant microflora promoting the anaerobe to increase to the ultimate microecological balance.

It is paramount of power to support this hypothesis by the wide application of probiotics preparation such as *Bacillus cereus, Bacillus subtilis*, *etc.*

22.1.4 Immune Activating

As non-specific immunoregulation factors, probiotics stimulates the host immunocytes by itself or the component of the surface of bacteria, which can be activated to secrete mitogenic factors elevating activities of natural killer cells and microphagocytes or exerting effect as adjuvant [3]. In addition, probiotics also have specific immune function to promote the ability of host B cells producing antibodies [4, 5]. For example, the inclined level of total IgA and intestinal IgA can be observed in feces from healthy children administered by *bifidobacterium*, which also enhances the capability of anti-infection.

22.1.5 Nutritional Effect

Probiotics (*e.g. Lactobacillus, Bifidobacterium*) are able to synthesize various kinds of vitamins such as niacin, folacin, vitamin B_1, B_2, B_6, B_{12}, *etc.*; they promote the digestion and absorption of protein, enhance uptake of calcium, iron and vitamin D, and function as assisting digestion and enhancing appetite. After fermentation in the gut, *Bifidobacterium* can produce lactic acid and acetic acid, raise the availability of calcium, phosphorus, iron, and enhance absorption of iron and vitamin D. In the processing of *Bifidobacterium* lactose fermentation, galactose is one component of the galactosyl ceramide in the nervous system that is closely related to the growth of infant brain. Simultaneously, *bifidobacterium* also can produce various vitamins and amino acids. With the capability of synthesizing vitamin and protein, *bifidobacterium* can promote digestion and absorption. It may enhance the protein digestibility by secreting phosphoprotein phosphatase which can degrade α-casein in milk. It is significant that the *Bifidobacterium* can also make lactase- deficiencies to be able to degrade lactose in the prevention of side-effects brought by the lactose intolerance.

22.1.6 Three Circulations Theory

The three circulations theory is one of the essences in macroecology and microecology. The three circulations are energy circulation, metabolite circulation and gene circulation. Energy circulation holds the relationship between normal microflora and host in maintaining the energy exchanging and cycling, especially in the field of microecology between host and microbes and/or between normal microorganisms. Metabolite circulation demonstrates that energy and substance of the normal microflora depend on the host that performs the necessary degradation metabolism and anabolism in microecology. By degradation and synthesis metabolites, normal microflora and host cell exchange substances with each other. In such exchange circulation, microbes utilize the lytic cells and perienzyme, otherwise the enzyme, vitamins and the cell component that stimulates and degrades are used by the host. Histamine, hydrogen sulfide, nitrous acid produced by microorganisms can harm the host. Gene circulation is called because gene transition generally occurred in normal microbes by the substances exchange, such as the resistance factor (R factor) and toxin-produced factor.

22.1.7 Continuous Succession Effect

Non-intestinal bacteria served as the manufacture of microecological modulator, besides the *Lactobacillus bulgaricus, Streptococcus thermophilus* and *Bacillus subtilis* previously utilized in France. Application of *Sporolactobacillus inulinus* and *Clostridium butyricum* in the Republic South Korea, and *Lactobacillus casei* and *Bnfillus licheniformis* in China, have been reported. These bacteria mentioned above are microaerophilic bacterium, facultative aerobe and aerobe respectively. The *Bacillus subtilis* is a special case, an obligate aerobe, it can grow quickly in tubes and has a strong oxygen tolerance. Oppositely, it can hardly be seen on the complex medium under anaerobic conditions. *Bacillus subtilis* is frequently found in the human intestinal tract; although, multiplication is limited results from its small count. Therefore, used as a modulator, it must be taken a large dose for the effect to be achieved. How does it exert effects when it cannot multiplicate or colonize in the intestine under a low Eh? It is suggested that the mechanism is concerned with glucoprotein in the bacteria or protodyne. Lallouette proved that the glycoprotein extracted from the *Bacillus subtilis* inoculates the mice thus preventing an attack from *E. coli*. Berger claimed that the non-specific immunity occurred attributed to protodyne, a component bacteria decomposed, uptake stimulating the immune mechanism. Nevertheless, the relation between *Bacillus subtilis* and intestinal microflora remains to be investigated.

Until now, no probiotics could perfectly fit in all types of environment, but only have effect in the corresponding ecosystem and microcommunity. Although under a lot of conditions, the bacteria cannot colonize, it should achieve effects after a flushing dose is taken resulting from the certain density (or level) of microcommunity maintained for continuous administration in a suitable environment. In history, *Lactobacillus bulgaricus* is regarded as a fermentative strain of yoghurt for a long time, and its effect has been confirmed. This strain cannot colonize in the human intestinal tract, the result arises from the continuous recapitulation, which means that the live organism count of probiotics decreased constantly as well as being elevated by adding continuously. In this process, probiotics products still could make the inherent eco-effect. The "continuous recapitulation" lays a foundation of probiotics preparation for allochthonous organisms.

22.2 Microecological Modulators

Microecological modulator is a class of biological products rising rapidly in the world, with applications in medical science, health protection, food science, agriculture, husbandry as well as fisheries. The appearance of microecological modulator has a close relationship with the development of ecological medicine. In the view of ecological medicine, the human body is not only a organism, but also a member of the earth's ecosystem. Environmental factors, technology of modern medicine, pesticide and veterinary drugs could break the organisms' microecological balance. Maintaining ecological balance and regulating microecological disturbance are new topics in modern medical science.

Microecological modulators include viable bacteria, dead bacteria, bacterial composition, metabolites and growth promoters. Currently, microecological modulators are divided into three general categories: probiotics, prebiotics and biogenetics.

22.2.1 Probiotics Definitions

The word 'probiotics' originates from Greek and means 'for life'. In 1965, Lilley and Stillwell defined such health-promoter microbes excrete as 'probiotics', the opposite word is antibiotics.

In 1989, Fuller defined probiotics as a preparation containing viable microorganism, which promotes the microflora in gut eubiosis, beneficially affecting the host. He emphasized that the probiotics must be living microbes, thereby excluding dead organisms and metabolites. In 1992, a more detailed description was made also by Fuller, as preparation criteria should be fulfilled as follows: (i) Probiotics must be viable in application of industry; (ii) Must be viable during storage and administering; (iii) Must be viable in gut or other ecoenvironment (multiplicity not necessary); (iv) Must have a beneficial effect on the host; (v) No toxin excretion and side-effects, safe and harmless.

In 1996, Arameo *et al.* further defined the probiotics as a microbial preparation which contained viable physiological microbes or dead ones (including the components and metabolites) , which administered orally or in other pathways, either improved the eubiosis between microflora and enzymes, or stimulated the mechanism of body specific or non-specific immunity, enhancing the capability of colonization or immunization.

There is a large number of probiotic products on the market. It is reported by many studies that the dead body, components and metabolites of the probiotics can also adjust the dysbiosis, keeping the eco-balanced and boosting the level of host health status. Therefore, it is coherent with Arameo and his definition is accepted by many scholars domestic and abroad.

22.2.2 Characteristics and Classification of Probiotics

The strains used as probiotics are mainly obtained from physiological predominant bacteria, non-resident bacteria and physiological eumycete of normal flora in host. The majority of physiological bacteria are lactic acid-producing bacteria, including hundreds of species from the 7 genus; non-resident bacteria has a low density of settlement in host, most of which are facultative anaerobe or aerobe from the autochthonous flora and environmental flora such as *Bacillus* and *Clostridia*; several probiotics yeasts are included in physiological eumycete.

There are probiotics approved to be administered to humans by the Ministry of Health of China: most of which are *Bifidobacterium, Lactobacillus, Clostridium butyricum, Bacteroides fragilis, Bacillus subtilis, Streprococcus thermophilus, Enterococcus* and photosynthetic bacteria.

The Ministry of Agriculture announced that the probiotics used by animals mainly are *Lactobacillus, Enterococcus faecalis, Bifidobacterium*, yeast, M423 *Bacillus cereus* and SA38 *Bacillus cereus*.

Traditionally, the classification of probiotics primarily is based on phenotype characteristics like morphology, carbohydrate fermentation, growing temperature, lactic acid configuration and the utilization of sorts of carbohydrates. The introduction of modern molecular methods, *e.g.* polymerase chain reaction, restriction enzyme, 16S rRNA sequence analysis, provides important ways to detect and identify probiotics strains in screening and application procedures. This paragraph summarizes the classification of current widely used probiotics according to the 9th edition of *Bergey's Manual of Systemaic Bacteriology* in combination with some new advancement of the categorization.

(i) *Acidobacterium.* Microbial strains regularly used as probiotics include: *L. delbrueckii, Lactobacillus brevis, Lactobacillus acidophilus, Lactobacillus bulgaricus, Lactobacillus casei, Lacrobacillus fermenti, Lactobacillus plantarum* and *Lactobacillus rhamnosus, etc.*

(ii) *Bifidobacterium.* Common used strains include: *Bifidobacterium bifidum, Bifidobacterium infantis, Bifidobacterium longum, Bifidobacterium breve* and *Bifidobacterium thermophilum, etc.*

(iii) *Enterococcus.* Representative strain consists of *Enterococcus faecalis, etc.*

(iv) *Streptococcus*: Used as probiotics strains *like Streptococcus thermophilus, Streptococcus acidi lactici* and *Streptococcus cremoris, etc.*

(v) *Bacillus.* Now applied extensively, strains include: *Bacillus subtilis, Bacillus cereus, Bnfillus licheniformis* and in the field of husbandry, *Bacterium anthracoides, Bacillus coagulans, etc.*

(vi) *Clostridia.* Strains widely utilized as probiotics such as *Clostridium butyricum*, discovered by Japanese in 1935, generally applied in foods, medicine, feed supplements.

(vii) *Leuconostoc.* In this genus, *Leuconostoc mesenteroides* is applied extensively, and its glucoside transferase can utilize sucrose to produce glucose oligosaccharide, which is the component of stabilizer. The more essential effect of *Leuconostoc mesenteroides* is stimulating the growth of gut intestinal flora.

(viii) *Pediococcus.* Applied strains primarily include *Pediococcus acidilactici, Pediococcus cerevisiae* and *Pediococcus pentosaceus. Pediococcus acidilactici* can also be used in the storage of meat and botanic foods.

(ix) *Lactococcus. Lactococcus* has an crucial status in dairy products, consisting of 3 subspecies of *lactococcus*. Some of these yield bacteriocin such as nisin, *Lactococcin* and so on. As a biological antiseptic, *galactococcus* is widely used.

(x) *Propionibacterium.* It includes *Propionibacterium shermanii* and *P. Freudenreichii*, both of which can be utilized as probiotics in feed supplements.

(xi) *Bacteroides.* The ones that can be used in animal feed supplements are *Bacteroides suis, B. capillosus,B. ruminocola* and *B. amylophilus.*

(xii) Yeast fungus Bioflor and *Candida mycoderma* are applied as probiotics in treatment of disease.

(xiii) Eumycete. Eumycete is used as feed supplements. The strains are black mold and *Aspergillus oryzae*, producing types of enzymes like amylase, protease, lipase, xylanase, cellulose, cellase and pectase promoting digestion of feed.

22.2.3 Bio-Safety of Probiotics

The critical factors used to evaluate safety are mainly demonstrated by the following aspects.

22.2.3.1 Study on Probiotics Pathogenicity and Infected Ability

Probiotics preparation on the market is in the forms of food and drugs. Biosafety for humans is becoming more and more important. Some traditional probiotics strains biosafety have beenidentified for a long time in application; for example, certain species of *Acidobacterium*, *Leuconostoc*, and *Pediococcus* have been used extensively in food processing for a long time. From the view of ecology, as a predominant microflora, *Bifidobacterium* exerts effects on maintaining the intestinal eubiosis and protecting human health. Furthermore, it is one probiotics genus that has been widely used until now and is generally recognized as a safe probiotics genus. As yet, safety of probiotics has not been a major problem because there are few reports concerning its harmful effects. However, during recent years, some species of *Acidobacterium*, *Leuconostoc*, *Pediococcus*, *Ent*erococcus and *Bifidobacterium* are isolated from types of focus of infection incidentally. Gasser covered that strains obtained from bacterial enteritis are *Lactobacillus acidophilus*, *Lactobacillus casei*, *Lactobacillus plantarum*, *Lactobacillus rhamnosus*, *Lactobacillus salivarius*, *L. delbrueckii sub.*, *Leuconostoc mesenteroides* and from blood and endogenous infection collected *Lactobacillus plantarum, Lactobacillus rhamnosus*, *Leuconostoc mesenteroides*, *Pediococcus acidilactici*, *Bifidobacterium dentium* and *Bifidobacterium adolescentis*. Gasser excluded *Enterococcus* and *Streptococcus* as they were recognized as having pathogenicity to some extent. Since *Bifidobacterium adolescentis* and *Bifidobacterium dentium* have similar physiological properties, it is possible to presume the *Bifidobacterium dentium* for *Bifidobacterium adolescentis*. Also Brook reported *Bifidobacterium dentium* and *Bacterium lacticum* are isolated from children infections. The above reports initiate a controversy surrounding the safetly of probiotics. A report from a European Union workshop indicates the ratio of infection induced by lactic acid bacteria is low except the *Enterococcus*, but *Lactobacillus rhamnosus* is ranged in supervision. It is extremely crucial to identify the safety of probiotics in commercial and industrial production considering many strains were found in the focus of infection.

The assessment of probiotics safety must include the pathogenicity, transmissibility, toxicity, metabolic activity and strain emphytic character. Donohur, Salminen, *et al.* provided some methods of assessing lactic acid bacteria safety and means of investigating infection, pharmacokinetics and internal trait of strains [6, 7].

22.2.3.2 Probiotic Metabolic Activity and Toxicant Produced

One part of assessing probiotics safety is to detect whether or not it produces toxic metabolic products or harmful enzymes, such as amine, benzopyrrole, phenol, amino compound and hydrogen sulfide in the process of fermenting protein. *Bifidobacterium* and *Lactobacillus* have not been reported to generate toxicant. By examination, *Bifidobacterium* has a lower level of deaminase activity, oppositely a higher ability of assimilating ammonia as well as desaturation of bile acid.

The normal rule methods of reviewing the probiotic toxicity are acute toxicity test and chronic toxicity test.

22.2.3.3 Platelet Agglutination Activity and Mucosa Desquamate Amotic Activity

The platelet agglutination raised from bacteria is considered as a factor promoting infection of endocarditis. Harry found the performance of platelet agglutination of all the 5 strains of *Lactobacillus rhamnosus* were isolated from the focus of endocarditis, but half of the 16 ones available from the laboratory. Agglutination is associated with the structure of outer protein of bacteria cells. The strains isolated from the focus of infection have stronger agglutination ability than the laboratory ones. It is certain to detect whether glucosidase, the shedding the glucoprotein on the surface of mucosa cells, has the toxicity or transmissibility. For example, the enzymes in *Lactobacillus rhamnosus* separated from the focus of endocarditis, but not in *Bifidobacteria* and *Lactobacillus.*

22.2.3.4 Antibiotics Resistance

The resistance factor is carried by some strains, between which the factor transferred. However, it is actually undesired in probiotics. Although some specially utilized probiotics strains could be resistant or multi resistanct, it is better not to be applied routinely.

Commonly, the resistance in *Enterococcus* is concerned, especially with whether or not the resistance genes combined by plasmid will be transferred to the normal flora or pathogen. It has been proven that *L. reuteri* transferred resistance factor to *Enterococcus Faecom* or *Enterococcus faecalis*, but resistance plasmid transfer exists under the choice of antibiotics opposite to the situation without drugs.

Both in *Enterococcus* and other probiotics strains, resistance transfer happened. All over the world, *Lactobacillus, Bifidobacterium* and *Streptococcus* used in majority of milk products have resistance, which are normal microflora having developed for a long term, and therefore can transfer initially. If there is no selective pressure, no probiotics resistance transfer will be considered. Until now, few reviews have reported on it.

22.2.4 *Bifidobacterium* Products

Bifidobacterium is one of the earliest discovered physiological bacteria, a probiotic colonized in healthy intestinal tract and accounting for 92% in breast feeding infants' gut, which is considered to be a sign for health. It is widely concerned that in the GI tract of the healthy adults and the longevity, *Bifidobacterium* is predominant; however, fewer in the diseases or apolexis and the number of which are dynamically changing related with the age as well as the physiological and pathological phenomena. Many microecological colloquiums both domestically and throughout the world assembled together at an international *Bifidobacterium* symposium held in Japan in 1990. The symposium widely discussedaspects of physiology, biochemistry, classification, mechanism and application, in addition, including its functions of biobarrier, anti-tumor, anti-aging and anti-infection. In a word, since the first *Bifidobacteria* strain was discovered, research on it has never stopped and now it is the focus in the field of microecology.

Until now, there are more than 30 *Bifidobacterium* species, 9 of which are isolated from human's bowel. *Bifidobacterium* establishes itself in the gut several hours after birth . It is one of the most common microflora in humans, the existence and the number of *Bifidobacterium* are crucial to person's physiology and health for the reasons that it can exert the effect of antagonism, nutrition, metabolism regulation, immune regulation and tumor prevention [8].

22.2.5 *Lactobacillus Products*

Lactobacillus is the predominant microflora in both human and animals' alimentary tract, whose physiological effect is next to that of *Bifidobacterium*. As a probiotics *Lactobacillus* colonized a host's digestive tract has the advantageous effects, which includes maintaining healthy probiotics floras, decreasing the pH of gastrointestine to inhibit the multiplication of pathogens by generating metabolites resembling bacteriocin, improving lactose tolerance, reducing serum cholesterol, diminishing concentration of blood ammonia, producing vitamins, stimulating immune system and possessing the ability of cancer prevention and anti-mutagenesis.

22.2.5.1 *Lactobacillus Acidophilus* Preparations

Lactobacillus acidophilus NCFM is a widely used commercial strain produced by France Rhodia Company, applied in liquid milk, yogurt, solid food, infant food and fruit juice, *etc*.. Next to the action of physiology, *Lactobacillus acidophilus* immediately participates in the composition of biological barriers to enhance the colonization in order to antagonize the invasion of pathogens or conditioned pathogens; they are also included in the synthesis of vitamins such as thiamin, lactoflavin, niacin, pantothen and folacin; it facilitates GI motility and absorption as well as improves mineral metabolism. For example, elevating blood concentration of calcium, magnesium and decreasing the potassium; activating the immune system generating more phagocytes and B cell antibodies to increase the organism general immunity; simultaneously, producing lactic acid, acetic acid resulted in low pH and bacteriocin and antibiotics like acidophilin, diplococcin, lactocin, which exert the effect of restriction between the species diminishing the excessive multiplication of gram-negative conditioned bacteria and production and releasing of the endotoxins; boost the tolerance of radioactive ray; reducing the ammonia emerged and decreasing the blood concentration of ammonia and phenol, improving liver function, lessening the blood fat and delaying arteriosclerosis; antagonizing the faculty of toxin-produced, decomposing the nitrite and represses the growth of tumor cells; depleting the creating of oxyradical and delaying the cell-aging.

Since *Lactobacillus acidophilus* was isolated from human bodies in 1990, the preparation of which had developed quickly and also the killed *L. acidophilus* preparations Lactocin killed *L. acidophilus* in heat treatment mixed with its metabolites by inoculating the strains in fresh milk. It contains effective components including lactic acid, vitamins and LAB-promoting substances (for example, lactose, galactose, casein hydrolysis and bifidus factors), which may regulate intestinal microflora disturbance, treat the acute and chronic enteritis and also have the functions of nutrition and immunity-raised.

Killed *Lactobacillus acidophilus* preparations (*e.g.*, Menarini (Belgium), Lacteo (France), Ramon Sala (Spanish), Lacteol (Switzerland), Lactobacillin (the Netherlands)) have already been widely applied in the treatment of diarrhea, and its trade name is Lacteol and included in Martindale the Extra Pharmacopoeia.

22.2.5.2 *L. delbrueckii* preparations

In the early1980s, Albert Doederlein from Germany presented the investigation on bacteriology in a gravid women's vagina in which *L. delbrueckii* produces the lactic acid. Shown on the medium or in the vagina, *L. delbrueckii*, firstly named as Doederlein, is able to suppress growth of pathogens. A subspecies strain *Lactobacillus debrueckii* Subsp. *lactis* (DM8909) is isolated by Dalian Medical University and proved it has a critical effect on regulation mechanism of vaginal microecology. It has been manifested that *Lactobacillus* is the predominant microflora in vagina and the *Enterobacter, Staphylococcus, Streptococcus*, yeast fungus are not. In early days, mid and late trimester of pregnancy, the amount of *Lactobacillus* increased gradually. The number of vaginal *lactobacillus* is one important physiological index during pregnancy and also a means for prophylaxis of infection. Dysbacteriosis in vagina occurs in bacterial vaginosis, non-specific vaginosis and dysbacterial vaginosis performed *Lactobacillus* decreased or disappeared and the other floras increased or disproportioned.

It is represented that *Lactobacillus* DM8909 viable organism preparation is innocuity, harmless, safe and effective to dysbacterial vaginosis, vaginal fungus infection and trichomoniasis. Compared with Metronidazole, the effect of this preparation is better, the recurrence rate is low and there is no side effect such as antibiotics intrinsic resistance, dysbacteriosis as well as toxic reactions.

22.2.5.3 *Lactobacillus plantarum* Preparations

Lactobacillus plantarum usually can be isolated from many different foods such as sauerkraut, fermentative vegetables, yogurt, air-dried sausages and beverages. It is common to use *L. plantarum* to ensure the microbiological safety of seafood products as it effectively represses the putrefying bacteria and pathogens by secretion of organic acid, H_2O_2 and bacteriocins. *L. plantarum* has the ability of metabolizing the lactose for producing lactic acid. Compared with the other strains, *L. plantarum* has a considerable superiority in translation of the lactose from the milk serum. It has a high efficiency in converting the lactose in milk serum into lactic acid and also in which can utilize the proteins. As the need of amino acids during the growth, *L. plantarum* can produce some ectoprotease to hydrolyse proteins in order to obtain the amino acids and short peptides for compensating those in its growth medium. Otherwise, *L. plantarum* can also emerge the lipase, phosphatase, amylase, peroxidase and bacteriocins applied in the field of antiseptic as well as the food refreshing.

22.2.5.4 *Lactobacillus rhamnosus* GG Preparations

Lactobacillus rhamnosus GG (LGG) is the product of Finland Vilio Co., Ltd, which is a probiotic strain studied in details for the considerable effects in adjustment of gastrointestine, prophylaxis and reduction of diarrhea, elevate the immunity and prevention of saprontia. As possessing a strong gastric acid and bile tolerance, LGG passages into the GI tract in the form of viability and colonize. The main functions of LGG are facilitating the growth of *Bifidobacteria* and *Lactobacillus acidophilus*, preventing and curing diarrhea, prophylaxis of respiratory tract infection and allergy, eduction toxins.

The products of LGG are predominantly fermentative milk goods like yogurt, milk, GHT milk, cheese, infant food, juice, beverage and medicine. Almost 30 countries and regions are now manufacturing and selling LGG products or the ones containing LGG. Yili Corporation has already purchased 5 years of sole right of use.

22.2.6 *Bacillus Products*

Bacillus is widely concerned with its special biological characteristics like strong anti-reversion force, high temperature and pressure tolerance, easy storage and so on, which has a large range of types applied in the fields of food industry, pharmaceuticals and agriculture. For example, *Bacillus thuringiensis* is utilized as a broad-spectrum insecticide resulting from the parasporal crystal, released during the metabolism, which has a strong killing effect on insects. A certain bacillus also can secrete some toxic and detrimental substances to elicit a human or animal disease or contamination of food influencing health. The probiotic strains allowed to use in every country is different. Now bacillus applied in feed includes *Bnfillus licheniformis*, *Bacillus coagulans*, *B. pumilus* Meyer and Gottheil, *B. subtilis*, *B. cereus, B. megaterium, B. firmus* and *B. toyoi*. Human Bacillus preparations are mainly utilized in the form of medicine, such as the *B. cereus* viable preparations produced by Shouyuan Pharmaceutical Corporation in Henan, *B. licheniformis* viable preparations by Shenyang Pharmaceutical Factory, *B. subtilis* viable preparations by Harbin White Swan Pharmaceutical Corporation and *B. coagulans* by Qingdao Donghai Pharmaceutical Corporation, those of which are all employed for intestinal prevention and treatment.

In China, there is a long history of producing black curd beans using Bacillus. In Japan, it is found that a *Bafillus natto* grow on the soya beans, which is different from the common one, secreting much more stickums, specific phages and some biotins sufficing the needs of growth. However, it is still called *Bacillus subtilis* worldwide.

22.2.7 Saccharomyces Preparations

The desirable growth environment of Yeast (aerobic bacteria) is an acid one resembling the intestines. *Saccharomycetaceae* includes 22 genus and 139 species and in those *Saccharomyces cerevisiae* shows its importance. Its cell is round or orbicular-ovate; not strictly required with nutrients; grows well on the beerwort or glucose medium; has an optimal growth temperature of 25 – 28 ℃; has the capability of fermenting glucose, sucrose, maltose, galactose and raffinose with CO_2 and ethanol brought out; resistanct to sunlight and dry; metabolizes actively in anaerobic surroundings. *Saccharomyces cerevisiae* spreads extensively not on the surface of all kinds of plants, but in guts of vertebrate and invertebrate. It is more critical to animal bowels that yeast supplements the vitamin and protein and enhances the activity of digestive enzyme as a food additive than the biological effects provided by its growth. In the application of medicine, *Saccharomyces cerevisiae* is used as a digestant, and the yeast for probiotics treatment strains include *Saccharomyces boulardii* and *Candide pintotopesii*.

Yeast takes up a crucial status in the industry of brewage, food and medicine. China has begun utilizing yeast to brew for 4,000 years. Yeast is rich in vitamin and protein, which can be used as food, medicine and feed, the originals for extracting biochemical products like nucleotide, CoA, cytochrome C, glutathione and ATP, and also can produce the vitamins, amino acids and organic acids. *C. lipolytica* is used for petroleum deparaffinage. Several species of yeast could lead to the corruption, for example, *Saccharomyces mellis* deteriorates honey and jam, *Hansenula anomala* contaminates the spirits and is detrimental in alcohol fermentation industry. *Monilia albicans* may elicit skin, mucosa, respiratory tract and urinary system diseases.

Recently in China, it is presented some health-care food and special restoratives with yeasts as the carriers supplementing some micronutrients like iron-enriched yeast, selenium-enriched yeast and zinc-enriched yeast. When culturing these yeasts, the concentration of iron, zinc and selenium are increased in order to obtain more of those edible substances. It is also feasible to mix the feed with nutrients and then convert into rich-milk, egg and meat. It is a question that whether it could include nutrients as the transferring functional substances, although it has been limited in nutrients so far. Physiological eumycete Bioflor preparations is probiotic ones. It is predominantly a therapy for the adult or child infectious or nonspecific diarrhea, prevents and treats the colonists and diarrhea induced by antibiotics, and cures the irritable bowel syndrome.

22.2.8 Enterococcus Preparations

Enterococcus was established in 1984. It includes 18 species, some of which are transformed from *Streptococcus* and some are newly defined. On the blood or nutrient agar, it forms round, integrity, smooth colonies, most strains perform no hemolytic reaction and a minority is observed α- or β-hemolytic reaction. A predominant part of the *Enterococcus* does not produce pigment. However, few strains emerge the xanthein. The bacteria cells are orbicular-ovate, and single, in pairs or short chains formed as well as prolonged in this direction. To be negative in gram's staining, no inner-spore formed and dynamic; facultative anaerobic and the end product of glucose fermentation is L-lactic.

As a regular microbial flora in human and animal's intestine, *Entercoccus* exists in all detected feces samples (Unsworth) and in most feces of mammals and birds, a large amount of *Enterococcus faecalis* and *Enterococcus faecium* could be obtained, and its isolation rate reaches nearly 100%. This bacteria is a resident in oral cavity, colon, vagina and perineal region. The proportion of the *Enterococcus* composition is determinate at a certain time and in a specified eubiosis environment, the performance of symbiosis maintains the normal physiological function of host. Once the situation is disturbed, superinfection will occur. Some species of *Enterococcus* is extensively applied in the preparations for prophylaxis and improvement of human and livestock diseases, the representatives are *E. facalis* and *E. faecium*.

Intestinal microflora are viewed as normal organs in humans, involved directly in host metabolism, and play a crucial role in regular physiological activity. *Enterococcus* is associated with infectious disease, cancers, immunity, the endogenous ones, as the human normal microflora, inhabited in the ileum and colon participating in the process of host substance metabolism. Reports abroad claimed that some members of intestinal floras anticipated the cholesterol metabolism immediately and adjusted the abnormal metabolism into a common one. In many covers, the *E. faecium* and *E. faecalis* draw the focus. The earliest usage of *E. faecalis* is Lactasin, first developing in Japan in 1930s, mainly treat all kinds of intestinal diseases; the other Japanese companies also have resembling products like the Laspan (キツセイ), Shin-Laclone (Asahi) and neo*bifidobacterium* preparations (Takeda) as well as in America, both Feed-mate68 and Lactiferm have *E. faecalis* get involved. Until now, this strain is still studied in China and a few patents of bioproducts are established regarding *Enterococcus* decreasing the blood cholesterol concentrate to cure arteriosclerosis.

In China, it is successful to develops *Enterococcus faecium* into a probiotic preparation to regulate the blood fat and obtained validation to be a health care product by Department of Health. The product could adjust the level of human blood fat but not the normal. Through the tests *in vivo* and *in vitro*, it is approved that this preparation has no toxicity, nor harm and is safe and effective.

22.2.9 Clostridium Butyricum Preparation

Clostridium butyricum, also named Miyarisan, is a species of *Clostridium spp.*, which was first founded by Dr. Miyairichikaji (Chiba Medical University, 1933). In 1935, Dr. Kingimiyairi isolated *Clostridium butyricum* from human discharge and soil. Later, he checked out its filters after anaerobic culture, which had a slight level of fatty acid and played a strong role in intestine adjustment effect that inhibited the gut pathogens and promoted the growth of *Bifidobacteria* and LAB [9].

The major metabolite of *Clostridium butyricum* is butanoic acid, the most essential material in the rebirth and repairment of entero-eptheliums. The functions of butanoic acid are: The energy resource, the ground substance of cytomembrane synthesis, promoting alimentary cell growth, inducing cell differentiation and apoptosis and enhancing protein acetylation [10].

The trade name of *Clostridium butyricum* is Mia granules. There are many types of *C. butyricum* made in China for sale, like the mixed preparation (*C. butyricum* and *B. coagulans*) producted by Qingdao Eastsea Parmaceutical Co. Ltd..

22.3 Prebiotics Preparation

Microecology theory concerning prevention and cure presents that the purpose of adjusting microflora was achieved by growth promoter in the choice of nutrition or environment to inhibit the non-physiological bacteria as well as increasing the amount of the physiologicals. As early as the beginning of this century, lactose had been administrated to enhance *E. coli* grow and inhibit the growth of dysentery bacillus in Germany because *E. coli* could utilize the lactose but the dysentery bacillus could not. *Enterococcus* grow with the help of folacin and vitamin B. *Bifidus* factor is the substance which advances the growth of *Bifidobacteria*, like oligosaccharide, could promote the growth of normal physiological bacteria as growth-promoters which restrain the non-physiologicals. These substances are called prebiotics throughout the world.

22.3.1 Definition

"Probiotics" was first introduced by Gibson GR and Roberforid in 1995. Probiotics, a sort of food ingredient or preparation, can promote the activity or growth of one or more than one species of resident bacteria in order to keep the host healthy. The four qualifications of probiotics are as following.

(i) In upper GI tract, neither can be hydrolyzed nor absorbed;

(ii) Only can promote beneficial bacteria growth or stimulate its metabolism selectively;

(iii) Can increase the amount of intestinal beneficial bacteria;

(iv) Strengthen host's health

Some food ingredients, like non-digestible carbohydrate (hardness starch, botan-cell wall polysaccharides, hemicellulose, pectin and gum, non-digestible oligosaccharides, some polypeptides and lipids) , can be prebiotics candidates for their structures. These ingredients are also called colon-digestion food for they are not absorbed and hydrolyzed in the upper alimentary, but utilized in the colon and provide host energy, metabolites and necessary nutrition ingredients.

In colon-digestion food, some non-digestible carbohydrates are consistent with prebiotics qualifications completely, but some peptides and proteins in milk and vegetables are not for their fermentation, because production of noxious substances (like ammonia and amine) in colon, although they are not absorbed in the upper alimentary tract as well as playing some beneficial effects (*e.g.* facilitating certain cation absorption, stimulating immune system). Non-digestible lipids are not prebiotics either because of its unclear metabolic mechanism. Moreover, non-digestible carbohydrates including all kinds of starch 1,500, non-linear chain polysaccharides, hemicellulose, pectin, non-digestible oligosaccharides are also not prebiotics for they are lack the bacteria growth-promoting effect in the colon (Table 22.1).

Table 22.1 Categorization of food carbohydrates as colon-digestion food and prebiotics

Saccharide	Colon-digestion food	Prebiotics
Hardiness starch	Yes	No
Non-starch polysaccharide		
Botan-cellwall	Yes	No
Hemicellulose	Yes	No
Pectin	Yes	No
Gum	Yes	No
Non-digestible polysaccharide		
Fructo-oligosaccharide	Yes	Yes
Xylo-oligosaccharide	Yes	Yes
Glacto-oligosaccharide	Yes	Yes
Soy oligosaccharide	Yes	Yes
Glucose oligosaccharide	Yes	No

22.3.2 Oligosaccharide as Prebiotics

As you can see from Table 22.1, most non-digestible oligosaccharides are both colon-digestion food and prebiotics. Oligosaccharide is the first choice as prebiotics among many natural substances, it has a clear effective factor, a rich resource and stable chemical characteristics, and is easily added into various kinds of food and beverage as well as could be produced in a large scale. Many countries are now devoted to oligosaccharide production.

For a long time, because of the complexity and specificity of oligosaccharide, it was so difficult to isolate, analyze and synthesize it that it restrained saccharide's development. Research on saccharide was limited to its metabolism and structure, which was falling behind the studies on proteins and nuclear. People recognized wrongly that functions of saccharide are structure materials (*e.g.* cellulose, chitin) and energy substances, not the life activity regulators. However, it has now been discovered that glycosyl- complexities are necessary to all proteins and in almost every biochemical process as well as natural life activities. Since the 1970s, there have been many outbreaks in the field of saccharide isolating and structure analyzing, which promoted the development of investment in saccharide and a new subfield-glycobiology was born. This subfield mainly conducts research on several aspects of oligosaccharides, through which we have already got preliminary recognitions of oligosaccharides' effects and significance.

According to the results of hydrolysis, saccharide is divided into monosaccharide, oligosaccharide and polysaccharide. Monosaccharide cannot be hydrolyzed into a smaller molecule, and its reducible hydroxyl is formed into glycosidic bond with hydroxyl of other monosaccharide, so the liner or branched chain oligosaccharide or polysaccharide is composite. Since 1996, the functional oligosaccharides have been defined as:

(i) Polymerized with 2 – 10 monosaccharide (same or not);

(ii) Possess the common properties of saccharides, could replace sucrose as a sweetener, cannot be degraded by gastric acid or enzyme, not absorbed in small intestine, able to reach large intestine;

(iii) Promote *Bifidobacteria* generation in human body.

The University of Georgia Complex Carbohydrate Research Center (CCRC) specializes in oligosaccharide structure collection. Till now, 49,897 different structures have been recorded. This complicated structure diversity determined that isolation, analysis and synthesis of oligosaccharides have an extreme difficulty, and it also explains functional diversity of oligosaccharides in life activities. Oligosaccharides may be a type of vector carrying bioinformation, in their large amount of structures a titanic information arsenal is hidden there providing prolificness besides limited gene translation in life activities.

According to the prebiotics definition, now, only oligosaccharides that cannot be digested by host are considered as prebiotics. Oligosaccharides are consisted of 2 – 10 straight or branched chain monosaccharides, which are the pentose and hexose such as glucose, fructose, galactose, xylose, arabinose, mannose. Some oligosaccharides can be modified with chemical groups (*e.g.* amino, carboxy, sulfate radical and phosphate radical), and the modified oligosaccharides still serve as prebiotics. Oligosaccharides are non-digestible saccharides, which can only can be used by a few bacteria (*Bifidobacteria* or *lactobacillus*) *in vivo*. They play the same result as probiotics and overcome the shortened shelf time of viable probiotics preparations.

These days countries all over the world have developed various kinds of oligosaccharide products, especially Japan. Oligosaccharides generally exist in every variety of natural food like fruit, vegetables, milk and honey, *etc.* In the past 10 years, it has been widely used as a low-calories sweetener, especially in Japan and Europe. In 1991, the Japanese Government legislate "Foods of Special Health Use, FOSHU", in which fructo-oligosaccharide, galacto-oligosaccharide, soy oligosaccharide and palatinose are listed, 450 products include oligosaccharides. Till 1996, FOSHU had ratified 58 foods, and 34 kinds of oligosaccharide are identified as functional food additives including Lactulose, xylo-oligosaccharide and isomaltose. Oligosaccharides, a type of food cellulose, multiplicate intestinal *Bifidobacteria* and maintain a good entero-environment. Japanese experts pointed out oligosaccharides products had the brightest future. Japan is the major country producing oligosaccharides, and utilization of oligosaccharides are world-wide.

22.3.3 Physiological Function

Prebiotics are different from other oligosaccharides in digestion. It cannot be digested, but can be utilized by beneficial bacteria (like *Bifidobacteria*) to exert a special physiological effect. Digestion of neotype oligosaccharides induces inherent beneficial bacteria *bifidobacteria* to generate, inhibit other pathogens from growing and decrease the formation of poisonous fermentation products as well. Neotype oligosaccharides have the same effects on *Lactobacillus*. Because *Bifidobacteria* are sensitive to oxygen, fever, and acid, their characteristics and functions are different *in vivo* and *in vitro*, and the degradation of gastric acid and gall, the quantities of viable *Bifidobacteria* decreased dramatically. Intestinal *Bifidobacteria* and *lactobacillus* multiplicated after administrating oligosaccharides, making beneficial bacteria occupy the predominant situation in the gut. Physiological functions of neotype oligosaccharides attribute to the contributions of their digestion and fermentation by beneficial bacteria.

22.3.3.1 Promote Bifidobacteria to Grow in Selectivity

Most prebiotics stimulate the growth of *Bifidobacteria* selectively, but are notutilized by other bacterium. That is mainly because these oligosaccharides could be selective growth substrates for fermentation; secondly, *Bifidobacteria* prohibit other bacteria by its acid produced in fermentation and other metabolites; thirdly, Anand proved that Bifidin secreted by *Bifidobacterium bifidum* had a strong inhibition effect on *Salmonella*, *Shigella*, *Staphylococcus aureus* and other bacteria. Studies indicated that administrating 2 – 10 g oligosaccharides for several weeks can raise the amount of *bifidobacterium* as much as 7.5 times as it was originally, *Clostridium* decreased 81% and *Lactobacillus* increased 2 – 3 times.

Ingesting fructo-oligosaccharides with *Bifidobacteria* together may reinforce the resistance to enterohaemorrhagic *E. coli* O157. Animal experiments indicated that mice colonized by *Bifidobacteria* had a low death rate when they are infected with O157 and fed with oligosaccharides together.

22.3.3.2 Short-Chained Fatty Acids (SCFA) Produced by Prebiotics in *Bifidobacteria* Fermentation

Fifty percent of the products of enterobacteria fermentation are organic acid including acetic acid, propanoic acid, butanoic acid and lactic acid. In the acidity environment, acetic acid could not be hydrolyzed and has a better effect on prohibiting putrefying bacteria from growing than *lactobacillus*. Lactic acid, the intermediate product of propanoic acid and butanoic acid, is absorbed slowly *in vivo*. The nutritional source of alimentary tract epithelium is butanoic acid, which can enhance the cell reborn, inhibit the *Clostridium difficile* and prevent cancer. Different organic acids have different physiological effects.

SCFA is produced by prebiotics in *bifidobacterium* fermentation. On one hand, they decrease intestine pH and suppress the growth of aerobes and facultative anaerobes; on the other hand, they reduce the production amount of harmful enzymes. If taking 3 – 6 g/d oligosaccharides, the toxicants and harmful enzymes will be cut down by 44.6% and 40.9% respectively, and then follow a low colon carcinoma incidence. SCFA may also be absorbed directly by intestinal mucosa epithelial cells and transported to liver, then turned to energies.

Edible fructo-oligosaccharides can inhibit spoilage in the gut, so the smell gets lighter. Research reported that after ingesting fructo-oligosaccharides, the concentration of smell ingredient methylmercaptan decreased significantly (from $1.9 \pm 2.36 \times 10^6$ to $1.02 \pm 1.92 \times 10^6$). The limit of methylmercaptan checked out is only 0.99×10^6.

22.3.3.3 Saprondontia Prevention

S. mutans is the major bacteria causing saprondontia, whose output is glucose transferase that cannot decompose the oligosaccharides into glucose, fructose and galactose. The metabolite of oligosaccharide lactose is less than that of monosaccharide. Therefore, oligosaccharides may replace monosaccharides to prevent saprondontia.

22.3.3.4 Blood Ammonia-Lower Effect

Administrating Lactose could increase the amount of *bifidobacterium*, reduce the quantities of urea-decomposed bacteria, decrease the level of ammonia down effectively in portal veins and blood. It has a positive significance for treating hepatic encephalopathy. Now, lactulose and oligosaccharides have been applied in adjunctive therapies of hepatic diseases and hepatic encephalopathy.

22.3.3.5 Prevention of Diarrhea and Constipation

Through probiotics fermentation *in vivo*, oligosaccharides produce various physiological acidic substances, which build up a barrier with probiotics to defend against pathogens causing diarrhea. Moreover, acetic acid and lactic acid generated by *Bifidobacteria* may stimulate the intestinal peristalsis. All these make sense in the prevention and treatment of constipation. Oligosaccharide is a sort of water-soluble dietary fiber, promoting intestinal peristalsis and increasing water content in dejecture as well as reducing the incidence of constipation.

22.3.3.6 Serum Cholesterol-Lower Effect

Prebiotics influence host metabolism by elevating the numbers of *bifidobacterium*, like the fat and cholesterol metabolism. Administrating oligosaccharides 6 – 12 g/d for 2 weeks to 3 mon can lower 20 – 50 g serum cholesterol.

22.3.3.7 Nutrition-Absorbed Promotion and Nutriment Production

Bifidobacterium generates various Vitamin B(VB), such as VB_1, VB_2, VB_6, VB_{12}, niacin, folacin, *etc.* Japanese experts did a rabbit calcium-absorption study utilizing fucto-oligosaccharides, galacto-oligosaccharides, isomerism oligomeric maltose, and it presented that oligosaccharides could enhance the absorption of calcium. The amount of calcium-absorption is correlated with that of L-lactic acid in the appendix. After oligosaccharides are transformed into L-lactic acid, it absorbs calcium compound to increase the dissolubility, so the calcium-absorbed capability of organism is strengthened. It was reported that compared with the controls, oligosaccharides elevated the calcium-absorbed rate by 26% – 28% and increased the bone density and improved the bone structure.

22.3.3.8 Blood Glucose Levels Improvement

Yamashita performed an oral administration test of functional oligosaccharides with non-insulin-dependent diabetes mellitus (NIDDM) patients. The test showed that oligosaccharides had a significant glucose-lower effect on NIDDM patients.

22.3.3.9 Immune Response Stimulation and Organism Immunity Elevation

Immunostimulation consists of adjuvant and immunological regulation effect. Oligosaccharides, as adjuvants, are combined with certain toxins, viruses and eukaryocytes. It reduces the antigen absorption, augments antigen potencies and enhances the function of cytoimmunity as well as humoral immunity.

22.4 Application of Probiotics and Prebiotics

With the further studies on probiotics and the relationship between human and animals, physiological senses probiotics have been revealed, which expand the range of treatment in the medical field. In general, probiotics preparation is studied mainly by high colonization to host of probiotics, influence on metabolism activity related with host health and stimulation to host immune response.

Based on the three effects presented above, through a great deal of research, it proved that the probiotics effect including reducing the generation of intestinal carcinogen and mutagenic material, antitumor effect, lowering cholesterols, relieving constipation, stimulating organism immunological activity, enhancing phygocytic activity of phygocytes, promoting peristalsis, treating antibiotics-associated diarrhea as well as preventing and curing intestinal infection diseases.

Applications of probiotics in the field of medical treatment are as follows.

22.4.1 Gastrointestinal Tract Infection Diseases Prevention

Through mechanisms of biological antagonism, immunity and microflora regulation, and probiotics preparations improve gut microecology as well as serve to prevent from intestinal infection diseases including pathogen transmissibility enteritis or virus transmissibility enteritis, colonitis, contaminated small bowel syndrome, ulcerative colitis, dysentery, rotavirus diarrhea and irritable bowel syndrome. Probiotics preparations have a good treatment effect on chronic diarrhea caused by gastrogenic, pancreatogenous or hepatobiliary diseases, lactose intolerance and despepsia.

22.4.1.1 Prevention of Antibiotics-Associated Dysbacteriosis

Antibiotics-associated diarrhea (AAD) belongs to hospital diseases. When using antibiotics or chemical medicines to treat infectious diseases in clinic, 3.2% – 29% inpatients will have antibiotics-associated diarrhea, especially among the oral administrators. This is also called superinfection which happens after taking antibiotics for 20 d continuously. Its incidence rate is higher when broad-spectrum anbiotics or multi-antibiotics are administered. For pathogens causing superinfection is resistant to general antibiotics and the resistance of patients is low, so it is difficult to control superinfection and its death rate (reaches as high as 45%) .

Probiotics preparations have an obvious effect on preventing and treating intestinal diseases caused by using large doses of antibiotics. The diseases include intestinal microflora disorders, amount of anaerobics reduction, AAD, antibiotics-associated enteritis and pseudomem-branous colitis *etc.* Supplementing large doses of *lactobacillus* preparation, *bifidobacterium* preparation and prebiotics may prevent and cure the antibiotics-associated dysbacteriosis.

22.4.1.2 Acute Diarrhea and Its Associated Diseases Treatment

In the view of microecology, dysbacteriosis is a combined reason as well as an essential reason causing diarrhea onset, which has already been recognized. The performances of dysbacteriosis are the changes of bacteria species, quantities and proportions. The general changing rule is the amount of aerobes or facultive anaerobes increase and anaerobes decrease especially the ones keeping gut flora balance.

Acute diarrhea may be caused by many reasons, like acute intestinal infection, bacterial food poisoning, acute systemic infection, allergic diseases, endocrine diseases and drug side effects. The microecological treatment for acute diarrhea is to analyze the microflora at first, mainly take *bifidobacterium* and *lactobacillus* in the meanwhile to antagonize the excessive growed *enterobacter*. Supplement of nutrition and dehydration prevention also needed to be paid attention to. The serious patients may take *bacillus, bifidobacterium* and *lactobacillus* mixed preparations. Otherwise, a single preparation is sufficient.

22.4.1.3 Treatment of Chronic Diarrhea and Its Associated Diseases

In general, chronic diarrhea lasts more than 2 mon. The defecation frequency of patients are more than that of the healthy, and it alternates with constipation. Chronic diarrhea includes infectious diarrhea, enterogenous diarrhea and irritable bowel syndrome, *etc.* Its cause is related with abnormal constitution of microbial population. Some cases indicated that when *bifidobacterium* quantities were under detection limit, the populations of *Klebsiella*, Proteus had an obvious increase. The treatment for chronic diarrhea sticks to the rules of 'microecology balance', supplementing both the *bifidobacterium, lactobacillus, bacillus* preparation and prebiotics preparations.

22.4.2 Constipation Treatment

Modern medicine studies proved that long-term constipation may induce different diseases which sped up the aging process. Stools stayed a long time in the gut, generating harmful bacteria, these harmful materials were produced and absorbed through the blood into the liver, which could damage the hypatocytes. In addition, constipation leads to difficulty in defecating; with people having to hold their breath in order to make the effort. In this process, the contraction of the abdominal muscle strongly leads to an elevation of abdominal cavity pressure. Ata certain pressure which the abdominal cavity blood vessles reach, a large volume of blood refluxes into heart, thereby burdening the heart and increasing the blood pressure. Constipation also induces chronic ulcers, haemorrhoids, carcinoma of rectum and breast cancer. Therefore, constipation is not only the evidence of intestinal contamination, but also the signals of body abnormalities.

As early as the 1920s, there were some studies on the relationship between constipation and intestinal bacteria, which pointed out that if the quantities of beneficial bacteria increased, the constipation would be effectively improved . Researchers found that bacterial flora structure in the body of constipation patients is not balanced, among which the beneficial bacteria is less than 25% of the healthy. Other reports uncovered in 50% constipationers' dejectures, there was no beneficial bacteria present.

The functional constipation takes microecological preparations to treat. Through administrating viable preparations, on one hand, it may supplement a large amount of physiological bacteria, correct the change of bacteria population and promote food digestion; on the other hand, in the metabolism process, probiotics generate many kinds of organic acids, which could lower the pH value down in enteric cavity, regulate intestinal peristalsis and relieve constipation. In addition, prebiotics preparations also can be taken by the constipationers. The acetic acid and lactic acid may get the environment weak tart, increase the volume of water penetrating into the body and stimulate intestinal peristalsis in order to relieve constipation.

22.4.3 Prevention and Treatment of Hepatic Diseases

Many chronic hepatic disease patients are accompanied with intestinal dysbacteriosis [11]. Normal physiological bacteria *in vivo* are inhibited, so a lot of proteins are not digested completely. Absorption and metabolism of organisms are blocked and lead to the accumulation of toxins in the blood. In addition, as using large doses of antibiotics eliminating or prohibiting sensitive physiological anaerobes in gut, the aerobic gram-negative bacillus multiplicate, which is the premise of endotoxins release.

22.4.3.1 Endotoxin and Intestinal Flora

Endotoxin is the lipopolysaccharide of gram-negative cell out membrane in intestines, which is detoxicated by liver. In the stage of cirrhosis or even chronic hepatitis, the intestinal dysbacteriosis has already existed showing that the number of *bifidobacterium* decreased and gram-negative bacilli grow excessively. Although the pathogens in the gut may produce a large amount of endotoxins, integrated intestinal mucosa can play an effective barrier function. The normal intestinal flora consisting mainly of obligate anaerobes also limit the contact between endotoxin and intestinal mucosa so as to prevent endotoxin from getting into blood circulation. Intestinal dysbacteriosis is one of the main reasons that cause the endotoxin level to rise [12].

22.4.3.2 Reduction of Enterogenous Endotoxin

Administrating *bifidobacterium* preparations may obviously improve intestinal function disorders, inhibit the number of endotoxin-producing gram-negative bacteria. It exerts a satisfied treatment effect on hepatopaths. By prohibiting harmful bacteria from growing, *Bifidobacterium* changes intestinal pH value, diminishes production and absorption of enterogenous ammonia to decrease the blood ammonia concentration so as to treat the hepatic coma. The Affiliated Hospital of Zhejiang University adopted *Bifidobacteria* preparations to cure hepatic coma, one week later, blood ammonia concentrations of most patients decreased obviously, the degree of the coma was relieved, and some of the patients came back into consciousness. *Bifidobacterium* is also *include*d in cytoimmunity mechanism. Wang *et al.* treated cirrhosis patients with *Bifidobacterium* and lactulose, they found the concentration of blood ammonia and endotoxin decreased, the intestinal microecological state improved and the ratios of cocci and bacilli rose.

22.4.3.3 Treatment of Hepatitis and Cirrhosis

As Miiting reported, probiotics preparations applied for treating chronic hepatitis and cirrhosis increased the level of ammonia in patients' dejecture as well as thrombin reduction time, decreased blood ammonia concentration, improved liver function indicators and hepatic protein metabolism as well as recovering intestinal flora and liver detoxicating function. In 1999, Li *et al.* utilized microecological regulators for treating severe liver diseases, which had already received crucial research achievements.

Lactulose and stachyose are both oligosaccharides prebiotics preparations, which cannot be absorbed in the gut, they are conducive to the growth of *Lactobacillus* and *Bifidobacterium.* Lactic acid and acetic acid, which the probiotics produce, may inhibit pathogens from growing and prevent people from getting infected. After administrating lactulose and stachyose, intestinal *Bafidobacterium* and *lactobacillus* in cirrhosis patients augmented. Lactulose and stachyose can inhibit the harmful bacteria, decrease toxins produced by putrefying bacteria as well as the formation and release of toxins generated by gram-negative bacillus, control endotoxemia and ease the detrimental effect to liver.

22.4.4 Prevention of Hypercholesterolemia

Intestinal microflora may be recognized as a normal human organ. These flora take part in the host metabolism activity directly, which is the indispensable main section in host normal physiological activities. As one of the normal floras, *Endogenous enterococci* predominantly colonized in the ileum and colon. However, some strains participate in the process of host substance metabolism immediately. Some reports presented some intestinal bacteria were included in cholesterol metabolism and regulated abnormal metabolism. Gilliand covered that a *lactobacillus* strain can suppress the high blood cholesterol concentration caused by cholesterol diet. Japanese studies showed drinking yogurt lowered the level of human serum cholesterol. Investigation indicated *bifidobacterium* preparations and *lactobacillus acidopilus* preparations could transform cholesterol into fecosterol, which the human body did not absorb. Human trials manifested taking *Lactobacillus bulgaricus* and *Streptococcus thermophilus* fermented yogurt may decrease blood cholesterol 5% – 10%. It is effective in treating andrelieving cholesterolemia. *Fu et al.* isolated a viable *Enterococcus Faecium* strain, which had an obvious prevention and treatment effect on hypercholesterolemia.

22.4.5 Treatment of Irritable Bowel Syndrome

Irritable bowel syndrome (IBS) is the common intestinal functional disease of the digestive system in clinics, whose symptoms are colonic abdominal pain, diarrhea, constipation, constipation and diarrhea alternated. IBS is also accompanied with intestinal dysbacteriosis, the numbers of bacteroides, *Bifidobacterium* and enteric bacilli decreased but *Clostridium* augmented. The reason leading to IBS is still uncovered. The onset mechanism of IBS is complicated, which is related with consciousness and psychological factors, intestinal infection, diet, internal organ paresthesia, GI dynamics disorders, GI hormone and neuropeptide as well as genetic factors. Now it is thought that the consciousness and psychological factors play an essential role in the development of IBS. Nowadays, doctors utilize entro-spasmolysant, probiotics preparations, antidiarrhea agent, *etc.*, to ease symptoms. Microecological preparations are applicated widely in treating various GI diseases, like inflammatory bowel disease, tourism diarrhea, *Clostridium difficile* diarrhea, AAD, which gets a good evaluation. However, it still needs large-scale prospective research to investigate microecological preparation whether it can colonize in gut for a long period, their effectiveness and safety for IBS patients also need to be checked. As the diversity of IBS causes, IBS future treatment trend is diversification and the microecological regulation may be one method in therapying IBS [13].

Zhang *et al.* used *Bifidobacterium adolescentis* to treat IBS patients whose course of disease had already lasted for 6 years. Among 35 cases, there were 32 patients with improved status. The efficient rate reached 91.4%.

22.4.6 Prevention and Treatment of Vaginal Infection Diseases

For normal healthy woman, the vagina has a natural defense function because of the anatomical organization characteristics it possesses: Closure of vaginal orifice, vaginal anterior and posterior wall close sticking, vaginal epithelium hyperplasy influenced by estrin, and superficial cell keratinization. Vaginal pH value is 4 – 5, which can inhibit the pathogen adapting alkaline environment from multiplication. When vaginal natural defense function is broken down, pathogen is easy to invoke vaginitis. In normal vaginal floras, *Lactobacillus* are predominant, which maintain the vaginal microecological balance. *Lactobacillus* decompose the vaginal epithelial glycogen into lactic acid to prohibit the other parasitic bacterium from excessively growing. Moreover, a part of *Lactobacillus* synthesized peroxide, which may inhibit other bacteria with other oxides. Furthermore, the acidic circumstance *Lactobacillus* caused is in favor of decreasing negative charges on cell surfaces and removing saccharide groups on surfaces of receptors in order to expose the site for bacteria adhering. It will induce vaginal ecological system imbalanced resulting in gynecological infections when host defense barriers are broken down, overies function lowered down, having systemic diseases, applicating large doses of antibiotics, antitumor drugs and immune inhibitor, getting infections or influenced by sexual hormones.

Bacterial vaginosis (BV) is one of the most common diseases in gynecology, which is treated by antibiotics in common now, but the curative effect is not satisfied. The application of antibiotics may induce the intestinal dysbacteriosis leading to a long recovery time, thus a lot of pains will be suffered. Now it is thought that BV is caused by vaginal dysbacteriosis that the *Lactobacillus* is suppressed but anaerobes and vaginal *Gardner coli* excessively grow. The amount of *Lactobacillus* in vagina of BV patient is more than several thousand times lower than that of the healthy, with sometimes none existing.

22.4.7 Application in Pediatric Diseases and Infant Care

GI tract is the biggest immune tissue and organ. The status of microecological balance in gut of children, especially the newborns, is different from that of adults. When an infant is born, there is no bacteria in its gastroinstestine, and the immune system is not developed. Bacteria colonization and microflora establishment in newborns are related closely with normal immune system development and maturity. Normal microflora in children is not only non-stable, and poorly balanced, but also easily influenced by many factors. The applications of probiotics in infection prevention and treatment mainly focus on AAD and infectious diarrhea. Supplementing probiotics may treat children's diarrhea, thus the accidence of children diarrhea is decreased.

National diarrhea control program proposed that supplementing probiotics for rebuilding microecological barrier conduced to children diarrhea recovery. Wu's studyproved that no matter the diarrhea was acute, persisting or chronic, there was no significant difference between adopting monoxenie preparation, dixenie preparation or even the multibacteria preparation to treat them.

Some investigation found that 80% children who broke out asthmatic bronchitis frequently were accompanied with instestinal dysbacteriosis. In the treating course besides the antiasthmatic medicine was used, the additional utilization of probiotics preparation obtained a pretty good effect. The method also works with children with pneumonia, typhoid, septicaemia or genitourinary tract infection. Application of probiotics could enhance the anti-infection effect, promoting recovery from infection and shorten the course of treatment .

22.4.8 Application in Antitumor

Intestinal microfloras have a close relationship with certain tumors. The probiotics antitumor effection is a hot study topic nowadays. The main mechanisms of probiotics antitumor effection are carcinogen detoxication and immune competence elevation [14].

Probiotics antitumor effect is concerned with organism immune activation. The significance of immune activation lies in many aspects, especially in antitumor effect. The effective components of *Lactobacillus* antitumor effect are Muramyl dipeptide (MDP) and lipoteichoicacid existing in cell walls. They activate macrophages, NK cells and B cells *etc.* in order to make them excrete cytotoxicity factors, such as the IL-1, IL-6,TNF-α,NO and other antibodies.

Goldin *et al.* observed that in the feces of healthy people drinking *Lactobacillus acidophilus* N_2 and *Lactobacillus acidophilus* NCFM fermented milk, β-glucuronidase, nitroreductase and azoreductase, which possess the catalytic capability of transforming precarcinogen into carcinogen, are reduced by 50% at least. But after stopping the administration of the milk, the enzymic activities are as what they were previously. Such things do not happen with low fat milk.

Rao *et al.* did a colon cancer prevention experiment by administrating mice with *Lactobacillus acidophilus* NCFM. The result showed that NCFM did not inhibit the total amount of abnormal colon follicular (ACF), but also suppress the follicular diversity and the number of ACF per unit area. ACF number is the representative data of colon cancer. The higher it is, the greater accidence of colon cancer happens.

Flavacin is one of the fungal toxins which contaminates food frequently, and may induce liver cancer. Experiments proved *bifidobacterium* is capable to combine flavacin. Otherwise, *bifidobacterium* could bind with inductor (like ammonium nitrite, nitrosoguanidine) to cover active genes of inductor or degrade the genes so as to inactivate them. Some studies also indicated that *Bifidobacteria* had a high absorption of BBQ meat or fried food keeping organism cells away from the carcinogens. *Bifidobacterium* restrain many putrefying bacteria from growing by regulating intestinal flora to reduce the carcinogen production, so the accidences of digestive tract cancers are lowered.

References

[1] Backhed F, Ley R E, Sonnenburg J L, *et al.* Host-bacterial mutualism in the human intestine. Science, 2005, 307: 1915-1920.

[2] Sullivan A, Edlund C, Nord C E. Effect of antimicrobial agents on the ecological balance of human microflora. Lancet Infect Dis, 2001, 1: 101.

[3] Haghighi H R, Gong J, Gyles C L, *et al.* Modulation of antibody-mediated immune response by probiotics in chickens. Clin Diagn Lab Immunol, 2005, 12: 1387-1392.

[4] Della R F, Criniti V, Della P V, *et al.* Genes modulated by histone acetylation as new effectors of butyrate activity. FEBS Lett, 2001, 499: 199-204.

[5] Kelly D, Conway S. Bacterial modulation of mucosal innate immunity. Mol Immunol, 2005, 42: 895-901.

[6] Donohur D C, Salminen S. Safety of probiotic lactic acid bacteria. Asia Pac J Clin Nutr, 1996, 5: 25-28.

[7] Gasser F. Safety of lactic acid bacteria and their occurrence in human clinical infections. Bull Inst Pasteur, 1994, 92: 45-67.

[8] Reuter G. Present and future of probiotics in Germany and in central Europe. Bioscience Microflora, 1997, 16: 43-51.

[9] Uncan S H, Louis P, Flint J H. Lactate-utilizing bacteria,isolated from human feces, that produce butyrate as a major fermentation product. Appl Environ Microbiol, 2004, 70: 5810-5817.

[10] Kalliomaki M, Isolauri E. Role of intestinal flora in the development of allergy. Curr Opin Allergy Clin Immunol, 2003, 30: 15-20.

[11] Wu Z W, Li L J, Ma W H, *et al.* Study on the intestinal microbial colonization resistance in patients with chronic severe hepatitis. Chin J Hepatol, 2001, 9: 329-330.

[12] Han D W. Intestinal endotoxemia as a pathogenetic mechanism in liver failure. World J Gastroenterol, 2002, 8: 961-965.

[13] Floch M H. Irritable bowel syndrome,and inflammatory bowel disease. Curr Treat Options Gastroenterol, 2003, 6: 283-288.

[14] Zhang Q P, Duan L F. Current progress in the study of subhealth. Journal of health care and medicine in Chinese PLA, 2003, 5: 67.

23

Future Development of Infectious Microecology

Lanjuan Li *, **Yanfei Chen**
The State Key Laboratory for Diagnosis and Treatment of Infectious Diseases, the First Affiliated Hospital, School of Medicine, Zhejiang University; Collaborative Innovation Center for Diagnosis and Treatment of Infectious Diseases, Hangzhou, 310003, China
* E-mail: ljli@zju.edu.cn

More and more studies indicate interactions between infectious diseases and microbiota. Advances in molecular techniques have led to a greater appreciation of the diversity of human microbiota, the extent of interactions with the human host, and how that relates to inter-individual variation. Realization of the interaction between infectious agents and the microbiota will definitely deepen our understanding of infectious diseases.

23.1 Evolving View of Infectious Disease

Microbial ecology is the relationship of microorganisms with one another, and with their environment [1]. A microbial ecosystem is defined as a system that consists of all the microorganisms that live in a certain area or niche, which function together in the context of other biotic (plants and animals) and abiotic (temperature, chemical composition, and structure of the surroundings) factors of the niche [2]. The human body is home to many indigenous microorganisms, with distinct communities at different anatomical sites. Within the body of a healthy adult, microbial cells are estimated to outnumber human cells by a factor of one or two orders of magnitude. All the microbial cells constitute a small microecology, which is the simplest ecology on the earth. The composition of this microbial community is host specific, evolving throughout an individual's lifetime and susceptible to both exogenous and endogenous modifications [3]. One of the main functions of this high density commensal microbiota that inhabits the intestine is to shield from infection. Colonization resistance is the mechanism whereby the host microflora protects itself against incursion by new and often harmful microorganisms [4]. An infectious disease is a clinically evident illness resulting from the presence of pathogenic microbial agents, including pathogenic viruses, pathogenic bacteria, fungi, protozoa, multicellular parasites, and aberrant proteins known as prions. Infectious disease results from the interplay between those few pathogens and the defenses of the hosts they infect. For a long time, much attention has been focused on identifying the bacteria which cause disease. It is of equal importance that bacteria associated with health also be researched, so that a microbiological view of infectious disease should be established.

23.2 Advances in Molecular Ecological Techniques

Due to long neglect, the complexity of the indigenous microbiota itself and the fact that many of its members resist cultivation, human microbiota is in fact new to science. Several developments have recently converged to renew interest in studying the normal human microbiota. Efforts to characterize microbial diversity increasingly rely on cultivation-independent, molecular techniques, since the vast majority of bacteria have yet to be cultivated. Advances in molecular techniques have led to a greater appreciation of the diversity of human microbiota, the extent of interactions with the human host, and how that relates to inter-individual variation.

Many of the tools developed for environmental microecology studies have recently been used on human samples, providing a more comprehensive view of our microbial symbionts. New technologies, such as metagenomics and metaproteomics, are shedding light on the wide diversity and function of the microbial consortium. Cell and animal experiments point to the intimate relationship between the immune system and the bacteria. Human microecology science is making rapid progress.

Genomics or metagenomics approaches have a tremendous capability to generate composition data and measure the metabolic potential encoded by the combined genomes of the gut microbiota. Because of its universal presence in cellular organisms, the presence of conserved regions, and its reliability, most molecular studies are based on the small subunit 16S rRNA gene for phylogenetic analysis. Several next-generation sequencing technologies, such as 454 pyrosequencing, have been introduced in recent years that dramatically outperform the traditional Sanger technology in terms of throughput and cost. These instruments, with the ability to produce millions of DNA sequence reads in a single run, are making great progress in human microecology research. For clinical diagnostic purposes, some simple and rapid methods for analysis of the composition of the gut microbiota such as the phylogeny chip and function chip have been developed [5].

Another post-genomics approach, metabonomics, has the capacity to measure the metabolic kinetics or flux of metabolites through an ecosystem at a particular point in time or over a course of time. Metabonomics thus derives data on the function of the gut microbiota *in situ* and how it responds to different environmental stimuli, *e.g.* substrates like prebiotics, antibiotics and other drugs and in response to disease.

Metagenomic sequencing has revealed information about the composition of genes in the gut microbiota. A major limitation of DNA-based approaches is that they predict potential functions, but it is not known whether the predicted genes are expressed at all or, if so, under what conditions and to what extent. In addition, it is not possible to determine whether the DNA is from cells that are active and viable, dormant or even dead. Verberkmoes *et al.* developed a novel high-throughput, non-targeted mass spectrometry (MS) approach, to determine the identities of thousands of microbial proteins in the faeces. This is the first step to developing an approach to obtain a first large-scale glimpse of the functional activities of the microbial community residing in the human gut [6].

Although in its infancy, the application of culture-independent tools has dramatically improved our ability to interrogate the vast diversity of unculturable microbial species. In the future, these three culture independent, high resolution approaches will be combined into a single "trans-genomic" approach which allows correlation of changes in metabolite profiles within human biofluids with microbiota composition metagenomic data. Such approaches are providing novel insight into the composition, function and evolution of our gut microbiota [7].

23.3 Normal Human Microbiota

As an ever-increasing body of evidence implicates the microbiota in defining states of health and disease, it is quite important to characterize the extent of normal microbiota [8]. With advanced technologies, we are able to learn more about the normal microbiota inside ourselves. Gut microbiota is the most complicated microbiota inside our body. Gut microbiota is an assortment of microorganisms inhabiting the length and width of the mammalian GI tract, which closely co-evolved with the human genome and diet. The distal GI tract houses up to 1,000 distinct bacterial species and an estimated excess of 1×10^{14} microorganisms. The composition of the microbiota in healthy people is not only unique, but also appears to be quite stable over time. There seems to be a vast "core" of approximately 300 phylotypes that are likely to exist in all individuals. The human oral microbiome is comprised of hundreds of microorganisms that mainly colonize on tooth surfaces as a biofilm. Dental plaque is a dynamic and extremely complex oral biofilm ecosystem. Oral bacterial researches have suggested that the oral cavity and intestinal tract harbor distinct sets of bacteria [9]. Two recent 16S rRNA gene tag pyrosequencing-based studies have suggested that there are approximately 250 – 300 species-level phylotypes in the mouth of any given individual, and that they segregate based on mucosal versus dental surfaces [10]. The predominant taxa belongs to *Firmicutes* (genus *Streptococcus*, family *Veillonellaceae*, genus *Granulicatella*), *Proteobacteria* (genus *Neisseria*, *Haemophilus*), *Actinobacteria* (genus *Corynebacterium*, *Rothia*, *Actinomyces*), *Bacteroidetes* (genus *Prevotella*, *Capnocytophaga*, *Porphyromonas*) and *Fusobacteria* (genus *Fusobacterium*) [11]. The flora of the vagina and the urinary tract consist of a well-balanced system of about 50 bacterial strains. *Lacto bacilli* dominate the healthy flora of premenopausal women. The balance can be disturbed by the overgrowth of indigenous bacteria of the vagina like *Gardnerella*, *Bacteroides*, *Peptostreptococcus*, *Prevotella spp.* or *aerobic cocci*, or by the invasion of foreign microorganisms, such as *Escherichia coli*, *Enterococcus faecalis*, *Enterobacteriaceae*, *Staphylococci* or *Candida*.

Viral diversity and life cycles are poorly understood in the human gut and other body habitats. Reyes *et al.* sequenced the viromes (metagenomes) of virus-like particles isolated from faecal samples collected from healthy adult female monozygotic twins and their mothers at three time points over a one-year period and found that viromes are unique to individuals regardless of their degree of genetic relatedness [12].

23.4 Interactions between Infectious Diseases and Microbiota

For a long time, much attention has been focused on identifying the bacteria which cause infectious disease. Nowadays, more and more studies indicate interactions between infectious diseases and microbiota. Antibiotics, often used to cure infections, are causing more and more problems to the normal balance of microbiota. Not only bacterial infection but also virus infections are relevant to microbiota dysbiosis.

23.4.1 Disturbance of Normal Microbiota by Therapy

The human microbiota helps to protect the GI tract from enteric infections. A healthy microbiota is important in the host response to intestinal pathogens. Perturbations in the intestinal microbiota significantly impact the clinical incidence and severity of enteric infections. Disruption of the gut microbiome, termed dysbiosis, is frequently accompanied by overgrowth of pathogenic bacteria or fungi, in conjunction with significant loss of microbial diversity or key functional groups and an inflammatory response by the host, which contributes to disease development.

Mucositis, also referred to as mucosal barrier injury, is one of the most debilitating side effects of radiotherapy and chemotherapy treatment. Clinically, mucositis is associated with pain, bacteremia and malnutrition. Gut microbiota was found to participate in the development and severity of chemotherapy-induced mucositis. Recently, it has been shown that chemotherapy treatment is associated with a decrease in the number of anaerobic bacteria and a decrease in microbial diversity [13]. The disappearance of commensal intestinal microbiota will minimize their protection of enterocytes against potential pathogens. Different initial microbial colonization may protect and predispose the pathophysiology of acute postradiotherapy diarrhea [14]. Stecher *et al.* have found that the presence of closely related species can increase the chance of invasion of newly incoming species into the gut ecosystem, and this principle might be of general validity for invasion of bacteria in preformed gut ecosystems [15]. However, research concerning the relationship between intestinal bacteria and chemotherapy-induced mucositis is still rare. Further research is needed to verify the mechanism of gut microbiota in chemotherapy-induced mucositis.

More and more evidence has shown that there is a new conceptual framework of the microbiota-gut-brain axis. Stressor exposure disrupts commensal microbial populations in the intestines, and leads to increased colonization by Citrobacter rodentium [16].

Antibiotics are the main, and often only, clinical intervention for prophylactic and active treatment of bacterial infections in humans. However, these drugs also shift the composition of commensal bacteria inside our bodies, especially those within the gut microbial community. More and more evidence show that antibiotic use can increase host susceptibility to pathogen infection. A number of opportunistic pathogens can cause disease during antibiotic therapy, including *Salmonella spp.*, *Clostridium perfringens*, *Klebsiella oxytoca*, *S. aureus*, *Candida albicans*, and *C. difficile*. Of these, *C. difficile* is the most common cause of pathogen-associated antibiotic-associated diarrhea (15% – 25%), the most common cause of severe disease, and it causes nearly all cases of nosocomial pseudomembranous colitis [17]. Animal (hamster and mouse) and *in vitro* models show antagonism between conventional microbiota and *C. difficile* population growth [18].

Antibiotic-associated diarrhea (AAD) and Clostridium difficile infections (CDI) are associated with altered intestinal microflora and other symptoms that may possibly lead to death. The association between AAD or CDI and perturbations of the gut microbiota is well established but poorly understood.

Chemotherapy treatment in pediatric patients with acute myeloid leukemia receiving antimicrobial prophylaxis leads to a relative increase in colonization with potentially pathogenic bacteria in the gut [19]. This disturbance in the balance between anaerobic and aerobic bacteria will further increase the risk of gram-positive aerobic infections.

23.4.2 Microbiota and Viral Infection

Infectious diseases caused by a virus may also have relationships with the bacterial community of the host. Structural responses of gut microbiota were found on children with rotavirus infection [20].

Alteration in gut microbiota was also found in patients with HIV infection, and was supposed to be a key factor in HIV pathogenesis [21].

Indigenous microbiota was found to play a crucial role in the expansion and maintenance of viral-specific CD8 memory T cells in mice infected with murine cytomegalovirus [22].

23.4.3 Microbiota and Autoimmunity Disease

Dental caries, once thought to be a simple disease caused by *S. mutans*, is now unraveling as an extremely complex disease, which reflects the response of the tooth to a microbial challenge. Normally, low populations of acido-genic and aciduric bacterial species will increase following high-frequency carbohydrate exposure. The metabolism of carbohydrate by these microbiota results in the acidification of plaque (pH < 5), and acid-induced demineralization of the enamel and dentin occurs, eventually resulting in cavitation. Cariogenic plaques are comprised of numerous different microbial species, including *S. mutans* and other low-pH streptococci (*Streptococcus oralis*, *Streptococcus mitis*, *Streptococcus anginosus*), *Rothia*, *Actinomyces*, *Lactobacilli* and *Bifidobacterium spp.*, and *Candida albicans* [23].

Studies of bacterial infections in developed countries suggest 75% of adults fully recover. However, around 25% have long lasting changes in bowel habits and a smaller number develop the irritable bowel syndrome (IBS). Post-infectious IBS (PI-IBS) usually follows bacterial infection with *Salmonella spp.*, *C. jejuni* and *Shigella spp.* [24]. Why this inflammation persists in some, but not others still needs further research.

23.5 Therapy

Improved understanding of the normal gut microbiota has made the therapeutic manipulation of the gut ecosystem a valid and realistic future prospect.

It is well known that antibiotic use can cause short-term ecological disturbance in the microbiota, while another disturbing consequence of antibiotic treatment is long-term persistence of antibiotic resistance genes. Prudence in the administration of antibiotics could partially alleviate the emergence of antibiotic resistant pathogenic strains. Emphasis should be placed on alternative therapeutic options such as probiotics, immunotherapy, *etc.*

23.5.1 Probiotics or Prebiotics

In response to problems caused by antibiotic use, new biological treatment for infectious disease is needed. The use of probiotics, prebiotics and synbiotics is increasing in popularity for both the prevention and treatment of a variety of diseases. Prebiotics are a group of non-digestible food ingredients including inulin, oligosaccharides, lactulose and resistant starch that are fermented by colonic commensal microbiota to potentially improve the health of the host by selectively stimulating the growth of certain gut bacteria. Lactic acid bacteria are thought to provide positive health effects for the host, and are usually referred to as probiotic bacteria. Bohle *et al.* found that the mucus adhesion promoting protein of Lactobacillus reuteri can be specially degraded to an antimicrobial peptide [25]. This finding gives some new perspectives on how probiotic bacteria may successfully contribute to a healthy microbiota. *Escherichia coli* O157:H7 is a food-borne pathogen causing hemorrhagic colitis and hemolytic-uremic syndrome, especially in children. The probiotic bacterium *Lactobacillus acidophilus* strain La-5 can secrete molecules, which is effective against enterohemorrhagic *Escherichia coli* (EHEC) O157:H7 infection [26]. Meta-analysis of probiotics trials for the prevention of AAD showed an overall reduction in the risk of AAD when probiotics were co-administered with antibiotics [27].

Many probiotic strains have been tested for CDI. *Saccharomyces boulardii* is the best studied one among these strains. In a phase 3 trail, adult patients with CDI were randomized in a combination treatment of oral vancomycin and *S. boulardii* or vancomycin and placebo. Patients treated with vancomycin and the probiotic had significantly decreased recurrence rates compared with vancomycin with placebo [28].

Probiotics have shown effects on HIV-infected patients. In a randomized, placebo-controlled study of 24 subjects with HIV infection or AIDS, daily ingestion of probiotic *Lactobacillus rhamnosus* GR-1 and *Lactobacillus reuteri* RC-14, supplemented in yogurt, led to rapid resolution of diarrhea, flatulence and nausea, as well as a small increase in CD4 cell counts [29].

Oral or local application of yogurt in bacteria vaginosis or candida-vaginitis has shown promising results for decades. In a human study with 49 subjects, the local application of *L. rhamnosus* reduced the rate of urinary tract infections by 73% [30]. The strain showed tight adhesion to epithelial surfaces, hydrogen peroxide production, or the release of biosurfactants.

Probiotics also exert effects against viral infection such as rotavirus infection or viruses whose target organ is not the intestine. The mechanism is considered to be by immunostimulation and not by direct competition with the infectious agent.

In vitro models with complex and complete gut microbiota are required to accurately assess the potential and efficacy of probiotics on pathogen colonisation in the gut.

23.5.2 Other Bacteriologic Therapy

Most of the drugs used in clinics today were derived from living matter in the external environment. It's time to mine new drugs from microbial-derived signaling molecules in the inner environment of the gut. Bacteriocins are a family of anti-microbial peptides by which the producer organism can inhibit the growth of other organisms. The broad-spectrum bacteriocin, lacticin 3147, has been shown to have resistance to *C. difficile* [31]. Some narrow-spectrum bacteriocins with relative specificity against specific organisms are being searched for now.

The gut microbiota is a source of immunomodulatory signals, and the microbiota composition has a profound impact on immunological differentiation. Some microbial-derived molecules with immunomodulatory potential have been well researched, which include bacterial nucleic acids or oligonucleotides containing hypomethylated CpG dinucloetides and cytoprotective or anti-inflammatory peptides [32]. The therapeutic potential of these molecules on infectious diseases is still under research. The exploration of the inner world of human microbiota for drug discovery or other bioactive development is in its infancy, but is very promising.

An alternative approach for shaping overall microbiota composition by probiotics or prebiotics aimed at reducing detrimental and potentially pathogenic bacteria is by means of specific bacteriophages. Bacteriophages are viruses that attach to their specific bacterial hosts and kill them by sequential internal replication and lysis. Bacteriophages administration is safe and can reduce the concentration of Listeria monocytogenes in the GI tract. It can also translocate to the spleen and liver in experimentally infected mice [33].

23.5.3 The Role of Microbiota in Drug Metabolism

Nowadays, it is clear that the complex microbial ecosystem in our intestines should be considered as a separate organ within the body, with a metabolic capacity which exceeds the liver, with a factor of 100. The intestinal microbiome is therefore closely involved in the first-pass metabolism of dietary compounds [34].

23.6 Summary and Prospects

This field is very much in its infancy, and considerable work still needs to be done to better understand the relationship between the microbiota and infectious diseases. Realization of the interaction between infectious agents and commensals in disease will require greater understanding of the normal microbiota, and the mechanisms of microbiota-host interactions. Technological advances in molecular analysis will definitely speed up the exploration.

There are still many concerns about the safety of probiotics use. Future work should define the possibly related molecular factors that promote probiotic functions, fitness, and facultative pathogenicity. Then we can give a definite answer as to when and how to use probiotic strains against infections.

Although many probiotic strains have been tested, the quality of evidence is still poor. Much of the mechanism remains unproven, *e.g.*, how probiotics work, which strains are effective, what can be expected to be achieved, and what dosage is required for effectiveness.

References

[1] Konopka A. What is microbial community ecology? ISME J, 2009, 3: 1223-1230.

[2] Raes J, Bork P. Molecular eco-systems biology: towards an understanding of community function. Nat Rev Microbiol, 2008, 6: 693-699.

[3] Xu J, Mahowald M A, Ley R E, *et al.* Evolution of symbiotic bacteria in the distal human intestine. PLoS Biol, 2007, 5: e156.

[4] Gorbach S L, Barza M, Giuliano M, *et al.* Colonization resistance of the human intestinal microflora: Testing the hypothesis in normal volunteers. Eur J Clin Microbiol Infect Dis, 1988, 7: 98-102.

[5] Bjerketorp J, Ng Tze Chiang A, Hjort K, *et al.* Rapid lab-on-a-chip profiling of human gut bacteria. J Microbiol Methods, 2008, 72: 82-90.

[6] Verberkmoes N C, Russell A L, Shah M, *et al.* Shotgun metaproteomics of the human distal gut microbiota. ISME J, 2009, 3: 179-189.

[7] Tuohy K M, Gougoulias C, Shen Q, *et al.* Studying the human gut microbiota in the trans-omics era — focus on metagenomics and metabonomics. Curr Pharm Des, 2009, 15: 1415-1427.

[8] Fujimura K E, Slusher NA, Cabana M D, *et al.* Role of the gut microbiota in defining human health. Expert Rev Anti Infect Ther, 2010, 8:435-454.

[9] Bik E M, Long C D, Armitage G C, *et al.* Bacterial diversity in the oral cavity of 10 healthy individuals. ISME J, 2010, 4: 962-974.

[10] Keijser B J, Zaura E, Huse S M, *et al.* Pyrosequencing analysis of the oral microflora of healthy adults. J Dent Res, 2008, 87: 1016-1020.

[11] Zaura E, Keijser B J, Huse S M, *et al.* Defining the healthy 'core microbiome' of oral microbial communities. BMC Microbiol, 2009, 9: 259.

[12] Reyes A, Haynes M, Hanson N, *et al.* Viruses in the faecal microbiota of monozygotic twins and their mothers. Nature, 2010, 466:334-338.

[13] van Vliet M J, Harmsen H J, de Bont E S, *et al.* The role of intestinal microbiota in the development and severity of chemotherapy-induced mucositis. PLoS Pathog, 2010, 62: 1223-1236.

[14] Manichanh C, Varela E, Martinez C, *et al.* The gut microbiota predispose to the pathophysiology of acute postradiotherapy diarrhea. Am J Gastroenterol, 2008, 103:1754-1761.

[15] Stecher B, Chaffron S, Kappeli R, *et al.* Like will to like: Abundances of closely related species can predict susceptibility to intestinal colonization by pathogenic and commensal bacteria. PLoS Pathog, 2010, 6: e1000711.

[16] Bailey M T, Dowd S E, Parry N M, *et al.* Stressor exposure disrupts commensal microbial populations in the intestines and leads to increased colonization by Citrobacter rodentium. Infect Immun, 2010, 78:1509-1519.

[17] Walk S T, Young V B. Emerging insights into antibiotic-associated diarrhea and clostridium difficile infection through the lens of microbial ecology. Interdiscip Perspect Infect Dis, 2008, 125081.

[18] Wilson K H. The microecology of Clostridium difficile. Clin Infect Dis, 1993, 16: S214-S218.

[19] van Vliet M J, Tissing W J, Dun C A, *et al.* Chemotherapy treatment in

pediatric patients with acute myeloid leukemia receiving antimicrobial prophylaxis leads to a relative increase of colonization with potentially pathogenic bacteria in the gut. Clin Infect Dis, 2009, 49: 262-270.

[20] Zhang M, Zhang C, Du H, *et al.* Pattern extraction of structural responses of gut microbiota to rotavirus infection *via* multivariate statistical analysis of clone library data. FEMS Microbiol Ecol, 2009, 70: 21-29.

[21] Gori A, Tincati C, Rizzardini G, *et al.* Early impairment of gut function and gut flora supporting a role for alteration of gastrointestinal mucosa in human immunodeficiency virus pathogenesis. J Clin Microbiol, 2008, 46:757-758.

[22] Tanaka K, Sawamura S, Satoh T, *et al.* Role of the indigenous microbiota in maintaining the virus-specific CD8 memory T cells in the lung of mice infected with murine cytomegalovirus. J Immunol, 2007, 178: 5209-5216.

[23] Filoche S, Wong L, Sissons C H. Oral biofilms: Emerging concepts in microbial ecology. J Dent Res, 2010, 89: 8-18.

[24] Spiller R, Garsed K. Infection, inflammation, and the irritable bowel syndrome. Dig Liver Dis, 2009, 41: 844-849.

[25] Bohle L A, Brede D A, Diep D B, *et al.* The mucus adhesion promoting protein (MapA) of Lactobacillus reuteri is specifically degraded to an antimicrobial peptide. Appl Environ Microbiol, 2010, 76:7306-7309.

[26] Medellin-Pena M J, Griffiths M W. Effect of molecules secreted by Lactobacillus acidophilus strain La-5 on Escherichia coli O157: H7 colonization. Appl Environ Microbiol, 2009, 75: 1165-1172.

[27] Doron S I, Hibberd P L, Gorbach S L. Probiotics for prevention of antibiotic-associated diarrhea. J Clin Gastroenterol, 2008, 42: S58-S63.

[28] Surawicz C M, McFarland L V, Greenberg R N, *et al.* The search for a better treatment for recurrent Clostridium difficile disease: use of high-dose vancomycin combined with Saccharomyces boulardii. Clin Infect Dis, 2000, 31: 1012-1017.

[29] Anukam K C, Osazuwa E O, Osadolor H B, *et al.* Yogurt containing probiotic *Lactobacillus* rhamnosus GR-1 and *L. reuteri* RC-14 helps resolve moderate diarrhea and increases CD4 count in HIV/AIDS patients. J Clin Gastroenterol, 2008, 42: 239-243.

[30]Reid G, Bruce A W. Low vaginal pH and urinary-tract infection. Lancet, 1995, 346: 1704.

[31] Rea M C, Clayton E, O'Connor PM, *et al.* Antimicrobial activity of lacticin 3,147 against clinical Clostridium difficile strains. J Med Microbiol, 2007, 56: 940-946.

[32] Shanahan F. 99th Dahlem conference on infection, inflammation and chronic inflammatory disorders: Host-microbe interactions in the gut: Target for drug therapy, opportunity for drug discovery. Clin Exp Immunol, 2010, 160: 92-97.

[33] Mai V, Ukhanova M, Visone L, *et al.* Bacteriophage administration reduces the concentration of listeria monocytogenes in the gastrointestinal tract and its translocation to spleen and liver in experimentally infected mice. Int J Microbiol, 2010, 624234.

[34] Possemiers S, Bolca S, Verstraete W, *et al.* The intestinal microbiome: A separate organ inside the body with the metabolic potential to influence the bioactivity of botanicals. Fitoterapia, 2010, 82: 53-66.

Index

Kolibri 360 GmbH
Ferdinand-Jühlke-Straße 7
99095 Erfurt, Deutschland
produktsicherheit@kolibri360.de